Solid-State Physics

James D. Patterson
Bernard C. Bailey

Solid-State Physics

Introduction to the Theory

With 202 Figures

Springer

Professor Emeritus, James D. Patterson, Ph.D.
3504 Parkview Drive
Rapid City, SD 57701
USA
e-mail: *marluce@rap.midco.net*

Dr. Bernard C. Bailey, Ph.D.
310 Taylor Ave. # C18
Cape Canaveral, FL 32920
USA
e-mail: *bbailey@brevard.net*

ISBN-10 3-540-24115-9 Springer Berlin Heidelberg New York
ISBN-13 978-3-540-24115-7 Springer Berlin Heidelberg New York

Library of Congress Control Number: 2006938043

Springer is part of Springer Science+Business Media

springer.com

Production and Typesetting: LE-TEX Jelonek, Schmidt & Vöckler GbR, Leipzig
Cover design: WMX Design GmbH, Heidelberg

SPIN 10963821 57/3100YL - 5 4 3 2 1 0 Printed on acid-free paper

Preface

Learning solid-state physics requires a certain degree of maturity, since it involves tying together diverse concepts from many areas of physics. The objective is to understand, in a basic way, how solid materials behave. To do this one needs both a good physical and mathematical background. One definition of solid-state physics is that it is the study of the physical (e.g. the electrical, dielectric, magnetic, elastic, and thermal) properties of solids in terms of basic physical laws. In one sense, solid-state physics is more like chemistry than some other branches of physics because it focuses on common properties of large classes of materials. It is typical that solid-state physics emphasizes how physical properties link to the electronic structure. In this book we will emphasize crystalline solids (which are periodic 3D arrays of atoms).

We have retained the term solid-state physics, even though condensed-matter physics is more commonly used. Condensed-matter physics includes liquids and non -crystalline solids such as glass, about which we have little to say. We have also included only a little material concerning soft condensed matter (which includes polymers, membranes and liquid crystals – it also includes wood and gelatins).

Modern solid-state physics came of age in the late 1930s and early 1940s (see Seitz [82]), and had its most extensive expansion with the development of the transistor, integrated circuits, and microelectronics. Most of microelectronics, however, is limited to the properties of inhomogeneously doped semiconductors. Solid-state physics includes many other areas of course; among the largest of these are ferromagnetic materials, and superconductors. Just a little less than half of all working physicists are engaged in condensed matter work, including solid-state.

One earlier version of this book was first published 30 years ago (J.D. Patterson, *Introduction to the Theory of Solid State Physics*, Addison-Wesley Publishing Company, Reading, Massachusetts, 1971, copyright reassigned to JDP 13 December, 1977), and bringing out a new modernized and expanded version has been a prodigious task. Sticking to the original idea of presenting basics has meant that the early parts are relatively unchanged (although they contain new and reworked material), dealing as they do with structure (Chap. 1), phonons (2), electrons (3), and interactions (4). Of course, the scope of solid-state physics has greatly expanded during the past 30 years. Consequently, separate chapters are now devoted to metals and the Fermi surface (5), semiconductors (6), magnetism (7, expanded and reorganized), superconductors (8), dielectrics and ferroelectrics (9), optical properties (10), defects (11), and a final chapter (12) that includes surfaces, and brief mention of modern topics (nanostructures, the quantum Hall effect, carbon nanotubes, amorphous materials, and soft condensed matter). The reference list has been brought up to date, and several relevant topics are further discussed in the appendices. The table of contents can be consulted for a full list of what is now included.

The fact that one of us (JDP) has taught solid-state physics over the course of this 30 years has helped define the scope of this book, which is intended as a textbook. Like golf, teaching is a humbling experience. One finds not only that the students don't understand as much as one hopes, but one constantly discovers limits to his own understanding. We hope this book will help students to begin a life-long learning experience, for only in that way can they gain a deep understanding of solid-state physics.

Discoveries continue in solid-state physics. Some of the more obvious ones during the last thirty years are: quasicrystals, the quantum Hall effect (both integer and fractional – where one must finally confront new aspects of electron–electron interactions), high-temperature superconductivity, and heavy fermions. We have included these, at least to some extent, as well as several others. New experimental techniques, such as scanning probe microscopy, LEED, and EXAFS, among others have revolutionized the study of solids. Since this is an introductory book on solid-state theory, we have only included brief summaries of these techniques. New ways of growing crystals and new "designer" materials on the nanophysics scale (superlattices, quantum dots, etc.) have also kept solid-state physics vibrant, and we have introduced these topics. There have also been numerous areas in which applications have played a driving role. These include semiconductor technology, spin-polarized tunneling, and giant magnetoresistance (GMR). We have at least briefly discussed these as well as other topics.

Greatly increased computing power has allowed many ab initio methods of calculations to become practical. Most of these require specialized discussions beyond the scope of this book. However, we continue to discuss pseudopotentials, and have added a Section on density functional techniques.

Problems are given at the end of each chapter (many new problems have been added). Occasionally they are quite long and have different approximate solutions. This may be frustrating, but it appears to be necessary to work problems in solid-state physics in order to gain a physical feeling for the subject. In this respect, solid-state physics is no different from many other branches of physics.

We should discuss what level of students for which this book is intended. One could perhaps more appropriately ask what degree of maturity of the students is assumed? Obviously, some introduction to quantum mechanics, solid-state physics, thermodynamics, statistical mechanics, mathematical physics, as well as basic mechanics and electrodynamics is necessary. In our experience, this is most commonly encountered in graduate students, although certain mature undergraduates will be able to handle much of the material in this book.

Although it is well to briefly mention a wide variety of topics, so that students will not be "blind sided" later, and we have done this in places, in general it is better to understand one topic relatively completely, than to scan over several. We caution professors to be realistic as to what their students can really grasp. If the students have a good start, they have their whole careers to fill in the details.

The method of presentation of the topics draws heavily on many other solid-state books listed in the bibliography. Acknowledgment due the authors of these books is made here. The selection of topics was also influenced by discussion with colleagues and former teachers, some of whom are mentioned later.

We think that solid-state physics abundantly proves that more is different, as has been attributed to P. W. Anderson. There really are emergent properties at higher levels of complexity. Seeking them, including applications, is what keeps solid-state physics alive.

In this day and age, no one book can hope to cover all of solid-state physics. We would like to particularly single out the following books for reference and or further study. Terms in brackets refer to references listed in the Bibliography.

1. Kittel – 7th edition – remains unsurpassed for what it does [23, 1996]. Also Kittel's book on advanced solid-state physics [60, 1963] is very good.

2. Ashcroft and Mermin, *Solid State Physics* – has some of the best explanations of many topics I have found anywhere [21, 1976].

3. Jones and March – a comprehensive two-volume work [22, 1973].

4. J.M. Ziman – many extremely clear physical explanation [25, 1972], see also Ziman's classic *Electrons and Phonons* [99, 1960].

5. O. Madelung, *Introduction to Solid-State Theory* – Complete with a very transparent and physical presentation [4.25].

6. M.P. Marder, *Condensed Matter Physics* – A modern presentation, including modern density functional methods with references [3.29].

7. P. Phillips, *Advanced Solid State Physics* – A modern Frontiers in Physics book, bearing the imprimatur of David Pines [A.20].

8. Dalven – a good start on applied solid-state physics [32, 1990].

9. Also Oxford University Press has recently put out a "Master Series in Condensed Matter Physics." There are six books which we recommend.
   a) Martin T. Dove, *Structure and Dynamics* – An atomic view of Materials [2.14].
   b) John Singleton, *Band Theory and Electronic Properties of Solids* [3.46].
   c) Mark Fox, *Optical Properties of Solids* [10.12].
   d) Stephen Blundell, *Magnetism in Condensed Matter* [7.9].
   e) James F. Annett, *Superconductivity, Superfluids, and Condensates* [8.3].
   f) Richard A. L. Jones, *Soft Condensed Matter* [12.30].

A word about notation is in order. We have mostly used SI units (although gaussian is occasionally used when convenient); thus $\boldsymbol{E}$ is the electric field, $\boldsymbol{D}$ is the electric displacement vector, $\boldsymbol{P}$ is the polarization vector, $\boldsymbol{H}$ is the magnetic field, $\boldsymbol{B}$ is the magnetic induction, and $\boldsymbol{M}$ is the magnetization. Note that the above quantities are in boldface. The boldface notation is used to indicate a vector. The magnitude of a vector $\boldsymbol{V}$ is denoted by V. In the SI system μ is the permeability (μ also represents other quantities). μ_0 is the permeability of free space, ε is the permittivity, and ε_0 the permittivity of free space. In this notation μ_0 should not be confused with μ_B, which is the Bohr magneton [$= |e|\ \hbar/2m$, where e = magnitude of electronic charge (i.e. e means $+|e|$ unless otherwise noted), $\hbar$ = Planck's constant divided by 2π, and m = electronic mass]. We generally prefer to write $\int A \mathrm{d}^3 r$ or $\int A \mathrm{d}\boldsymbol{r}$ instead of $\int A\ \mathrm{d}x\ \mathrm{d}y\ \mathrm{d}z$, but they all mean the same thing. Both $\langle i|H|j\rangle$ and $(i|H|j)$ are used for the matrix elements of an operator H. Both mean $\int \psi^* H \psi \mathrm{d}\tau$ where the integral over τ means to integrate over whatever space is appropriate (e.g., it could mean an integral over real space and a sum over spin space). By $\sum$ a summation is indicated and by $\prod$ a product. The Kronecker delta δ_{ij} is 1 when $i = j$ and zero when $i \neq j$. We have not used covariant and contravariant spaces; thus δ_{ij} and δ_i^j, for example, mean the same thing. We have labeled sections by A for advanced, B for basic, and EE for material that might be especially interesting for electrical engineers, and similarly MS for materials science, and MET for metallurgy. Also by [number], we refer to a reference at the end of the book.

There are too many colleagues to thank, to include a complete list. JDP wishes to specifically thank several. A beautifully prepared solid-state course by Professor W. R Wright at the University of Kansas gave him his first exposure to a logical presentation of solid-state physics, while also at Kansas, Dr. R.J. Friauf, was very helpful in introducing JDP to the solid-state. Discussions with Dr. R.D. Redin, Dr. R.G. Morris, Dr. D.C. Hopkins, Dr. J. Weyland, Dr. R.C. Weger and others who were at the South Dakota School of Mines and Technology were always useful. Sabbaticals were spent at Notre Dame and the University of Nebraska, where working with Dr. G.L. Jones (Notre Dame) and D.J. Sellmyer (Nebraska) deepened JDP's understanding. At the Florida Institute of Technology, Drs. J. Burns, and J. Mantovani have read parts of this book, and discussions with Dr. R. Raffaelle and Dr. J. Blatt were useful. Over the course of JDP's career, a variety of summer jobs were held that bore on solid-state physics; these included positions at Hughes Semiconductor Laboratory, North American Science Center, Argonne National Laboratory, Ames Laboratory of Iowa State University, the Federal University of Pernambuco in Recife, Brazil, Sandia National Laboratory, and the Marshal Space Flight Center. Dr. P. Richards of Sandia, and Dr. S.L. Lehoczky of Marshall, were particularly helpful to JDP. Brief, but very pithy conversations of JDP with Dr. M. L. Cohen of the University of Califonia/ Berkeley, over the years, have also been uncommonly useful.

Dr. B.C. Bailey would like particularly to thank Drs. J. Burns and J. Blatt for the many years of academic preparation, mentorship, and care they provided at Florida Institute of Technology. A special thanks to Dr. J.D. Patterson who, while Physics Department Head at Florida Institute of Technology, made a conscious decision to take on a coauthor for this extraordinary project.

All mistakes, misconceptions and failures to communicate ideas are our own. No doubt some sign errors, misprints, incorrect shading of meanings, and perhaps more serious errors have crept in, but hopefully their frequency decreases with their gravity.

Most of the figures, for the first version of this book, were prepared in preliminary form by Mr. R.F. Thomas. However, for this book, the figures are either new or reworked by the coauthor (BCB).

We gratefully acknowledge the cooperation and kind support of Dr. C. Asheron, Ms. E. Sauer, and Ms. A. Duhm of Springer. Finally, and most importantly, JDP would like to note that without the constant encouragement and patience of his wife Marluce, this book would never have been completed.

J.D. Patterson, Rapid City, South Dakota
B.C. Bailey, Cape Canaveral, Florida

October 2005

Contents

1 Crystal Binding and Structure

It has been argued that solid-state physics was born, as a separate field, with the publication, in 1940, of Fredrick Seitz's book, *Modern Theory of Solids* [82]. In that book parts of many fields such as metallurgy, crystallography, magnetism, and electronic conduction in solids were in a sense coalesced into the new field of solid-state physics. About twenty years later, the term condensed-matter physics, which included the solid-state but also discussed liquids and related topics, gained prominent usage (see, e.g., Chaikin and Lubensky [26]). In this book we will focus on the traditional topics of solid-state physics, but particularly in the last chapter consider also some more general areas. The term "solid-state" is often restricted to mean only crystalline (periodic) materials. However, we will also consider, at least briefly, amorphous solids (e.g., glass that is sometimes called a supercooled viscous liquid),[1] as well as liquid crystals, something about polymers, and other aspects of a new subfield that has come to be called soft condensed-matter physics (see Chap. 12).

The physical definition of a solid has several ingredients. We start by defining a solid as a large collection (of the order of Avogadro's number) of atoms that attract one another so as to confine the atoms to a definite volume of space. Additionally, in this chapter, the term *solid* will mostly be restricted to crystalline solids. A *crystalline solid* is a material whose atoms have a regular arrangement that exhibits translational symmetry. The exact meaning of translational symmetry will be given in Sect. 1.2.2. When we say that the atoms have a regular arrangement, what we mean is that the equilibrium positions of the atoms have a regular arrangement. At any given temperature, the atoms may vibrate with small amplitudes about fixed equilibrium positions. For the most part, we will discuss only perfect crystalline solids, but defects will be considered later in Chap. 11.

Elements form solids because for some range of temperature and pressure, a solid has less free energy than other states of matter. It is generally supposed that at low enough temperature and with suitable external pressure (helium requires external pressure to solidify) everything becomes a solid. No one has ever proved that this must happen. We cannot, in general, prove from first principles that the crystalline state is the lowest free-energy state.

---

[1] The viscosity of glass is typically greater than 10^{13} poise and it is disordered.

P.W. Anderson has made the point[2] that just because a solid is complex does not mean the study of solids is less basic than other areas of physics. More is different. For example, crystalline symmetry, perhaps the most important property discussed in this book, cannot be understood by considering only a single atom or molecule. It is an emergent property at a higher level of complexity. Many other examples of emergent properties will be discussed as the topics of this book are elaborated.

The goal of this chapter is three-fold. All three parts will help to define the universe of crystalline solids. We start by discussing why solids form (the binding), then we exhibit how they bind together (their symmetries and crystal structure), and finally we describe one way we can experimentally determine their structure (X-rays).

Section 1.1 is concerned with chemical bonding. There are approximately four different forms of bonds. A bond in an actual crystal may be predominantly of one type and still show characteristics related to others, and there is really no sharp separation between the types of bonds.

1.1 Classification of Solids by Binding Forces (B)

A complete discussion of crystal binding cannot be given this early because it depends in an essential way on the electronic structure of the solid. In this Section, we merely hope to make the reader believe that it is not unreasonable for atoms to bind themselves into solids.

1.1.1 Molecular Crystals and the van der Waals Forces (B)

Examples of molecular crystals are crystals formed by nitrogen (N_2) and rare-gas crystals formed by argon (Ar). Molecular crystals consist of chemically inert atoms (atoms with a rare-gas electronic configuration) or chemically inert molecules (neutral molecules that have little or no affinity for adding or sharing additional electrons and that have affinity for the electrons already within the molecule). We shall call such atoms or molecules *chemically saturated units.* These interact weakly, and therefore their interaction can be treated by quantum-mechanical perturbation theory.

The interaction between chemically saturated units is described by the van der Waals forces. Quantum mechanics describes these forces as being due to correlations in the fluctuating distributions of charge on the chemically saturated units. The appearance of virtual excited states causes transitory dipole moments to appear on adjacent atoms, and if these dipole moments have the right directions, then the atoms can be attracted to one another. The quantum-mechanical description of these forces is discussed in more detail in the example below. The van der

[2] See Anderson [1.1].

Waals forces are weak, short-range forces, and hence molecular crystals are characterized by low melting and boiling points. The forces in molecular crystals are almost central forces (central forces act along a line joining the atoms), and they make efficient use of their binding in close-packed crystal structures. However, the force between two atoms is somewhat changed by bringing up a third atom (i.e. the van der Waals forces are not exactly two-body forces). We should mention that there is also a repulsive force that keeps the lattice from collapsing. This force is similar to the repulsive force for ionic crystals that is discussed in the next Section. A sketch of the interatomic potential energy (including the contributions from the van der Waals forces and repulsive forces) is shown in Fig. 1.1.

A relatively simple model [14, p. 438] that gives a qualitative feeling for the nature of the van der Waals forces consists of two one-dimensional harmonic oscillators separated by a distance R (see Fig. 1.2). Each oscillator is electrically neutral, but has a time-varying electric dipole moment caused by a fixed $+e$ charge and a vibrating $-e$ charge that vibrates along a line joining the two oscillators. The displacements from equilibrium of the $-e$ charges are labeled d_1 and d_2. When $d_i = 0$, the $-e$ charges will be assumed to be separated exactly by the distance R. Each charge has a mass M, a momentum P_i, and hence a kinetic energy $P_i^2/2M$.

The spring constant for each charge will be denoted by k and hence each oscillator will have a potential energy $kd_i^2/2$. There will also be a Coulomb coupling energy between the two oscillators. We shall neglect the interaction between the $-e$ and the $+e$ charges on the same oscillator. This is not necessarily physically reasonable. It is just the way we choose to build our model. The attraction between these charges is taken care of by the spring.

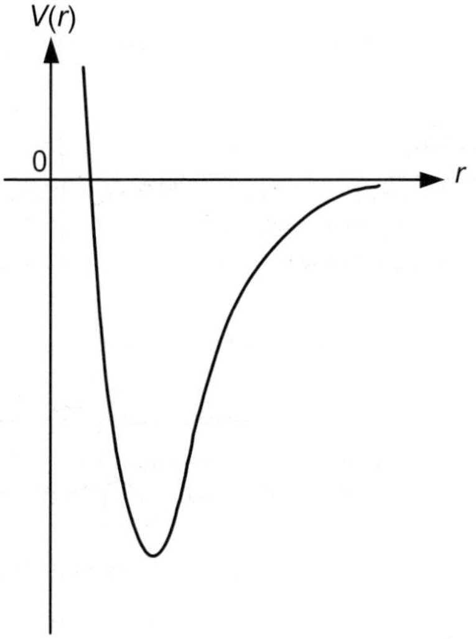

Fig. 1.1. The interatomic potential $V(r)$ of a rare-gas crystal. The interatomic spacing is r

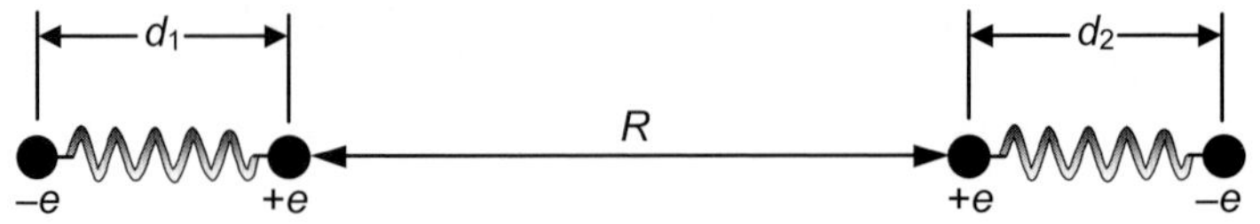

Fig. 1.2. Simple model for the van der Waals forces

The total energy of the vibrating dipoles may be written

$$E = \frac{1}{2M}(P_1^2 + P_2^2) + \frac{1}{2}k(d_1^2 + d_2^2) + \frac{e^2}{4\pi\varepsilon_0(R + d_1 + d_2)} + \frac{e^2}{4\pi\varepsilon_0 R} - \frac{e^2}{4\pi\varepsilon_0(R + d_1)} - \frac{e^2}{4\pi\varepsilon_0(R + d_2)}, \quad (1.1)$$

where ε_0 is the permittivity of free space. In (1.1) and throughout this book for the most part, mks units are used (see Appendix A). Assuming that $R >> d$ and using

$$\frac{1}{1+\eta} \cong 1 - \eta + \eta^2, \quad (1.2)$$

if $|\eta| << 1$, we find a simplified form for (1.1):

$$E \cong \frac{1}{2M}(P_1^2 + P_2^2) + \frac{1}{2}k(d_1^2 + d_2^2) + \frac{2e^2 d_1 d_2}{4\pi\varepsilon_0 R^3}. \quad (1.3)$$

If there were no coupling term, (1.3) would just be the energy of two independent oscillators each with frequency (in radians per second)

$$\omega_0 = \sqrt{k/M}. \quad (1.4)$$

The coupling splits this single frequency into two frequencies that are slightly displaced (or alternatively, the coupling acts as a perturbation that removes a two-fold degeneracy).

By defining new coordinates (making a normal coordinate transformation) it is easily possible to find these two frequencies. We define

$$Y_+ = \frac{1}{\sqrt{2}}(d_1 + d_2), \quad Y_- = \frac{1}{\sqrt{2}}(d_1 - d_2),$$
$$P_+ = \frac{1}{\sqrt{2}}(P_1 + P_2), \quad P_- = \frac{1}{\sqrt{2}}(P_1 - P_2). \quad (1.5)$$

By use of this transformation, the energy of the two oscillators can be written

$$E \cong \left[\frac{1}{2M}P_+^2 + \left(\frac{k}{2} + \frac{e^2}{4\pi\varepsilon_0 R^3}\right)Y_+^2\right] + \left[\frac{1}{2M}P_-^2 + \left(\frac{k}{2} - \frac{e^2}{4\pi\varepsilon_0 R^3}\right)Y_-^2\right]. \quad (1.6)$$

Note that (1.6) is just the energy of two uncoupled harmonic oscillators with frequencies ω_+ and ω_- given by

$$\omega_{\pm} = \sqrt{\frac{1}{M}\left(k \pm \frac{e^2}{2\pi\varepsilon_0 R^3}\right)} . \tag{1.7}$$

The lowest possible quantum-mechanical energy of this system is the zero-point energy given by

$$E \cong \frac{\hbar}{2}(\omega_+ + \omega_-) , \tag{1.8}$$

where $\hbar$ is Planck's constant divided by 2π.

A more instructive form for the ground-state energy is obtained by making an assumption that brings a little more physics into the model. The elastic restoring force should be of the same order of magnitude as the Coulomb forces so that

$$\frac{e^2}{4\pi\varepsilon_0 R^2} \cong kd_i .$$

This expression can be cast into the form

$$\frac{e^2}{4\pi\varepsilon_0 R^3}\frac{R}{d_i} \cong k .$$

It has already been assumed that $R >> d_i$ so that the above implies $e^2/4\pi\varepsilon_0 R^3 << k$. Combining this last inequality with (1.7), making an obvious expansion of the square root, and combining the result with (1.8), one readily finds for the approximate ground-state energy

$$E \cong \hbar\omega_0(1 - C/R^6) , \tag{1.9}$$

where

$$C = \frac{e^4}{32\pi^2 k^2 \varepsilon_0^2} .$$

From (1.9), the additional energy due to coupling is approximately $-C\hbar\omega_0/R^6$. The negative sign tells us that the two dipoles attract each other. The R^{-6} tells us that the attractive force (proportional to the gradient of energy) is an inverse seventh power force. This is a short-range force. Note that without the quantum-mechanical zero-point energy (which one can think of as arising from the uncertainty principle) there would be no binding (at least in this simple model).

While this model gives one a useful picture of the van der Waals forces, it is only qualitative because for real solids:

1. More than one dimension must be considered,
2. The binding of electrons is not a harmonic oscillator binding, and
3. The approximation $R >> d$ (or its analog) is not well satisfied.
4. In addition, due to overlap of the core wave functions and the Pauli principle there is a repulsive force (often modeled with an R^{-12} potential). The totality of R^{-12} linearly combined with the $-R^{-6}$ attraction is called a Lennard–Jones potential.

1.1.2 Ionic Crystals and Born–Mayer Theory (B)

Examples of ionic crystals are sodium chloride (NaCl) and lithium fluoride (LiF). Ionic crystals also consist of chemically saturated units (the ions that form their basic units are in rare-gas configurations). The ionic bond is due mostly to Coulomb attractions, but there must be a repulsive contribution to prevent the lattice from collapsing. The Coulomb attraction is easily understood from an electron-transfer point of view. For example, we view LiF as composed of $Li^{+}(1s^2)$ and $F^{-}(1s^2 2s^2 2p^6)$, using the usual notation for configuration of electrons. It requires about one electron volt of energy to transfer the electron, but this energy is more than compensated by the energy produced by the Coulomb attraction of the charged ions. In general, alkali and halogen atoms bind as singly charged ions. The core repulsion between the ions is due to an overlapping of electron clouds (as constrained by the Pauli principle).

Since the Coulomb forces of attraction are strong, long-range, nearly two-body, central forces, ionic crystals are characterized by close packing and rather tight binding. These crystals also show good ionic conductivity at high temperatures, good cleavage, and strong infrared absorption.

A good description of both the attractive and repulsive aspects of the ionic bond is provided by the semi-empirical theory due to Born and Mayer. To describe this theory, we will need a picture of an ionic crystal such as NaCl. NaCl-like crystals are composed of stacked planes, similar to the plane in Fig. 1.3. The theory below will be valid only for ionic crystals that have the same structure as NaCl.

Let N be the number of positive or negative ions. Let $\boldsymbol{r}_{ij}$ (a symbol in boldface type means a vector quantity) be the vector connecting ions i and j so that $|\boldsymbol{r}_{ij}|$ is the distance between ions i and j. Let E_{ij} be $(+1)$ if the i and j ions have the same signs and (-1) if the i and j ions have opposite signs. With this notation the potential energy of ion i is given by

$$U_i = \sum\nolimits_{\text{all } j(\neq i)} E_{ij} \frac{e^2}{4\pi\varepsilon_0 |\boldsymbol{r}_{ij}|}, \tag{1.10}$$

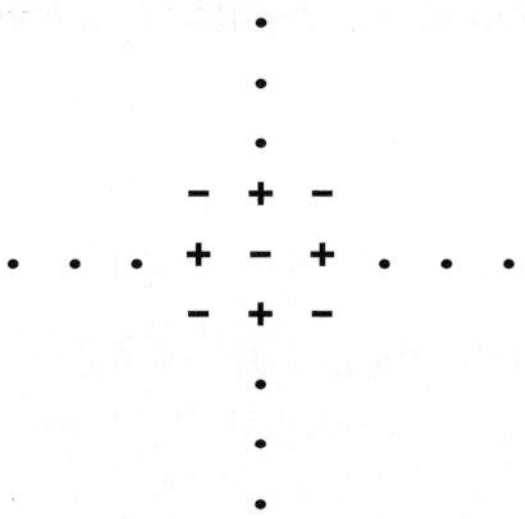

Fig. 1.3. NaCl-like ionic crystals

where e is, of course, the magnitude of the charge on any ion. For the whole crystal, the total potential energy is $U = NU_i$. If N_1, N_2 and N_3 are integers, and a is the distance between adjacent positive and negative ions, then (1.10) can be written as

$$U_i = {\sum}'_{(N_1,N_2,N_3)} \frac{(-)^{N_1+N_2+N_3}}{\sqrt{N_1^2+N_2^2+N_3^2}} \frac{e^2}{4\pi\varepsilon_0 a} . \tag{1.11}$$

In (1.11), the term $N_1 = 0$, $N_2 = 0$, and $N_3 = 0$ is omitted (this is what the prime on the sum means). If we assume that the lattice is almost infinite, the N_i in (1.11) can be summed over an infinite range. The result for the total Coulomb potential energy is

$$U = -N \frac{M_{\mathrm{NaCl}} e^2}{4\pi\varepsilon_0 a} , \tag{1.12}$$

where

$$M_{\mathrm{NaCl}} = -{\sum}'^{\infty}_{N_1,N_2,N_3=-\infty} \frac{(-)^{N_1+N_2+N_3}}{\sqrt{N_1^2+N_2^2+N_3^2}} \tag{1.13}$$

is called the Madelung constant for a NaCl-type lattice. Evaluation of (1.13) yields $M_{\mathrm{NaCl}} = 1.7476$. The value for M depends only on geometrical arrangements. The series for M given by (1.13) is very slowly converging. Special techniques are usually used to obtain good results [46].

As already mentioned, the stability of the lattice requires a repulsive potential, and hence a repulsive potential energy. Quantum mechanics suggests (basically from the Pauli principle) that the form of this repulsive potential energy between ions i and j is

$$U_{ij}^R = X_{ij} \exp\left(-\frac{|\boldsymbol{r}_{ij}|}{R_{ij}} \right) , \tag{1.14}$$

where X_{ij} and R_{ij} depend, as indicated, on the pair of ions labeled by i and j. "Common sense" suggests that the repulsion be of short-range. In fact, one usually assumes that only nearest-neighbor repulsive interactions need be considered. There are six nearest neighbors for each ion, so that the total repulsive potential energy is

$$U^R = 6NX\exp(-a/R)\,. \tag{1.15}$$

This usually amounts to only about 10% of the magnitude of the total cohesive energy. In (1.15), X_{ij} and R_{ij} are assumed to be the same for all six interactions (and equal to the X and R). That this should be so is easily seen by symmetry.

Combining the above, we have for the total potential energy for the lattice

$$U = N\left(-\frac{\mathrm{M}_{\mathrm{NaCl}}e^2}{4\pi\varepsilon_0 a}\right) + 6NX\exp\left(-\frac{a}{R}\right). \tag{1.16}$$

The cohesive energy for free ions equals U plus the kinetic energy of the ions in the solid. However, the magnitude of the kinetic energy of the ions (especially at low temperature) is much smaller than U, and so we simply use U in our computations of the cohesive energy. Even if we refer U to zero temperature, there would be, however, a small correction due to zero-point motion. In addition, we have neglected a very weak attraction due to the van der Waals forces.

Equation (1.16) shows that the Born–Mayer theory is a two-parameter theory. Certain thermodynamic considerations are needed to see how to feed in the results of experiment.

The combined first and second laws for reversible processes is

$$T\mathrm{d}S = \mathrm{d}U + p\,\mathrm{d}V\,, \tag{1.17}$$

where S is the entropy, U is the internal energy, p is the pressure, V is the volume, and T is the temperature. We want to derive an expression for the isothermal compressibility k that is defined by

$$\frac{1}{kV} = -\left(\frac{\partial p}{\partial V}\right)_T. \tag{1.18}$$

The isothermal compressibility is not very sensitive to temperature, so we will evaluate k for $T = 0$. Combining (1.17) and (1.18) at $T = 0$, we obtain

$$\left(\frac{1}{kV}\right)_{T=0} = \left(\frac{\partial^2 U}{\partial V^2}\right)_{T=0}. \tag{1.19}$$

There is one more relationship between R, X, and experiment. At the equilibrium spacing $a = A$ (determined by experiment using X-rays), there must be no net force on an ion so that

$$\left(\frac{\partial U}{\partial a}\right)_{a=A} = 0\,. \tag{1.20}$$

Thus, a measurement of the compressibility and the lattice constant serves to fix the two parameters *R* and *X*. When we know *R* and *X*, it is possible to give a theoretical value for the cohesive energy per molecule (*U*/*N*). This quantity can also be independently measured by the Born–Haber cycle [46].[3] Comparing these two quantities gives a measure of the accuracy of the Born–Mayer theory. Table 1.1 shows that the Born–Mayer theory gives a good estimate of the cohesive energy. (For some types of complex solid-state calculations, an accuracy of 10 to 20% can be achieved.)

Table 1.1. Cohesive energy in kcal mole^{-1}

Solid	Born–Mayer theory	Experiment
LiCl	196.3	201.5
NaCl	182.0	184.7
NaBr	172.7	175.9
NaI	159.3	166.3

Adapted from Born M and Huang K, *Dynamical Theory of Crystal Lattices,* selected parts of Table 9 (p.26) Clarendon Press, Oxford, 1954. By permission of Oxford University Press.

1.1.3 Metals and Wigner–Seitz Theory (B)

Examples of metals are sodium (Na) and copper (Cu). A metal such as Na is viewed as being composed of positive ion cores (Na^+) immersed in a "sea" of free conduction electrons that come from the removal of the 3s electron from atomic Na. Metallic binding can be partly understood within the context of the Wigner-Seitz theory. In a full treatment, it would be necessary to confront the problem of electrons in a periodic lattice. (A discussion of the Wigner–Seitz theory will be deferred until Chap. 3.) One reason for the binding is the lowering of the kinetic energy of the "free" electrons relative to their kinetic energy in the atomic 3s state [41]. In a metallic crystal, the valence electrons are free (within the constraints of the Pauli principle) to wander throughout the crystal, causing them to have a smoother wave function and hence less $\nabla^2\psi$. Generally speaking this spreading of the electrons wave function also allows the electrons to make better use of the attractive potential. Lowering of the kinetic and/or potential

[3] The Born–Haber cycle starts with (say) NaCl solid. Let *U* be the energy needed to break this up into Na^+ gas and Cl^- gas. Suppose it takes E_F units of energy to go from Cl^- gas to Cl gas plus electrons, and E_I units of energy are gained in going from Na^+ gas plus electrons to Na gas. The Na gas gives up heat of sublimation energy *S* in going to Na solid, and the Cl gas gives up heat of dissociation *D* in going to Cl_2 gas. Finally, let the Na solid and Cl_2 gas go back to NaCl solid in its original state with a resultant energy *W*. We are back where we started and so the energies must add to zero: $U - E_I + E_F - S - D - W = 0$. This equation can be used to determine *U* from other experimental quantities.

energy implies binding. However, the electron–electron Coulomb repulsions cannot be neglected (see, e.g., Sect. 3.1.4), and the whole subject of binding in metals is not on so good a quantitative basis as it is in crystals involving the interactions of atoms or molecules which do not have free electrons. One reason why the metallic crystal is prevented from collapsing is the kinetic energy of the electrons. Compressing the solid causes the wave functions of the electrons to "wiggle" more and hence raises their kinetic energy.

A very simple picture[4] suffices to give part of the idea of metallic binding. The ground-state energy of an electron of mass M in a box of volume V is [19]

$$E = \frac{\hbar^2 \pi^2}{2M} V^{-2/3} .$$

Thus the energy of N electrons in N separate boxes is

$$E_A = N \frac{\hbar^2 \pi^2}{2M} V^{-2/3} . \tag{1.21}$$

The energy of N electrons in a box of volume NV is (neglecting electron–electron interaction that would tend to increase the energy)

$$E_M = N \frac{\hbar^2 \pi^2}{2M} V^{-2/3} N^{-2/3} . \tag{1.22}$$

Therefore $E_M/E_A = N^{-2/3} << 1$ for large N and hence the total energy is lowered considerably by letting the electrons spread out. This model of binding is, of course, not adequate for a real metal, since the model neglects not only electron–electron interactions but also the potential energy of interaction between electrons and ions and between ions and other ions. It also ignores the fact that electrons fill up states by satisfying the Pauli principle. That is, they fill up in increasing energy. But it does clearly show how the energy can be lowered by allowing the electronic wave functions to spread out.

In modern times, considerable progress has been made in understanding the cohesion of metals by the density functional method, see Chap. 3. We mention in particular, Daw [1.6].

Due to the important role of the free electrons in binding, metals are good electrical and thermal conductors. They have moderate to fairly strong binding. We do not think of the binding forces in metals as being two-body, central, or short-range.

---

4 A much more sophisticated approach to the binding of metals is contained in the pedagogical article by Tran and Perdew [1.26]. This article shows how exchange and correlation effects are important and discusses modern density functional methods (see Chap. 3).

1.1.4 Valence Crystals and Heitler–London Theory (B)

An example of a valence crystal is carbon in diamond form. One can think of the whole valence crystal as being a huge chemically saturated molecule. As in the case of metals, it is not possible to understand completely the binding of valence crystals without considerable quantum-mechanical calculations, and even then the results are likely to be only qualitative. The quantum-mechanical considerations (Heitler–London theory) will be deferred until Chap. 3.

Some insight into covalent bonds (also called homopolar bonds) of valence crystals can be gained by considering them as being caused by sharing electrons between atoms with unfilled shells. Sharing of electrons can lower the energy because the electrons can get into lower energy states without violating the Pauli principle. In carbon, each atom has four electrons that participate in the valence bond. These are the electrons in the 2s2p shell, which has eight available states.[5] The idea of the valence bond in carbon is (*very* schematically) indicated in Fig. 1.4. In this figure each line symbolizes an electron bond. The idea that the eight 2s2p states participate in the valence bond is related to the fact that we have drawn each carbon atom with eight bonds.

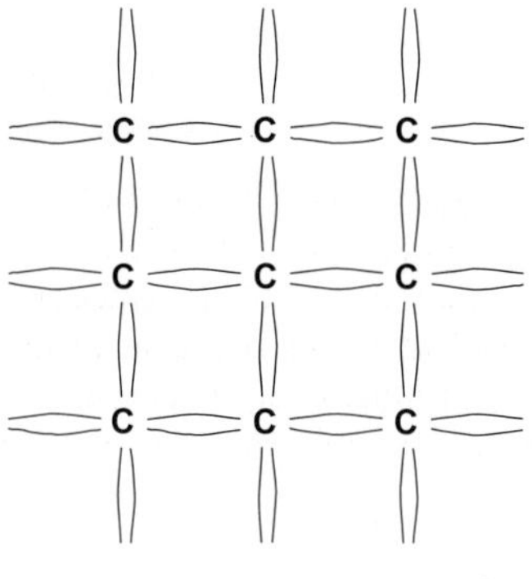

Fig. 1.4. The valence bond of diamond

Valence crystals are characterized by hardness, poor cleavage, strong bonds, poor electronic conductivity, and poor ionic conductivity. The forces in covalent bonds can be thought of as short-range, two-body, but not central forces. The covalent bond is very directional, and the crystals tend to be loosely packed.

[5] More accurately, one thinks of the electron states as being combinations formed from s and p states to form sp^3 *hybrids*. A very simple discussion of this process as well as the details of other types of bonds is given by Moffatt et al [1.17].

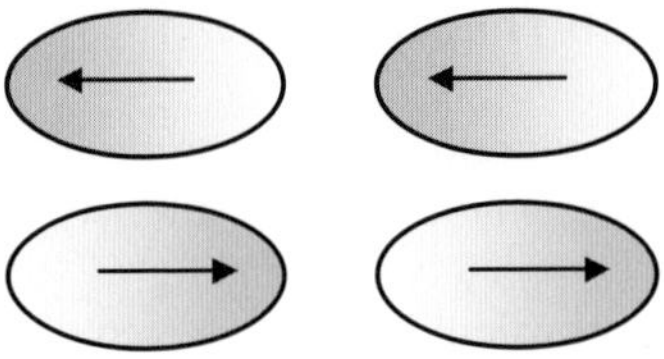

Molecular crystals are bound by the van der Waals forces caused by fluctuating dipoles in each molecule. A "snap-shot" of the fluctuations.
Example: argon

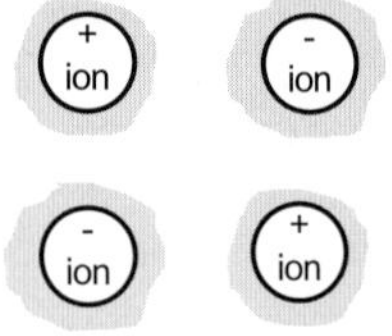

Ionic crystals are bound by ionic forces as described by the Born–Mayer theory.
Example: NaCl

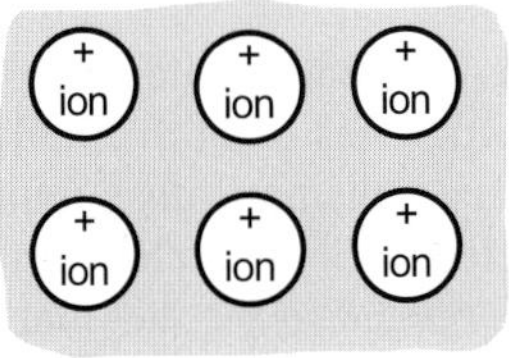

Metallic crystalline binding is described by quantum-mechanical means. One simple theory which does this is the Wigner–Seitz theory.
Example: sodium

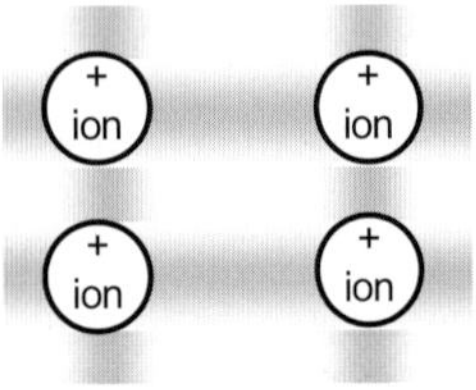

Valence crystalline binding is described by quantum-mechanical means. One simple theory that does this is the Heitler–London theory.
Example: carbon in diamond form

Fig. 1.5. Schematic view of the four major types of crystal bonds. All binding is due to the Coulomb forces and quantum mechanics is needed for a complete description, but some idea of the binding of molecular and ionic crystals can be given without quantum mechanics. The density of electrons is indicated by the shading. Note that the outer atomic electrons are progressively smeared out as one goes from an ionic crystal to a valence crystal to a metal

1.1.5 Comment on Hydrogen-Bonded Crystals (B)

Many authors prefer to add a fifth classification of crystal bonding: hydrogen-bonded crystals [1.18]. The hydrogen bond is a bond between two atoms due to the presence of a hydrogen atom between them. Its main characteristics are caused by the small size of the proton of the hydrogen atom, the ease with which the electron of the hydrogen atom can be removed, and the mobility of the proton.

The presence of the hydrogen bond results in the possibility of high dielectric constant, and some hydrogen-bonded crystals become ferroelectric. A typical example of a crystal in which hydrogen bonds are important is ice. One generally thinks of hydrogen-bonded crystals as having fairly weak bonds. Since the hydrogen atom often loses its electron to one of the atoms in the hydrogen-bonded molecule, the hydrogen bond is considered to be largely ionic in character. For this reason we have not made a separate classification for hydrogen-bonded crystals. Of course, other types of bonding may be important in the total binding together of a crystal with hydrogen bonds. Figure 1.5 schematically reviews the four major types of crystal bonds.

1.2 Group Theory and Crystallography

We start crystallography by giving a short history [1.14].

1. In 1669 Steno gave the law of constancy of angle between like crystal faces. This of course was a key idea needed to postulate there was some underlying microscopic symmetry inherent in crystals.
2. In 1784 Abbe Hauy proposed the idea of unit cells.
3. In 1826 Naumann originated the idea of 7 crystal systems.
4. In 1830 Hessel said there were 32 crystal classes because only 32 point groups were consistent with the idea of translational symmetry.
5. In 1845 Bravais noted there were only 14 distinct lattices, now called Bravais lattices, which were consistent with the 32 point groups.
6. By 1894 several groups had enumerated the 230 space groups consistent with only 230 distinct kinds of crystalline symmetry.
7. By 1912 von Laue started X-ray experiments that could delineate the space groups.
8. In 1936 Seitz started deriving the irreducible representations of the space groups.
9. In 1984 Shectmann, Steinhardt et al found quasi-crystals, substances that were neither crystalline nor glassy but nevertheless ordered in a quasi periodic way.

The symmetries of crystals determine many of their properties as well as simplify many calculations. To discuss the symmetry properties of solids, one needs an appropriate formalism. The most concise formalism for this is group theory. Group theory can actually provide deep insight into the classification by quantum numbers of quantum-mechanical states. However, we shall be interested at this stage in crystal symmetry. This means (among other things) that finite groups will be of interest, and this is a simplification. We will not use group theory to discuss crystal symmetry in this Section. However, it is convenient to introduce some group-theory notation in order to use the crystal symmetry operations as examples of groups and to help in organizing in one's mind the various sorts of symmetries that are presented to us by crystals. We will use some of the concepts (presented here) in parts of the chapter on magnetism (Chap. 7) and also in a derivation of Bloch's theorem in Appendix C.

1.2.1 Definition and Simple Properties of Groups (AB)

There are two basic ingredients of a group: a set of elements $G = \{g_1, g_2, \ldots\}$ and an operation ($*$) that can be used to combine the elements of the set. In order that the set form a group, there are four rules that must be satisfied by the operation of combining set elements:

1. *Closure*. If g_i and g_j are arbitrary elements of G, then

$$g_i * g_j \in G$$

   ($\in$ means "included in").

2. *Associative Law*. If g_i, g_j and g_k are arbitrary elements of G, then

$$(g_i * g_j) * g_k = g_i * (g_j * g_k) .$$

3. *Existence of the identity*. There must exist a $g_e \in G$ with the property that for any

$$g_k \in G , \quad g_e * g_k = g_k * g_e = g_k .$$

   Such a g_e is called E, the identity.

4. *Existence of the inverse*. For each $g_i \in G$ there exists a $g_i^{-1} \in G$ such that

$$g_i * g_i^{-1} = g_i^{-1} * g_i = E ,$$

   Where g_i^{-1} is called the inverse of g_i.
   From now on the $*$ will be omitted and $g_i * g_j$ will simply be written $g_i g_j$.

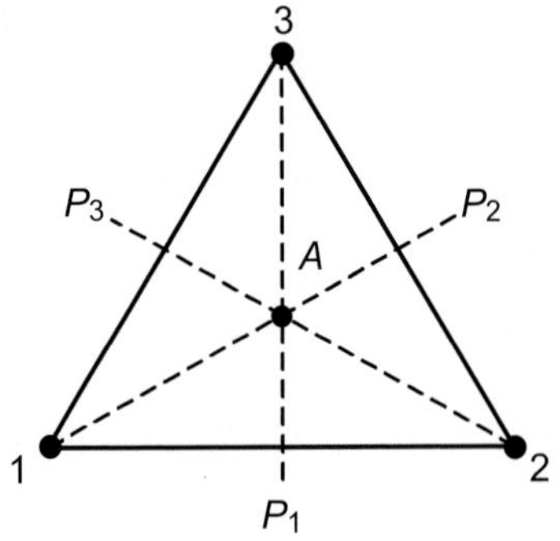

Fig. 1.6. The equilateral triangle

An example of a group that is small enough to be easily handled and yet large enough to have many features of interest is the group of rotations in three dimensions that bring the equilateral triangle into itself. This group, denoted by D_3, has six elements. One thus says its *order* is 6.

In Fig. 1.6, let A be an axis through the center of the triangle and perpendicular to the plane of the paper. Let g_1, g_2, and g_3 be rotations of 0, $2\pi/3$, and $4\pi/3$ about A. Let g_4, g_5, and g_6 be rotations of π about the axes P_1, P_2, and P_3. The group multiplication table of D_3 can now be constructed. See Table 1.2.

Table 1.2. Group multiplication table of D_3

D_3	g_1	g_2	g_3	g_4	g_5	g_6
g_1	g_1	g_2	g_3	g_4	g_5	g_6
g_2	g_2	g_3	g_1	g_6	g_4	g_5
g_3	g_3	g_1	g_2	g_5	g_6	g_4
g_4	g_4	g_5	g_6	g_1	g_2	g_3
g_5	g_5	g_6	g_4	g_3	g_1	g_2
g_6	g_6	g_4	g_5	g_2	g_3	g_1

The group elements can be very easily described by indicating how the vertices are mapped. Below, arrows are placed in the definition of g_1 to define the notation. After g_1, the arrows are omitted:

$$g_1 = \begin{pmatrix} 1 & 2 & 3 \\ \downarrow & \downarrow & \downarrow \\ 1 & 2 & 3 \end{pmatrix}, \quad g_2 = \begin{pmatrix} 1 & 2 & 3 \\ 2 & 3 & 1 \end{pmatrix}, \quad g_3 = \begin{pmatrix} 1 & 2 & 3 \\ 3 & 1 & 2 \end{pmatrix},$$
$$g_4 = \begin{pmatrix} 1 & 2 & 3 \\ 2 & 1 & 3 \end{pmatrix}, \quad g_5 = \begin{pmatrix} 1 & 2 & 3 \\ 1 & 3 & 2 \end{pmatrix}, \quad g_6 = \begin{pmatrix} 1 & 2 & 3 \\ 3 & 2 & 1 \end{pmatrix}.$$

Using this notation we can see why the group multiplication table indicates that $g_4 g_2 = g_5$:[6]

$$g_4 g_2 = \begin{pmatrix} 1 & 2 & 3 \\ 2 & 1 & 3 \end{pmatrix} \begin{pmatrix} 1 & 2 & 3 \\ 2 & 3 & 1 \end{pmatrix} = \begin{pmatrix} 1 & 2 & 3 \\ 1 & 3 & 2 \end{pmatrix} = g_5 .$$

The table also says that $g_2 g_4 = g_6$. Let us check this:

$$g_2 g_4 = \begin{pmatrix} 1 & 2 & 3 \\ 2 & 3 & 1 \end{pmatrix} \begin{pmatrix} 1 & 2 & 3 \\ 2 & 1 & 3 \end{pmatrix} = \begin{pmatrix} 1 & 2 & 3 \\ 3 & 2 & 1 \end{pmatrix} = g_6 .$$

In a similar way, the rest of the group multiplication table was easily derived.

Certain other definitions are worth noting [61]. A is a *proper subgroup* of G if A is a group contained in G and not equal to E (E is the identity that forms a trivial group of order 1) or G. In D_3, $\{g_1, g_2, g_3\}$, $\{g_1, g_4\}$, $\{g_1, g_5\}$, $\{g_1, g_6\}$ are proper subgroups. The *class* of an element $g \in G$ is the set of elements $\{g_i^{-1} g g_i\}$ for all $g_i \in G$. Mathematically this can be written for $g \in G$, $Cl(g) = \{g_i^{-1} g g_i |$ for all $g_i \in G\}$.

[6] Note that the application starts on the right so $3 \to 1 \to 2$, for example.

Two operations belong to the same class if they perform the same sort of geometrical operation. For example, in the group D_3 there are three classes:

$$\{g_1\},\quad \{g_2, g_3\},\quad \text{and}\quad \{g_4, g_5, g_6\}.$$

Two very simple sorts of groups are often encountered. One of these is the *cyclic* group. A cyclic group can be generated by a single element. That is, in a cyclic group there exists a $g \in G$, such that all $g_k \in G$ are given by $g_k = g^k$ (of course one must name the group elements suitably). For a cyclic group of order N with *generator* g, $g^N \equiv E$. Incidentally, the *order* of a group element is the smallest power to which the element can be raised and still yield E. Thus the order of the generator (g) is N.

The other simple group is the *abelian* group. In the abelian group, the order of the elements is unimportant ($g_i g_j = g_j g_i$ for all g_i, $g_j \in G$). The elements are said to *commute*. Obviously all cyclic groups are abelian. The group D_3 is not abelian but all of its subgroups are.

In the abstract study of groups, all *isomorphic* groups are equivalent. Two groups are said to be isomorphic if there is a one-to-one correspondence between the elements of the group that preserves group "multiplication." Two isomorphic groups are identical except for notation. For example, the three subgroups of D_3 that are of order 2 are isomorphic.

An interesting theorem, called *Lagrange's theorem*, states that the order of a group divided by the order of a subgroup is always an integer. From this it can immediately be concluded that the only possible proper subgroups of D_3 have order 2 or 3. This, of course, checks with what we actually found for D_3.

Lagrange's theorem is proved by using the concept of a *coset*. If A is a subgroup of G, the *right cosets* are of the form Ag_i for all $g_i \in G$ (cosets with identical elements are not listed twice) – each g_i generates a coset. For example, the right cosets of $\{g_1, g_6\}$ are $\{g_1, g_6\}$, $\{g_2, g_4\}$, and $\{g_3, g_5\}$. A similar definition can be made of the term *left coset*.

A subgroup is *normal* or *invariant* if its right and left cosets are identical. In D_3, $\{g_1, g_2, g_3\}$ form a normal subgroup. The *factor group* of a normal subgroup is the normal subgroup plus all its cosets. In D_3, the factor group of $\{g_1, g_2, g_3\}$ has elements $\{g_1, g_2, g_3\}$ and $\{g_4, g_5, g_6\}$. It can be shown that the order of the factor group is the order of the group divided by the order of the normal subgroup. The factor group forms a group under the operation of taking the *inner product*. The inner product of two sets is the set of all possible distinct products of the elements, taking one element from each set. For example, the inner product of $\{g_1, g_2, g_3\}$ and $\{g_4, g_5, g_6\}$ is $\{g_4, g_5, g_6\}$. The arrangement of the elements in each set does not matter.

It is often useful to form a larger group from two smaller groups by taking the direct product. Such a group is naturally enough called a *direct product group*. Let $G = \{g_1 \ldots g_n\}$ be a group of order n, and $H = \{h_1 \ldots h_m\}$ be a group of order m. Then the direct product $G \times H$ is the group formed by all products of the form $g_i h_j$. The order of the direct product group is nm. In making this definition, it has been assumed that the group operations of G and H are independent. When this is not so, the definition of the direct product group becomes more complicated (and less interesting – at least to the physicist). See Sect. 7.4.4 and Appendix C.

1.2.2 Examples of Solid-State Symmetry Properties (B)

All real crystals have defects (see Chap. 11) and in all crystals the atoms vibrate about their equilibrium positions. Let us define ideal crystals as real crystals in which these complications are not present. This chapter deals with ideal crystals. In particular we will neglect boundaries. In other words, we will assume that the crystals are infinite. Ideal crystals exhibit many types of symmetry, one of the most important of which is *translational* symmetry. Let m_1, m_2, and m_3 be arbitrary integers. A crystal is said to be translationally symmetric or periodic if there exist three linearly independent vectors $(\boldsymbol{a}_1, \boldsymbol{a}_2, \boldsymbol{a}_3)$ such that a translation by $m_1\boldsymbol{a}_1 + m_2\boldsymbol{a}_2 + m_3\boldsymbol{a}_3$ brings one back to an equivalent point in the crystal. We summarize several definitions and facts related to the $\boldsymbol{a}_i$:

1. The $\boldsymbol{a}_i$ are called basis vectors. Usually, they are not orthogonal.
2. The set $(\boldsymbol{a}_1, \boldsymbol{a}_2, \boldsymbol{a}_3)$ is not unique. Any linear combination with integer coefficients gives another set.
3. By parallel extensions, the $\boldsymbol{a}_i$ form a parallelepiped whose volume is $V = \boldsymbol{a}_1 \cdot (\boldsymbol{a}_2 \times \boldsymbol{a}_3)$. This parallelepiped is called a unit cell.
4. Unit cells have two principal properties:
   a) It is possible by stacking unit cells to fill all space.
   b) Corresponding points in different unit cells are equivalent.
5. The smallest possible unit cells that satisfy properties a) and b) above are called *primitive cells* (primitive cells are not unique). The corresponding basis vectors $(\boldsymbol{a}_1, \boldsymbol{a}_2, \boldsymbol{a}_3)$ are then called *primitive translations*.
6. The set of all translations $\boldsymbol{T} = m_1\boldsymbol{a}_1 + m_2\boldsymbol{a}_2 + m_3\boldsymbol{a}_3$ form a group. The group is of infinite order, since the crystal is assumed to be infinite in size.[7]

The symmetry operations of a crystal are those operations that bring the crystal back onto itself. Translations are one example of this sort of operation. One can find other examples by realizing that any operation that maps three noncoplanar points on equivalent points will map the whole crystal back on itself. Other types of symmetry transformations are *rotations* and *reflections*. These transformations are called *point* transformations because they leave at least one point fixed. For example, D_3 is a *point group* because all its operations leave the center of the equilateral triangle fixed.

---

[7] One can get around the requirement of having an infinite crystal and still preserve translational symmetry by using periodic boundary conditions. These will be described later.

We say we have an axis of symmetry of the nth order if a rotation by $2\pi/n$ about the axis maps the body back onto itself. C_n is often used as a symbol to represent the $2\pi/n$ rotations about a given axis. Note that $(C_n)^n = C_1 = E$, the identity.

A unit cell is mapped onto itself when reflected in a plane of reflection symmetry. The operation of reflecting in a plane is called σ. Note that $\sigma^2 = E$.

Another symmetry element that unit cells may have is a *rotary reflection* axis. If a body is mapped onto itself by a rotation of $2\pi/n$ about an axis and a simultaneous reflection through a plane normal to this axis, then the body has a rotary reflection axis of nth order.

If $f(x, y, z)$ is any function of the Cartesian coordinates (x, y, z), then the *inversion* I through the origin is defined by $I[f(x, y, z)] = f(-x, -y, -z)$. If $f(x, y, z) = f(-x, -y, -z)$, then the origin is said to be a *center of symmetry for f*. Denote an nth order rotary reflection by S_n, a reflection in a plane perpendicular to the axis of the rotary reflection by σ_h, and the operation of rotating $2\pi/n$ about the axis by C_n. Then $S_n = C_n\sigma_h$. In particular, $S_2 = C_2\sigma_h = I$. A second-order *rotary reflection* is the same as an inversion.

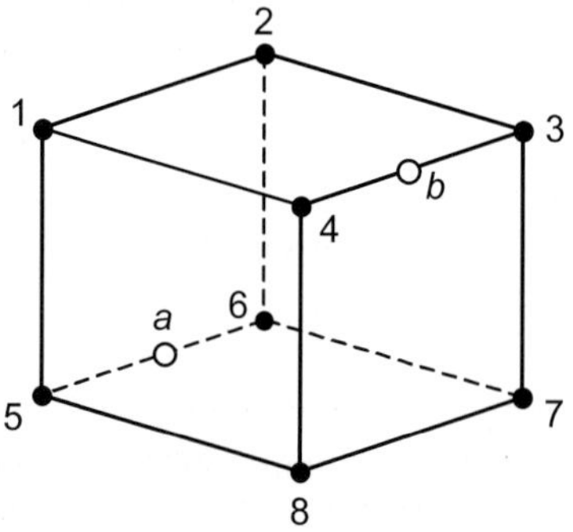

Fig. 1.7. The cubic unit cell

To illustrate some of the point symmetry operations, use will be made of the example of the unit cell being a cube. The cubic unit cell is shown in Fig. 1.7. It is obvious from the figure that the cube has rotational symmetry. For example,

$$C_2 = \begin{pmatrix} 1 & 2 & 3 & 4 & 5 & 6 & 7 & 8 \\ 8 & 7 & 6 & 5 & 4 & 3 & 2 & 1 \end{pmatrix}$$

obviously maps the cube back on itself. The rotation represented by C_2 is about a horizontal axis. There are two other axes that also show two-fold symmetry. It turns out that all three rotations belong to the same class (in the mathematical sense already defined) of the 48-element cubic point group O_h (the group of operations that leave the center point of the cube fixed and otherwise map the cube onto itself or leave the figure invariant).

The cube has many other rotational symmetry operations. There are six four-fold rotations that belong to the class of

$$C_4 = \begin{pmatrix} 1 & 2 & 3 & 4 & 5 & 6 & 7 & 8 \\ 4 & 3 & 7 & 8 & 1 & 2 & 6 & 5 \end{pmatrix}.$$

There are six two-fold rotations that belong to the class of the π rotation about the axis *ab*. There are eight three-fold rotation elements that belong to the class of $2\pi/3$ rotations about the body diagonal. Counting the identity, $(1 + 3 + 6 + 6 + 8) = 24$ elements of the cubic point group have been listed.

It is possible to find the other 24 elements of the cubic point group by taking the product of the 24 rotation elements with the inversion element. For the cube,

$$I = \begin{pmatrix} 1 & 2 & 3 & 4 & 5 & 6 & 7 & 8 \\ 7 & 8 & 5 & 6 & 3 & 4 & 1 & 2 \end{pmatrix}.$$

The use of the inversion element on the cube also introduces the reflection symmetry. A mirror reflection can always be constructed from a rotation and an inversion. This can be seen explicitly for the cube by direct computation.

$$\begin{aligned} IC_2 &= \begin{pmatrix} 1 & 2 & 3 & 4 & 5 & 6 & 7 & 8 \\ 7 & 8 & 5 & 6 & 3 & 4 & 1 & 2 \end{pmatrix}\begin{pmatrix} 1 & 2 & 3 & 4 & 5 & 6 & 7 & 8 \\ 8 & 7 & 6 & 5 & 4 & 3 & 2 & 1 \end{pmatrix} \\ &= \begin{pmatrix} 1 & 2 & 3 & 4 & 5 & 6 & 7 & 8 \\ 2 & 1 & 4 & 3 & 6 & 5 & 8 & 7 \end{pmatrix} = \sigma_h. \end{aligned}$$

It has already been pointed out that rotations about equivalent axes belong to the same class. Perhaps it is worthwhile to make this statement somewhat more explicit. *If in the group there is an element that carries one axis into another, then rotations about the axes through the same angle belong to the same class.*

A crystalline solid may also contain symmetry elements that are not simply group products of its rotation, inversion, and translational symmetry elements. There are two possible types of symmetry of this type. One of these types is called a *screw-axis* symmetry, an example of which is shown in Fig. 1.8.

The symmetry operation (which maps each point on an equivalent point) for Fig. 1.8 is to simultaneously rotate by $2\pi/3$ and translate by *d*. In general a screw axis is the combination of a rotation about an axis with a displacement parallel to

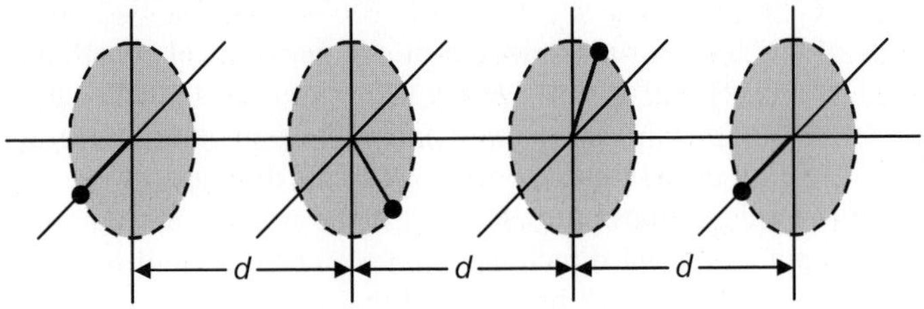

Fig. 1.8. Screw-axis symmetry

the axis. Suppose one has an n-fold screw axis with a displacement distance d. Let a be the smallest period (translational symmetry distance) in the direction of the axis. Then it is clear that $nd = pa$, where $p = 1, 2, \ldots, n - 1$. This is a restriction on the allowed types of screw-axis symmetry.

An example of *glide plane symmetry* is shown in Fig. 1.9. The line beneath the d represents a plane perpendicular to the page. The symmetry element for Fig. 1.9 is to simultaneously reflect through the plane and translate by d. In general, a glide plane is a reflection with a displacement parallel to the reflection plane. Let d be the translation operation involved in the glide-plane symmetry operation. Let a be the length of the period of the lattice in the direction of the translation. Only those glide-reflection planes are possible for which $2d = a$.

When one has a geometrical entity with several types of symmetry, the various symmetry elements must be consistent. For example, a three-fold axis cannot have only one mirror plane that contains it. The fact that we have a three-fold axis automatically requires that if we have one mirror plane that contains the axis, then we must have three such planes. The three-fold axis implies that every physical property must be repeated three times as one goes around the axis. A particularly interesting consistency condition is examined in the next Section.

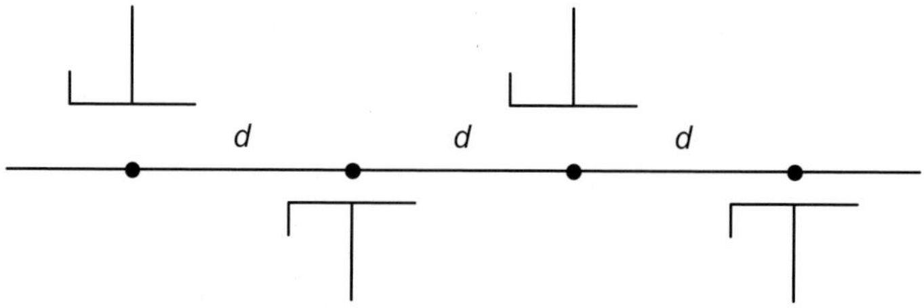

Fig. 1.9. Glide-plane symmetry

1.2.3 Theorem: No Five-fold Symmetry (B)

Any real crystal exhibits both translational and rotational symmetry. The mere fact that a crystal must have translational symmetry places restrictions on the types of rotational symmetry that one can have.

The theorem is:

A crystal can have only one-, two-, three-, four-, and six-fold axes of symmetry.

The proof of this theorem is facilitated by the geometrical construction shown in Fig. 1.10 [1.5, p. 32]. In Fig. 1.10, $\boldsymbol{R}$ is a vector drawn to a lattice point (one of the points defined by $m_1\boldsymbol{a}_1 + m_2\boldsymbol{a}_2 + m_3\boldsymbol{a}_3$), and $\boldsymbol{R}_1$ is another lattice point. $\boldsymbol{R}_1$ is chosen so as to be the closest lattice point to $\boldsymbol{R}$ in the direction of one of the translations in the (x,z)-plane; thus $|\boldsymbol{a}| = |\boldsymbol{R} - \boldsymbol{R}_1|$ is the minimum separation distance between lattice points in that direction. The coordinate system is chosen so that the z-axis is parallel to $\boldsymbol{a}$. It will be assumed that a line parallel to the y-axis and passing through the lattice point defined by $\boldsymbol{R}$ is an n-fold axis of symmetry.

Strictly speaking, one would need to prove one can always find a lattice plane perpendicular to an n-fold axis. Another way to look at it is that our argument is really in two dimensions, but one can show that three-dimensional Bravais lattices do not exist unless two-dimensional ones do. These points are discussed by Ashcroft and Mermin in two problems [21, p. 129]. Since all lattice points are equivalent, there must be a similar axis through the tip of $\boldsymbol{R}_1$. If $\theta = 2\pi/n$, then a counterclockwise rotation of $\boldsymbol{a}$ about $\boldsymbol{R}$ by θ produces a new lattice vector $\boldsymbol{R}^r$. Similarly a clockwise rotation by the same angle of **a** about $\boldsymbol{R}_1$ produces a new lattice point $\boldsymbol{R}_1{}^r$. From Fig. 1.10, $\boldsymbol{R}^r - \boldsymbol{R}_1{}^r$ is parallel to the z-axis and $\boldsymbol{R}^r - \boldsymbol{R}_1{}^r = p|\boldsymbol{a}|$. Further, $|p\boldsymbol{a}| = |\boldsymbol{a}| + 2|\boldsymbol{a}| \sin(\theta - \pi/2) = |\boldsymbol{a}|\,(1 - 2\cos\theta)$. Therefore $p = 1 - 2\cos\theta$ or $|\cos\theta| = |(p-1)/2| \le 1$. This equation can be satisfied only for $p = 3, 2, 1, 0, -1$ or $\theta = \pm(2\pi/1, 2\pi/2, 2\pi/3, 2\pi/4, 2\pi/6)$. This is the result that was to be proved.

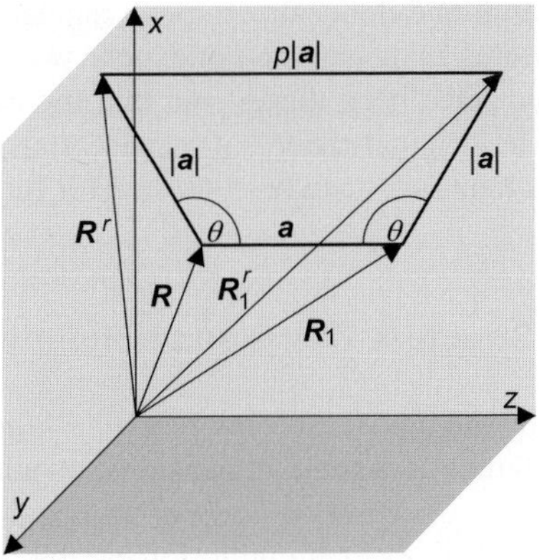

Fig. 1.10. The impossibility of five-fold symmetry. All vectors are in the (x,z)-plane

The requirement of translational symmetry and symmetry about a point, when combined with the formalism of group theory (or other appropriate means), allows one to classify all possible symmetry types of solids. Deriving all the results is far beyond the scope of this chapter. For details, the book by Buerger [1.5] can be consulted. The following sections (1.2.4 and following) give some of the results of this analysis.

Quasiperiodic Crystals or Quasicrystals (A)

These materials represented a surprise. When they were discovered in 1984, crystallography was supposed to be a long dead field, at least for new fundamental results. We have just proved a fundamental theorem for crystalline materials that forbids, among other symmetries, a 5-fold one. In 1984, materials that showed relatively sharp Bragg peaks and that had 5-fold symmetry were discovered. It was soon realized that the tacit assumption that the presence of Bragg peaks implied crystalline structure was false.

It is true that purely crystalline materials, which by definition have translational periodicity, cannot have 5-fold symmetry and will have sharp Bragg peaks. However, quasicrystals that are not crystalline, that is not translationally periodic, can have perfect (that is well-defined) long-range order. This can occur, for example, by having a symmetry that arises from the sum of noncommensurate periodic functions, and such materials will have sharp (although perhaps dense) Bragg peaks (see Problems 1.10 and 1.12). If the amplitude of most peaks is very small the denseness of the peaks does not obscure a finite number of diffraction peaks being observed. Quasiperiodic crystals will also have a long-range orientational order that may be 5-fold.

The first quasicrystals that were discovered (Shechtman and coworkers)[8] were grains of AlMn intermetallic alloys with icosahedral symmetry (which has 5-fold axes). An icosahedron is one of the five regular polyhedrons (the others being tetrahedron, cube, octahedron and dodecahedron). A regular polyhedron has identical faces (triangles, squares or pentagons) and only two faces meet at an edge. Other quasicrystals have since been discovered that include AlCuCo alloys with decagonal symmetry. The original theory of quasicrystals is attributed to Levine and Steinhardt.[9] The book by Janot can be consulted for further details [1.12].

1.2.4 Some Crystal Structure Terms and Nonderived Facts (B)

A set of points defined by the tips of the vectors $m_1\boldsymbol{a}_1 + m_2\boldsymbol{a}_2 + m_3\boldsymbol{a}_3$ is called a *lattice*. In other words, a lattice is a three-dimensional regular net-like structure. If one places at each point a collection or *basis* of atoms, the resulting structure is called a *crystal structure*. Due to interatomic forces, the basis will have no symmetry not contained in the lattice. The points that define the lattice are not necessarily at the location of the atoms. Each collection or basis of atoms is to be identical in structure and composition.

Point groups are collections of crystal symmetry operations that form a group and also leave one point fixed. From the above, the point group of the basis must be a point group of the associated lattice. There are only 32 different point groups allowed by crystalline solids. An explicit list of point groups will be given later in this chapter.

Crystals have only 14 different possible parallelepiped networks of points. These are the 14 *Bravais* lattices. All lattice points in a Bravais lattice are equivalent. The Bravais lattice must have at least as much point symmetry as its basis. For any given crystal, there can be no translational symmetry except that specified by its Bravais lattice. In other words, there are only 14 basically different types of translational symmetry. This result can be stated another way. The requirement that a lattice be invariant under one of the 32 point groups leads to symmetrically specialized types of lattices. These are the Bravais lattices. The types of symmetry

[8] See Shechtman et al [1.21].

[9] See Levine and Steinhardt [1.15]. See also Steinhardt and Ostlund [1.22].

of the Bravais lattices with respect to rotations and reflections specify the crystal *systems*. There are seven crystal systems. The meaning of Bravais lattice and crystal system will be clearer after the next Section, where unit cells for each Bravais lattice will be given and each Bravais lattice will be classified according to its crystal system.

Associating bases of atoms with the 14 Bravais lattices gives a total of 230 three-dimensional periodic patterns. (Loosely speaking, there are 230 different kinds of "three-dimensional wall paper.") That is, there are 230 possible *space groups*. Each one of these space groups must have a group of primitive translations as a subgroup. As a matter of fact, this subgroup must be an invariant subgroup. Of these space groups, 73 are simple group products of point groups and translation groups. These are the so-called *symmorphic* space groups. The rest of the space groups have screw or glide symmetries. In all cases, the factor group of the group of primitive translations is isomorphic to the point group that makes up the (proper and improper – an improper rotation has a proper rotation plus an inversion or a reflection) rotational parts of the symmetry operations of the space group. The above very brief summary of the symmetry properties of crystalline solids is by no means obvious and it was not produced very quickly. A brief review of the history of crystallography can be found in the article by Koster [1.14].

1.2.5 List of Crystal Systems and Bravais Lattices (B)

The seven crystal systems and the Bravais lattice for each type of crystal system are described below. The crystal systems are discussed in order of increasing symmetry.

1. *Triclinic Symmetry*. For each unit cell, $\alpha \neq \beta$, $\beta \neq \gamma$, $\alpha \neq \gamma$, $a \neq b$, $b \neq c$, and $a \neq c$, and there is only one Bravais lattice. Refer to Fig. 1.11 for nomenclature.

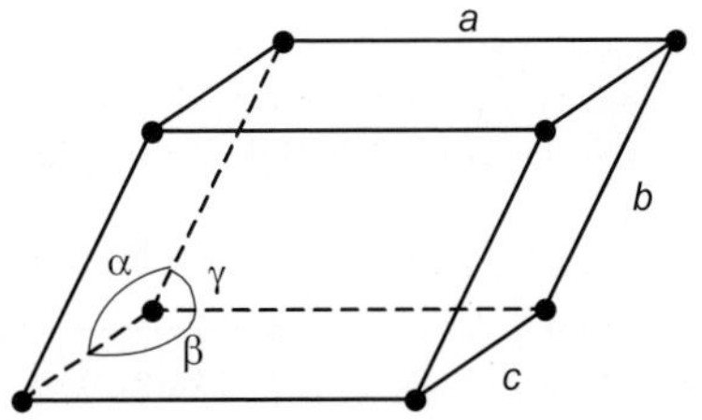

Fig. 1.11. A general unit cell (triclinic)

2. *Monoclinic Symmetry*. For each unit cell, $\alpha = \gamma = \pi/2$, $\beta \neq \alpha$, $a \neq b$, $b \neq c$, and $a \neq c$. The two Bravais lattices are shown in Fig. 1.12.

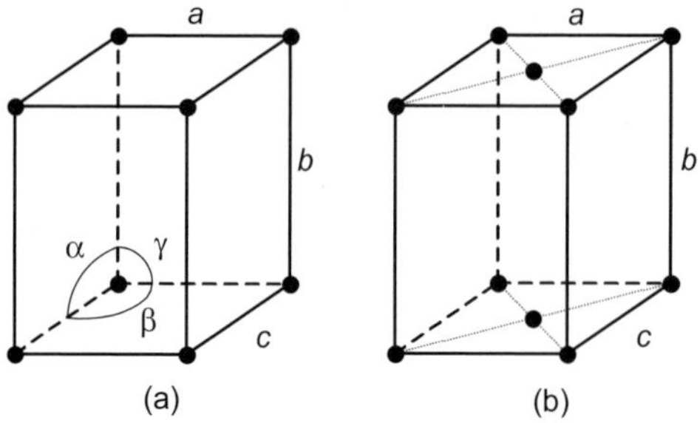

Fig. 1.12. (**a**) The simple monoclinic cell, and (**b**) the base-centered monoclinic cell

3. *Orthorhombic Symmetry*. For each unit cell, $\alpha = \beta = \gamma = \pi/2$, $a \neq b$, $b \neq c$, and $a \neq c$. The four Bravais lattices are shown in Fig. 1.13.

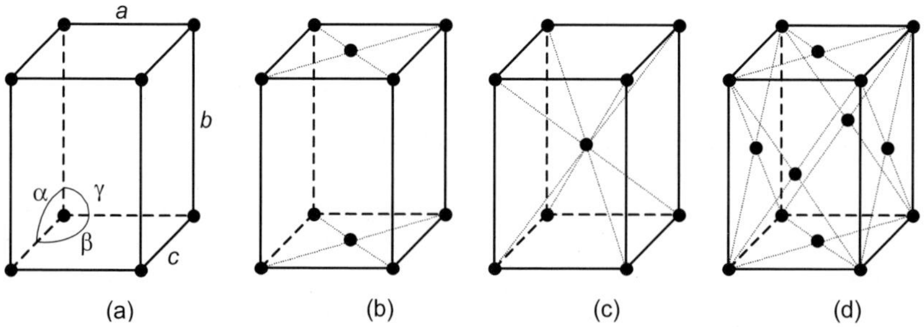

Fig. 1.13. (**a**) The simple orthorhombic cell, (**b**) the base-centered orthorhombic cell, (**c**) the body-centered orthorhombic cell, and (**d**) the face-centered orthorhombic cell

4. *Tetragonal Symmetry*. For each unit cell, $\alpha = \beta = \gamma = \pi/2$ and $a = b \neq c$. The two unit cells are shown in Fig. 1.14.

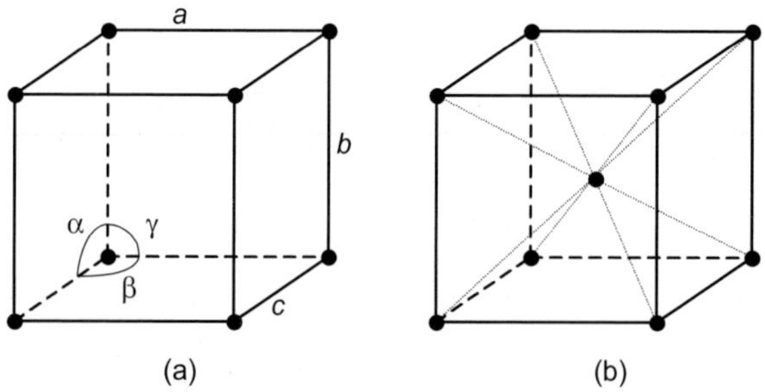

Fig. 1.14. (**a**) The simple tetragonal cell, and (**b**) the body-centered tetragonal cell

5. *Trigonal Symmetry*. For each unit cell, $\alpha = \beta = \gamma \neq \pi/2, < 2\pi/3$ and a = b = c. There is only one Bravais lattice, whose unit cell is shown in Fig. 1.15.

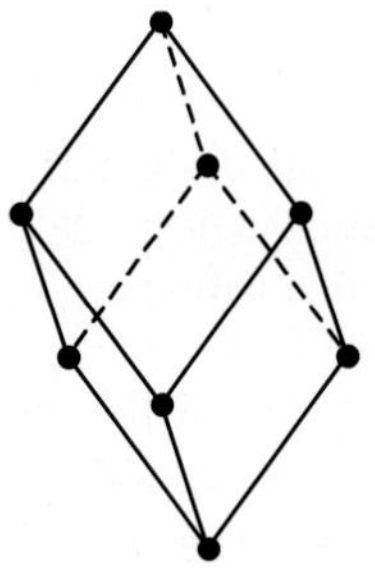

Fig. 1.15. Trigonal unit cell

6. *Hexagonal Symmetry*. For each unit cell, $\alpha = \beta = \pi/2$, $\gamma = 2\pi/3$, $a = b$, and $a \neq c$. There is only one Bravais lattice, whose unit cell is shown in Fig. 1.16.

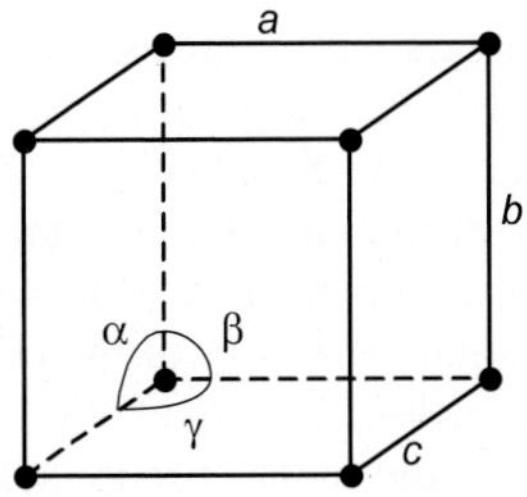

Fig. 1.16. Hexagonal unit cell

7. *Cubic Symmetry*. For each unit cell, $\alpha = \beta = \gamma = \pi/2$ and $a = b = c$. The unit cells for the three Bravais lattices are shown in Fig. 1.17.

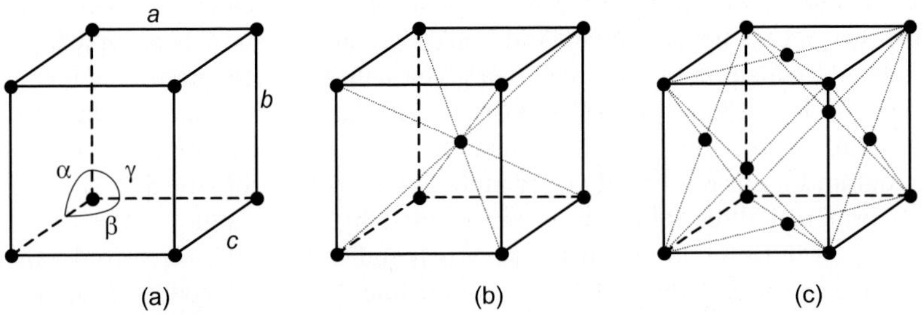

Fig. 1.17. (**a**) The simple cubic cell, (**b**) the body-centered cubic cell, and (**c**) the face-centered cubic cell

1.2.6 Schoenflies and International Notation for Point Groups (A)

There are only 32 point group symmetries that are consistent with translational symmetry. In this Section a descriptive list of the point groups will be given, but first a certain amount of notation is necessary.

The international (sometimes called Hermann–Mauguin) notation will be defined first. The Schoenflies notation will be defined in terms of the international notation. This will be done in a table listing the various groups that are compatible with the crystal systems (see Table 1.3).

An f-fold axis of rotational symmetry will be specified by f. Also, f will stand for the group of f-fold rotations. For example, 2 means a two-fold axis of symmetry (previously called C_2), and it can also mean the group of two-fold rotations. $\bar{f}$ will denote a rotation inversion axis. For example, $\bar{2}$ means that the crystal is brought back into itself by a rotation of π followed by an inversion. f/m means a rotation axis with a perpendicular mirror plane. $f2$ means a rotation axis with a perpendicular two-fold axis (or axes). fm means a rotation axis with a parallel mirror plane (or planes) ($m = \bar{2}$). $\bar{f}2$ means a rotation inversion axis with a perpendicular two-fold axis (or axes). $\bar{f}m$ means that the mirror plane m (or planes) is parallel to the rotation inversion axis. A rotation axis with a mirror plane normal and mirror planes parallel is denoted by f/mm or $(f/m)m$. Larger groups are compounded out of these smaller groups in a fairly obvious way. Note that 32 point groups are listed.

A very useful pictorial way of thinking about point group symmetries is by the use of *stereograms* (*or stereographic projections*). Stereograms provide a way of representing the three-dimensional symmetry of the crystal in two dimensions. To construct a stereographic projection, a lattice point (or any other point about which one wishes to examine the point group symmetry) is surrounded by a sphere. Symmetry axes extending from the center of the sphere intersect the sphere at points. These points are joined to the south pole (for points above the equator) by straight lines. Where the straight lines intersect a plane through the equator, a geometrical symbol may be placed to indicate the symmetry of the appropriate symmetry axis. The stereogram is to be considered as viewed by someone at the north pole. Symmetry points below the equator can be characterized by turning the process upside down. Additional diagrams to show how typical points are mapped by the point group are often given with the stereogram. The idea is illustrated in Fig. 1.18. Wood [98] and Brown [49] have stereograms of the 32 point groups. Rather than going into great detail in describing stereograms, let us look at a stereogram for our old friend D_3 (or in the *international* notation 32).

The principal three-fold axis is represented by the triangle in the center of Fig. 1.19b. The two-fold symmetry axes perpendicular to the three-fold axis are represented by the dark ovals at the ends of the line through the center of the circle.

In Fig. 1.19a, the dot represents a point above the plane of the paper and the open circle represents a point below the plane of the paper. Starting from any given point, it is possible to get to any other point by using the appropriate symmetry operations. D_3 has no reflection planes. Reflection planes are represented by dark lines. If there had been a reflection plane in the plane of the paper, then the outer boundary of the circle in Fig. 1.19b would have been dark.

Table 1.3. Schoenflies[1] and international[2] symbols for point groups, and permissible point groups for each crystal system

Crystal system	International symbol	Schoenflies symbol
Triclinic	1	C_1
	$\bar{1}$	C_i
Monoclinic	2	C_2
	m	C_{1h}
	$(2/m)$	C_{2h}
Orthorhombic	222	D_2
	$2mm$	C_{2v}
	$(2/m)(2/m)(2/m)$	D_{2h}
Tetragonal	4	C_4
	$\bar{4}$	S_4
	$(4/m)$	C_{4h}
	422	D_4
	$4mm$	C_{4v}
	$\bar{4}2m$	D_{2d}
	$(4/m)(2/m)(2/m)$	D_{4h}
Trigonal	3	C_3
	$\bar{3}$	C_{3i}
	32	D_3
	$3m$	C_{3v}
	$\bar{3}(2/m)$	D_{3d}
Hexagonal	6	C_6
	$\bar{6}$	C_{3h}
	$(6/m)$	C_{6h}
	622	D_6
	$6mm$	C_{6v}
	$\bar{6}m2$	D_{3h}
	$(6/m)(2/m)(2/m)$	D_{6h}
Cubic	23	T
	$(2/m)\bar{3}$	T_h
	432	O
	$\bar{4}3m$	T_d
	$(4/m)(\bar{3})(2/m)$	O_h

[1] A. Schoenflies, *Krystallsysteme und Krystallstruktur*, Leipzig, 1891.

[2] C. Hermann, Z. *Krist.*, **76**, 559 (1931); C. Mauguin, Z. *Krist.*, **76**, 542 (1931).

At this stage it might be logical to go ahead with lists, descriptions, and names of the 230 space groups. This will not be done for the simple reason that it would be much too confusing in a short time and would require most of the book otherwise. For details, Buerger [1.5] can always be consulted. A large part of the theory of solids can be carried out without reference to any particular symmetry type. For the rest, a research worker is usually working with one crystal and hence one space group and facts about that group are best learned when they are needed (unless one wants to specialize in crystal structure).

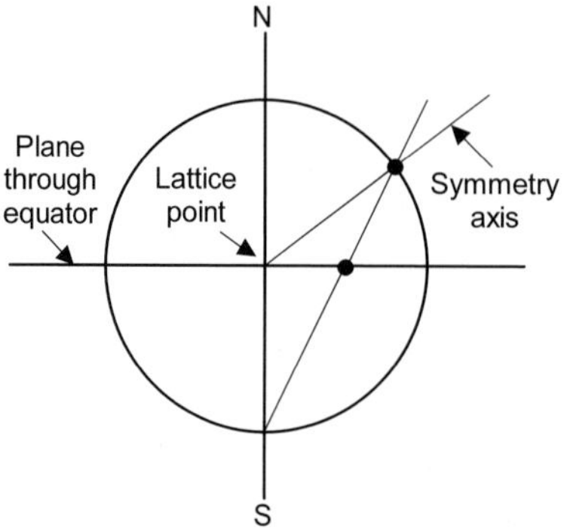

Fig. 1.18. Illustration of the way a stereogram is constructed

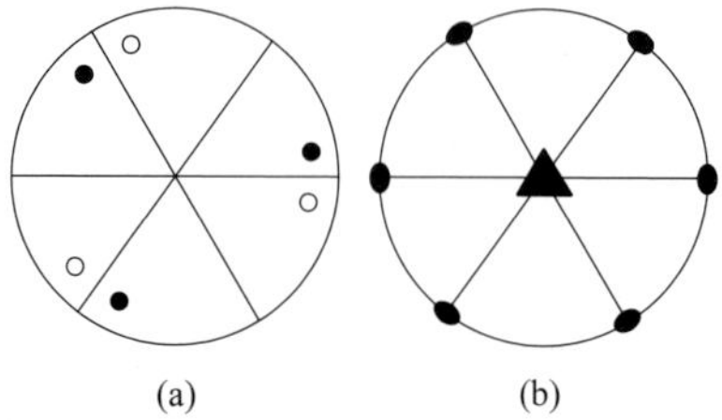

Fig. 1.19. Stereogram for D_3

1.2.7 Some Typical Crystal Structures (B)

The Sodium Chloride Structure. The sodium chloride structure, shown in Fig. 1.20, is one of the simplest and most familiar. In addition to NaCl, PbS and MgO are examples of crystals that have the NaCl arrangement. The space lattice is fcc (face-centered cubic). Each ion (Na^+ or Cl^-) is surrounded by six nearest-neighbor ions of the opposite sign. We can think of the basis of the space lattice as being a NaCl molecule.

Table 1.4. Packing fractions (PF) and coordination numbers (CN)

Crystal Structure	PF	CN
fcc	$\frac{\sqrt{2}\pi}{6} = 0.74$	12
bcc	$\frac{\sqrt{3}\pi}{8} = 0.68$	8
sc	$\frac{\pi}{6} = 0.52$	6
diamond	$\frac{\sqrt{3}\pi}{16} = 0.34$	4

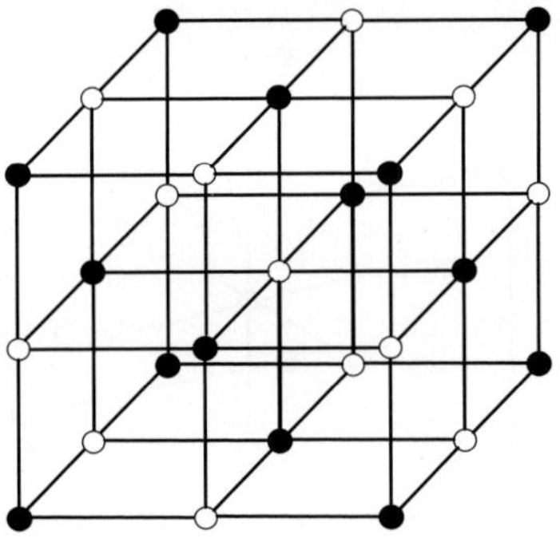

Fig. 1.20. The sodium chloride structure

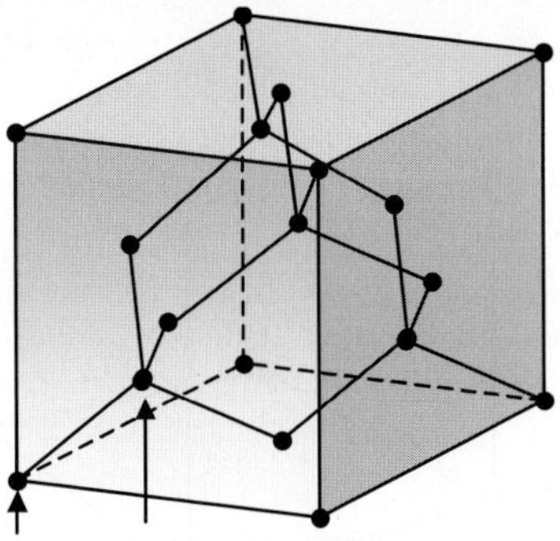

Fig. 1.21. The diamond structure

The Diamond Structure. The crystal structure of diamond is somewhat more complicated to draw than that of NaCl. The diamond structure has a space lattice that is fcc. There is a basis of two atoms associated with each point of the fcc lattice. If the lower left-hand side of Fig. 1.21 is a point of the fcc lattice, then the basis places atoms at this point [labeled (0, 0, 0)] and at (a/4, a/4, a/4). By placing bases at each point in the fcc lattice in this way, Fig. 1.21 is obtained. The characteristic feature of the diamond structure is that each atom has four nearest neighbors or each atom has tetrahedral bonding. Carbon (in the form of diamond), silicon, and germanium are examples of crystals that have the diamond structure. We compare sc, fcc, bcc, and diamond structures in Table 1.4.

The packing fraction is the fraction of space filled by spheres on each lattice point that are as large as they can be so as to touch but not overlap. The coordination number is the number of nearest neighbors to each lattice point.

The Cesium Chloride Structure. The cesium chloride structure, shown in Fig. 1.22, is one of the simplest structures to draw. Each atom has eight nearest neighbors. Besides CsCl, CuZn (β-brass) and AlNi have the CsCl structure. The Bravais lattice is simple cubic (sc) with a basis of (0,0,0) and (a/2)(1,1,1). If all the atoms were identical this would be a body-centered cubic (bcc) unit cell.

The Perovskite Structure. Perovskite is calcium titanate. Perhaps the most familiar crystal with the perovskite structure is barium titanate, $BaTiO_3$. Its structure is shown in Fig. 1.23. This crystal is ferroelectric. It can be described with a sc lattice with basis vectors of (0,0,0), (a/2)(0,1,1), (a/2)(1,0,1), (a/2)(1,1,0), and (a/2)(1,1,1).

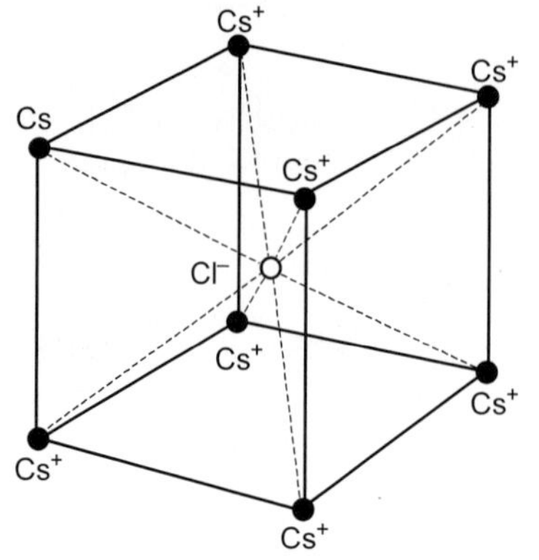

Fig. 1.22. The cesium chloride structure

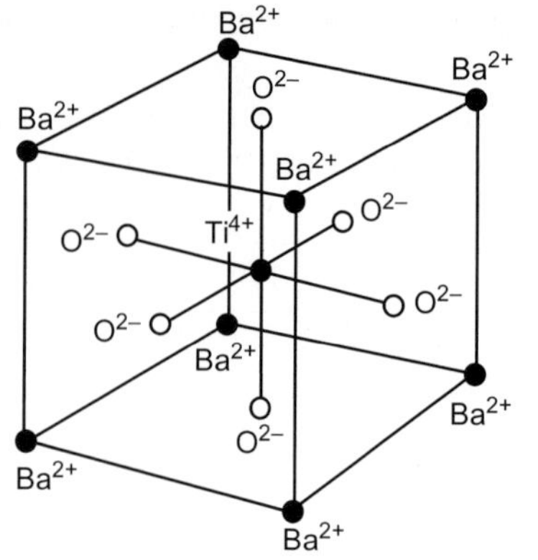

Fig. 1.23. The barium titanate ($BaTiO_3$) structure

Crystal Structure Determination (B)

How do we know that these are the structures of actual crystals? The best way is by the use of diffraction methods (X-ray, electron, or neutron). See Sect. 1.2.9 for more details about X-ray diffraction. Briefly, X-rays, neutrons and electrons can all be diffracted from a crystal lattice. In each case, the wavelength of the diffracted entity must be comparable to the spacing of the lattice planes. For X-rays to have a wavelength of order Angstroms, the energy needs to be of order keV, neutrons need to have energy of order fractions of an eV (thermal neutrons), and electrons should have energy of order eV. Because they carry a magnetic moment and hence interact magnetically, neutrons are particularly useful for determining magnetic structure.[10] Neutrons also interact by the nuclear interaction, rather than with electrons, so they are used to located hydrogen atoms (which in a solid have few or no electrons around them to scatter X-rays). We are concerned here with elastic scattering. Inelastic scattering of neutrons can be used to study lattice vibrations (see the end of Sect. 4.3.1). Since electrons interact very strongly with other electrons their diffraction is mainly useful to elucidate surface structure.[11]

Ultrabright X-rays: Synchrotron radiation from a storage ring provides a major increase in X-ray intensity. X-ray fluorescence can be used to study bonds on the surface because of the high intensity.

[10] For example, Shull and Smart in 1949 used elastic neutron diffraction to directly demonstrate the existence of two magnetic sublattices on an antiferromagnet.

[11] Diffraction of electrons was originally demonstrated by Davisson and Germer in an experiment clearly showing the wave nature of electrons.

1.2.8 Miller Indices (B)

In a Bravais lattice we often need to describe a plane or a set of planes, or a direction or a set of directions. The Miller indices are a notation for doing this. They are also convenient in X-ray work.

To describe a plane:

1. Find the intercepts of the plane on the three axes defined by the basis vectors $(\boldsymbol{a}_1, \boldsymbol{a}_2, \boldsymbol{a}_3)$.
2. Step 1 gives three numbers. Take the reciprocal of the three numbers.
3. Divide the reciprocals by their greatest common divisor (which yields a set of integers). The resulting set of three numbers (h, k, l) is called the Miller indices for the plane. $\{h, k, l\}$ means all planes equivalent (by symmetry) to (h, k, l).

To find the Miller indices for a direction:

1. Find any vector in the desired direction.
2. Express this vector in terms of the basis $(\boldsymbol{a}_1, \boldsymbol{a}_2, \boldsymbol{a}_3)$.
3. Divide the coefficients of $(\boldsymbol{a}_1, \boldsymbol{a}_2, \boldsymbol{a}_3)$ by their greatest common divisor. The resulting set of three integers $[h, k, l]$ defines a direction. $\langle h, k, l\rangle$ means all vectors equivalent to $[h, k, l]$. Negative signs in any of the numbers are indicated by placing a bar over the number (thus $\bar{h}$).

1.2.9 Bragg and von Laue Diffraction (AB)[12]

By discussing crystal diffraction, we accomplish two things: (1) We make clear how we know actual crystal structures exist, and (2) We introduce the concept of the reciprocal lattice, which will be used throughout the book

The simplest approach to Bragg diffraction is illustrated in Fig. 1.24. We assume specular reflection with angle of incidence equal to angle of reflection. We also assume the radiation is elastically scattered so that incident and reflected waves have the same wavelength.

For constructive interference we must have the path difference between reflected rays equal to an integral (n) number of wavelengths (λ). Using Fig. 1.24, the condition for diffraction peaks is then

$$n\lambda = 2d\sin\theta\,, \tag{1.23}$$

which is the famous Bragg law. Note that peaks in the diffraction only occur if λ is less than $2d$, and we will only resolve the peaks if λ and d are comparable.

---

[12] A particularly clear discussion of these topics is found in Brown and Forsyth [1.4]. See also Kittel [1.13, Chaps. 2 and 19].

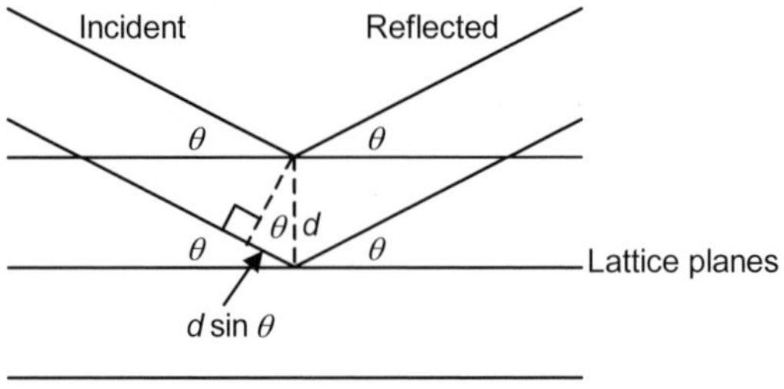

Fig. 1.24. Bragg diffraction

The Bragg approach gives a simple approach to X-ray diffraction. However, it is not easily generalized to include the effects of a basis of atoms, of the distribution of electrons, and of temperature. For that we need the von Laue approach.

We will begin our discussion in a fairly general way. X-rays are electromagnetic waves and so are governed by the Maxwell equations. In SI and with no charges or currents (i.e. neglecting the interaction of the X-rays with the electron distribution except for scattering), we have for the electric field $\boldsymbol{E}$ and the magnetic field $\boldsymbol{H}$ (with the magnetic induction $\boldsymbol{B} = \mu_0\boldsymbol{H}$)

$$\nabla \cdot \boldsymbol{E} = 0\,, \quad \nabla \times \boldsymbol{H} = \varepsilon_0 \frac{\partial \boldsymbol{E}}{\partial t}\,, \quad \nabla \times \boldsymbol{E} = -\frac{\partial \boldsymbol{B}}{\partial t}\,, \quad \nabla \cdot \boldsymbol{B} = 0.$$

Taking the curl of the third equation, using $\boldsymbol{B} = \mu_0\boldsymbol{H}$ and using the first and second of the Maxwell equations we find the usual wave equation:

$$\nabla^2 \boldsymbol{E} = \frac{1}{c^2}\frac{\partial^2 \boldsymbol{E}}{\partial t^2}\,, \tag{1.24}$$

where $c = (\mu_0\varepsilon_0)^{-1/2}$ is the speed of light. There is also a similar wave equation for the magnetic field. For simplicity we will focus on the electric field for this discussion. We assume plane-wave X-rays are incident on an atom and are scattered as shown in Fig. 1.25.

In Fig. 1.25 we use the center of the atom as the origin and $\boldsymbol{r}_s$ locates the electron that scatters the X-ray. As mentioned earlier, we will first specialize to the case of the lattice of point scatterers, but the present setup is useful for generalizations.

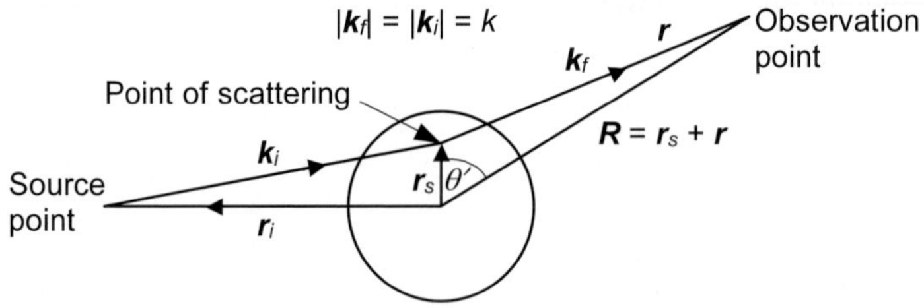

Fig. 1.25. Plane-wave scattering

The solution of the wave equation for the incident plane wave is

$$\boldsymbol{E}_i(\boldsymbol{r}) = \boldsymbol{E}_0 \exp[\mathrm{i}(\boldsymbol{k}_i \cdot \boldsymbol{r}_i - \omega t)], \tag{1.25}$$

where $\boldsymbol{E}_0$ is the amplitude and $\omega = kc$. If the wave equation is written in spherical coordinates, one can find a solution for the spherically scattered wave (retaining only dominant terms far from the scattering location)

$$\boldsymbol{E}_s = K_1 \boldsymbol{E}(\boldsymbol{r}_s) \frac{\mathrm{e}^{\mathrm{i}kr}}{r}, \tag{1.26}$$

where K_1 is a constant, with the scattered wave having the same frequency and wavelength as the incident wave. Spherically scattered waves are important ones since the wavelength being scattered is much greater than the size of the atom. Also, we assume the source and observation points are very far from the point of scattering. From the diagram $\boldsymbol{r} = \boldsymbol{R} - \boldsymbol{r}_s$, so by squaring, taking the square root, and using that $r_s/R \ll 1$ (i.e. far from the scattering center), we have

$$r = R\left(1 - \frac{r_s}{R}\cos\theta'\right), \tag{1.27}$$

from which since $kr_s \cos\theta \cong \boldsymbol{k}_f \cdot \boldsymbol{r}_s$;

$$kr \cong kR - \boldsymbol{k}_f \cdot \boldsymbol{r}_s. \tag{1.28}$$

Therefore

$$\boldsymbol{E}_s = K_1 \boldsymbol{E}_0 \frac{\mathrm{e}^{\mathrm{i}kR}}{R} \mathrm{e}^{\mathrm{i}(\boldsymbol{k}_i - \boldsymbol{k}_f)\cdot \boldsymbol{r}_s} \mathrm{e}^{-\mathrm{i}\omega t}, \tag{1.29}$$

where we have used (1.28), (1.26), and (1.25) and also assumed $r^{-1} \cong R^{-1}$ to sufficient accuracy. Note that $(\boldsymbol{k}_i - \boldsymbol{k}_f) \cdot \boldsymbol{r}_s$, as we will see, can be viewed as the phase difference between the wave scattered from the origin and that scattered from $\boldsymbol{r}_s$ in the approximation we are using. Thus, the scattering intensity is proportional to $|P|^2$ (given by (1.32)) that, as we will see, could have been written down immediately. Thus, we can write the scattered wave as

$$E_{\mathrm{sc}} = FP, \tag{1.30}$$

where the magnitude of F^2 is proportional to the incident intensity $\boldsymbol{E}_0$ and

$$|F| = \left|\frac{K_1 \boldsymbol{E}_0}{R}\right|, \tag{1.31}$$

$$P = \sum\nolimits_s \mathrm{e}^{-\mathrm{i}\Delta \boldsymbol{k}\cdot \boldsymbol{r}_s}, \tag{1.32}$$

summed over all scatterers, and

$$\Delta \boldsymbol{k} = \boldsymbol{k}_f - \boldsymbol{k}_i. \tag{1.33}$$

P can be called the (relative) scattering amplitude.

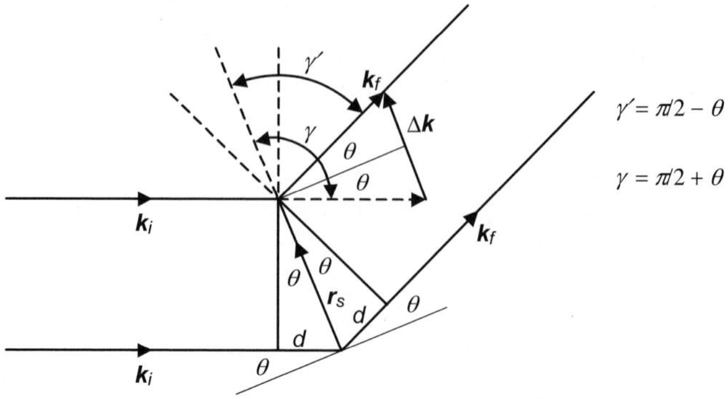

Fig. 1.26. Schematic for simpler discussion of scattering

It is useful to follow up on the comment made above and give a simpler discussion of scattering. Looking at Fig. 1.26, we see the path difference between the two beams is $2d = 2\, r_s \sin\theta$. So the phase difference is

$$\Delta\varphi = \frac{4\pi}{\lambda} r_s \sin\theta = 2kr_s \sin\theta ,$$

since $|\boldsymbol{k}_f| = |\boldsymbol{k}_i| = k$. Note also

$$\Delta\boldsymbol{k}\cdot\boldsymbol{r}_s = kr_s\left[\cos\left(\frac{\pi}{2}-\theta\right) - \cos\left(\frac{\pi}{2}+\theta\right)\right] = 2kr_s \sin\theta ,$$

which is the phase difference. We obtain for a continuous distribution of scatterers

$$P = \int \exp(-\mathrm{i}\Delta\boldsymbol{k}\cdot\boldsymbol{r}_s)\rho(r_s)\mathrm{d}V , \tag{1.34}$$

where we have assumed each scatterer scatters proportionally to its density.

We assume now the general case of a lattice with a basis of atoms, each atom with a distribution of electrons. The lattice points are located at

$$\boldsymbol{R}_{pmn} = p\boldsymbol{a}_1 + m\boldsymbol{a}_2 + n\boldsymbol{a}_3 , \tag{1.35}$$

where p, m and n are integers and $\boldsymbol{a}_1$, $\boldsymbol{a}_2$, $\boldsymbol{a}_3$ are the fundamental translation vectors of the lattice. For each $\boldsymbol{R}_{pmn}$ there will be a basis at

$$\boldsymbol{R}_j = a_j\boldsymbol{a}_1 + b_j\boldsymbol{a}_2 + c_j\boldsymbol{a}_3 , \tag{1.36}$$

where $j = 1$ to q for q atoms per unit cell and a_j, b_j, c_j are numbers that are generally not integers. Starting at $\boldsymbol{R}_j$ we can assume the electrons are located at $\boldsymbol{r}_s$ so the electron locations are specified by

$$\boldsymbol{r} = \boldsymbol{R}_{pmn} + \boldsymbol{R}_j + \boldsymbol{r}_s , \tag{1.37}$$

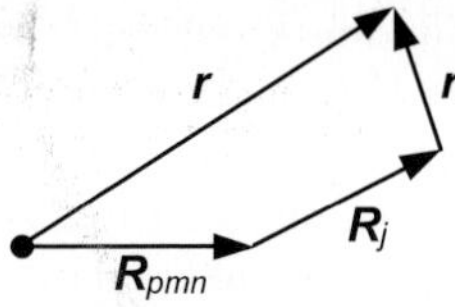

Fig. 1.27. Vector diagram of electron positions for X-ray scattering

as shown in Fig. 1.27. Relative to $\boldsymbol{R}_j$ then the electron's position is

$$\boldsymbol{r}_s = \boldsymbol{r} - \boldsymbol{R}_{pmn} - \boldsymbol{R}_j .$$

If we let $\rho_j(\boldsymbol{r})$ be the density of electrons of atom j then the total density of electrons is

$$\rho(\boldsymbol{r}) = \sum_{pmn} \sum_{j=1}^{q} \rho_j(\boldsymbol{r} - \boldsymbol{R}_j - \boldsymbol{R}_{pmn}) . \tag{1.38}$$

By a generalization of (1.34) we can write the scattering amplitude as

$$P = \sum_{pmn} \sum_j \int \rho_j(\boldsymbol{r} - \boldsymbol{R}_j - \boldsymbol{R}_{pmn}) \mathrm{e}^{-\mathrm{i}\Delta\boldsymbol{k}\cdot\boldsymbol{r}} \,\mathrm{d}V. \tag{1.39}$$

Making a dummy change of integration variable and using (1.37) (dropping s on $\boldsymbol{r}_s$) we write

$$P = \sum_{pmn} \mathrm{e}^{-\mathrm{i}\Delta\boldsymbol{k}\cdot\boldsymbol{R}_{pmn}} \left(\sum_j \mathrm{e}^{-\mathrm{i}\Delta\boldsymbol{k}\cdot\boldsymbol{R}_j} \int \rho_j(\boldsymbol{r}) \mathrm{e}^{\mathrm{i}\Delta\boldsymbol{k}\cdot\boldsymbol{r}} \mathrm{d}V \right).$$

For N^3 unit cells the lattice factor separates out and we will show below that

$$\sum_{pmn} \exp(-\mathrm{i}\Delta\boldsymbol{k} \cdot \boldsymbol{R}_{pmn}) = N^3 \delta_{\boldsymbol{G}_{hkl}}^{\Delta\boldsymbol{k}} ,$$

where as defined below, the $\boldsymbol{G}$ are reciprocal lattice vectors. So we find

$$P = N^3 \delta_{\boldsymbol{G}_{hkl}}^{\Delta\boldsymbol{k}} S_{hkl} , \tag{1.40}$$

where S_{hkl} is the structure factor defined by

$$S_{hkl} = \sum_j \mathrm{e}^{-\mathrm{i}\boldsymbol{G}_{hkl}\cdot\boldsymbol{R}_j} f_j^{hkl} , \tag{1.41}$$

and f_j is the atomic form factor defined by

$$f_j^{hkl} = \int \rho_j(\boldsymbol{r}) \mathrm{e}^{-\mathrm{i}\boldsymbol{G}_{hkl}\cdot\boldsymbol{r}} \mathrm{d}V . \tag{1.42}$$

Since nuclei do not interact appreciably with X-rays, $\rho_j(\boldsymbol{r})$ is only determined by the density of electrons as we have assumed. Equation (1.42) can be further simplified for $\rho_j(\boldsymbol{r})$ representing a spherical distribution of electrons and can be worked out if its functional form is known, such as $\rho_j(\boldsymbol{r})$ = (constant) $\exp(-\lambda r)$.

This is the general case. Let us work out the special case of a lattice of point scatterers where $f_j = 1$ and $R_j = 0$. For this case, as in a three-dimension diffraction grating (crystal lattice), it is useful to introduce the concept of a *reciprocal lattice*. This concept will be used throughout the book in many different contexts. The basis vectors $\boldsymbol{b}_j$ for the reciprocal lattice are defined by the set of equations

$$\boldsymbol{a}_i \cdot \boldsymbol{b}_j = \delta_{ij}, \tag{1.43}$$

where $i, j \rightarrow 1$ to 3 and δ_{ij} is the Kronecker delta. The reciprocal lattice is then defined by

$$\boldsymbol{G}_{hkl} = 2\pi(h\boldsymbol{b}_1 + k\boldsymbol{b}_2 + l\boldsymbol{b}_3), \tag{1.44}$$

where h, k, l are integers.[13] As an aside, we mention that we can show that

$$\boldsymbol{b}_1 = \frac{1}{\Omega}\boldsymbol{a}_2 \times \boldsymbol{a}_3 \tag{1.45}$$

plus cyclic changes where $\Omega = \boldsymbol{a}_1 \cdot (\boldsymbol{a}_2 \times \boldsymbol{a}_3)$ is the volume of a unit cell in direct space. It is then easy to show that the volume of a unit cell in reciprocal space is

$$\Omega_{\mathrm{RL}} = \boldsymbol{b}_1 \cdot (\boldsymbol{b}_2 \times \boldsymbol{b}_3) = \frac{1}{\Omega}. \tag{1.46}$$

The vectors $\boldsymbol{b}_1$, $\boldsymbol{b}_2$, and $\boldsymbol{b}_3$ span three-dimensional space, so $\Delta\boldsymbol{k}$ can be expanded in terms of them,

$$\Delta\boldsymbol{k} = 2\pi(h\boldsymbol{b}_1 + k\boldsymbol{b}_2 + l\boldsymbol{b}_3), \tag{1.47}$$

where now h, k, l are not necessarily integers. Due to (1.43) we can write

$$\boldsymbol{R}_{pmn} \cdot \Delta\boldsymbol{k} = 2\pi(ph + mk + ln), \tag{1.48}$$

with p, m, n still being integers. Using (1.32) with $\boldsymbol{r}_s = \boldsymbol{R}_{pmn}$, (1.48), and assuming a lattice of N^3 atoms, the structure factor can be written:

$$P = \sum_{p=0}^{N-1} \mathrm{e}^{-\mathrm{i}2\pi ph} \sum_{m=0}^{N-1} \mathrm{e}^{-\mathrm{i}2\pi mk} \sum_{n=0}^{N-1} \mathrm{e}^{-\mathrm{i}2\pi nl}. \tag{1.49}$$

This can be evaluated by the law of geometric progressions. We find:

$$|P|^2 = \left(\frac{\sin^2 \pi hN}{\sin^2 \pi h}\right)\left(\frac{\sin^2 \pi kN}{\sin^2 \pi k}\right)\left(\frac{\sin^2 \pi lN}{\sin^2 \pi l}\right). \tag{1.50}$$

For a real lattice N is very large, so we assume $N \rightarrow \infty$ and then if h, k, l are not integers $|P|$ is negligible. If they are integers, each factor is N^2 so

$$|P|^2 = N^6 \delta_{h,k,l}^{\mathrm{integers}}. \tag{1.51}$$

[13] Alternatively, as is often done, we could include a 2π in (1.43) and remove the multiplicative factor on the right-hand side of (1.44).

Thus for a lattice of point ions then, the diffraction peaks occur for

$$\Delta \boldsymbol{k} = \boldsymbol{k}_f - \boldsymbol{k}_i = \boldsymbol{G}_{hkl} = 2\pi(h\boldsymbol{b}_1 + k\boldsymbol{b}_2 + l\boldsymbol{b}_3), \tag{1.52}$$

where h, k, and l are now integers.

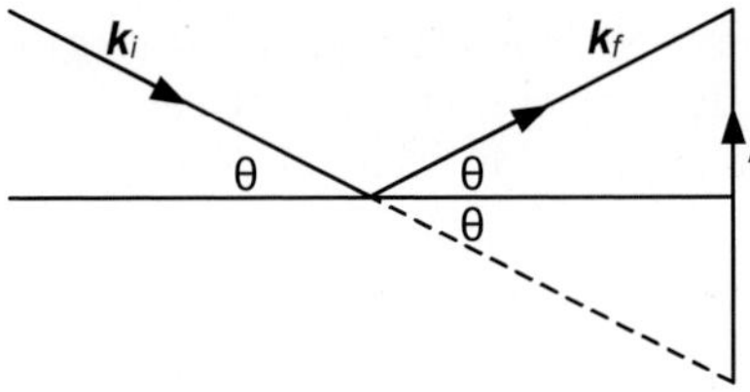

Fig. 1.28. Wave vector–reciprocal lattice relation for diffraction peaks

Thus the X-ray diffraction peaks directly determine the reciprocal lattice that in turn determines the direct lattice. For diffraction peaks (1.51) is valid. Let $\boldsymbol{G}_{hkl} = n\boldsymbol{G}'_{h'k'l'}$, where now h', k', l' are Miller indices and $\boldsymbol{G}'_{h'k'l'}$ is the shortest vector in the direction of $\boldsymbol{G}_{hkl}$. $\boldsymbol{G}_{hkl}$ is perpendicular to (h, k, l) plane, and we show in Problem 1.10 that the distance between adjacent such planes is

$$d_{hkl} = \frac{2\pi}{G'_{h'k'l'}}. \tag{1.53}$$

Thus

$$|\boldsymbol{G}| = 2k\sin\theta = n\,|\boldsymbol{G}'_{h'k'l'}| = n\frac{2\pi}{d_{hkl}}, \tag{1.54}$$

so since $k = 2\pi/\lambda$,

$$n\lambda = 2d_{hkl}\sin\theta, \tag{1.55}$$

which is Bragg's equation.

So far our discussion has assumed a rigid fixed lattice. The effect of temperature on the lattice can be described by the Debye–Waller factor. We state some results but do not derive them as they involve lattice-vibration concepts discussed in Chap. 2.[14] The results for intensity are:

$$I = I_{T=0}\mathrm{e}^{-2W}, \tag{1.56}$$

where $D(T) = \mathrm{e}^{-2W}$, and W is known as the Debye–Waller factor. If $\boldsymbol{K} = \boldsymbol{k} - \boldsymbol{k}'$, where $|\boldsymbol{k}| = |\boldsymbol{k}'|$ are the incident and scattered wave vectors of the X-rays, and if

[14] See, e.g., Ghatak and Kothari [1.9].

$\boldsymbol{e}(\boldsymbol{q}, j)$ is the polarization vector of the phonons (see Chap. 2) in the mode j with wave vector $\boldsymbol{q}$, then one can show[14, 15] that the Debye–Waller factor is

$$2W = \frac{\hbar^2}{2MN} \sum_{\boldsymbol{q},j} \frac{\boldsymbol{K} \cdot \boldsymbol{e}(\boldsymbol{q}, j)}{\hbar \omega_j(\boldsymbol{q})} \coth \frac{\hbar \omega_j(\boldsymbol{q})}{2kT}, \tag{1.57}$$

where N is the number of atoms, M is their mass and $\omega_j(\boldsymbol{q})$ is the frequency of vibration of phonons in mode j, wave vector $\boldsymbol{q}$. One can further show that in the Debye approximation (again discussed in Chap. 2): At low temperature ($T << \theta_D$)

$$2W = \frac{3}{4M} \frac{\hbar^2 K^2}{k\theta_D} = \text{constant}, \tag{1.58}$$

and at high temperature ($T >> \theta_D$)

$$2W = \frac{3}{M\hbar\theta_D} \frac{T}{\theta_D} K^2 \propto T, \tag{1.59}$$

where θ_D is the Debye Temperature defined from the cutoff frequency in the Debye approximation (see Sect. 2.3.3). The effect of temperature is to reduce intensity but not broaden lines. Even at $T = 0$ the Debye–Waller factor is not unity so there is always some "diffuse" scattering, in addition to the diffraction.

As an example of the use of the structure factor, we represent the bcc lattice as a sc lattice with a basis. Let the simple cubic unit cell have side a. Consider a basis at $R_0 = (0,0,0)a$, $R_1 = (1,1,1)a/2$. The structure factor is

$$S_{hkl} = f_0 + f_1 \mathrm{e}^{-\mathrm{i}2\pi(h+k+l)a/2} = f_0 + f_1(-1)^{h+k+l}. \tag{1.60}$$

Suppose also the atoms at R_0 and R_1 are identical, then $f_0 = f_1 = f$ so

$$\begin{aligned} S_{hkl} &= f(1+(-)^{h+k+l}), \\ &= 0 \quad \text{if } h+k+l \text{ is odd}, \\ &= 2f \quad \text{if } h+k+l \text{ is even}. \end{aligned} \tag{1.61}$$

The nonvanishing structure factor ends up giving results identical to a bcc lattice.

Problems

1.1. Show by construction that stacked regular pentagons do not fill all two-dimensional space. What do you conclude from this? Give an example of a geometrical figure that when stacked will fill all two-dimensional space.

1.2. Find the Madelung constant for a one-dimensional lattice of alternating, equally spaced positive and negative charged ions.

---

[15] See Maradudin et al [1.16]

1.3. Use the Evjen counting scheme [1.19] to evaluate approximately the Madelung constant for crystals with the NaCl structure.

1.4. Show that the set of all rational numbers (without zero) forms a group under the operation of multiplication. Show that the set of all rational numbers (with zero) forms a group under the operation of addition.

1.5. Construct the group multiplication table of D_4 (the group of three dimensional rotations that map a square into itself).

1.6. Show that the set of elements (1, –1, i, –i) forms a group when combined under the operation of multiplication of complex numbers. Find a geometric group that is isomorphic to this group. Find a subgroup of this group. Is the whole group cyclic? Is the subgroup cyclic? Is the whole group abelian?

1.7. Construct the stereograms for the point groups 4(C_4) and 4mm(C_{4v}). Explain how all elements of each group are represented in the stereogram (see Table 1.3).

1.8. Draw a bcc (body-centered cubic) crystal and draw in three crystal planes that are neither parallel nor perpendicular. Name these planes by the use of Miller indices. Write down the Miller indices of three directions, which are neither parallel nor perpendicular. Draw in these directions with arrows.

1.9. Argue that electrons should have energy of order electron volts to be diffracted by a crystal lattice.

1.10. Consider lattice planes specified by Miller indices (h, k, l) with lattice spacing determined by $d(h,k,l)$. Show that the reciprocal lattice vectors $\mathbf{G}(h,k,l)$ are orthogonal to the lattice plane (h,k,l) and if $\mathbf{G}(h,k,l)$ is the shortest such reciprocal lattice vector then

$$d(h,k,l) = \frac{2\pi}{|\mathbf{G}(h,k,l)|} .$$

1.11. Suppose a one-dimensional crystal has atoms located at nb and αmb where n and m are integers and α is an irrational number. Show that sharp Bragg peaks are still obtained.

1.12. Find the Bragg peaks for a grating with a modulated spacing. Assume the grating has a spacing

$$d_n = nb + \varepsilon b \sin(2\pi knb) ,$$

where ε is small and kb is irrational. Carry your results to first order in ε and assume that all scattered waves have the same geometry. You can use the geometry shown in the figure of this problem. The phase φ_n of scattered wave n at angle θ is

$$\varphi_n = \frac{2\pi}{\lambda} d_n \sin\theta ,$$

where λ is the wavelength. The scattered intensity is proportional to the square of the scattered amplitude, which in turn is proportional to

$$E \equiv \left|\sum_0^N \exp(\mathrm{i}\varphi_n)\right|$$

for N+1 scattered wavelets of equal amplitude.

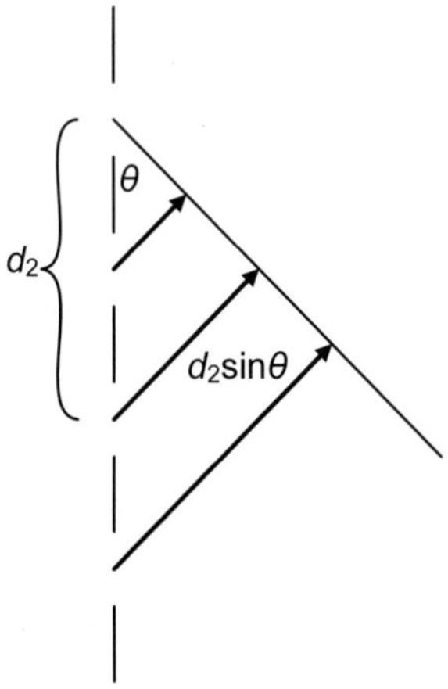

1.13. Find all Bragg angles less than 50 degrees for diffraction of X-rays with wavelength 1.5 angstroms from the (100) planes in potassium. Use a conventional unit cell with structure factor.

2 Lattice Vibrations and Thermal Properties

Chapter 1 was concerned with the binding forces in crystals and with the manner in which atoms were arranged. Chapter 1 defined, in effect, the universe with which we will be concerned. We now begin discussing the elements of this universe with which we interact. Perhaps the most interesting of these elements are the internal energy excitation modes of the crystals. The quanta of these modes are the "particles" of the solid. This chapter is primarily devoted to a particular type of internal mode – the lattice vibrations.

The lattice introduced in Chap. 1, as we already mentioned, is not a static structure. At any finite temperature there will be thermal vibrations. Even at absolute zero, according to quantum mechanics, there will be zero-point vibrations. As we will discuss, these lattice vibrations can be described in terms of normal modes describing the collective vibration of atoms. The quanta of these normal modes are called phonons.

The phonons are important in their own right as, e.g., they contribute both to the specific heat and the thermal conduction of the crystal, and they are also important because of their interaction with other energy excitations. For example, the phonons scatter electrons and hence cause electrical resistivity. Scattering of phonons, by whatever mode, in general also limits thermal conductivity. In addition, phonon–phonon interactions are related to thermal expansion. Interactions are the subject of Chap. 4.

We should also mention that the study of phonons will introduce us to wave propagation in periodic structures, allowed energy bands of elementary excitations propagating in a crystal, and the concept of Brillouin zones that will be defined later in this chapter.

There are actually two main reservoirs that can store energy in a solid. Besides the phonons or lattice vibrations, there are the electrons. Generally, we start out by discussing these two independently, but this is an approximation. This approximation is reasonably clear-cut in insulators, but in metals it is much harder to justify. Its intellectual framework goes by the name of the *Born–Oppenheimer* approximation. This approximation paves the way for a systematic study of solids in which the electron–phonon interactions can later be put in, often by perturbation theory. In this chapter we will discuss a wide variety of lattice vibrations in one and three dimensions. In three dimensions we will also discuss the vibration problem in the elastic continuum approximation. Related topics will follow: in Chap. 3 electrons moving in a static lattice will be considered, and in Chap. 4 electron–phonon interactions (and other topics).

2.1 The Born–Oppenheimer Approximation (A)

The most fundamental problem in solid-state physics is to solve the many-particle Schrödinger wave equation,

$$\mathcal{H}_{\mathrm{c}}\psi = \mathrm{i}\hbar\frac{\partial\psi}{\partial t}, \tag{2.1}$$

where $\mathcal{H}_{\mathrm{c}}$ is the crystal Hamiltonian defined by (2.3). In a sense, this equation is the "Theory of Everything" for solid-state physics. However, because of the many-body problem, solutions can only be obtained after numerous approximations. As mentioned in Chap. 1, P. W. Anderson has reminded us, "more is different!" There are usually emergent properties at higher levels of complexity [2.1]. In general, the wave function ψ is a function of all electronic and nuclear coordinates and of the time t. That is,

$$\psi = \psi(\boldsymbol{r}_i, \boldsymbol{R}_l, t), \tag{2.2}$$

where the $\boldsymbol{r}_i$ are the electronic coordinates and the $\boldsymbol{R}_l$ are the nuclear coordinates. The Hamiltonian $\mathcal{H}_{\mathrm{c}}$ of the crystal is

$$\begin{aligned}\mathcal{H}_{\mathrm{c}} = &-\sum_i \frac{\hbar^2}{2m}\nabla_i^2 - \sum_l \frac{\hbar^2}{2M_l}\nabla_l^2 + \frac{1}{2}\sum_{i,j}{}' \frac{e^2}{4\pi\varepsilon_0 |\boldsymbol{r}_i - \boldsymbol{r}_j|} \\ &- \sum_{i,l} \frac{e^2 Z_l}{4\pi\varepsilon_0 |\boldsymbol{r}_i - \boldsymbol{R}_l|} + \frac{1}{2}\sum_{l,l'}{}' \frac{e^2 Z_l Z_{l'}}{4\pi\varepsilon_0 |\boldsymbol{R}_l - \boldsymbol{R}_{l'}|}.\end{aligned} \tag{2.3}$$

In (2.3), m is the electronic mass, M_l is the mass of the nucleus located at $\boldsymbol{R}_l$, Z_l is the atomic number of the nucleus at $\boldsymbol{R}_l$, and e has the magnitude of the electronic charge. The sums over i and j run over all electrons.[1] The prime on the third term on the right-hand side of (2.3) means the terms $i = j$ are omitted. The sums over l and l' run over all nuclear coordinates and the prime on the sum over l and l' means that the $l = l'$ terms are omitted. The various terms all have a physical interpretation. The first term is the operator representing the kinetic energy of the electrons. The second term is the operator representing the kinetic energy of the nuclei. The third term is the Coulomb potential energy of interaction between the electrons. The fourth term is the Coulomb potential energy of interaction between the electrons and the nuclei. The fifth term is the Coulomb potential energy of interaction between the nuclei.

---

[1] Had we chosen the sum to run over only the outer electrons associated with each atom, then we would have to replace the last term in (2.3) by an ion–ion interaction term. This term could have three and higher body interactions as well as two-body forces. Such a procedure would be appropriate [51, p. 3] for the practical discussion of lattice vibrations. However, we shall consider only two-body forces.

In (2.3) internal magnetic interactions are left out because of their assumed smallness. This corresponds to neglecting relativistic effects. In solid-state physics, it is seldom necessary to assign a structure to the nucleus. It is never necessary (or possible) to assign a structure to the electron. Thus in (2.3) both electrons and nuclei are treated as point charges. Sometimes it will be necessary to allow for the fact that the nucleus can have nonzero spin, but this is only when much smaller energy differences are being considered than are of interest now. Because of statistics, as will be evident later, it is usually necessary to keep in mind that the electron is a spin 1/2 particle. For the moment, it is necessary to realize only that the wave function of (2.2) is a function of the spin degrees of freedom as well as of the space degrees of freedom. If we prefer, we can think of $\boldsymbol{r}_i$ in the wave function as symbolically labeling all the coordinates of the electron. That is, $\boldsymbol{r}_i$ gives both the position and the spin. However, ∇_i^2 is just the ordinary spatial Laplacian.

For purposes of shortening the notation it is convenient to let T_{E} be the kinetic energy of the electrons, T_{N} be the kinetic energy of the nuclei, and U be the total Coulomb energy of interaction of the nuclei and the electrons. Then (2.3) becomes

$$\mathcal{H}_{\mathrm{c}} = T_{\mathrm{E}} + U + T_{\mathrm{N}} . \tag{2.4}$$

It is also convenient to define

$$\mathcal{H}_0 = T_{\mathrm{E}} + U . \tag{2.5}$$

Nuclei have large masses and hence in general (cf. the classical equipartition theorem) they have small kinetic energies. Thus in the expression $\mathcal{H}_{\mathrm{c}} = \mathcal{H}_0 + T_{\mathrm{N}}$, it makes some sense to regard T_{N} as a perturbation on $\mathcal{H}_0$. However, for metals, where the electrons have no energy gap between their ground and excited states, it is by no means clear that T_{N} should be regarded as a small perturbation on $\mathcal{H}_0$. At any rate, one can proceed to make expansions just as if a perturbation sequence would converge.

Let M_0 be a mean nuclear mass and define

$$K = \left(\frac{m}{M_0}\right)^{1/4} .$$

If we define

$$\mathcal{H}_{\mathrm{L}} = -\sum\nolimits_l \frac{M_0}{M_l}\frac{\hbar^2}{2m}\nabla_l^2 , \tag{2.6}$$

then

$$T_{\mathrm{N}} = K^4 \mathcal{H}_{\mathrm{L}} . \tag{2.7}$$

The total Hamiltonian then has the form

$$\mathcal{H}_{\mathrm{c}} = \mathcal{H}_0 + K^4 \mathcal{H}_{\mathrm{L}} , \tag{2.8}$$

and the time-independent Schrödinger wave equation that we wish to solve is

$$\mathcal{H}_{\mathrm{c}}\psi(\boldsymbol{r}_i,\boldsymbol{R}_l)=E\psi(\boldsymbol{r}_i,\boldsymbol{R}_l)\,. \tag{2.9}$$

The time-independent Schrödinger wave equation for the electrons, if one assumes the nuclei are at fixed positions $\boldsymbol{R}_l$, is

$$\mathcal{H}_0\phi(\boldsymbol{r}_i,\boldsymbol{R}_l)=E^0\phi(\boldsymbol{r}_i,\boldsymbol{R}_l)\,. \tag{2.10}$$

Born and Huang [46] have made a perturbation expansion of the solution of (2.9) in powers of K. They have shown that if the wave function is evaluated to second order in K, then a product separation of the form $\psi_n(\boldsymbol{r}_i,\boldsymbol{R}_l)=\phi_n(\boldsymbol{r}_i)X(\boldsymbol{R}_l)$ where n labels an electronic state, is possible. The assertion that the total wave function can be written as a product of the electronic wave function (depending only on electronic coordinates with the nuclei at fixed positions) times the nuclear wave function (depending only on nuclear coordinates with the electrons in some fixed state) is the physical content of the Born–Oppenheimer approximation (1927). In this approximation the electrons provide a potential energy for the motion of the nuclei while the moving nuclei *continuously* deform the wave function of the electrons (rather than causing any sudden changes). Thus this idea is also called the adiabatic approximation.

It turns out when the wave function is evaluated to second order in K that the effective potential energy of the nuclei involves nuclear displacements to fourth order and lower. Expanding the nuclear potential energy to second order in the nuclear displacements yields the harmonic approximation. Terms higher than second order are called *anharmonic* terms. Thus it is possible to treat anharmonic terms and still stay within the Born–Oppenheimer approximation.

If we evaluate the wave function to third order in K, it turns out that a simple product separation of the wave function is no longer possible. Thus the Born–Oppenheimer approximation breaks down. This case corresponds to an effective potential energy for the nuclei of fifth order. Thus it really does not appear to be correct to assume that there exists a nuclear potential function that includes fifth or higher power terms in the nuclear displacement, at least from the viewpoint of the perturbation expansion.

Apparently, in actual practice the adiabatic approximation does not break down quite so quickly as the above discussion suggests. To see that this might be so a somewhat simpler development of the Born–Oppenheimer approximation [46] is sometimes useful. In this development, we attempt to find a solution for ψ in (2.9) of the form

$$\psi(\boldsymbol{r}_i,\boldsymbol{R}_l)=\textstyle\sum_n\psi_n(\boldsymbol{R}_l)\phi_n(\boldsymbol{r}_i,\boldsymbol{R}_l)\,. \tag{2.11}$$

The ϕ_n are eigenfunctions of (2.10). Substituting into (2.9) gives

$$\textstyle\sum_n\mathcal{H}_{\mathrm{c}}\psi_n\phi_n=E\sum_n\psi_n\phi_n\,,$$

or

$$\sum_n \mathcal{H}_0 \psi_n \phi_n + \sum_n T_\mathrm{N} \psi_n \phi_n = E \sum_n \psi_n \phi_n ,$$

or using (2.10) gives

$$\sum_n E_n^0 \psi_n \phi_n + \sum_n T_\mathrm{N} (\psi_n \phi_n) = E \sum_n \psi_n \phi_n .$$

Noting that

$$T_\mathrm{N}(\psi_n \phi_n) = (T_\mathrm{N} \psi_n)\phi_n + \psi_n (T_\mathrm{N} \phi_n) + \sum_l \frac{1}{M_l} (\boldsymbol{P}_l \phi_n) \cdot (\boldsymbol{P}_l \psi_n) ,$$

where

$$T_\mathrm{N} = \sum_l \frac{1}{2M_l} P_l^2 = -\hbar \sum_l \frac{1}{2M_l} \nabla_{\boldsymbol{R}_l}^2 ,$$

we can write the above as

$$\begin{aligned} &\sum_{n^1} \phi_{n^1} (T_N + E_n^0 - E)\psi_{n^1} + \sum_{n^1} \psi_{n^1} T_N \phi_{n^1} \\ &+ \sum_{n^1} \sum_l \frac{1}{M_l} (\boldsymbol{P}_l, \phi_{n^1}) \cdot (\boldsymbol{P}_l, \psi_{n^1}) = 0 \end{aligned} .$$

Multiplying the above equation by ϕ_n^* and integrating over the electronic coordinates gives

$$(T_\mathrm{N} + E_n^0 - E)\psi_n + \sum_{n^1} C_{nn^1}(\boldsymbol{R}_l, \boldsymbol{P}_l)\psi_{n^1} = 0 , \tag{2.12}$$

where

$$C_{nn^1} = \sum_{li} \frac{1}{M_l} \left(Q_{nn^1}^{li} P_{li} + R_{nn^1}^{li} \right) \tag{2.13}$$

(the sum over i goes from 1 to 3, labeling the x, y, and z components) and

$$Q_{nn^1}^{li} = \int \phi_n^* P_{li} \phi_{n^1} \mathrm{d}\tau , \tag{2.14}$$

$$R_{nn^1}^{li} = \tfrac{1}{2} \int \phi_n^* P_{li}^2 \phi_{n^1} \mathrm{d}\tau . \tag{2.15}$$

The integration is over electronic coordinates.

For stationary states, the ϕs can be chosen to be real and so it is easily seen that the diagonal elements of Q vanish:

$$Q_{nn^1}^{li} = \int \phi_n P_{li} \phi_n \mathrm{d}\tau = \frac{\hbar}{2i} \frac{\partial}{\partial X_{li}} \int \phi_n^2 \mathrm{d}\tau = 0 .$$

From this we see that the effect of the diagonal elements of C is a multiplication effect and not an operator effect. Therefore the diagonal elements of C can be added to E_n^0 to give an effective potential energy U_{eff}.[2] Equation (2.12) can be written as

$$(T_{\text{N}} + U_{\text{eff}} - E)\psi_n + \sum\nolimits_{n^1(\neq n)} C_{nn^1}\psi_{n^1} = 0 . \tag{2.16}$$

If the C_{nn^1} vanish, then we can split the discussion of the electronic and nuclear motions apart as in the adiabatic approximation. Otherwise, of course, we cannot. For metals there appears to be no reason to suppose that the effect of the C is negligible. This is because the excited states are continuous in energy with the ground state, and so the sum in (2.16) goes over into an integral. Perhaps the best way to approach this problem would be to just go ahead and make the Born–Oppenheimer approximation. Then wave functions could be evaluated so that the C_{nn^1} could be evaluated. One could then see if the calculations were consistent, by seeing if the C were actually negligible in (2.16).

In general, perturbation theory indicates that if there is a large energy gap between the ground and excited electronic states, then an adiabatic approximation may be valid.

Can we even speak of lattice vibrations in metals without explicitly also discussing the electrons? The above discussion might lead one to suspect that the answer is no. However, for completely free electrons (whose wave functions do not depend at all on the $\boldsymbol{R}_l$) it is clear that all the C vanish. Thus the presence of free electrons does not make the Born–Oppenheimer approximation invalid (using the concept of *completely free* electrons to represent any of the electrons in a solid is, of course, unrealistic). In metals, when the electrons can be thought of as almost free, perhaps the net effect of the C is small enough to be neglected in zeroth-order approximation. We shall suppose this is so and suppose that the Born–Oppenheimer approximation can be applied to conduction electrons in metals. But we should also realize that strange effects may appear in metals due to the fact that the coupling between electrons and lattice vibrations is not negligible. In fact, as we shall see in a later chapter, the mere presence of electrical resistivity means that the Born–Oppenheimer approximation is breaking down. The phenomenon of superconductivity is also due to this coupling. At any rate, we can always write the Hamiltonian as $\mathcal{H} = \mathcal{H}$(electrons) + $\mathcal{H}$(lattice vibrations) + $\mathcal{H}$(coupling). It just may be that in metals, $\mathcal{H}$(coupling) cannot always be regarded as a small perturbation.

Finally, it is well to note that the perturbation expansion results depend on K being fairly small. If nature had not made the mass of the proton much larger than the mass of the electron, it is not clear that there would be any valid Born–Oppenheimer approximation.[3]

---

[2] We have used the terms Born–Oppenheimer approximation and adiabatic approximation interchangeably. More exactly, Born–Oppenheimer corresponds to neglecting C_{nn}, whereas in the adiabatic approximation C_{nn} is retained.

[3] For further details of the Born–Oppenheimer approximation, references [46], [82], [22, Vol 1, pp 611-613] and the references cited therein can be consulted.

2.2 One-Dimensional Lattices (B)

Perhaps it would be most logical at this stage to plunge directly into the problem of solving quantum-mechanical three-dimensional lattice vibration problems either in the harmonic or in a more general adiabatic approximation. But many of the interesting features of lattice vibrations are not quantum-mechanical and do not depend on three-dimensional motion. Since our aim is to take a fairly easy path to the understanding of lattice vibrations, it is perhaps best to start with some simple classical one-dimensional problems. The classical theory of lattice vibrations is due to M. Born, and Born and Huang [2.5] contains a very complete treatment.

Even for the simple problems, we have a choice as to whether to use the harmonic approximation or the general adiabatic approximation. Since the latter involves quartic powers of the nuclear displacements while the former involves only quadratic powers, it is clear that the former will be the simplest starting place. For many purposes the harmonic approximation gives an adequate description of lattice vibrations. This chapter will be devoted almost entirely to a description of lattice vibrations in the harmonic approximation.

A very simple physical model of this approximation exists. It involves a potential with quadratic displacements of the nuclei. We could get the same potential by connecting suitable springs (which obey Hooke's law) between appropriate atoms. This in fact is an often-used picture.

Even with the harmonic approximation there is still a problem as to what value we should assign to the "spring constants" or force constants. No one can answer this question from first principles (for a real solid). To do this we would have to know the electronic energy eigenvalues as a function of nuclear position ($\boldsymbol{R}_l$). This is usually too complicated a many-body problem to have a solution in any useful approximation. So the "spring constants" have to be left as unknown parameters, which are determined from experiment or from a model that involves certain approximations.

It should be mentioned that our approach (which we could call the *unrestricted force constants approach*) to discussing lattice vibration is probably as straightforward as any and it also is probably as good a way to begin discussing the lattice vibration problem as any. However, there has been a considerable amount of progress in discussing lattice vibration problems beyond that of our approach. In large part this progress has to do with the way the interaction between atoms is viewed. In particular, the shell model[4] has been applied with good results to ionic and covalent crystals.[5] The shell model consists in regarding each atom as consisting of a core (the nucleus and inner electrons) plus a shell. The core and shell are coupled together on each atom. The shells of nearest-neighbor atoms are coupled. Since the cores can move relative to the shells, it is possible to polarize the atoms. Electric dipole interactions can then be included in neighbor interactions.

---

[4] See Dick and Overhauser [2.12].

[5] See, for example, Cochran [2.9].

Lattice vibrations in metals can be particularly difficult to treat by starting from the standpoint of force constants as we do. A special way of looking at lattice vibrations in metals has been given.[6] Some metals can apparently be described by a model in which the restoring forces between ions are either of the bond-stretching or axially symmetric bond-bending variety.[7]

We have listed some other methods for looking at the vibrational problems in Table 2.1. Methods, besides the Debye approximation (Sect. 2.3.3), for approximating the frequency distribution include root sampling and others [2.26, Chap. 3]. Montroll[8] has given an elegant way for estimating the frequency distribution, at least away from singularities. This method involves taking a trace of the Dynamical Matrix (2.3.2) and is called the moment-trace method. Some later references for lattice dynamics calculations are summarized in Table 2.1.

Table 2.1. References for Lattice vibration calculations

Lattice vibrational calculations	Reference
Einstein	Kittel [23, Chap. 5]
Debye	Chap. 2, this book
Rigid Ion Models	Bilz and Kress [2.3]
Shell Model	Jones and March [2.20, Chap. 3]. Also footnotes 4 and 5.
ab initio models	Kunc et al [2.22]. Strauch et al [2.33]. Density Functional Techniques are used (see Chap. 3).
General reference	Maradudin et al [2.26]. See also Born and Huang [46]

2.2.1 Classical Two-Atom Lattice with Periodic Boundary Conditions (B)

We start our discussion of lattice vibrations by considering the simplest problem that has any connection with real lattice vibrations. Periodic boundary conditions will be used on the two-atom lattice because these are the boundary conditions that are used on large lattices where the effects of the surface are relatively unimportant. Periodic boundary conditions mean that when we come to the end of the lattice we assume that the lattice (including its motion) identically repeats itself. It will be assumed that adjacent atoms are coupled with springs of spring constant γ. Only nearest-neighbor coupling will be assumed (for a two-atom lattice, you couldn't assume anything else).

6 See Toya [2.34].

7 See Lehman et al [2.23]. For a more general discussion, see Srivastava [2.32].

8 See Montroll [2.28].

As should already be clear from the Born–Oppenheimer approximation, in a lattice all motions of sufficiently small amplitude are describable by Hooke's law forces. This is true no matter what the physical origin (ionic, van der Waals, etc.) of the forces. This follows directly from a Taylor series expansion of the potential energy using the fact that the first derivative of the potential evaluated at the equilibrium position must vanish.

The two-atom lattice is shown in Fig. 2.1, where a is the equilibrium separation of atoms, x_1 and x_2 are coordinates measuring the displacement of atoms 1 and 2 from equilibrium, and m is the mass of atom 1 or 2. The idea of periodic boundary conditions is shown by repeating the structure outside the vertical dashed lines.

With periodic boundary conditions, Newton's second law for each of the two atoms is

$$\begin{aligned} m\ddot{x}_1 &= \gamma(x_2 - x_1) - \gamma(x_1 - x_2), \\ m\ddot{x}_2 &= \gamma(x_1 - x_2) - \gamma(x_2 - x_1). \end{aligned} \tag{2.17}$$

In (2.17), each dot means a derivative with respect to time.

Solutions of (2.17) will be sought in which both atoms vibrate with the same frequency. Such solutions are called *normal mode solutions* (see Appendix B). Substituting

$$x_n = u_n \exp(\mathrm{i}\omega t) \tag{2.18}$$

in (2.17) gives

$$\begin{aligned} -\omega^2 m u_1 &= \gamma(u_2 - u_1) - \gamma(u_1 - u_2), \\ -\omega^2 m u_2 &= \gamma(u_1 - u_2) - \gamma(u_2 - u_1). \end{aligned} \tag{2.19}$$

Equation (2.19) can be written in matrix form as

$$\begin{pmatrix} 2\gamma - \omega^2 m & -2\gamma \\ -2\gamma & 2\gamma - \omega^2 m \end{pmatrix} \begin{pmatrix} u_1 \\ u_2 \end{pmatrix} = 0. \tag{2.20}$$

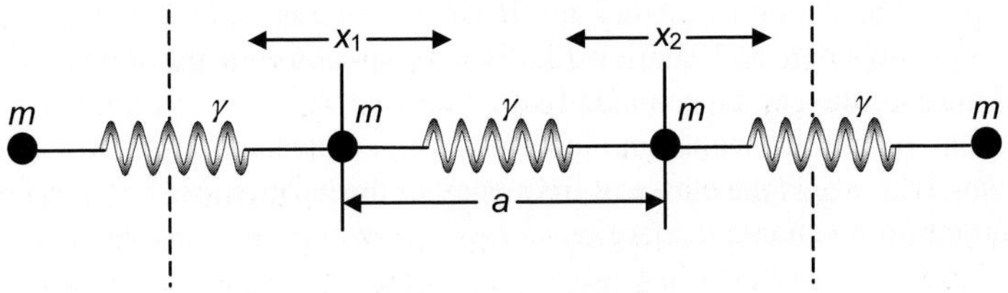

Fig. 2.1. The two-atom lattice (with periodic boundary conditions schematically indicated)

For nontrivial solutions (u_1 and u_2 not both equal to zero) of (2.20) the determinant (written det below) of the matrix of coefficients must be zero or

$$\det\begin{bmatrix} 2\gamma - \omega^2 m & -2\gamma \\ -2\gamma & 2\gamma - \omega^2 m \end{bmatrix} = 0 . \tag{2.21}$$

Equation (2.21) is known as the *secular equation,* and the two frequencies that satisfy (2.21) are known as *eigenfrequencies*.

These two eigenfrequencies are

$$\omega_1^2 = 0 , \tag{2.22}$$

and

$$\omega_2^2 = 4\gamma / m . \tag{2.23}$$

For (2.22), $u_1 = u_2$ and for (2.23),

$$(2\gamma - 4\gamma)u_1 = 2\gamma u_2 \quad \text{or} \quad u_1 = -u_2 .$$

Thus, according to Appendix B, the normalized eigenvectors corresponding to the frequencies ω_1 and ω_2 are

$$E_1 = \frac{(1,1)}{\sqrt{2}} , \tag{2.24}$$

and

$$E_2 = \frac{(1,-1)}{\sqrt{2}} . \tag{2.25}$$

The first term in the row matrix of (2.24) or (2.25) gives the relative amplitude of u_1 and the second term gives the relative amplitude of u_2. Equation (2.25) says that in mode 2, $u_2/u_1 = -1$, which checks our previous results. Equation (2.24) describes a pure translation of the crystal. If we are interested in a fixed crystal, this solution is of no interest. Equation (2.25) corresponds to a motion in which the center of mass of the crystal remains fixed.

Since the quantum-mechanical energies of a harmonic oscillator are $E_n = (n + 1/2)\hbar\omega$, where ω is the classical frequency of the harmonic oscillator, it follows that the quantum-mechanical energies of the *fixed* two-atom crystal are given by

$$E_n = \left(n + \frac{1}{2}\right)\hbar\sqrt{\frac{4\gamma}{m}} . \tag{2.26}$$

This is our first encounter with normal modes, and since we shall encounter them continually throughout this chapter, it is perhaps worthwhile to make a few more comments. The sets E_1 and E_2 determine the normal coordinates of the normal mode. They do this by defining a transformation. In this simple example, the theory of small oscillations tells us that the normal coordinates are

$$X_1 = \frac{u_1}{\sqrt{2}} + \frac{u_2}{\sqrt{2}} \quad \text{and} \quad X_2 = \frac{u_1}{\sqrt{2}} - \frac{u_2}{\sqrt{2}}.$$

Note that X_1, X_2 are given by

$$\begin{pmatrix} X_1 \\ X_2 \end{pmatrix} = \begin{pmatrix} E_1 \\ E_2 \end{pmatrix} \begin{pmatrix} u_1 \\ u_2 \end{pmatrix} = \frac{1}{\sqrt{2}} \begin{pmatrix} 1 & 1 \\ 1 & -1 \end{pmatrix} \begin{pmatrix} u_1 \\ u_2 \end{pmatrix}.$$

X_1 and X_2 are the amplitudes of the normal modes. If we want the time-dependent normal coordinates, we would multiply the first set by $\exp(i\omega_1 t)$ and the second set by $\exp(i\omega_2 t)$. In most applications when we say *normal coordinates* it should be obvious which set (time-dependent or otherwise) we are talking about.

The following comments are also relevant:

1. In an n-dimensional problem with m atoms, there are $(n{\cdot}m)$ normal coordinates corresponding to nm different independent motions.
2. In the harmonic approximation, each normal coordinate describes an independent mode of vibration with a single frequency.
3. In a normal mode, all atoms vibrate with the same frequency.
4. Any vibration in the crystal is a superposition of normal modes.

2.2.2 Classical, Large, Perfect Monatomic Lattice, and Introduction to Brillouin Zones (B)

Our calculation will still be classical and one-dimensional but we shall assume that our chain of atoms is long. Further, we shall give brief consideration to the possibility that the forces are not harmonic or nearest-neighbor. By a long crystal will be meant a crystal in which it is not very important what happens at the boundaries. However, since the crystal is finite, some choice of boundary conditions must be made. Periodic boundary conditions (sometimes called Born–von Kárman or cyclic boundary conditions) will be used. These boundary conditions can be viewed as the large line of atoms being bent around to form a ring (although it is not topologically possible analogously to represent periodic boundary conditions in three dimensions). A perfect crystal will mean here that the forces between any two atoms depend only on the separation of the atoms and that there are no defect atoms. Perfect monatomic further implies that all atoms are identical.

N atoms of mass M will be assumed. The equilibrium spacing of the atoms will be a. x_n will be the displacement of the nth atom from equilibrium. V will be the potential energy of the interacting atoms, so that $V = V(x_1,\ldots,x_n)$. By the Born–Oppenheimer approximation it makes sense to expand the potential energy to fourth order in displacements:

$$\begin{aligned} V(x_1,\ldots,x_N) = \\ & V(0,\ldots,0) + \frac{1}{2}\sum\nolimits_{n,n'}\left(\frac{\partial^2 V}{\partial x_n \partial x_{n'}}\right)_{(x_1,\ldots,x_N)=0} x_n x_{n'} \\ & + \frac{1}{6}\sum\nolimits_{n,n',n''}\left(\frac{\partial^3 V}{\partial x_n \partial x_{n'} \partial x_{n''}}\right)_{(x_1,\ldots,x_N)=0} x_n x_{n'} x_{n''} \\ & + \frac{1}{24}\sum\nolimits_{n,n',n'',n'''}\left(\frac{\partial^4 V}{\partial x_n \partial x_{n'} \partial x_{n''} \partial x_{n'''}}\right)_{(x_1,\ldots,x_N)=0} x_n x_{n'} x_{n''} x_{n'''} . \end{aligned} \tag{2.27}$$

In (2.27), $V(0,\ldots,0)$ is just a constant and the zero of the potential energy can be chosen so that this constant is zero. The first-order term $(\partial V/\partial x)_{(x1,\ldots,xN)=0}$ is the negative of the force acting on atom n in equilibrium; hence it is zero and was left out of (2.27). The second-order terms are the terms that one would use in the harmonic approximation. The last two terms are the anharmonic terms.

Note in the summations that there is no restriction that says that n' and n must refer to adjacent atoms. Hence (2.27), as it stands, includes the possibility of forces between all pairs of atoms.

The dynamical problem that (2.27) gives rise to is only exactly solvable in closed form if the anharmonic terms are neglected. For small oscillations, their effect is presumably much smaller than the harmonic terms. The cubic and higher-order terms are responsible for certain effects that completely vanish if they are left out. Whether or not one can neglect them depends on what one wants to describe. We need anharmonic terms to explain thermal expansion, a small correction (linear in temperature) to the specific heat of an insulator at high temperatures, and the thermal resistivity of insulators at high temperatures. The effect of the anharmonic terms is to introduce interactions between the various normal modes of the lattice vibrations. A separate chapter is devoted to interactions and so they will be neglected here. This still leaves us with the possibility of forces of greater range than nearest-neighbors.

It is convenient to define

$$V_{n,n'} = \left(\frac{\partial^2 V}{\partial x_n \partial x_{n'}}\right)_{(x_1,\ldots,x_N)=0} . \tag{2.28}$$

$V_{n,n'}$ has several properties. The order of taking partial derivatives doesn't matter, so that

$$V_{n,n'} = V_{n',n} . \tag{2.29}$$

Two further restrictions on the V may be obtained from the equations of motion. These equations are simply obtained by Lagrangian mechanics [2]. From our model, the Lagrangian is

$$L = (M/2)\sum_n \dot{x}_n^2 - \tfrac{1}{2}\sum_{n,n'} V_{n,n'} x_n x_{n'} \,. \tag{2.30}$$

The sums extend over the one-dimensional crystal. The Lagrange equations are

$$\frac{\mathrm{d}}{\mathrm{d}t}\frac{\partial L}{\partial \dot{x}_n} - \frac{\partial L}{\partial x_n} = 0 \,. \tag{2.31}$$

The equation of motion is easily found by combining (2.30) and (2.31):

$$M\ddot{x}_n = -\sum_{n'} V_{n,n'} x_{n'} \,. \tag{2.32}$$

If all atoms are displaced a constant amount, this corresponds to a translation of the crystal, and in this case the resulting force on each atom must be zero. Therefore

$$\sum_{n'} V_{n,n'} = 0 \,. \tag{2.33}$$

If all atoms except the kth are at their equilibrium position, then the force on the nth atom is the force acting between the kth and nth atoms,

$$F = M\ddot{x}_n = -V_{nk} x_k \,.$$

But because of periodic boundary conditions and translational symmetry, this force can depend only on the relative positions of n and k, and hence on their difference, so that

$$V_{n,k} = V(n-k) \,. \tag{2.34}$$

With these restrictions on the V in mind, the next step is to solve (2.32).

Normal mode solutions of the form

$$x_n = u_n \mathrm{e}^{\mathrm{i}\omega t} \tag{2.35}$$

will be sought. The u_n are assumed to be time independent. Substituting (2.35) into (2.32) gives

$$pu_n \equiv M\omega^2 u_n - \sum_{n'} V(n'-n) u_{n'} = 0 \,. \tag{2.36}$$

Equation (2.36) is a difference equation with constant coefficients. Note that a new operator p is defined by (2.36).

This difference equation has a nice property due to its translational symmetry. Let n go to $n+1$ in (2.36). We obtain

$$M\omega^2 u_{n+1} - \sum_{n'} V(n'-n-1) u_{n'} = 0 \,. \tag{2.37}$$

Then make the change $n' \to n' + 1$ in the dummy variable of summation. Because of periodic boundary conditions, no change is necessary in the limits of summation. We obtain

$$M\omega^2 u_{n+1} - \sum_{n'} V(n' - n) u_{n'+1} = 0 \,. \tag{2.38}$$

Comparing (2.36) and (2.38) we see that if $pu_n = 0$, then $pu_{n+1} = 0$. If $pf = 0$ had only one solution, then it follows that

$$u_{n+1} = \mathrm{e}^{iqa} u_n \,, \tag{2.39}$$

where e^{iqa} is some arbitrary constant K, that is, $q = \ln(K/ia)$. Equation (2.39) is an expression of a very important theorem by Bloch that we will have occasion to discuss in great detail later. The fact that we get all solutions by this assumption follows from the fact that if $pf = 0$ has N solutions, then N linearly independent linear combinations of solutions can always be constructed so that each satisfies an equation of the form (2.39) [75].

By applying (2.39) n times starting with $n = 0$ it is readily seen that

$$u_n = \mathrm{e}^{iqna} u_0 \,. \tag{2.40}$$

If we wish to keep u_n finite as $n \to \pm\infty$, then it is evident that q must be real. Further, if there are N atoms, it is clear by periodic boundary conditions that $u_n = u_0$, so that

$$qNa = 2\pi m \,, \tag{2.41}$$

where m is an integer.

Over a restricted range, each different value of m labels a different normal mode solution. We will show later that the modes corresponding to m and $m + N$ are in fact the same mode. Therefore, all physically interesting modes are obtained by restricting m to be any N consecutive integers. A common such range is (supposing N to be even)

$$-(N/2) + 1 \le m \le N/2 \,.$$

For this range of m, q is restricted to

$$-\pi / a < q \le \pi / a \,. \tag{2.42}$$

This range of q is called the *first Brillouin zone.*

Substituting (2.40) into (2.36) shows that (2.40) is indeed a solution, provided that ω_q satisfies

$$M\omega_q^2 = \sum_{n'} V(n' - n) \mathrm{e}^{iqa(n'-n)} \,,$$

or

$$\omega_q^2 = \frac{1}{M} \sum_{l=-\infty}^{\infty} V(l) \mathrm{e}^{iqal} \,, \tag{2.43}$$

or

$$\omega_q^2 = \frac{1}{M}\sum_{l=-\infty}^{\infty} V(l)\cos(qla),$$

for an infinite crystal (otherwise the sum can run over appropriate limits specifying the crystal). In getting the *dispersion relation* (2.43), use has been made of (2.29).

Equation (2.43) directly shows one general property of the dispersion relation for lattice vibrations:

$$\omega^2(-q) = \omega^2(q). \tag{2.44}$$

Another general property is obtained by expanding $\omega^2(q)$ in a Taylor series:

$$\omega^2(q) = \omega^2(0) + (\omega^2)'_{q=0}q + \tfrac{1}{2}(\omega^2)''_{q=0}q^2 + \cdots. \tag{2.45}$$

From (2.43), (2.33), and (2.34),

$$\omega^2(0) \propto \sum_l V(l) = 0.$$

From (2.44), $\omega^2(q)$ is an even function of q and hence $(\omega^2)'_{q=0} = 0$. Thus for sufficiently small q,

$$\omega^2(q) = (\text{constant})q^2 \quad \text{or} \quad \omega(q) = (\text{constant})q. \tag{2.46}$$

Equation (2.46) is a dispersion relation for waves propagating without dispersion (that is, their group velocity $\mathrm{d}\omega/\mathrm{d}q$ equals their phase velocity ω/q). This is the type of relation that is valid for vibrations in a continuum. It is not surprising that it is found here. The small q approximation is a low-frequency or long-wavelength approximation; hence the discrete nature of the lattice is unimportant.

That small q can be thought of as indicating a long-wavelength is perhaps not evident. q (which is often called the *wave vector*) can be given the interpretation of $2\pi/\lambda$, where λ is a wavelength, This is easily seen from the fact that the amplitude of the vibration for the nth atom should equal the amplitude of vibration for the zeroth atom provided $na = \lambda$.

In that case

$$u_n = \mathrm{e}^{\mathrm{i}qna}u_0 = \mathrm{e}^{\mathrm{i}q\lambda}u_0 = u_0,$$

so that $q = 2\pi/\lambda$. This equation for q also indicates why there is no unique q to describe a vibration. In a discrete (not continuous) lattice there are several wavelengths that will describe the same physical vibration. The point is that in order to describe the vibrations, we have to know only the value of a function at a discrete set of points and we do not care what values it takes on in between. There are obviously many distinct functions that have the same value at many discrete points. The idea is illustrated in Fig. 2.2.

Restricting $q = 2\pi/\lambda$ to the first Brillouin zone is equivalent to selecting the range of q to have as small a $|q|$ or as large a wavelength as possible. Letting q become negative just means that the direction of propagation of the wave is reversed. In Fig. 2.2, (a) is a first Brillouin zone description of the wave, whereas (b) is not.

It is worthwhile to get an explicit solution to this problem in the case where only nearest-neighbor forces are involved. This means that

$$V(l) = 0 \quad (\text{if } l \neq 0 \text{ or } 1)\,.$$

By (2.29) and (2.34),

$$V(+l) = V(-l)\,.$$

By (2.33) and the nearest-neighbor assumption,

$$V(+l) + V(0) + V(-l) = 0\,.$$

Thus

$$V(+l) = V(-l) = -\tfrac{1}{2}V(0)\,. \tag{2.47}$$

By combining (2.47) with (2.43), we find that

$$\omega^2 = \frac{V(0)}{M}(1 - \cos qa)\,,$$

or that

$$\omega = \sqrt{\frac{2V(0)}{M}}\left|\sin\frac{qa}{2}\right|. \tag{2.48}$$

This is the dispersion relation for our problem. The largest value that ω can have is

$$\omega_c = \sqrt{\frac{2V(0)}{M}}\,. \tag{2.49}$$

By (2.48) it is obvious that increasing q by $2\pi/a$ leaves the value of ω unchanged. By (2.35), (2.40), (2.41), and (2.48), the displacement of the nth atom in the mth normal mode is given by

$$x_n^{(m)} = u_0 \exp\left[ina\left(\frac{2\pi m}{Na}\right)\right]\exp\left[it\left(\frac{2V(0)}{M}\right)\left|\sin\left(\frac{a}{2}\cdot\frac{2\pi m}{Na}\right)\right|\right]. \tag{2.50}$$

This is also invariant to increasing $q = 2\pi m/Na$ by $2\pi/a$.

A plot of the dispersion relation (ω versus q) as given by (2.48) looks something like Fig. 2.3. In Fig. 2.3, we imagine $N \to \infty$ so that the curve is defined by an almost continuous set of points.

For the two-atom case, the theory of small oscillations tells us that the normal coordinates (X_1, X_2) are found from the transformation

$$\begin{pmatrix} X_1 \\ X_2 \end{pmatrix} = \begin{pmatrix} \frac{1}{\sqrt{2}} & \frac{1}{\sqrt{2}} \\ \frac{1}{\sqrt{2}} & -\frac{1}{\sqrt{2}} \end{pmatrix} \begin{pmatrix} x_1 \\ x_2 \end{pmatrix}. \tag{2.51}$$

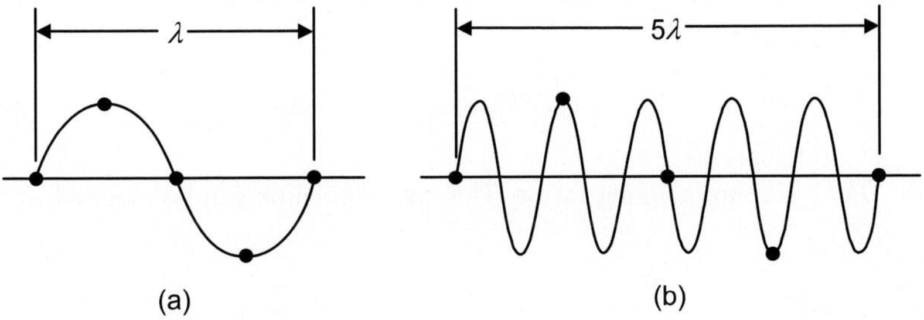

Fig. 2.2. Different wavelengths describe the same vibration in a discrete lattice. (The dots represent atoms. Their displacement is indicated by the distance of the dots from the horizontal axis.) (a) $q = \pi/2a$, (b) $q = 5\pi/2a$

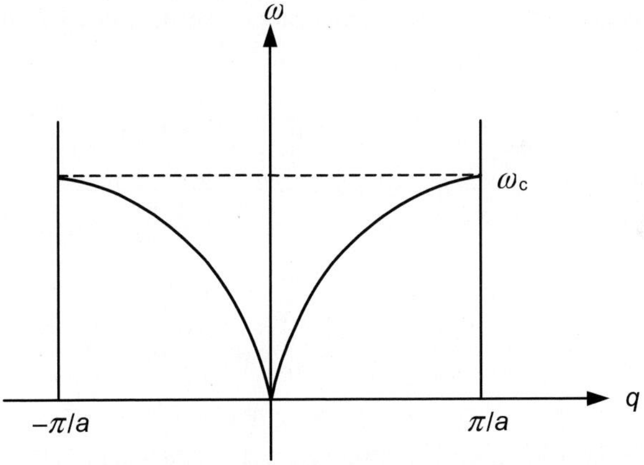

Fig. 2.3. Frequency versus wave vector for a large one-dimensional crystal

If we label the various components of the eigenvectors (E_i) by adding a subscript, we find that

$$X_i = \sum_j E_{ij} x_j \, . \tag{2.52}$$

The equations of motion of each X_i are harmonic oscillator equations of motion. The normal coordinate transformation reduced the two-atom problem to the problem of two decoupled harmonic oscillators.

We also want to investigate if the normal coordinate transformation reduces the N-atom problem to a set of N decoupled harmonic oscillators. The normal coordinates each vibrate with a time factor $e^{i\omega t}$ and so they must describe some sort of harmonic oscillators. However, it is useful for later purposes to demonstrate this explicitly.

By analogy with the two-atom case, we expect that the normal coordinates in the N-atom case are given by

$$X_{m'} = \frac{1}{\sqrt{N}} \sum\nolimits_{n'} \exp\left(\frac{i2\pi m'n'}{N}\right) x_{n'} , \tag{2.53}$$

where $1/N^{1/2}$ is a normalizing factor. This transformation can be inverted as follows:

$$\begin{aligned} \frac{1}{\sqrt{N}} \sum\nolimits_{m'} \exp\left(-\frac{2\pi i m'n}{N}\right) X_{m'} &= \frac{1}{N} \sum\nolimits_{m',n'} \exp\left[\frac{2\pi i}{N}(n'-n)m'\right] x_{n'} \\ &= \frac{1}{N} \sum\nolimits_{n'} x_{n'} \sum\nolimits_{m'} \exp\left[\frac{2\pi i}{N}(n'-n)m'\right]. \end{aligned} \tag{2.54}$$

In (2.54), the sum over m' runs over any continuous range in m' equivalent to one Brillouin zone. For convenience, this range can be chosen from 0 to $N - 1$. Then

$$\begin{aligned} \sum\nolimits_{m'=0}^{N-1} \exp\left[\frac{2\pi i}{N}(n'-n)m'\right] &= \frac{1-\left\{\exp\left[\frac{2\pi i}{N}(n'-n)\right]\right\}^N}{1-\exp\left[\frac{2\pi i}{N}(n'-n)\right]} \\ &= \frac{1-1}{1-\exp\left[\frac{2\pi i}{N}(n'-n)\right]} \\ &= 0 \quad \text{unless } n' = n. \end{aligned}$$

If $n' = n$, then $\sum_{m'}$ just gives N. Therefore we can say in general that

$$\frac{1}{N} \sum\nolimits_{m'=0}^{N-1} \exp\left[\frac{2\pi i}{N}(n'-n)m'\right] = \delta_{n'}^{n} . \tag{2.55}$$

Equations (2.54) and (2.55) together give

$$x_n = \frac{1}{\sqrt{N}} \sum\nolimits_{m'} \exp\left(-\frac{2\pi i}{N} m'n\right) X_{m'} , \tag{2.56}$$

which is the desired inversion of the transformation defined by (2.53).

We wish to show now that this normal coordinate transformation reduces the Hamiltonian for the N interacting atoms to a Hamiltonian representing a set of N decoupled harmonic oscillators. The reason for the emphasis on the Hamiltonian is that this is the important quantity to consider in nonrelativistic quantum-mechanical problems. This reduction not only shows that the ω are harmonic oscillator frequencies, but it gives an example of an N-body problem that can be exactly solved because it reduces to N one-body problems.

First, we must construct the Hamiltonian. If the Lagrangian $L(q_k, \dot{q}_k, t)$ is expressed in terms of generalized coordinates q_k and velocities $\dot{q}_k$, then the canonically conjugate generalized momenta are defined by

$$p_k = \frac{\partial L(q_k, \dot{q}_k, t)}{\partial \dot{q}_k}. \tag{2.57}$$

$\mathcal{H}$ is defined by

$$\mathcal{H}(p_k, q_k, t) = \sum_j \dot{q}_j p_j - L(q_k, \dot{q}_k, t). \tag{2.58}$$

The equations of motion of the system can be obtained by Hamilton's canonical equations,

$$\dot{q}_k = \frac{\partial \mathcal{H}}{\partial p}, \tag{2.59}$$

$$\dot{p}_k = -\frac{\partial \mathcal{H}}{\partial q_k}. \tag{2.60}$$

If the constraints are independent of the time and if the potential V is independent of the velocity, then the Hamiltonian is just the total energy, $T + V$ ($T \equiv$ kinetic energy), and is constant. In this case we really do not need to use (2.58) to construct the Hamiltonian.

From the above, the Hamiltonian of our system is

$$\mathcal{H} = \frac{M}{2}\sum_n \dot{x}_n^2 + \frac{1}{2}\sum_{n,n'} V_{n,n'} x_n x_{n'}. \tag{2.61}$$

As yet, no conditions requiring x_n to be real have been inserted in the normal coordinate definitions. Since the x_n are real, the normal coordinates, defined by (2.56), must satisfy

$$X_{-m} = X_m^*. \tag{2.62}$$

Similarly $\dot{x}_n$ is real, and this implies that

$$\dot{X}_{-m} = \dot{X}_m^*. \tag{2.63}$$

Substituting (2.56) into (2.61) yields

$$\mathcal{H} = \frac{M}{2}\sum_n \frac{1}{N}\sum_{m,m'} \exp\left[-\frac{2\pi i}{N} n(m+m')\right]\dot{X}_m \dot{X}_{m'}$$
$$+\frac{1}{2}\sum_{n,n'} V_{n,n'} \sum_{m,m'} \frac{1}{N}\exp\left[-\frac{2\pi i}{N}(nm+n'm')\right] X_m X_{m'} .$$

The last equation can be written

$$\mathcal{H} = \frac{M}{2N}\sum_{m,m'} \dot{X}_m \dot{X}_{m'} \sum_n \exp\left[-\frac{2\pi i}{N} n(m+m')\right]$$
$$+\frac{1}{2N}\sum_{m,m'} X_m X_{m'} \sum_{n-n'} V(n-n') \exp\left[-\frac{2\pi i}{N}(n-n')m\right] \tag{2.64}$$
$$\times \sum_{n'} \exp\left[-\frac{2\pi i}{N} n'(m+m')\right].$$

Using the results of Problem 2.2, we can write (2.64) as

$$\mathcal{H} = \frac{M}{2}\sum_m \dot{X}_m \dot{X}_{-m} + \frac{1}{2}\sum_m X_m X_{-m} \sum_l V(l) \exp\left(-\frac{2\pi i}{N} lm\right),$$

or by (2.43), (2.62), and (2.63),

$$\mathcal{H} = \sum_m \left(\frac{M}{2}\left|\dot{X}_m^2\right| + \frac{1}{2} M\omega_m^2 \left|X_m\right|^2\right). \tag{2.65}$$

Equation (2.65) is practically the correct form. What is needed is an equation similar to (2.65) but with the X real. It is possible to find such an expression by making the following transformation: Define u and v so that

$$X_m = u_m + iv_m . \tag{2.66}$$

Since $X_m^* = X_{-m}$, it is seen that $u_m = u_{-m}$ and $v_m = -v_{-m}$. The second condition implies that $v_0 = 0$, and also because $X_m = X_{m+N}$ that $v_{N/2} = 0$ (we are assuming that N is even). Therefore the number of independent u and v is $1+2(N/2-1)+1=N$, as it should be.

If the definitions

$$\begin{aligned} z_0 &= u_0 \\ z_1 &= \sqrt{2}u_1, \ldots, z_{(N/2)-1} = \sqrt{2}u_{(N/2)-1}, \quad z_{N/2} = u_{N/2}, \\ z_{-1} &= \sqrt{2}v_1, \ldots, z_{-(N/2)+1} = \sqrt{2}v_{(N/2)-1} \end{aligned} \tag{2.67}$$

are made, then the z are real, there are N of them, and the Hamiltonian may be written, by (2.65), (2.66), and (2.67),

$$\mathcal{H} = \frac{M}{2}\sum_{m=-(N/2)+1}^{N/2} (\dot{z}_m^2 + \omega_m^2 z_m^2) . \tag{2.68}$$

Equation (2.68) is explicitly the Hamiltonian for N uncoupled harmonic oscillators. This is what was to be proved. The allowed quantum-mechanical energies are then

$$E = \sum_{m=-(N/2)+1}^{N/2} (N_m + \tfrac{1}{2})\hbar\omega_m \,. \tag{2.69}$$

By relabeling, the sum in (2.69) could just as well go from 0 to $N - 1$. The N_m are integers.

2.2.3 Specific Heat of Linear Lattice (B)

We will use the canonical ensemble to derive the specific heat of the one-dimensional crystal.[9] A good reference for the use of the canonical ensemble is Huang [11]. In a canonical ensemble calculation, we first calculate the partition function. The partition function and the Helmholtz free energy are related, and by use of this relation we can calculate all thermodynamic properties once the partition function is known.

If the allowed quantum-mechanical states of the system are labeled by E_M, then the partition function Z is given by

$$Z = \sum_M \exp(-E_M/kT)\,.$$

If there are N atoms in the linear lattice, and if we are interested only in the harmonic approximation, then

$$E_M = E_{m_1,m_2,\ldots,m_n} = \hbar\sum_{n=1}^{N} m_n\omega_n + \frac{\hbar}{2}\sum_{n=1}^{N}\omega_n \,,$$

where the m_n are integers. The partition function is then given by

$$Z = \exp\left(-\frac{\hbar}{2kT}\sum_{n=1}^{N}\omega_n\right)\sum_{(m_1,m_2,\ldots,m_N)=0}^{\infty}\exp\left(-\frac{\hbar}{kT}\sum_{n=1}^{N}\omega_n m_n\right). \tag{2.70}$$

Equation (2.70) can be rewritten as

$$Z = \exp\left(-\frac{\hbar}{2kT}\sum_{n=1}^{N}\omega_n\right)\prod_{n=1}^{N}\sum_{m_n=0}^{\infty}\exp\left(-\frac{\hbar}{kT}\omega_n m_n\right). \tag{2.71}$$

---

9 The discussion of 1D (and 2D) lattices is perhaps mainly of interest because it sets up a formalism that is useful in 3D. One can show that the mean square displacement of atoms in 1D (and 2D) diverges in the phonon approximation. Such lattices are apparently inherently unstable. Fortunately, the mean energy does not diverge, and so the calculation of it in 1D (and 2D) perhaps makes some sense. However, in view of the divergence, things are not as simple as implied in the text. Also see a related comment on the Mermin–Wagner theorem in Chap. 7 (Sect. 7.2.5 under *Two Dimensional Structures*).

The result (2.71) is a consequence of a general property. Whenever we have a set of independent systems, the partition function can be represented as a product of partition functions (one for each independent system). In our case, the independent systems are the independent harmonic oscillators that describe the normal modes of the lattice vibrations.

Since $1/(1-\alpha) = \sum_0^\infty \alpha^n$ if $|\alpha| < 1$, we can write (2.71) as

$$Z = \exp\left(-\frac{\hbar}{2kT}\sum_{n=1}^{N}\omega_n\right)\prod_{n=1}^{N}\frac{1}{1-\exp(-\hbar\omega_n/kT)} . \tag{2.72}$$

The relation between the Helmholtz free energy F and the partition function Z is given by

$$F = -kT \ln Z . \tag{2.73}$$

Combining (2.72) and (2.73) we easily find

$$F = \frac{\hbar}{2}\sum_{n=1}^{N}\omega_n + kT\sum_{n=1}^{N}\ln\left[1-\exp\left(-\frac{\hbar\omega}{kT}\right)\right] . \tag{2.74}$$

Using the thermodynamic formulas for the entropy S,

$$S = -(\partial F/\partial T)_V , \tag{2.75}$$

and the internal energy U,

$$U = F + TS , \tag{2.76}$$

we easily find an expression for U,

$$U = \frac{\hbar}{2}\sum_{n=1}^{N}\omega_n + \sum_{n=1}^{N}\frac{\hbar\omega_n}{\exp(\hbar\omega/kT)-1} . \tag{2.77}$$

Equation (2.77) without the zero-point energy can be arrived at by much more intuitive reasoning. In this formulation, the zero-point energy ($\hbar/2 \sum_{n=1}^{N} \omega_n$) does not contribute anything to the specific heat anyway, so let us neglect it. Call each energy excitation of frequency ω_n and energy $\hbar\omega_n$ a *phonon*. Assume that the phonons are *bosons*, which can be created and destroyed. We shall suppose that the chemical potential is zero so that the number of phonons is not conserved. In this situation, the mean number of phonons of energy $\hbar\omega_n$ (when the system has a temperature T) is given by $1/[\exp(\hbar\omega_n/kT - 1)]$. Except for the zero-point energy, (2.77) now follows directly. Since (2.77) follows so easily, we might wonder if the use of the canonical ensemble is really worthwhile in this problem. In the first place, we need an argument for why phonons act like bosons of zero chemical potential. In the second place, if we had included higher-order terms (than the second-order terms) in the potential, then the phonons would interact and hence have an interaction energy. The canonical ensemble provides a straightforward method of including this interaction energy (for practical cases, approximations would be necessary). The simpler method does not.

The zero-point energy has zero temperature derivative, and so need not be considered for the specific heat. The indicated sum in (2.77) is easily done if $N \to \infty$. Then the modes become infinitesimally close together, and the sum can be replaced by an integral. We can then write

$$U = 2\int_0^{\omega_c} \frac{1}{\exp(\hbar\omega/kT) - 1} \hbar\omega n(\omega) \mathrm{d}\omega , \tag{2.78}$$

where $n(\omega)\mathrm{d}\omega$ is the number of modes (with $q > 0$) between ω and $\omega + \mathrm{d}\omega$. The factor 2 arises from the fact that for every (q) mode there is a ($-q$) mode of the same frequency.

$n(\omega)$ is called the density of states and it can be evaluated from the appropriate dispersion relation, which is $\omega_n = \omega_c\,|\sin(\pi n/N)|$ for the nearest-neighbor approximation. To obtain the density of states, we differentiate the dispersion relation

$$\begin{aligned} \mathrm{d}\omega_n &= \pi\omega_c \cos(\pi n / N)\,\mathrm{d}(n/N), \\ &= \sqrt{\omega_c^2 - \omega_n^2}\,\pi\,\mathrm{d}(n/N). \end{aligned}$$

Therefore

$$N\mathrm{d}(n/N) = (N/\pi)(\omega_c^2 - \omega_n^2)^{-1/2}\mathrm{d}\omega_n \equiv n(\omega_n)\mathrm{d}\omega_n ,$$

or

$$n(\omega_n) = (N/\pi)(\omega_c^2 - \omega_n^2)^{-1/2} . \tag{2.79}$$

Combining (2.78), (2.79), and the definition of specific heat at constant volume, we have

$$\begin{aligned} C_v &= \left(\frac{\partial U}{\partial T}\right)_v \\ &= \frac{2N\hbar}{\pi}\int_0^{\omega_c}\left\{\frac{\omega}{\sqrt{\omega_c^2-\omega^2}}\left[\exp\left(\frac{\hbar\omega}{kT}\right)-1\right]^{-2}\exp\left(\frac{\hbar\omega}{kT}\right)\frac{\hbar\omega}{kT^2}\right\}\mathrm{d}\omega . \end{aligned} \tag{2.80}$$

In the high-temperature limit this gives

$$C_v = \frac{2Nk}{\pi}\int_0^{\omega_c}\left(\sqrt{\omega_c^2-\omega^2}\right)^{-1}\mathrm{d}\omega = \frac{2Nk}{\pi}\sin^{-1}(\omega/\omega_c)\Big|_0^{\omega_c} = Nk . \tag{2.81}$$

Equation (2.81) is just a one-dimensional expression of the law of Dulong and Petit, which is also the classical limit.

2.2.4 Classical Diatomic Lattices: Optic and Acoustic Modes (B)

So far we have considered only linear lattices in which all atoms are identical. There exist, of course, crystals that have more than one type of atom. In this Section we will discuss the case of a linear lattice with two types of atoms in alternating positions. We will consider only the harmonic approximation with nearest-neighbor interactions. By symmetry, the force between each pair of atoms is described by the same spring constant. In the diatomic linear lattice we can think of each unit cell as containing two atoms of differing mass. It is characteristic of crystals with two atoms per unit cell that two types of mode occur. One of these modes is called the *acoustic mode*. In an acoustic mode, we think of adjacent atoms as vibrating almost in phase. The other mode is called the *optic mode*. In an optic mode, we think of adjacent atoms as vibrating out of phase. As we shall show, these descriptions of optic and acoustic modes are valid only in the long-wavelength limit. In three dimensions we would also have to distinguish between longitudinal and transverse modes. Except for special crystallographic directions, these modes would not have the simple physical interpretation that their names suggest. The longitudinal mode is, usually, the mode with highest frequency for each wave vector in the three optic modes and also in the three acoustic modes.

A picture of the diatomic linear lattice is shown in Fig. 2.4. Atoms of mass m are at $x = (2n + 1)a$ for $n = 0, \pm 1, \pm 2, \ldots$, and atoms of mass M are at $x = 2na$ for $n = 0, \pm 1, \ldots$ The displacements from equilibrium of the atoms of mass m are labeled d_n^m and the displacements from equilibrium of the atoms of mass M are labeled d_n^M. The spring constant is k.

From Newton's laws[10]

$$m\ddot{d}_n^m = k(d_{n+1}^M - d_n^m) + k(d_n^M - d_n^m)\,, \tag{2.82a}$$

and

$$M\ddot{d}_n^M = k(d_n^m - d_n^M) + k(d_{n-1}^m - d_n^M)\,. \tag{2.82b}$$

It is convenient to define $K_1 = k/m$ and $K_2 = k/M$. Then (2.82) can be written

$$\ddot{d}_n^m = -K_1(2d_n^m - d_n^M - d_{n+1}^M) \tag{2.83a}$$

and

$$\ddot{d}_n^M = -K_2(d_n^M - d_n^m - d_{n-1}^m)\,. \tag{2.83b}$$

[10] When we discuss lattice vibrations in three dimensions we give a more general technique for handling the case of two atoms per unit cell. Using the dynamical matrix defined in that section (or its one-dimensional analog), it is a worthwhile exercise to obtain (2.87a) and (2.87b).

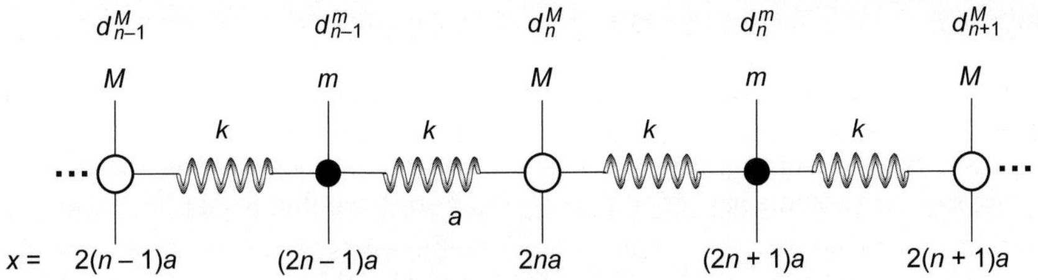

Fig. 2.4. The diatomic linear lattice

Consistent with previous work, normal mode solutions of the form

$$d_n^m = A\exp[\mathrm{i}(qx_n^m - \omega t)], \tag{2.84a}$$

and

$$d_n^M = B\exp[\mathrm{i}(qx_n^M - \omega t)] \tag{2.84b}$$

will be sought. Substituting (2.84) into (2.83) and finding the coordinates of the atoms (x_n) from Fig. 2.4, we have

$$\begin{aligned}
-\omega^2 A\exp\{\mathrm{i}[q(2n+1)a - \omega t]\} &= -K_1(2A\exp\{\mathrm{i}[q(2n+1)a - \omega t]\} \\
&\quad - B\exp\{\mathrm{i}[q(2na) - \omega t]\} \\
&\quad - B\exp\{\mathrm{i}[q(n+1)2a - \omega t]\}) \\
-\omega^2 B\exp\{\mathrm{i}[q(2na) - \omega t]\} &= -K_2(2B\exp\{\mathrm{i}[q(2na) - \omega t]\} \\
&\quad - A\exp\{\mathrm{i}[q(2n+1)a - \omega t]\} \\
&\quad - A\exp\{\mathrm{i}[q(2n-1)a - \omega t]\})
\end{aligned},$$

or

$$\omega^2 A = K_1(2A - B\mathrm{e}^{-\mathrm{i}qa} - B\mathrm{e}^{+\mathrm{i}qa}), \tag{2.85a}$$

and

$$\omega^2 B = K_2(2B - A\mathrm{e}^{-\mathrm{i}qa} - A\mathrm{e}^{+\mathrm{i}qa}). \tag{2.85b}$$

Equations (2.85) can be written in the form

$$\begin{bmatrix} \omega^2 - 2K_1 & 2K_1\cos qa \\ 2K_2\cos qa & \omega^2 - 2K_2 \end{bmatrix}\begin{bmatrix} A \\ B \end{bmatrix} = 0. \tag{2.86}$$

Equation (2.86) has nontrivial solutions only if the determinant of the coefficient matrix is zero. This yields the two roots

$$\omega_1^2 = (K_1 + K_2) - \sqrt{(K_1 + K_2)^2 - 4K_1K_2\sin^2 qa}, \tag{2.87a}$$

and

$$\omega_2^2 = (K_1 + K_2) + \sqrt{(K_1 + K_2)^2 - 4K_1K_2 \sin^2 qa} \,. \tag{2.87b}$$

In (2.87) the symbol √ means the positive square root. In figuring the positive square root, we assume $m < M$ or $K_1 > K_2$. As $q \to 0$, we find from (2.87) that

$$\omega_1 = 0 \quad \text{and} \quad \omega_2 = \sqrt{2(K_1 + K_2)} \,.$$

As $q \to (\pi/2a)$ we find from (2.87) that

$$\omega_1 = \sqrt{2K_2} \quad \text{and} \quad \omega_2 = \sqrt{2K_1} \,.$$

Plots of (2.87) look similar to Fig. 2.5. In Fig. 2.5, ω_1 is called the acoustic mode and ω_2 is called the optic mode. The reason for naming ω_1 and ω_2 in this manner will be given later. The first Brillouin zone has $-\pi/2a \le q \le \pi/2a$. This is only half the size that we had in the monatomic case. The reason for this is readily apparent. In the diatomic case (with the same total number of atoms as in the monatomic case) there are two modes for every q in the first Brillouin zone, whereas in the monatomic case there is only one. For a fixed number of atoms and a fixed number of dimensions, the number of modes is constant.

In fact it can be shown that the diatomic case reduces to the monatomic case when $m = M$. In this case $K_1 = K_2 = k/m$ and

$$\omega_1^2 = 2k/m - (2k/m)\cos qa = (2k/m)(1 - \cos qa),$$
$$\omega_2^2 = 2k/m + (2k/m)\cos qa = (2k/m)(1 + \cos qa).$$

But note that cos qa for $-\pi/2 < qa < 0$ is the same as $-$cos qa for $\pi/2 < qa < \pi$, so that we can just as well combine ω_1^2 and ω_2^2 to give

$$\omega = (2k/m)(1 - \cos qa) = (4k/m)\sin^2(qa/2)$$

for $-\pi < qa < \pi$. This is the same as the dispersion relation that we found for the linear lattice.

The reason for the names optic and acoustic modes becomes clear if we examine the motions for small qa. We can write (2.87a) as

$$\omega_1 \cong \sqrt{\frac{2K_1K_2}{(K_1 + K_2)}} qa \tag{2.88}$$

for small qa. Substituting (2.88) into $(\omega^2 - 2K_1)A + 2K_1 \cos(qa)B = 0$, we find

$$\frac{B}{A} = -\left[\frac{2K_1K_2q^2a^2/(K_1 + K_2) - 2K_1}{2K_1 \cos qa}\right]^{qa \to 0} \to +1 \,. \tag{2.89}$$

Therefore in the long-wavelength limit of the ω_1 mode, adjacent atoms vibrate in phase. This means that the mode is an acoustic mode.

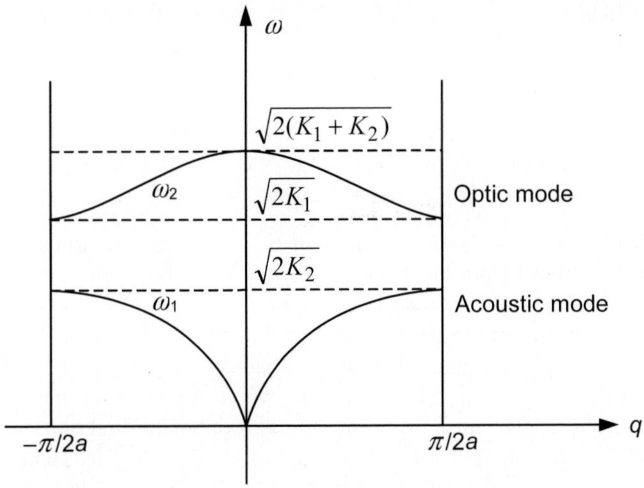

Fig. 2.5. The dispersion relation for the optic and acoustic modes of a diatomic linear lattice

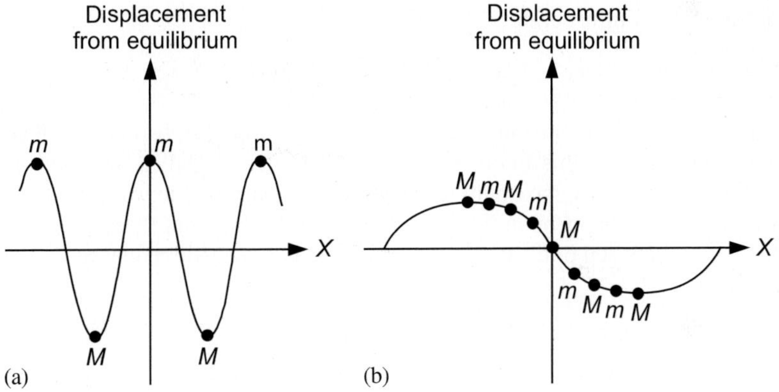

Fig. 2.6. (**a**) Optic and (**b**) acoustic modes for qa very small (the long-wavelength limit)

It is instructive to examine the ω_1 solution (for small qa) still further:

$$\omega_1 = \sqrt{\frac{2K_1K_2}{(K_1+K_2)}}qa = \sqrt{\frac{2k^2/(mM)}{k/m+k/M}}qa = \sqrt{\frac{ka}{(m+M)/2a}}q\,. \tag{2.90}$$

For (2.90), $\omega_1/q = \mathrm{d}\omega/\mathrm{d}q$, the phase and group velocities are the same, and so there is no dispersion. This is just what we would expect in the long-wavelength limit.

Let us examine the ω_2 modes in the $qa \to 0$ limit. It is clear that

$$\omega_2^2 \cong 2(K_1+K_2) + \frac{2K_1K_2}{(K_1+K_2)}q^2a^2 \quad \text{as } qa \to 0. \tag{2.91}$$

Substituting (2.91) into $(\omega^2 - 2K_1)A + 2K_1 \cos(qa)B = 0$ and letting $qa = 0$, we have

$$2K_2 A + 2K_1 B = 0 ,$$

or

$$mA + MB = 0 . \tag{2.92}$$

Equation (2.92) corresponds to the center of mass of adjacent atoms being fixed. Thus in the long-wavelength limit, the atoms in the ω_2 mode vibrate with a phase difference of π. Thus the ω_2 mode is the optic mode. Suppose we shine electromagnetic radiation of visible frequencies on the crystal. The wavelength of this radiation is much greater than the lattice spacing. Thus, due to the opposite charges on adjacent atoms in a polar crystal (which we assume), the electromagnetic wave would tend to push adjacent atoms in opposite directions just as they move in the long-wavelength limit of a (transverse) optic mode. Hence the electromagnetic waves would interact strongly with the optic modes. Thus we see where the name optic mode came from. The long-wavelength limits of optic and acoustic modes are sketched in Fig. 2.6

In the small qa limit for optic modes by (2.91),

$$\omega_2 = \sqrt{2k(1/m + 1/M)} . \tag{2.93}$$

Electromagnetic waves in ionic crystals are very strongly absorbed at this frequency. Very close to this frequency, there is a frequency called the *restrahl* frequency where there is a maximum reflection of electromagnetic waves [93].

A curious thing happens in the $q \to \pi/2a$ limit. In this limit there is essentially no distinction between optic and acoustic modes. For acoustic modes as $q \to \pi/2a$, from (2.86),

$$(\omega^2 - 2K_1)A = -2K_1 B \cos qa ,$$

or as $qa \to \pi/2$,

$$\frac{A}{B} = K_1 \frac{\cos qa}{K_1 - K_2} = 0 ,$$

so that only M moves. In the same limit $\omega_2 \to (2K_1)^{1/2}$, so by (2.86)

$$2K_2 (\cos qa) A + (2K_1 - 2K_2) B = 0 ,$$

or

$$\frac{B}{A} = 2K_2 \frac{\cos qa}{K_2 - K_1} = 0 ,$$

so that only m moves. The two modes are sketched in Fig. 2.7. Table 2.2 collects some one-dimensional results.

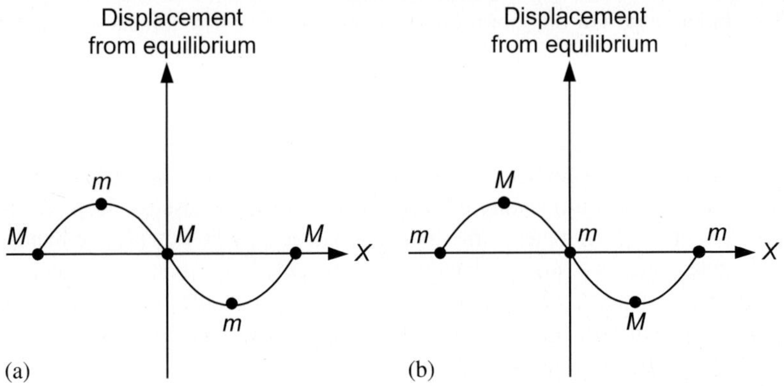

Fig. 2.7. (**a**) Optic and (**b**) acoustic modes in the limit $qa \to \pi/2$

Table 2.2. One-dimensional dispersion relations and density of states†

Model	Dispersion relation	Density of states
Monatomic	$\omega = \omega_0 \left\| \sin \frac{qa}{2} \right\|$	$D(\omega) \propto \frac{1}{\sqrt{\omega_0^2 - \omega^2}}$
Diatomic $[M > m,$ $\mu = Mm/(M+m)]$		Small q
– acoustic	$\omega^2 \propto \left(\frac{1}{\mu} - \sqrt{\frac{1}{\mu^2} - \frac{4}{Mm} \sin^2 qa} \right)$	$D(\omega) \propto$ constant
– optical	$\omega^2 \propto \left(\frac{1}{\mu} + \sqrt{\frac{1}{\mu^2} - \frac{4}{Mm} \sin^2 qa} \right)$	$D(\omega) \propto \|q\|^{-1}$

† q = wave vector, ω = frequency, a = distance between atoms.

2.2.5 Classical Lattice with Defects (B)

Most of the material in this Section was derived by Lord Rayleigh many years ago. However, we use more modern techniques (Green's functions). The calculation will be done in one dimension, but the technique can be generalized to three dimensions. Much of the present formulation is due to A. A. Maradudin and co-workers.[11]

---

[11] See [2.39].

The modern study of the vibration of a crystal lattice with defects was begun by Lifshitz in about 1942 [2.25] and Schaefer [2.29] has shown experimentally that local modes do exist. Schaefer examined the infrared absorption of H^- ions (impurities) in KCl.

Point defects can cause the appearance of localized states. Here we consider lattice vibrations and later (in Sect. 3.2.4) electronic states. Strong as well as weak perturbations can lead to interesting effects. For example, we discuss deep electronic effects in Sect. 11.2. In general, the localized states may be outside the bands and have discrete energies or inside a band with transiently bound resonant levels.

In this Section the word *defect* is used in a rather specialized manner. The only defects considered will be substitutional atoms with different masses from those of the atoms of the host crystal.

We define an operator p such that (compare (2.36))

$$pu_n = \omega^2 Mu_n + \gamma(u_{n+1} - 2u_n + u_{n+1}) , \tag{2.94}$$

where u_n is the amplitude of vibration of atom n, with mass M and frequency ω. For a perfect lattice (in the harmonic nearest-neighbor approximation with $\gamma = M\omega_c^2/4$ = spring constant),

$$pu_n = 0 .$$

This result readily follows from the material in Sect. 2.2.2. If the crystal has one or more defects, the equations describing the resulting vibrations can always be written in the form

$$pu_n = \sum\nolimits_k d_{nk} u_k . \tag{2.95}$$

For example, if there is a defect atom of mass M^1 at $n = 0$ and if the spring constants are not changed, then

$$d_{nk} = (M - M^1)\omega^2 \delta_n^0 \delta_k^0 . \tag{2.96}$$

Equation (2.95) will be solved with the aid of Green's functions. Green's functions (G_{mn}) for this problem are defined by

$$pG_{mn} = \delta_{mn} . \tag{2.97}$$

To motivate the introduction of the G_{mn}, it is useful to prove that a solution to (2.95) is given by

$$u_n = \sum\nolimits_{l,k} G_{nl} d_{lk} u_k . \tag{2.98}$$

Since p operates on index n in pu_n, we have

$$pu_n = \sum\nolimits_{l,k} pG_{nl} d_{lk} u_k = \sum\nolimits_{l,k} \delta_{nl} d_{lk} u_k = \sum\nolimits_k d_{nk} u_k ,$$

and hence (2.98) is a formal solution of (2.95).

The next step is to find an explicit expression for the G_{mn}. By the arguments of Sect. 2.2.2, we know that (we are supposing that there are N atoms, where N is an even number)

$$\delta_{mn} = \frac{1}{N}\sum_{s=0}^{N-1} \exp\left[\frac{2\pi i s}{N}(m-n)\right]. \tag{2.99}$$

Since G_{mn} is determined by the lattice, and since periodic boundary conditions are being used, it should be possible to make a *Fourier* analysis of G_{mn}:

$$G_{mn} = \frac{1}{N}\sum_{s=0}^{N-1} g_s \exp\left[\frac{2\pi i s}{N}(m-n)\right]. \tag{2.100}$$

From the definition of p, we can write

$$\begin{aligned} p\exp\left[2\pi i\frac{s}{N}(m-n)\right] &= \omega^2 M \exp\left[2\pi i\frac{s}{N}(m-n)\right] \\ &+ \gamma\left\{\exp\left[2\pi i\frac{s}{N}(m-n-1)\right] - 2\exp\left[2\pi i\frac{s}{N}(m-n)\right]\right. \\ &\left.+ \exp\left[2\pi i\frac{s}{N}(m-n+1)\right]\right\}. \end{aligned} \tag{2.101}$$

To prove that we can find solutions of the form (2.100), we need only substitute (2.100) and (2.99) into (2.97). We obtain

$$\begin{aligned} \frac{1}{N}\sum_{s=0}^{N-1} g_s \Bigg(\omega^2 M \exp\left[2\pi i\frac{s}{N}(m-n)\right] &+ \gamma\left\{\exp\left[2\pi i\frac{s}{N}(m-n-1)\right]\right. \\ -2\exp\left[2\pi i\frac{s}{N}(m-n)\right] &+ \left.\exp\left[2\pi i\frac{s}{N}(m-n+1)\right]\right\}\Bigg) \\ &= \frac{1}{N}\sum_{s=0}^{N-1} \exp\left[2\pi i\frac{s}{N}(m-n)\right]. \end{aligned} \tag{2.102}$$

Operating on both sides of the resulting equation with

$$\sum_{m-n} \exp\left[-\frac{2\pi i}{N}(m-n)s'\right],$$

we find

$$\sum_s g_s\{\omega^2 M\delta_s^{s'} - 2\gamma\delta_s^{s'}[1-\cos(2\pi s/N)]\} = \sum_s \delta_s^{s'}. \tag{2.103}$$

Thus a G of the form (2.100) has been found provided that

$$g_s = \frac{1}{M\omega^2 - 2\gamma(1-\cos 2\pi s/N)} = \frac{1}{M\omega^2 - 4\gamma\sin^2(\pi s/N)}. \tag{2.104}$$

By (2.100), G_{mn} is a function only of $m - n$, and, further by Problem 2.4, G_{mn} is a function only of $|m - n|$. Thus it is convenient to define

$$G_{mn} = G_l \,, \tag{2.105}$$

where $l = |m - n| \geq 0$.

It is possible to find a more convenient expression for G. First, define

$$\cos\phi = 1 - \frac{M\omega^2}{2\gamma} \,. \tag{2.106}$$

Then for a perfect lattice

$$0 < \omega^2 \leq \omega_c^2 = \frac{4\gamma}{M} \,,$$

so

$$1 \geq 1 - \frac{M\omega^2}{2\gamma} \geq -1 \,. \tag{2.107}$$

Thus when ϕ is real in (2.106), ω^2 is restricted to the range defined by (2.107). With this definition, we can prove that a general expression for the G_n is[12]

$$G_n = \frac{1}{2\gamma \sin\phi}\left(\cot\frac{N\phi}{2} \cos n\phi + \sin|n|\phi \right) . \tag{2.108}$$

The problem of a mass defect in a linear chain can now be solved. We define the relative change in mass e by

$$e = (M - M^1)/M \,, \tag{2.109}$$

with the defect mass M^1 assumed to be less than M for the most interesting case. Using (2.96) and (2.98), we have

$$u_n = G_n M e \omega^2 u_0 \,. \tag{2.110}$$

Setting $n = 0$ in (2.110), using (2.108) and (2.106), we have (assuming $u_0 \neq 0$, this limits us to modes that are not antisymmetric)

$$\frac{1}{G_n} = 2\frac{\gamma \sin\phi}{\cot(N\phi/2)} = eM\omega^2 = 2e\gamma(1 - \cos\phi) \,,$$

or

$$\frac{\sin\phi}{\cot(N\phi/2)} = e(1 - \cos\phi) \,,$$

[12] For the derivation of (2.108), see the article by Maradudin op cit (and references cited therein).

or

$$\tan\frac{N\phi}{2} = e\tan\frac{\phi}{2}\,. \tag{2.111}$$

We would like to solve for ω^2 as a function of e. This can be found from ϕ as a function of e by use of (2.111). For small e, we have

$$\phi(e) \cong \phi(0) + \left.\frac{\partial\phi}{\partial e}\right|_{e=0} e\,. \tag{2.112}$$

From (2.111),

$$\phi(0) = 2\pi s / N\,. \tag{2.113}$$

Differentiating (2.111), we find

$$\frac{\mathrm{d}}{\mathrm{d}e}\tan\frac{N\phi}{2} = \frac{\mathrm{d}}{\mathrm{d}e}\left(e\tan\frac{\phi}{2}\right),$$

or

$$\frac{N}{2}\sec^2\frac{N\phi}{2}\frac{\partial\phi}{\partial e} = \tan\frac{\phi}{2} + \frac{e}{2}\sec^2\frac{\phi}{2}\frac{\partial\phi}{\partial e}\,,$$

or

$$\left.\frac{\partial\phi}{\partial e}\right|_{e=0} = \left.\frac{\tan\phi/2}{(N/2)\sec^2(N\phi/2)}\right|_{e=0}\,. \tag{2.114}$$

Combining (2.112), (2.113), and (2.114), we find

$$\phi \cong \frac{2\pi s}{N} + \frac{2e}{N}\tan\frac{\pi s}{N}\,. \tag{2.115}$$

Therefore, for small e, we can write

$$\begin{aligned}
\cos\phi &\cong \cos\left(\frac{2\pi s}{N} + \frac{2e}{N}\tan\frac{\pi s}{N}\right) \\
&= \cos\frac{2\pi s}{N}\cos\left(\frac{2e}{N}\tan\frac{\pi s}{N}\right) - \sin\frac{2\pi s}{N}\sin\frac{2e}{N}\cdot\tan\frac{\pi s}{N} \\
&\cong \cos\frac{2\pi s}{N} - \frac{2e}{N}\tan\frac{\pi s}{N}\sin\frac{2\pi s}{N} \\
&= \cos\frac{2\pi s}{N} - \frac{4e}{N}\sin^2\frac{\pi s}{N}\,.
\end{aligned} \tag{2.116}$$

Using (2.106), we have

$$\omega^2 \cong \frac{2\gamma}{M}\left(1 - \cos\frac{2\pi s}{N} + \frac{4e}{N}\sin^2\frac{\pi s}{N}\right). \tag{2.117}$$

Using the half-angle formula $\sin^2 \theta/2 = (1 - \cos \theta)/2$, we can recast (2.117) into the form

$$\omega \cong \omega_c \left|\sin\frac{\pi s}{N}\right| \left(1 + \frac{e}{N}\right). \tag{2.118}$$

We can make several physical comments about (2.118). As noted earlier, if the description of the lattice modes is given by symmetric (about the impurity) and antisymmetric modes, then our development is valid for symmetric modes. Antisymmetric modes cannot be affected because $u_0 = 0$ for them anyway and it cannot matter then what the mass of the atom described by u_0 is. When $M > M^1$, then $e > 0$ and all frequencies (of the symmetric modes) are shifted upward. When $M < M^1$, then $e < 0$ and all frequencies (of the symmetric modes) are shifted downward. There are no local modes here, but one does speak of resonant modes.[13] When $N \to \infty$, then the frequency shift of all modes given by (2.118) is negligible. Actually when $N \to \infty$, there is one mode for the $e > 0$ case that is shifted in frequency by a non-negligible amount. This mode is the impurity mode. The reason we have not yet found the impurity mode is that we have not allowed the ϕ defined by (2.106) to be complex. Remember, real ϕ corresponds only to modes whose amplitude does not diminish. With impurities, it is reasonable to seek modes whose amplitude does change. Therefore, assume $\phi = \pi + \mathrm{i}z$ ($\phi = \pi$ corresponds to the highest frequency unperturbed mode). Then from (2.111),

$$\tan\left[\frac{N}{2}(\pi + \mathrm{i}z)\right] = e\tan\tfrac{1}{2}(\pi + \mathrm{i}z). \tag{2.119}$$

Since $\tan(A + B) = (\tan A + \tan B)/(1 - \tan A \tan B)$, then as $N \to \infty$ (and remains an even number), we have

$$\tan\left(\frac{N\pi}{2} + \frac{\mathrm{i}Nz}{2}\right) = \tan\frac{\mathrm{i}Nz}{2} = \mathrm{i}. \tag{2.120}$$

Also

$$\begin{aligned}\tan\left(\frac{\pi + \mathrm{i}z}{2}\right) &= \frac{\sin(\pi/2 + \mathrm{i}z/2)}{\cos(\pi/2 + \mathrm{i}z/2)} = -\frac{\sin(\pi/2)\cos(\mathrm{i}z/2)}{\sin(\pi/2)\sin(\mathrm{i}z/2)} \\ &= -\cot\frac{\mathrm{i}z}{2} = +\mathrm{i}\coth\frac{z}{2}.\end{aligned} \tag{2.121}$$

[13] Elliott and Dawber [2.15].

Combining (2.119), (2.120), and (2.121), we have

$$e\coth\frac{z}{2}=1\,. \tag{2.122}$$

Equation (2.122) can be solved for z to yield

$$z=\ln\frac{1+e}{1-e}\,. \tag{2.123}$$

But

$$\begin{aligned}\cos\phi &= \cos(\pi+\mathrm{i}z)=\cos\pi\cos\mathrm{i}z\\ &= -\tfrac{1}{2}(\exp z+\exp -z)\\ &= -\frac{1+e^2}{1-e^2}\end{aligned} \tag{2.124}$$

by (2.122). Combining (2.124) and (2.106), we find

$$\omega^2=\omega_{\mathrm{c}}^2/(1-e^2)\,. \tag{2.125}$$

The mode with frequency given by (2.125) can be considerably shifted even if $N \to \infty$. The amplitude of the motion can also be estimated. Combining previous results and letting $N \to \infty$, we find

$$u_n=(-)^{|n|}\frac{M-M^1}{2\gamma}\frac{\omega_{\mathrm{c}}^2}{2e}\left(\frac{1-e}{1+e}\right)^{|n|}u_0=(-1)^n\left(\frac{1-e}{1+e}\right)^{|n|}u_0\,. \tag{2.126}$$

This is truly an impurity mode. The amplitude dies away as we go away from the impurity. No new modes have been gained, of course. In order to gain a mode with frequency described by (2.125), we had to give up a mode with frequency described by (2.118). For further details see Maradudin et al [2.26 Sect. 5.5]

2.2.6 Quantum-Mechanical Linear Lattice (B)

In a previous Section we found the quantum-mechanical energies of a linear lattice by first reducing the classical problem to a set of classical harmonic oscillators. We then quantized the harmonic oscillators. Another approach would be initially to consider the lattice from a quantum viewpoint. Then we transform to a set of independent quantum-mechanical harmonic oscillators. As we demonstrate below, the two procedures amount to the same thing. However, it is not always true that we can get correct results by quantizing the Hamiltonian in any set of generalized coordinates [2.27].

With our usual assumptions of nearest-neighbor interactions and harmonic forces, the classical Hamiltonian of the linear chain can be written

$$\mathcal{H}(p_l, x_l) = \frac{1}{2M}\sum_l p_l^2 + \frac{\gamma}{2}\sum_l (2x_l^2 - x_l x_{l+1} - x_l x_{l-1}) . \tag{2.127}$$

In (2.127), $p_1 = M\dot{x}_1$, and in the potential energy term use can always be made of periodic boundary conditions in rearranging the terms without rearranging the limits of summation (for N atoms, $x_l = x_{l+N}$). The sum in (2.127) runs over the crystal, the equilibrium position of the lth atom being at la. The displacement from equilibrium of the lth atom is x_l and γ is the spring constant.

To quantize (2.127) we associate operators with dynamical quantities. For (2.127), the method is clear because p_l and x_l are canonically conjugate. The momentum p_l was defined as the derivative of the Lagrangian with respect to $\dot{x}_l$. This implies that Poisson bracket relations are guaranteed to be satisfied. Therefore, when operators are associated with p_l and x_l, they must be associated in such a way that the commutation relations (analog of Poisson bracket relations)

$$[x_l, p_{l'}] = i\hbar\delta_l^{l'} \tag{2.128}$$

are satisfied. One way to do this is to let

$$p_l \to \frac{\hbar}{i}\frac{\partial}{\partial x_i}, \quad \text{and} \quad x_l \to x_l . \tag{2.129}$$

This is the choice that will usually be made in this book.

The quantum-mechanical problem that must be solved is

$$\mathcal{H}\left(\frac{\hbar}{i}\frac{\partial}{\partial x_l}, x_l\right)\psi(x_1 \cdots x_n) = E(x_1 \cdots x_n) . \tag{2.130}$$

In (2.130), $\psi(x_1 \cdots x_n)$ is the wave function describing the lattice vibrational state with energy E.

How can (2.130) be solved? A good way to start would be to use normal coordinates just as in the Section on vibrations of a classical lattice. Define

$$X_q = \frac{1}{\sqrt{N}}\sum_l \mathrm{e}^{iqla} x_l , \tag{2.131}$$

where $q = 2\pi m/Na$ and m is an integer, so that

$$x_l = \frac{1}{\sqrt{N}}\sum_q \mathrm{e}^{-iqla} X_q . \tag{2.132}$$

The next quantities that are needed are a set of new momentum operators that are canonically conjugate to the new coordinate operators. The simplest way to get these operators is to write down the correct ones and show they are correct by the fact that they satisfy the correct commutation relations:

$$P_{q'} = \frac{1}{\sqrt{N}}\sum_l p_l \mathrm{e}^{-iq'la} , \tag{2.133}$$

or

$$p_l = \frac{1}{\sqrt{N}} \sum_{q''} P_{q''} e^{iq''la} . \tag{2.134}$$

The fact that the commutation relations are still satisfied is easily shown:

$$\begin{aligned}
[X_q, P_{q'}] &= \frac{1}{N} \sum_{l,l'} [x_{l'}, p_l] \exp[ia(ql' - q'l)] \\
&= \frac{1}{N} \sum_{l,l'} i\hbar \delta_l^{l'} \exp[ia(ql' - q'l)] \\
&= i\hbar \delta_q^{q'} .
\end{aligned} \tag{2.135}$$

Substituting (2.134) and (2.132) into (2.127), we find in the usual way that the Hamiltonian reduces to a fairly simple form:

$$\mathcal{H} = \frac{1}{2M} \sum_q P_q P_{-q} + \gamma \sum_q X_q X_{-q} (1 - \cos qa) . \tag{2.136}$$

Thus, the normal coordinate transformation does the same thing quantum-mechanically as it does classically.

The quantities X_q and X_{-q} are related. Let † (dagger) represent the Hermitian conjugate operation. Then for all operators A that represent physical observables (e.g. p_l), $A^\dagger = A$. The † of a scalar is equivalent to complex conjugation (*).

Note that

$$P_q^\dagger = \frac{1}{\sqrt{N}} \sum_l p_l e^{iqla} = P_{-q} ,$$

and similarly that

$$X_q^\dagger = X_{-q} .$$

From the above, we can write the Hamiltonian in a Hermitian form:

$$\mathcal{H} = \sum_q \left[\frac{1}{2M} P_q P_q^\dagger + \gamma (1 - \cos qa) X_q X_q^\dagger \right] . \tag{2.137}$$

From the previous work on the classical lattice, it is already known that (2.137) represents a set of independent simple harmonic oscillators whose classical frequencies are given by

$$\omega_q = \sqrt{2\gamma(1 - \cos qa)/M} = \sqrt{2\gamma/M} \left| \sin(qa/2) \right| . \tag{2.138}$$

However, if we like, we can regard (2.138) as a useful definition. Its physical interpretation will become clear later on. With ω_q defined by (2.138), (2.137) becomes

$$\mathcal{H} = \sum_q \left(\frac{1}{2M} P_q P_q^\dagger + \tfrac{1}{2} M \omega^2 X_q X_q^\dagger \right) . \tag{2.139}$$

The Hamiltonian can be further simplified by introducing the two variables [99]

$$a_q = \frac{1}{\sqrt{2M\hbar\omega_q}} P_q - \mathrm{i}\sqrt{\frac{M\omega_q}{2\hbar}} X_q^\dagger, \tag{2.140}$$

$$a_q^\dagger = \frac{1}{\sqrt{2M\hbar\omega_q}} P_q^\dagger + \mathrm{i}\sqrt{\frac{M\omega_q}{2\hbar}} X_q . \tag{2.141}$$

Let us compute $[a_q, a_{q^1}^\dagger]$. By (2.140) and (2.141),

$$\begin{aligned}
[a_q, a_{q^1}^\dagger] &= \frac{\mathrm{i}}{\sqrt{2M\hbar\omega_q}} \sqrt{\frac{M\omega_q}{2\hbar}} \{[P_q, X_{q^1}] - [X_q^\dagger, P_{q^1}^\dagger]\} \\
&= \frac{\mathrm{i}}{2\hbar}(-\mathrm{i}\hbar\delta_q^{q^1} - -\mathrm{i}\hbar\delta_q^{q^1}) \\
&= \delta_q^{q^1},
\end{aligned}$$

or in summary,

$$[a_q, a_{q^1}^\dagger] = \delta_q^{q^1} . \tag{2.142}$$

It is also interesting to compute $½\Sigma_q \hbar\omega_q\{a_q, a_q^\dagger\}$, where $\{a_q, a_q^\dagger\}$ stands for the anticommutator; i.e. it represents $a_q a_q^\dagger + a_q^\dagger a_q$:

$$\begin{aligned}
&\tfrac{1}{2}\sum_q \hbar\omega_q \{a_q, a_q^\dagger\} = \\
&\quad \tfrac{1}{2}\sum_q \hbar\omega_q \left(\frac{1}{\sqrt{2M\hbar\omega_q}} P_q - \mathrm{i}\sqrt{\frac{M\omega_q}{2\hbar}} X_q^\dagger\right)\left(\frac{1}{\sqrt{2M\hbar\omega_q}} P_q^\dagger + \mathrm{i}\sqrt{\frac{M\omega_q}{2\hbar}} X_q\right) \\
&\quad + \tfrac{1}{2}\sum_q \hbar\omega_q \left(\frac{1}{\sqrt{2M\hbar\omega_q}} P_q^\dagger + \mathrm{i}\sqrt{\frac{M\omega_q}{2\hbar}} X_q\right)\left(\frac{1}{\sqrt{2M\hbar\omega_q}} P_q - \mathrm{i}\sqrt{\frac{M\omega_q}{2\hbar}} X_q^\dagger\right) \\
&= \tfrac{1}{2}\sum_q \hbar\omega_q \left(\frac{1}{2M\hbar\omega_q} P_q P_q^\dagger + \frac{M\omega_q}{2\hbar} X_q^\dagger X_q - \frac{\mathrm{i}}{2\hbar} X_q^\dagger P_q^\dagger + \frac{\mathrm{i}}{2\hbar} P_q X_q \right. \\
&\qquad \left. + \frac{1}{2M\hbar\omega_q} P_q^\dagger P_q + \frac{M\omega_q}{2\hbar} X_q X_q^\dagger + \frac{\mathrm{i}}{2\hbar} X_q P_q - \frac{\mathrm{i}}{2\hbar} P_q^\dagger X_q^\dagger \right).
\end{aligned}$$

Observing that

$$\begin{aligned}
X_q P_q + P_q X_q - X_q^\dagger P_q^\dagger - P_q^\dagger X_q^\dagger &= 2(P_q X_q - P_q^\dagger X_q^\dagger), \\
P_q^\dagger = P_{-q}, X_q^\dagger &= X_{-q},
\end{aligned}$$

and $\omega_q = \omega_{-q}$, we see that

$$\sum_q \hbar\omega_q (P_q X_q - P_q^\dagger X_q^\dagger) = 0 .$$

Also $[X_q^\dagger, X_q] = 0$ and $[P_q^\dagger, P_q] = 0$, so that we obtain

$$\tfrac{1}{2}\sum_q \hbar\omega_q \{a_q, a_q^\dagger\} = \sum_q \left(\frac{1}{2M} P_q P_q^\dagger + \tfrac{1}{2} M\omega_q^2 X_q X_q^\dagger \right) = H . \quad (2.143)$$

Since the a_q operators obey the commutation relations of (2.142) and by Problem 2.6, they are isomorphic (can be set in one-to-one correspondence) to the step-up and step-down operators of the harmonic oscillator [18, p349ff]. Since the harmonic oscillator is a solved problem so is (2.143).

By (2.142) and (2.143) we can write

$$\mathcal{H} = \sum_q \hbar\omega_q (a_q^\dagger a_q + \tfrac{1}{2}) . \quad (2.144)$$

But from the quantum mechanics of the harmonic oscillator, we know that

$$a_q^\dagger \left| n_q \right\rangle = \sqrt{(n_q + 1)} \left| n_q + 1 \right\rangle , \quad (2.145)$$

$$a_q \left| n_q \right\rangle = \sqrt{n_q} \left| n_q - 1 \right\rangle . \quad (2.146)$$

Where $|n_q\rangle$ is the eigenket of a single harmonic oscillator in a state with energy $(n_q + ½)\hbar\omega_q$, ω_q is the classical frequency and n_q is an integer. Equations (2.145) and (2.146) imply that

$$\mathrm{a}_\mathrm{q}^\dagger \mathrm{a}_\mathrm{q} \left| \mathrm{n}_\mathrm{q} \right\rangle = \mathrm{n}_\mathrm{q} \left| \mathrm{n}_\mathrm{q} \right\rangle . \quad (2.147)$$

Equation (2.144) is just an operator representing a sum of decoupled harmonic oscillators with classical frequency ω_q. Using (2.147), we find that the energy eigenvalues of (2.143) are

$$E = \sum_q \hbar\omega_q (n_q + \tfrac{1}{2}) . \quad (2.148)$$

This is the same result as was previously obtained.

From relations (2.145) and (2.146) it is easy to see why $a_q^\dagger$ is often called a *creation* operator and a_q is often called an *annihilation* operator. We say that $a_q^\dagger$ creates a *phonon* in the mode q. The quantities n_q are said to be the number of phonons in the mode q. Since n_q can be any integer from 0 to ∞, the phonons are said to be *bosons*. In fact, the commutation relations of the a_q operators are typical commutation relations for boson annihilation and creation operators. The Hamiltonian in the form (2.144) is said to be written in *second quantization notation.* (See Appendix G for a discussion of this notation.) The eigenkets $|n_q\rangle$ are said to be *kets in occupation number space*.

With the Hamiltonian written in the form (2.144), we never really need to say much about eigenkets. All eigenkets are of the form

$$\left|m_q\right\rangle = \frac{1}{\sqrt{m_q!}}(a_q^\dagger)^{m_q}\left|0\right\rangle,$$

where $|0\rangle$ is the vacuum eigenket. More complex eigenkets are built up by taking a product. For example, $|m_1,m_2\rangle = |m_1\rangle\,|m_2\rangle$. States of the $|m_q\rangle$, which are eigenkets of both the annihilation and creation operators, are often called coherent states.

Let us briefly review what we have done in this section. We have found the eigenvalues and eigenkets of the Hamiltonian representing one-dimensional lattice vibrations in the harmonic and nearest-neighbor approximations. We have introduced the concept of the phonon, but some more discussion of the term may well be in order. We also need to give some more meaning to the subscript q that has been used. For both of these purposes it is useful to consider the symmetry properties of the crystal as they are reflected in the Hamiltonian.

The energy eigenvalue equation has been written

$$\mathcal{H}\psi(x_1 \cdots x_N) = E\psi(x_1 \cdots x_N).$$

Now suppose we define a translation operator T_m that translates the coordinates by ma. Since the Hamiltonian is invariant to such translations, we have

$$[\mathcal{H}, T_m] = 0. \tag{2.149}$$

By quantum mechanics [18] we know that it is possible to find a set of functions that are simultaneous eigenfunctions of both T_m and $\mathcal{H}$. In particular, consider the case $m = 1$. Then there exists an eigenket $|E\rangle$ such that

$$\mathcal{H}\left|E\right\rangle = E\left|E\right\rangle, \tag{2.150}$$

and

$$T_1\left|E\right\rangle = t^1\left|E\right\rangle. \tag{2.151}$$

Clearly $|t^1| = 1$ for $(T_1)^N\,|E\rangle = |E\rangle$ by periodic boundary conditions, and this implies $(t_1)^N = 1$ or $|t^1| = 1$. Therefore let

$$t^1 = \exp(\mathrm{i}k_q a), \tag{2.152}$$

where k_q is real. Since $|t_1| = 1$ we know that $k_q aN = p\pi$, where p is an integer. Thus

$$k_q = \frac{2\pi}{Na}\cdot p, \tag{2.153}$$

and hence k_q is of the same form as our old friend q. Statements (2.150) to (2.153) are equivalent to the already-mentioned *Bloch's theorem*, which is a general theorem for waves propagating in periodic media. For further proofs of Bloch's theorem and a discussion of its significance see Appendix C.

What is the q then? It is a quantum number labeling a state of vibration of the system. Because of translational symmetry (in discrete translations by a) the system naturally vibrates in certain states. These states are labeled by the q quantum number. There is nothing unfamiliar here. The hydrogen atom has rotational symmetry and hence its states are labeled by the quantum numbers characterizing the eigenfunctions of the rotational operators (which are related to the angular momentum operators). Thus it might be better to write (2.150) and (2.151) as

$$\mathcal{H}|E,q\rangle = E_q|E,q\rangle \tag{2.154}$$

$$T_1|E,q\rangle = \mathrm{e}^{\mathrm{i}k_q a}|E,q\rangle . \tag{2.155}$$

Incidentally, since $|E,q\rangle$ is an eigenket of T_1 it is also an eigenket of T_m. This is easily seen from the fact that $(T_1)^m = T_m$.

We now have a little better insight into the meaning of q. Several questions remain. What is the relation of the eigenkets $|E,q\rangle$ to the eigenkets $|n_q\rangle$? They, in fact, can be chosen to be the same.[14] This is seen if we utilize the fact that T_1 can be represented by

$$T_1 = \exp(\mathrm{i}a\textstyle\sum_{q'} q' a_{q'}^{\dagger} a_{q'}) . \tag{2.156}$$

Then it is seen that

$$\begin{aligned} T_1|n_q\rangle &= \exp(\mathrm{i}a\textstyle\sum_{q'} q' a_{q'}^{\dagger} a_{q'})|n_q\rangle \\ &= \exp(\mathrm{i}a\textstyle\sum_{q'} q' n_{q'} \delta_q^{q'})|n_q\rangle = \exp(\mathrm{i}aqn_q)|n_q\rangle . \end{aligned} \tag{2.157}$$

Let us now choose the set of eigenkets that simultaneously diagonalize both the Hamiltonian and the translation operator (the $|E,q\rangle$) to be the $|n_q\rangle$. Then we see that

$$k_q = q \cdot n_q . \tag{2.158}$$

This makes physical sense. If we say we have one phonon in mode q (which state we characterize by $|1_q\rangle$) then

$$T_1|1_q\rangle = \mathrm{e}^{\mathrm{i}qa}|1_q\rangle ,$$

and we get the typical factor $\mathrm{e}^{\mathrm{i}qa}$ for Bloch's theorem. However, if we have two phonons in mode q, then

$$T_1|2_q\rangle = \mathrm{e}^{\mathrm{i}qa(2)}|2_q\rangle ,$$

and the typical factor of Bloch's theorem appears twice.

---

[14] See, for example, Jensen [2.19]

The above should make clear what we mean when we say that a phonon is a quantum of a lattice vibrational state.

Further insight into the nature of the q can be gained by taking the expectation value of x_1 in a time-dependent state of fixed q. Define

$$|q\rangle \equiv \sum_{n_q} C_{n_q} \exp[-(\mathrm{i}/\hbar)(E_{n_q})t]\,|\,n_q\rangle\,. \tag{2.159}$$

We choose this state in order that the classical limit will represent a wave of fixed wavelength. Then we wish to compute

$$\langle \mathrm{q}\,|\,\mathrm{x_p}\,|\,q\rangle = \sum_{n_q,n_q^1} C_{n_q}^* C_{n_q^1} \exp[+(\mathrm{i}/\hbar)(E_{n_q} - E_{n_q^1})t]\cdot\langle n_q\,|\,\mathrm{x_p}\,|\,n_q^1\rangle\,. \tag{2.160}$$

By previous work we know that

$$x_p = (1/\sqrt{N})\sum_{q^1} \exp(-\mathrm{i}paq^1)X_{q^1}\,, \tag{2.161}$$

where the X_q can be written in terms of creation and annihilation operators as

$$X_q = \frac{1}{2\mathrm{i}}\sqrt{\frac{2\hbar}{M\omega_q}}(a_q^\dagger - a_{-q})\,. \tag{2.162}$$

Therefore,

$$x_p = \frac{1}{2\mathrm{i}}\sqrt{\frac{2\hbar}{NM}}\sum_{q^1} \exp(-\mathrm{i}paq^1)(a_{q^1}^\dagger - a_{-q^1})\frac{1}{\sqrt{\omega_{q^1}}}\,. \tag{2.163}$$

Thus

$$\begin{aligned}\left\langle n_q\middle|x_p\middle|n_q^1\right\rangle = \frac{1}{2\mathrm{i}}\sqrt{\frac{2\hbar}{NM}}\Big[&\sum_{q^1}(\omega_{q^1})^{-1/2}\exp(-\mathrm{i}paq^1)\left\langle n_q\middle|a_{q^1}^\dagger\middle|n_q^1\right\rangle \\ &-\sum_{q^1}\exp(-\mathrm{i}paq^1)\left\langle n_q\middle|a_{-q^1}\middle|n_q^1\right\rangle\Big]\,.\end{aligned} \tag{2.164}$$

By (2.145) and (2.146), we can write (2.164) as

$$\left\langle n_q\middle|x_p\middle|n_q^1\right\rangle = \frac{1}{2\mathrm{i}}\sqrt{\frac{2\hbar}{NM\omega_q}}\,\mathrm{e}^{-\mathrm{i}paq}\sqrt{n_q^1+1}\,\delta_{n_q^1+1}^{n_q} - \mathrm{e}^{+\mathrm{i}paq}\sqrt{n_q^1}\,\delta_{n_q}^{n_q^1-1}\,. \tag{2.165}$$

Then by (2.160) we can write

$$\begin{aligned}\left\langle q\middle|x_p\middle|q\right\rangle = \frac{1}{2\mathrm{i}}\sqrt{\frac{2\hbar}{NM\omega_q}}\Big[&\sum_{n_q} C_{n_q}^* C_{n_q-1}\sqrt{n_q}\,\mathrm{e}^{-\mathrm{i}paq}\mathrm{e}^{+\mathrm{i}\omega_q t} \\ &-\sum_{n_q} C_{n_q} C_{n_q+1}\sqrt{n_q+1}\,\mathrm{e}^{+\mathrm{i}paq}\mathrm{e}^{-\mathrm{i}\omega_q t}\Big]\,.\end{aligned} \tag{2.166}$$

In (2.166) we have used that

$$E_{n_q} = (n_q + \tfrac{1}{2})\hbar\omega_q .$$

Now let us go to the classical limit. In the classical limit only those C_n for which n_q is large are important. Further, let us suppose that C_n are very slowly varying functions of n_q. Since for large n_q we can write

$$\sqrt{n_q} \cong \sqrt{n_q + 1} ,$$

$$\langle q|x_p|q\rangle = \sqrt{\frac{2\hbar}{NM\omega_q}}(\textstyle\sum_{n_q=0}^{\infty}\sqrt{n_q}\,|C_{n_q}|^2)\sin[\omega_q t - q(pa)] . \quad (2.167)$$

Equation (2.167) is similar to the equation of a running wave on a classical lattice where pa serves as the coordinate (it locates the equilibrium position of the vibrating atom), and the displacement from equilibrium is given by x_p. In this classical limit then it is clear that q can be interpreted as 2π over the wavelength.

In view of the similarity of (2.167) to a plane wave, it might be tempting to call $\hbar q$ the momentum of the phonons. Actually, this should not be done because phonons do not carry momentum (except for the $q = 0$ phonon, which corresponds to a translation of the crystal as a whole). The q do obey a conservation law (as will be seen in the chapter on interactions), but this conservation law is somewhat different from the conservation of momentum.

To see that phonons do not carry momentum, it suffices to show that

$$\langle n_q|P_{\text{tot}}|n_q\rangle = 0 , \quad (2.168)$$

where

$$P_{\text{tot}} = \textstyle\sum_l p_l . \quad (2.169)$$

By previous work

$$p_l = (1/\sqrt{N})\textstyle\sum_{q^1} P_{q^1} \exp(\mathrm{i}q^1 la) ,$$

and

$$P_{q^1} = \sqrt{2M\hbar\omega_{q^1}}(a_{q^1} + a^{\dagger}_{-q^1}) .$$

Then

$$\langle n_q|P_{tot}|n_q\rangle = \sqrt{\frac{M\hbar}{2N}}\textstyle\sum_l\sum_{q^1}\sqrt{\omega_{q^1}}\exp(\mathrm{i}q^1 la)\langle n_q|(a_{q^1} + a^{\dagger}_{-q^1})|n_q\rangle = 0 \quad (2.170)$$

by (2.145) and (2.146). The $q^1 \to 0$ mode can be treated by a limiting process. However, it is simpler to realize it corresponds to all the atoms moving together so it obviously can carry momentum. Anybody who has been hit by a thrown rock knows that.

2.3 Three-Dimensional Lattices

Up to now only one-dimensional lattice vibration problems have been considered. They have the very great advantage of requiring only simple notation. The prolixity of symbols is what makes the three-dimensional problems somewhat more cumbersome. Not too many new ideas come with the added dimensions, but numerous superscripts and subscripts do.

2.3.1 Direct and Reciprocal Lattices and Pertinent Relations (B)

Let ($\boldsymbol{a}_1$, $\boldsymbol{a}_2$, $\boldsymbol{a}_3$) be the primitive translation vectors of the lattice. All points defined by

$$\boldsymbol{R}_l = l_1\boldsymbol{a}_1 + l_2\boldsymbol{a}_2 + l_3\boldsymbol{a}_3 , \tag{2.171}$$

where (l_1, l_2, l_3,) are integers, define the *direct lattice*. This vector will often be written as simply $\boldsymbol{l}$. Let ($\boldsymbol{b}_1$, $\boldsymbol{b}_2$, $\boldsymbol{b}_3$) be three vectors chosen so that

$$\boldsymbol{a}_i \cdot \boldsymbol{b}_j = \delta_{ij} . \tag{2.172}$$

Compare (2.172) to (1.38). The 2π could be inserted in (2.172) and left out of (2.173), which should be compared to (1.44). Except for notation, they are the same. There are two alternative ways of defining the reciprocal lattice. All points described by

$$\boldsymbol{G}_n = 2\pi(n_1\boldsymbol{b}_1 + n_2\boldsymbol{b}_2 + n_3\boldsymbol{b}_3) , \tag{2.173}$$

where (n_1, n_2, n_3) are integers, define the *reciprocal lattice* (we will sometimes use $\boldsymbol{K}$ for $\boldsymbol{G}_n$ type vectors). Cyclic boundary conditions are defined on a fundamental parallelepiped of volume

$$V_{\text{f.p.p.}} = N_1\boldsymbol{a}_1 \cdot (N_2\boldsymbol{a}_2 \times N_3\boldsymbol{a}_3) , \tag{2.174}$$

where N_1, N_2, N_3 are very large integers such that $(N_1)(N_2)(N_3)$ is of the order of Avogadro's number.

With cyclic boundary conditions, all wave vectors $\boldsymbol{q}$ (generalizations of the old q) in one dimension are given by

$$\boldsymbol{q} = 2\pi[(n_1/N_1)\boldsymbol{b}_1 + (n_2/N_2)\boldsymbol{b}_2 + (n_3/N_3)\boldsymbol{b}_3] . \tag{2.175}$$

The $\boldsymbol{q}$ are said to be restricted to a fundamental range when the n_i in (2.175) are restricted to the range

$$-N_i/2 < n_i < N_i/2 . \tag{2.176}$$

We can always add a $\boldsymbol{G}_n$ type vector to a $\boldsymbol{q}$ vector and obtain an equivalent vector. When the $\boldsymbol{q}$ in a fundamental range are modified (if necessary) by this technique to give a complete set of $\boldsymbol{q}$ that are closer to the origin than any other lattice point, then the $\boldsymbol{q}$ are said to be in the first Brillouin zone. Any general vector in direct space is given by

$$\boldsymbol{r} = \eta_1 \boldsymbol{a}_1 + \eta_2 \boldsymbol{a}_2 + \eta_3 \boldsymbol{a}_3 , \tag{2.177}$$

where the η_i are arbitrary real numbers.

Several properties of the quantities defined by (2.171) to (2.177) can now be derived. These properties are results of what can be called crystal mathematics. They are useful for three-dimensional lattice vibrations, the motion of electrons in crystals, and any type of wave motion in a periodic medium. Since most of the results follow either from the one-dimensional calculations or from Fourier series or integrals, they will not be derived here but will be presented as problems (Problem 2.11). However, most of these results are of great importance and are constantly used.

The most important results are summarized below:

1.
$$\frac{1}{N_1 N_2 N_3} \sum\nolimits_{\boldsymbol{R}_l} \exp(\mathrm{i}\boldsymbol{q} \cdot \boldsymbol{R}_l) = \sum\nolimits_{\boldsymbol{G}_n} \delta_{\boldsymbol{q},\boldsymbol{G}_n} . \tag{2.178}$$

2.
$$\frac{1}{N_1 N_2 N_3} \sum\nolimits_{\boldsymbol{q}} \exp(\mathrm{i}\boldsymbol{q} \cdot \boldsymbol{R}_l) = \delta_{\boldsymbol{R}_l,0} \tag{2.179}$$

(summed over one Brillouin zone).

3. In the limit as $V_{\text{f.p.p}} \to \infty$, one can replace

$$\sum\nolimits_{\boldsymbol{q}} \text{ by } \frac{V_{\text{f.p.p.}}}{(2\pi)^3} \int \mathrm{d}^3 q . \tag{2.180}$$

Whenever we speak of an integral over q space, we have such a limit in mind.

4.
$$\frac{\Omega_a}{(2\pi)^3} \int_{\text{one Brillouin zone}} \exp(\mathrm{i}\boldsymbol{q} \cdot \boldsymbol{R}_l) \mathrm{d}^3 q = \delta_{\boldsymbol{R}_l,0} , \tag{2.181}$$

where $\Omega_\mathrm{a} = a_1 \cdot a_2 \times a_3$ is the volume of a unit cell.

5.
$$\frac{1}{\Omega_a} \int_{\Omega_a} \exp[\mathrm{i}(\boldsymbol{G}_{l^1} - \boldsymbol{G}_l) \cdot \boldsymbol{r}] \mathrm{d}^3 r = \delta_{l^1,l} . \tag{2.182}$$

6.
$$\frac{1}{(2\pi)^3} \int_{\text{all } \boldsymbol{q} \text{ space}} \exp[\mathrm{i}\boldsymbol{q} \cdot (\boldsymbol{r} - \boldsymbol{r}^1)] \mathrm{d}^3 q = \delta(\boldsymbol{r} - \boldsymbol{r}^1) , \tag{2.183}$$

where $\delta(\boldsymbol{r} - \boldsymbol{r}^1)$ is the Dirac delta function.

7.
$$\frac{1}{(2\pi)^3} \int_{V_{\text{f.p.p.}\to\infty}} \exp[\mathrm{i}(\boldsymbol{q} - \boldsymbol{q}^1) \cdot \boldsymbol{r}] \mathrm{d}^3 r = \delta(\boldsymbol{q} - \boldsymbol{q}^1) . \tag{2.184}$$

2.3.2 Quantum-Mechanical Treatment and Classical Calculation of the Dispersion Relation (B)

This Section is similar to Sect. 2.2.6 on one-dimensional lattices but differs in three ways. It is three-dimensional. More than one atom per unit cell is allowed. Also, we indicate that so far as calculating the dispersion relation goes, we may as well stick to the notation of classical calculations. The use of $\boldsymbol{R}_l$ will be dropped in this section, and $\boldsymbol{l}$ will be used instead. It is better not to have subscripts of subscripts of...etc.

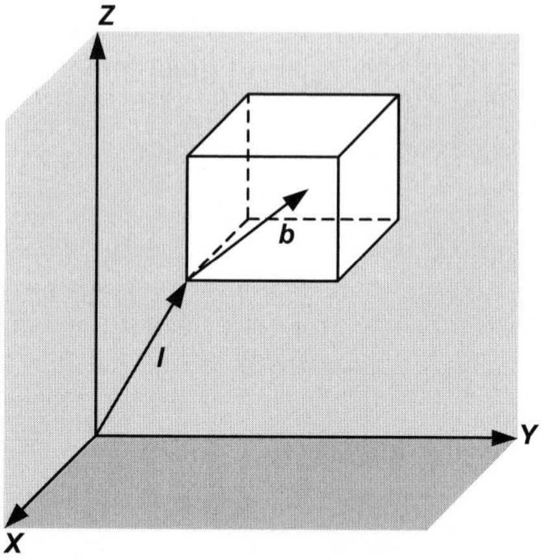

Fig. 2.8. Notation for three-dimensional lattices

In Fig. 2.8, $\boldsymbol{l}$ specifies the location of the unit cell and $\boldsymbol{b}$ specifies the location of the atoms in the unit cell (there may be several $\boldsymbol{b}$ for each unit cell).

The actual coordinates of the atoms will be $\boldsymbol{d}_{l,b}$ and

$$\boldsymbol{x}_{l,b} = \boldsymbol{d}_{l,b} - (\boldsymbol{l} + \boldsymbol{b}) \tag{2.185}$$

will be the coordinates that specify the deviation of the position of an atom from equilibrium.

The potential energy function will be $V(\boldsymbol{x}_{l,b})$. In the equilibrium state, by definition,

$$(\nabla_{\boldsymbol{x}_{l,b}} V)_{\text{all } \boldsymbol{x}_{l,b}=0} = 0 . \tag{2.186}$$

Expanding the potential energy in a Taylor series, and neglecting the anharmonic terms, we have

$$V(\boldsymbol{x}_{l,b}) = V_0 + \tfrac{1}{2}\sum\nolimits_{l,b,l^1,b^1(\alpha,\beta)} x^{\alpha}_{lb} J^{\alpha\beta}_{lbl^1b^1} x^{\beta}_{l^1b^1} . \tag{2.187}$$

In (2.187), $x^{\alpha}_{l,b}$ is the αth component of $\boldsymbol{x}_{l,b}$. V_0 can be chosen to be zero, and this choice then fixes the zero of the potential energy. If $\boldsymbol{p}_{lb}$ is the momentum (operator) of the atom located at $\boldsymbol{l} + \boldsymbol{b}$ with mass m_b, the Hamiltonian can be written

$$\mathcal{H} = \frac{1}{2}\sum\nolimits_{\substack{\boldsymbol{l}\ (\text{all unit cells}),\ \alpha=1 \\ \boldsymbol{b}\ (\text{all atoms} \\ \text{within a cell})}}^{\alpha=3} \frac{1}{m_b} p^{\alpha}_{\boldsymbol{l}\boldsymbol{b}} p^{\alpha}_{\boldsymbol{l}\boldsymbol{b}} + \frac{1}{2}\sum\nolimits_{\boldsymbol{l},\boldsymbol{b},\boldsymbol{l}^1,\boldsymbol{b}^1,\alpha=1,\beta=1}^{\alpha=3,\beta=3} J^{\alpha\beta}_{\boldsymbol{l}\boldsymbol{b}\boldsymbol{l}^1\boldsymbol{b}^1} x^{\alpha}_{\boldsymbol{l}\boldsymbol{b}} x^{\beta}_{\boldsymbol{l}^1\boldsymbol{b}^1}. \tag{2.188}$$

In (2.188), summing over α or β corresponds to summing over three Cartesian coordinates, and

$$J^{\alpha\beta}_{\boldsymbol{l}\boldsymbol{b}\boldsymbol{l}^1\boldsymbol{b}^1} = \left(\frac{\partial^2 V}{\partial x^{\alpha}_{\boldsymbol{l}\boldsymbol{b}} \partial x^{\beta}_{\boldsymbol{l}^1\boldsymbol{b}^1}}\right)_{\text{all } \boldsymbol{x}_{lb}=0}. \tag{2.189}$$

The Hamiltonian simplifies much as in the one-dimensional case. We make a normal coordinate transformation or a Fourier analysis of the coordinate and momentum variables. The transformation is canonical, and so the new variables obey the same commutation relations as the old:

$$\boldsymbol{x}_{\boldsymbol{l},\boldsymbol{b}} = \frac{1}{\sqrt{N}}\sum\nolimits_{\boldsymbol{q}} \boldsymbol{X}^1_{\boldsymbol{q},\boldsymbol{b}} \mathrm{e}^{-\mathrm{i}\boldsymbol{q}\cdot\boldsymbol{l}}, \tag{2.190}$$

$$\boldsymbol{p}_{\boldsymbol{l},\boldsymbol{b}} = \frac{1}{\sqrt{N}}\sum\nolimits_{\boldsymbol{q}} \boldsymbol{P}^1_{\boldsymbol{q},\boldsymbol{b}} \mathrm{e}^{+\mathrm{i}\boldsymbol{q}\cdot\boldsymbol{l}}, \tag{2.191}$$

where $N = N_1N_2N_3$. Since $\boldsymbol{x}_{l,b}$ and $\boldsymbol{p}_{l,b}$ are Hermitian, we must have

$$\boldsymbol{X}^1_{-\boldsymbol{q},\boldsymbol{b}} = \boldsymbol{X}^{1\dagger}_{\boldsymbol{q},\boldsymbol{b}}, \tag{2.192}$$

and

$$\boldsymbol{P}^1_{-\boldsymbol{q},\boldsymbol{b}} = \boldsymbol{P}^{1\dagger}_{\boldsymbol{q},\boldsymbol{b}}. \tag{2.193}$$

Substituting (2.190) and (2.191) into (2.188) gives

$$\mathcal{H} = \frac{1}{2}\sum\nolimits_{\boldsymbol{l},\boldsymbol{b}} \frac{1}{m_b}\frac{1}{N}\sum\nolimits_{\boldsymbol{q},\boldsymbol{q}^1} \boldsymbol{P}^1_{\boldsymbol{q},\boldsymbol{b}} \cdot \boldsymbol{P}^1_{\boldsymbol{q}^1,\boldsymbol{b}} \mathrm{e}^{\mathrm{i}(\boldsymbol{q}+\boldsymbol{q}^1)\cdot\boldsymbol{l}} + \frac{1}{2}\sum\nolimits_{\boldsymbol{l},\boldsymbol{b},\boldsymbol{l}^1,\boldsymbol{b}^1,\alpha,\beta} \frac{1}{N}\sum\nolimits_{\boldsymbol{q},\boldsymbol{q}^1} J^{\alpha\beta}_{\boldsymbol{l},\boldsymbol{b},\boldsymbol{l}^1,\boldsymbol{b}^1} X^{1\alpha}_{\boldsymbol{q},\boldsymbol{b}} X^{1\beta}_{\boldsymbol{q}^1,\boldsymbol{b}^1} \mathrm{e}^{-\mathrm{i}(\boldsymbol{q}\cdot\boldsymbol{l}+\boldsymbol{q}^1\cdot\boldsymbol{l}^1)}. \tag{2.194}$$

Using (2.178) on the first term of the right-hand side of (2.194) we can write

$$\mathcal{H} = \frac{1}{2}\sum_{q,b}\frac{1}{m_b}\boldsymbol{P}^1_{q,b}\cdot\boldsymbol{P}^{1\dagger}_{q,b} + \frac{1}{2N}\sum_{\substack{q,q^1,b,b^1\\ \alpha,\beta}}\left\{\sum_{l,l^1} J^{\alpha\beta}_{l,b,l^1,b^1}\, e^{-iq^1\cdot(l-l^1)}\, \mathrm{e}^{-i(q+q^1)}\right\} X^{1\alpha}_{q,b}\, X^{1\beta}_{q^1,b^1}\,. \tag{2.195}$$

The force between any two atoms in our perfect crystal cannot depend on the position of the atoms but only on the vector separation of the atoms. Therefore, we must have that

$$J^{\alpha\beta}_{l,b,l^1,b^1} = J^{\alpha\beta}_{b,b^1}(\boldsymbol{l}-\boldsymbol{l}^1)\,. \tag{2.196}$$

Letting $\boldsymbol{m} = (\boldsymbol{l} - \boldsymbol{l}^1)$, defining

$$K_{bb^1}(\boldsymbol{q}) = \sum_{\boldsymbol{m}} J_{bb^1}(\boldsymbol{m})\mathrm{e}^{-i\boldsymbol{q}\cdot\boldsymbol{m}}\,, \tag{2.197}$$

and again using (2.178), we find that the Hamiltonian becomes

$$\mathcal{H} = \sum_{q}\mathcal{H}_q\,, \tag{2.198a}$$

where

$$\mathcal{H}_q = \frac{1}{2}\sum_b \frac{1}{m_b}\boldsymbol{P}^1_{q,b}\cdot\boldsymbol{P}^{1\dagger}_{q,b} + \frac{1}{2}\sum_{\substack{b,b^1\\ \alpha,\beta}} K^{\alpha\beta}_{b,b^1}\, X^{1\alpha}_{q,b}\, X^{1\beta\dagger}_{q^1,b^1}\,. \tag{2.198b}$$

The transformation has used translational symmetry in decoupling terms in the Hamiltonian. The rest of the transformation depends on the crystal structure and is found by straightforward small vibration theory applied to each unit cell. If there are K particles per unit cell, then there are $3K$ normal modes described by (2.198). Let $\omega_{q,p}$, where p goes from 1 to $3K$, represent the eigenfrequencies of the normal modes, and let $\boldsymbol{e}_{q,p,b}$ be the components of the eigenvectors of the normal modes. The quantities $\boldsymbol{e}_{q,p,b}$ allow us to calculate[15] the magnitude and direction of vibration of the atom at $\boldsymbol{b}$ in the mode labeled by $(\boldsymbol{q}, p)$. The eigenvectors can be chosen to obey the usual orthogonality relation

$$\sum_b \boldsymbol{e}^*_{qpb}\cdot\boldsymbol{e}_{qp^1b} = \delta_{p,p^1}\,. \tag{2.199}$$

It is convenient to allow for the possibility that $\boldsymbol{e}_{qpb}$ is complex due to the fact that all we readily know about $\mathcal{H}_q$ is that it is Hermitian. A Hermitian matrix can always be diagonalized by a unitary transformation. A real symmetric matrix can always be diagonalized by a real orthogonal transformation. It can be shown that

[15] The way to do this is explained later when we discuss the classical calculation of the dispersion relation.

with only one atom per unit cell the polarization vectors $\boldsymbol{e}_{qpb}$ are real. We can choose $\boldsymbol{e}_{-q,p,b} = \boldsymbol{e}^{*}_{q,p,b}$ in more general cases.

Once the eigenvectors are known, we can make a normal coordinate transformation and hence diagonalize the Hamiltonian [99]:

$$X^{11}_{q,p} = \sum_b \sqrt{m_b}\, \boldsymbol{e}_{qpb} \cdot \boldsymbol{X}^{1}_{qb} \,. \tag{2.200}$$

The momentum $\boldsymbol{P}^{11}_{q,p}$, which is canonically conjugate to (2.200), is

$$\boldsymbol{P}^{11}_{q,p} = \sum_b (1/\sqrt{m_b}) \boldsymbol{e}^{*}_{qpb} \cdot \boldsymbol{P}^{1}_{qb} \,. \tag{2.201}$$

Equations (2.200) and (2.201) can be inverted by use of the closure notation

$$\sum_p e^{\alpha *}_{qpb} e^{\beta}_{qpb^1} = \delta^{\beta}_{\alpha} \delta^{b^1}_{b} \,. \tag{2.202}$$

Finally, define

$$a_{q,p} = 1/\sqrt{2\hbar\omega_{q,p}}\, P^{11}_{q,p} - i\sqrt{(\omega_{q,p}/2\hbar)}\, X^{11\dagger}_{q,p} \,, \tag{2.203}$$

and a similar expression for $a^{\dagger}_{q,p}$. In the same manner as was done in the one-dimensional case, we can show that

$$[a_{q,p}, a^{\dagger}_{q,p}] = \delta^{q^1}_{q} \delta^{p^1}_{p} \,, \tag{2.204}$$

and that the other commutators vanish. Therefore the as are boson annihilation operators, and the $a^{\dagger}$ are boson creation operators. In this second quantization notation, the Hamiltonian reduces to a set of decoupled harmonic oscillators:

$$\mathcal{H} = \sum_{q,p} \hbar\omega_{q,p} (a^{\dagger}_{q,p} a_{q,p} + \tfrac{1}{2}) \,. \tag{2.205}$$

By (2.205) we have seen that the Hamiltonian can be represented by $3NK$ decoupled harmonic oscillators. This decomposition has been shown to be formally possible within the context of quantum mechanics. However, the only thing that we do not know is the dispersion relationship that gives ω as a function of $\boldsymbol{q}$ for each p. The dispersion relation is the same in quantum mechanics and classical mechanics because the calculation is the same. Hence, we may as well stay with classical mechanics to calculate the dispersion relation (except for estimating the forces), as this will generally keep us in a simpler notation. In addition, we do not know what the potential V is and hence the J and K ((2.189), (2.197)) are unknown also.

This last fact emphasizes what we mean when we say we have obtained a formal solution to the lattice-vibration problem. In actual practice the calculation of the dispersion relation would be somewhat cruder than the above might lead one to suspect. We gave some references to actual calculations in the introduction to Sect. 2.2. One approach to the problem might be to imagine the various atoms hooked together by springs. We would try to choose the spring constants so that the elastic constants, sound velocity, and the specific heat were given correctly. Perhaps not all the spring constants would be determined by this method. We might like to try to select the rest so that they gave a dispersion relation that agreed with the dispersion relation provided by neutron diffraction data (if available). The details of such a program would vary from solid to solid.

Let us briefly indicate how we would calculate the dispersion relation for a crystal lattice if we were interested in doing it for an actual case. We suppose we have some combination of model, experiment, and general principles so the

$$J^{\alpha\beta}_{l,b,l^1,b^1}$$

can be determined. We would start with the Hamiltonian (2.188) except that we would have in mind staying with classical mechanics:

$$\mathcal{H} = \frac{1}{2}\sum_{l,b,\alpha=1}^{\alpha=3} \frac{1}{m_b}(p^{\alpha}_{l,b})^2 + \frac{1}{2}\sum_{l,b,l^1,b^1,\alpha=1,\beta=1}^{\alpha=3,\beta=3} J^{\alpha\beta}_{l,b,l^1,b^1} x^{\alpha}_{lb} x^{\beta}_{l^1b^1} . \tag{2.206}$$

We would use the known symmetry in J:

$$J^{\alpha\beta}_{l,b,l^1,b^1} = J^{\alpha\beta}_{l^1,b^1,l,b} , \quad J^{\alpha\beta}_{l,b,l^1,b^1} = J^{\alpha\beta}_{(l-l^1),b,b^1} . \tag{2.207}$$

It is also possible to show by translational symmetry (similarly to the way (2.33) was derived) that

$$\sum_{l^1,b^1} J^{\alpha\beta}_{l,b,l^1,b^1} = 0 . \tag{2.208}$$

Other restrictions follow from the rotational symmetry of the crystal.[16]

The equations of motion of the lattice are readily obtained from the Hamiltonian in the usual way. They are

$$m_b \ddot{x}^{\alpha}_{lb} = -\sum_{l^1,b^1,\beta} J^{\alpha\beta}_{l,b,l^1,b^1} x^{\beta}_{l^1,b^1} . \tag{2.209}$$

If we seek normal mode solutions of the form (whose real part corresponds to the physical solutions)[17]

$$x^{\alpha}_{l,b} = \frac{1}{\sqrt{m_b}} x^{\alpha}_{b} \, \mathrm{e}^{-\mathrm{i}\omega t + \boldsymbol{q}\cdot\boldsymbol{l}} , \tag{2.210}$$

---

[16] Maradudin et. al. [2.26].

[17] Note that this substitution assumes the results of Bloch's theorem as discussed after (2.39).

we find (using the periodicity of the lattice) that the equations of motion reduce to

$$\omega^2 x_b^\alpha = \sum_{b^1,\beta} M_{q,b,b^1}^{\alpha\beta} x_{b^1}^\beta , \tag{2.211}$$

where

$$M_{q,b,b^1}^{\alpha\beta}$$

is called the *dynamical matrix* and is defined by

$$M_{q,b,b^1}^{\alpha\beta} = \frac{1}{\sqrt{m_b m_{b^1}}} \sum_{(l-l^1)} J_{(l-l^1)b,b^1}^{\alpha\beta} \mathrm{e}^{-\mathrm{i}\boldsymbol{q}\cdot(\boldsymbol{l}-\boldsymbol{l}^1)} . \tag{2.212}$$

These equations have nontrivial solutions provided that

$$\det(M_{q,b,b^1}^{\alpha\beta} - \omega^2 \delta_{\alpha\beta} \delta_{b,b^1}) = 0 . \tag{2.213}$$

If there are K atoms per unit cell, the determinant condition yields $3K$ values of ω^2 for each $\boldsymbol{q}$. These correspond to the $3K$ branches of the dispersion relation. There will always be three branches for which $\omega = 0$ if $\boldsymbol{q} = 0$. These branches are called the *acoustic modes*. Higher branches, if present, are called the *optic modes*.

Suppose we let the solutions of the determinantal condition be defined by $\omega_p^2(\boldsymbol{q})$, where $p = 1$ to $3K$. Then we can define the polarization vectors by

$$\omega_p^2(\boldsymbol{q}) e_{q,p,b}^\alpha = \sum_{b^1,\beta} M_{q,b,b^1}^{\alpha\beta} e_{q,p,b}^\beta . \tag{2.214}$$

It is seen that these polarization vectors are just the eigenvectors. In evaluating the determinantal equation, it will probably save time to make full use of the symmetry properties of J via M. The physical meaning of complex polarization vectors is obtained when they are substituted for x_b^α and then the resulting real part of $x_{l,b}^\alpha$ is calculated.

The central problem in lattice-vibration dynamics is to determine the dispersion relation. As we have seen, this is a purely classical calculation. Once the dispersion relation is known (and it never is fully known exactly – either from calculation or experiment), quantum mechanics can be used in the formalism already developed (see, for example, (2.205) and preceding equations).

2.3.3 The Debye Theory of Specific Heat (B)

In this Section an exact expression for the specific heat will be written down. This expression will then be approximated by something that can actually be evaluated. The method of approximation used is called the *Debye approximation*. Note that in three dimensions (unlike one dimension), the form of the dispersion relation and hence the density of states is not exactly known [2.11]. Since the Debye model works so well, for many years after it was formulated nobody tried very hard to do better. Actually, it is always a surprise that the approximation does work well because the assumptions, on first glance, do not appear to be completely reasonable. Before Debye's work, Einstein showed (see Problem 2.24) that a simple model in which each mode had the same frequency, led with quantum mechanics to a specific heat that vanished at absolute zero. However, the Einstein model predicted an exponential temperature decrease at low temperatures rather than the correct T^3 dependence.

The average number of phonons in mode ($\boldsymbol{q}$, p) is

$$\bar{n}_{\boldsymbol{q},p} = \frac{1}{\exp(\hbar\omega_{\boldsymbol{q},p}/kT) - 1}. \tag{2.215}$$

The average energy per mode is

$$\hbar\omega_{\boldsymbol{q},p}\bar{n}_{\boldsymbol{q},p},$$

so that the thermodynamic average energy is (neglecting a constant zero-point correction, cf. (2.77))

$$U = \sum\nolimits_{\boldsymbol{q},p} \frac{\hbar\omega_{\boldsymbol{q},p}}{\exp(\hbar\omega_{\boldsymbol{q},p}/kT) - 1}. \tag{2.216}$$

The specific heat at constant volume is then given by

$$C_v = \left(\frac{\partial U}{\partial T}\right)_v = \frac{1}{kT^2}\sum\nolimits_{\boldsymbol{q},p} \frac{(\hbar\omega_{\boldsymbol{q},p})^2 \exp(\hbar\omega_{\boldsymbol{q},p}/kT)}{[\exp(\hbar\omega_{\boldsymbol{q},p}/kT) - 1]^2}. \tag{2.217}$$

Incidentally, when we say we are differentiating at constant volume it may not be in the least evident where there could be any volume dependence. However, the $\omega_{\boldsymbol{q},p}$ may well depend on the volume. Since we are interested only in a crystal with a fixed volume, this effect is not relevant. The student may object that this is not realistic as there is a thermal expansion of the solids. It would not be consistent to include anything about thermal expansion here. Thermal expansion is due to the anharmonic terms in the potential and we are consistently neglecting these. Furthermore, the Debye theory works fairly well in its present form without refinements.

The Debye model is a model based on the exact expression (2.217) in which the sum is evaluated by replacing it by an integral in which there is a density of states. Let the total density of states $D(\omega)$ be represented by

$$D(\omega) = \sum\nolimits_p D_p(\omega), \tag{2.218}$$

where $D_p(\omega)$ is the number of modes of type p per unit frequency at frequency ω. The Debye approximation consists in assuming that the lattice vibrates as if it were an elastic continuum. This should work at low temperatures because at low temperatures only long-wavelength (low $\boldsymbol{q}$) acoustic modes should be important. At high temperatures the cutoff procedure that we will introduce for $D(\omega)$ will assure that we get the results of the classical equipartition theorem whether or not we use the elastic continuum model. We choose the cutoff frequency so that we have only $3NK$ (where N is the number of unit cells and K is the number of atoms per unit cell) distinct continuum frequencies corresponding to the $3NK$ normal modes. The details of choosing this cutoff frequency will be discussed in more detail shortly.

In a box with length L_x, width L_y, and height L_z, classical elastic isotropic continuum waves have frequencies given by

$$\omega_j^2 = \pi^2 c^2 \left(\frac{k_j^2}{L_x^2} + \frac{l_j^2}{L_y^2} + \frac{m_j^2}{L_z^2} \right), \tag{2.219}$$

where c is the velocity of the wave (it may differ for different types of waves), and (k_j, l_j and m_j) are positive integers.

We can use the dispersion relation given by (2.219) to derive the density of states $D_p(\omega)$.[18] For this purpose, it is convenient to define an ω space with base vectors

$$\hat{\boldsymbol{e}}_1 = \frac{\pi c}{L_x}\hat{\boldsymbol{i}}, \quad \hat{\boldsymbol{e}}_2 = \frac{\pi c}{L_y}\hat{\boldsymbol{j}}, \quad \text{and} \quad \hat{\boldsymbol{e}}_3 = \frac{\pi c}{L_z}\hat{\boldsymbol{k}}. \tag{2.220}$$

Note that

$$\omega_j^2 = k_j^2 \hat{e}_1^2 + l_j^2 \hat{e}_2^2 + m_j^2 \hat{e}_3^2. \tag{2.221}$$

Since the (k_i, l_i, m_i) are positive integers, for each state ω_j, there is an associated cell in ω space with volume

$$\hat{\boldsymbol{e}}_1 \cdot (\hat{\boldsymbol{e}}_2 \times \hat{\boldsymbol{e}}_3) = \frac{(\pi c)^3}{L_x L_y L_z}. \tag{2.222}$$

The volume of the crystals is $V = L_x L_y L_z$, so that the number of states per unit volume of ω space is $V/(\pi c)^3$. If n is the number of states in a sphere of radius ω in ω space, then

$$n = \frac{1}{8}\frac{4\pi}{3}\omega^3 \frac{V}{(\pi c)^3}.$$

The factor $^1/_8$ enters because only positive k_j, l_j, and m_j are allowed. Simplifying, we obtain

$$n = \frac{\pi}{6}\omega^3 \frac{V}{(\pi c)^3}. \tag{2.223}$$

[18] We will later introduce more general ways of deducing the density of states from the dispersion relation, see (2.258).

The density of states for mode p (which is the number of modes of type p per unit frequency) is

$$D_p(\omega) = \frac{\mathrm{d}n}{\mathrm{d}\omega} = \frac{\omega^2 V}{(2\pi^2 c_p^3)} . \tag{2.224}$$

In (2.224), c_p means the velocity of the wave in mode p.

Debye assumed (consistent with the isotropic continuum limit) that there were two transverse modes and one longitudinal mode. Thus for the total density of states, we have $D(\omega) = (\omega^2 V/2\pi^2)\,(1/c_l^3 + 2/c_t^3)$, where c_l and c_t are the velocities of the longitudinal and transverse modes. However, the total number of modes must be $3NK$. Thus, we have

$$3NK = \int_0^{\omega_D} D(\omega)\mathrm{d}\omega .$$

Note that when $K = 2$ = the number of atoms per unit cell, the assumptions we have made push the optic modes into the high-frequency part of the density of states. We thus have

$$3NK = \int_0^{\omega_D} \frac{V}{2\pi^2}\left[\frac{1}{c_l^3} + \frac{1}{c_t^3}\right]\omega^2 \mathrm{d}\omega . \tag{2.225}$$

We have assumed only one cutoff frequency ω_D. This was not necessary. We could just as well have defined a set of cutoff frequencies by the set of equations

$$\begin{aligned} 2NK &= \int_0^{\omega_D^t} D(\omega)_t \mathrm{d}\omega, \\ NK &= \int_0^{\omega_D^l} D(\omega)_l \mathrm{d}\omega. \end{aligned} \tag{2.226}$$

There are yet further alternatives. But we are already dealing with a phenomenological treatment. Such modifications may improve the agreement of our results with experiment, but they hardly increase our understanding from a fundamental point of view. Thus for simplicity let us also assume that $c_p = c$ = constant. We can regard c as some sort of average of the c_p.

Equation (2.225) then gives us

$$\omega_D = \left(\frac{6\pi^2 N c^3}{V} K\right)^{1/3} . \tag{2.227}$$

The *Debye temperature* θ_D is defined as

$$\theta_D = \frac{\hbar\omega_D}{k} = \frac{\hbar}{k}\left(\frac{6\pi^2 NKc^3}{V}\right)^{1/3} . \tag{2.228}$$

Combining previous results, we have for the specific heat

$$C_v = \frac{3}{kT^2}\int_0^{\omega_D} \frac{(\hbar\omega)^2 \exp(\hbar\omega/kT)}{[\exp(\hbar\omega/kT)-1]^2}\frac{V}{2\pi^2 c^3}\omega^2 \mathrm{d}\omega ,$$

which gives for the specific heat per unit volume (after a little manipulation)

$$\frac{C_v}{V} = 9k(NK/V)D(\theta_D/T), \tag{2.229}$$

where $D(\theta_D/T)$ is the *Debye function* defined by

$$D(\theta_D/T) = (T/\theta_D)^3 \int_0^{\theta_D/T} \frac{z^4 e^z dz}{(e^z - 1)^2}. \tag{2.230}$$

In Problem 2.13, you are asked to show that (2.230) predicts a T^3 dependence for C_v at low temperature and the classical limit of $3k(NK)$ at high temperature. Table 2.3 gives some typical Debye temperatures. For metals θ_D in K for Al is about 394, Fe about 420, and Pb about 88. See, e.g., Parker [24, p 104].

Table 2.3. Approximate Debye temperature for alkali halides at 0 K

Alkali halide	Debye temperature (K)
LiF	734
NaCl	321
KBr	173
RbI	103

Adapted with permission from Lewis JT et al. *Phys Rev* **161**, 877, 1967. Copyright 1967 by the American Physical Society.

In discussing specific heats there is, as mentioned, one big difference between the one-dimensional case and the three-dimensional case. In the one-dimensional case, the dispersion relation is known exactly (for nearest-neighbor interactions) and from it the density of states can be exactly computed. In the three-dimensional case, the dispersion relation is not known, and so the dispersion relation of a classical isotropic elastic continuum is often used instead. From this dispersion relation, a density of states is derived. As already mentioned, in recent years it has been possible to determine the dispersion relation directly by the technique of neutron diffraction (which will be discussed in a later chapter). Somewhat less accurate methods are also available. From the dispersion relation we can (rather laboriously) get a fairly accurate density of states curve. Generally speaking, this density of states curve does not compare very well with the density of states used in the Debye approximation. The reason the error is not serious is that the specific heat uses only an integral over the density of states.

In Fig. 2.9 and Fig. 2.10 we have some results of dispersion curves and density of states curves that have been obtained from neutron work. Note that only in the crudest sense can we say that Debye theory fits a dispersion curve as represented by Fig. 2.10. The vibrational frequency spectrum can also be studied by other methods such as for example by X-ray scattering. See Maradudin et al [2.26, Chap. VII] and Table 2.4.

Table 2.4. Experimental methods of studying phonon spectra

Method	Reference
Inelastic scattering of neutrons by phonons. See the end of Sect. 4.3.1	Brockhouse and Stewart [2.6]. Shull and Wollan [2.31]
Inelastic scattering of X-rays by phonons (in which the diffuse background away from Bragg peaks is measured). Synchrotron radiation with high photon flux has greatly facilitated this technique.	Dorner et al [2.13]
Raman scattering (off optic modes) and Brillouin scattering (off acoustic modes). See Sect. 10.11.	Vogelgesang et al [2.36].

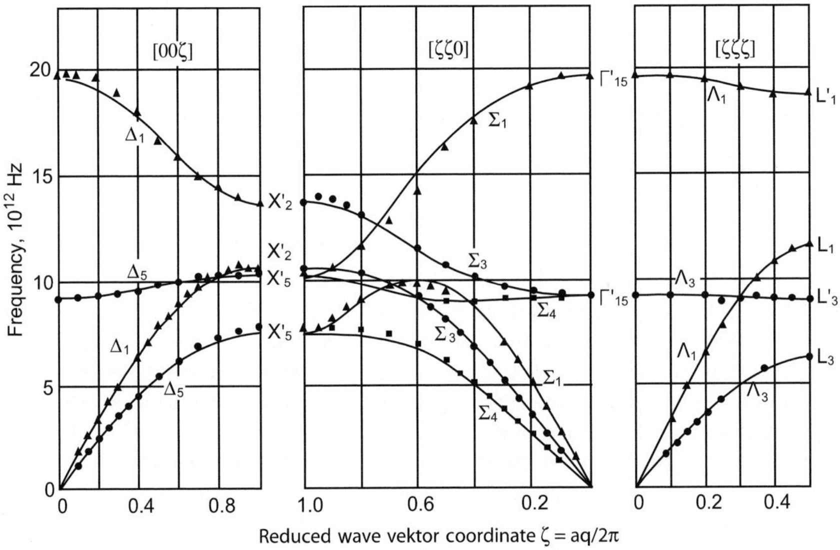

Fig. 2.9. Measured dispersion curves. The dispersion curves are for Li^7F at 298 °K. The results are presented along three directions of high symmetry. Note the existence of both optic and acoustic modes. The solid lines are a best least-squares fit for a seven-parameter model. [Reprinted with permission from Dolling G, Smith HG, Nicklow RM, Vijayaraghavan PR, and Wilkinson MK, *Physical Review*, **168**(3), 970 (1968). Copyright 1968 by the American Physical Society.] For a complete definition of all terms, reference can be made to the original paper

The Debye theory is often phenomenologically improved by letting $\theta_D = \theta_D(T)$ in (2.229). Again this seems to be a curve-fitting procedure, rather than a procedure that leads to better understanding of the fundamentals. It is, however, a good

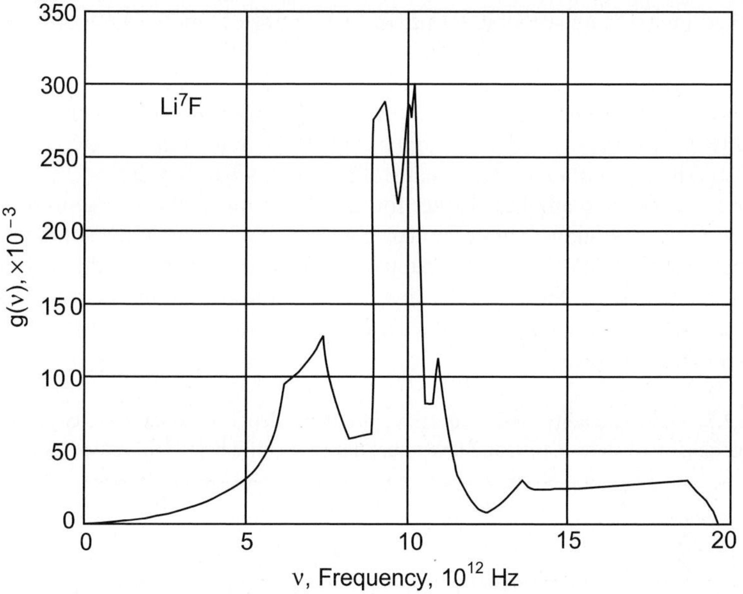

Fig. 2.10. Density of states $g(\nu)$ for Li^7F at 298 °K. [Reprinted with permission from Dolling G, Smith HG, Nicklow RM, Vijayaraghavan PR, and Wilkinson MK, *Physical Review*, **168**(3), 970 (1968). Copyright 1968 by the American Physical Society.]

way of measuring the consistency of the Debye approximation. That is, the more θ_D varies with temperature, the less accurate the Debye density of states is in representing the true density of states.

We should mention that from a purely theoretical point we know that the Debye model must, in general, be wrong. This is because of the existence of Van Hove singularities [2.35]. A general expression for the density of states involves one over the k space gradient of the frequency (see (3.258)). Thus, Van Hove has shown that the translational symmetry of a lattice causes critical points [values of $\boldsymbol{k}$ for which $\nabla_k \omega_p(\boldsymbol{k}) = 0$] and that these critical points (which are maxima, minima, or saddle points) in general cause singularities (e.g. a discontinuity of slope) in the density of states. See Fig. 2.10. It is interesting to note that the approximate Debye theory has no singularities except that due to the cutoff procedure.

The experimental curve for the specific heat of insulators looks very much like Fig. 2.11. The Debye expression fits this type of curve fairly well at all temperatures. Kohn has shown that there is another cause of singularities in the phonon spectrum that can occur in metals. These occur when the phonon wave vector is twice the Fermi wave vector. Related comments are made in Sects. 5.3, 6.6, and 9.5.3.

In this chapter we have set up a large mathematical apparatus for defining phonons and trying to understand what a phonon is. The only thing we have calculated that could be compared to experiment is the specific heat. Even the specific heat was not exactly evaluated. First, we made the Debye approximation. Second, if we had included anharmonic terms, we would have found a small term linear in T at high T. For the experimentally minded student, this is not very satisfactory. He would want to see calculations and comparisons to experiment for a wide variety of cases. However, our plan is to defer such considerations. Phonons are one of the two most important basic energy excitations in a solid (electrons being the other) and it is important to understand, at first, just what they are.

We have reserved another chapter for the discussion of the interactions of phonons with other phonons, with other basic energy excitations of the solid, and with external probes such as light. This subject of interactions contains the real meat of solid-state physics. One topic in this area is introduced in the next section. Table 2.5 summarizes simple results for density of states and specific heat in one, two, and three dimensions.

Table 2.5. Dimensionality and frequency (ω) dependence of long-wavelength acoustic phonon density of states $D(\omega)$, and low-temperature specific heat C_v of lattice vibrations

	$D(\omega)$	C_v
One dimension	A_1	$B_1 T$
Two dimensions	$A_2\,\omega$	$B_2\,T^2$
Three dimensions	$A_3\,\omega^2$	$B_3\,T^3$

Note that the A_i and B_i are constants.

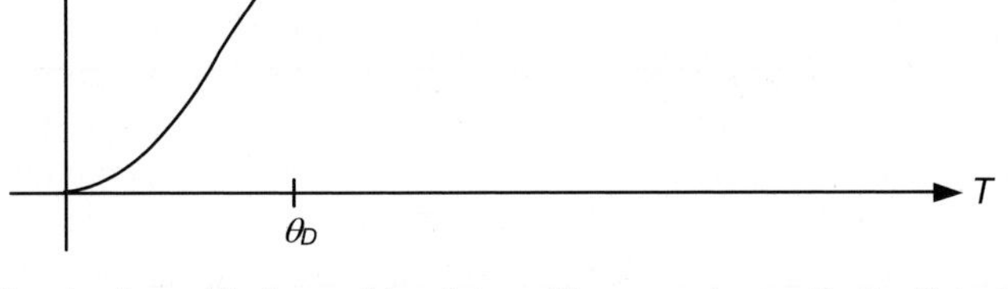

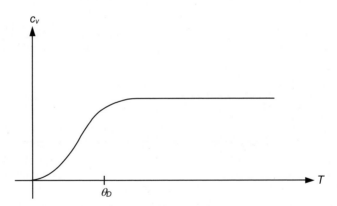

Fig. 2.11. Sketch of specific heat of insulators. The curve is practically flat when the temperature is well above the Debye temperature

2.3.4 Anharmonic Terms in The Potential / The Gruneisen Parameter (A)[19]

We wish to address the topic of thermal expansion, which would not exist without anharmonic terms in the potential (for then the average position of the atoms would be independent of their amplitude of vibration). Other effects of the anharmonic terms are the existence of finite thermal conductivity (which we will discuss later in Sect. 4.2) and the increase of the specific heat beyond the classical Dulong and Petit value at high temperature. Here we wish to obtain an approximate expression for the coefficient of thermal expansion (which would vanish if there were no anharmonic terms).

We first derive an expression for the free energy of the lattice due to thermal vibrations. The free energy is given by

$$F_L = -k_B T \ln Z\,, \tag{2.231}$$

where Z is the partition function. The partition function is given by

$$Z = \textstyle\sum_{\{n\}} \exp(-\beta E_{\{n\}})\,,\ \beta = \dfrac{1}{k_B T}\,, \tag{2.232}$$

where

$$E_{\{n\}} = \textstyle\sum_{\boldsymbol{k},j} (n_{\boldsymbol{k}} + \frac{1}{2})\hbar\omega_j(\boldsymbol{k}) \tag{2.233}$$

in the harmonic approximation and $\omega_j(\boldsymbol{k})$ labels the frequency of the different modes at wave vector $\boldsymbol{k}$. Each $n_{\boldsymbol{k}}$ can vary from 0 to ∞. The partition function can be rewritten as

$$\begin{aligned} Z &= \textstyle\sum_{n_1}\sum_{n_2}\cdots \exp(-\beta E_{\{n_{\boldsymbol{k}}\}}) \\ &= \prod_{\boldsymbol{k},j}\prod_{n_{\boldsymbol{k}}} \exp[-\beta(n_{\boldsymbol{k}} + \tfrac{1}{2})\hbar\omega_j(\boldsymbol{k})] \\ &= \prod_{\boldsymbol{k},j} \exp[-\hbar\omega_j(\boldsymbol{k})/2] \prod_{n_{\boldsymbol{k}}} \exp[-\beta n_{\boldsymbol{k}}\hbar\omega_j(\boldsymbol{k})], \end{aligned}$$

which readily leads to

$$F_L = k_B T \textstyle\sum_{\boldsymbol{k},j} \ln\left[2\sinh\left(\dfrac{\hbar\omega_j(\boldsymbol{k})}{2k_B T}\right)\right]. \tag{2.234}$$

[19] [2.10, 1973, Chap. 8]

Equation (2.234) could have been obtained by rewriting and generalizing (2.74). We must add to this the free energy at absolute zero due to the increase in elastic energy if the crystal changes its volume by ΔV. We call this term U_0.[20]

$$F = k_B T \sum\nolimits_{\boldsymbol{k},j} \ln\left[2\sinh\left(\frac{\hbar\omega_j(\boldsymbol{k})}{2k_B T}\right)\right] + U_0 . \tag{2.235}$$

We calculate the volume coefficient of thermal expansion α

$$\alpha = \frac{1}{V}\left(\frac{\partial V}{\partial T}\right)_P . \tag{2.236}$$

But,

$$\left(\frac{\partial V}{\partial T}\right)_P \left(\frac{\partial P}{\partial V}\right)_T \left(\frac{\partial T}{\partial P}\right)_V = -1 .$$

The isothermal compressibility is defined as

$$\kappa = -\frac{1}{V}\left(\frac{\partial V}{\partial P}\right)_T , \tag{2.237}$$

then we have

$$\alpha = \kappa\left(\frac{\partial P}{\partial T}\right)_V . \tag{2.238}$$

But

$$P = -\left(\frac{\partial F}{\partial V}\right)_T ,$$

so

$$P = -\frac{\partial U_0}{\partial V} - k_B T \sum\nolimits_{\boldsymbol{k},j} \coth\left(\frac{\hbar\omega_j(\boldsymbol{k})}{2k_B T}\right)\frac{\hbar}{2k_B T}\frac{\partial\omega_j(\boldsymbol{k})}{\partial V} . \tag{2.239}$$

The anharmonic terms come into play by assuming the $\omega_j(\boldsymbol{k})$ depend on volume. Since the average number of phonons in the mode $\boldsymbol{k}, j$ is

$$\bar{n}_j(\boldsymbol{k}) = \frac{1}{\exp\left(\frac{\hbar\omega_j(\boldsymbol{k})}{k_B T}\right) - 1} = \frac{1}{2}\left[\coth\left(\frac{\hbar\omega_j(\boldsymbol{k})}{2k_B T}\right) - 1\right] . \tag{2.240}$$

---

[20] U_0 is included for completeness, but we end up only using a vanishing temperature derivative so it could be left out.

Thus

$$P = -\frac{\partial U_0}{\partial V} - \sum_{\boldsymbol{k},j} (\bar{n}_j(\boldsymbol{k}) + \tfrac{1}{2})\hbar \frac{\partial \omega_j(\boldsymbol{k})}{\partial V}. \tag{2.241}$$

We define the Gruneisen parameter for the mode $\boldsymbol{k}, j$ as

$$\gamma_j(\boldsymbol{k}) = -\frac{V}{\omega_j(\boldsymbol{q})} \frac{\partial \omega_j(\boldsymbol{k})}{\partial V} = -\frac{\partial \ln \omega_j(\boldsymbol{k})}{\partial \ln V}. \tag{2.242}$$

Thus

$$P = -\frac{\partial}{\partial V}[U_0 + \sum_{\boldsymbol{k},j} \tfrac{1}{2}\hbar\omega_j(\boldsymbol{k})] + \sum_{\boldsymbol{k},j} \bar{n}_j(\boldsymbol{k}) \frac{\hbar\omega_j(\boldsymbol{k})\gamma_j}{V}. \tag{2.243}$$

However, the lattice internal energy is (in the harmonic approximation)

$$U = \sum_{\boldsymbol{k},j} (\bar{n}_j(\boldsymbol{k}) + \tfrac{1}{2})\hbar\omega_j(\boldsymbol{k}). \tag{2.244}$$

So

$$\frac{\partial U}{\partial T} = \sum_{\boldsymbol{k},j} \hbar\omega_j(\boldsymbol{k}) \frac{\partial \bar{n}_j(\boldsymbol{k})}{\partial T}, \tag{2.245}$$

$$c_V = \frac{1}{V}\frac{\partial U}{\partial T} = \frac{1}{V}\sum_{\mathbf{k},j} \hbar\omega_j(\mathbf{k}) \frac{\partial \bar{n}_j(\mathbf{k})}{\partial T} = \sum c_{V_j}(\mathbf{k}), \tag{2.246}$$

which defines a specific heat for each mode. Since the first term of P in (2.243) is independent of T at constant V, and using

$$\alpha = \kappa \left.\frac{\partial P}{\partial T}\right)_V,$$

we have

$$\alpha = \kappa \frac{1}{V} \sum_{\boldsymbol{k},j} \hbar\omega_j(\boldsymbol{k})\gamma_j(\boldsymbol{k}) \frac{\partial \bar{n}_j(\boldsymbol{k})}{\partial T}. \tag{2.247}$$

Thus

$$\alpha = \kappa \sum_{\boldsymbol{k},j} \gamma_j(\boldsymbol{k}) c_{V_j}(\boldsymbol{k}). \tag{2.248}$$

Let us define the overall Gruneisen parameter γ_T as the average Gruneisen parameter for mode $\boldsymbol{k}, j$ weighted by the specific heat for that mode. Then by (2.242) and (2.246) we have

$$c_V\gamma_T = \sum_{\boldsymbol{k},j} \gamma_j(\boldsymbol{k}) c_{V_j}(\boldsymbol{k}). \tag{2.249}$$

We then find

$$\alpha = \kappa \gamma_T c_V . \tag{2.250}$$

If γ_T (the Gruneisen parameter) were actually a constant α would tend to follow the changes of c_V, which happens for some materials.

From thermodynamics

$$c_P = c_V + \frac{\alpha^2 T}{\kappa}, \tag{2.251}$$

so $c_p = c_v(1 + \gamma\alpha T)$ and γ is often between 1 and 2.

Table 2.6. Gruneisen constants

Temperature	LiF	NaCl	KBr	KI
0 K	1.7 ± 0.05	0.9 ± 0.03	0.29 ± 0.03	0.28 ± 0.02
283 K	1.58	1.57	1.49	1.47

Adaptation of Table 3 from White GK, *Proc Roy Soc London* **A286**, 204, 1965. By permission of The Royal Society.

2.3.5 Wave Propagation in an Elastic Crystalline Continuum[21] (MET, MS)

In the limit of long waves, classical mechanics can be used for the discussion of elastic waves in a crystal. The relevant wave equations can be derived from Newton's second law and a form of Hooke's law. The appropriate generalized form of Hooke's law says the stress and strain are linearly related. Thus we start by defining the stress and strain tensors.

The Stress Tensor (σ_{ij}) (MET, MS)

We define the stress tensor $\boldsymbol{\sigma}_{ij}$ in such a way that

$$\sigma_{yx} = \frac{\Delta F_y}{\Delta y \Delta z} \tag{2.252}$$

for an infinitesimal cube. See Fig. 2.12. Thus i labels the force (positive for tension) per unit area in the i direction and j indicates which face the force acts on (the face is normal to the j direction). The stress tensor is symmetric in the absence of body torques, and it transforms as the products of vectors so it truly is a tensor.

[21] See, e.g., Ghatak and Kothari [2.16, Chap. 4] or Brown [2.7, Chap. 5].

By considering Fig. 2.13, we derive a useful expression for the stress that we will use later. The normal to dS is $\boldsymbol{n}$ and $\boldsymbol{\sigma}_{in}$dS is the force on dS in the ith direction. Thus for equilibrium

$$\boldsymbol{\sigma}_{in}\mathrm{d}S = \boldsymbol{\sigma}_{ix}n_x\mathrm{d}S + \boldsymbol{\sigma}_{iy}n_y\mathrm{d}S + \boldsymbol{\sigma}_{iz}n_z\mathrm{d}S\,,$$

so that

$$\boldsymbol{\sigma}_{in} = \sum\nolimits_j \boldsymbol{\sigma}_{ij}n_j\,. \tag{2.253}$$

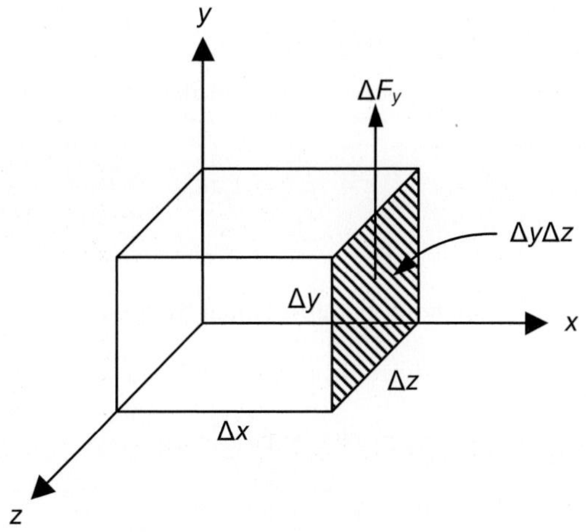

Fig. 2.12. Schematic definition of stress tensor σ_{ij}

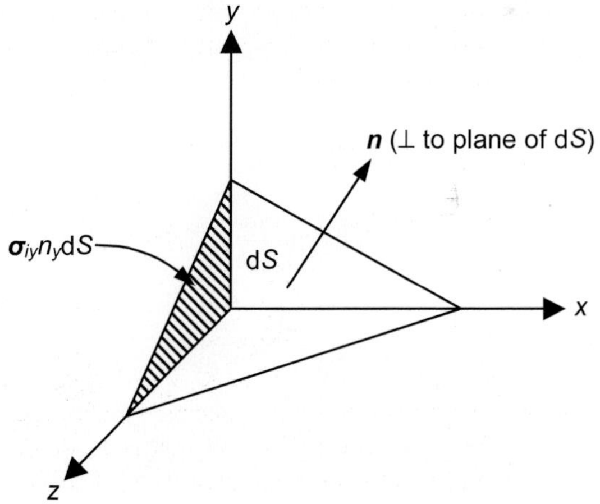

Fig. 2.13. Useful pictorial of stress tensor σ_{ij}

The Strain Tensor (ε_{ij}) (MET, MS)

Consider infinitesimal and uniform strains and let $\boldsymbol{i}$, $\boldsymbol{j}$, $\boldsymbol{k}$ be a set of orthogonal axes in the unstrained crystal. Under strain, they will go to a not necessarily orthogonal set $\boldsymbol{i}'$, $\boldsymbol{j}'$, $\boldsymbol{k}'$. We define ε_{ij} so

$$\boldsymbol{i}' = (1+\varepsilon_{xx})\boldsymbol{i} + \varepsilon_{xy}\boldsymbol{j} + \varepsilon_{xz}\boldsymbol{k}\ , \tag{2.254a}$$

$$\boldsymbol{j}' = \varepsilon_{yx}\boldsymbol{i} + (1+\varepsilon_{yy})\boldsymbol{j} + \varepsilon_{yz}\boldsymbol{k}\ , \tag{2.254b}$$

$$\boldsymbol{k}' = \varepsilon_{zx}\boldsymbol{i} + \varepsilon_{zy}\boldsymbol{j} + (1+\varepsilon_{zz})\boldsymbol{k}\ . \tag{2.254c}$$

Let $\boldsymbol{r}$ represent a point in an unstrained crystal that becomes $\boldsymbol{r}'$ under uniform infinitesimal strain.

$$\boldsymbol{r} = x\boldsymbol{i} + y\boldsymbol{j} + z\boldsymbol{k}\ , \tag{2.255a}$$

$$\boldsymbol{r}' = x\boldsymbol{i}' + y\boldsymbol{j}' + z\boldsymbol{k}'\ . \tag{2.255b}$$

Let the displacement of the point be represented by $\boldsymbol{u} = \boldsymbol{r}' - \boldsymbol{r}$, so

$$u_x = x\varepsilon_{xx} + y\varepsilon_{yx} + z\varepsilon_{zx}\ , \tag{2.256a}$$

$$u_y = x\varepsilon_{xy} + y\varepsilon_{yy} + z\varepsilon_{zy}\ , \tag{2.256b}$$

$$u_z = x\varepsilon_{xz} + y\varepsilon_{yz} + z\varepsilon_{zz}\ . \tag{2.256c}$$

We define the strain components in the following way

$$e_{xx} = \frac{\partial u_x}{\partial x}\ , \tag{2.257a}$$

$$e_{yy} = \frac{\partial u_y}{\partial y}\ , \tag{2.257b}$$

$$e_{zz} = \frac{\partial u_z}{\partial z}\ , \tag{2.257c}$$

$$e_{xy} = \frac{1}{2}\left(\frac{\partial u_x}{\partial y} + \frac{\partial u_y}{\partial x}\right), \tag{2.257d}$$

$$e_{yz} = \frac{1}{2}\left(\frac{\partial u_y}{\partial z} + \frac{\partial u_z}{\partial y}\right), \tag{2.257e}$$

$$e_{zx} = \frac{1}{2}\left(\frac{\partial u_z}{\partial x} + \frac{\partial u_x}{\partial z}\right). \tag{2.257f}$$

The diagonal components are the normal strain and the off-diagonal components are the shear strain. Pure rotations have not been considered, and the strain tensor (e_{ij}) is symmetric. It is a tensor as it transforms like one. The dilation, or change in volume per unit volume is,

$$\theta = \frac{\delta V}{V} = \boldsymbol{i}' \cdot (\boldsymbol{j}' \times \boldsymbol{k}') = e_{xx} + e_{yy} + e_{zz} . \tag{2.258}$$

Due to symmetry there are only 6 independent stress, and 6 independent strain components. The six component stresses and strains may be defined by:

$$\sigma_1 = \sigma_{xx} , \tag{2.259a}$$

$$\sigma_2 = \sigma_{yy} , \tag{2.259b}$$

$$\sigma_3 = \sigma_{zz} , \tag{2.259c}$$

$$\sigma_4 = \sigma_{yz} = \sigma_{zy} , \tag{2.259d}$$

$$\sigma_5 = \sigma_{xz} = \sigma_{zx} , \tag{2.259e}$$

$$\sigma_6 = \sigma_{xy} = \sigma_{yx} , \tag{2.259f}$$

$$\varepsilon_1 = e_{xx} , \tag{2.260a}$$

$$\varepsilon_2 = e_{yy} , \tag{2.260b}$$

$$\varepsilon_3 = e_{zz} , \tag{2.260c}$$

$$\varepsilon_4 = 2e_{yz} = 2e_{zy} , \tag{2.260d}$$

$$\varepsilon_5 = 2e_{xz} = 2e_{zx} , \tag{2.260e}$$

$$\varepsilon_6 = 2e_{xy} = 2e_{yx} . \tag{2.260f}$$

(The introduction of the 2 in (2.260 d, e, f) is convenient for later purposes).

Hooke's Law (MET, MS)

The generalized Hooke's law says stress is proportional to strain or in terms of the six-component representation:

$$\sigma_i = \sum\nolimits_{j=1}^{6} c_{ij} \varepsilon_j , \tag{2.261}$$

where the c_{ij} are the elastic constants of the crystal.

General Equation of Motion (MET, MS)

It is fairly easy, using Newton's second law, to derive an expression relating the displacements u_i and the stresses σ_{ij}. Reference can be made to Ghatak and Kothari [2.16, pp 59-62] for details. If $\boldsymbol{\sigma}_i^B$ denotes body force per unit mass in the direction i and if ρ is the density of the material, the result is

$$\rho \frac{\partial^2 u_i}{\partial t^2} = \rho \sigma_i^B + \sum_j \frac{\partial \sigma_{ij}}{\partial x_j} . \tag{2.262}$$

In the absence of external body forces the term $\boldsymbol{\sigma}_i^B$, of course, drops out.

Strain Energy (MET, MS)

Equation (2.262) seems rather complicated because there are 36 c_{ij}. However, by looking at an expression for the strain energy [2.16, p 63-65] and by using (2.261) it is possible to show

$$c_{ij} = \frac{\partial \sigma_i}{\partial \varepsilon_j} = \frac{\partial^2 u_V}{\partial \varepsilon_j \partial \varepsilon_i} , \tag{2.263}$$

where u_V is the potential energy per unit volume. Thus c_{ij} is a symmetric matrix and of the 36 c_{ij}, only 21 are independent.

Now consider only cubic crystals. Since the x-, y-, z-axes are equivalent,

$$c_{11} = c_{22} = c_{33} \tag{2.264a}$$

and

$$c_{44} = c_{55} = c_{66} \, . \tag{2.264b}$$

By considering inversion symmetry, we can show all the other off-diagonal elastic constants are zero except for

$$c_{12} = c_{13} = c_{23} = c_{21} = c_{31} = c_{32} \, .$$

Thus there are only three independent elastic constants,[22] which can be represented as:

$$c_{ij} = \begin{pmatrix} c_{11} & c_{12} & c_{12} & 0 & 0 & 0 \\ c_{12} & c_{11} & c_{12} & 0 & 0 & 0 \\ c_{12} & c_{12} & c_{11} & 0 & 0 & 0 \\ 0 & 0 & 0 & c_{44} & 0 & 0 \\ 0 & 0 & 0 & 0 & c_{44} & 0 \\ 0 & 0 & 0 & 0 & 0 & c_{44} \end{pmatrix} . \tag{2.265}$$

[22] If one can assume central forces Cauchy proved that $c_{12} = c_{44}$, however, this is not a good approximation in real materials.

Equations of Motion for Cubic Crystals (MET, MS)

From (2.262) (with no external body forces)

$$\rho \frac{\partial^2 u_i}{\partial t^2} = \sum_j \frac{\partial \sigma_{ij}}{\partial x_j} = \frac{\partial \sigma_{xx}}{\partial x} + \frac{\partial \sigma_{xy}}{\partial y} + \frac{\partial \sigma_{xz}}{\partial x}, \tag{2.266}$$

but

$$\begin{aligned} \sigma_{xx} &= \sigma_1 = c_{11}\varepsilon_1 + c_{12}\varepsilon_2 + c_{13}\varepsilon_3 \\ &= (c_{11} - c_{12})\varepsilon_1 + c_{12}(\varepsilon_1 + \varepsilon_2 + \varepsilon_3), \end{aligned} \tag{2.267a}$$

$$\sigma_{xy} = \sigma_6 = c_{44}\varepsilon_6, \tag{2.267b}$$

$$\sigma_{xz} = \sigma_5 = c_{44}\varepsilon_5. \tag{2.267c}$$

Using also (2.257), and combining with the above we get an equation for $\partial^2 u_x/\partial t^2$. Following a similar procedure we can also get equations for $\partial^2 u_y/\partial t^2$ and $\partial^2 u_z/\partial t^2$. Seeking solutions of the form

$$u_j = K_j \mathrm{e}^{\mathrm{i}(\boldsymbol{k}\cdot\boldsymbol{v} - \omega t)} \tag{2.268}$$

for $j = 1, 2, 3$ or x, y, z, we find nontrivial solutions only if

$$\begin{vmatrix} \begin{Bmatrix} (c_{11}-c_{44})k_x^2 \\ + c_{44}k^2 - \rho\omega^2 \end{Bmatrix} & (c_{12}+c_{44})k_x k_y & (c_{12}+c_{44})k_x k_z \\ (c_{12}+c_{44})k_y k_x & \begin{Bmatrix} (c_{11}-c_{44})k_y^2 \\ + c_{44}k^2 - \rho\omega^2 \end{Bmatrix} & (c_{12}+c_{44})k_y k_z \\ (c_{12}+c_{44})k_z k_x & (c_{12}+c_{44})k_z k_y & \begin{Bmatrix} (c_{11}-c_{44})k_z^2 \\ + c_{44}k^2 - \rho\omega^2 \end{Bmatrix} \end{vmatrix} = 0. \tag{2.269}$$

Suppose the wave travels along the x direction so $k_y = k_z = 0$. We then find the three wave velocities:

$$v_1 = \sqrt{\frac{c_{11}}{\rho}}, \quad v_2 = v_3 = \sqrt{\frac{c_{44}}{\rho}} \text{ (degenerate)}. \tag{2.270}$$

v_1 is a longitudinal wave and v_2, v_3 are the two transverse waves. Thus, one way of determining these elastic constants is by measuring appropriate wave velocities. Note that for an isotropic material $c_{11} = c_{12} + 2c_{44}$ so $v_1 > v_2$ and v_3. The longitudinal sound wave is greater than the transverse sound velocity.

Problems

2.1 Find the normal modes and normal-mode frequencies for a three-atom "lattice" (assume the atoms are of equal mass). Use periodic boundary conditions.

2.2 Show when m and m' are restricted to a range consistent with the first Brillouin zone that

$$\frac{1}{N}\sum_n \exp\left(\frac{2\pi i}{N}(m-m')n\right) = \delta_m^{m'},$$

where $\delta_m^{m'}$ is the Kronecker delta.

2.3 Evaluate the specific heat of the linear lattice (given by (2.80)) in the low temperature limit.

2.4 Show that $G_{mn} = G_{nm}$, where G is given by (2.100).

2.5 This is an essay length problem. It should clarify many points about impurity modes. Solve the five-atom lattice problem shown in Fig. 2.14. Use periodic boundary conditions. To solve this problem define $A = \beta/\alpha$ and $\delta = m/M$ (α and β are the spring constants) and find the normal modes and eigenfrequencies. For each eigenfrequency, plot $m\omega^2/\alpha$ versus δ for $A = 1$ and $m\omega^2/\alpha$ versus A for $\delta = 1$. For the first plot: (a) The degeneracy at $\delta = 1$ is split by the presence of the impurity. (b) No frequency is changed by more than the distance to the next unperturbed frequency. This is a general property. (c) The frequencies that are unchanged by changing δ correspond to modes with a node at the impurity (M). (d) Identify the mode corresponding to a pure translation of the crystal. (e) Identify the impurity mode(s). (f) Note that as we reduce the mass of M, the frequency of the impurity mode increases. For the second plot: (a) The degeneracy at $A = 1$ is split by the presence of an impurity. (b) No frequency is changed more than the distance to the next unperturbed frequency. (c) Identify the pure translation mode. (d) Identify the impurity modes. (e) Note that the frequencies of the impurity mode(s) increase with β.

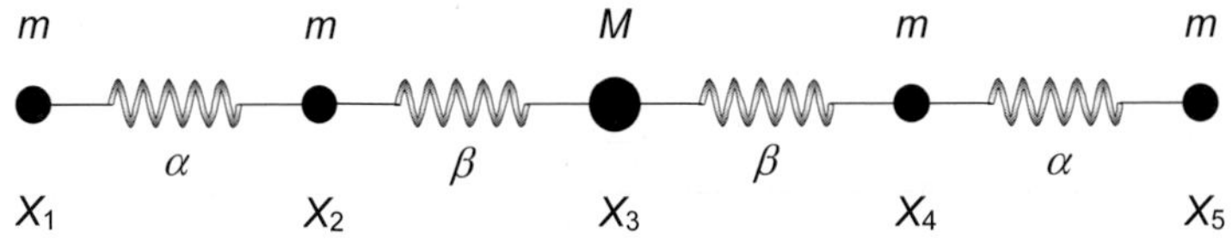

Fig. 2.14. The five-atom lattice

2.6 Let a_q and $a_q^\dagger$ be the phonon annihilation and creation operators. Show that

$$[a_q, a_{q^1}] = 0 \quad \text{and} \quad [a_q^\dagger, a_{q^1}^\dagger] = 0.$$

2.7 From the phonon annihilation and creation operator commutation relations derive that

$$a_q^\dagger|n_q\rangle = \sqrt{n_q+1}\,|n_q+1\rangle,$$

and

$$a_q|n_q\rangle = \sqrt{n_q}\,|n_q-1\rangle.$$

2.8 If $\boldsymbol{a}_1$, $\boldsymbol{a}_2$, and $\boldsymbol{a}_3$ are the primitive translation vectors and if $\Omega_a \equiv \boldsymbol{a}_1\cdot(\boldsymbol{a}_2\times\boldsymbol{a}_3)$, use the method of Jacobians to show that $\mathrm{d}x\,\mathrm{d}y\,\mathrm{d}z = \Omega_a\,\mathrm{d}\eta_1\,\mathrm{d}\eta_2\,\mathrm{d}\eta_3$, where x, y, z are the Cartesian coordinates and η_1, η_2, and η_3 are defined by $\boldsymbol{r} = \eta_1\boldsymbol{a}_1 + \eta_2\boldsymbol{a}_2 + \eta_3\boldsymbol{a}_3$.

2.9 Show that the b_i vectors defined by (2.172) satisfy

$$\Omega_a\boldsymbol{b}_1 = \boldsymbol{a}_2\times\boldsymbol{a}_3,\quad \Omega_a\boldsymbol{b}_2 = \boldsymbol{a}_3\times\boldsymbol{a}_1,\quad \Omega_a\boldsymbol{b}_3 = \boldsymbol{a}_1\times\boldsymbol{a}_2,$$

where $\Omega_a = a_1\cdot(a_2\times a_3)$.

2.10 If $\Omega_b = \boldsymbol{b}_1\cdot(\boldsymbol{b}_2\times\boldsymbol{b}_3)$, $\Omega_a = \boldsymbol{a}_1\cdot(\boldsymbol{a}_2\times\boldsymbol{a}_3)$, the $\boldsymbol{b}_i$ are defined by (2.172), and the a_i are the primitive translation vectors, show that $\Omega_b = 1/\Omega_a$.

2.11 This is a long problem whose results are very important for crystal mathematics. (See (2.178)–(2.184)). Show that

a)
$$\frac{1}{N_1N_2N_3}\sum\nolimits_{\boldsymbol{R}_l}\exp(\mathrm{i}\boldsymbol{q}\cdot\boldsymbol{R}_l) = \sum\nolimits_{G_n}\delta_{q,G_n},$$

where the sum over R_l is a sum over the lattice.

b)
$$\frac{1}{N_1N_2N_3}\sum\nolimits_{q}\exp(\mathrm{i}\boldsymbol{q}\cdot\boldsymbol{R}_l) = \delta_{\boldsymbol{R}_l,0},$$

where the sum over $\boldsymbol{q}$ is a sum over one Brillouin zone.

c) In the limit as $V_{\text{f.p.p.}}\to\infty$ ($V_{\text{f.p.p.}}$ means the volume of the parallelepiped representing the actual crystal), one can replace

$$\sum\nolimits_q f(\boldsymbol{q}) \quad\text{by}\quad \frac{V_{\text{f.p.p.}}}{(2\pi)^3}\int f(\boldsymbol{q})\mathrm{d}^3\boldsymbol{q}.$$

d)
$$\frac{\Omega_a}{(2\pi)^3}\int_{\text{B.Z.}}\exp(\mathrm{i}\boldsymbol{q}\cdot\boldsymbol{R}_l)\mathrm{d}^3q = \delta_{\boldsymbol{R}_l,0},$$

where the integral is over one Brillouin zone.

e)
$$\frac{1}{\Omega_a}\int\exp[\mathrm{i}(\boldsymbol{G}_{l'}-\boldsymbol{G}_l)\cdot\boldsymbol{r}]d^3r = \delta_{l',l},$$

where the integral is over a unit cell.

f)
$$\frac{1}{(2\pi)^3}\int \exp[\mathrm{i}\boldsymbol{q}\cdot(\boldsymbol{r}-\boldsymbol{r}')]\mathrm{d}^3q = \delta(\boldsymbol{r}-\boldsymbol{r}'),$$

where the integral is over all of reciprocal space and $\delta(\boldsymbol{r} - \boldsymbol{r}')$ is the Dirac delta function.

g)
$$\frac{1}{(2\pi)^3}\int_{V_{\text{f.p.p.}\to\infty}} \exp[\mathrm{i}(\boldsymbol{q}-\boldsymbol{q}')\cdot\boldsymbol{r}]\mathrm{d}^3r = \delta(\boldsymbol{q}-\boldsymbol{q}').$$

In this problem, the $\boldsymbol{a}_i$ are the primitive translation vectors. $N_1\boldsymbol{a}_1$, $N_2\boldsymbol{a}_2$, and $N_3\boldsymbol{a}_3$ are vectors along the edges of the fundamental parallelepiped. $\boldsymbol{R}_l$ defines lattice points in the direct lattice by (2.171). $\boldsymbol{q}$ are vectors in reciprocal space defined by (2.175). The $\boldsymbol{G}_l$ define the lattice points in the reciprocal lattice by (2.173). $\Omega_a = \boldsymbol{a}_1 \cdot (\boldsymbol{a}_2 \times \boldsymbol{a}_3)$, and the $\boldsymbol{r}$ are vectors in direct space.

2.12 This problem should clarify the discussion of diagonalizing H_q (defined by 2.198). Find the normal mode eigenvalues and eigenvectors associated with

$$m_i\ddot{x}_i = -\sum_{j=1}^{3}\gamma_{ij}x_j\,,$$

$$m_1 = m_3 = m, \quad m_2 = M, \quad \text{and} \quad (\gamma_{ij}) = \begin{pmatrix} k, & -k, & 0 \\ -k, & 2k, & -k \\ 0, & -k, & k \end{pmatrix}.$$

A convenient substitution for this purpose is

$$x_i = u_i\frac{\mathrm{e}^{\mathrm{i}\omega t}}{\sqrt{m_i}}.$$

2.13 By use of the Debye model, show that

$$c_v \propto T^3 \quad \text{for} \quad T << \theta_D$$

and

$$c_v \propto 3k(NK) \quad \text{for} \quad T >> \theta_D.$$

Here, k = the Boltzmann gas constant, N = the number of unit cells in the fundamental parallelepiped, and K = the number of atoms per unit cell. Show that this result is independent of the Debye model.

2.14 The nearest-neighbor one-dimensional lattice vibration problem (compare Sect. 2.2.2) can be exactly solved. For this lattice: (a) Plot the average number (per atom) of phonons (with energies between ω and $\omega + \mathrm{d}\omega$) versus ω for several temperatures. (b) Plot the internal energy per atom versus temperature. (c) Plot the entropy per atom versus temperature. (d) Plot the specific heat per atom versus temperature. [Hint: Try to use convenient dimensionless quantities for both ordinates and abscissa in the plots.]

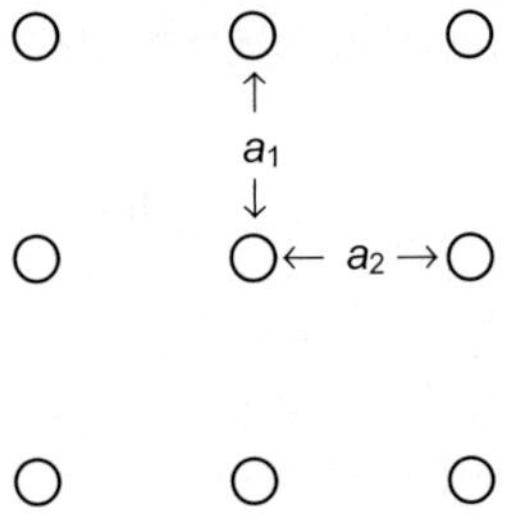

2.15 Find the reciprocal lattice of the two-dimensional square lattice shown above.

2.16 Find the reciprocal lattice of the three-dimensional body-centered cubic lattice. Use for primitive lattice vectors

$$\boldsymbol{a}_1 = \frac{a}{2}(\hat{x} + \hat{y} - \hat{z}), \quad \boldsymbol{a}_2 = \frac{a}{2}(-\hat{x} + \hat{y} + \hat{z}), \quad \boldsymbol{a}_3 = \frac{a}{2}(\hat{x} - \hat{y} + \hat{z}) .$$

2.17 Find the reciprocal lattice of the three-dimensional face-centered cubic lattice. Use as primitive lattice vectors

$$\boldsymbol{a}_1 = \frac{a}{2}(\hat{x} + \hat{y}), \quad \boldsymbol{a}_2 = \frac{a}{2}(\hat{y} + \hat{z}), \quad \boldsymbol{a}_3 = \frac{a}{2}(\hat{y} + \hat{x}) .$$

2.18 Sketch the first Brillouin zone in the reciprocal lattice of the fcc lattice. The easiest way to do this is to draw planes that perpendicularly bisect vectors (in reciprocal space) from the origin to other reciprocal lattice points. The volume contained by all planes is the first Brillouin zone. This definition is equivalent to the definition just after (2.176).

2.19 Sketch the first Brillouin zone in the reciprocal lattice of the bcc lattice. Problem 2.18 gives a definition of the first Brillouin zone.

2.20 Find the dispersion relation for the two-dimensional monatomic square lattice in the harmonic approximation. Assume nearest-neighbor interactions.

2.21 Write an exact expression for the heat capacity (at constant area) of the two-dimensional square lattice in the nearest-neighbor harmonic approximation. Evaluate this expression in an approximation that is analogous to the Debye approximation, which is used in three dimensions. Find the exact high- and low-temperature limits of the specific heat.

2.22 Use (2.200) and (2.203), the fact that the polarization vectors satisfy

$$\sum_p e^{*\alpha}_{\boldsymbol{q}p\boldsymbol{b}} e^{\beta}_{\boldsymbol{q}p\boldsymbol{b}'} = \delta^{\beta}_{\alpha} \delta^{\boldsymbol{b}'}_{\boldsymbol{b}}$$

(the α and β refer to Cartesian components), and

$$X^{11\dagger}_{-\boldsymbol{q},p} = X^{11}_{\boldsymbol{q},p}, P^{11\dagger}_{-\boldsymbol{q},p} = P^{11}_{\boldsymbol{q},p} .$$

(you should convince yourself that these last two relations are valid) to establish that

$$\boldsymbol{X}^1_{\boldsymbol{q},p} = -\mathrm{i}\sum\nolimits_p \sqrt{\frac{\hbar}{2m_b\omega_{\boldsymbol{q},p}}}\boldsymbol{e}^*_{\boldsymbol{q},p,b}(a^\dagger_{\boldsymbol{q},p} - a_{-\boldsymbol{q},p}) \,.$$

2.23 Show that the specific heat of a lattice at low temperatures goes as the temperature to the power of the dimension of the lattice as in Table 2.5.

2.24 Discuss the Einstein theory of specific heat of a crystal in which only one lattice vibrational frequency is considered. Show that this leads to a vanishing of the specific heat at absolute zero, but not as T cubed.

2.25 In (2.270) show v_1 is longitudinal and v_2, v_3 are transverse.

2.26 Derive wave velocities and physically describe the waves that propagate along the [110] directions in a cubic crystal. Use (2.269).

3 Electrons in Periodic Potentials

As we have said, the universe of traditional solid-state physics is defined by the crystalline lattice. The principal actors are the elementary excitations in this lattice. In the previous chapter we discussed one of these, the phonons that are the quanta of lattice vibration. Another is the electron that is perhaps the principal actor in all of solid-state physics. By an electron in a solid we will mean something a little different from a free electron. We will mean a dressed electron or an electron plus certain of its interactions. Thus we will find that it is often convenient to assign an electron in a solid an effective mass.

There is more to discuss on lattice vibrations than was covered in Chap. 2. In particular, we need to analyze anharmonic terms in the potential and see how these terms cause phonon–phonon interactions. This will be done in the next chapter. Electron–phonon interactions are also included in Chap. 4 and before we get there we obviously need to discuss electrons in solids. After making the Born–Oppenheimer approximation (Chap. 2), we still have to deal with a many-electron problem (as well as the behavior of the lattice). A way to reduce the many-electron problem approximately to an equivalent one-electron problem[1] is given by the Hartree and Hartree–Fock methods. The density functional method, which allows at least in principle, the exact evaluation of some ground-state properties is also important. In a certain sense, it can be regarded as an extension of the Hartree–Fock method and it has been much used in recent years.

After justifying the one-electron approximation by discussing the Hartree, Hartree–Fock, and density functional methods, we consider several applications of the elementary quasifree-electron approximation.

We then present the nearly free and tight binding approximations for electrons in a crystalline lattice. After that we discuss various band structure approximations. Finally we discuss some electronic properties of lattice defects. We begin with the variational principle, which is used in several of our developments.

---

[1] A much more sophisticated approach than we wish to use is contained in Negele and Orland [3.36]. In general, with the hope that this book may be useful to all who are entering solid-state physics, we have stayed away from most abstract methods of quantum field theory.

3.1 Reduction to One-Electron Problem

3.1.1 The Variational Principle (B)

The variational principle that will be derived in this Section is often called the *Rayleigh–Ritz variational principle.* The principle in itself is extremely simple. For this reason, we might be surprised to learn that it is of great practical importance. It gives us a way of constructing energies that have a value greater than or equal to the ground-state energy of the system. In other words, it gives us a way of constructing upper bounds for the energy. There are also techniques for constructing lower bounds for the energy, but these techniques are more complicated and perhaps not so useful.[2] The variational technique derived in this Section will be used to derive both the Hartree and Hartree–Fock equations. A variational procedure will also be used with the density functional method to develop the Kohn–Sham equations.

Let $\mathcal{H}$ be a positive definite Hermitian operator with eigenvalues E_μ and eigenkets $|\mu\rangle$. Since $\mathcal{H}$ is positive definite and Hermitian it has a lowest E_μ and the E_μ are real. Let the E_μ be labeled so that E_0 is the lowest. Let $|\psi\rangle$ be an arbitrary ket (not necessarily normalized) in the space of interest and define a quantity $Q(\psi)$ such that

$$Q(\psi) = \frac{\langle\psi|\mathcal{H}|\psi\rangle}{\langle\psi|\psi\rangle}. \tag{3.1}$$

The eigenkets $|\mu\rangle$ are assumed to form a complete set so that

$$|\psi\rangle = \sum\nolimits_\mu a_\mu |\mu\rangle. \tag{3.2}$$

Since $\mathcal{H}$ is Hermitian, we can assume that the $|\mu\rangle$ are orthonormal, and we find

$$\langle\psi|\psi\rangle = \sum\nolimits_{\mu^1,\mu} a^*_{\mu^1} a_\mu \langle\mu^1|\mu\rangle = \sum\nolimits_\mu |a_\mu|^2, \tag{3.3}$$

and

$$\langle\psi|\mathcal{H}|\psi\rangle = \sum\nolimits_{\mu^1,\mu} a^*_{\mu^1} a_\mu \langle\mu^1|\mathcal{H}|\mu\rangle = \sum\nolimits_\mu |a_\mu|^2 E_\mu. \tag{3.4}$$

Q can then be written as

$$Q(\psi) = \frac{\sum_\mu E_\mu |a_\mu|^2}{\sum_\mu |a_\mu|^2} = \frac{\sum_\mu E_0 |a_\mu|^2}{\sum_\mu |a_\mu|^2} + \frac{\sum_\mu (E_\mu - E_0)|a_\mu|^2}{\sum_\mu |a_\mu|^2},$$

---

[2] See, for example, Friedman [3.18].

or

$$Q(\psi) = E_0 + \frac{\sum_\mu (E_\mu - E_0)|a_\mu|^2}{\sum_\mu |a_\mu|^2}. \tag{3.5}$$

Since $E_\mu > E_0$ and $|a_\mu|^2 \geq 0$, we can immediately conclude from (3.5) that

$$Q(\psi) \geq E_0 . \tag{3.6}$$

Summarizing, we have

$$\frac{\langle \psi | \mathcal{H} | \psi \rangle}{\langle \psi | \psi \rangle} \geq E_0 . \tag{3.7}$$

Equation (3.7) is the basic equation of the variational principle. Suppose ψ is a trial wave function with a variable parameter η. Then the η that are the best if $Q(\psi)$ is to be as close to the lowest eigenvalue as possible (or as close to the ground-state energy if $\mathcal{H}$ is the Hamiltonian) are among the η for which

$$\frac{\partial Q}{\partial \eta} = 0 . \tag{3.8}$$

For the $\eta = \eta_b$ that solves (3.8) and minimizes $Q(\psi)$, $Q(\psi(\eta_b))$ is an approximation to E_0. By using successively more sophisticated trial wave functions with more and more variable parameters (this is where the hard work comes in), we can get as close to E_0 as desired. $Q(\psi) = E_0$ exactly only if ψ is an exact wave function corresponding to E_0.

3.1.2 The Hartree Approximation (B)

When applied to electrons, the Hartree method neglects the effects of antisymmetry of many electron wave functions. It also neglects *correlations* (this term will be defined precisely later). Despite these deficiencies, the Hartree approximation can be very useful, e.g. when applied to many-electron atoms. The fact that we have a shell structure in atoms appears to make the deficiencies of the Hartree approximation not very serious (strictly speaking even here we have to use some of the ideas of the Pauli principle in order that all electrons are not in the same lowest-energy shell). The Hartree approximation is also useful for gaining a crude understanding of why the quasifree-electron picture of metals has some validity. Finally, it is easier to understand the Hartree–Fock method as well as the density functional method by slowly building up the requisite ideas. The Hartree approximation is a first step.

For a solid, the many-electron Hamiltonian whose Schrödinger wave equation must be solved is

$$\begin{aligned}\mathcal{H} = &-\frac{\hbar^2}{2m}\sum\nolimits_{i\,(\text{electrons})}\nabla_i^2 - \sum\nolimits_{\substack{a\,(\text{nuclei})\\ i\,(\text{electrons})}}\frac{e^2}{4\pi\varepsilon_0 r_{ai}} \\ &+\frac{1}{2}\sum\nolimits'_{a,b\,(\text{nuclei})}\frac{Z_a Z_b e^2}{4\pi\varepsilon_0 R_{ab}} + \frac{1}{2}\sum\nolimits'_{i,j\,(\text{electron})}\frac{e^2}{4\pi\varepsilon_0 r_{ij}}.\end{aligned} \tag{3.9}$$

This equals $\mathcal{H}_0$ of (2.10).

The first term in the Hamiltonian is the operator representing the kinetic energy of all the electrons. Each different i corresponds to a different electron The second term is the potential energy of interaction of all of the electrons with all of the nuclei, and r_{ai} is the distance from the ath nucleus to the ith electron. This potential energy of interaction is due to the Coulomb forces. Z_a is the atomic number of the nucleus at a. The third term is the Coulomb potential energy of interaction between the nuclei. R_{ab} is the distance between nucleus a and nucleus b. The prime on the sum as usual means omission of those terms for which $a = b$. The fourth term is the Coulomb potential energy of interaction between the electrons, and r_{ij} is the distance between the ith and jth electrons. For electronic calculations, the internuclear distances are treated as constant parameters, and so the third term can be omitted. This is in accord with the Born–Oppenheimer approximation as discussed at the beginning of Chap. 2. Magnetic interactions are relativistic corrections to the electrical interactions, and so are often small. They are omitted in (3.9).

For the purpose of deriving the Hartree approximation, this N-electron Hamiltonian is unnecessarily cumbersome. It is more convenient to write it in the more abstract form

$$\mathcal{H}(x_1 \cdots x_n) = \sum\nolimits_{i=1}^N \mathcal{H}(i) + \tfrac{1}{2}\sum\nolimits'_{i,j} V(ij), \tag{3.10a}$$

where

$$V(ij) = V(ji). \tag{3.10b}$$

In (3.10a), $\mathcal{H}(i)$ is a one-particle operator (e.g. the kinetic energy), $V(ij)$ is a two-particle operator (e.g. the fourth term in (3.9)), and i refers to the electron with coordinate x_i (or $\boldsymbol{r}_i$ if you prefer). Spin does not need to be discussed for a while, but again we can regard x_i in a wave function as including the spin of electron i if we so desire.

Eigenfunctions of the many-electron Hamiltonian defined by (3.10a) will be sought by use of the variational principle. If there were no interaction between electrons and if the indistinguishability of electrons is forgotten, then the eigenfunction can be a product of N functions, each function being a function of the coordinates of only one electron. So even though we have interactions, let us try a trial wave function that is a simple product of one-electron wave functions:

$$\psi(x_1 \cdots x_n) = u_1(x_1)u_2(x_2)\cdots u_n(x_n)\,. \tag{3.11}$$

The u will be assumed to be normalized, but not necessarily orthogonal. Since the u are normalized, it is easy to show that the ψ are normalized:

$$\begin{aligned}\int \psi^*(x_1,\cdots,x_N)\psi(x_1,\cdots,x_N)\mathrm{d}\tau &= \int u_1^*(x_1)u(x_1)\mathrm{d}\tau_1 \cdots \int u_N^*(x_N)u(x_N)\mathrm{d}\tau_N \\ &= 1.\end{aligned}$$

Combining (3.10) and (3.11), we can easily calculate

$$\begin{aligned}\langle\psi|\mathcal{H}|\psi\rangle &\equiv \int \psi^* \mathcal{H}\psi \mathrm{d}\tau \\ &= \int u_1^*(x_1)\cdots u_N^*(x_N)[\textstyle\sum \mathcal{H}(i) + \frac{1}{2}\sum_{i,j}' \mathrm{V(ij)}]u_1(x_1)\cdots u_N(x_N)\mathrm{d}\tau \\ &= \textstyle\sum_i \int u_i^*(x_i)\mathcal{H}(i)u_i(x_i)\mathrm{d}\tau_i \\ &\quad + \textstyle\frac{1}{2}\sum_{i,j}' \int u_i^*(x_i)u_j^*(x_j)\mathrm{V(ij)}u_i(x_i)u_j(x_j)\mathrm{d}\tau_i\mathrm{d}\tau_j \\ &= \textstyle\sum_i \int u_i^*(x_1)\mathcal{H}(1)u_i(x_1)\mathrm{d}\tau_1 \\ &\quad + \textstyle\frac{1}{2}\sum_{i,j}' \int u_i^*(x_1)u_j^*(x_2)\mathrm{V}(1,2)u_i(x_1)u_j(x_2)\mathrm{d}\tau_1\mathrm{d}\tau_2,\end{aligned} \tag{3.12}$$

where the last equation comes from making changes of dummy integration variables.

By (3.7) we need to find an extremum (hopefully a minimum) for $\langle\psi|\mathcal{H}|\psi\rangle$ while at the same time taking into account the constraint of normalization. The convenient way to do this is by the use of Lagrange multipliers [2]. The variational principle then tells us that the best choice of u is determined from

$$\delta[\langle\psi|\mathcal{H}|\psi\rangle - \textstyle\sum_i \lambda_i \int u_i^*(x_i)u_i(x_i)\mathrm{d}\tau_i] = 0\,. \tag{3.13}$$

In (3.13), δ is an arbitrary variation of the u. u_i and u_j can be treated independently (since Lagrange multipliers λ_i are being used) as can u_i and u_j^*. Thus it is convenient to choose $\delta = \delta_k$, where $\delta_k u_k^*$ and $\delta_k u_k$ are independent and arbitrary, $\delta_k u_{i\,(\neq k)} = 0$, and $\delta_k u_{i\,(\neq k)}^* = 0$.

By (3.10b), (3.12), (3.13), $\delta = \delta_k$, and a little manipulation we easily find

$$\begin{aligned}\int \delta_k u_k^*(x_1)\{[\mathcal{H}(1)u_k(x_1) + (\textstyle\sum_{j(\neq \mathrm{k})} \int u_j^*(x_2)V(1,2)u_j(x_2)\mathrm{d}\tau)u_k(x_1)] \\ - \lambda_k u_k(x_1)\}\mathrm{d}\tau + C.C. = 0.\end{aligned} \tag{3.14}$$

In (3.14), C.C. means the complex conjugate of the terms that have already been written on the left-hand side of (3.14). The second term is easily seen to be the complex conjugate of the first term because

$$\delta\langle\psi|\mathcal{H}|\psi\rangle = \langle\delta\psi|\mathcal{H}|\psi\rangle + \langle\psi|\mathcal{H}|\delta\psi\rangle = \langle\delta\psi|\mathcal{H}|\psi\rangle + \langle\delta\psi|\mathcal{H}|\psi\rangle^*,$$

since $\mathcal{H}$ is Hermitian.

In (3.14), two terms have been combined by making changes of dummy summation and integration variables, and by using the fact that $V(1,2) = V(2,1)$. In (3.14), $\delta_k u_k^*(x_1)$ and $\delta_k u_k(x_1)$ are independent and arbitrary, so that the integrands involved in the coefficients of either $\delta_k u_k$ or $\delta_k u_k^*$ must be zero. The latter fact gives the Hartree equations

$$\mathcal{H}(x_1)u_k(x_1) + [\sum_{j(\neq k)} \int u_j^*(x_2)V(1,2)u_j(x_2)\mathrm{d}\tau_2]u_k(x_1) = \lambda_k u_k(x_1) \cdot \quad (3.15)$$

Because we will have to do the same sort of manipulation when we derive the Hartree–Fock equations, we will add a few comments on the derivation of (3.15). Allowing for the possibility that the λ_k may be complex, the most general form of (3.14) is

$$\int \delta_k u_k^*(x_1)\{F(1)u_k(1) - \lambda_k u_k(x_1)\}\mathrm{d}\tau_1$$
$$+ \int \delta_k u_k(x_1)\{F(1)u_k(1) - \lambda_k^* u_k(x_1)\}^* \mathrm{d}\tau_1 = 0,$$

where $F(1)$ is defined by (3.14). Since $\delta_k u_k(x_1)$ and $\delta_k u_k(x_1)^*$ are independent (which we will argue in a moment), we have

$$F(1)u_k(1) = \lambda_k u_k(1) \quad \text{and} \quad F(1)u_k(1) = \lambda_k^* u_k(1) \, .$$

F is Hermitian so that these equations are consistent because then $\lambda_k = \lambda_k^*$ and is real. The independence of $\delta_k u_k$ and $\delta_k u_k^*$ is easily seen by the fact that if $\delta_k u_k = \alpha + \mathrm{i}\beta$ then α and β are real and independent. Therefore if

$$(C_1 + C_2)\alpha + (C_1 - C_2)\mathrm{i}\beta = 0, \quad \text{then} \quad C_1 = C_2 \quad \text{and} \quad C_1 = -C_2,$$

or $C_1 = C_2 = 0$ because this is what we mean by independence. But this implies $C_1(\alpha + \mathrm{i}\beta) + C_2(\alpha - \mathrm{i}\beta) = 0$ implies $C_1 = C_2 = 0$ so $\alpha + \mathrm{i}\beta = \delta_k u_k$ and $\alpha - \mathrm{i}\beta = \delta_k u_k^*$ are independent.

Several comments can be made about these equations. The Hartree approximation takes us from one Schrödinger equation for N electrons to N Schrödinger equations each for one electron. The way to solve the Hartree equations is to guess a set of u_i and then use (3.15) to calculate a new set. This process is to be continued until the u we calculate are similar to the u we guess. When this stage is reached, we say we have a *consistent set* of equations. In the Hartree approximation, the state u_i is not determined by the instantaneous positions of the electrons in state j, but only by their average positions. That is, the sum $-e\sum_{j(\neq k)} u_j^*(x_2)u_j(x_2)$ serves as a time-independent density $\rho(2)$ of electrons for calculating $u_k(x_1)$. If $V(1,2)$ is the Coulomb repulsion between electrons, the second term on the left-hand side corresponds to

$$-\int \rho(2)\frac{1}{4\pi\varepsilon_0 r_{12}}\mathrm{d}\tau_2 \, .$$

Thus this term has a classical and intuitive meaning. The u_i, obtained by solving the Hartree equations in a self-consistent manner, are the best set of one-electron orbitals in the sense that for these orbitals $Q(\psi) = \langle\psi|\mathcal{H}|\psi\rangle/\langle\psi|\psi\rangle$ (with $\psi = u_1,\ldots,u_N$) is a minimum. The physical interpretation of the Lagrange multipliers λ_k has not yet been given. Their values are determined by the eigenvalue condition as expressed by (3.15). From the form of the Hartree equations we might expect that the λ_k correspond to "the energy of an electron in state k." This will be further discussed and made precise within the more general context of the Hartree–Fock approximation.

3.1.3 The Hartree–Fock Approximation (A)

The derivation of the Hartree–Fock equations is similar to the derivation of the Hartree equations. The difference in the two methods lies in the form of the trial wave function that is used. In the Hartree–Fock approximation the fact that electrons are fermions and must have antisymmetric wave functions is explicitly taken into account. If we introduce a "spin coordinate" for each electron, and let this spin coordinate take on two possible values (say ± ½), then the general way we put into the Pauli principle is to require that the many-particle wave function be antisymmetric in the interchange of *all* the coordinates of any two electrons. If we form the antisymmetric many-particle wave functions out of one-particle wave functions, then we are led to the idea of the Slater determinant for the trial wave function. Applying the ideas of the variational principle, we are then led to the Hartree–Fock equations. The details of this program are given below. First, we shall derive the Hartree–Fock equations using the same notation as was used for the Hartree equations. We will then repeat the derivation using the more convenient second quantization notation. The second quantization notation often shortens the algebra of such derivations. Since much of the current literature is presented in the second quantization notation, some familiarity with this method is necessary.

Derivation of Hartree–Fock Equations in Old Notation (A)[3]

Given N one-particle wave functions $u_i(x_i)$, where x_i in the wave functions represents all the coordinates (space and spin) of particle i, there is only one antisymmetric combination that can be formed (this is a theorem that we will not prove). This antisymmetric combination is a determinant. Thus the trial wave function that will be used takes the form

$$\psi(x_1,\ldots,x_N) = M\begin{vmatrix} u_1(x_1) & u_2(x_1) & \cdots & u_N(x_1) \\ u_1(x_2) & u_2(x_2) & \cdots & u_N(x_2) \\ \vdots & \vdots & & \vdots \\ u_1(x_N) & u_2(x_N) & \cdots & u_N(x_N) \end{vmatrix}. \tag{3.16}$$

In (3.16), M is a normalizing factor to be chosen so that $\int|\psi|^2 \mathrm{d}\tau = 1$.

[3] Actually, for the most part we assume restricted Hartree–Fock Equations where there are an even number of electrons divided into sets of 2 with the same spatial wave functions paired with either a spin-up or spin-down function. In unrestricted Hartree–Fock we do not make these assumptions. See, e.g., Marder [3.34, p. 209].

It is easy to see why the use of a determinant automatically takes into account the Pauli principle. If two electrons are in the same state, then for some i and j, $u_i = u_j$. But then two columns of the determinant would be equal and hence $\psi = 0$, or in other words $u_i = u_j$ is physically impossible. For the same reason, two electrons with the same spin cannot occupy the same point in space. The antisymmetry property is also easy to see. If we interchange x_i and x_j, then two rows of the determinant are interchanged so that ψ changes sign. All physical properties of the system in state ψ depend only quadratically on ψ, so the physical properties are unaffected by the change of sign caused by the interchange of the two electrons. This is an example of the indistinguishability of electrons. Rather than using (3.16) directly, it is more convenient to write the determinant in terms of its definition that uses permutation operators:

$$\psi(x_1 \cdots x_n) = M\sum\nolimits_p (-)^p Pu_1(x_1)\cdots u_N(x_N) . \tag{3.17}$$

In (3.17), P is the permutation operator and it acts either on the subscripts of u (in pairs) or on the coordinates x_i (in pairs). $(-)^P$ is ± 1, depending on whether P is an even or an odd permutation. A permutation of a set is even (odd), if it takes an even (odd) number of interchanges of pairs of the set to get the set from its original order to its permuted order.

In (3.17) it will be assumed that the single-particle wave functions are orthonormal:

$$\int u_i^*(x_1)u_j(x_1)\mathrm{d}x_1 = \delta_i^j . \tag{3.18}$$

In (3.18) the symbol $\int$ means to integrate over the spatial coordinates and to sum over the spin coordinates. For the purposes of this calculation, however, the symbol can be regarded as an ordinary integral (most of the time) and things will come out satisfactorily.

From Problem 3.2, the correct normalizing factor for the ψ is $(N!)^{-1/2}$, and so the normalized ψ have the form

$$\psi(x_1 \cdots x_n) = (1/\sqrt{N!})\sum\nolimits_p (-)^p Pu_1(x_1)\cdots u_N(x_N) . \tag{3.19}$$

Functions of the form (3.19) are called *Slater determinants*.

The next obvious step is to apply the variational principle. Using Lagrange multipliers λ_{ij} to take into account the orthonormality constraint, we have

$$\delta\left(\langle\psi|\mathcal{H}|\psi\rangle - \sum\nolimits_{i,j}\lambda_{i,j}\langle u_i|u_j\rangle\right) = 0 . \tag{3.20}$$

Using the same Hamiltonian as was used in the Hartree problem, we have

$$\langle\psi|\mathcal{H}|\psi\rangle = \langle\psi|\sum\mathcal{H}(i)|\psi\rangle + \langle\psi|\tfrac{1}{2}\sum\nolimits_{i,j}' V(ij)|\psi\rangle . \tag{3.21}$$

The first term can be evaluated as follows:

$$\left\langle \psi \middle| \sum \mathcal{H}(i) \middle| \psi \right\rangle$$
$$= \frac{1}{N!} \sum\nolimits_{p,p'} (-)^{p+p'} \int [P u_1^*(x_1) \cdots u_N^*(x_N)] \sum \mathcal{H}(i) [P' u_1(x_1) \cdots u_N(x_N)] \mathrm{d}\tau$$
$$= \frac{1}{N!} \sum\nolimits_{p,p'} (-)^{p+p'} P \int [u_1^*(x_1) \cdots u_N^*(x_N)] \sum \mathcal{H}(i) P^{-1} P' [u_1(x_1) \cdots u_N(x_N)] \mathrm{d}\tau,$$

since P commutes with $\sum \mathcal{H}(i)$. Defining $Q = P^{-1}P'$, we have

$$\left\langle \psi \middle| \sum \mathcal{H}(i) \middle| \psi \right\rangle$$
$$= \frac{1}{N!} \sum\nolimits_{p,q} (-)^q P \int [u_1^*(x_1) \cdots u_N^*(x_N)] \sum \mathcal{H}(i) Q [u_1(x_1) \cdots u_N(x_N)] \mathrm{d}\tau,$$

where $Q \equiv P^{-1}P'$ is also a permutation,

$$= \sum\nolimits_q (-)^q \int [u_1^*(x_1) \cdots u_N^*(x_N)] \sum \mathcal{H}(i) Q [u_1(x_1) \cdots u_N(x_N)] \mathrm{d}\tau,$$

where P is regarded as acting on the coordinates, and by dummy changes of integration variables, the $N!$ integrals are identical,

$$= \sum\nolimits_q (-)^q \int [u_1^*(x_1) \cdots u_N^*(x_N)] \sum \mathcal{H}(i) [u_{q_1}(x_1) \cdots u_{q_N}(x_N)] \mathrm{d}\tau,$$

where $q_1 \ldots q_N$ is the permutation of $1 \ldots N$ generated by Q,

$$= \sum\nolimits_q (-)^q \sum\nolimits_i \int u_i^* \mathcal{H}(i) u_{q_i} \delta_{q_1}^1 \delta_{q_2}^2 \cdots \delta_{q_{i-1}}^{i-1} \delta_{q_{i+1}}^{i+1} \cdots \delta_{q_N}^N \mathrm{d}\tau_i ,$$

where use has been made of the orthonormality of the u_i,

$$= \sum\nolimits_i \int u_i^*(x_1) \mathcal{H}(1) u_1(x_1) \mathrm{d}\tau_1 , \qquad (3.22)$$

where the delta functions allow only $Q = I$ (the identity) and a dummy change of integration variables has been made.

The derivation of an expression for the matrix element of the two-particle operator is somewhat longer:

$$\left\langle \psi \middle| \tfrac{1}{2} \sum\nolimits_{i,j}' V(i,j) \middle| \psi \right\rangle$$
$$= \frac{1}{2N!} \sum\nolimits_{p,p'} (-)^{p+p'} \int [P u_1^*(x_1) \cdots u_N^*(x_N)]$$
$$\times \sum\nolimits_{i,j}' V(i,j) [P' u_1(x_1) \cdots u_N(x_N)] \mathrm{d}\tau$$
$$= \frac{1}{2N!} \sum\nolimits_{p,p'} (-)^{p+p'} P \Big\{ \int [u_1^*(x_1) \cdots u_N^*(x_N)]$$
$$\times \sum\nolimits_{i,j}' V(i,j) P^{-1} P' [u_1(x_1) \cdots u_N(x_N)] \mathrm{d}\tau \Big\},$$

since P commutes with $\sum'_{i,j}V(i,j)$,

$$= \frac{1}{2N!}\sum_{p,q}(-)^q P\left[\int u_1^*(x_1)\cdots u_N^*(x_N)\sum_{i,j}{}' V(i,j)Qu_1(x_1)\cdots u_N(x_N)\mathrm{d}\tau\right],$$

where $Q \equiv P^{-1}P'$ is also a permutation,

$$= \frac{1}{2N!}\sum_q(-)^q \int[u_1^*(x_1)\cdots u_N^*(x_N)]\sum_{i,j}{}' V(i,j)[u_{q_1}(x_1)\cdots u_{q_N}(x_N)]\mathrm{d}\tau\,,$$

since all $N!$ integrals generated by P can be shown to be identical and $q_1 \ldots q_N$ is the permutation of $1 \ldots N$ generated by Q,

$$= \frac{1}{2}\sum_q(-)^q\sum_{i,j}{}' \int u_i^*(x_i)u_j^*(x_j)V(i,j)u_{q_i}(x_i)u_{q_j}(x_j)\mathrm{d}\tau_i\mathrm{d}\tau_j\delta^1_{q_1}\cdots\delta^{i-1}_{q_{i-1}}$$
$$\times\delta^{i+1}_{q_{i+1}}\cdots\delta^{j-1}_{q_{j-1}}\delta^{j+1}_{q_{j+1}}\cdots\delta^N_{q_N}\,,$$

where use has been made of the orthonormality of the u_i,

$$\begin{aligned} = \frac{1}{2}\sum_{i,j}{}' \int[&u_i^*(x_1)u_j^*(x_2)V(1,2)u_i(x_1)u_j(x_2) \\ &- u_i^*(x_1)u_j^*(x_2)V(1,2)u_j(x_1)u_i(x_2)]\mathrm{d}\tau_1\mathrm{d}\tau_2\,, \end{aligned} \tag{3.23}$$

where the delta function allows only $q_i = i$, $q_j = j$ or $q_i = j$, $q_j = i$, and these permutations differ in the sign of $(-1)^q$ and a change in the dummy variables of integration has been made.

Combining (3.20), (3.21), (3.22), (3.23), and choosing $\delta = \delta_k$ in the same way as was done in the Hartree approximation, we find

$$\begin{aligned} \int \mathrm{d}\tau_1\delta_k u_k^*(x_1)\Big\{&\mathcal{H}(1)u_k(x_1) + \sum_{j(\neq k)}\int d\tau_2 u_j^*(x_2)V(1,2)u_j(x_2)u_k(x_2) \\ &- \sum_{j(\neq k)}\int d\tau_2 u_j^*(x_2)V(1,2)u_k(x_2)u_j(x_1) - \sum_j u_j(x_1)\lambda_{kj}\Big\} \\ &+ C.C. = 0. \end{aligned}$$

Since $\delta_k u_k^*$ is completely arbitrary, the part of the integrand inside the brackets must vanish. There is some arbitrariness in the λ just because the u are not unique (there are several sets of us that yield the same determinant). The arbitrariness is sufficient that we can choose $\lambda_{k\neq j} = 0$ without loss in generality. Also note that we can let the sums run over $j = k$ as the $j = k$ terms cancel one another. The following equations are thus obtained:

$$\begin{aligned} \mathcal{H}(1)u_k(x_1) + \sum_j[&\int \mathrm{d}\tau_2 u_j^*(x_2)V(1,2)u_j(x_2)u_k(x_1) \\ &- \int \mathrm{d}\tau_2 u_j^*(x_2)V(1,2)u_k(x_2)u_j(x_1)] = \varepsilon_k u_k\,, \end{aligned} \tag{3.24}$$

where $\varepsilon_k \equiv \lambda_{kk}$.

Equation (3.24) gives the set of equations known as the *Hartree–Fock* equations. The derivation is not complete until the ε_k are interpreted. From (3.24) we can write

$$\varepsilon_k = \langle u_k(1)|\mathcal{H}(1)|u_k(1)\rangle + \sum_j \{\langle u_k(1)u_j(2)|V(1,2)|u_k(1)u_j(2)\rangle - \langle u_k(1)u_j(2)|V(1,2)|u_j(1)u_k(2)\rangle\}, \tag{3.25}$$

where 1 and 2 are a notation for x_1 and x_2. It is convenient at this point to be explicit about what we mean by this notation. We must realize that

$$u_k(x_1) \equiv \psi_k(\boldsymbol{r}_1)\xi_k(s_1), \tag{3.26}$$

where ψ_k is the spatial part of the wave function, and ξ_k is the spin part.

Integrals mean integration over space and summation over spins. The spin functions refer to either "+1/2" or "−1/2" spin states, where ±1/2 refers to the eigenvalues of $s_z/\hbar$ for the spin in question. Two spin functions have inner product equal to one when they are both in the same spin state. They have inner product equal to zero when one is in a +1/2 spin state and one is in a −1/2 spin state. Let us rewrite (3.25) where the summation over the spin part of the inner product has already been done. The inner products now refer only to integration over space:

$$\varepsilon_k = \langle \psi_k(1)|\mathcal{H}(1)|\psi_k(1)\rangle + \sum_j \langle \psi_k(1)\psi_j(2)|V(1,2)|\psi_k(1)\psi_j(2)\rangle - \sum_{\mathrm{j}(\|\mathrm{k})} \langle \psi_k(1)\psi_j(2)|V(1,2)|\psi_j(1)\psi_k(2)\rangle. \tag{3.27}$$

In (3.27), $j(\| k)$ means to sum only over states j that have spins that are in the same state as those states labeled by k.

Equation (3.27), of course, does not tell us what the ε_k are. A theorem due to Koopmans gives the desired interpretation. Koopmans' theorem states that ε_k is the negative of the energy required to remove an electron in state k from the solid. The proof is fairly simple. From (3.22) and (3.23) we can write (using the same notation as in (3.27))

$$E = \sum_i \langle \psi_i(1)|\mathcal{H}(1)|\psi_i(1)\rangle + \tfrac{1}{2}\sum_{i,j} \langle \psi_i(1)\psi_j(2)|V(1,2)|\psi_i(1)\psi_j(2)\rangle - \tfrac{1}{2}\sum_{i,j(\|)} \langle \psi_i(1)\psi_j(2)|V(1,2)|\psi_j(1)\psi_i(2)\rangle. \tag{3.28}$$

Denoting E(w.o.k.) as (3.28) in which terms for which $i = k, j = k$ are omitted from the sums we have

$$E(\text{w.o.k.}) - E = -\langle \psi_k(1)|\mathcal{H}(1)|\psi_k(1)\rangle - \sum_j \langle \psi_k(1)\psi_j(2)|V(1,2)|\psi_k(1)\psi_j(2)\rangle + \sum_{\mathrm{i,j}(\|)} \langle \psi_k(1)\psi_j(2)|V(1,2)|\psi_j(1)\psi_k(2)\rangle. \tag{3.29}$$

Combining (3.27) and (3.29), we have

$$\varepsilon_k = -[E(\text{w.o.k.}) - E], \tag{3.30}$$

which is the precise mathematical statement of Koopmans' theorem. A similar theorem holds for the Hartree method.

Note that the statement that ε_k is the negative of the energy required to remove an electron in state k is valid only in the approximation that the other states are unmodified by removal of an electron in state k. For a metal with many electrons, this is a good approximation. It is also interesting to note that

$$\begin{aligned}\sum\nolimits_1^N \varepsilon_k &= E + \tfrac{1}{2}\sum\nolimits_{i,j}\left\langle \psi_i(1)\psi_j(2)\middle| V(1,2)\middle|\psi_i(1)\psi_j(2)\right\rangle \\ &\quad -\tfrac{1}{2}\sum\nolimits_{i,j(\parallel)}\left\langle \psi_i(1)\psi_j(2)\middle| V(1,2)\middle|\psi_j(1)\psi_i(2)\right\rangle .\end{aligned} \tag{3.31}$$

Derivation of Hartree–Fock Equations in Second Quantization Notation (A)

There really aren't many new ideas introduced in this section. Its purpose is to gain some familiarity with the second quantization notation for fermions. Of course, the idea of the variational principle will still have to be used.[4]

According to Appendix G, if the Hamiltonian is of the form (3.10), then we can write it as

$$\mathcal{H} = \sum\nolimits_{i,j}\mathcal{H}_{ij}a_i^{\dagger}a_j + \tfrac{1}{2}\sum\nolimits_{i,j,k,l}V_{ij,kl}a_j^{\dagger}a_i^{\dagger}a_k a_l \,, \tag{3.32}$$

where the $\mathcal{H}_{ij}$ and the $V_{ij,kl}$ are matrix elements of the one- and two-body operators,

$$V_{ij,kl} = V_{ji,lk} \quad \text{and} \quad a_i a_j^{\dagger} + a_j^{\dagger}a_i = \delta_{ij} \,. \tag{3.33}$$

The rest of the anticommutators of the a are zero.

We shall assume that the occupied states for the normalized ground state Φ (which is a Slater determinant) that minimizes $\langle\Phi|\mathcal{H}|\Phi\rangle$ are labeled from 1 to N. For Φ giving a true extremum, as we saw in the Section on the Hartree approximation, we need require only that

$$\left\langle \delta\Phi\middle|\mathcal{H}\middle|\Phi\right\rangle = 0 \,. \tag{3.34}$$

It is easy to see that if $\langle\Phi|\Phi\rangle = 1$, then $|\Phi\rangle + |\delta\Phi\rangle$ is still normalized to first order in the variation. For example, let us assume that

$$\left|\delta\Phi\right\rangle = (\delta s)a_{k^1}^{\dagger}a_{i^1}\left|\Phi\right\rangle \quad \text{for} \quad k^1 > N,\; i^1 \le N \,, \tag{3.35}$$

[4] For additional comments, see Thouless [3.54].

where δs is a small number and where all one-electron states up to the Nth are occupied in the ground state of the electron system. That is, $|\delta\Phi\rangle$ differs from $|\Phi\rangle$ by having the electron in state Φ_{i^1} go to state Φ_{k^1}. Then

$$\begin{aligned}
&\left(\langle\Phi| + \langle\delta\Phi|\right)\left(|\Phi\rangle + |\delta\Phi\rangle\right) \\
&= \left(\langle\Phi| + \langle\Phi|a^{\dagger}_{i^1}a_{k^1}\delta s^*\right)\left(|\Phi\rangle + a^{\dagger}_{k^1}a_{i^1}\delta s|\Phi\rangle\right) \\
&= 1 + (\delta s)^*\langle\Phi|a^{\dagger}_{i^1}a_{k^1}|\Phi\rangle + \delta s\langle\Phi|a^{\dagger}_{k^1}a_{i^1}|\Phi\rangle + O(\delta s)^2 \\
&= 1 + O(\delta s)^2.
\end{aligned} \tag{3.36}$$

According to the variational principle, we have as a basic condition

$$0 = \langle\delta\Phi|\mathcal{H}|\Phi\rangle = (\delta s)^*\langle\Phi|\mathcal{H}a^{\dagger}_{i^1}a_{k^1}|\Phi\rangle\,. \tag{3.37}$$

Combining (3.32) and (3.37) yields

$$0 = \sum\nolimits_{i,j}\mathcal{H}_{i,j}\langle\Phi|a^{\dagger}_{i^1}a_{k^1}a^{\dagger}_i a_j|\Phi\rangle + \tfrac{1}{2}\sum\nolimits_{i,j,k,l}V_{ij,kl}\langle\Phi|a^{\dagger}_{i^1}a_{k^1}a^{\dagger}_j a^{\dagger}_i a_k a_l|\Phi\rangle \tag{3.38}$$

where the summation is over all values of i, j, k, l (both occupied and unoccupied).

There are two basically different matrix elements to consider. To evaluate them we can make use of the anticommutation relations. Let us do the simplest one first. Φ has been assumed to be the Slater determinant approximation to the ground state, so:

$$\begin{aligned}
\langle\Phi|a^{\dagger}_{i^1}a_{k^1}a^{\dagger}_i a_j|\Phi\rangle &= \langle\Phi|a^{\dagger}_{i^1}(\delta^i_{k^1} - a^{\dagger}_i a_{k^1})a_j|\Phi\rangle \\
&= \langle\Phi|a^{\dagger}_{i^1}a_j|\Phi\rangle\delta^{i}_{k^1} - \langle\Phi|a^{\dagger}_{i^1}a^{\dagger}_i a_{k^1}a_j|\Phi\rangle.
\end{aligned}$$

In the second term a_{k^1} operating to the right gives zero (the only possible result of annihilating a state that isn't there). Since $a_j|\Phi\rangle$ is orthogonal to $a_{i^1}|\Phi\rangle$ unless $i^1 = j$, the first term is just $\delta^j{}_{i^1}$. Thus we obtain

$$\langle\Phi|a^{\dagger}_{i^1}a_{k^1}a^{\dagger}_i a_j|\Phi\rangle = \delta^j_{i^1}\delta^i_{k^1}\,. \tag{3.39}$$

The second matrix element in (3.38) requires a little more manipulation to evaluate

$$\begin{aligned}
&\langle\Phi|a^{\dagger}_{i^1}a_{k^1}a^{\dagger}_j a^{\dagger}_i a_k a_l|\Phi\rangle \\
&= \langle\Phi|a^{\dagger}_{i^1}(\delta^j_{k^1} - a^{\dagger}_j a_{k^1})a^{\dagger}_i a_k a_l|\Phi\rangle \\
&= \delta^j_{k^1}\langle\Phi|a^{\dagger}_{i^1}a^{\dagger}_i a_k a_l|\Phi\rangle - \langle\Phi|a^{\dagger}_{i^1}a^{\dagger}_j a_{k^1}a^{\dagger}_i a_k a_l|\Phi\rangle \\
&= \delta^j_{k^1}\langle\Phi|a^{\dagger}_{i^1}a^{\dagger}_i a_k a_l|\Phi\rangle - \langle\Phi|a^{\dagger}_{i^1}a^{\dagger}_j(\delta^i_{k^1} - a^{\dagger}_i a_{k^1})a_k a_l|\Phi\rangle \\
&= \delta^j_{k^1}\langle\Phi|a^{\dagger}_{i^1}a^{\dagger}_i a_k a_l|\Phi\rangle - \delta^i_{k^1}\langle\Phi|a^{\dagger}_{i^1}a^{\dagger}_j a_k a_l|\Phi\rangle \\
&\quad + \langle\Phi|a^{\dagger}_{i^1}a^{\dagger}_j a^{\dagger}_i a_{k^1}a_k a_l|\Phi\rangle.
\end{aligned}$$

Since $a_{k^1}|\Phi\rangle = 0$, the last matrix element is zero. The first two matrix elements are both of the same form, so we need evaluate only one of them:

$$
\begin{aligned}
\langle\Phi|a^\dagger_{i^1} a^\dagger_i a_k a_l|\Phi\rangle &= -\langle\Phi|a^\dagger_i a^\dagger_{i^1} a_k a_l|\Phi\rangle \\
&= -\langle\Phi|a^\dagger_i(\delta^k_{i^1} - a_k a^\dagger_{i^1})a_l|\Phi\rangle \\
&= -\langle\Phi|a^\dagger_i a_l|\Phi\rangle\delta^k_{i^1} + \langle\Phi|a^\dagger_i a_k a^\dagger_{i^1} a_l|\Phi\rangle \\
&= -\delta_i^{l\le N}\delta^k_{i^1} - \langle\Phi|a^\dagger_i a_k(\delta^l_{i^1} - a_l a^\dagger_{i^1})|\Phi\rangle .
\end{aligned}
$$

$a^\dagger_{i^1}|\Phi\rangle$ is zero since this tries to create a fermion in an already occupied state. So

$$\langle\Phi|a^\dagger_{i^1} a^\dagger_i a_k a_l|\Phi\rangle = -\delta_i^{l\le N}\delta^k_{i^1} + \delta^l_{i^1}\delta_i^{k\le N} .$$

Combining with previous results, we finally find

$$
\begin{aligned}
\langle\Phi|a^\dagger_{i^1} a_{k^1} a^\dagger_j a^\dagger_i a_k a_l|\Phi\rangle &= \delta^j_{k^1}\delta^l_{i^1}\delta_i^{k\le N} - \delta^j_{k^1}\delta_i^{l\le N}\delta^k_{i^1} \\
&\quad - \delta^i_{k^1}\delta^l_{i^1}\delta_j^{k\le N} + \delta^i_{k^1}\delta_j^{l\le N}\delta^k_{i^1} .
\end{aligned}
\tag{3.40}
$$

Combining (3.38), (3.39), and (3.40), we have

$$
\begin{aligned}
0 &= \sum\nolimits_{i,j} \mathcal{H}_{i,j}\delta^j_{i^1}\delta^i_{k^1} \\
&\quad + \tfrac{1}{2}\sum\nolimits_{ijkl} V_{ij,kl}\left(\delta^j_{k^1}\delta^l_{i^1}\delta^k_i + \delta^i_{k^1}\delta^l_j\delta^k_{i^1} - \delta^j_{k^1}\delta^l_i\delta^k_{i^1} - \delta^i_{k^1}\delta^l_{i^1}\delta^k_j\right),
\end{aligned}
$$

or

$$
\begin{aligned}
0 &= \mathcal{H}_{k^1 i^1} \\
&\quad + \tfrac{1}{2}\left(\sum\nolimits_{i=1}^N V_{ik^1,ii^1} + \sum\nolimits_{j=1}^N V_{k^1 j,i^1 j} - \sum\nolimits_{i=1}^N V_{ik^1,i^1 i} - \sum\nolimits_{j=1}^N V_{k^1 j,ji^1}\right).
\end{aligned}
$$

By using the symmetry in the V and making dummy changes in summation variables this can be written as

$$0 = \mathcal{H}_{k^1 i^1} + \sum\nolimits_{j=1}^N\left(V_{k^1 j,i^1 j} - V_{k^1 j,ji^1}\right). \tag{3.41}$$

Equation (3.41) suggests a definition of a one-particle operator called the self-consistent one-particle Hamiltonian:

$$\mathcal{H}_C = \sum\nolimits_{ki}[\mathcal{H}_{ki} + \sum\nolimits_{j=1}^N(V_{kj,ij} - V_{kj,ji})]a^\dagger_k a_i . \tag{3.42}$$

At first glance we might think that this operator is identically zero by comparing it to (3.41). But in (3.41) $k^1 > N$ and $i^1 < N$, whereas in (3.42) there is no such restriction.

An important property of $\mathcal{H}_C$ is that it has no matrix elements between occupied (i^1) and normally unoccupied (k^1) levels. Letting $\mathcal{H}_C = \sum_{ki} f_{ki} a_k^{\dagger} a_i$, we have

$$\begin{aligned}\left\langle k^1 \middle| \mathcal{H}_C \middle| i^1 \right\rangle &= \sum_{ki} f_{ki} \left\langle k^1 \middle| a_k^{\dagger} a_i \middle| i^1 \right\rangle \\ &= \sum_{ki} f_{ki} \left\langle 0 \middle| a_{k^1} a_k^{\dagger} a_i a_{i^1}^{\dagger} \middle| 0 \right\rangle \\ &= \sum_{ki} f_{ki} \left\langle 0 \middle| (a_k^{\dagger} a_{k^1} - \delta_{k^1}^{k})(a_{i^1}^{\dagger} a_{i^1} - \delta_{i^1}^{i}) \middle| 0 \right\rangle .\end{aligned}$$

Since $a_i |0\rangle = 0$, we have

$$\left\langle k^1 \middle| \mathcal{H}_C \middle| i^1 \right\rangle = + f_{k^1 i^1} = 0$$

by the definition of f_{ki} and (3.41).

We have shown that $\langle \delta\Phi | \mathcal{H} | \Phi \rangle = 0$ (for Φ constructed by Slater determinants) if, and only if, (3.41) is satisfied, which is true if, and only if, $\mathcal{H}_C$ has no matrix elements between occupied (i^1) and unoccupied (k^1) levels. Thus in a matrix representation $\mathcal{H}_C$ is in block diagonal form since all $\langle i^1 | \mathcal{H} | k^1 \rangle = \langle k^1 | \mathcal{H} | i^1 \rangle = 0$. H_C is Hermitian, so that it can be diagonalized. Since it is already in block diagonal form, each block can be separately diagonalized. This means that the new occupied levels are linear combinations of the old occupied levels only and the new occupied levels are linear combinations of the old unoccupied levels only. By new levels we mean those levels that have wave functions $\langle i |$, $\langle j |$ such that $\langle i | \mathcal{H}_C | j \rangle$ vanishes unless $i = j$.

Using this new set of levels, we can say

$$\mathcal{H}_C = \sum_i \varepsilon_i a_i^{\dagger} a_i \ . \tag{3.43}$$

In order that (3.43) and (3.42) are equivalent, we have

$$\mathcal{H}_{ki} + \sum_{j=1}^{N} (V_{kj,ij} - V_{kj,ji}) = \varepsilon_i \delta_{ki} \, . \tag{3.44}$$

These equations are the Hartree–Fock equations. Compare (3.44) and (3.24). That is, we have established that $\langle \delta\Phi | \mathcal{H} | \Phi \rangle = 0$ (for Φ a Slater determinant) implies (3.44). It is also true that the set of one-electron wave functions for which (3.44) is true minimizes $\langle \Phi | \mathcal{H} | \Phi \rangle$, where Φ is restricted to be a Slater determinant of the one-electron functions.

Hermitian Nature of the Exchange Operator (A)

In this section, the Hartree–Fock "Hamiltonian" will be proved to be Hermitian. If the Hartree–Fock Hamiltonian, in addition, has nondegenerate eigenfunctions, then we are guaranteed that the eigenfunctions will be orthogonal. Regardless of degeneracy, the orthogonality of the eigenfunctions was built into the Hartree–Fock equations from the very beginning. More importantly, perhaps, the Hermitian nature of the Hartree–Fock Hamiltonian guarantees that its eigenvalues are real. They have to be real. Otherwise Koopmans' theorem would not make sense.

The Hartree–Fock Hamiltonian is defined as that operator $\mathcal{H}^{\mathrm{F}}$ for which

$$\mathcal{H}^{\mathrm{F}} u_k = \varepsilon_k u_k \,. \tag{3.45}$$

$\mathcal{H}^{\mathrm{F}}$ is then defined by comparing (3.24) and (3.45). Taking care of the spin summations as has already been explained, we can write

$$\mathcal{H}^{\mathrm{F}} = \mathcal{H}_1 + \sum\nolimits_j \int \psi_j^*(\boldsymbol{r}_2) V(1,2) \psi_j(\boldsymbol{r}_2) d\tau_2 + A_1 \,, \tag{3.46}$$

where

$$A_1 \psi_{\boldsymbol{k}}(\boldsymbol{r}_1) = -\sum\nolimits_{j(\| k)} \int \psi_j^*(\boldsymbol{r}_2) V(1,2) \psi_{\boldsymbol{k}}(\boldsymbol{r}_2) \mathrm{d}\tau_2 \psi_j(\boldsymbol{r}_1) \,,$$

and A_1 is called the *exchange operator*.

For the Hartree–Fock Hamiltonian to be Hermitian we have to prove that

$$\langle i | \mathcal{H}^F | j \rangle = \langle j | \mathcal{H}^F | i \rangle^* \,. \tag{3.47}$$

This property is obvious for the first two terms on the right-hand side of (3.46) and so needs only to be proved for A_1:

$$\begin{aligned}
\langle l | A_1 | m \rangle^* &= -\Big(\sum\nolimits_{j(\| m)} \int \psi_l^*(\boldsymbol{r}_1) \int \psi_j^*(\boldsymbol{r}_2) V(1,2) \psi_m(\boldsymbol{r}_2) \psi_j(\boldsymbol{r}_1) d\tau_2 d\tau_1\Big)^* \\
&= -\Big(\sum\nolimits_{j(\| m)} \int \psi_l^*(\boldsymbol{r}_1) \psi_j(\boldsymbol{r}_1) \int \psi_j^*(\boldsymbol{r}_2) V(1,2) \psi_m(\boldsymbol{r}_2) d\tau_2 d\tau_1\Big)^* \\
&= -\Big(\sum\nolimits_{j(\| m)} \int \psi_m(\boldsymbol{r}_1) \psi_j^*(\boldsymbol{r}_1) \int \psi_j(\boldsymbol{r}_2) V(1,2) \psi_l^*(\boldsymbol{r}_2) d\tau_2 d\tau_1\Big)^* \\
&= \langle m | A_1 | l \rangle.
\end{aligned}$$

[5] In the proof, use has been made of changes of dummy integration variable and of the relation $V(1,2) = V(2,1)$.

The Fermi Hole (A)

The exchange term (when the interaction is the Coulomb interaction energy and e is the magnitude of the charge on the electron) is

$$\begin{aligned}
A_1 \psi_i(\boldsymbol{r}_1) &\equiv -\sum\nolimits_{j(\| i)} \int \frac{e^2}{4\pi\varepsilon_0 r_{12}} \psi_j^*(\boldsymbol{r}_2) \psi_i(\boldsymbol{r}_2) \mathrm{d}\tau_2 \cdot \psi_i(\boldsymbol{r}_1) \\
&= -\sum\nolimits_{j(\| i)} \int \frac{e}{4\pi\varepsilon_0 r_{12}} \left(\frac{e \psi_j^*(\boldsymbol{r}_2) \psi_i(\boldsymbol{r}_2) \psi_j(\boldsymbol{r}_1)}{\psi_i(\boldsymbol{r}_1)} \right) \psi_i(\boldsymbol{r}_1) \mathrm{d}\tau_2
\end{aligned}$$

$$A_1 \psi_i(\boldsymbol{r}_1) = \int \frac{(-e)}{4\pi\varepsilon_0 r_{12}} \rho(\boldsymbol{r}_1, \boldsymbol{r}_2) \psi_i(\boldsymbol{r}_1) \mathrm{d}\tau_2 \,, \tag{3.48}$$

[5] The matrix elements in (3.47) would vanish if i and j did not refer to spin states which were parallel.

where

$$\rho(\boldsymbol{r}_1, \boldsymbol{r}_2) = \frac{e\sum_{j(\|i)} \psi_j^*(\boldsymbol{r}_2)\psi_i(\boldsymbol{r}_2)\psi_j(\boldsymbol{r}_1)}{\psi_i(\boldsymbol{r}_1)}.$$

From (3.48) and (3.49) we see that exchange can be interpreted as the potential energy of interaction of an electron at $\boldsymbol{r}_1$ with a charge distribution with charge density $\rho(\boldsymbol{r}_1, \boldsymbol{r}_2)$. This charge distribution is a mathematical rather than a physical charge distribution.

Several comments can be made about the exchange charge density $\rho(\boldsymbol{r}_1, \boldsymbol{r}_2)$:

1.
$$\begin{aligned}\int \rho(\boldsymbol{r}_1, \boldsymbol{r}_2)\mathrm{d}\tau_2 &= +e\int \sum_{j(\|i)} \psi_j^*(\boldsymbol{r}_2)\psi_i(\boldsymbol{r}_2)\mathrm{d}\tau_2 \cdot \frac{\psi_j(\boldsymbol{r}_1)}{\psi_i(\boldsymbol{r}_1)} \\ &= e\sum_{j(\|i)} \delta_i^j \cdot \frac{\psi_j(\boldsymbol{r}_1)}{\psi_i(\boldsymbol{r}_1)} = +e.\end{aligned} \tag{3.49}$$

Thus we can think of the total exchange charge as being of magnitude $+e$.

2. $\rho(\boldsymbol{r}_1, \boldsymbol{r}_1) = e\sum_{j(\|i)}|\psi_j(\boldsymbol{r}_1)|^2$, which has the same magnitude and opposite sign of the charge density of parallel spin electrons.

3. From (1) and (2) we can conclude that $|\rho|$ must decrease as r_{12} increases. This will be made quantitative in the section below on Two Free Electrons and Exchange.

4. It is convenient to think of the Fermi hole and exchange charge density in the following way: in $\mathcal{H}^{\mathrm{F}}$, neglecting for the moment A_1, the potential energy of the electron is the potential energy due to the ion cores and all the electrons. Thus the electron interacts with itself in the sense that it interacts with a charge density constructed from its own wave function. The exchange term cancels out this unwanted interaction in a sense, but it cancels it out locally. That is, the exchange term A_1 cancels the potential energy of interaction of electrons with parallel spin in the neighborhood of the electron with given spin. Pictorially we say that the electron with given spin is surrounded by an exchange charge hole (or Fermi hole of charge +e).

The idea of the Fermi hole still does not include the description of the *Coulomb correlations* between electrons due to their mutual repulsion. In this respect the Hartree–Fock method is no better than the Hartree method. In the Hartree method, the electrons move in a field that depends only on the average charge distribution of all other electrons. In the Hartree–Fock method, the only correlations included are those that arise because of the Fermi hole, and these are simply due to the fact that the Pauli principle does not allow two electrons with parallel spin to have the same spatial coordinates. We could call these kinematic correlations (due to constraints) rather than dynamic correlations (due to forces). For further comments on Coulomb correlations see Sect. 3.1.4.

The Hartree–Fock Method Applied to the Free-Electron Gas (A)

To make the above concepts clearer, the Hartree–Fock method will be applied to a free-electron gas. This discussion may actually have some physical content. This is because the Hartree–Fock equations applied to a monovalent metal can be written

$$
\begin{aligned}
&\left[-\frac{\hbar^2}{2m}\nabla_1^2+\sum\nolimits_{I=1}^{N}V_I(\boldsymbol{r}_1)+e^2\sum\nolimits_{j=1}^{N}\int\frac{|\psi_j(\boldsymbol{r}_2)|^2}{4\pi\varepsilon_0 r_{12}}\mathrm{d}\tau_2\right]\psi_i(\boldsymbol{r}_1)\\
&\qquad -e\sum\nolimits_{j(\|i)}\left[\int\frac{\psi_j^*(\boldsymbol{r}_2)\psi_i(\boldsymbol{r}_2)\psi_j(\boldsymbol{r}_1)}{4\pi\varepsilon_0 r_{12}\psi_i(\boldsymbol{r}_1)}\mathrm{d}\tau_2\right]\psi_i(\boldsymbol{r}_1)=E_i\psi_i(\boldsymbol{r}_1).
\end{aligned}
\tag{3.50}
$$

The $V_I(\boldsymbol{r}_1)$ are the ion core potential energies. Let us smear out the net positive charge of the ion cores to make a uniform positive background charge. We will find that the eigenfunctions of (3.50) are plane waves. This means that the electronic charge distribution is a uniform smear as well. For this situation it is clear that the second and third terms on the left-hand side of (3.50) must cancel. This is because the second term represents the negative potential energy of interaction between smeared out positive charge and an equal amount of smeared out negative electronic charge. The third term equals the positive potential energy of interaction between equal amounts of smeared out negative electronic charge. We will, therefore, drop the second and third terms in what follows.

With such a drastic assumption about the ion core potentials, we might also be tempted to throw out the exchange term as well. If we do this we are left with just a set of *one-electron, free-electron* equations. That even this crude model has some physical validity is shown in several following sections. In this section, the exchange term will be retained, and the Hartree–Fock equations for a free-electron gas will later be considered as approximately valid for a monovalent metal.

The equations we are going to solve are

$$
\begin{aligned}
-\frac{\hbar^2}{2m}\nabla_1^2\psi_{\boldsymbol{k}}(\boldsymbol{r}_1)-e\sum\nolimits_{\boldsymbol{k}'}\left[\int\frac{\psi_{\boldsymbol{k}'}^*(\boldsymbol{r}_2)\psi_{\boldsymbol{k}}(\boldsymbol{r}_2)\psi_{\boldsymbol{k}'}(\boldsymbol{r}_1)}{4\pi\varepsilon_0 r_{12}\psi_{\boldsymbol{k}}(\boldsymbol{r}_1)}\mathrm{d}\tau_2\right]\psi_{\boldsymbol{k}}(\boldsymbol{r}_1)\\
=E_{\boldsymbol{k}}\psi_{\boldsymbol{k}}(\boldsymbol{r}_1).
\end{aligned}
\tag{3.51}
$$

Dropping the Coulomb terms is not consistent unless we can show that the solutions of (3.51) are of the form of plane waves

$$
\psi_{\boldsymbol{k}}(\boldsymbol{r}_1)=\frac{1}{\sqrt{V}}\mathrm{e}^{\mathrm{i}\boldsymbol{k}\cdot\boldsymbol{r}_1},
\tag{3.52}
$$

where V is the volume of the crystal.

In (3.51) all integrals are over V. Since $\hbar\boldsymbol{k}$ refers just to linear momentum, it is clear that there is no reference to spin in (3.51). When we sum over $\boldsymbol{k}'$, we sum over distinct spatial states. If we assume each spatial state is doubly occupied with one spin 1/2 electron and one spin −1/2 electron, then a sum over $\boldsymbol{k}'$ sums over all electronic states with spin parallel to the electron in $\boldsymbol{k}$.

To establish that (3.52) is a solution of (3.51) we have only to substitute. The kinetic energy is readily disposed of:

$$-\frac{\hbar^2}{2m}\nabla_1^2\psi_{\boldsymbol{k}}(\boldsymbol{r}_1)=\frac{\hbar^2 k^2}{2m}\psi_{\boldsymbol{k}}(\boldsymbol{r}_1)\,. \tag{3.53}$$

The exchange term requires a little more thought. Using (3.52), we obtain

$$\begin{aligned}A_1\psi_{\boldsymbol{k}}(\boldsymbol{r}_1) &= -\frac{e^2}{4\pi\varepsilon_0 V}\sum\nolimits_{\boldsymbol{k}'}\left[\int\frac{\psi_{\boldsymbol{k}'}^*(\boldsymbol{r}_2)\psi_{\boldsymbol{k}}(\boldsymbol{r}_2)\psi_{\boldsymbol{k}'}(\boldsymbol{r}_1)}{r_{12}\psi_{\boldsymbol{k}}(\boldsymbol{r}_1)}\mathrm{d}\tau_2\right]\psi_{\boldsymbol{k}}(\boldsymbol{r}_1)\\ &= -\frac{e^2}{4\pi\varepsilon_0 V}\sum\nolimits_{\boldsymbol{k}'}\left[\int\frac{\mathrm{e}^{\mathrm{i}(\boldsymbol{k}-\boldsymbol{k}')\cdot(\boldsymbol{r}_2-\boldsymbol{r}_1)}}{r_{12}}\mathrm{d}\tau_2\right]\psi_{\boldsymbol{k}}(\boldsymbol{r}_1)\\ &= -\frac{e^2}{4\pi\varepsilon_0 V}\sum\nolimits_{\boldsymbol{k}'}\mathrm{e}^{-\mathrm{i}(\boldsymbol{k}-\boldsymbol{k}')\cdot\boldsymbol{r}_1}\left[\int\frac{\mathrm{e}^{\mathrm{i}(\boldsymbol{k}-\boldsymbol{k}')\cdot\boldsymbol{r}_2}}{r_{12}}\mathrm{d}\tau_2\right]\psi_{\boldsymbol{k}}(\boldsymbol{r}_1).\end{aligned} \tag{3.54}$$

The last integral in (3.54) can be evaluated by making an analogy to a similar problem in electrostatics. Suppose we have a collection of charges that have a charge density $\rho(\boldsymbol{r}_2) = \exp[\mathrm{i}(\boldsymbol{k}-\boldsymbol{k}')\cdot\boldsymbol{r}_2]$. Let $\phi(\boldsymbol{r}_1)$ be the potential at the point $\boldsymbol{r}_1$ due to these charges. Let us further suppose that we can treat $\rho(\boldsymbol{r}_2)$ as if it is a collection of real charges. Then Coulomb's law would tell us that the potential and the charge distribution are related in the following way:

$$\phi(\boldsymbol{r}_1)=\int\frac{\mathrm{e}^{\mathrm{i}(\boldsymbol{k}-\boldsymbol{k}')\cdot\boldsymbol{r}_2}}{4\pi\varepsilon_0 r_{12}}\mathrm{d}\tau_2\,. \tag{3.55}$$

However, since we are regarding $\rho(\boldsymbol{r}_2)$ as if it were a real distribution of charge, we know that $\phi(\boldsymbol{r}_1)$ must satisfy Poisson's equation. That is,

$$\nabla_1^2\phi(\boldsymbol{r}_1)=-\frac{1}{\varepsilon_0}\mathrm{e}^{\mathrm{i}(\boldsymbol{k}-\boldsymbol{k}')\cdot\boldsymbol{r}_1}\,. \tag{3.56}$$

By substitution, we see that a solution of this equation is

$$\phi(\boldsymbol{r}_1)=\frac{\mathrm{e}^{\mathrm{i}(\boldsymbol{k}-\boldsymbol{k}')\cdot\boldsymbol{r}_1}}{\varepsilon_0\,|\boldsymbol{k}-\boldsymbol{k}'|^2}\,. \tag{3.57}$$

Comparing (3.55) with (3.57), we find

$$\int\frac{\mathrm{e}^{\mathrm{i}(\boldsymbol{k}-\boldsymbol{k}')\cdot\boldsymbol{r}_2}}{4\pi\varepsilon_0 r_{12}}\mathrm{d}\tau_2=\frac{\mathrm{e}^{\mathrm{i}(\boldsymbol{k}-\boldsymbol{k}')\cdot\boldsymbol{r}_1}}{\varepsilon_0\,|\boldsymbol{k}-\boldsymbol{k}'|^2}\,. \tag{3.58}$$

We can therefore write the exchange operator defined in (3.54) as

$$A_1\psi_{\boldsymbol{k}}(\boldsymbol{r}_1) = -\frac{e^2}{\varepsilon_0 V}\sum\nolimits_{\boldsymbol{k}'}\frac{1}{|\boldsymbol{k}-\boldsymbol{k}'|^2}\psi_{\boldsymbol{k}}(\boldsymbol{r}_1). \tag{3.59}$$

If we define $A_1(\boldsymbol{k})$ as the eigenvalue of the operator defined by (3.59), then we find that we have plane-wave solutions of (3.51), provided that the energy eigenvalues are given by

$$E_{\boldsymbol{k}} = \frac{\hbar^2\boldsymbol{k}^2}{2m} + A_1(\boldsymbol{k}). \tag{3.60}$$

If we propose that the above be valid for monovalent metals, then we can make a comparison with experiment. If we imagine that we have a very large crystal, then we can evaluate the sum in (3.59) by replacing it by an integral. We have

$$A_1(\boldsymbol{k}) = -\frac{e^2}{\varepsilon_0 V}\frac{V}{8\pi^3}\int\frac{1}{|\boldsymbol{k}-\boldsymbol{k}'|^2}\mathrm{d}^3k'. \tag{3.61}$$

We assume that the energy of the electrons depends only on $|\boldsymbol{k}|$ and that the maximum energy electrons have $|\boldsymbol{k}| = k_M$. If we use spherical polar coordinates (in $\boldsymbol{k}'$-space) with the k'_z-axis chosen to be parallel to the $\boldsymbol{k}$-axis, we can write

$$\begin{aligned}
A_1(\boldsymbol{k}) &= -\frac{e^2}{8\pi^3\varepsilon_0}\int_0^{k_M}\left[\int_0^{\pi}\left(\int_0^{2\pi}\frac{k'^2\sin\theta}{k^2+k'^2-2kk'\cos\theta}\mathrm{d}\phi\right)\mathrm{d}\theta\right]\mathrm{d}k' \\
&= -\frac{e^2}{4\pi^2\varepsilon_0}\int_0^{k_M}\left[\int_{-1}^{1}\frac{k'^2}{k^2+k'^2-2kk'\cos\theta}\mathrm{d}(\cos\theta)\right]\mathrm{d}k' \\
&= -\frac{e^2}{4\pi\varepsilon_0}\int_0^{k_M}k'^2\left[\int_{-1}^{1}\frac{\ln(k^2+k'^2-2kk'f)}{-2kk'}\right]_{f=-1}^{f=+1}\mathrm{d}k' \\
&= \frac{e^2}{8\pi^2\varepsilon_0 k}\int_0^{k_M}k'\ln\left(\frac{k^2+k'^2-2kk'}{k^2+k'^2+2kk'}\right)\mathrm{d}k' \\
&= -\frac{e^2}{4\pi^2\varepsilon_0 k}\int_0^{k_M}k'\ln\left|\frac{k+k'}{k-k'}\right|\mathrm{d}k'.
\end{aligned} \tag{3.62}$$

But $\int x(\ln x)\mathrm{d}x = (x^2/2)\ln x - x^2/4$, so we can evaluate this last integral and finally find

$$A_1(\boldsymbol{k}) = -\frac{e^2k_M}{4\pi^2\varepsilon_0}\left(2+\frac{k_M^2-k^2}{kk_M}\ln\left|\frac{k+k'}{k-k'}\right|\right). \tag{3.63}$$

The results of Problem 3.5 combined with (3.60) and (3.63) tell us on the Hartree–Fock free-electron model for the monovalent metals that the lowest energy in the conduction band should be given by

$$E(0) = -\frac{e^2}{2\pi^2}\frac{k_M}{\varepsilon_0}, \tag{3.64}$$

while the energy of the highest filled electronic state in the conduction band should be given by

$$E(k_M) = \frac{\hbar^2 k_M^2}{2m} - \frac{e^2 k_M}{4\pi^2 \varepsilon_0}. \tag{3.65}$$

Therefore, the width of the filled part of the conduction band is readily obtained as a simple function of k_M:

$$\left[E(k_M) - E(0)\right] = \frac{\hbar^2 k_M^2}{2m} + \frac{e^2 k_M}{4\pi^2 \varepsilon_0}. \tag{3.66}$$

To complete the calculation we need only express k_M in terms of the number of electrons N in the conduction band:

$$N = \sum\nolimits_{\boldsymbol{k}}(1) = 2\frac{V}{8\pi^3}\int_0^{k_M} \mathrm{d}^3\boldsymbol{k} = \frac{2V}{8\pi^3}\cdot\frac{4\pi}{3}k_M^3. \tag{3.67}$$

The factor of 2 in (3.67) comes from having two spin states per $\boldsymbol{k}$-state. Equation (3.67) determines k_M only for absolute zero temperature. However, we only have an upper limit on the electron energy at absolute zero anyway. We do not introduce much error by using these expressions at finite temperature, however, because the preponderance of electrons always has $|\boldsymbol{k}| < k_M$ for any reasonable temperature.

The first term on the right-hand side of (3.66) is the Hartree result for the bandwidth (for occupied states). If we run out the numbers, we find that the Hartree–Fock bandwidth is typically more than twice as large as the Hartree bandwidth. If we compare this to experiment for sodium, we find that the Hartree result is much closer to the experimental value. The reason for this is that the Hartree theory makes two errors (neglect of the Pauli principle and neglect of Coulomb correlations), but these errors tend to cancel. In the Hartree–Fock theory, Coulomb correlations are left out and there is no other error to cancel this omission. In atoms, however, the Hartree–Fock method usually gives better energies than the Hartree method. For further discussion of the topics in these last two sections as well as in the next section, see the book by Raimes [78].

Two Free Electrons and Exchange (A)

To give further insight into the nature of exchange and to the meaning of the Fermi hole, it is useful to consider the two free-electron model. A direct derivation of the charge density of electrons (with the same spin state as a given electron) will be made for this model. This charge density will be found as a function of the distance from the given electron. If we have two free electrons with the same spin in states $\boldsymbol{k}$ and $\boldsymbol{k}'$, the spatial wave function is

$$\psi_{k,k'}(\boldsymbol{r}_1,\boldsymbol{r}_2) = \frac{1}{\sqrt{2V^2}} \begin{vmatrix} e^{i\boldsymbol{k}\cdot\boldsymbol{r}_1} & e^{i\boldsymbol{k}\cdot\boldsymbol{r}_2} \\ e^{i\boldsymbol{k}'\cdot\boldsymbol{r}_1} & e^{i\boldsymbol{k}'\cdot\boldsymbol{r}_2} \end{vmatrix} . \tag{3.68}$$

By quantum mechanics, the probability $P(\boldsymbol{r}_1,\boldsymbol{r}_2)$ that $\boldsymbol{r}_1$ lies in the volume element $d\boldsymbol{r}_1$, and $\boldsymbol{r}_2$ lies in the volume element $d\boldsymbol{r}_2$ is

$$\begin{aligned} P(\boldsymbol{r}_1,\boldsymbol{r}_2)d^3r_1d^3r_2 &= |\psi_{k,k'}(\boldsymbol{r}_1,\boldsymbol{r}_2)|^2\, d^3r_1d^3r_2 \\ &= \frac{1}{V^2}\{1-\cos[(\boldsymbol{k}'-\boldsymbol{k})\cdot(\boldsymbol{r}_1-\boldsymbol{r}_2)]\}d^3r_1d^3r_2 . \end{aligned} \tag{3.69}$$

The last term in (3.69) is obtained by using (3.68) and a little manipulation.

If we now assume that there are N electrons (half with spin 1/2 and half with spin −1/2), then there are $(N/2)(N/2 - 1) \cong N^2/4$ pairs with parallel spins. Averaging over all pairs, we have for the average probability of parallel spin electron at $\boldsymbol{r}_1$ and $\boldsymbol{r}_2$

$$\bar{P}(\boldsymbol{r}_1,\boldsymbol{r}_2)d^3r_1d^3r_2 = \frac{4}{V^2N^2}\sum\nolimits_{k,k'}\int\!\!\int\{1-\cos[(\boldsymbol{k}'-\boldsymbol{k})\cdot(\boldsymbol{r}_1-\boldsymbol{r}_2)]\}d^3r_1d^3r_2 ,$$

and after considerable manipulation we can recast this into the form

$$\begin{aligned} \bar{P}(\boldsymbol{r}_1,\boldsymbol{r}_2) &= \left(\frac{4}{N^2}\right)\left(\frac{1}{8\pi^3}\right)^2\left(\frac{4\pi}{3}k_M^3\right)^2 \\ &\quad\times\left\{1-9\left[\frac{\sin(k_Mr_{12})-k_Mr_{12}\cos(k_Mr_{12})}{k_M^3r_{12}^3}\right]^2\right\} \\ &\equiv \frac{1}{V^2}\rho(k_Mr_{12}). \end{aligned} \tag{3.70}$$

If there were no exchange (i.e. if we use a simple product wave function rather than a determinantal wave function), then ρ would be 1 everywhere. This means that parallel spin electrons would have no tendency to avoid each other. But as Fig. 3.1 shows, exchange tends to "correlate" the motion of parallel spin electrons in such a way that they tend to not come too close. This is, of course, just an example of the Pauli principle applied to a particular situation. This result should be compared to the Fermi hole concept introduced in a previous section. These oscillations are related to the Rudermann–Kittel oscillations of Sect. 7.2.1 and the Friedel oscillations mentioned in Sect. 9.5.3.

In later sections, the Hartree approximation on a free-electron gas with a uniform positive background charge will be used. It is surprising how many experiments can be interpreted with this model. The main use that is made of this model is in estimating a density of states of electrons. (We will see how to do this in the Section on the specific heat of an electron gas.) Since the final results usually depend only on an integral over the density of states, we can begin to see why this model does not introduce such serious errors. More comments need to be made about the progress in understanding Coulomb correlations. These comments are made in the next section.

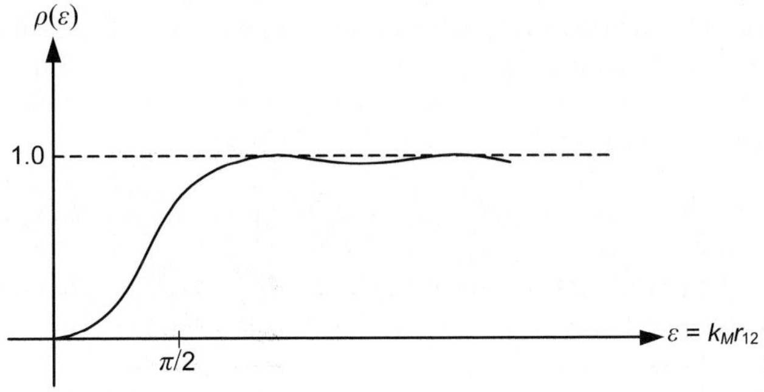

Fig. 3.1. Sketch of density of electrons within a distance r_{12} of a parallel spin electron

3.1.4 Coulomb Correlations and the Many-Electron Problem (A)

We often assume that the Coulomb interactions of electrons (and hence Coulomb correlations) can be neglected. The Coulomb force between electrons (especially at metallic densities) is not a weak force. However, many phenomena (such as Pauli paramagnetism and thermionic emission, which we will discuss later) can be fairly well explained by theories that ignore Coulomb correlations.

This apparent contradiction is explained by admitting that the electrons do interact strongly. We believe that the strongly interacting electrons in a metal form a (normal) Fermi liquid.[6] The elementary energy excitations in the Fermi liquid are called Landau[7] quasiparticles or quasielectrons. *For every electron there is a quasielectron.* The Landau theory of the Fermi liquid is discussed a little more in Sect. 4.1.

Not all quasielectrons are important. Only those that are near the Fermi level in energy are detected in most experiments. This is fortunate because it is only these quasielectrons that have fairly long lifetimes.

---

[6] A normal Fermi liquid can be thought to evolve adiabatically from a Fermi liquid in which the electrons do not interact and in which there is a 1 to 1 correspondence between noninteracting electrons and the quasiparticles. This excludes the formation of "bound" states as in superconductivity (Chap. 8).

[7] See Landau [3.31].

We may think of the quasielectrons as being weakly interacting. Thus our discussion of the N-electron problem in terms of N one-electron problems is approximately valid if we realize we are talking about quasielectrons and not electrons.

Further work on interacting electron systems has been done by Bohm, Pines, and others. Their calculations show two types of fundamental energy excitations: quasielectrons and *plasmons*.[8] The plasmons are collective energy excitations somewhat like a wave in the electron "sea." Since plasmons require many electron volts of energy for their creation, we may often ignore them. This leaves us with the quasielectrons that interact by shielded Coulomb forces and so interact weakly. Again we see why a free-electron picture of an interacting electron system has some validity.

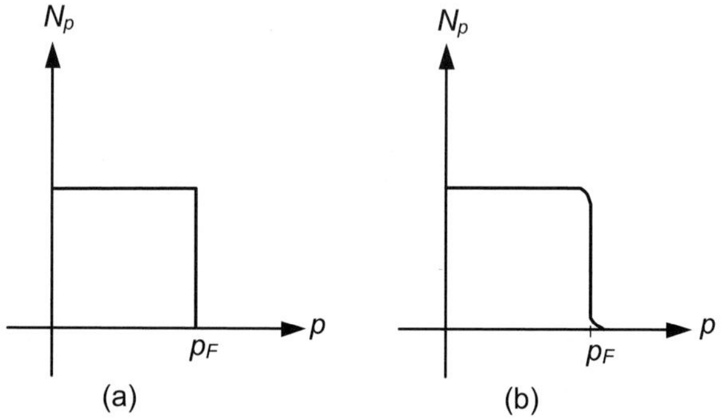

Fig. 3.2. The Fermi distribution at absolute zero (**a**) with no interactions, and (**b**) with interactions (sketched)

We should also mention that Kohn, Luttinger, and others have indicated that electron–electron interactions may change (slightly) the Fermi–Dirac distribution.[8] Their results indicate that the interactions introduce a tail in the Fermi distribution as sketched in Fig. 3.2. N_p is the probability per state for an electron to be in a state with momentum p. Even with interactions there is a discontinuity in the slope of N_p at the Fermi momentum. However, we expect for all calculations in this book that we can use the Fermi–Dirac distribution without corrections and still achieve little error.

The study of many-electron systems is fundamental to solid-state physics. Much research remains to be done in this area. Further related comments are made in Sect. 3.2.2 and in Sect. 4.4.

[8] See Pines [3.41].

3.1.5 Density Functional Approximation[9] (A)

We have discussed the Hartree–Fock method in detail, but, of course, it has its difficulties. For example, a true, self-consistent Hartree–Fock approximation is very complex, and the correlations between electrons due to Coulomb repulsions are not properly treated. The density functional approximation provides another starting point for treating many-body systems, and it provides a better way of teaching electron correlations, at least for ground-state properties. One can regard the density functional method as a generalization of the much older Thomas–Fermi method discussed in Sect. 9.5.2. Sometimes density functional theory is said to be a part of *The Standard Model* for periodic solids [3.27].

There are really two parts to density functional theory (DFT). The first part, upon which the whole theory is based, derives from a basic theorem of P. Hohenberg and W. Kohn. This theorem reduces the solution of the many body ground state to the solution of a one-particle Schrödinger-like equation for the electron density. The electron density contains all needed information. In principle, this equation contains the Hartree potential, exchange and correlation.

In practice, an approximation is needed to make a problem treatable. This is the second part. The most common approximation is known as the local density approximation (LDA). The approximation involves treating the effective potential at a point as depending on the electron density in the same way as it would be for jellium (an electron gas neutralized by a uniform background charge). The approach can also be regarded as a generalization of the Thomas–Fermi–Dirac method.

The density functional method has met with considerable success for calculating the binding energies, lattice parameters, and bulk moduli of metals. It has been applied to a variety of other systems, including atoms, molecules, semiconductors, insulators, surfaces, and defects. It has also been used for certain properties of itinerant electron magnetism. Predicted energy gap energies in semiconductors and insulators can be too small, and the DFT has difficulty predicting excitation energies. DFT-LDA also has difficulty in predicting the ground states of open-shell, 3d, transition element atoms. In 1998, Walter Kohn was awarded a Nobel prize in chemistry for his central role in developing the density functional method [3.27].

Hohenberg–Kohn Theorem (HK Theorem) (A)

As the previous discussion indicates, the most important difficulty associated with the Hartree–Fock approximation is that electrons with opposite spin are left uncorrelated. However, it does provide a rational self-consistent calculation that is more or less practical, and it does clearly indicate the exchange effect. It is a useful starting point for improved calculations. In one sense, density functional theory can be regarded as a modern improved and generalized Hartree–Fock calculation, at least for ground-state properties. This is discussed below.

---

[9] See Kohn [3.27] and Callaway and March [3.8].

We start by deriving the basic theorem for DFT for N identical spinless fermions with a nondegenerate ground state. This theorem is: The ground-state energy E_0 is a unique functional of the electron density $n(\boldsymbol{r})$, i.e. $E_0 = E_0[n(\boldsymbol{r})]$. Further, $E_0[n(\boldsymbol{r})]$ has a minimum value for $n(\boldsymbol{r})$ having its correct value. In all variables, n is constrained, so $N = \int n(\boldsymbol{r})\mathrm{d}\boldsymbol{r}$.

In deriving this theorem, the concept of an external (local) field with a local external potential plays an important role. We will basically show that the external potential $v(\boldsymbol{r})$, and thus, all properties of the many-electron systems will be determined by the ground-state electron distribution function $n(\boldsymbol{r})$. Let $\varphi = \varphi_0(\boldsymbol{r}_1,\boldsymbol{r}_2,\ldots\boldsymbol{r}_N)$ be the normalized wave function for the nondegenerate ground state. The electron density can then be calculated from

$$n(\boldsymbol{r}_1) = N\int \varphi_0^* \varphi_0 d\boldsymbol{r}_2 \ldots d\boldsymbol{r}_n ,$$

where $\mathrm{d}\boldsymbol{r}_i = \mathrm{d}x_i\mathrm{d}y_i\mathrm{d}z_i$. Assuming the same potential for each electron $\upsilon(\boldsymbol{r})$, the potential energy of all electrons in the external field is

$$V(\boldsymbol{r}_1 \ldots \boldsymbol{r}_N) = \sum_{i=1}^{N} \upsilon(\boldsymbol{r}_i) . \tag{3.71}$$

The proof of the theorem starts by showing that $n(\boldsymbol{r})$ determines $\upsilon(\boldsymbol{r})$, (up to an additive constant, of course, changing the overall potential by a constant amount does not affect the ground state). More technically, we say that $\upsilon(\boldsymbol{r})$ is a unique functional of $n(\boldsymbol{r})$. We prove this by a reductio ad absurdum argument.

We suppose υ' determines the Hamiltonian $\mathcal{H}'$ and hence the ground state φ'_0, similarly, υ determines $\mathcal{H}$ and hence, φ_0. We further assume $\upsilon' \neq \upsilon$ but the ground-state wave functions have $n' = n$. By the variational principle for nondegenerate ground states (the proof can be generalized for degenerate ground states):

$$E_0' < \int \varphi_0^* \mathcal{H}' \varphi_0 \mathrm{d}\tau , \tag{3.72}$$

where $\mathrm{d}\tau = \mathrm{d}\boldsymbol{r}_1 \ldots \mathrm{d}\boldsymbol{r}_N$, so

$$E_0' < \int \varphi_0^* (\mathcal{H} - V + V')\varphi_0 \mathrm{d}\tau ,$$

or

$$\begin{aligned} E_0' &< E_0 + \int \varphi_0^* (V' - V)\varphi_0 \mathrm{d}\tau \\ &< E_0 + \sum_{i=1}^{N} \int \varphi_0^* (1\ldots N)[\upsilon'(\boldsymbol{r}_i) - \upsilon(\boldsymbol{r}_i)]\varphi_0(1\ldots N)\mathrm{d}\tau , \\ &< E_0 + N\int \varphi_0^* (1\ldots N)[\upsilon'(\boldsymbol{r}_i) - \upsilon(\boldsymbol{r}_i)]\varphi_0(1\ldots N)\mathrm{d}\tau \end{aligned} \tag{3.73}$$

by the symmetry of $|\varphi_0|^2$ under exchange of electrons. Thus, using the definitions of $n(\boldsymbol{r})$, we can write

$$E_0' < E_0 + N\int [\upsilon'(\boldsymbol{r}_i) - \upsilon(\boldsymbol{r}_i)](\int \varphi_0^* (1\ldots N)\varphi_0(1\ldots N) \cdot \mathrm{d}\boldsymbol{r}_2 \ldots \mathrm{d}\boldsymbol{r}_N)\mathrm{d}\boldsymbol{r}_1 ,$$

or

$$E_0' < E_0 + \int n(\boldsymbol{r}_1)[\upsilon'(\boldsymbol{r}_1) - \upsilon(\boldsymbol{r}_1)]\mathrm{d}\boldsymbol{r}_1 \ . \tag{3.74}$$

Now, $n(\boldsymbol{r})$ is assumed to be the same for υ and υ', so interchanging the primed and unprimed terms leads to

$$E_0 < E_0' + \int n(\boldsymbol{r}_1)[\upsilon(\boldsymbol{r}_1) - \upsilon'(\boldsymbol{r}_1)]\mathrm{d}\boldsymbol{r}_1 \ . \tag{3.75}$$

Adding the last two results, we find

$$E_0 + E_0' < E_0' + E_0 \, , \tag{3.76}$$

which is, of course, a contradiction. Thus, our original assumption that n and n' are the same must be false. Thus $\upsilon(\boldsymbol{r})$ is a unique functional (up to an additive constant) of $n(\boldsymbol{r})$.

Let the Hamiltonian for all the electrons be represented by $\mathcal{H}$. This Hamiltonian will include the total kinetic energy T, the total interaction energy U between electrons, and the total interaction with the external field $V = \sum \upsilon(\boldsymbol{r}_i)$. So,

$$\mathcal{H} = T + U + \sum \upsilon(\boldsymbol{r}_i) \ . \tag{3.77}$$

We have shown $n(\mathbf{r})$ determines $\upsilon(\mathbf{r})$, and hence, $\mathcal{H}$, which determines the ground-state wave function φ_0. Therefore, we can define the functional

$$F[n(\boldsymbol{r})] = \int \varphi_0^*(T + U)\varphi_0 \mathrm{d}\tau \ . \tag{3.78}$$

We can also write

$$\int \varphi_0 \sum \upsilon(r)\varphi_0 \mathrm{d}\tau = \sum \int \varphi_0^*(1 \ldots N)\upsilon(r_i)\varphi_0(1 \ldots N)\mathrm{d}\tau \, , \tag{3.79}$$

by the symmetry of the wave function,

$$\begin{aligned} \int \varphi_0 \sum \upsilon(r)\varphi_0 \mathrm{d}\tau &= N \int \varphi_0^*(1 \ldots N)\upsilon(r_i)\varphi_0(1 \ldots N)\mathrm{d}\tau \\ &= \int \upsilon(\boldsymbol{r})n(\boldsymbol{r})\mathrm{d}\boldsymbol{r} \end{aligned} \tag{3.80}$$

by definition of $n(\boldsymbol{r})$. Thus the total energy functional can be written

$$E_0[n] = \int \varphi_0^* \mathcal{H} \varphi_0 \mathrm{d}\tau = F[n] + \int n(\boldsymbol{r})\upsilon(\boldsymbol{r})\mathrm{d}\boldsymbol{r} \ . \tag{3.81}$$

The ground-state energy E_0 is a unique functional of the ground-state electron density. We now need to show that E_0 is a minimum when $n(\boldsymbol{r})$ assumes the correct electron density. Let n be the correct density function, and let us vary $n \to n'$, so $\upsilon \to \upsilon'$ and $\varphi \to \varphi'$ (the ground-state wave function). All variations are subject to $N = \int n(\boldsymbol{r})\mathrm{d}\boldsymbol{r} = \int n'(\boldsymbol{r})\mathrm{d}\boldsymbol{r}$ being constant. We have

$$\begin{aligned} E_0[n'] &= \int \varphi_0' \mathcal{H} \varphi_0' \mathrm{d}\tau \\ &= \int \varphi_0'(T + U)\varphi_0' \mathrm{d}\tau + \int \varphi_0' \sum \upsilon(\boldsymbol{r}_i)\varphi_0' \mathrm{d}\tau \\ &= F[n'] + \int \upsilon n' \mathrm{d}\boldsymbol{r}. \end{aligned} \tag{3.82}$$

By the principle $\int \varphi'_0 \mathcal{H} \varphi'_0 \mathrm{d}\tau > \int \varphi_0 \mathcal{H} \varphi_0 \mathrm{d}\tau$, we have

$$E_0[n'] > E_0[n] , \tag{3.83}$$

as desired. Thus, the HK Theorem is proved.

The HK Theorem can be extended to the more realistic case of electrons with spin and also to finite temperature. To include spin, one must consider both a spin density $s(\boldsymbol{r})$, as well as a particle density $n(\boldsymbol{r})$. The HK Theorem then states that the ground state is a unique functional of both these densities.

Variational Procedure (A)

Just as the single particle Hartree–Fock equations can be derived from a variational procedure, analogous single-particle equations can be derived from the density functional expressions. In DFT, the energy functional is the sum of $\int \upsilon n \mathrm{d}\tau$ and $F[n]$. In turn, $F[n]$ can be split into a kinetic energy term, an exchange-correlation term and an electrostatic energy term. We may formally write (using Gaussian units so $1/4\pi\varepsilon_0$ can be left out)

$$F[n] = F_{\mathrm{KE}}[n] + E_{\mathrm{xc}}[n] + \frac{e^2}{2} \int \frac{n(\boldsymbol{r}) n(\boldsymbol{r}') \mathrm{d}\tau \mathrm{d}\tau'}{|\boldsymbol{r} - \boldsymbol{r}'|} . \tag{3.84}$$

Equation (3.84), in fact, serves as the definition of $E_{xc}[n]$. The variational principle then states that

$$\delta E_0[n] = 0 , \tag{3.85}$$

subject to $\delta \int n(\boldsymbol{r}) \mathrm{d}\tau = \delta N = 0$, where

$$E_0[n] = F_{\mathrm{KE}}[n] + E_{\mathrm{xc}}[n] + \frac{e^2}{2} \int \frac{n(\boldsymbol{r}) n(\boldsymbol{r}') \mathrm{d}\tau \mathrm{d}\tau'}{|\boldsymbol{r} - \boldsymbol{r}'|} + \int \upsilon(\boldsymbol{r}) n(\boldsymbol{r}) \mathrm{d}\tau . \tag{3.86}$$

Using a Lagrange multiplier μ to build in the constraint of a constant number of particles, and making

$$\delta \left[\frac{e^2}{2} \int \frac{n(\boldsymbol{r}) n(\boldsymbol{r}') \mathrm{d}\tau \mathrm{d}\tau'}{|\boldsymbol{r} - \boldsymbol{r}'|} \right] = e^2 \int \delta n(\boldsymbol{r}) \int \frac{n(\boldsymbol{r}') \mathrm{d}\tau' \mathrm{d}\tau}{|\boldsymbol{r} - \boldsymbol{r}'|} , \tag{3.87}$$

we can write

$$\int \delta n(\boldsymbol{r}) \left[\frac{\delta F_{\mathrm{KE}}[n]}{\delta n(\boldsymbol{r})} + \upsilon(\boldsymbol{r}) + e^2 \int \frac{n(\boldsymbol{r}') \mathrm{d}\tau'}{|\boldsymbol{r} - \boldsymbol{r}'|} + \frac{\delta E_{\mathrm{xc}}[n]}{\delta n(\boldsymbol{r})} \right] d\tau - \mu \int \delta n \mathrm{d}\tau = 0 . \tag{3.88}$$

Defining

$$\upsilon_{\mathrm{xc}}(\boldsymbol{r}) = \frac{\delta E_{\mathrm{xc}}[n]}{\delta n(\boldsymbol{r})} \tag{3.89}$$

(an exchange correlation potential which, in general may be nonlocal), we can then define an effective potential as

$$v_{\mathrm{eff}}(\boldsymbol{r}) = \upsilon(\boldsymbol{r}) + \upsilon_{\mathrm{xc}}(\boldsymbol{r}) + e^2 \int \frac{n(\boldsymbol{r}')\mathrm{d}\tau'}{|\boldsymbol{r}-\boldsymbol{r}'|} . \tag{3.90}$$

The Euler–Lagrange equations can now be written as

$$\frac{\delta F_{\mathrm{KE}}[n]}{\delta n(\boldsymbol{r})} + v_{\mathrm{eff}}(\boldsymbol{r}) = \mu . \tag{3.91}$$

Kohn–Sham Equations (A)

We need to find usable expressions for the kinetic energy and the exchange correlation potential. Kohn and Sham assumed that there existed some N single-particle wave functions $u_i(r)$, which could be used to determine the electron density. They assumed that if this made an error in calculating the kinetic energy, then this error could be lumped into the exchange correlation potential. Thus,

$$n(\boldsymbol{r}) = \sum_{i=1}^{N} |u_i(\boldsymbol{r})|^2 , \tag{3.92}$$

and assume the kinetic energy can be written as

$$\begin{aligned} F_{\mathrm{KE}}(n) &= \frac{1}{2}\sum_{i=1}^{N} \int \nabla u_i^* \cdot \nabla u_i \mathrm{d}\tau \\ &= \sum_{i=1}^{N} \int u_i^* \left(-\frac{1}{2}\nabla^2\right) u_i \mathrm{d}\tau \end{aligned} , \tag{3.93}$$

where units are used so $\hbar^2/m = 1$. Notice this is a kinetic energy for non interacting particles In order for F_{KE} to represent the kinetic energy, the u_i must be orthogonal. Now, without loss in generality, we can write

$$\delta n = \sum_{i=1}^{N} (\delta u_i^*) u_i , \tag{3.94}$$

with the u_i constrained to be orthogonal so $\int u_i^* u_i = \delta_{ij}$. The energy functional $E_0[n]$ is now given by

$$\begin{aligned} E_0[n] = &\sum_{i=1}^{N} \int u_i^* \left(-\frac{1}{2}\nabla^2\right) u_i \mathrm{d}\tau + E_{xc}[n] \\ &+ \frac{e^2}{2} \int \frac{n(\boldsymbol{r})n(\boldsymbol{r}')\mathrm{d}\tau\mathrm{d}\tau'}{|\boldsymbol{r}-\boldsymbol{r}'|} + \int \upsilon(\boldsymbol{r})n(\boldsymbol{r})\mathrm{d}\tau. \end{aligned} \tag{3.95}$$

Using Lagrange multipliers ε_{ij} to put in the orthogonality constraints, the variational principle becomes

$$\delta E_0[n] - \sum_{i=1}^{N} \varepsilon_{ij} \int \delta u_i^* u_i \mathrm{d}\tau = 0 . \tag{3.96}$$

This leads to

$$\sum_{i=1}^{N} \int \delta u_i^* \left[\left(-\frac{1}{2}\nabla^2 + v_{\mathrm{eff}}(\boldsymbol{r}) \right) u_i - \sum_j \varepsilon_{ij} u_i \right] \mathrm{d}\tau = 0 . \tag{3.97}$$

Since the u_i^* can be treated as independent, the terms in the bracket can be set equal to zero. Further, since ε_{ij} is Hermitian, it can be diagonalized without affecting the Hamiltonian or the density. We finally obtain one form of the Kohn–Sham equations

$$\left(-\frac{1}{2}\nabla^2 + v_{\mathrm{eff}}(\boldsymbol{r}) \right) u_i = \varepsilon_i u_i , \tag{3.98}$$

where $v_{\mathrm{eff}}(\boldsymbol{r})$ has already been defined. There is no Koopmans' Theorem in DFT and care is necessary in the interpretation of ε_i. In general, for DFT results for excited states, the literature should be consulted. We can further derive an expression for the ground state energy. Just as for the Hartree–Fock case, the ground-state energy does not equal $\sum \varepsilon_i$. However, using the definition of n,

$$\begin{aligned} \sum_i \varepsilon_i &= \sum_i \int u_i^* \left[-\frac{1}{2}\nabla^2 + \upsilon(\boldsymbol{r}) + e^2 \int \frac{n(\boldsymbol{r}')d\tau'}{|\boldsymbol{r}-\boldsymbol{r}'|} + \upsilon_{xc}(\boldsymbol{r}) \right] u_i \mathrm{d}\tau \\ &= F_{\mathrm{KE}}[n] + \int n\upsilon \mathrm{d}\tau + \int n\upsilon_{\mathrm{xc}} \mathrm{d}\tau + e^2 \int \frac{n(\boldsymbol{r}')n(\boldsymbol{r})\mathrm{d}\tau\mathrm{d}\tau'}{|\boldsymbol{r}-\boldsymbol{r}'|} . \end{aligned} \tag{3.99}$$

Equations (3.90), (3.92), and (3.98) are the Kohn–Sham equations. If υ_{xc} were zero these would just be the Hartree equations. Substituting the expression into the equation for the ground-state energy, we find

$$E_0[n] = \sum \varepsilon_i - \frac{e^2}{2} \int \frac{n(\boldsymbol{r})n(\boldsymbol{r}')\mathrm{d}\tau\mathrm{d}\tau'}{|\boldsymbol{r}-\boldsymbol{r}'|} - \int \upsilon_{\mathrm{xc}}(\boldsymbol{r})n(\boldsymbol{r})\mathrm{d}\tau + E_{\mathrm{xc}}[n] . \tag{3.100}$$

We now want to look at what happens when we include spin. We must define both spin-up and spin-down densities, $n_\uparrow$ and $n_\downarrow$. The total density n would then be a sum of these two, and the exchange correlation energy would be a functional of both. This is shown as follows:

$$E_{\mathrm{xc}} = E_{\mathrm{xc}}[n_\uparrow, n_\downarrow] . \tag{3.101}$$

We also assume single-particle states exist, so

$$n_\uparrow(\boldsymbol{r}) = \sum_{i=1}^{N_\uparrow} |u_{i\uparrow}(\boldsymbol{r})|^2 , \tag{3.102}$$

and

$$n_{\downarrow}(\boldsymbol{r}) = \sum_{i=1}^{N_{\downarrow}} | u_{i\downarrow}(\boldsymbol{r}) |^2 . \tag{3.103}$$

Similarly, there would be both spin-up and spin-down exchange correlation energy as follows:

$$\upsilon_{\text{xc}\uparrow} = \frac{\delta E_{\text{xc}}[n_{\uparrow}, n_{\downarrow}]}{\delta n_{\uparrow}} , \tag{3.104}$$

and

$$\upsilon_{\text{xc}\downarrow} = \frac{\delta E_{\text{xc}}[n_{\uparrow}, n_{\downarrow}]}{\delta n_{\downarrow}} . \tag{3.105}$$

Using σ to represent either $\uparrow$ or $\downarrow$, we can find both the single-particle equations and the expression for the ground-state energy

$$\left[-\frac{1}{2}\nabla^2 + \upsilon(\boldsymbol{r}) + e^2 \int \frac{n(\boldsymbol{r}')\mathrm{d}\tau'}{|\boldsymbol{r}-\boldsymbol{r}'|} + \upsilon_{\text{xc}\sigma}(\boldsymbol{r}) \right] u_{i\sigma} = \varepsilon_{i\sigma} u_{i\sigma} , \tag{3.106}$$

$$\begin{aligned} E_0[n] = \textstyle\sum_{i,\sigma} \varepsilon_{i\sigma} - \frac{e^2}{2} \int \frac{n(\boldsymbol{r})n(\boldsymbol{r}')\mathrm{d}\tau\mathrm{d}\tau'}{|\boldsymbol{r}-\boldsymbol{r}'|} \\ - \textstyle\sum_{\sigma} \int \upsilon_{\text{xc}\sigma}(\boldsymbol{r}) n_{\sigma}(\boldsymbol{r}) \mathrm{d}\tau + E_{\text{xc}}[n], \end{aligned} \tag{3.107}$$

over N lowest $\varepsilon_{i\sigma}$.

Local Density Approximation (LDA) to υ_{xc} (A)

The equations are still not in a tractable form because we have no expression for υ_{xc}. We assume the local density approximation of Kohn and Sham, in which we assume that locally E_{xc} can be calculated as if it were a uniform electron gas. That is, we assume for the spinless case

$$E_{\text{xc}}^{\text{LDA}} = \int n \varepsilon_{\text{xc}}^{\text{uniform}}[n(\boldsymbol{r})] \mathrm{d}\tau ,$$

and for the spin $^1/_2$ case,

$$E_{\text{xc}}^{\text{LDA}} = \int n \varepsilon_{\text{xc}}^{u}[n_{\uparrow}(\boldsymbol{r}), n_{\downarrow}(\boldsymbol{r})] \mathrm{d}\tau ,$$

where ε_{xc} represents the energy per electron. For the spinless case, the exchange-correlation potential can be written

$$\upsilon_{\text{xc}}^{\text{LDA}}(\boldsymbol{r}) = \frac{\delta E_{\text{xc}}^{\text{LDA}}}{\delta n(\boldsymbol{r})} , \tag{3.108}$$

and

$$\delta E_{\text{xc}}^{\text{LDA}} = \int \delta n \varepsilon_{\text{xc}}^{u} \cdot \text{d}\tau + \int n \frac{\delta \varepsilon_{\text{xc}}^{u}}{\delta n} \delta n \cdot \text{d}\tau \tag{3.109}$$

by the chain rule. So,

$$\delta E_{\text{xc}}^{\text{LDA}} = \int \frac{\delta E_{\text{xc}}^{\text{LDA}}}{\delta n} \delta n \cdot \text{d}\tau = \int \left(\varepsilon_{\text{xc}}^{u} + n \frac{\delta \varepsilon_{\text{xc}}^{u}}{\delta n} \right) \delta n \cdot \text{d}\tau . \tag{3.110}$$

Thus,

$$\frac{\delta E_{\text{xc}}^{\text{LDA}}}{\delta n} = \varepsilon_{\text{xc}}^{u}(n) + n \frac{\delta \varepsilon_{\text{xc}}^{u}(n)}{\delta n} . \tag{3.111}$$

The exchange correlation energy per particle can be written as a sum of exchange and correlation energies, $\varepsilon_{\text{xc}}(n) = \varepsilon_{\text{x}}(n) + \varepsilon_{\text{c}}(n)$. The exchange part can be calculated from the equations

$$E_x = \frac{1}{2} \frac{V}{\pi^2} \int_0^{k_M} A_1(k) k^2 \text{d}k , \tag{3.112}$$

and

$$A_1(k) = -\frac{e^2 k_M}{2\pi} \left[2 + \frac{k_M^2 - k^2}{k k_M} \ln \left| \frac{k_M + k}{k_M - k} \right| \right] , \tag{3.113}$$

see (3.63), where 1/2 in E_x is inserted so as not to count interactions twice. Since

$$N = \frac{V}{\pi^2} \frac{k_M^3}{3} ,$$

we obtain by doing all the integrals,

$$\frac{E_x}{N} = -\frac{3}{4} \left(\frac{3}{\pi} \cdot \frac{N}{V} \right)^{1/3} . \tag{3.114}$$

By applying this equation locally, we obtain the Dirac exchange energy functional

$$\varepsilon_x(n) = -c_x [n(r)]^{1/3} , \tag{3.115}$$

where

$$c_x = \frac{3}{4} \left(\frac{3}{\pi} \right)^{1/3} . \tag{3.116}$$

The calculation of ε_c is lengthy and difficult. Defining r_s so

$$\frac{4}{3}\pi r_s^3 = \frac{1}{n}, \tag{3.117}$$

one can derive exact expressions for ε_c at large and small r_s. An often-used expression in atomic units (see Appendix A) is

$$\varepsilon_c = 0.0252F\left(\frac{r_s}{30}\right), \tag{3.118}$$

where

$$F(x) = (1 + x^3)\ln\left(1 + \frac{1}{x}\right) + \frac{x}{2} - x^2 - \frac{1}{3}. \tag{3.119}$$

Other expressions are often given. See, e.g., Ceperley and Alder [3.9] and Pewdew and Zunger [3.39]. More complicated expressions are necessary for the nonspin compensated case (odd number of electrons and/or spin-dependent potentials).

Reminder: Functions and Functional Derivatives A function assigns a number $g(x)$ to a variable x, while a functional assigns a number $F[g]$ to a function whose values are specified over a whole domain of x. If we had a function $F(g_1, g_2, \ldots, g_n)$ of the function evaluated at a finite number of x_i, so that $g_1 = g(x_1)$, etc., the differential of the function would be

$$\mathrm{d}F = \sum_{i=1}^{N} \frac{\partial F}{\partial g_i}\mathrm{d}g_i \,. \tag{3.120}$$

Since we are dealing with a continuous domain D of the x-values over a whole domain, we define a functional derivative in a similar way. But now, the sum becomes an integral and the functional derivative should really probably be called a functional derivative density. However, we follow current notation and determine the variation in F (δF) in the following way:

$$\delta F = \int_{x \in D} \frac{\delta F}{\delta g(x)} \delta g(x)\mathrm{d}x \,. \tag{3.121}$$

This relates to more familiar ideas often encountered with, say, Lagrangians. Suppose

$$F[x] = \int_D L(x, \dot{x})\mathrm{d}t \,;\ \dot{x} = \mathrm{d}x/\mathrm{d}t \,,$$

and assume $\delta x = 0$ at the boundary of D, then

$$\delta F = \int \frac{\delta F}{\delta x(t)} \delta x(t)\mathrm{d}t \,,$$

but

$$\delta L(x, \dot{x}) = \frac{\partial L}{\partial x}\delta x + \frac{\partial L}{\partial \dot{x}}\delta \dot{x} .$$

If

$$\int \frac{\partial L}{\partial \dot{x}}\delta \dot{x} \mathrm{d}t = \int \frac{\partial L}{\partial \dot{x}}\frac{\mathrm{d}}{\mathrm{d}t}\delta x \mathrm{d}t = \underbrace{\left.\frac{\partial L}{\partial \dot{x}}\right|_{Boundary}}_{\to 0}\delta x - \int \frac{\mathrm{d}}{\mathrm{d}t}\frac{\partial L}{\partial \dot{x}}\delta x \mathrm{d}t ,$$

then

$$\int_D \frac{\delta F}{\delta x(t)}\delta x(t)\mathrm{d}t = \int_D \left(\frac{\partial L}{\partial x} - \frac{\mathrm{d}}{\mathrm{d}t}\frac{\partial L}{\partial \dot{x}}\right)\delta x(t)\mathrm{d}t .$$

So

$$\frac{\delta F}{\delta x(t)} = \frac{\partial L}{\partial x} - \frac{\mathrm{d}}{\mathrm{d}t}\frac{\partial L}{\partial \dot{x}} ,$$

which is the typical result of Lagrangian mechanics. For example,

$$E_{\mathrm{X}}^{\mathrm{LDA}} = \int n(\boldsymbol{r})\varepsilon_x \mathrm{d}\tau , \tag{3.122}$$

where $\varepsilon_x = -c_x n(\boldsymbol{r})^{1/3}$, as given by the Dirac exchange. Thus,

$$\begin{aligned} E_{\mathrm{X}}^{\mathrm{LDA}} &= -c_x \int n(\boldsymbol{r})^{4/3}\mathrm{d}\tau \\ \delta E_{\mathrm{X}}^{\mathrm{LDA}} &= -c_x \frac{4}{3}\int n(\boldsymbol{r})^{1/3}\delta n \mathrm{d}\tau , \\ &= \int \frac{\delta E_{\mathrm{X}}^{\mathrm{LDA}}}{\delta n}\delta n \mathrm{d}\tau \end{aligned} \tag{3.123}$$

so,

$$\frac{\delta E_{\mathrm{X}}^{\mathrm{LDA}}}{\delta n} = -\frac{4}{3}c_x n(\boldsymbol{r})^{1/3} . \tag{3.124}$$

Further results may easily be found in the functional analysis literature (see, e.g., Parr and Yang [3.38].

We summarize in Table 3.1 the one-electron approximations we have discussed thus far.

Table 3.1. One-electron approximations

Approximation	Equations defining	Comments
Free electrons	$\mathcal{H} = -\frac{\hbar^2}{2m^*}\nabla^2 + V$ $V = \text{constant}$ $m^* = \text{effective mass}$ $\mathcal{H}\psi_k = E\psi$ $E_k = \frac{\hbar^2 k^2}{2m^*} + V$ $\psi_{\mathbf{k}} = A e^{\mathrm{i}\mathbf{k}\cdot\mathbf{r}}$ $A = \text{constant}$	Populate energy levels with Fermi–Dirac statistics useful for simple metals.
Hartree	$[\mathcal{H} + V(\mathbf{r})]u_k(\mathbf{r}) = E_k u_k(\mathbf{r})$ $V(\mathbf{r}) = V_{\text{nucl}} + V_{\text{coul}}$ $V_{\text{nucl}} = -\sum_{\substack{a(\text{nuclei})\\ i(\text{electrons})}} \frac{e^2}{4\pi\varepsilon_0 r_{ai}} + \text{const}$ $V_{\text{coul}} = \sum_{j(\neq k)} \int u_j^*(x_2)V(1,2)u_j(x_2)\mathrm{d}\tau_2$ V_{coul} arises from Coulomb interactions of electrons	See (3.9), (3.15)
Hartree–Fock	$[\mathcal{H} + V(\mathbf{r}) + V_{\text{exch}}]u_k(\mathbf{r}) = E_k u_k(\mathbf{r})$ $V_{\text{exch}}u_k(\mathbf{r}) =$ $-\sum_j \int \mathrm{d}\tau_2 u_j^*(x_2)V(1,2)u_k(x_2)u_j(x_1)$ and $V(\mathbf{r})$ as for Hartree (without the $j \neq k$ restriction in the sum).	E_k is defined by Koopmans' Theorem (3.30).
Hohenberg–Kohn Theorem	An external potential $v(\mathbf{r})$ is uniquely determined by the ground-state density of electrons in a band system. This local electronic charge density is the basic quantity in density functional theory, rather than the wave function.	No Koopmans' theorem.

Table 3.1. (cont)

Approximation	Equations defining	Comments
Kohn–Sham equations	$\left(-\frac{1}{2}\nabla^2 + v_{\text{eff}}(\boldsymbol{r}) - \varepsilon_j\right)\varphi_j(\boldsymbol{r}) = 0$ where $n(\boldsymbol{r}) = \sum_{j=1}^{N}\left\|\varphi_j(\boldsymbol{r})\right\|^2$	Related to Slater's earlier ideas (see Marder op cit p. 219)
Local density approximation	$v_{\text{eff}}(\boldsymbol{r}) = v(\boldsymbol{r}) + \int \frac{n(\boldsymbol{r}')}{\left\|\boldsymbol{r}-\boldsymbol{r}'\right\|}\mathrm{d}\boldsymbol{r}' + v_{\text{xc}}(\boldsymbol{r})$ $E_{\text{xc}}^{\text{LDA}} = \int n\varepsilon_{\text{xc}}^{\text{u}}[n(\boldsymbol{r})]\mathrm{d}\boldsymbol{r}$, exchange correlation energy ε_{xc} per particle $v_{\text{xc}}(\boldsymbol{r}) = \frac{\delta E_{\text{xc}}[n]}{\delta n(\boldsymbol{r})}$ and see (3.111) and following	See (3.90).

3.2 One-Electron Models

We now have some feeling about the approximation in which an N-electron system can be treated as N one-electron systems. The problem we are now confronted with is how to treat the motion of one electron in a three-dimensional periodic potential. Before we try to solve this problem it is useful to consider the problem of one electron in a spatially infinite one-dimensional periodic potential. This is the Kronig–Penney model.[10] Since it is exactly solvable, the Kronig–Penney model is very useful for giving some feeling for electronic energy bands, Brillouin zones, and the concept of effective mass. For some further details see also Jones [58], as well as Wilson [97, p26ff].

3.2.1 The Kronig–Penney Model (B)

The potential for the Kronig–Penney model is shown schematically in Fig. 3.3. A good reference for this Section is Jones [58, Chap. 1, Sect. 6].

Rather than using a finite potential as shown in Fig. 3.3, it is mathematically convenient to let the widths a of the potential become vanishingly narrow and the

[10] See Kronig and Penny [3.30].

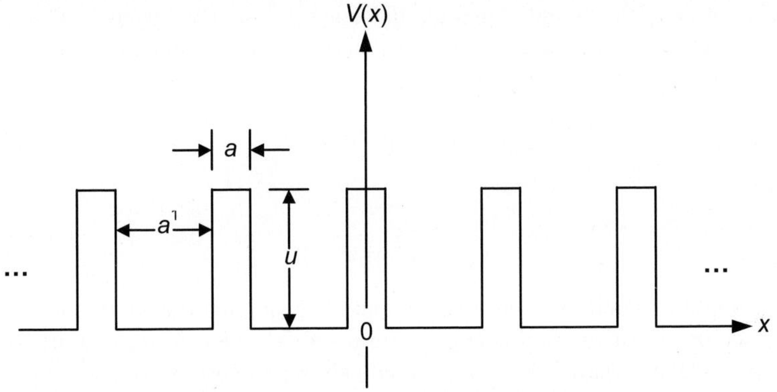

Fig. 3.3. The Kronig–Penney potential

heights u become infinitely high so that their product au remains a constant. In this case, we can write the potential in terms of Dirac delta functions

$$V(x) = au\sum_{n=-\infty}^{n=\infty} \delta(x - na^1)\,, \tag{3.125}$$

where $\delta(x)$ is Dirac's delta function.

With delta function singularities in the potential, the boundary conditions on the wave functions must be discussed rather carefully. In the vicinity of the origin, the wave function must satisfy

$$-\frac{\hbar^2}{2m}\frac{d^2\psi}{dx^2} + au\delta(x)\psi = E\psi\,. \tag{3.126}$$

Integrating across the origin, we find

$$\frac{\hbar^2}{2m}\int_{-\varepsilon}^{\varepsilon}\frac{\mathrm{d}^2\psi}{\mathrm{d}x^2}\mathrm{d}x - au\int_{-\varepsilon}^{\varepsilon}\delta(x)\psi(x)\mathrm{d}x = -E\int_{-\varepsilon}^{\varepsilon}\psi\mathrm{d}x\,,$$

or

$$\frac{\hbar^2}{2m}\frac{\mathrm{d}\psi}{\mathrm{d}x}\bigg|_{-\varepsilon}^{\varepsilon} - au\psi(0) = -E\int_{-\varepsilon}^{\varepsilon}\psi\mathrm{d}x\,.$$

Taking the limit as $\varepsilon \to 0$, we find

$$\left.\frac{\mathrm{d}\psi}{\mathrm{d}x}\right)_+ - \left.\frac{\mathrm{d}\psi}{\mathrm{d}x}\right)_- = \frac{2m(au)}{\hbar^2}\psi(0)\,. \tag{3.127}$$

Equation (3.127) is the appropriate boundary condition to apply across the Dirac delta function potential.

Our problem now is to solve the Schrödinger equation with periodic Dirac delta function potentials with the aid of the boundary condition given by (3.127). The periodic nature of the potential greatly aids our solution. By Appendix C we know that Bloch's theorem can be applied. This theorem states, for our case, that the wave equation has stationary-state solutions that can always be chosen to be of the form

$$\psi_k(x) = \mathrm{e}^{ikx} u_k(x), \tag{3.128}$$

where

$$u_k(x + a^1) = u_k(x). \tag{3.129}$$

Knowing the boundary conditions to apply at a singular potential, and knowing the consequences of the periodicity of the potential, we can make short work of the Kronig–Penney model. We have already chosen the origin so that the potential is symmetric in x, i.e. $V(x) = V(-x)$. This implies that $\mathcal{H}(x) = \mathcal{H}(-x)$. Thus if $\psi(x)$ is a stationary-state wave function,

$$\mathcal{H}(x)\psi(x) = E\psi(x).$$

By a dummy variable change

$$\mathcal{H}(-x)\psi(-x) = E\psi(-x),$$

so that

$$\mathcal{H}(x)\psi(-x) = E\psi(-x).$$

This little argument says that if $\psi(x)$ is a solution, then so is $\psi(-x)$. In fact, any linear combination of $\psi(x)$ and $\psi(-x)$ is then a solution. In particular, we can always choose the stationary-state solutions to be even $z_s(x)$ or odd $z_a(x)$:

$$z_s(x) = \tfrac{1}{2}[\psi(x) + \psi(-x)], \tag{3.130}$$

$$z_s(x) = \tfrac{1}{2}[\psi(x) - \psi(-x)]. \tag{3.131}$$

To avoid confusion, it should be pointed out that this result does not necessarily imply that there is always a two-fold degeneracy in the solutions; $z_s(x)$ or $z_a(x)$ could vanish. In this problem, however, there always is a two-fold degeneracy.

It is always possible to write a solution as

$$\psi(x) = Az_s(x) + Bz_a(x). \tag{3.132}$$

From Bloch's theorem

$$\psi(a^1/2) = \mathrm{e}^{ika^1}\psi(-a^1/2), \tag{3.133}$$

and

$$\psi'(a^1/2) = \mathrm{e}^{ika^1}\psi'(-a^1/2), \tag{3.134}$$

where the prime means the derivative of the wave function.

Combining (3.132), (3.133), and (3.134), we find that

$$A[z_s(a^1/2) - \mathrm{e}^{ika^1} z_s(-a^1/2)] = B[\mathrm{e}^{ika^1} z_a(-a^1/2) - z_a(a^1/2)], \quad (3.135)$$

and

$$A[z_s'(a^1/2) - \mathrm{e}^{ika^1} z_s'(-a^1/2)] = B[\mathrm{e}^{ika^1} z_a'(-a^1/2) - z_a'(a^1/2)]. \quad (3.136)$$

Recalling that z_s, z_a' are even, and z_a, z_s' are odd, we can combine (3.135) and (3.136) to find that

$$\left(\frac{1-\mathrm{e}^{ika^1}}{1+\mathrm{e}^{ika^1}}\right)^2 = \frac{z_s'(a^1/2)z_a(a^1/2)}{z_s(a^1/2)z_a'(a^1/2)}. \quad (3.137)$$

Using the fact that the left-hand side is

$$-\tan^2\frac{ka^1}{2} = -\tan^2\frac{\theta}{2} = 1 - \frac{1}{\cos^2(\theta/2)},$$

and $\cos^2(\theta/2) = (1 + \cos\theta)/2$, we can write (3.137) as

$$\cos(ka^1) = -1 + \frac{2z_s(a^1/2)z_a'(a^1/2)}{W}, \quad (3.138)$$

where

$$W = \begin{vmatrix} z_s & z_a \\ z_s' & z_a' \end{vmatrix}. \quad (3.139)$$

The solutions of the Schrödinger equation for this problem will have to be sinusoidal solutions. The odd solutions will be of the form

$$z_a(x) = \sin(rx), \quad -a^1/2 \le x \le a^1/2, \quad (3.140)$$

and the even solution can be chosen to be of the form [58]

$$z_s(x) = \cos r(x + K), \qquad 0 \le x \le a^1/2, \quad (3.141)$$

$$z_s(x) = \cos r(-x + K), \quad -a^1/2 \le x \le 0. \quad (3.142)$$

At first glance, we might be tempted to chose the even solution to be of the form $\cos(rx)$. However, we would quickly find that it is impossible to satisfy the boundary condition (3.127). Applying the boundary condition to the odd solution, we simply find the identity $0 = 0$. Applying the boundary condition to the even solution, we find

$$-2r\sin rK = (\cos rK)\cdot 2mau/\hbar^2,$$

or in other words, K is determined from

$$\tan rK = -\frac{m(au)}{r\hbar^2}. \quad (3.143)$$

Putting (3.140) and (3.141) into (3.139), we find

$$W = r\cos rK\,. \tag{3.144}$$

Combining (3.138), (3.140), (3.141), and (3.144), we find

$$\cos ka^1 = -1 + \frac{2r\cos[r(a^1/2+K)]\cos(ra^1/2)}{r\cos(rK)}\,. \tag{3.145}$$

Using (3.143), this last result can be written

$$\cos ka^1 = \cos ra^1 + \frac{m(au)}{\hbar^2}a^1\frac{\sin ra^1}{ra^1}\,. \tag{3.146}$$

Note the fundamental 2π periodicity of ka^1. This is the usual Brillouin zone periodicity.

Equation (3.146) is the basic equation describing the energy eigenvalues of the Kronig–Penney model. The reason that (3.146) gives the energy eigenvalue relation is that r is proportional to the square root of the energy. If we substitute (3.141) into the Schrödinger equation, we find that

$$r = \frac{\sqrt{2mE}}{\hbar}\,. \tag{3.147}$$

Thus (3.146) and (3.147) explicitly determine the energy eigenvalue relation (E versus k; this is also called the dispersion relationship) for electrons propagating in a periodic crystal.

The easiest thing to get out of this dispersion relation is that there are allowed and disallowed energy bands. If we plot the right-hand side of (3.146) versus ra, the results are somewhat as sketched in Fig. 3.4.

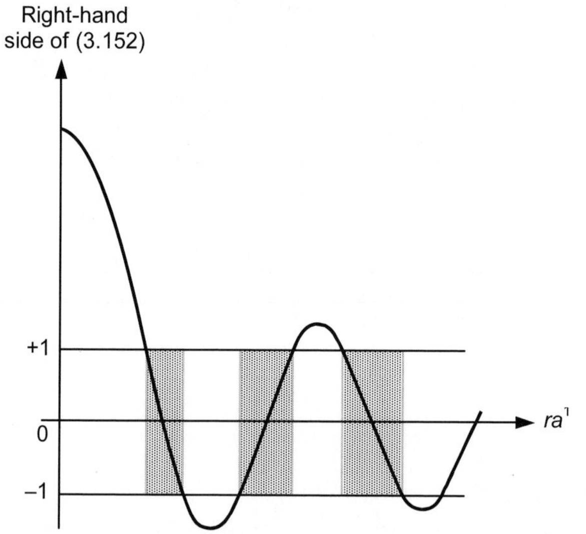

Fig. 3.4. Sketch showing how to get energy bands from the Kronig–Penney model

From (3.146), however, we see we have a solution only when the right-hand side is between +1 and −1 (because these are the bounds of cos ka^1, with real k). Hence the only allowed values of ra^1 are those values in the shaded regions of Fig. 3.4. But by (3.147) this leads to the concept of energy bands.

Detailed numerical analysis of (3.146) and (3.147) will yield a plot similar to Fig. 3.5 for the first band of energies as plotted in the first Brillouin zone. Other bands could be similarly obtained.

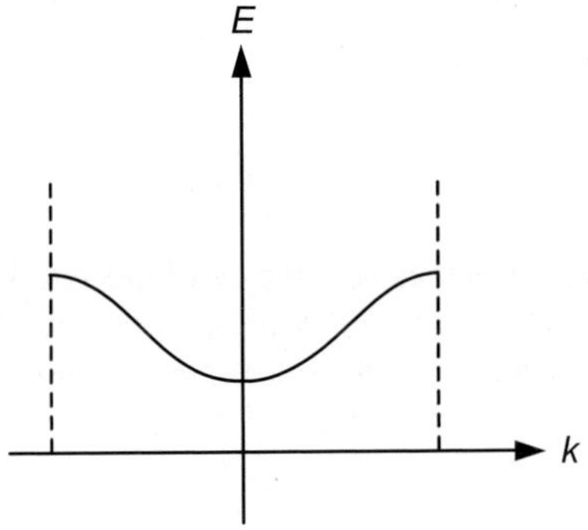

Fig. 3.5. Sketch of the first band of energies in the Kronig–Penney model (an arbitrary $k = 0$ energy is added in)

Figure 3.5 looks somewhat like the plot of the dispersion relation for a one-dimensional lattice vibration. This is no accident. In both cases we have waves propagating through periodic media. There are significant differences that distinguish the dispersion relation for electrons from the dispersion relation for lattice vibrations. For electrons in the lowest band as $k \to 0$, $E \propto k^2$, whereas for phonons we found $E \propto |k|$. Also, for lattice vibrations there is only a finite number of energy bands (equal to the number of atoms per unit cell times 3). For electrons, there are infinitely many bands of allowed electronic energies (however, for realistic models the bands eventually overlap and so form a continuum).

We can easily check the results of the Kronig–Penney model in two limiting cases. To do this, the equation will be rewritten slightly:

$$\cos(ka^1) = \cos(ra^1) + \mu \frac{\sin ra^1}{ra^1} \equiv P(ra^1)\,, \tag{3.148}$$

where

$$\mu \equiv \frac{ma^1(au)}{\hbar^2}\,. \tag{3.149}$$

In the limit as the potential becomes extremely weak, $\mu \to 0$, so that $ka^1 \equiv ra^1$. Using (3.147), one easily sees that the energies are given by

$$E = \frac{\hbar^2 k^2}{2m}. \tag{3.150}$$

Equation (3.150) is just what one would expect. It is the free-particle solution.

In the limit as the potential becomes extremely strong, $\mu \to \infty$, we can have solutions of (3.148) only if $\sin ra^1 = 0$. Thus $ra^1 = n\pi$, where n is an integer, so that the energy is given by

$$E = \frac{n^2 \pi^2 \hbar^2}{2m(a^1)^2}. \tag{3.151}$$

Equation (3.151) is expected as these are the "particle-in-a-box" solutions.

It is also interesting to study how the widths of the energy bands vary with the strength of the potential. From (3.148), the edges of the bands of allowed energy occur when $P(ra^1) = \pm 1$. This can certainly occur when $ra^1 = n\pi$. The other values of ra^1 at the band edges are determined in the argument below. At the band edges,

$$\pm 1 = \cos ra^1 + \frac{\mu}{ra^1} \sin(ra^1).$$

This equation can be recast into the form,

$$0 = 1 + \frac{\mu}{ra^1} \frac{\sin(ra^1)}{\mp 1 + \cos(ra^1)}. \tag{3.152}$$

From trigonometric identities

$$\tan \frac{ra^1}{2} = \frac{\sin(ra^1)}{1 + \cos(ra^1)}, \tag{3.153}$$

and

$$\cot \frac{ra^1}{2} = \frac{\sin(ra^1)}{1 - \cos(ra^1)}. \tag{3.154}$$

Combining the last three equations gives

$$0 = 1 + \frac{\mu}{ra^1} \tan \frac{ra^1}{2} \quad \text{or} \quad 0 = 1 - \frac{\mu}{ra^1} \cot \frac{ra^1}{2},$$

or

$$\tan(ra^1/2) = -(ra^1)/\mu, \quad \cot(ra^1/2) = +(ra^1)/\mu.$$

Since $1/\tan\theta = \cot\theta$, these last two equations can be written

$$\cot(ra^1/2) = -\mu/(ra^1),$$
$$\tan(ra^1/2) = +\mu/(ra^1),$$

or

$$(ra^1/2)\cot(ra^1/2) = -ma^1(au)/2\hbar^2, \quad (3.155)$$

and

$$(ra^1/2)\tan(ra^1/2) = +ma^1(au)/2\hbar^2. \quad (3.156)$$

Figure 3.6 uses $ra^1 = n\pi$, (3.155), and (3.156) (which determine the upper and lower ends of the energy bands) to illustrate the variation of bandwidth with the strength of the potential.

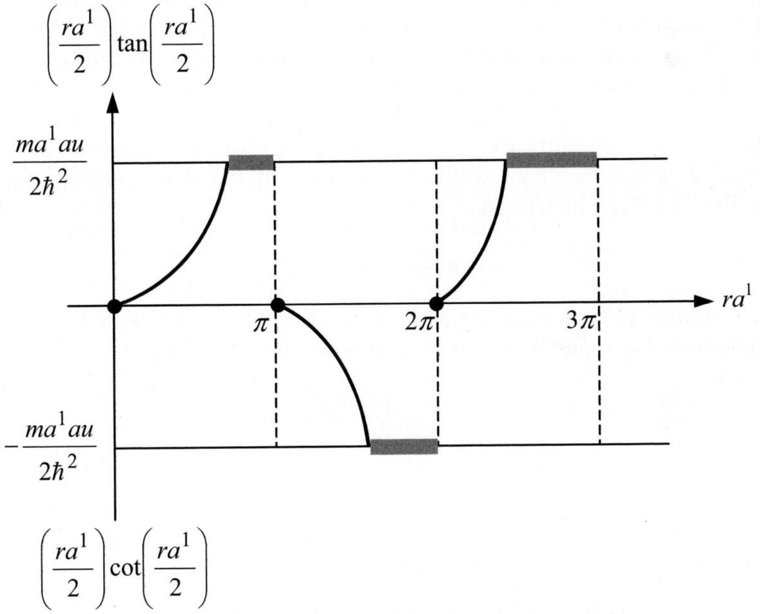

Fig. 3.6. Variation of bandwidth with strength of the potential

Note that increasing u decreases the bandwidth of any given band. For a fixed u, the higher r (or the energy) is, the larger is the bandwidth. By careful analysis it can be shown that the bandwidth increases as a^1 decreases. The fact that the bandwidth increases as the lattice spacing decreases has many important consequences as it is valid in the more important three-dimensional case. For example, Fig. 3.7 sketches the variation of the 3s and 3p bonds for solid sodium. Note that at the equilibrium spacing a_0, the 3s and 3p bands form one continuous band.

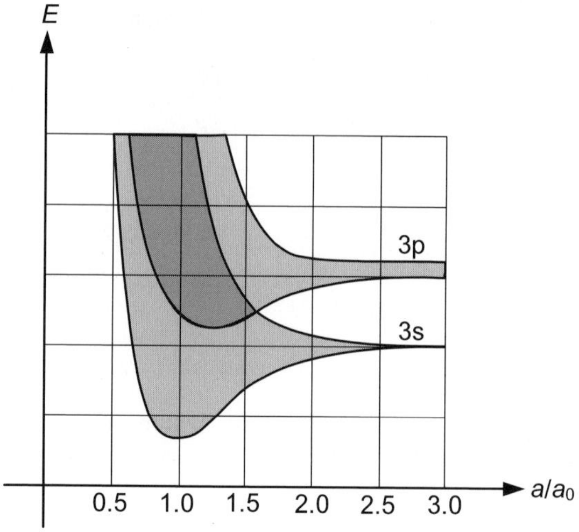

Fig. 3.7. Sketch of variation (with distance between atoms) of bandwidths of Na. Each energy unit represents 2 eV. The equilibrium lattice spacing is a_0. Higher bands such as the 4s and 3d are left out

The concept of the effective mass of an electron is very important. A simple example of it can be given within the context of the Kronig–Penney model. Equation (3.148) can be written as

$$\cos ka^1 = P(ra^1) .$$

Let us examine this equation for small k and for r near r_0 (= r at k = 0). By a Taylor series expansion for both sides of this equation, we have

$$1 - \tfrac{1}{2}(ka^1)^2 = 1 + P_0' a^1 (r - r_0) ,$$

or

$$r_0 - \frac{1}{2}\frac{k^2 a^1}{P_0'} = r .$$

Squaring both sides and neglecting terms in k^4, we have

$$r^2 = r_0^2 - r_0 \frac{k^2 a^1}{P_0'} .$$

Defining an *effective mass* m^* as

$$m^* = -\frac{mP_0'}{r_0 a^1} ,$$

we have by (3.147) that

$$E = \frac{\hbar^2 r^2}{2m} = E_0 + \frac{\hbar^2 k^2}{2m^*}, \tag{3.157}$$

where $E_0 = \hbar^2 r_0^2/2m$. Except for the definition of mass, this equation is just like an equation for a free particle. Thus for small k we may think of m^* as acting as a mass; hence it is called an *effective mass*. For small k, at any rate, we see that the only effect of the periodic potential is to modify the apparent mass of the particle.

The appearances of allowed energy bands for waves propagating in periodic lattices (as exhibited by the Kronig–Penney model) is a general feature. The physical reasons for this phenomenon are fairly easy to find.

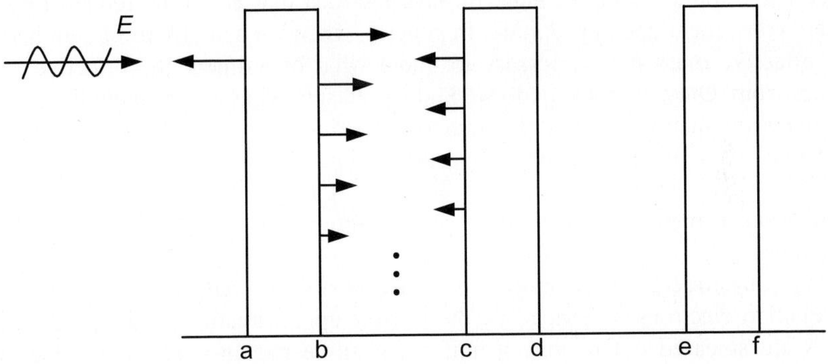

Fig. 3.8. Wave propagating through periodic potential. E is the kinetic energy of the particle with which there is associated a wave with de Broglie wavelength $\lambda = h/(2mE)^{1/2}$ (internal reflections omitted for clarity)

Consider a quantum-mechanical particle moving along with energy E as shown in Fig. 3.8. Associated with the particle is a wave of de Broglie wavelength λ. In regions a–b, c–d, e–f, etc., the potential energy is nonzero. These regions of "hills" in the potential cause the wave to be partially reflected and partially transmitted. After several partial reflections and partial transmissions at a–b, c–d, e–f, etc., it is clear that the situation will be very complex. However, there are two possibilities. The reflections and transmissions may or may not result in destructive interference of the propagating wave. Destructive interference will result in attenuation of the wave. Whether or not we have destructive interference depends clearly on the wavelength of the wave (and of course on the spacings of the "hills" of the potential) and hence on the energy of the particle. Hence we see qualitatively, at any rate, that for some energies the wave will not propagate because of attenuation. This is what we mean by a disallowed band of energy. For other energies, there will be no net attenuation and the wave will propagate. This is what we mean by an allowed band of energy. The Kronig–Penney model calculations were just a way of expressing these qualitative ideas in precise quantum-mechanical form.

3.2.2 The Free-Electron or Quasifree-Electron Approximation (B)

The Kronig–Penney model indicates that for small $|ka^1|$ we can take the periodic nature of the solid into account by using an effective mass rather than an actual mass for the electrons. In fact we can always treat independent electrons in a periodic potential in this way so long as we are interested only in a group of electrons that have energy clustered about minima in an E versus $\boldsymbol{k}$ plot (in general this would lead to a tensor effective mass, but let us restrict ourselves to minima such that $E \propto k^2$ + constant near the minima). Let us agree to call the electrons with effective mass *quasifree electrons*. Perhaps we should also include Landau's ideas here and say that what we mean by quasifree electrons are Landau quasiparticles with an effective mass enhanced by the periodic potential. We will often use m rather than m^*, but will have the idea that m can be replaced by m^* where convenient and appropriate. In general, when we actually use a number for the effective mass it is necessary to quote what experiment the effective mass comes from. Only in this way do we know precisely what we are including. There are many interactions beyond that due to the periodic lattice that can influence the effective mass of an electron. Any sort of interaction is liable to change the effective mass (or "renormalize it"). It is now thought that the electron–phonon interaction in metals can be important in determining the effective mass of the electrons.

The quasifree-electron model is most easily arrived at by treating the conduction electrons in a metal by the Hartree approximation. If the positive ion cores are smeared out to give a uniform positive background charge, then the interaction of the ion cores with the electrons exactly cancels the interactions of the electrons with each other (in the Hartree approximation). We are left with just a one-electron, free-electron Schrödinger equation. Of course, we really need additional ideas (such as discussed in Sect. 3.1.4 and in Sect. 4.4 as well as the introduction of Chap. 4) to see why the electrons can be thought of as rather weakly interacting, as seems to be required by the "uncorrelated" nature of the Hartree approximation. Also, if we smear out the positive ion cores, we may then have a hard time justifying the use of an effective mass for the electrons or indeed the use of a periodic potential. At any rate, before we start examining in detail the effect of a three-dimensional lattice on the motion of electrons in a crystal, it is worthwhile to pursue the quasifree-electron picture to see what can be learned. The picture appears to be useful (with some modifications) to describe the motions of electrons in simple monovalent metals. It is also useful for describing the motion of charge carriers in semiconductors. At worst it can be regarded as a useful phenomenological picture.[11]

Density of States in the Quasifree-Electron Model (B)

Probably the most useful prediction made by the quasifree-electron approximation is a prediction regarding the number of quantum states per unit energy. This

[11] See also Kittel C [59, 60].

quantity is called the density of states. For a quasifree electron with effective mass m^*,

$$-\frac{\hbar^2}{2m^*}\nabla^2\psi = E\psi\,. \tag{3.158}$$

This equation has the solution (normalized in a volume V)

$$\psi = \frac{1}{\sqrt{V}}\exp(\mathrm{i}\boldsymbol{k}\cdot\boldsymbol{r})\,, \tag{3.159}$$

provided that

$$E = \frac{\hbar^2}{2m^*}(k_1^2 + k_2^2 + k_3^2)\,. \tag{3.160}$$

If periodic boundary conditions are applied on a parallelepiped of sides $N_i\boldsymbol{a}_i$ and volume V, then $\boldsymbol{k}$ is of the form

$$\boldsymbol{k} = 2\pi\left(\frac{n_1}{N_1}\boldsymbol{b}_1 + \frac{n_2}{N_2}\boldsymbol{b}_2 + \frac{n_3}{N_3}\boldsymbol{b}_3\right), \tag{3.161}$$

where the n_i are integers and the b_i are the customary reciprocal lattice vectors that are defined from the a_i. (For the case of quasifree electrons, we really do not need the concept of reciprocal lattice, but it is convenient for later purposes to carry it along.) There are thus $N_1N_2N_3$ k-type states in a volume $(2\pi)^3\boldsymbol{b}_1\cdot(\boldsymbol{b}_2\times\boldsymbol{b}_3)$ of $\boldsymbol{k}$ space. Thus the number of states per unit volume of $\boldsymbol{k}$ space is

$$\frac{N_1N_2N_3}{(2\pi)^3\boldsymbol{b}_1\cdot(\boldsymbol{b}_2\times\boldsymbol{b}_3)} = \frac{N_1N_2N_3\Omega_a}{(2\pi)^3} = \frac{V}{(2\pi)^3}\,, \tag{3.162}$$

where $\Omega = \boldsymbol{a}_1\cdot(\boldsymbol{a}_2\times\boldsymbol{a}_3)$. Since the states in $\boldsymbol{k}$ space are uniformly distributed, the number of states per unit volume of real space in d^3k is

$$\mathrm{d}^3k/(2\pi)^3\,. \tag{3.163}$$

If $E = \hbar^2\boldsymbol{k}^2/2m^*$, the number of states with energy less than E (with $\boldsymbol{k}$ defined by this equation) is

$$\frac{4\pi}{3}|\boldsymbol{k}|^3\frac{V}{(2\pi)^3} = \frac{Vk^3}{6\pi^2}\,,$$

where $|\boldsymbol{k}| = k$, of course. Thus, if $N(E)$ is the number of states in E to $E + \mathrm{d}E$, and $N(k)$ is the number of states in k to $k + \mathrm{d}k$, we have

$$N(E)\mathrm{d}E = N(k)\mathrm{d}k = \frac{\mathrm{d}}{\mathrm{d}k}\left(\frac{Vk^3}{6\pi^2}\right)\mathrm{d}k = \frac{Vk^2}{2\pi^2}\mathrm{d}k\,.$$

Table 3.2. Dependence of density of states of free electrons $D(E)$ on dimension and energy E.

	$D(E)$
One Dimension	$A_1 E^{-1/2}$
Two Dimensions	A_2
Three Dimensions	$A_3 E^{1/2}$

Note that the A_i are constants, and in all cases the dispersion relation is of the form $E_{\boldsymbol{k}} = \hbar^2 k^2/(2m^*)$.

But

$$\mathrm{d}E = \frac{\hbar^2}{m^*} k\mathrm{d}k, \quad \text{so} \quad \mathrm{d}k = \frac{m^*}{\hbar^2}\frac{\mathrm{d}E}{k},$$

or

$$N(E)\mathrm{d}E = \frac{V}{2\pi^2}\sqrt{\frac{2m^* E}{\hbar^2}}\frac{m^*}{\hbar^2}\mathrm{d}E,$$

or

$$N(E)\mathrm{d}E = \frac{V}{4\pi^2}\left(\frac{2m^*}{\hbar^2}\right)^{3/2} E^{1/2}\mathrm{d}E. \tag{3.164}$$

Equation (3.164) is the basic equation for the density of states in the quasifree-electron approximation. If we include spin, there are two spin states for each $\boldsymbol{k}$, so (3.164) must be multiplied by 2.

Equation (3.164) is most often used with *Fermi–Dirac* statistics. The Fermi function $f(E)$ tells us the average number of electrons per state at a given temperature, $0 \leq f(E) \leq 1$. With Fermi–Dirac statistics, the number of electrons per unit volume with energy between E and $E + \mathrm{d}E$ and at temperature T is

$$\mathrm{d}n = f(E)K\sqrt{E}\mathrm{d}E = \frac{K\sqrt{E}\mathrm{d}E}{\exp[(E-E_F)/kT]+1}, \tag{3.165}$$

where $K = (1/2\pi^2)(2m^*/\hbar^2)^{3/2}$ and E_F is the Fermi energy.

If there are N electrons per unit volume, then E_F is determined from

$$N = \int_0^\infty K\sqrt{E}f(E)\mathrm{d}E. \tag{3.166}$$

Once the Fermi energy E_F is obtained, the mean energy of an electron gas is determined from

$$E = \int_0^\infty K f(E) \sqrt{E} E \mathrm{d}E . \tag{3.167}$$

We shall find (3.166) and (3.167) particularly useful in the next Section where we evaluate the specific heat of an electron gas. We summarize the density of states for free electrons in one, two, and three dimensions in Table 3.2.

Specific Heat of an Electron Gas (B)

This Section and the next one follow the early ground-breaking work of Pauli and Sommerfeld. In this Section all we have to do is to find the Fermi energy from (3.166), perform the indicated integral in (3.167), and then take the temperature derivative. However, to perform these operations exactly is impossible in closed form and so it is useful to develop an approximate way of evaluating the integrals in (3.166) and (3.167). The approximation we will use will be an excellent approximation for metals at all ordinary temperatures.

We first develop a general formula (the Sommerfeld expansion) for the evaluation of integrals of the needed form for "low" temperatures (room temperature qualifies as a very low temperature for the approximation that we will use).

Let $f(E)$ be the Fermi distribution function, and $R(E)$ be a function that vanishes when E vanishes. Define

$$S = +\int_0^\infty f(E) \frac{\mathrm{d}R(E)}{\mathrm{d}E} \mathrm{d}E \tag{3.168}$$

$$= -\int_0^\infty R(E) \frac{\mathrm{d}f(E)}{\mathrm{d}E} \mathrm{d}E. \tag{3.169}$$

At low temperature, $f'(E)$ has an appreciable value only where E is near the Fermi energy E_F. Thus we make a Taylor series expansion of $R(E)$ about the Fermi energy:

$$R(E) = R(E_F) + (E - E_F) R'(E_F) + \tfrac{1}{2}(E - E_F)^2 R''(E_F) + \cdots . \tag{3.170}$$

In (3.170) $R''(E_F)$ means

$$\left(\frac{\mathrm{d}^2 R(E)}{\mathrm{d}E^2} \right)_{E=E_F} .$$

Combining (3.169) and (3.170), we can write

$$S \cong aR(E_F) + bR'(E_F) + cR''(E_F) , \tag{3.171}$$

where

$$a = -\int_0^\infty f'(E)\mathrm{d}E = 1,$$

$$b = -\int_0^\infty (E - E_F) f'(E)\mathrm{d}E = 0,$$

$$c = -\frac{1}{2}\int_0^\infty (E - E_F)^2 f'(E)\mathrm{d}E \cong \frac{kT^2}{2}\int_{-\infty}^\infty \frac{x^2 \mathrm{e}^x dx}{(\mathrm{e}^x + 1)^2} = \frac{\pi^2}{6}(kT)^2.$$

Thus we can write

$$\int_0^\infty f(E)\frac{\mathrm{d}R(E)}{\mathrm{d}E}\mathrm{d}E = R(E_F) + \frac{\pi^2}{6}(kT)^2 R''(E_F) + \cdots. \tag{3.172}$$

By (3.166),

$$N = \int_0^\infty K\frac{\mathrm{d}}{\mathrm{d}E}\frac{2}{3}E^{3/2} f(E)\mathrm{d}E \cong \frac{2}{3}KE_F^{3/2} + \frac{\pi^2}{6}(kT)^2\frac{K}{2}\frac{1}{\sqrt{E_F}}. \tag{3.173}$$

At absolute zero temperature, the Fermi function $f(E)$ is 1 for $0 \le E \le E_F(0)$ and zero otherwise. Therefore we can also write

$$N = \int_0^{E_F(0)} KE^{1/2}\mathrm{d}E = \frac{2}{3}K[E_F(0)]^{3/2}. \tag{3.174}$$

Equating (3.173) and (3.174), we obtain

$$[E_F(0)]^{3/2} \cong E_F^{3/2} + \frac{\pi^2}{8}\frac{(kT)^2}{\sqrt{E_F}}.$$

Since the second term is a small correction to the first, we can let $E_F = E_F(0)$ in the second term:

$$[E_F(0)]^{3/2}\left[1 - \frac{\pi^2}{8}\frac{(kT)^2}{[E_F(0)]^2}\right] \cong E_F^{3/2}.$$

Again, since the second term is a small correction to the first term, we can use $(1 - \varepsilon)^{3/2} \cong 1 - 3/2\varepsilon$ to obtain

$$E_F = E_F(0)\left\{1 - \frac{\pi^2}{12}\left[\frac{kT}{E_F(0)}\right]^2\right\}. \tag{3.175}$$

For all temperatures that are normally of interest, (3.175) is a good approximation for the variation of the Fermi energy with temperature. We shall need this expression in our calculation of the specific heat.

The mean energy $\bar{E}$ is given by (3.167) or

$$\bar{E} = \int_0^\infty f(E)\frac{d}{dE}[\tfrac{2}{5}K(E)^{5/2}]dE \cong \frac{2K}{5}E_F^{5/2} + \frac{\pi^2}{6}(kT)^2\frac{3K}{2}\sqrt{E_F}\,. \quad (3.176)$$

Combining (3.176) and (3.175), we obtain

$$\bar{E} \cong \frac{2K}{5}[E_F(0)]^{5/2} + [E_F(0)]^{5/2}\frac{\pi^2}{6}K\left[\frac{kT}{E_F(0)}\right]^2 .$$

The specific heat of the electron gas is then the temperature derivative of $\bar{E}$:

$$C_V = \frac{\partial \bar{E}}{\partial T} = \frac{\pi^2}{3}k^2 K\sqrt{E_F(0)}\,T .$$

This is commonly written as

$$C_V = \gamma T , \quad (3.177)$$

where

$$\gamma = \frac{\pi^2}{3}k^2 K\sqrt{E_F(0)} . \quad (3.178)$$

There are more convenient forms for γ. From (3.174),

$$K = \tfrac{3}{2}N[E_F(0)]^{-3/2} ,$$

so that

$$\gamma = \frac{\pi^2}{2}Nk\frac{k}{E_F(0)} .$$

The Fermi temperature T_F is defined as $T_F = E_F(0)/k$ so that

$$\gamma \cong \frac{\pi^2}{2}\frac{Nk}{T_F} . \quad (3.179)$$

The expansions for $\bar{E}$ and E_F are expansions in powers of $kT/E_F(0)$. Clearly our results (such as (3.177)) are valid only when $kT \ll E_F(0)$. But as we already mentioned, this does not limit us to very low temperatures. If 1/40 eV corresponds to 300° K, then $E_F(0) \cong 1$ eV (as for metals) corresponds to approximately 12 000° K. So for temperatures well below 12 000° K, our results are certainly valid.

A similar calculation for the specific heat of a free electron gas using Hartree–Fock theory yields $C_v \propto (T/\ln T)$, which is not even qualitatively correct. This shows that Coulomb correlations really do have some importance, and our free-electron theory does well only because the errors (involved in neglecting both Coulomb corrections and exchange) approximately cancel.

Pauli Spin Paramagnetism (B)

The quasifree electrons in metals show both a paramagnetic and diamagnetic effect. Paramagnetism is a fairly weak induced magnetization in the direction of the applied field. Diamagnetism is a very weak induced magnetization opposite the direction of the applied field. The paramagnetism of quasifree electrons is called *Pauli spin paramagnetism*. This phenomenon will be discussed now because it is a simple application of Fermi–Dirac statistics to electrons.

For Pauli spin paramagnetism we must consider the effect of an external magnetic field on the spins and hence magnetic moments of the electrons. If the magnetic moment of an electron is parallel to the magnetic field, the energy of the electron is lowered by the magnetic field. If the magnetic moment of the electron is in the opposite direction to the magnetic field, the energy of the electron is raised by the magnetic field. In equilibrium at absolute zero, all of the electrons are in as low an energy state as they can get into without violating the Pauli principle. Consequently, in the presence of the magnetic field there will be more electrons with magnetic moment parallel to the magnetic field than antiparallel. In other words there will be a net magnetization of the electrons in the presence of a magnetic field. The idea is illustrated in Fig. 3.9, where μ is the magnetic moment of the electron and H is the magnetic field.

Using (3.165), Fig. 3.9, and the definition of magnetization, we see that for absolute zero and for a small magnetic field the net magnetization is given approximately by

$$M = \tfrac{1}{2} K \sqrt{E_F(0)}\, 2\mu^2 \mu_0 H \,. \tag{3.180}$$

The factor of 1/2 arises because D_a and D_p (in Fig. 3.9) refer only to half the total number of electrons. In (3.180), K is given by $(1/2\pi^2)(2m^*/\hbar^2)^{3/2}$.

Equations (3.180) and (3.174) give the following results for the magnetic susceptibility:

$$\chi = \frac{\partial M}{\partial H} = \mu_0 \mu^2 \sqrt{E_F(0)}\, \frac{3N}{2} [E_F(0)]^{-3/2} = \frac{3N\mu_0\mu^2}{2E_F(0)} \,,$$

or, if we substitute for E_F,

$$\chi = \frac{3N\mu_0\mu^2}{2kT_F(0)} \,. \tag{3.181}$$

This result was derived for absolute zero, it is fairly good for all $T \ll T_F(0)$. The only trouble with the result is that it is hard to compare to experiment. Experiment measures the total magnetic susceptibility. Thus the above must be corrected for the diamagnetism of the ion cores and the diamagnetism of the conduction electrons if it is to be compared to experiment. Better agreement with experiment is obtained if we use an appropriate effective mass, in the evaluation of $T_F(0)$, and if we try to make some corrections for exchange and Coulomb correlation.

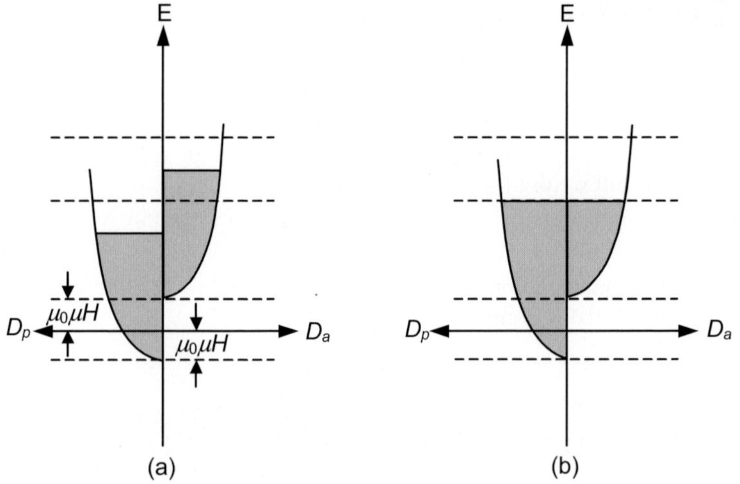

Fig. 3.9. A magnetic field is applied to a free-electron gas. (a) Instantaneous situation, and (b) equilibrium situation. Both (a) and (b) are at absolute zero. D_p is the density of states of parallel (magnetic moment parallel to field) electrons. D_a is the density of states of antiparallel electrons. The shaded areas indicate occupied states

Landau Diamagnetism (B)

It has already been mentioned that quasifree electrons show a diamagnetic effect. This diamagnetic effect is referred to as *Landau diamagnetism*. This Section will not be a complete discussion of Landau diamagnetism. The main part will be devoted to solving exactly the quantum-mechanical problem of a free electron moving in a region in which there is a constant magnetic field. We will find that this situation yields a particularly simple set of energy levels. Standard statistical-mechanical calculations can then be made, and it is from these calculations that a prediction of the magnetic susceptibility of the electron gas can be made. The statistical-mechanical analysis is rather complicated, and it will only be outlined. The analysis here is also closely related to the analysis of the de Haas–van Alphen effect (oscillations of magnetic susceptibility in a magnetic field). The de Haas–van Alphen effect will be discussed in Chap. 5. This Section is also related to the quantum Hall effect, see Sect. 12.7.2. In SI units, neglecting spin effects, the Hamiltonian of an electron in a constant magnetic field described by a vector potential $\boldsymbol{A}$ is (here $e > 0$)

$$\mathcal{H} = \frac{1}{2m}(\boldsymbol{p} + e\boldsymbol{A})^2 = -\frac{\hbar^2}{2m}\nabla^2 + \frac{e\hbar}{2mi}\nabla\cdot\boldsymbol{A} + \frac{e\hbar}{2mi}\boldsymbol{A}\cdot\nabla + \frac{e^2}{2m}\boldsymbol{A}^2. \quad (3.182)$$

Using $\nabla\cdot(\boldsymbol{A}\psi) = \boldsymbol{A}\cdot\nabla\psi + \psi\nabla\cdot\boldsymbol{A}$, we can formally write the Hamiltonian as

$$\mathcal{H} = -\frac{\hbar^2}{2m}\nabla^2 + \frac{e\hbar}{2mi}\nabla\cdot\boldsymbol{A} + \frac{e\hbar}{mi}\boldsymbol{A}\cdot\nabla + \frac{e^2}{2m}\boldsymbol{A}^2\,. \quad (3.183)$$

A constant magnetic field in the z direction is described by the nonunique vector potential

$$\boldsymbol{A} = -\frac{\mu_0 H y}{2}\hat{\boldsymbol{i}} + \frac{\mu_0 H x}{2}\hat{\boldsymbol{j}}. \tag{3.184}$$

To check this result we use the defining relation

$$\mu_0 \boldsymbol{H} = \nabla \times \boldsymbol{A}. \tag{3.185}$$

and after a little manipulation it is clear that (3.184) and (3.185) imply $H = H\hat{\boldsymbol{k}}$. It is also easy to see that $\boldsymbol{A}$ defined by (3.184) implies

$$\nabla \cdot \boldsymbol{A} = 0, \tag{3.186}$$

Combining (3.183), (3.184), and (3.186), we find that the Hamiltonian for an electron in a constant magnetic field is given by

$$\mathcal{H} = -\frac{\hbar^2}{2m}\nabla^2 + \frac{e\hbar\mu_0 H}{2mi}\left(x\frac{\partial}{\partial y} - y\frac{\partial}{\partial x}\right) + \frac{e^2\mu_0^2 H^2}{8m}(x^2 + y^2). \tag{3.187}$$

It is perhaps worth pointing out that (3.187) plus a central potential is a Hamiltonian often used for atoms. In the atomic case, the term $(x\,\partial/\partial y - y\,\partial/\partial x)$ gives rise to paramagnetism (orbital), while the term $(x^2 + y^2)$ gives rise to diamagnetism. For free electrons, however, we will retain both terms as it is possible to obtain an exact energy eigenvalue spectrum of (3.187).

The exact energy eigenvalue spectrum of (3.187) can readily be found by making three transformations. The first transformation that it is convenient to make is

$$\psi(x, y, z) = \phi(x, y, z)\exp\left(\frac{\mathrm{i}e\mu_0 H}{2}\frac{xy}{\hbar}\right). \tag{3.188}$$

Substituting (3.188) into $\mathcal{H}\psi = E\psi$ with $\mathcal{H}$ given by (3.187), we see that ϕ satisfies the differential equation

$$-\frac{\hbar^2}{2m}\nabla^2\phi - \frac{e\hbar\mu_0 H}{im}x\frac{\partial\phi}{\partial y} + \frac{H^2\mu_0^2 e^2}{2m}x^2\phi = E\phi. \tag{3.189}$$

A further transformation is suggested by the fact that the effective Hamiltonian of (3.189) does not involve y or z so p_y and p_z are conserved:

$$\phi(x, y, z) = F(x)\exp[-\mathrm{i}(k_y y + k_z z)]. \tag{3.190}$$

This transformation reduces the differential equation to

$$\frac{\mathrm{d}^2 F}{\mathrm{d}x^2} + (A + Bx)^2 F = CF, \tag{3.191}$$

or more explicitly

$$-\frac{\hbar^2}{2m}\frac{\mathrm{d}^2F}{\mathrm{d}x^2}+\frac{1}{2m}[\hbar k_y-(H\mu_0)(ex)]^2F=\left(E-\frac{\hbar^2k_z^2}{2m}\right)F\,. \tag{3.192}$$

Finally, if we make a transformation of the dependent variable x,

$$x^1=x-\frac{\hbar k_y}{eH\mu_0}, \tag{3.193}$$

then we find

$$-\frac{\hbar^2}{2m}\frac{\mathrm{d}^2F}{\mathrm{d}(x^1)^2}+\frac{e^2H^2\mu_0^2}{2m}(x^1)^2F=\left(E-\frac{\hbar^2k_z^2}{2m}\right)F\,. \tag{3.194}$$

Equation (3.194) is the equation of a harmonic oscillator. Thus the allowed energy eigenvalues are

$$E_{n,k_z}=\frac{\hbar^2k_z^2}{2m}+\hbar\omega_c(n+\tfrac{1}{2}), \tag{3.195}$$

where n is an integer and

$$\omega_c\equiv\left|\frac{eH\mu_0}{m}\right| \tag{3.196}$$

is just the *cyclotron* frequency.

This quantum-mechanical result can be given quite a simple classical meaning. We think of the electron as describing a helix about the magnetic field. The helical motion comes from the fact that, in general, the electron may have a velocity parallel to the magnetic field (which velocity is unaffected by the magnetic field) in addition to the component of velocity that is perpendicular to the magnetic field. The linear motion has the kinetic energy $p^2/2m = \hbar^2k_z^2/2m$, while the circular motion is quantized and is mathematically described by harmonic oscillator wave functions.

It is at this stage that the rather complex statistical-mechanical analysis must be made. Landau diamagnetism for electrons in a periodic lattice requires a still more complicated analysis. The general method is to compute the free energy and concentrate on the terms that are monotonic in H. Then thermodynamics tells us how to relate the free energy to the magnetic susceptibility.

A beginning is made by calculating the partition function for a canonical ensemble,

$$Z=\textstyle\sum_i\exp(-E_i\,/\,kT), \tag{3.197}$$

where E_i is the energy of the whole system in state i, and i may represent several quantum numbers. (Proper account of the Pauli principle must be taken in calculating E_i from (3.195).) The Helmholtz free energy F is then obtained from

$$F = -kT \ln Z\,, \tag{3.198}$$

and from this the magnetization is determined:

$$M = \frac{\partial F}{\mu_0 \partial H}\,. \tag{3.199}$$

Finally the magnetic susceptibility is determined from

$$\chi = \left(\frac{\partial M}{\partial H}\right)_{H=0}\,. \tag{3.200}$$

The approximate result obtained for free electrons is

$$\chi_{Landau} = -\tfrac{1}{3}\chi_{Pauli} = -N\mu_0\mu^2 / 2kT_F\,. \tag{3.201}$$

Physically, Landau diamagnetism (negative χ) arises because the coalescing of energy levels (described by (3.195)) increases the total energy of the system. Fermi–Dirac statistics play an essential role in making the average energy increase. Seitz [82] is a basic reference for this section.

Soft X-ray Emission Spectra (B)

So far we have discussed the concept of density of states but we have given no direct experimental way of measuring this concept for the quasifree electrons. Soft X-ray emission spectra give a way of measuring the density of states. They are even more directly related to the concept of the bandwidth. If a metal is exposed to a beam of electrons, electrons may be knocked out of the inner or bound levels. The conduction-band electrons tend to drop into the inner or bound levels and they emit an X-ray photon in the process. If E_1 is the energy of a conduction-band electron and E_2 is the energy of a bound level, the conduction-band electron emits a photon of angular frequency

$$\omega = (E_1 - E_2)/\hbar\,.$$

Because these X-ray photons have, in general, low frequency compared to other X-rays, they are called soft X-rays. Compare Fig. 3.10. The conduction-band width is determined by the spread in frequency of all the X-rays. The intensities of the X-rays for the various frequencies are (at least approximately) proportional to the density of states in the conduction band. It should be mentioned that the measured bandwidths so obtained are only the width of the occupied portion of the band. This may be less than the actual bandwidth.

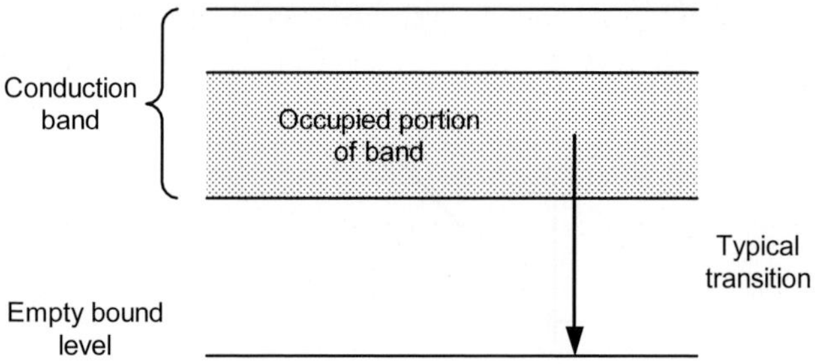

Fig. 3.10. Soft X-ray emission

The results of some soft X-ray measurements have been compared with Hartree calculations.[12] Hartree–Fock theory does not yield nearly so accurate agreement unless one somehow fixes the omission of Coulomb correlation. With the advent of synchrotron radiation, soft X-rays have found application in a wide variety of areas. See Smith [3.51].

The Wiedeman–Franz Law (B)

This law applies to metals where the main carriers of both heat and charge are electrons. It states that the thermal conductivity is proportional to the electrical conductivity times the absolute temperature. Good conductors seem to obey this law quite well if the temperature is not too low.

The straightforward way to derive this law is to derive simple expressions for the electrical and thermal conductivity of quasifree electrons, and to divide the two expressions. Simple expressions may be obtained by kinetic theory arguments that treat the electrons as classical particles. The thermal conductivity will be derived first.

Suppose one has a homogeneous rod in which there is a temperature gradient of $\partial T/\partial z$ along its length. Suppose $\dot{Q}$ units of energy cross any cross-sectional area (perpendicular to the axis of the rod) of the rod per unit area per unit time. Then the thermal *conductivity* k of the rod is defined as

$$k = \left| \frac{\dot{Q}}{\partial T / \partial z} \right| . \tag{3.202}$$

Figure 3.11 sets the notation for our calculation of the thermal conductivity.

[12] See Raimes [3.42, Table I, p 190].

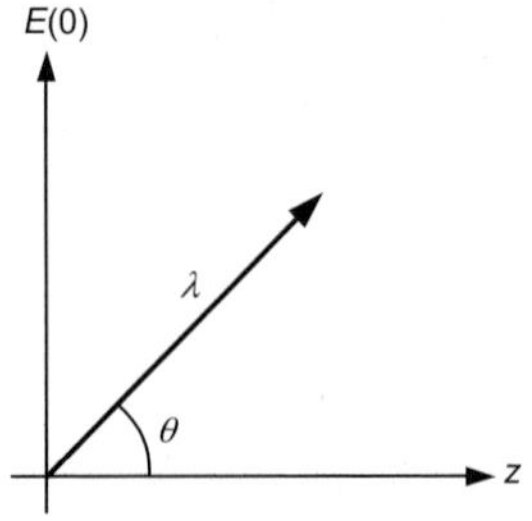

Fig. 3.11. Picture used for a simple kinetic theory calculation of the thermal conductivity. $E(0)$ is the mean energy of an electron in the (x,y)-plane, and λ is the mean free path of an electron. A temperature gradient exists in the z direction

If an electron travels a distance equal to the mean free path λ after leaving the (x,y)-plane at an angle θ, then it has a mean energy

$$E(0) + \lambda \cos\theta \frac{\partial E}{\partial z} . \tag{3.203}$$

Note that θ going from 0 to π takes care of both forward and backward motion. If N is the number of electrons per unit volume and u is their average velocity, then the number of electrons that cross unit area of the (x,y)-plane in unit time and that make an angle between θ and $\theta + \mathrm{d}\theta$ with the z-axis is

$$\frac{2\pi \sin\theta \mathrm{d}\theta}{4\pi} Nu\cos\theta = \tfrac{1}{2} Nu \cos\theta \sin\theta \mathrm{d}\theta . \tag{3.204}$$

From (3.203) and (3.204) it can be seen that the net energy flux is

$$\begin{aligned}
\dot{Q} = \left| k \frac{\partial T}{\partial z} \right| &= \tfrac{1}{2} Nu \int_0^{\pi} \cos\theta \sin\theta \left(E(0) + \lambda \cos\theta \frac{\partial E}{\partial z} \right) \mathrm{d}\theta \\
&= \tfrac{1}{2} Nu \int_0^{\pi} \lambda \cos^2\theta \sin\theta \frac{\partial E}{\partial z} \mathrm{d}\theta \\
&= \tfrac{1}{3} Nu\lambda \frac{\partial E}{\partial z} = \tfrac{1}{3} Nu\lambda \frac{\partial E}{\partial T} \frac{\partial T}{\partial z} ,
\end{aligned}$$

but since the heat capacity is $C = N(\partial E/\partial T)$, we can write the thermal conductivity as

$$k = \tfrac{1}{3} Cu\lambda . \tag{3.205}$$

Equation (3.205) is a basic equation for the thermal conductivity. Fermi–Dirac statistics can somewhat belatedly be put in by letting $u \to u_{\mathrm{F}}$ (the Fermi velocity) where

$$\tfrac{1}{2} m u_F^2 = kT_F , \tag{3.206}$$

and by using the correct (by Fermi–Dirac statistics) expression for the heat capacity,

$$C = \frac{\pi^2 N k^2 T}{m u_F^2} . \qquad (3.207)$$

It is also convenient to define a relaxation time τ:

$$\tau \equiv \lambda / u_F . \qquad (3.208)$$

The expression for the thermal conductivity of an electron gas is then

$$k = \frac{\pi^2}{3} \frac{N k^2 \tau T}{m} . \qquad (3.209)$$

If we replace m by a suitable m^* in (3.209), then (3.209) would probably give more reliable results.

An expression is also needed for the electrical conductivity of a gas of electrons. We follow here essentially the classical Drude–Lorentz theory. If v_i is the velocity of electron i, we define the average drift velocity of N electrons to be

$$\bar{v} = \frac{1}{N} \sum_{i=1}^{N} v_i . \qquad (3.210)$$

If τ is the relaxation time for the electrons (or the mean time between collisions) and a constant external field E is applied to the gas of the electrons, then the equation of motion of the drift velocity is

$$m \frac{d\bar{v}}{dt} + \frac{\bar{v}}{\tau} = -eE . \qquad (3.211)$$

The steady-state solution of (3.211) is

$$\bar{v} = -e\tau E / m . \qquad (3.212)$$

Thus the electric current density j is given by

$$j = -Ne\bar{v} = Ne^2 (\tau / m) E . \qquad (3.213)$$

Therefore, the electrical conductivity is given by

$$\sigma = Ne^2 \tau / m . \qquad (3.214)$$

Equation (3.214) is a basic equation for the electrical conductivity. Again, (3.214) agrees with experiment more closely if m is replaced by a suitable m^*.

Dividing (3.209) by (3.214), we obtain the law of Wiedeman and Franz:

$$\frac{k}{\sigma} = \frac{\pi^2}{3} \left(\frac{k}{e} \right)^2 T = LT , \qquad (3.215)$$

where L is by definition the *Lorentz number* and has a value of 2.45×10^{-8} $\mathrm{w}\cdot\Omega\cdot\mathrm{K}^{-2}$. At room temperature, most metals do obey (3.215); however, the experimental value of $k/\sigma T$ may easily differ from L by 20% or so. Of course, we should not be surprised as, for example, our derivation assumed that the relaxation times for both electrical and thermal conductivity were the same. This perhaps is a reasonable first approximation when electrons are the main carriers of both heat and electricity. However, it clearly is not good when the phonons carry an appreciable portion of the thermal energy.

We might also note in the derivation of the Wiedeman–Franz law that the electrons are treated as partly classical and more or less noninteracting, but it is absolutely essential to assume that the electrons collide with something. Without this assumption, $\tau \to \infty$ and our equations obviously make no sense. We also see why the Wiedeman–Franz law may be good even though the expressions for k and σ were only qualitative. The phenomenological and unknown τ simply cancelled out on division. For further discussion of the conditions for the validity of Weideman–Franz law see Berman [3.4].

There are several other applications of the quasifree electron model as it is often used in some metals and semiconductors. Some of these will be treated in later chapters. These include thermionic and cold field electron emission (Chap. 11), the plasma edge and transparency of metals in the ultraviolet (Chap. 10), and the Hall effect (Chap. 6).

Angle-resolved Photoemission Spectroscopy (ARPES) (B)

Starting with Spicer [3.52], a very effective technique for learning about band structure has been developed by looking at the angular dependence of the photoelectric effect. When light of suitable wavelength impinges on a metal, electrons are emitted and this is the photoelectric effect. Einstein explained this by saying the light consisted of quanta called photons of energy $E = \hbar\omega$ where ω is the frequency. For emission of electrons the light has to be above a cutoff frequency, in order that the electrons have sufficient energy to surmount the energy barrier at the surface.

The idea of angle-resolved photoemission is based on the fact that the component of the electron's wave vector k parallel to the surface is conserved in the emission process. Thus there are three conserved quantities in this process: the two components of k parallel to the surface, and the total energy. Various experimental techniques are then used to unravel the energy band structure for the band in which the electron originally resided (say the valence band $E_v(k)$). One technique considers photoemission from differently oriented surfaces. Another uses high enough photon energies that the final state of the electron is free-electron like. If one assumes high energies so there is ballistic transport near the surface then k perpendicular to the surface is also conserved. Energy conservation and experiment will then yield both k perpendicular and $E_v(k)$, and k parallel to the surface can also by obtained from experiment—thus $E_v(k)$ is obtained. In most cases, the photon momentum can be neglected compared to the electron's $\hbar \boldsymbol{k}$.[13]

3.2.3 The Problem of One Electron in a Three-Dimensional Periodic Potential

There are two easy problems in this Section and one difficult problem. The easy problems are the limiting cases where the periodic potential is very strong or where it is very weak. When the periodic potential is very weak, we can treat it as a perturbation and we say we have the nearly free-electron approximation. When the periodic potential is very strong, each electron is almost bound to a minimum in the potential and so one can think of the rest of the lattice as being a perturbation on what is going on in this minimum. This is known as the *tight binding approximation*. For the interesting bands in most real solids neither of these methods is adequate. In this intermediate range we must use much more complex methods such as, for example, orthogonalized plane wave (OPW), augmented plane wave (APW), or in recent years more sophisticated methods. Many methods are applicable only at high symmetry points in the Brillouin zone. For other places we must use more sophisticated methods or some sort of interpolation procedure. Thus this Section breaks down to discussing easy limiting cases, harder realistic cases, and interpolation methods.

Metals, Insulators, and Semiconductors (B)

From the band structure and the number of electrons filling the bands, one can predict the type of material one has. If the highest filled band is full of electrons and there is a sizeable gap (3 eV or so) to the next band, then one has an insulator. Semiconductors result in the same way except the bandgap is smaller (1 eV or so). When the highest band is only partially filled, one has a metal. There are other issues, however. Band overlapping can complicate matters and cause elements to form metals, as can the Mott transition (*qv*) due to electron–electron interactions. The simple picture of solids with noninteracting electrons in a periodic potential was exhaustively considered by Bloch and Wilson [97].

The Easy Limiting Cases in Band Structure Calculations (B)

The Nearly Free-Electron Approximation (B) Except for the one-dimensional calculation, we have not yet considered the effects of the lattice structure. Obviously, the smeared out positive ion core approximation is rather poor, and the free-electron model does not explain all experiments. In this section, the effects of the periodic potential are considered as a perturbation. As in the one-dimensional Kronig–Penny calculation, it will be found that a periodic potential has the effect of splitting the allowed energies into bands. It might be thought that the nearly

[13] A longer discussion is given by Marder [3.34 footnote 3, p. 654].

free-electron approximation would have little validity. In recent years, by the method of pseudopotentials, it has been shown that the assumptions of the nearly free-electron model make more sense than one might suppose.

In this Section it will be assumed that a one-electron approximation (such as the Hartree approximation) is valid. The equation that must be solved is

$$\left[-\frac{\hbar^2}{2m}\nabla^2 + V(\boldsymbol{r})\right]\psi_{\boldsymbol{k}}(\boldsymbol{r}) = E_{\boldsymbol{k}}\psi_{\boldsymbol{k}}(\boldsymbol{r}). \tag{3.216}$$

Let $\boldsymbol{R}$ be any direct lattice vector that connects equivalent points in two unit cells. Since $V(\boldsymbol{r}) = V(\boldsymbol{r} + \boldsymbol{R})$, we know by Bloch's theorem that we can always choose the wave functions to be of the form

$$\psi_{\boldsymbol{k}}(\boldsymbol{r}) = \mathrm{e}^{\mathrm{i}\boldsymbol{k}\cdot\boldsymbol{r}}U_{\boldsymbol{k}}(\boldsymbol{r}),$$

where $U_k(\boldsymbol{r}) = U_k(\boldsymbol{r} + \boldsymbol{R})$.

Since both U_k and V have the fundamental translational symmetry of the crystal, we can make a Fourier analysis [71] of them in the form

$$V(\boldsymbol{r}) = \sum\nolimits_{\boldsymbol{K}} V(\boldsymbol{K})\mathrm{e}^{\mathrm{i}\boldsymbol{K}\cdot\boldsymbol{r}} \tag{3.217}$$

$$U_{\boldsymbol{k}}(\boldsymbol{r}) = \sum\nolimits_{\boldsymbol{K}} U(\boldsymbol{K})\mathrm{e}^{\mathrm{i}\boldsymbol{K}\cdot\boldsymbol{r}}. \tag{3.218}$$

In the above equations, the sum over $\boldsymbol{K}$ means to sum over all the lattice points in the reciprocal lattice. Substituting (3.217) and (3.218) into (3.216) with the Bloch condition on the wave function, we find that

$$\begin{aligned}\frac{\hbar^2}{2m}\sum\nolimits_{\boldsymbol{K}} U(\boldsymbol{K})|\boldsymbol{k}+\boldsymbol{K}|^2\,\mathrm{e}^{\mathrm{i}\boldsymbol{K}\cdot\boldsymbol{r}} + \sum\nolimits_{\boldsymbol{K}^1,\boldsymbol{K}^{11}} V(\boldsymbol{K}^1)U(\boldsymbol{K}^{11})\mathrm{e}^{\mathrm{i}(\boldsymbol{K}^1+\boldsymbol{K}^{11})\cdot\boldsymbol{r}} \\ = E_{\boldsymbol{k}}\sum\nolimits_{\boldsymbol{K}} U(\boldsymbol{K})\mathrm{e}^{\mathrm{i}\boldsymbol{K}\cdot\boldsymbol{r}}.\end{aligned} \tag{3.219}$$

By equating the coefficients of $\mathrm{e}^{\mathrm{i}\boldsymbol{K}\cdot\boldsymbol{r}}$, we find that

$$\left(\frac{\hbar^2}{2m}|\boldsymbol{k}+\boldsymbol{K}|^2 - E_{\boldsymbol{k}}\right)U(\boldsymbol{K}) = -\sum\nolimits_{\boldsymbol{K}^1} V(\boldsymbol{K}^1)U(\boldsymbol{K}-\boldsymbol{K}^1). \tag{3.220}$$

If we had a constant potential, then all $V(\boldsymbol{K})$ with $K \neq 0$ would equal zero. Thus it makes sense to assume in the nearly free-electron approximation (in other words in the approximation that the potential is almost constant) that $V(\boldsymbol{K}) \ll V(0)$. As we will see, this also implies that $U(\boldsymbol{K}) \ll U(0)$.

Therefore (3.220) can be approximately written

$$\left[E_{\boldsymbol{k}} - V(0) - \frac{\hbar^2}{2m}|\boldsymbol{k}+\boldsymbol{K}|^2\right]U(\boldsymbol{K}) = V(\boldsymbol{K})U(0)(1-\delta_{\boldsymbol{K}}^0). \tag{3.221}$$

Note that the part of the sum in (3.220) involving $V(0)$ has already been placed in the left-hand side of (3.221). Thus (3.221) with $\boldsymbol{K} = 0$ yields

$$E_{\boldsymbol{k}} \cong V(0) + \frac{\hbar^2 k^2}{2m} . \tag{3.222}$$

These are the free-particle eigenvalues. Using (3.222) and (3.221), we obtain for $\boldsymbol{K} \neq 0$ in the same approximation:

$$\frac{U(\boldsymbol{K})}{U(0)} = -\frac{m}{\hbar^2} \frac{V(\boldsymbol{K})}{\boldsymbol{k} \cdot \boldsymbol{K} + \frac{1}{2}\boldsymbol{K}^2} . \tag{3.223}$$

Note that the above approximation obviously fails when

$$\boldsymbol{k} \cdot \boldsymbol{K} + \tfrac{1}{2}\boldsymbol{K}^2 = 0 , \tag{3.224}$$

if $V(\boldsymbol{K})$ is not equal to zero.

The $\boldsymbol{k}$ that satisfy (3.224) (for each value of $\boldsymbol{K}$) span the surface of the Brillouin zones. If we construct all *Brillouin zones* except those for which $V(\boldsymbol{K}) = 0$ then we have the *Jones zones*.

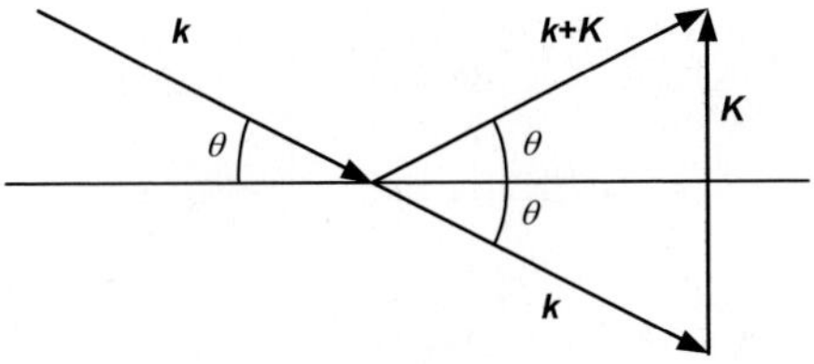

Fig. 3.12. Brillouin zones and Bragg reflection

Condition (3.224) can be given an interesting interpretation in terms of *Bragg reflection*. This situation is illustrated in Fig. 3.12. The $\boldsymbol{k}$ in the figure satisfy (3.224). From Fig. 3.12,

$$k \sin\theta = \tfrac{1}{2} K . \tag{3.225}$$

But $k = 2\pi/\lambda$, where λ is the de Broglie wavelength of the electron, and one can find K for which $K = n \cdot 2\pi/a$, where a is the distance between a given set of parallel lattice planes (see Sect. 1.2.9 where this is discussed in more detail in connection with X-ray diffraction). Thus we conclude that (3.225) implies that

$$\frac{2\pi}{\lambda}\sin\theta = \tfrac{1}{2} n \frac{2\pi}{a} , \tag{3.226}$$

or that

$$n\pi = 2a\sin\theta . \tag{3.227}$$

Since θ can be interpreted as an angle of incidence or reflection, (3.227) will be recognized as the familiar law describing Bragg reflection. It will presently be shown that at the Jones zone, there is a gap in the E versus k energy spectrum. This happens because the electron is Bragg reflected and does not propagate, and this is what we mean by having a gap in the energy. It will also be shown that when $V(\boldsymbol{K}) = 0$ there is no gap in the energy. This last fact is not obvious from the Bragg reflection picture. However, we now see why the Jones zones are the important physical zones. It is only at the Jones zones that the energy gaps appear. Note also that (3.225) indicates a simple way of defining the Brillouin zones by construction. We just draw reciprocal space. Starting from any point in reciprocal space, we draw straight lines connecting this point to all other points. We then bisect all these lines with planes perpendicular to the lines. Starting from the point of interest; these planes form the boundaries of the Brillouin zones. The first zone is the first enclosed volume. The second zone is the volume between the first set of planes and the second set. The idea should be clear from the two-dimensional representation in Fig. 3.13.

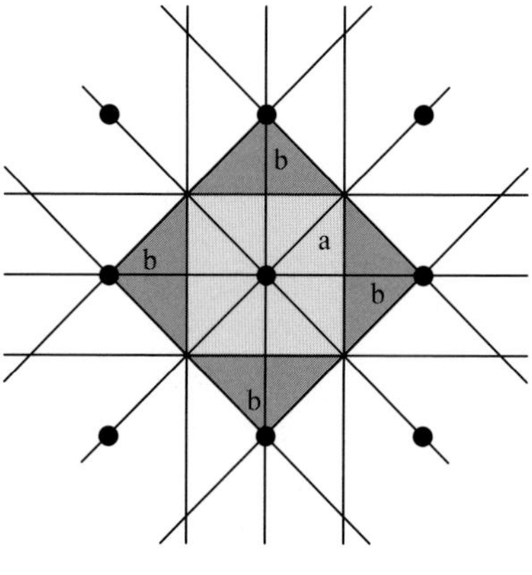

Fig. 3.13. Construction of Brillouin zones in reciprocal space: (a) the first Brillouin zone, and (b) the second Brillouin zone. The dots are lattice points in reciprocal space. Any vector joining two dots is a K-type reciprocal vector

To finish the calculation, let us treat the case when $\boldsymbol{k}$ is near a Brillouin zone boundary so that $U(\boldsymbol{K}^1)$ may be very large. Equation (3.220) then gives two equations that must be satisfied:

$$\left[E_{\boldsymbol{k}} - V(0) - \frac{\hbar^2}{2m}|\boldsymbol{k}+\boldsymbol{K}^1|^2\right]U(\boldsymbol{K}^1) = V(\boldsymbol{K}^1)U(0), \quad \boldsymbol{K}^1 \neq 0, \tag{3.228}$$

$$\left[E_{\boldsymbol{k}} - V(0) - \frac{\hbar^2}{2m}\boldsymbol{k}^2\right]U(0) = V(-\boldsymbol{K}^1)U(\boldsymbol{K}^1). \tag{3.229}$$

The equations have a nontrivial solution only if the following secular equation is satisfied:

$$\begin{vmatrix} E_{\boldsymbol{k}} - V(0) - \dfrac{\hbar^2}{2m}(\boldsymbol{k}+\boldsymbol{K}^1)^2 & -V(\boldsymbol{K}^1) \\ -V(-\boldsymbol{K}^1) & E_{\boldsymbol{k}} - V(0) - \dfrac{\hbar^2}{2m}\boldsymbol{k}^2 \end{vmatrix} = 0. \tag{3.230}$$

By problem 3.7 we know that (3.230) is equivalent to

$$E_{\boldsymbol{k}} = \tfrac{1}{2}(E^0_{\boldsymbol{k}} + E^0_{\boldsymbol{k}^1}) \pm \tfrac{1}{2}[4|V(\boldsymbol{K}^1)|^2 + (E^0_{\boldsymbol{k}} - E^0_{\boldsymbol{k}^1})^2]^{1/2}, \tag{3.231}$$

where

$$E^0_{\boldsymbol{k}} = V(0) + \frac{\hbar^2}{2m}\boldsymbol{k}^2, \tag{3.232}$$

and

$$E^0_{\boldsymbol{k}^1} = V(0) + \frac{\hbar^2}{2m}(\boldsymbol{k}+\boldsymbol{K}^1)^2. \tag{3.233}$$

For k on the Brillouin zone surface of interest, i.e. for $k^2 = (k + K^1)^2$, we see that there is an energy gap of magnitude

$$E^+_{\boldsymbol{k}} - E^-_{\boldsymbol{k}} = 2|V(\boldsymbol{K}^1)|. \tag{3.234}$$

This proves our point that the gaps in energy appear whenever $|V(K^1)| \neq 0$.

The next question that naturally arises is: "When does $V(K^1) = 0$?" This question leads to a discussion of the concept of the *structure factor*. The structure factor arises whenever there is more than one atom per unit cell in the Bravais lattice.

If there are m atoms located at the coordinates $\boldsymbol{r}_b$ in each unit cell, if we assume each atom contributes $U(\boldsymbol{r})$ (with the coordinate system centered at the center of the atom) to the potential, and if we assume the potential is additive, then with a fixed origin the potential in any cell can be written

$$V(\boldsymbol{r}) = \textstyle\sum_{b=1}^{m} U(\boldsymbol{r} - \boldsymbol{r}_b). \tag{3.235}$$

Since $V(\boldsymbol{r})$ is periodic in a unit cube, we can write

$$V(\boldsymbol{r}) = \textstyle\sum_{\boldsymbol{K}} V(\boldsymbol{K}) \mathrm{e}^{\mathrm{i}\boldsymbol{k}\cdot\boldsymbol{r}}, \tag{3.236}$$

where

$$V(\boldsymbol{K}) = \frac{1}{\Omega}\int_{\Omega} V(r) \mathrm{e}^{-\mathrm{i}\boldsymbol{K}\cdot\boldsymbol{r}} \mathrm{d}^3 r, \tag{3.237}$$

and Ω is the volume of a unit cell. Combining (3.235) and (3.237), we can write the Fourier coefficient

$$\begin{aligned} V(\boldsymbol{K}) &= \frac{1}{\Omega}\sum_{b=1}^{m}\int_{\Omega} U(\boldsymbol{r}-\boldsymbol{r}_b)\mathrm{e}^{-\mathrm{i}\boldsymbol{K}\cdot\boldsymbol{r}}\mathrm{d}^3 r \\ &= \frac{1}{\Omega}\sum_{b=1}^{m}\int_{\Omega} U(\boldsymbol{r}')\mathrm{e}^{-\mathrm{i}\boldsymbol{K}\cdot(\boldsymbol{r}'+\boldsymbol{r}_b)}\mathrm{d}^3 r' \\ &= \frac{1}{\Omega}\sum_{b=1}^{m}\mathrm{e}^{-\mathrm{i}\boldsymbol{K}\cdot\boldsymbol{r}_b}\int_{\Omega} U(\boldsymbol{r}')\mathrm{e}^{-\mathrm{i}\boldsymbol{K}\cdot\boldsymbol{r}'}\mathrm{d}^3 r', \end{aligned}$$

or

$$V(\boldsymbol{K}) \equiv S_{\boldsymbol{K}} v(\boldsymbol{K}), \tag{3.238}$$

where

$$S_{\boldsymbol{K}} \equiv \sum_{b=1}^{m}\mathrm{e}^{-\mathrm{i}\boldsymbol{K}\cdot\boldsymbol{r}_b}, \tag{3.239}$$

(structure factors are also discussed in Sect. 1.2.9) and

$$v(\boldsymbol{K}) \equiv \frac{1}{\Omega}\int_{\Omega} U(\boldsymbol{r}^1)\mathrm{e}^{-\mathrm{i}\boldsymbol{K}\cdot\boldsymbol{r}^1}\mathrm{d}^3 r^1 . \tag{3.240}$$

$S_{\boldsymbol{K}}$ is the structure factor, and if it vanishes, then so does $V(\boldsymbol{K})$. If there is only one atom per unit cell, then $|S_{\boldsymbol{K}}| = 1$. With the use of the structure factor, we can summarize how the first Jones zone can be constructed:

1. Determine all planes from

$$\boldsymbol{k}\cdot\boldsymbol{K} + \tfrac{1}{2}\boldsymbol{K}^2 = 0 .$$

2. Retain those planes for which $S_{\boldsymbol{K}} \neq 0$, and that enclose the smallest volume in $\boldsymbol{k}$ space.

To complete the discussion of the nearly free-electron approximation, the pseudopotential needs to be mentioned. However, the pseudopotential is also used as a practical technique for band-structure calculations, especially in semiconductors. Thus we discuss it in a later section.

The Tight Binding Approximation (B)[14]

This method is often called by the more descriptive name *linear combination of atomic orbitals* (LCAO). It was proposed by Bloch, and was one of the first types of band-structure calculation. The tight binding approximation is valid for the inner or core electrons of most solids and approximately valid for all electrons in an insulator.

All solids with periodic potentials have allowed and forbidden regions of energy. Thus it is no great surprise that the tight binding approximation predicts

[14] For further details see Mott and Jones [71].

a band structure in the energy. In order to keep things simple, the tight binding approximation will be done only for the s-band (the band of energy formed by s-electron states).

To find the energy bands one must solve the Schrödinger equation

$$\mathcal{H}\psi_0 = E_0\psi_0 , \tag{3.241}$$

where the subscript zero refers to s-state wave functions. In the spirit of the tight binding approximation, we attempt to construct the crystalline wave functions by using a superposition of atomic wave functions

$$\psi_0(\boldsymbol{r}) = \sum_{i=1}^{N} d_i \phi_0(\boldsymbol{r} - \boldsymbol{R}_i) . \tag{3.242}$$

In (3.242), N is the number of the lattice ions, ϕ_0 is an atomic s-state wave function, and the $\boldsymbol{R}_i$ are the vectors labeling the location of the atoms.

If the d_i are chosen to be of the form

$$d_i = \mathrm{e}^{\mathrm{i}\boldsymbol{k}\cdot\boldsymbol{R}_i} , \tag{3.243}$$

then $\psi_0(\boldsymbol{r})$ satisfies the Bloch condition. This is easily proved:

$$\begin{aligned}\psi(\boldsymbol{r} + \boldsymbol{R}_k) &= \sum_i \mathrm{e}^{\mathrm{i}\boldsymbol{k}\cdot\boldsymbol{R}_i} \phi_0(\boldsymbol{r} + \boldsymbol{R}_k - \boldsymbol{R}_i) \\ &= \mathrm{e}^{\mathrm{i}\boldsymbol{k}\cdot\boldsymbol{R}_k} \sum_i \mathrm{e}^{\mathrm{i}\boldsymbol{k}\cdot(\boldsymbol{R}_i - \boldsymbol{R}_k)} \phi_0[\boldsymbol{r} - (\boldsymbol{R}_i - \boldsymbol{R}_k)] \\ &= \mathrm{e}^{\mathrm{i}\boldsymbol{k}\cdot\boldsymbol{R}_k} \psi(\boldsymbol{r}).\end{aligned}$$

Note that this argument assumes only one atom per unit cell. Actually a much more rigorous argument for

$$\psi_0(\boldsymbol{r}) = \sum_{i=1}^{N} \mathrm{e}^{\mathrm{i}\boldsymbol{k}\cdot\boldsymbol{R}_i} \phi_0(\boldsymbol{r} - \boldsymbol{R}_i) \tag{3.244}$$

can be given by the use of projection operators.[15] Equation (3.244) is only an approximate equation for $\psi_0(\boldsymbol{r})$.

Using (3.244), the energy eigenvalues are given approximately by

$$E_0 \cong \frac{\int \psi_0^* \mathcal{H} \psi_0 \mathrm{d}\tau}{\int \psi_0^* \psi_0 \mathrm{d}\tau} , \tag{3.245}$$

where $\mathcal{H}$ is the crystal Hamiltonian.

We define an atomic Hamiltonian

$$\mathcal{H}_i = -(\hbar^2 / 2m)\nabla^2 + V_0(\boldsymbol{r} - \boldsymbol{R}_i) , \tag{3.246}$$

where $V_0(\boldsymbol{r} - \boldsymbol{R}_i)$ is the atomic potential. Then

$$\mathcal{H}_i \phi_0(\boldsymbol{r} - \boldsymbol{R}_i) = E_0^0 \phi_0(\boldsymbol{r} - \boldsymbol{R}_i) , \tag{3.247}$$

[15] See Löwdin [3.33].

and

$$\mathcal{H} - \mathcal{H}_i = V(\boldsymbol{r}) - V_0(\boldsymbol{r} - \boldsymbol{R}_i)\,, \tag{3.248}$$

where E_0^0 and ϕ_0 are atomic eigenvalues and eigenfunctions, and V is the crystal potential energy.

Using (3.244), we can now write

$$\mathcal{H}\psi_0 = \sum_{i=1}^{N} \mathrm{e}^{\mathrm{i}\boldsymbol{k}\cdot\boldsymbol{R}_i}[\mathcal{H}_i + (\mathcal{H} - \mathcal{H}_i)]\phi_0(\boldsymbol{r} - \boldsymbol{R}_i)\,,$$

or

$$\mathcal{H}\psi_0 = E_0^0\psi_0 + \sum_{i=1}^{N} \mathrm{e}^{\mathrm{i}\boldsymbol{k}\cdot\boldsymbol{R}_i}[V(\boldsymbol{r}) - V_0(\boldsymbol{r} - \boldsymbol{R}_i)]\phi_0(\boldsymbol{r} - \boldsymbol{R}_i)\,. \tag{3.249}$$

Combining (3.245) and (3.249), we readily find

$$E_0 - E_0^0 \cong \frac{\sum_{i=1}^{N} \mathrm{e}^{\mathrm{i}\boldsymbol{k}\cdot\boldsymbol{R}_i} \int \psi_0^*[V(\boldsymbol{r}) - V_0(\boldsymbol{r} - \boldsymbol{R}_i)]\phi_0(\boldsymbol{r} - \boldsymbol{R}_i)\mathrm{d}\tau}{\int \psi_0^*\psi_0 \mathrm{d}\tau}\,. \tag{3.250}$$

Using (3.244) once more, this last equation becomes

$$E_0 - E_0^0 \cong \frac{\sum_{i,j} \mathrm{e}^{\mathrm{i}\boldsymbol{k}\cdot(\boldsymbol{R}_i - \boldsymbol{R}_j)} \int \phi_0^*(\boldsymbol{r} - \boldsymbol{R}_j)[V(\boldsymbol{r}) - V_0(\boldsymbol{r} - \boldsymbol{R}_i)]\phi_0(\boldsymbol{r} - \boldsymbol{R}_i)\mathrm{d}\tau}{\sum_{i,j} \mathrm{e}^{\mathrm{i}\boldsymbol{k}\cdot(\boldsymbol{R}_i - \boldsymbol{R}_j)} \int \phi_0^*(\boldsymbol{r} - \boldsymbol{R}_j)\phi_0(\boldsymbol{r} - \boldsymbol{R}_i)\mathrm{d}\tau}\,. \tag{3.251}$$

Neglecting overlap, we have approximately

$$\int \phi_0^*(\boldsymbol{r} - \boldsymbol{R}_j)\phi_0(\boldsymbol{r} - \boldsymbol{R}_i)\mathrm{d}\tau \cong \delta_{i,j}\,.$$

Combining (3.250) and (3.251) and using the periodicity of $V(\boldsymbol{r})$, we have

$$E_0 - E_0^0 \cong \frac{1}{N}\sum_{i,j} \mathrm{e}^{\mathrm{i}\boldsymbol{k}\cdot(\boldsymbol{R}_i - \boldsymbol{R}_j)} \int \phi_0^*[\boldsymbol{r} - (\boldsymbol{R}_j - \boldsymbol{R}_i)][V(\boldsymbol{r}) - V_0(\boldsymbol{r})]\phi_0(\boldsymbol{r})\mathrm{d}\tau\,,$$

or

$$E_0 - E_0^0 \cong \frac{1}{N}\sum_{l} \mathrm{e}^{-\mathrm{i}\boldsymbol{k}\cdot\boldsymbol{R}_l} \int \phi_0^*(\boldsymbol{r} - \boldsymbol{R}_l)[V(\boldsymbol{r}) - V_0(\boldsymbol{r})]\phi_0(\boldsymbol{r})\mathrm{d}\tau\,. \tag{3.252}$$

Assuming that the terms in the sum of (3.252) are very small beyond nearest neighbors, and realizing that only s-wave functions (which are isotropic) are involved, then it is useful to define two parameters:

$$\int \phi_0^*(\boldsymbol{r})[V(\boldsymbol{r}) - V_0(\boldsymbol{r})]\phi_0(\boldsymbol{r})\mathrm{d}\tau = -\alpha\,, \tag{3.253}$$

$$\int \phi_0^*(\boldsymbol{r} + \boldsymbol{R}_l')[V(\boldsymbol{r}) - V_0(\boldsymbol{r})]\phi_0(\boldsymbol{r})\mathrm{d}\tau = -\gamma\,, \tag{3.254}$$

where $\boldsymbol{R}_l'$ is a vector of the form $\boldsymbol{R}_l$ for nearest neighbors.

Thus the tight binding approximation reduces to a two-parameter (α, γ) theory with the dispersion relationship (i.e. the E versus $\boldsymbol{k}$ relationship) for the s-band given by

$$E_0 - (E_0^0 - \alpha) = -\gamma \sum\nolimits_{j(\text{n.n.})} \mathrm{e}^{\mathrm{i}\boldsymbol{k}\cdot\boldsymbol{R}'_j} . \tag{3.255}$$

Explicit expressions for (3.255) are easily obtained in three cases

1. The simple cubic lattice. Here

$$\boldsymbol{R}'_j = (\pm a,0,0),(0,\pm a,0),(0,0,\pm a) ,$$

and

$$E_0 - (E_0^0 - \alpha) = -2\gamma(\cos k_x a + \cos k_y a + \cos k_z a) .$$

The bandwidth in this case is given by 12γ.

2. The body-centered cubic lattice. Here there are eight nearest neighbors at

$$\boldsymbol{R}'_j = \tfrac{1}{2}(\pm a,\pm a,\pm a) .$$

Equation (3.255) and a little algebra gives

$$E_0 - (E_0^0 - \alpha) = -8\gamma\left(\cos\frac{k_x a}{2}\right)\left(\cos\frac{k_y a}{2}\right)\left(\cos\frac{k_z a}{2}\right) .$$

The bandwidth in this case is 16γ.

3. The face-centered cubic lattice. Here the 12 nearest neighbors are at

$$\boldsymbol{R}'_j = \frac{1}{2}(0,\pm a,\pm a),\frac{1}{2}(\pm a,0,\pm a),\frac{1}{2}(\pm a,\pm a,0) .$$

A little algebra gives

$$E_0 - (E_0^0 - \alpha) = -4\gamma\left[\cos\left(\frac{k_y a}{2}\right)\cos\left(\frac{k_z a}{2}\right) + \cos\left(\frac{k_z a}{2}\right)\cos\left(\frac{k_x a}{2}\right) + \cos\left(\frac{k_x a}{2}\right)\cos\left(\frac{k_y a}{2}\right)\right] .$$

The bandwidth for this case is 16γ. The tight binding approximation is valid when γ is small, i.e., when the bands are narrow.

As must be fairly obvious by now, one of the most important results that we get out of an electronic energy calculation is the density of states. It was fairly easy to get the density of states in the free-electron approximation (or more generally when E is a quadratic function $|\boldsymbol{k}|$). The question that now arises is how we can get a density of states from a general dispersion relation similar to (3.255).

Since the $\boldsymbol{k}$ in reciprocal space are uniformly distributed, the number of states in a small volume $\mathrm{d}\boldsymbol{k}$ of phase space (per unit volume of real space) is

$$2\frac{\mathrm{d}^3k}{(2\pi)^3}.$$

Now look at Fig. 3.14 that shows a small volume between two constant electronic energy surfaces in $\boldsymbol{k}$-space.

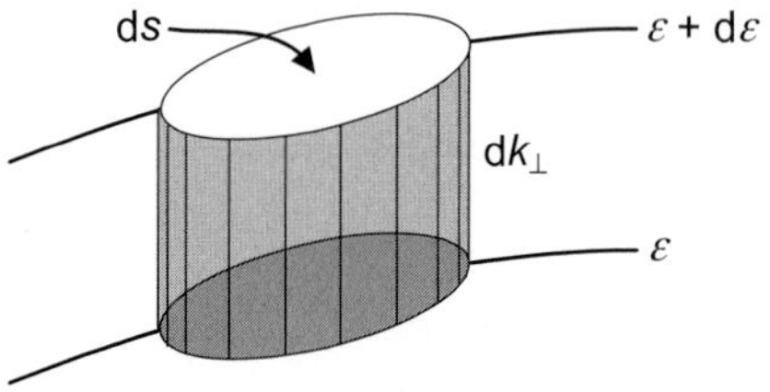

Fig. 3.14. Infinitesimal volume between constant energy surfaces in $\boldsymbol{k}$-space

From the figure we can write

$$\mathrm{d}^3k = \mathrm{d}s\mathrm{d}k_{\perp}.$$

But

$$\mathrm{d}\varepsilon = |\nabla_{\boldsymbol{k}}\varepsilon(\boldsymbol{k})|\,\mathrm{d}k_{\perp},$$

so that if $D(\varepsilon)$ is the number of states between ε and $\varepsilon + \mathrm{d}\varepsilon$, we have

$$D(\varepsilon) = \frac{2}{(2\pi)^3}\int_s \frac{\mathrm{d}s}{|\nabla_{\boldsymbol{k}}\varepsilon(\boldsymbol{k})|}. \tag{3.256}$$

Equation (3.256) can always be used to calculate a density of states when a dispersion relation is known. As must be obvious from the derivation, (3.256) applies also to lattice vibrations when we take into account that phonons have different polarizations (rather than the different spin directions that we must consider for the case of electrons).

Table 3.3. Simple models of electronic bands

Model	Energies
Nearly free electron near Brillouin zone boundary on surface where $\boldsymbol{k}\cdot\boldsymbol{K}+\frac{1}{2}K^2=0$	$E_k=\frac{1}{2}(E_k^0+E_{k'}^0)\pm\frac{1}{2}\sqrt{(E_k^0-E_{k'}^0)^2+4\lvert V(K)\rvert^2}$ $E_k^0=V(0)+\frac{\hbar^2k^2}{2m}$ $E_{k'}^0=V(0)+\frac{\hbar^2}{2m}(k+K)^2$ $V(K)=\frac{1}{\Omega}\int_\Omega V(\boldsymbol{r})e^{-i\boldsymbol{K}\cdot\boldsymbol{r}}dV$ Ω = unit cell volume
Tight binding	A, B appropriately chosen parameters. a = cell side.
simple cube	$E_k=A-B(\cos k_xa+\cos k_ya+\cos k_za)$
body-centered cubic	$E_k=A-4B\cos\frac{k_xa}{2}\cos\frac{k_ya}{2}\cos\frac{k_za}{2}$
face-centered cubic	$E_k=A-2B\left(\cos\frac{k_xa}{2}\cos\frac{k_ya}{2}+\cos\frac{k_ya}{2}\cos\frac{k_za}{2}+\cos\frac{k_za}{2}\cos\frac{k_xa}{2}\right)$
Kronig–Penny $r=\sqrt{\frac{2mE}{\hbar^2}}$ $P=\frac{mub}{\hbar^2}a$ a – distance between barriers u – height of barriers b – width of barrier	$\cos ka=\cos ra+P\frac{\sin ka}{ra}$ determines energies in $b\to 0$, $ua\to$ constant limit

Tight binding approximation calculations are more complicated for p, d., etc., bands, and also when there is an overlapping of bands. When things get too complicated, it may be easier to use another method such as one of those that will be discussed in the next section.

The tight binding method and its generalizations are often subsumed under the name linear combination of atomic orbital (LCAO) methods. The tight binding method here gave the energy of an s-band as a function of $\boldsymbol{k}$. This energy depended on the interpolation parameters α and γ. The method can be generalized to include other interpolation parameters. For example, the overlap integrals that were neglected could be treated as interpolation parameters. Similarly, the integrals for the energy involved only nearest neighbors in the sum. If we summed to next-nearest neighbors, more interpolation parameters would be introduced and hence greater accuracy would be achieved.

Results for the nearly free-electron approximation, the tight binding approximation, and the Kronig–Penny model are summarized in Table 3.3.

The Wigner–Seitz method (1933) (B)

The Wigner–Seitz method [3.57] was perhaps the first genuine effort to solve the Schrödinger wave equation and produce useful band-structure results for solids. This technique is generally applied to the valence electrons of alkali metals. It will also help us to understand their binding. We can partition space with polyhedra. These polyhedra are constructed by drawing planes that bisect the lines joining each atom to its nearest neighbors (or further neighbors if necessary). The polyhedra so constructed are called the *Wigner–Seitz* cells.

Sodium is a typical solid for which this construction has been used (as in the original Wigner–Seitz work, see [3.57]), and the Na^+ ions are located at the center of each polyhedron. In a reasonable approximation, the potential can be assumed to be spherically symmetric inside each polyhedron.

Let us first consider Bloch wave functions for which $\boldsymbol{k} = 0$ and deal with only s-band wave functions.

The symmetry and periodicity of this wave function imply that the normal derivative of it must vanish on the surface of each boundary plane. This boundary condition would be somewhat cumbersome to apply, so the atomic polyhedra are replaced by spheres of equal volume having radius r_0. In this case the boundary condition is simply written as

$$\left(\frac{\partial \psi_0}{\partial r}\right)_{r=r_0} = 0\,. \tag{3.257}$$

With $\boldsymbol{k} = 0$ and a spherically symmetric potential, the wave equation that must be solved is simply

$$\left[-\frac{\hbar^2}{2mr^2}\frac{\mathrm{d}}{\mathrm{d}r}\left(r^2\frac{\mathrm{d}}{\mathrm{d}r}\right) + V(r)\right]\psi_0 = E\psi_0\,, \tag{3.258}$$

subject to the boundary condition (3.257). The simultaneous solution of (3.257) and (3.258) gives both the eigenfunction ψ_0 and the eigenvalue E.

The biggest problem remaining is the usual problem that confronts one in making band-structure calculations. This is the problem of selecting the correct ion core potential in each polyhedra. We select $V(r)$ that gives a best fit to the electronic energy levels of the isolated atom or ion. Note that this does not imply that the eigenvalue E of (3.258) will be a free-ion eigenvalue, because we use boundary condition (3.257) on the wave function rather than the boundary condition that the wave function must vanish at infinity. The solution of (3.258) may be obtained by numerically integrating this radial equation.

Once ψ_0 has been obtained, higher k value wave functions may be approximated by

$$\psi_k(\boldsymbol{r}) \cong \mathrm{e}^{\mathrm{i}\boldsymbol{k}\cdot\boldsymbol{r}}\psi_0\,, \tag{3.259}$$

with $\psi_0 = \psi_0(\boldsymbol{r})$ being the same in each cell. This set of wave functions at least has the virtue of being nearly plane waves in most of the atomic volume, and of wiggling around in the vicinity of the ion cores as physically they should.

Finally, a Wigner–Seitz calculation can be used to explain, from the calculated eigenvalues, the cohesion of metals. Physically, the zero slope of the wave function causes less wiggling of the wave function in a region of nearly constant potential energy. Thus the kinetic and hence total energy of the conduction electrons is lowered. Lower energy means cohesion. The idea is shown schematically in Fig. 3.15.[16]

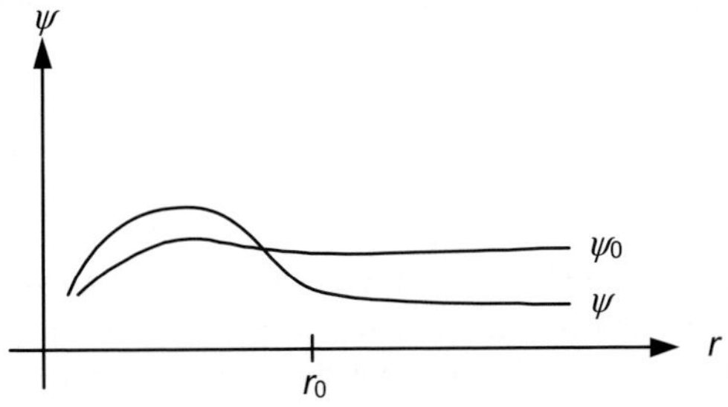

Fig. 3.15. The boundary condition on the wave function ψ_0 in the Wigner–Seitz model. The free-atom wave function is ψ

The Augmented Plane Wave Method (A)

The augmented plane wave method was developed by J. C. Slater in 1937, but continues in various forms as a very effective method. (Perhaps the best early reference is Slater [88] and also the references contained therein as well as Loucks [63] and Dimmock [3.16].) The basic assumption of the method is that the potential in a spherical region near an atom is spherically symmetric, whereas the potential in regions away from the atom is assumed constant. Thus one gets a "muffin tin" as shown in Fig. 3.16.

The Schrödinger equation can be solved exactly in both the spherical region and the region of constant potential. The solutions in the region of constant potential are plane waves. By choosing a linear combination of solutions (involving several l values) in the spherical region, it is possible to obtain a fit at the spherical surface (in value, not in normal derivative) of each plane wave to

---

[16] Of course there are much more sophisticated techniques nowadays using the density functional techniques. See, e.g., Schlüter and Sham [3.44] and Tran and Pewdew [3.55].

a linear combination of spherical solutions. Such a procedure gives an *augmented plane wave* for one Wigner–Seitz cell. (As already mentioned, Wigner–Seitz cells are constructed in direct space in the same way first Brillouin zones are constructed in reciprocal space.) We can extend the definition of the augmented plane wave to all points in space by requiring that the extension satisfy the Bloch condition. Then we use a linear combination of augmented plane waves in a variational calculation of the energy. The use of symmetry is quite useful in this calculation.

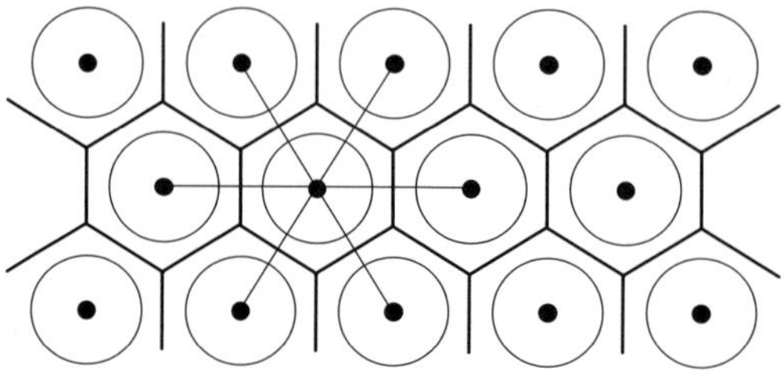

Fig. 3.16. The "muffin tin" potential of the augmented plane wave method

Before a small mathematical development of the augmented plane method is made, it is convenient to summarize a few more facts about it. First, the exact crystalline potential is never either exactly constant or precisely spherically symmetric in any region. Second, a real strength of early augmented plane wave methods lay in the fact that the boundary conditions are applied over a sphere (where it is relatively easy to satisfy them) rather than over the boundaries of the Wigner–Seitz cell where it is relatively hard to impose and satisfy reasonable boundary conditions. The best linear combination of augmented plane waves greatly reduces the discontinuity in normal derivative of any single plane wave. As will be indicated later, it is only at points of high symmetry in the Brillouin zone that the APW calculation goes through well. However, nowadays with huge computing power, this is not as big a problem as it used to be. The augmented plane wave has also shed light on why the nearly free-electron approximation appears to work for the alkali metals such as sodium. In those cases where the nearly free-electron approximation works, it turns out that just one augmented plane wave is a good approximation to the actual crystalline wave function.

The APW method has a strength that has not yet been emphasized. The potential is relatively flat in the region between ion cores and the augmented plane wave method takes this flatness into account. Furthermore, the crystalline potential is essentially identical to an atomic potential when one is near an atom. The augmented plane wave method takes this into account also.

The augmented plane wave method is not completely rigorous, since there are certain adjustable parameters (depending on the approximation) involved in its use. The radius R_0 of the spherically symmetric region can be such a parameter. The main constraint on R_0 is that it be smaller than r_0 of the Wigner–Seitz method. The value of the potential in the constant potential region is another adjustable parameter. The type of spherically symmetric potential in the spherical region is also adjustable, at least to some extent.

Let us now look at the augmented plane wave method in a little more detail. Inside a particular sphere of radius R_0, the Schrödinger wave equation has a solution

$$\phi^a(\boldsymbol{r}) = \sum_{l,m} d_{lm} R_l(r, E) Y_{lm}(\theta, \phi) . \tag{3.260}$$

For other spheres, $\phi^a(\boldsymbol{r})$ is constructed from (3.260) so as to satisfy the Bloch condition. In (3.260), $R_l(r, E)$ is a solution of the radial wave equation and it is a function of the energy parameter E. The d_{lm} are determined by fitting (3.260) to a plane wave of the form $e^{i\boldsymbol{k}\cdot\boldsymbol{r}}$. This gives a different $\phi^a = \phi_{\boldsymbol{k}}^a$ for each value of $\boldsymbol{k}$. The functions $\phi_{\boldsymbol{k}}^a$ that are either plane waves or linear combinations of spherical harmonics (according to the spatial region of interest) are the augmented plane waves $\phi_{\boldsymbol{k}}^a(r)$.

The most general function that can be constructed from augmented plane waves and that satisfies Bloch's theorem is

$$\psi_{\boldsymbol{k}}(\boldsymbol{r}) = \sum_{\boldsymbol{G}_n} K_{\boldsymbol{k}+\boldsymbol{G}_n} \phi_{\boldsymbol{k}+\boldsymbol{G}_n}^a(\boldsymbol{r}) . \tag{3.261}$$

The use of symmetry has already reduced the number of augmented plane waves that have to be considered in any given calculation. If we form a wave function that satisfies Bloch's theorem, we form a wave function that has all the symmetry that the translational symmetry of the crystal requires. Once we do this, we are not required to mix together wave functions with different *reduced* wave vectors $\boldsymbol{k}$ in (3.261).

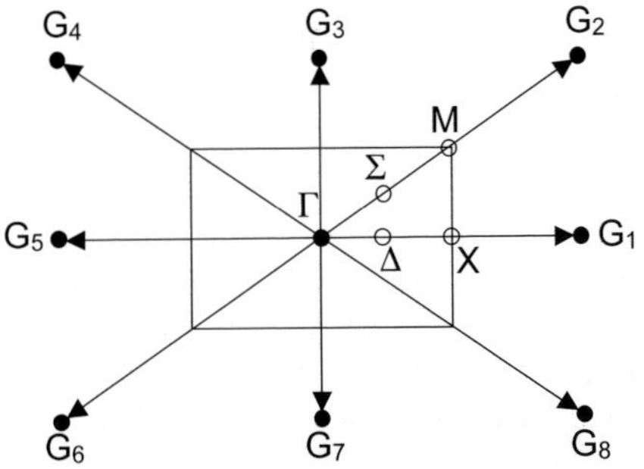

Fig. 3.17. Points of high symmetry (Γ, Δ, X, Σ, M) in the Brillouin zone. [Adapted from Ziman JM, *Principles of the Theory of Solids*, Cambridge University Press, New York, 1964, Fig. 53, p. 99. By permission of the publisher.]

The coefficients K_{k+G_n}, are determined by a variational calculation of the energy. This calculation also gives $E(\boldsymbol{k})$. The calculation is not completely straightforward, however. This is because of the $E(\boldsymbol{k})$ dependence that is implied in the $R_l(r,E)$ when the d_{lm} are determined by fitting spherical solutions to plane waves. Because of this, and other obvious complications, the augmented plane wave method is practical to use only with a digital computer, which nowadays is not much of a restriction. The great merit of the augmented plane wave method is that if one works hard enough on it, one gets good results.

There is yet another way in which symmetry can be used in the augmented plane wave method. By the use of group theory we can also take into account some rotational symmetry of the crystal. In the APW method (as well as the OPW method, which will be discussed) group theory may be used to find relations among the coefficients K_{k+G_n}. The most accurate values for $E(\boldsymbol{k})$ can be obtained at the points of highest symmetry in the zone. The ideas should be much clearer after reasoning from Fig. 3.17, which is a picture of a two-dimensional reciprocal space with a very simple symmetry.

For the APW (or OPW) expansions, the expansions are of the form

$$\psi_{\boldsymbol{k}} = \sum\nolimits_n K_{\boldsymbol{k}-\boldsymbol{G}_n} \psi_{\boldsymbol{k}-\boldsymbol{G}_n} \,.$$

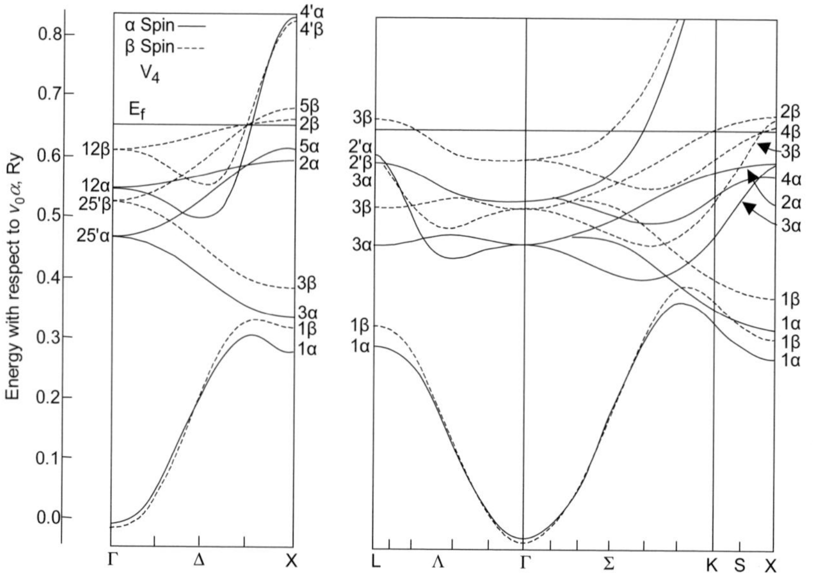

Fig. 3.18. Self-consistent energy bands in ferromagnetic Ni along the three principal symmetry directions. The letters along the horizontal axis refer to different symmetry points in the Brillouin zone [refer to Bouckaert LP, Smoluchowski R, and Wigner E, *Physical Review*, **50**, 58 (1936) for notation]. [Reprinted by permission from Connolly JWD, *Physical Review*, **159**(2), 415 (1967). Copyright 1967 by the American Physical Society.]

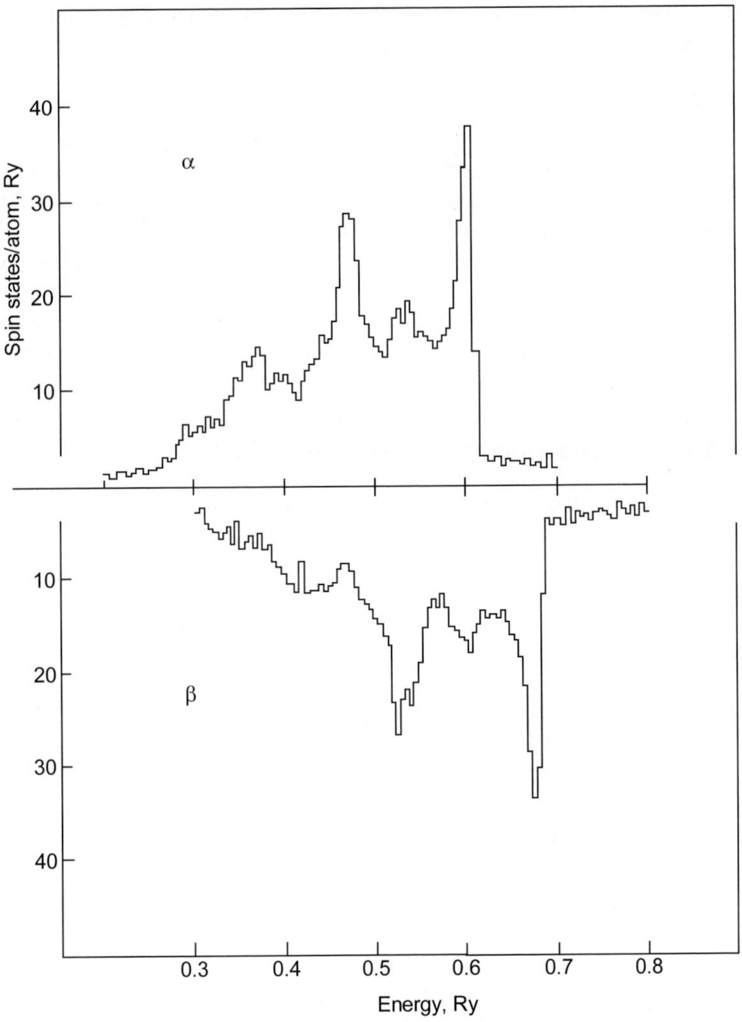

Fig. 3.19. Density of states for up (α) and down (β) spins in ferromagnetic Ni. [Reprinted by permission from Connolly JWD, *Physical Review*, **159**(2), 415 (1967). Copyright 1967 by the American Physical Society.]

Suppose it is assumed that only G_1 through G_8 need to be included in the expansions. Further assume we are interested in computing $E(\boldsymbol{k}_\Delta)$ for a $\boldsymbol{k}$ on the Δ symmetry axis. Then due to the fact that the calculation cannot be affected by appropriate rotations in reciprocal space, we must have

$$K_{\boldsymbol{k}-\boldsymbol{G}_2} = K_{\boldsymbol{k}-\boldsymbol{G}_8}, \quad K_{\boldsymbol{k}-\boldsymbol{G}_3} = K_{\boldsymbol{k}-\boldsymbol{G}_7}, \quad K_{\boldsymbol{k}-\boldsymbol{G}_4} = K_{\boldsymbol{k}-\boldsymbol{G}_6},$$

and so we have only five independent coefficients rather than eight (in three dimensions there would be more coefficients and more relations). Complete details for applying group theory in this way are available.[17] At a general point $\boldsymbol{k}$ in reciprocal space, there will be no relations among the coefficients.

Figure 3.18 illustrates the complexity of results obtained by an APW calculation of several electronic energy bands in Ni. The letters along the horizontal axis refer to different symmetry points in the Brillouin zone. For a more precise definition of terms, the paper by Connolly can be consulted. One rydberg (Ry) of energy equals approximately 13.6 eV. Results for the density of states (on Ni) using the APW method are shown in Fig. 3.19. Note that in Connolly's calculations, the fact that different spins may give different energies is taken into account. This leads to the concept of spin-dependent bands. This is tied directly to the fact that Ni is ferromagnetic.

The Orthogonalized Plane Wave Method (A)

The orthogonalized plane wave method was developed by C. Herring in 1940.[18]

The orthogonalized plane wave (OPW) method is fairly similar to the augmented plane wave method, but it does not seem to be as much used. Both methods address themselves to the same problem, namely, how to have wave functions wiggle like an atomic function near the cores but behave as a plane wave in regions far from the core. Both are improvements over the nearly free-electron method and the tight binding method. The nearly free-electron model will not work well when the wiggles of the wave function near the core are important because it requires too many plane waves to correctly reproduce these wiggles. Similarly, the tight binding method does not work when the plane-wave behavior far from the cores is important because it takes too many core wave functions to reproduce correctly the plane-wave behavior.

The basic assumption of the OPW method is that the wiggles of the conduction-band wave functions near the atomic cores can be represented by terms that cause the conduction-band wave function to be orthogonal to the core-band wave functions. We will see how (in the Section *The Pseudopotential Method*) this idea led to the idea of the pseudopotential. The OPW method can be stated fairly simply. To each plane wave we add on a sum of (Bloch sums of) atomic core wave functions. The functions formed in the previous sentence are orthogonal to Bloch sums of atomic wave functions. The resulting wave functions are called the OPWs and are used to construct trial wave functions in a variational calculation of the energy. The OPW method uses the tight binding approximation for the core wave functions.

Let us be a little more explicit about the technical details of the OPW method. Let $C_{tk}(\boldsymbol{r})$ be the crystalline atomic core wave functions (where t labels different core bands). The conduction band states ψ_k should look very much like plane waves between the atoms and like core wave functions near the atoms. A good

[17] See Bouckaert et al [3.7].

[18] See [3.21, 3.22].

choice for the base set of functions for the trial wave function for the conduction band states is

$$\psi_{\boldsymbol{k}} = \mathrm{e}^{\mathrm{i}\boldsymbol{k}\cdot\boldsymbol{r}} - \sum_t K_t C_{t\boldsymbol{k}}(\boldsymbol{r}) . \tag{3.262}$$

The Hamiltonian is Hermitian and so $\psi_{\boldsymbol{k}}$ and $C_{t\boldsymbol{k}}(\boldsymbol{r})$ must be orthogonal. With K_t chosen so that

$$(\psi_{\boldsymbol{k}}, C_{t\boldsymbol{k}}) = 0 , \tag{3.263}$$

where $(u, v) = \int u^* v \, \mathrm{d}\tau$, we obtain the orthogonalized plane waves

$$\psi_{\boldsymbol{k}} = \mathrm{e}^{\mathrm{i}\boldsymbol{k}\cdot\boldsymbol{r}} - \sum_t (C_{t\boldsymbol{k}}, \mathrm{e}^{\mathrm{i}\boldsymbol{k}\cdot\boldsymbol{r}}) C_{t\boldsymbol{k}}(\boldsymbol{r}) . \tag{3.264}$$

Linear combinations of OPWs satisfy the Bloch condition and are a good choice for the trial wave function $\psi_{\boldsymbol{k}}^T$.

$$\psi_{\boldsymbol{k}}^T = \sum_{l'} K_{\boldsymbol{k}-\boldsymbol{G}_{l'}} \psi_{\boldsymbol{k}-\boldsymbol{G}_{l'}} . \tag{3.265}$$

The choice for the core wave functions is easy. Let $\phi_t(\boldsymbol{r} - \boldsymbol{R}_l)$ be the atomic "core" states appropriate to the ion site $\boldsymbol{R}_l$. The Bloch wave functions constructed from atomic core wave functions are given by

$$C_{t\boldsymbol{k}} = \sum_l \mathrm{e}^{\mathrm{i}\boldsymbol{k}\cdot\boldsymbol{R}_l} \phi_t(\boldsymbol{r} - \boldsymbol{R}_l) . \tag{3.266}$$

We discuss in Appendix C how such a Bloch sum of atomic orbitals is guaranteed to have the symmetry appropriate for a crystal.

Usually only a few (at a point of high symmetry in the Brillouin zone) OPWs are needed to get a fairly good approximation to the crystal wave function. It has already been mentioned how the use of symmetry can help in reducing the number of variational parameters. The basic problem remaining is to choose the Hamiltonian (i.e. the potential) and then do a variational calculation with (3.265) as the trial wave function.

For a detailed list of references to actual OPW calculations (as well as other band-structure calculations) the book by Slater [89] can be consulted. Rather briefly, the OPW method was first applied to beryllium and has since been applied to diamond, germanium, silicon, potassium, and other crystals.

Better Ways of Calculating Electronic Energy Bands (A)

The process of calculating good electronic energy levels has been slow in reaching accuracy. Some claim that the day is not far off when computers can be programmed so that one only needs to push a few buttons to obtain good results for any solid. It would appear that this position is somewhat overoptimistic. The comments below should convince you that there are many remaining problems.

In an actual band-structure calculation there are many things that have to be decided. We may assume that the Born–Oppenheimer approximation and the density functional approximation (or Hartree–Fock or whatever) introduce little error. But we must always keep in mind that neglect of electron–phonon interactions and other interactions may importantly affect the electronic density of states. In particular this may lead to errors in predicting some of the optical properties. We should also remember that we do not do a completely *self-consistent* calculation.

The exchange-correlation term in the density functional approximation is difficult to treat exactly so it can be approximated by the free-electron-like *Slater* $\rho^{1/3}$ *term* [88] or the related local density approximation. However, density functional techniques suggest some factor[19] other than the one Slater suggests should multiply the $\rho^{1/3}$ term. In the treatment below we will not concern ourselves with this problem. We shall just assume that the effects of exchange (and correlation) are somehow lumped approximately into an ordinary crystalline potential.

This latter comment brings up what is perhaps the crux of an energy-band calculation. Just how is the "ordinary crystalline potential" selected? We don't want to do an energy-band calculation for all electrons in a solid. We want only to calculate the energy bands of the outer or valence electrons. The inner or core electrons are usually assumed to be the same in a free atom as in an atom that is in a solid. We never rigorously prove this assumption.

Not all electrons in a solid can be thought of as being nonrelativistic. For this reason it is sometimes necessary to put in relativistic corrections.[20]

Before we discuss other techniques of band-structure calculations, it is convenient to discuss a few features that would be common to any method.

For any crystal and for any method of energy-band calculation we always start with a Hamiltonian. The Hamiltonian may not be very well known but it always is invariant to all the symmetry operations of the crystal. In particular the crystal always has translational symmetry. The single-electron Hamiltonian satisfies the equation,

$$\mathcal{H}(\boldsymbol{p},\boldsymbol{r}) = \mathcal{H}(\boldsymbol{p},\boldsymbol{r}+\boldsymbol{R}_l)\,, \tag{3.267}$$

for any $\boldsymbol{R}_l$.

This property allows us to use Bloch's theorem that we have already discussed (see Appendix C). The eigenfunctions $\psi_{n\boldsymbol{k}}$ (n labeling *a* band, $\boldsymbol{k}$ labeling a wave vector) of $\mathcal{H}$ can always be chosen so that

$$\psi_{n\boldsymbol{k}}(\boldsymbol{r}) = \mathrm{e}^{\mathrm{i}\boldsymbol{k}\cdot\boldsymbol{r}} U_{n\boldsymbol{k}}(\boldsymbol{r})\,, \tag{3.268}$$

where

$$U_{n\boldsymbol{k}}(\boldsymbol{r}+\boldsymbol{R}_l) = U_{n\boldsymbol{k}}(\boldsymbol{r})\,. \tag{3.269}$$

Three possible Hamiltonians can be listed,[21] depending on whether we want to do (a) a completely nonrelativistic calculation, (b) a nonrelativistic calculation

[19] See Kohn and Sham [3.29].

[20] See Loucks [3.32].

[21] See Blount [3.6].

with some relativistic corrections, or (c) a completely relativistic calculation, or at least one with more relativistic corrections than (b) has.

a) Schrödinger Hamiltonian:

$$\mathcal{H} = \frac{p^2}{2m} + V(\boldsymbol{r}) . \tag{3.270}$$

b) Low-energy Dirac Hamiltonian:

$$\mathcal{H} = \frac{p^2}{2m_0} - \frac{p^4}{8m_0^3c^2} + V + \frac{\hbar^2}{4m_0^2c^2}[\boldsymbol{\sigma}\cdot(\nabla V \times \boldsymbol{p}) - \nabla V \cdot \nabla \psi] , \tag{3.271}$$

where m_0 is the rest mass and the third term is the spin-orbit coupling term (see Appendix F). (More comments will be made about spin-orbit coupling later in this chapter).

c) Dirac Hamiltonian:

$$\mathcal{H} = \beta m_0 c^2 + c\boldsymbol{\alpha}\cdot\boldsymbol{p} + V , \tag{3.272}$$

where $\boldsymbol{\alpha}$ and β are the Dirac matrices (see Appendix F).

Finally, two more general comments will be made on energy-band calculations. The first is in the frontier area of electron–electron interactions. Some related general comments have already been made in Sect. 3.1.4. Here we should note that no completely accurate method has been found for computing electronic correlations for metallic densities that actually occur [78], although the density functional technique [3.27] provides, at least in principle, an exact approach for dealing with ground-state many-body effects. Another comment has to do with Bloch's theorem and core electrons. There appears to be a paradox here. We think of core electrons as having well-localized wave functions but Bloch's theorem tells us that we can always choose the crystalline wave functions to be not localized. There is no paradox. It can be shown for infinitesimally narrow energy bands that either localized or nonlocalized wave functions are possible because a large energy degeneracy implies many possible descriptions [95, p. 160], [87, Vol. II, p. 154ff]. Core electrons have narrow energy bands and so core electronic wave functions can be thought of as approximately localized. This can always be done. For narrow energy bands, the localized wave functions are also good approximations to energy eigenfunctions.[22]

[22] For further details on band structure calculations, see Slater [88, 89, 90] and Jones and March [3.26, Chap. 1].

Interpolation and Pseudopotential Schemes (A)

An energy calculation is practical only at points of high symmetry in the Brillouin zone. This statement is almost true but, of course, as computers become more and more efficient, calculations at a general point in the Brillouin zone become more and more practical. Still, it will be a long time before the calculations are so "dense" in $\boldsymbol{k}$-space that no (nontrivial) interpolations between calculated values are necessary. Even if such calculations were available, interpolation methods would still be useful for many considerations in which their accuracy was sufficient. The interpolation methods are the LCAO method (already mentioned in the tight binding method section), the pseudopotential method (which is closely related to the OPW method and will be discussed), and the $\boldsymbol{k}\cdot\boldsymbol{p}$ method. Since the first two methods have other uses let us discuss the $\boldsymbol{k}\cdot\boldsymbol{p}$ method.

The $\boldsymbol{k}\cdot\boldsymbol{p}$ Method (A)[23] We let the index n label different bands. The solutions of

$$\mathcal{H}\psi_{n\boldsymbol{k}} = E_n(\boldsymbol{k})\psi_{n\boldsymbol{k}} \tag{3.273}$$

determine the energy band structure $E_n(\boldsymbol{k})$. By Bloch's theorem, the wave functions can be written as

$$\psi_{n\boldsymbol{k}} = \mathrm{e}^{\mathrm{i}\boldsymbol{k}\cdot\boldsymbol{r}}U_{n\boldsymbol{k}}\,.$$

Substituting this result into (3.273) and multiplying both sides of the resulting equation by $\mathrm{e}^{-\mathrm{i}\boldsymbol{k}\cdot\boldsymbol{r}}$ gives

$$(\mathrm{e}^{-\mathrm{i}\boldsymbol{k}\cdot\boldsymbol{r}}\mathcal{H}\mathrm{e}^{\mathrm{i}\boldsymbol{k}\cdot\boldsymbol{r}})U_{n\boldsymbol{k}} = E_n(\boldsymbol{k})U_{n\boldsymbol{k}}\,. \tag{3.274}$$

It is possible to define

$$\mathcal{H}(\boldsymbol{p}+\hbar\boldsymbol{k},\boldsymbol{r}) \equiv \mathrm{e}^{-\mathrm{i}\boldsymbol{k}\cdot\boldsymbol{r}}\mathcal{H}\mathrm{e}^{\mathrm{i}\boldsymbol{k}\cdot\boldsymbol{r}}\,. \tag{3.275}$$

It is not entirely obvious that such a definition is reasonable; let us check it for a simple example.

If $\mathcal{H} = \boldsymbol{p}^2/2m$, then $\mathcal{H}(\boldsymbol{p}+\hbar\boldsymbol{k}) = (1/2m)(\boldsymbol{p}^2 + 2\hbar\boldsymbol{k}\cdot\boldsymbol{p} + \hbar^2\boldsymbol{k}^2)$. Also

$$\begin{aligned}\mathrm{e}^{-\mathrm{i}\boldsymbol{k}\cdot\boldsymbol{r}}\mathcal{H}\mathrm{e}^{\mathrm{i}\boldsymbol{k}\cdot\boldsymbol{r}}F &= \frac{1}{2m}\mathrm{e}^{-\mathrm{i}\boldsymbol{k}\cdot\boldsymbol{r}}\left(-\frac{\hbar}{i}\nabla\right)^2\mathrm{e}^{\mathrm{i}\boldsymbol{k}\cdot\boldsymbol{r}}F \\ &= \frac{1}{2m}[\boldsymbol{p}^2 + 2\hbar\boldsymbol{k}\cdot\boldsymbol{p} + (\hbar\boldsymbol{k})^2]F\end{aligned},$$

which is the same as $[\mathcal{H}(\boldsymbol{p}+h\boldsymbol{k})]F$ for our example.

---

[23] See Blount [3.6].

By a series expansion

$$\mathcal{H}(\boldsymbol{p}+\hbar\boldsymbol{k},\boldsymbol{r}) = \mathcal{H} + \left(\frac{\partial \mathcal{H}}{\partial \boldsymbol{p}}\right)\cdot \hbar\boldsymbol{k} + \tfrac{1}{2}\sum_{i,j=1}^{3}\left(\frac{\partial^2 \mathcal{H}}{\partial p_i \partial p_j}\right)(\hbar k_i)(\hbar k_j)\,. \quad (3.276)$$

Note that if $\mathcal{H} = \boldsymbol{p}^2/2m$, where $\boldsymbol{p}$ is an operator, then

$$\nabla_{\boldsymbol{p}}\mathcal{H} \equiv \frac{\partial \mathcal{H}}{\partial p} = \frac{\boldsymbol{p}}{m} \equiv \boldsymbol{v}\,, \quad (3.277)$$

where $\boldsymbol{v}$ might be called a velocity operator. Further

$$\frac{\partial^2 \mathcal{H}}{\partial p_i \partial p_l} = \frac{1}{m}\delta_{il}\,, \quad (3.278)$$

so that (3.276) becomes

$$\mathcal{H}(\boldsymbol{p}+\hbar\boldsymbol{k},\boldsymbol{r}) \cong \mathcal{H} + \hbar\boldsymbol{k}\cdot\boldsymbol{v} + \frac{\hbar^2 \boldsymbol{k}^2}{2m}\,. \quad (3.279)$$

Then

$$\begin{aligned}\mathcal{H}(\boldsymbol{p}+\hbar\boldsymbol{k}+\hbar\boldsymbol{k}',\boldsymbol{r}) &= \mathcal{H} + \hbar(\boldsymbol{k}+\boldsymbol{k}')\cdot\boldsymbol{v} + \frac{\hbar^2}{2m}(\boldsymbol{k}+\boldsymbol{k}')^2 \\ &= \mathcal{H} + \hbar\boldsymbol{k}\cdot\boldsymbol{v} + \frac{\hbar^2}{2m}\boldsymbol{k}^2 + \hbar\boldsymbol{k}'\cdot\boldsymbol{v} + \frac{\hbar^2}{2m}\boldsymbol{k}\cdot\boldsymbol{k}' + \frac{\hbar^2}{2m}\boldsymbol{k}'^2 \\ &= \mathcal{H}(\boldsymbol{p}+\hbar\boldsymbol{k},\boldsymbol{r}) + \hbar\boldsymbol{k}'\cdot\left(\boldsymbol{v}+\frac{\hbar\boldsymbol{k}}{2m}\right) + \frac{\hbar^2}{2m}\boldsymbol{k}'^2.\end{aligned}$$

Defining

$$\boldsymbol{v}(\boldsymbol{k}) \equiv \boldsymbol{v} + \hbar\boldsymbol{k}/m\,, \quad (3.280)$$

and

$$\mathcal{H}' = \hbar\boldsymbol{k}'\cdot\boldsymbol{v}(\boldsymbol{k}) + \frac{\hbar^2 \boldsymbol{k}'^2}{2m}\,, \quad (3.281)$$

we see that

$$\mathcal{H}(\boldsymbol{p}+\hbar\boldsymbol{k}+\hbar\boldsymbol{k}') \cong \mathcal{H}(\boldsymbol{p}+\hbar\boldsymbol{k},\boldsymbol{r}) + \mathcal{H}'\,. \quad (3.282)$$

Thus comparing (3.274), (3.275), (3.280), (3.281), and (3.282), we see that if we know U_{nk}, E_{nk}, and $\boldsymbol{v}$ for a $\boldsymbol{k}$, we can find $E_{n,k+k'}$ for small $\boldsymbol{k}'$ by perturbation theory. Thus perturbation theory provides a means of *interpolating* to other energies in the vicinity of E_{nk}.

The Pseudopotential Method (A) The idea of the pseudopotential relates to the simple idea that electron wave functions corresponding to different energies are orthogonal. It is thus perhaps surprising that it has so many ramifications as we will indicate below. Before we give a somewhat detailed exposition of it, let us start with several specific comments that otherwise might be lost in the ensuing details.

1. In one form, the idea of a pseudopotential originated with Enrico Fermi [3.17].
2. The pseudopotential and OPW methods are focused on constructing valence wave functions that are orthogonal to the core wave functions. The pseudopotential method clearly relates to the orthogonalized plane wave method.
3. The pseudopotential as it is often used today was introduced by Phillips and Kleinman [3.40].
4. More general formalisms of the pseudopotential have been given by Cohen and Heine [3.14] and Austin et al [3.3].
5. In the hands of Marvin Cohen it has been used extensively for band-structure calculations of many materials – particularly semiconductors (Cohen [3.11], and also [3.12, 3.13]).
6. W. A. Harrison was another pioneer in relating pseudopotential calculations to the band structure of metals [3.19].
7. The use of the pseudopotential has not died away. Nowadays, e.g., people are using it in conjunction with the density functional method (for an introduction, see, e.g., Marder [3.34, p232ff].
8. Two complications of using the pseudopotential are that it is nonlocal and nonunique. We will show these below, as well as note that it is short range.
9. There are many aspects of the pseudopotential. There is the empirical pseudopotential method (EPM), ab initio calculations, and the pseudopotential can also be considered with other methods for broad discussions of solid-state properties [3.12].
10. As we will show below, the pseudopotential can be used as a way to assess the validity of the nearly free-electron approximation, using the so-called cancellation theorem.
11. Since the pseudopotential, for valence states, is positive it tends to cancel the attractive potential in the core leading to an empty-core method (ECM).

12. We will also note that the pseudopotential projects into the space of core wave functions, so its use will not change the valence eigenvalues.

13. Finally, the use of pseudopotentials has grown vastly and we can only give an introduction. For further details, one can start with a monograph like Singh [3.45].

We start with the original Phillips–Kleinman derivation of the pseudopotential because it is particularly transparent.

Using a one-electron picture, we write the Schrödinger equation as

$$\mathcal{H}|\psi\rangle = E|\psi\rangle, \tag{3.283}$$

where $\mathcal{H}$ is the Hamiltonian of the electron in energy state E with corresponding eigenket $|\psi\rangle$. For core eigenfunctions $|c\rangle$

$$\mathcal{H}|c\rangle = E_c|c\rangle. \tag{3.284}$$

If $|\psi\rangle$ is a valence wave function, we require that it be orthogonal to the core wave functions. Thus for appropriate $|\phi\rangle$ it can be written

$$|\psi\rangle = |\phi\rangle - \sum_{c'}|c'\rangle\langle c'|\phi\rangle, \tag{3.285}$$

so $\langle c|\psi\rangle = 0$ for all c, $c' \in$ the core wave functions. $|\phi\rangle$ will be a relatively smooth function as the "wiggles" of $|\psi\rangle$ in the core region that are necessary to make $\langle c|\psi\rangle = 0$ are included in the second term of (3.285) (This statement is complicated by the nonuniqueness of $|\phi\rangle$ as we will see below). See also Ziman [3.59, p. 53].

Substituting (3.285) in (3.283) and (3.284) yields, after rearrangement

$$(\mathcal{H} + V_R)|\phi\rangle = E|\phi\rangle, \tag{3.286}$$

where

$$V_R|\phi\rangle = \sum_c (E - E_c)|c\rangle\langle c|\phi\rangle. \tag{3.287}$$

Note V_R has several properties:

a. It is short range since the wave function ψ_c corresponds to $|c\rangle$ and is short range. This follows since if $\boldsymbol{r}|\boldsymbol{r}'\rangle = \boldsymbol{r}'|\boldsymbol{r}'\rangle$ is used to define $|\boldsymbol{r}\rangle$, then $\psi_c(\boldsymbol{r}) = \langle \boldsymbol{r}|c\rangle$.

b. It is nonlocal since

$$\langle \boldsymbol{r}'|V_R|\phi\rangle = \sum_c (E - E_c)\psi_c(\boldsymbol{r}')\int \psi_c^*(\boldsymbol{r})\phi(\boldsymbol{r})\mathrm{d}V,$$

or $V_R\phi(\boldsymbol{r}) \neq f(\boldsymbol{r})\phi(\boldsymbol{r})$ but rather the effect of V_R on ϕ involves values of $\phi(\boldsymbol{r})$ for all points in space.

c. The pseudopotential is not unique. This is most easily seen by letting $|\phi\rangle \to |\phi\rangle + \delta|\phi\rangle$ (provided $\delta|\phi\rangle$ can be expanded in core states). By substitution $\delta|\psi\rangle \to 0$ but

$$\delta V_R|\phi\rangle = \sum_c (E - E_c)\langle c|\delta\phi\rangle|c\rangle \neq 0.$$

d. Also note that $E > E_c$, when dealing with valence wave functions so $V_R > 0$ and since $V < 0$, $|V + V_R| < |V|$. This is an aspect of the cancellation theorem.

e. Note also, by (3.287) that since V_R projects $|\phi\rangle$ into the space of core wave functions it will not affect the valence eigenvalues as we have mentioned and will see in more detail later.

Since $\mathcal{H} = T + V$ where T is the kinetic energy operator and V is the potential energy, if we define the total pseudopotential V_p as

$$V_p = V + V_R, \tag{3.288}$$

then (3.286) can be written as

$$(T + V_p)|\phi\rangle = E|\phi\rangle. \tag{3.289}$$

To derive further properties of the pseudopotential it is useful to develop the formulation of Austin et al. We start with the following five equations:

$$\mathcal{H}\psi_n = E_n\psi_n \ (n = c \text{ or } v), \tag{3.290}$$

$$\mathcal{H}_p\phi_n = (\mathcal{H} + V_R)\phi_n = \bar{E}_n\phi_n \text{ (allowing for several } \phi), \tag{3.291}$$

$$V_R\phi = \sum\nolimits_c \langle F_c|\phi\rangle\psi_c, \tag{3.292}$$

where note F_c is arbitrary so V_R is not yet specified.

$$\phi_c = \sum\nolimits_{c'} \alpha_{c'}^c \psi_{c'} + \sum\nolimits_v \alpha_v^c \psi_v, \tag{3.293}$$

$$\phi_v = \sum\nolimits_c \alpha_c^v \psi_c + \sum\nolimits_{v'} \alpha_{v'}^v \psi_{v'}. \tag{3.294}$$

Combining (3.291) with $n = c$ and (3.293), we obtain

$$(\mathcal{H} + V_R)(\sum\nolimits_{c'} \alpha_{c'}^c \psi_{c'} + \sum\nolimits_v \alpha_v^c \psi_v) = \bar{E}_n(\sum\nolimits_{c'} \alpha_{c'}^c \psi_{c'} + \sum\nolimits_v \alpha_v^c \psi_v). \tag{3.295}$$

Using (3.283), we have

$$\begin{aligned} &\sum\nolimits_{c'} \alpha_{c'}^c E_{c'}\psi_{c'} + \sum\nolimits_v \alpha_v^c E_v\psi_v + \sum\nolimits_{c'} \alpha_{c'}^c V_R\psi_{c'} + \sum\nolimits_v \alpha_v^c V_R\psi_v \\ &\qquad = \bar{E}_c(\sum\nolimits_{c'} \alpha_{c'}^c \psi_{c'} + \sum\nolimits_v \alpha_v^c \psi_v). \end{aligned} \tag{3.296}$$

Using (3.292), this last equation becomes

$$\begin{aligned} &\sum\nolimits_{c'} \alpha_{c'}^c E_{c'}\psi_{c'} + \sum\nolimits_v \alpha_v^c E_v\psi_v + \sum\nolimits_{c'} \alpha_{c'}^c \sum\nolimits_c \langle F_c|\psi_{c'}\rangle\psi_c \\ &\quad + \sum\nolimits_v \alpha_v^c \sum\nolimits_c \langle F_c|\psi_v\rangle\psi_c = \bar{E}_c(\sum\nolimits_{c'} \alpha_{c'}^c \psi_{c'} + \sum\nolimits_v \alpha_v^c \psi_v). \end{aligned} \tag{3.297}$$

This can be recast as

$$\begin{aligned}&\sum_{c'c''}[(E_{c'}-\overline{E}_c)\delta_{c'}^{c''}+\langle F_{c'}|\psi_{c''}\rangle]\alpha_{c''}^{c}\psi_{c'}\\&\quad+\sum_{c'}\sum_{v}\alpha_v^c\langle F_{c'}|\psi_v\rangle\psi_{c'}+\sum_v\alpha_v^c(E_v-\overline{E}_c)\psi_v=0.\end{aligned}\tag{3.298}$$

Taking the inner product of (3.298) with $\psi_{v'}$ gives

$$\sum_v\alpha_v^c(E_v-\overline{E}_c)\delta_v^{v'}=0\quad\text{or}\quad\alpha_{v'}^c(E_{v'}-\overline{E}_c)=0\quad\text{or}\quad\alpha_{v'}^c=0\,.$$

unless there is some sort of strange accidental degeneracy. We shall ignore such degeneracies. This means by (3.293) that

$$\phi_c=\sum_{c'}\alpha_{c'}^c\psi_{c'}\,.\tag{3.299}$$

Equation (3.298) becomes

$$\sum_{c'c''}[(E_{c'}-\overline{E}_c)\delta_{c'}^{c''}+\langle F_{c'}|\psi_{c''}\rangle]\alpha_{c''}^c\psi_{c'}=0\,.\tag{3.300}$$

Taking the matrix element of (3.300) with the core state ψ_c and summing out a resulting Kronecker delta function, we have

$$\sum_{c''}[(E_c-\overline{E}_c)\delta_c^{c''}+\langle F_c|\psi_{c''}\rangle]\alpha_{c''}^{c'}=0\,.\tag{3.301}$$

For nontrivial solutions of (3.301), we must have

$$\det[(E_c-\overline{E}_c)\delta_c^{c''}+\langle F_c|\psi_{c''}\rangle]=0\,.\tag{3.302}$$

The point to (3.302) is that the "core" eigenvalues $\overline{E}_c$ are formally determined.

Combining (3.291) with $n = v$, and using ϕ_v from (3.294), we obtain

$$(H+V_R)(\sum_c\alpha_c^v\psi_c+\sum_{v'}\alpha_{v'}^v\psi_{v'})=\overline{E}_v(\sum_c\alpha_c^v\psi_c+\sum_{v'}\alpha_{v'}^v\psi_{v'})\,.$$

By (3.283) this becomes

$$\begin{aligned}&\sum_c\alpha_c^vE_c\psi_c+\sum_{v'}\alpha_{v'}^vE_{v'}\psi_{v'}+\sum_c\alpha_c^vV_R\psi_c+\sum_{v'}\alpha_{v'}^vV_R\psi_{v'}\\&\qquad=\overline{E}_v(\sum_c\alpha_c^v\psi_c+\sum_{v'}\alpha_{v'}^v\psi_{v'}).\end{aligned}$$

Using (3.292), this becomes

$$\begin{aligned}&\sum_c\alpha_c^v(E_c-\overline{E}_v)\psi_c+\sum_{v'}\alpha_{v'}^v(E_{v'}-\overline{E}_v)\psi_{v'}\\&\quad+\sum_c\alpha_c^v\sum_{c'}\langle F_{c'}|\psi_c\rangle\psi_{c'}+\sum_{v'}\alpha_{v'}^v\sum_c\langle F_c|\psi_{v'}\rangle\psi_c=0.\end{aligned}\tag{3.303}$$

With a little manipulation we can write (3.303) as

$$\begin{aligned}&\sum_{c,c'}[(E_c-\overline{E}_v)\delta_{cc'}+\langle F_c|\psi_{c'}\rangle]\alpha_{c'}^v\psi_c\\&+\sum_c\alpha_v^v\langle F_c|\psi_v\rangle\psi_c+\sum_{v'(\neq v),c}\alpha_{v'}^v\langle F_c|\psi_{v'}\rangle\psi_c\\&+(E_v-\overline{E}_v)\alpha_v^v\psi_v+\sum_{v'(\neq v)}(E_{v'}-\overline{E}_v)\alpha_{v'}^v\psi_{v'}=0.\end{aligned}\tag{3.304}$$

Taking the inner product of (3.304) with ψ_v and $\psi_{v''}$, we find

$$(E_v - \overline{E}_v)\alpha_v^v = 0\,, \tag{3.305}$$

and

$$(E_{v''} - \overline{E}_v)\alpha_{v''}^v = 0\,. \tag{3.306}$$

This implies that $E_v = \overline{E}_v$ and

$$\alpha_{v''}^v = 0\,.$$

The latter result is really true only in the absence of degeneracy in the set of E_v. Combining with (3.294), we have (if $\alpha_v^v = 1$)

$$\phi_v = \psi_v + \sum_c \alpha_c^v \psi_c\,. \tag{3.307}$$

Equation (3.304) can now be written

$$\sum_{c'}[(E_{c''} - E_v)\delta_{c''}^{c'} + \langle F_{c''}|\psi_{c'}\rangle]\alpha_{c'}^v = -\langle F_{c''}|\psi_v\rangle\,. \tag{3.308}$$

With these results we can understand the general pseudopotential theorem as given by Austin at al.:

The pseudo-Hamiltonian $\mathcal{H}_P = \mathcal{H} + V_R$, *where* $V_R\phi = \sum_c \langle F_c|\phi\rangle\psi_c$, *has the same valence eigenvalues* E_v *as* $\mathcal{H}$ *does. The eigenfunctions are given by* (3.299) *and* (3.307).

We get a particularly interesting form for the pseudopotential if we choose the arbitrary function to be

$$F_c = -V\psi_c\,. \tag{3.309}$$

In this case

$$V_R\phi = -\sum_c \langle\psi_c|V|\phi\rangle\psi_c\,, \tag{3.310}$$

and thus the pseudo-Hamiltonian can be written

$$\mathcal{H}_p\phi_n = (T + V + V_R)\phi_n = T\phi_n + V\phi_n - \sum_c \psi_c\langle\psi_c|V\phi_n\rangle\,. \tag{3.311}$$

Note that by completeness

$$\begin{aligned} V\phi_n &= \sum_m a_m\psi_m \\ &= \sum_m \psi_m\langle\psi_m|V\phi_n\rangle \\ &= \sum_c \psi_c\langle\psi_c|V\phi_n\rangle + \sum_v \psi_v\langle\psi_v|V\phi_n\rangle \end{aligned}\,,$$

so

$$V\phi_n - \sum_c \psi_c\langle\psi_c|V\phi_n\rangle = \sum_v \psi_v\langle\psi_v|V\phi_n\rangle\,. \tag{3.312}$$

If the ψ_c are almost a complete set for $V\phi_n$, then the right-hand side of (3.312) is very small and hence

$$\mathcal{H}_p\phi_n \cong T\phi_n\,. \tag{3.313}$$

This is another way of looking at the cancellation theorem. Notice this equation is just the free-electron approximation, and, furthermore, $\mathcal{H}_p$ has the same eigenvalues as $\mathcal{H}$. Thus we see how the nearly free-electron approximation is partially justified by the pseudopotential.

Physically, the use of a pseudopotential assures us that the valence wave functions are orthogonal to the core wave functions. Using (3.307) and the orthonormality of the core and valence eigenfunction, we can write

$$\left|\psi_v\right\rangle = \left|\phi_v\right\rangle - \sum_c \left|\psi_c\right\rangle \left\langle\psi_c\middle|\phi_v\right\rangle \tag{3.314}$$

$$\equiv (I - \sum_c \left|\psi_c\right\rangle \left\langle\psi_c\right|) \left|\phi_v\right\rangle . \tag{3.315}$$

The operator $(I - \sum_c|\psi_c\rangle\langle\psi_c|)$ simply projects out from $|\phi_v\rangle$ all components that are perpendicular to $|\psi_c\rangle$. We can crudely say that the valence electrons would have to wiggle a lot (and hence raise their energy) to be in the vicinity of the core and also be orthogonal to the core wave function. The valence electron wave functions have to be orthogonal to the core wave functions and so they tend to stay out of the core. This effect can be represented by an effective repulsive pseudopotential that tends to cancel out the attractive core potential when we use the effective equation for calculating volume wave functions.

Since V_R can be constructed so as to cause $V + V_R$ to be small in the core region, the following simplified form of the pseudopotential V_P is sometimes used.

$$\begin{aligned} V_P(r) &= -\frac{Ze}{4\pi\varepsilon_0 r} && \text{for } r > r_{core} \\ V_P(r) &= 0 && \text{for } r \le r_{core} \end{aligned} \tag{3.316}$$

This is sometimes called the empty-core pseudopotential or empty-core method (ECM).

Cohen [3.12, 3.13], has developed an empirical pseudopotential model (EPM) that has been very effective in relating band-structure calculations to optical properties. He expresses $V_p(\boldsymbol{r})$ in terms of Fourier components and structure factors (see [3.12, p. 21]). He finds that only a few Fourier components need be used and fitted from experiment to give useful results. If one uses the correct nonlocal version of the pseudopotential, things are more complicated but still doable [3.12, p. 23]. Even screening effects can be incorporated as discussed by Cohen and Heine [3.13].

Note that the pseudopotential can be broken up into different core angular momentum components (where the core wave functions are expressed in atomic form). To see this, write

$$\left|c\right\rangle = \left|N, L\right\rangle ,$$

where N is all the quantum number necessary to define c besides L. Thus

$$\begin{aligned} V_R &= \sum_c |c\rangle(E - E_c)\langle c| \\ &= \sum_L \left(\sum_N |N,L\rangle(E - E_{N,L})\langle N,L| \right) \end{aligned}.$$

This may help in finding simplified calculations.

For further details see Chelikowsky and Louie [3.10]. This is a Festschrift in honor of Marvin L. Cohen. This volume shows how the calculations of Cohen and his school intertwine with experiment: in many cases explaining experimental results, and in other cases predicting results with consequent experimental verification. We end this discussion of pseudopotentials with a qualitative roundup.

Table 3.4. Band structure and related references

Band-structure calculational techniques	Reference	Comments
Nearly free electron methods (NFEM)	3.2.3	Perturbed electron gas of free electrons
Tight binding/LCAO methods (TBM)	3.2.3	Starts from atomic nature of electron states.
Wigner–Seitz method	[3.57], 3.2.3	First approximate quantitative solution of wave equation in crystal.
Augmented plane wave and related methods (APW)	[3.16], [63], 3.2.3	Muffin tin potential with spherical wave functions inside and plane wave outside (Slater).
Orthogonalized plane wave methods (OPW)	Jones [58] Ch. 6, [3.58], 3.2.3	Basis functions are plane waves plus core wave functions (Herring). Related to pseudopotential.
Empirical pseudopotential methods (EPM) as well as Self-consistent and ab initio pseudopotential methods	[3.12, 3.20]	Builds in orthogonality to core with a pseudopotential.
Kohn–Korringa–Rostocker or KKR Green function methods	[3.26]	Related to APW.
Kohn–Sham density functional Techniques (for many-body properties)	[3.23, 3.25, 3.27, 3.28]	For calculating ground-state properties.
$\boldsymbol{k} \cdot \boldsymbol{p}$ Perturbation Theory	[3.5, 3.16, 3.26], 3.2.3	An interpolation scheme.
G. W. approximation	[3.2]	G is for Green's function, W for Coulomb interaction, Evaluates self-energy of quasi-particles.
General reference	[3.1, 3.37]	

As already mentioned, M. L. Cohen's early work (in the 1960s) was with the empirical pseudopotential. In brief review, the pseudopotential idea can be traced back to Fermi and is clearly based on the orthogonalized plane wave (OPW) method of Conyers Herring. In the pseudopotential method for a solid, one considers the ion cores as a background in which the valence electrons move. J. C. Phillips and L. Kleinman demonstrated how the requirement of orthogonality of the valence wave function to core atomic functions could be folded into the potential. M. L. Cohen found that the pseudopotentials converged rapidly in Fourier space, and so only a few were needed for practical calculations. These could be fitted from experiment (reflectivity for example), and then the resultant pseudopotential was very useful in determining the optical response – this method was particularly useful for several semiconductors. Band structures, and even electron–phonon interactions were usefully determined in this way. M. L. Cohen and his colleagues have continually expanded the utility of pseudopotentials. One of the earliest extensions was to an angular-momentum-dependent nonlocal pseudopotential, as discussed above. This was adopted early on in order to improve the accuracy, at the cost of more computation. Of course, with modern computers, this is not much of a drawback.

Nowadays, one often uses a pseudopotential-density functional method. One can thus develop ab initio pseudopotentials. The density functional method (in say the local density approximation – LDA) allows one to treat the electron–electron interaction in the core of the atom quite accurately. As we have already shown, the density functional method reduces a many-electron problem to a set of one-electron equations (the Kohn–Sham equations) in a rational way. Morrel Cohen (another pioneer in the elucidation of pseudopotentials, see Chap. 23 of Chelikowsky and Louie, op cit) has said, with considerable truth, that the Kohn–Sham equations taught us the real meaning of our one-electron calculations. One then uses the pseudopotential to treat the interaction between the valence electrons and the ion core. Again as noted, the pseudopotential allows us to understand why the electron–ion core interaction is apparently so small. This combined pseudopotential-density functional approach has facilitated good predictions of ground-state properties, phonon vibrations, and structural properties such as phase transitions caused by pressure.

There are still problems that need additional attention, such as the correct prediction of bandgaps, but it should not be overlooked that calculations on real materials, not "toy" models are being considered. In a certain sense, M. L. Cohen and his colleagues are developing a "Standard Model of Condensed Matter Physics." The Holy Grail is to feed in only information about the constituents, and from there, at a given temperature and pressure, to predict all solid-state properties. Perhaps at some stage one can even theoretically design materials with desired properties. Along this line, the pseudopotential-density functional method is now being applied to nanostructures such as arrays of quantum dots (nanophysics, quantum dots, etc. are considered in Chap. 12 of Chelikowsky and Louie).

We have now described in some detail the methods of calculating the $E(\boldsymbol{k})$ relation for electrons in a perfect crystal. Comparisons of actual calculations with experiment will not be made here. Later chapters give some details about the type of experimental results that need $E(\boldsymbol{k})$ information for their interpretation. In particular, the Section on the Fermi surface gives some details on experimental results that can be obtained for the conduction electrons in metals. Further references for band-structure calculations are in Table 3.4. See also Altman [3.1].

The Spin-Orbit Interaction (B)

As shown in Appendix F, the spin-orbit effect can be correctly derived from the Dirac equation. As mentioned there, perhaps the most familiar form of the spin-orbit interaction is the form that is appropriate for spherical symmetry. This form is

$$\mathcal{H}' = f(r)\boldsymbol{L} \cdot \boldsymbol{S} \,. \tag{3.317}$$

In (3.317), $\mathcal{H}'$ is the part of the Hamiltonian appropriate to the spin-orbit interaction and hence gives the energy shift for the spin-orbit interaction. In solids, spherical symmetry is not present and the contribution of the spin-orbit effect to the Hamiltonian is

$$\mathcal{H} = \frac{\hbar}{2m_0^2c^2}\boldsymbol{S} \cdot (\nabla V \times \boldsymbol{p}) \,. \tag{3.318}$$

There are other relativistic corrections that derive from approximating the Dirac equation but let us neglect these.

A relatively complete account of spin-orbit splitting will be found in Appendix 9 of the second volume of Slater's book on the quantum theory of molecules and solids [89]. Here, we shall content ourselves with making a few qualitative observations. If we look at the details of the spin-orbit interaction, we find that it usually has unimportant effects for states corresponding to a general point of the Brillouin zone. At symmetry points, however, it can have important effects because degeneracies that would otherwise be present may be lifted. This lifting of degeneracy is often similar to the lifting of degeneracy in the atomic case. Let us consider, for example, an atomic case where the $j = l \pm ½$ levels are degenerate in the absence of spin-orbit interaction. When we turn on a spin-orbit interaction, two levels arise with a splitting proportional to $\boldsymbol{L} \cdot \boldsymbol{S}$ (using $J^2 = L^2 + S^2 + 2\boldsymbol{L} \cdot \boldsymbol{S}$). The energy difference between the two levels is proportional to

$$\begin{aligned}&(l+\tfrac{1}{2})(l+\tfrac{3}{2}) - l(l+1) - \tfrac{1}{2}(\tfrac{3}{2}) - (l-\tfrac{1}{2})(l+\tfrac{1}{2}) + l(l+1) + \tfrac{1}{2}(\tfrac{3}{2})\\ &= (l+\tfrac{1}{2})[(l+\tfrac{3}{2}) - l + \tfrac{1}{2}] = (l+\tfrac{1}{2}) \cdot 2 = 2l+1 \,.\end{aligned}$$

This result is valid when $l > 0$. When $l = 0$, there is no splitting. Similar results are obtained in solids. A practical case is shown in Fig. 3.20. Note that we might have been able to guess (a) and (b) from the atomic consideration given above.

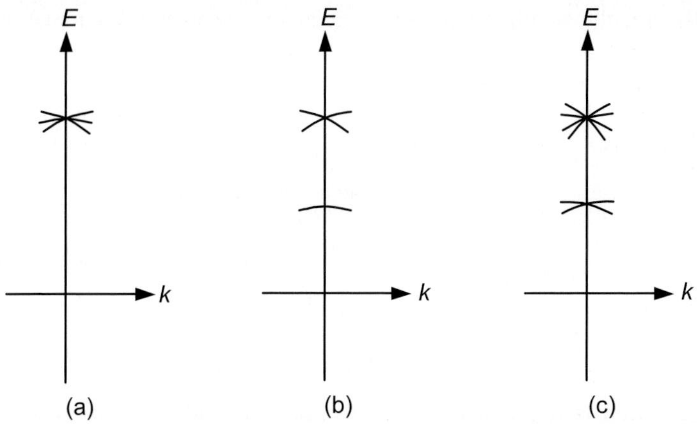

Fig. 3.20. Effect of spin-orbit interaction on the $l = 1$ level in solids: (**a**) no spin-orbit, six degenerate levels at $k = 0$ (a point of cubic symmetry), (**b**) spin-orbit with inversion symmetry (e.g. Ge), (c) spin-orbit without inversion symmetry (e.g. InSb). [Adapted from Ziman JM, *Principles of the Theory of Solids*, Cambridge University Press, New York, 1964, Fig. 54, p. 100. By permission of the publisher.]

3.2.4 Effect of Lattice Defects on Electronic States in Crystals (A)

The results that will be derived here are similar to the results that were derived for lattice vibrations with a defect (see Sect. 2.2.5). In fact, the two methods are abstractly equivalent; it is just that it is convenient to have a little different formalism for the two cases. Unified discussions of the impurity state in a crystal, including the possibility of localized spin waves, are available.[24] Only the case of one-dimensional motion will be considered here; however, the method is extendible to three dimensions.

The model of defects considered here is called the Slater–Koster model.[25] In the discussion below, no consideration will be given to the practical details of the calculation. The aim is to set up a general formalism that is useful in the understanding of the general features of electronic impurity states.[26] The Slater–Koster model is also useful for discussing deep levels in semiconductors (see Sect. 11.3).

In order to set the notation, the Schrödinger equation for stationary states will be rewritten:

$$\mathcal{H}\psi_{n,k}(x) = E_n(k)\psi_{n,k}(x)\,. \tag{3.319}$$

[24] See Izynmov [3.24].

[25] See [3.49, 3.50]

[26] Wannier [95, p181ff]

In (3.319), $\mathcal{H}$ is the Hamiltonian without defects, n labels the different bands, and k labels the states within each band. The solutions of (3.319) are assumed known.

We shall now suppose that there is a localized perturbation (described by V) on one of the lattice sites of the crystal. For the perturbed crystal, the equation that must be solved is

$$(\mathcal{H} + V)\psi = E\psi\,. \tag{3.320}$$

(This equation is true by definition; $\mathcal{H} + V$ is by definition the total Hamiltonian of the crystal with defect.)

Green's function for the problem is defined by

$$\mathcal{H}G_E(x, x_0) - EG_E(x, x_0) = -4\pi\delta(x - x_0)\,. \tag{3.321}$$

Green's function is required to satisfy the same boundary conditions as $\psi_{nk}(x)$. Writing $\psi_{nk} = \psi_m$, and using the fact that the ψ_m form a complete set, we can write

$$G_E(x, x_0) = \sum\nolimits_m A_m \psi_m(x)\,. \tag{3.322}$$

Substituting (3.322) into the equation defining Green's function, we obtain

$$\sum\nolimits_m A_m (E_m - E)\psi_m(x) = -4\pi\delta(x - x_0)\,. \tag{3.323}$$

Multiplying both sides of (3.323) by $\psi_n^*(x)$ and integrating, we find

$$A_n = -4\pi \frac{\psi_n^*(x_0)}{E_n - E}\,. \tag{3.324}$$

Combining (3.324) with (3.322) gives

$$G_E(x, x_0) = -4\pi \sum\nolimits_m \frac{\psi_m^*(x_0)\psi_n(x)}{E_m - E}\,. \tag{3.325}$$

Green's function has the property that it can be used to convert a differential equation into an integral equation. This property can be demonstrated. Multiply (3.320) by G_E^* and integrate:

$$\int G_E^* H\psi \mathrm{d}x - E\int G_E^* \psi \mathrm{d}x = -\int G_E^* V\psi \mathrm{d}x\,. \tag{3.326}$$

Multiply the complex conjugate of (3.321) by ψ and integrate:

$$\int \psi H G_E^* \mathrm{d}x - E\int G_E^* \psi \mathrm{d}x = -4\pi\psi(x_0)\,. \tag{3.327}$$

Since $\mathcal{H}$ is Hermitian,

$$\int G_E^* \mathcal{H}\psi dx = \int \psi \mathcal{H} G_E^* dx\,. \tag{3.328}$$

Thus subtracting (3.326) from (3.327), we obtain

$$\psi(x_0) = \frac{1}{4\pi} \int G_E^*(x, x_0) V(x) \psi(x) \mathrm{d}x \,. \tag{3.329}$$

Therefore the equation governing the impurity problem can be formally written as

$$\psi(x_0) = -\sum\nolimits_{n,k} \frac{\psi_{n,k}(x_0)}{E_n(k) - E} \int \psi_{n,k}^*(x) V(x) \psi(x) \mathrm{d}x \,. \tag{3.330}$$

Since the $\psi_{n,k}(x)$ form a complete orthonormal set of wave functions, we can define another complete orthonormal set of wave functions through the use of a unitary transformation. The unitary transformation most convenient to use in the present problem is

$$\psi_{n,k}(x) = \frac{1}{\sqrt{N}} \sum\nolimits_j \mathrm{e}^{\mathrm{i}k(ja)} A_n(x - ja) \,. \tag{3.331}$$

Equation (3.331) should be compared to (3.244), which was used in the tight binding approximation. We see the $\phi_0(\boldsymbol{r} - \boldsymbol{R}_i)$ are analogous to the $A_n(x - ja)$. The $\phi_0(\boldsymbol{r} - \boldsymbol{R}_i)$ are localized atomic wave functions, so that it is not hard to believe that the $A_n(x - ja)$ are localized. The $A_n(x - ja)$ are called Wannier functions.[27]

In (3.331), a is the spacing between atoms in a one-dimensional crystal (with N unit cells) and so the ja (for j an integer) labels the coordinates of the various atoms. The inverse of (3.331) is given by

$$A_n(x - ja) = \frac{1}{\sqrt{N}} \sum\nolimits_{k(\text{a Brillouin zone})} \mathrm{e}^{-\mathrm{i}k(ja)} \psi_{n,k}(x) \,. \tag{3.332}$$

If we write the $\psi_{n,k}$ as functions satisfying the Bloch condition, it is possible to give a somewhat simpler form for (3.332). However, for our purposes (3.332) is sufficient.

Since (3.332) form a complete set, we can expand the impurity-state wave function ψ in terms of them:

$$\psi(x) = \sum\nolimits_{l,i} U_l(ia) A_l(x - ia) \,. \tag{3.333}$$

Substituting (3.331) and (3.333) into (3.330) gives

$$\sum\nolimits_{l,i'} U_l(i'a) A_l(x - i'a)$$
$$= -\sum\nolimits_{\substack{l,i' \\ j,j'}}^{n,k} \frac{1}{N} \frac{\mathrm{e}^{ikja}}{E - E_n(k)} A_n(x_0 - ja) \int \mathrm{e}^{-\mathrm{i}kj'a} A_n^*(x - j'a) V U_l(i'a) A_l(x - i'a) dx. \tag{3.334}$$

[27] See Wannier [3.56].

Multiplying the above equation by $A_m^*(x_0 - pa)$, integrating over all space, using the orthonormality of the A_m, and defining

$$V_{n,l}(j',i) = \int A_n^*(x - j'a)VA_l(x - ia)\mathrm{d}x\,, \tag{3.335}$$

we find

$$\sum\nolimits_{l,i'} U_l(i'a)\left[\delta_l^m\delta_{i'}^p + \frac{1}{N}\sum\nolimits_{k,j'}\frac{\mathrm{e}^{\mathrm{i}k(pa-j'a)}}{E_m(k)-E}V_{m,l}(j',i')\right] = 0\,. \tag{3.336}$$

For a nontrivial solution, we must have

$$\det\left[\delta_l^m\delta_{i'}^p + \frac{1}{N}\sum\nolimits_{k,j'}\frac{\mathrm{e}^{\mathrm{i}k(p-j')a}}{E_m(k)-E}V_{m,l}(j',i')\right] = 0\,. \tag{3.337}$$

This appears to be a very difficult equation to solve, but if $V_{ml}(j', i) = 0$ for all but a finite number of terms, then the determinant would be drastically simplified.

Once the energy of a state has been found, the expansion coefficients may be found by going back to (3.334).

To show the type of information that can be obtained from the Slater–Koster model, the potential will be assumed to be short range (centered on $j = 0$), and it will be assumed that only one band is involved. Explicitly, it will be assumed that

$$V_{m,l}(j',i) = \delta_l^b\delta_m^b\delta_{j'}^0\delta_{i'}^0V_0\,. \tag{3.338}$$

Note that the local character of the functions defined by (3.332) is needed to make such an approximation.

From (3.337) and (3.338) we find that the condition on the energy is

$$f(E) \equiv \frac{N}{V_0} + \sum\nolimits_k \frac{1}{E_b(k)-E} = 0\,. \tag{3.339}$$

Equation (3.339) has N real roots. If $V_0 = 0$, the solutions are just the unperturbed energies $E_b(k)$. If $V_0 \neq 0$, then we can use graphical methods to find E such that $f(E)$ is zero. See Fig. 3.21. In the figure, V_0 is assumed to be negative.

The crosses in Fig. 3.21 are the perturbed energies; these are the roots of $f(E)$. The poles of $f(E)$ are the unperturbed levels. The roots are all smaller than the unperturbed roots if V_0 is negative and larger if V_0 is positive. The size of the shift in E due to V_0 is small (negligible for large N) for all roots but one. This is characterized by saying that all but one level is "pinned" in between two unperturbed levels. As expected, these results are similar to the lattice defect vibration problem. It should be intuitive, if not obvious, that the state that splits off from the band for V_0 negative is a localized state. We would get one such state for each band.

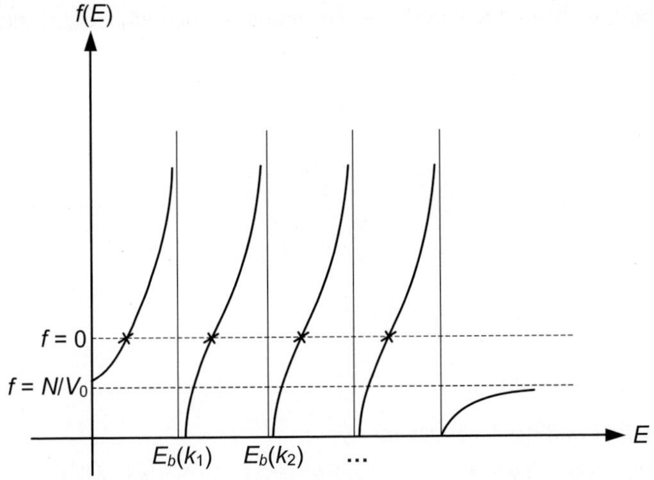

Fig. 3.21. A qualitative plot of $f(E)$ versus E for the Slater–Koster model. The crosses determine the energies that are solutions of (3.339)

This Section has discussed the effects of isolated impurities on electronic states. We have found, except for the formation of isolated localized states, that the Bloch view of a solid is basically unchanged. A related question is what happens to the concept of Bloch states and energy bands in a disordered alloy. Since we do not have periodicity here, we might expect these concepts to be meaningless. In fact, the destruction of periodicity may have much less effect on Bloch states than one might imagine. The changes caused by going from a periodic potential to a potential for a disordered lattice may tend to cancel one another out.[28] However, the entire subject is complex and incompletely understood. For example, sufficiently large disorder can cause localization of electron states.[29]

Problems

3.1 Use the variational principle to find the approximate ground-state energy of the helium atom (two electrons). Assume a trial wave function of the form $\exp[-\eta(r_1+r_2)]$, where r_1 and r_2 are the radial coordinates of the electron.

3.2 By use of (3.17) and (3.18) show that $\int |\psi|^2 \mathrm{d}\tau = N!\,|M|^2$.

3.3 Derive (3.31) and explain physically why $\sum_1^N \varepsilon_k \neq E$.

---

[28] For a discussion of these and related questions, see Stern [3.53], and references cited therein.

[29] See Cusack [3.15].

3.4 For singly charged ion cores whose charge is smeared out uniformly and for plane-wave solutions so that $|\psi_j| = 1$, show that the second and third terms on the left-hand side of (3.50) cancel.

3.5 Show that

$$\lim_{k\to 0}\frac{k_M^2 - k^2}{kk_M}\ln\left|\frac{k_M + k}{k_M - k}\right| = 2\,,$$

and

$$\lim_{k\to k_M}\frac{k_M^2 - k^2}{kk_M}\ln\left|\frac{k_M + k}{k_M - k}\right| = 0\,,$$

relate to (3.64) and (3.65).

3.6 Show that (3.230) is equivalent to

$$E_{\boldsymbol{k}} = \tfrac{1}{2}(E_{\boldsymbol{k}}^0 + E_{\boldsymbol{k}'}^0) \pm \tfrac{1}{4}[4|V(\boldsymbol{K}')|^2 + (E_{\boldsymbol{k}}^0 - E_{\boldsymbol{k}'}^0)^2]^{1/2}\,,$$

where

$$E_{\boldsymbol{k}}^0 = V(0) + \frac{\hbar^2 \boldsymbol{k}^2}{2m} \quad \text{and} \quad E_{\boldsymbol{k}'}^0 = V(0) + \frac{\hbar^2}{2m}(\boldsymbol{k}+\boldsymbol{K}')^2\,.$$

3.7 Construct the first Jones zone for the simple cubic lattice, face-centered cubic lattice, and body-centered cubic lattice. Describe the fcc and bcc with a sc lattice with basis. Assume identical atoms at each lattice point.

3.8 Use (3.255) to derive E_0 for the simple cubic lattice, the body-centered cubic lattice, and the face-centered cubic lattice.

3.9 Use (3.256) to derive the density of states for free electrons. Show that your results check (3.164).

3.10 For the one-dimensional potential well shown in Fig. 3.22 discuss either mathematically or physically the behavior of the low-lying energy levels as a function of V_0, b, and a. Do you see any analogies to band structure?

3.11 How does soft X-ray emission differ from the more ordinary type of X-ray emission?

3.12 Suppose the first Brillouin zone of a two-dimensional crystal is as shown in Fig. 3.23 (the shaded portion). Suppose that the surfaces of constant energy are either circles or pieces of circles as shown. Suppose also that where k is on a sphere or a spherical piece that $E = (\hbar^2/2m)k^2$. With all of these assumptions, compute the density of states.

3.13 Use Fermi–Dirac statistics to evaluate approximately the low-temperature specific heat of quasi free electrons in a two-dimensional crystal.

3.14 For a free-electron gas at absolute zero in one dimension, show the average energy per electron is one third of the Fermi energy.

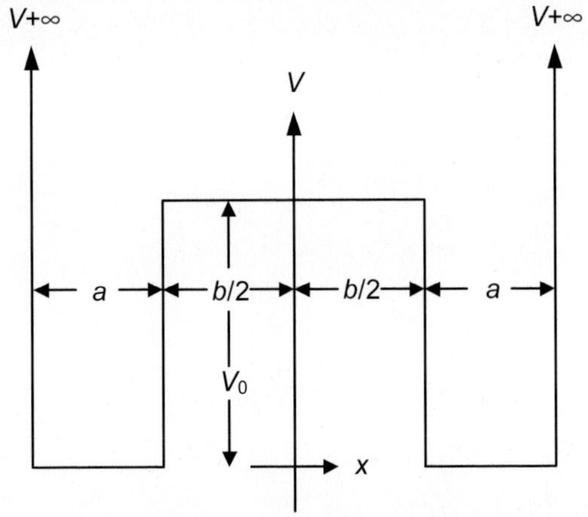

Fig. 3.22. A one-dimensional potential well

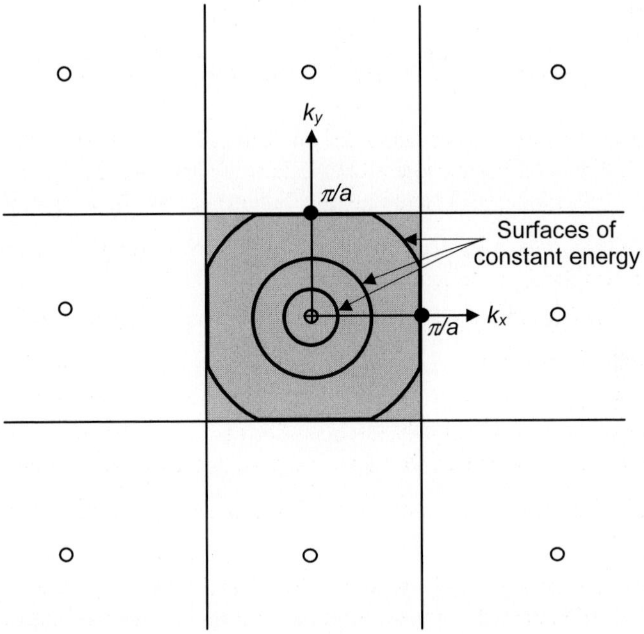

Fig. 3.23. First Brillouin zone and surfaces of constant energy in a simple two-dimensional reciprocal lattice

4 The Interaction of Electrons and Lattice Vibrations

4.1 Particles and Interactions of Solid-state Physics (B)

There are, in fact, two classes of types of interactions that are of interest. One type involves interactions of the solid with external probes (such as electrons, positrons, neutrons, and photons). Perhaps the prime example of this is the study of the structure of a solid by the use of X-rays as discussed in Chap. 1. In this chapter, however, we are more concerned with the other class of interactions; those that involve interactions of the elementary energy excitations among themselves.

So far the only energy excitations that we have discussed are phonons (Chap. 2) and electrons (Chap. 3). Thus the kinds of internal interactions that we consider at present are electron–phonon, phonon–phonon, and electron–electron. There are of course several othe kinds of elementary energy excitations in solids and thus there are many other examples of interaction. Several of these will be treated in later parts of this book. A summary of most kinds of possible pair wise interactions is given in Table 4.1.

The concept of the "particle" as an entity by itself makes sense only if its life time in a given state is fairly long even with the interactions. In fact interactions between particles may be of such character as to form new "particles." Only a limited number of these interactions will be important in discussing any given experiment. Most of them may be important in discussing all possible experiments. Some of them may not become important until entirely new types of solids have been formed. In view of the fact that only a few of these interactions have actually been treated in detail, it is easy to believe that the field of solid-state physics still has a considerable amount of growing to do.

We have not yet defined all of the fundamental energy excitations.[1] Several of the excitations given in Table 4.1 are defined in Table 4.2. Neutrons, positrons, and photons, while not solid-state particles, can be used as external probes. For some purposes, it may be useful to make the distinctions in terminology that are noted in Table 4.3. However, in this book, we hope the meaning of our terms will be clear from the context in which they are used.

---

[1] A simplified approach to these ideas is in Patterson [4.33]. See also Mattuck [17, Chap. 1].

Table 4.1. Possible sorts of interactions of interest in interpreting solid-state experiments*

	1 e^-	2 h	3 ph	4 m	5 pl	6 b	7 ex	8 ext	9 pe	10 he	11 n	12 e^+	13 ν
1. electrons (e^-)	e^-–e^-												
2. holes (h)	h–e^-	h–h											
3. phonons (ph)	ph–e^-	ph–h	ph–ph										
4. magnons (m)	m–e^-	m–h	m–ph	m–m									
5. plasmons (pl)	pl–e^-	pl–h	pl–ph	pl–m	pl–pl								
6. bogolons (b)	b–e^-	b–h	b–ph	b–m	b–pl	b–b							
7. excitons (ex)	ex–e^-	ex–h	ex–ph	ex–m	ex–pl	ex–b	ex–ex						
8. politarons (pn)	pn–e^-	pn–h	pn–ph	pn–m	pn–pl	pn–b	pn–ex	pn–pn					
9. polarons (po)	po–e^-	po–h	po–ph	po–m	po–pl	po–b	po–ex	po–pn	po–po				
10. helicons (he)	he–e^-	he–h	he–ph	he–m	he–pl	he–b	he–ex	he–pn	he–po	he–he			
11. neutrons (n)	n–e^-	n–h	n–ph	n–m	n–pl	n–b	n–ex	n–pn	n–po	n–he	n–n		
12. positrons (e^+)	e^+–e^-	e^+–h	e^+–ph	e^+–m	e^+–pl	e^+–b	e^+–ex	e^+–pn	e^+–po	e^+–he	e^+–n	e^+–e^+	
13. photons (ν)	ν–e^-	ν–h	ν–ph	ν–m	ν–pl	ν–b	ν–ex	ν–pn	ν–po	ν–he	ν–n	ν–e^+	ν–ν

* For actual use in a physical situation, each interaction would have to be carefully examined to make sure it did not violate some fundamental symmetry of the physical system and that a physical mechanism to give the necessary coupling was present. Each of these quantities are defined in Table 4.2.

Table 4.2. Solid-state particles and related quantities

Bogolon (or Bogoliubov quasiparticles)	Elementary energy excitations in a superconductor. Linear combinations of electrons in $(+k, +)$, and holes in $(-k, -)$ states. See Chap. 8. The + and − after the ks refer to "up" and "down" spin states.
Cooper pairs	Loosely coupled electrons in the states $(+k, +)$, $(-k, -)$. See Chap. 8.
Electrons	Electrons in a solid can have their masses dressed due to many interactions. The most familiar contribution to their effective mass is due to scattering from the periodic static lattice. See Chap. 3.
Mott–Wannier and Frenkel excitons	The Mott–Wannier excitons are weakly bound electron-hole pairs with energy less than the energy gap. Here we can think of the binding as hydrogen-like except that the electron–hole attraction is screened by the dielectric constant and the mass is the reduced mass of the effective electron and hole masses. The effective radius of this exciton is the Bohr radius modified by the dielectric constant and effective reduced mass of electron and hole. Since the static dielectric constant can only have meaning for dimensions large compared with atomic dimensions, strongly bound excitations as in, e.g., molecular crystals are given a different name Frenkel excitons. These are small and tightly bound electron–hole pairs. We describe Frenkel excitons with a hopping excited state model. Here we can think of the energy spectrum as like that given by tight binding. Excitons may give rise to absorption structure below the bandgap. See Chap. 10.
Helicons	Slow, low-frequency (much lower than the cyclotron frequency), circularly polarized propagating electromagnetic waves coupled to electrons in a metal that is in a uniform magnetic field that is in the direction of propagation of the electromagnetic waves. The frequency of helicons is given by (see Chap. 10) $$\omega_H = \frac{\omega_c (kc)^2}{\omega_p^2}.$$
Holes	Vacant states in a band normally filled with electrons. See Chap. 5.
Magnon	The low-lying collective states of spin systems, found in ferromagnets, ferrimagnets, antiferromagnets, canted, and helical spin arrays, whose spins are coupled by exchange interactions are called spin waves. Their quanta are called magnons. One can also say the spin waves are fluctuations in density in the spin angular momentum. At very long wavelength, the magnetostatic interaction can dominate exchange, and then one speaks of magnetostatic spin waves. The dispersion relation links the frequency with the reciprocal wavelength, which typically, for ordinary spin waves, at long wavelengths goes as the square of the wave vector for ferromagnets but is linear in the wave vector for antiferromagnets. The magnetization at low temperatures for ferromagnets can be described by spin-wave excitations that reduce it, as given by the famous Bloch $T^{3/2}$ law. See Chap. 7.

Table 4.2. (cont.)

Neutron	Basic neutral constituent of nucleus. Now thought to be a composite of two down quarks and one up quark whose charge adds to zero. Very useful as a scattering projectile in studying solids.
Acoustical phonons	Sinusoidal oscillating wave where the adjacent atoms vibrate in phase with the frequency, vanishing as the wavelength becomes infinite. See Chap. 2.
Optical phonons	Here the frequency does not vanish when the wavelength become infinite and adjacent atoms tend to vibrate out of phase. See Chap. 2.
Photon	Quanta of electromagnetic field.
Plasmons	Quanta of collective longitudinal excitation of an electron gas in a metal involving sinusoidal oscillations in the density of the electron gas. The alkali metals are transparent in the ultraviolet, that is for frequencies above the plasma frequency. In semiconductors, the plasma edge in absorption can occur in the infrared. Plasmons can be observed from the absorption of electrons (which excite the plasmons) incident on thin metallic films. See Chap. 9.
Polaritons	Waves due to the interaction of transverse optical phonons with transverse electromagnetic waves. Another way to say this is that they are coupled or mixed transverse electromagnetic and mechanical waves. There are two branches to these modes. At very low and very high wave vectors the branches can be identified as photons or phonons but in between the modes couple to produce polariton modes. The coupling of modes also produces a gap in frequency through which radiation cannot propagate. The upper and lower frequencies defining the gap are related by the Lyddane–Sachs–Teller relation. See Chap. 10.
Polarons	A polaron is an electron in the conduction band (or hole in the valence band) together with the surrounding lattice with which it is coupled. They occur in both insulators and semiconductors. The general idea is that an electron moving through a crystal interacts via its charge with the ions of the lattice. This electron–phonon interaction leads to a polarization field that accompanies the electron. In particle language, the electron is dressed by the phonons and the combined particle is called the polaron. When the coupling extends over many lattice spacings, one speaks of a large polaron. Large polarons are formed in polar crystals by electrons coulombically interacting with longitudinal optical phonons. One thinks of a large polaron as a particle moving in a band with a somewhat increased effective mass. A small polaron is localized and hops or tunnels from site to site with larger effective mass. An equation for the effective mass of a polaron is: $$m_{\text{polaron}} \cong m \frac{1}{1-\frac{\alpha}{6}},$$ where α is the polaron coupling constant. This equation applies both to small and large polarons.

Table 4.2. (cont.)

Polarons summary	(1) Small polarons: $\alpha > 6$. These are not band-like. The transport mechanism for the charge carrier is that of hopping. The electron associated with a small polaron spends most of its time near a particular ion. (2) Large polarons: $1 < \alpha < 6$. These are band-like but their mobility is low. See Chap. 4.
Positron	The antiparticle of an electron with positive charge.
Proton	A basic constituent of the nucleus thought to be a composite of two up and one down quarks whose charge total equals the negative of the charge on the electron. Protons and neutrons together form the nuclei of solids.

Table 4.3. Distinctions that are sometimes made between solid-state quasi particles (or "particles")

1. Landau quasi particles	Quasi electrons interact weakly and have a long lifetime provided their energies are near the Fermi energy. The Landau quasi electrons stand in one-to-one relation to the real electrons, where a real electron is a free electron in its measured state; i.e. the real electron is already "dressed" (see below for a partial definition) due to its interaction with virtual photons (in the sense of quantum electrodynamics), but it is not dressed in the sense of interactions of interest to solid-state physics. The term Fermi liquid is often applied to an electron gas in which correlations are strong, such as in a simple metal. The normal liquid, which is what is usually considered, means as the interaction is turned on adiabatically and forms the one-to-one correspondence, that there are no bound states formed. Superconducting electrons are not a Fermi liquid.
2. Fundamental energy excitations from ground state of a solid	Quasi particles (e.g. electrons): These may be "dressed" electrons where the "dressing" is caused by mutual electron–electron interaction or by the interaction of the electrons with other "particles." The dressed electron is the original electron surrounded by a "cloud" of other particles with which it is interacting and thus it may have a different effective mass from the real electron. The effective interaction between quasi electrons may be much less than the actual interaction between real electrons. The effective interaction between quasi electrons (or quasi holes) usually means their lifetime is short (in other words, the quasi electron picture is not a good description) unless their energies are near the Fermi energy and so if the quasi electron picture is to make sense, there must be many fewer quasi electrons than real electrons. Note that the term quasi electron as used here corresponds to a Landau quasi electron.

Table 4.3. (cont.)

2. (cont.)	Collective excitations (e.g. phonons, magnons, or plasmons): These may also be dressed due to their interaction with other "particles." In this book these are also called quasi particles but this practice is not followed everywhere. Note that collective excitations do not resemble a real particle because they involve wave-like motion of all particles in the system considered.
3. Excitons and bogolons	Note that *excitons* and *bogolons* do not correspond either to a simple quasi particle (as discussed above) or to a collective excitation. However, in this book we will also call these quasi particles or "particles."
4. Goldstone boson	Quanta of long-wavelength and low-frequency modes associated with conservation laws and broken symmetry. The existence of broken symmetry implies this mode. Broken symmetry (see Sect. 7.2.6) means quantum eigenstates with lower symmetry than the underlying Hamiltonian. Phonons and magnons are examples.

Once we know something about the interactions, the question arises as to what to do with them. A somewhat oversimplified viewpoint is that all solid-state properties can be discussed in terms of fundamental energy excitations and their interactions. Certainly, the interactions are the dominating feature of most transport processes. Thus we would like to know how to use the properties of the interactions to evaluate the various transport coefficients. One way (perhaps the most practical way) to do this is by the use of the Boltzmann equation. Thus in this chapter we will discuss the interactions, the Boltzmann equation, how the interactions fit into the Boltzmann equation, and how the solutions of the Boltzmann equation can be used to calculate transport coefficients. Typical transport coefficients that will be discussed are those for electrical and thermal conductivity.

The Boltzmann equation itself is not very rigorous, at least in the situations where it will be applied in this chapter, but it does yield some practical results that are helpful in interpreting experiments. In general, the development in this whole chapter will not be very rigorous. Many ideas are presented and the main aim will be to get the ideas across. If we treat any interaction with great care, and if we use the interaction to calculate a transport property, we will usually find that we are engaged in a sizeable research project.

In discussing the rigor of the Boltzmann equation, an attempt will be made to show how its predictions can be true, but no attempt will be made to discover the minimum number of assumptions that are necessary so that the predictions made by use of the Boltzmann equation must be true.

It should come as no surprise that the results in this chapter will not be rigorous. The systems considered are almost as complicated as they can be: they are interacting *many-body systems,* and *nonequilibrium* statistical properties are the properties of interest. Low-order perturbation theory will be used to discuss the interactions in the many-body system. An essentially classical technique (the Boltzmann equation) will be used to derive the statistical properties. No precise statement of the errors introduced by the approximations can be given. We start with the phonon–phonon interaction.

4.2 The Phonon–Phonon Interaction (B)

The mathematics is not always easy but we can see physically why phonons scatter phonons. Wave-like motions propagate through a periodic lattice without scattering only if there are no distortions from periodicity. One phonon in a lattice distorts the lattice from periodicity and hence scatters another phonon. This view is a little oversimplified because it is essential to have anharmonic terms in the lattice potential in order for phonon–phonon scattering to occur. These cause the first phonon to modify the original periodicity in the elastic properties.

4.2.1 Anharmonic Terms in the Hamiltonian (B)

From the *Golden rule of* perturbation theory (see for example, Appendix E), the basic quantity that determines the transition probability from one phonon state ($|i\rangle$) to another ($|f\rangle$) is the matrix element $|\langle i|\mathcal{H}^1|f\rangle|^2$, where $\mathcal{H}^1$ is that part of the Hamiltonian that causes phonon–phonon interactions.

For phonon–phonon interactions, the perturbing Hamiltonian $\mathcal{H}^1$ is the part containing the cubic (and higher if necessary) anharmonic terms.

$$\mathcal{H}^1 = \sum_{\substack{lbl'b'l''b'' \\ \alpha,\beta,\gamma}} U^{\alpha,\beta,\gamma}_{lbl'b'l''b''} x^{\alpha}_{lb} x^{\beta}_{l'b'} x^{\gamma}_{l''b''} , \tag{4.1}$$

where $\boldsymbol{x}^{\alpha}$ is the αth component of vector $\boldsymbol{x}$ and U is determined by Taylor's theorem,

$$U^{\alpha,\beta,\gamma}_{lbl'b'l''b''} \equiv \frac{1}{3!}\left(\frac{\partial^3 V}{\partial x^{\alpha}_{lb}\partial x^{\beta}_{l'b'}\partial x^{\gamma}_{l''b''}}\right)_{\text{all } x_{lb}=0} , \tag{4.2}$$

and the V is the potential energy of the atoms as a function of their position. In practice, we generally do not try to calculate the U from (4.2) but we carry them along as parameters to be determined from experiment.

As usual, the mathematics is easier to do if the Hamiltonian is expressed in terms of annihilation and creation operators. Thus it is useful to work toward this end by starting with the transformation (2.190). We find,

$$\begin{aligned}\mathcal{H}^1 = &\frac{1}{N^{3/2}} \sum_{\substack{q,b,q',b'q'',b'' \\ \alpha,\beta,\gamma}} \sum_{l,l',l''} \exp[-\mathrm{i}(\boldsymbol{q}\cdot\boldsymbol{l} + \boldsymbol{q}'\cdot\boldsymbol{l}' + \boldsymbol{q}''\cdot\boldsymbol{l}'')] \\ &\times U^{\alpha,\beta,\gamma}_{lbl'b'l''b''} X'^{\alpha}_{q,b} X'^{\beta}_{q',b'} X'^{\gamma}_{q'',b''} .\end{aligned} \tag{4.3}$$

In (4.3) it is convenient to make the substitutions $\boldsymbol{l}' = \boldsymbol{l} + \boldsymbol{m}'$, and $\boldsymbol{l}'' = \boldsymbol{l} + \boldsymbol{m}''$:

$$\begin{aligned}\mathcal{H}' = &\frac{1}{N^{3/2}} \sum_{\substack{q,b,q',b',q'',b'' \\ \alpha,\beta,\gamma}} \sum_{l} \exp[-\mathrm{i}(\boldsymbol{q} + \boldsymbol{q}' + \boldsymbol{q}'')\cdot\boldsymbol{l}] \\ &\times X'^{\alpha}_{q,b} X'^{\beta}_{q',b'} X'^{\gamma}_{q'',b''} D^{\alpha,\beta,\gamma}_{q,b,q',b',q'',b''} .\end{aligned} \tag{4.4}$$

where

$$D^{\alpha,\beta,\gamma}_{\boldsymbol{q},\boldsymbol{b},\boldsymbol{q}',\boldsymbol{b}',\boldsymbol{q}'',\boldsymbol{b}''}$$

could be expressed in terms of the U if necessary, but its fundamental property is that

$$D^{\alpha,\beta,\gamma}_{\boldsymbol{q},\boldsymbol{b},\boldsymbol{q}',\boldsymbol{b}',\boldsymbol{q}'',\boldsymbol{b}''} \neq f(\boldsymbol{l}) \,, \tag{4.5}$$

because there is no preferred lattice point.

We obtain

$$\mathcal{H}^1 = \frac{1}{N^{1/2}} \sum_{\substack{\boldsymbol{q},\boldsymbol{b},\boldsymbol{q}',\boldsymbol{b}',\boldsymbol{q}'',\boldsymbol{b}'' \\ \alpha,\beta,\gamma}} \delta^{\boldsymbol{G}_n}_{\boldsymbol{q}+\boldsymbol{q}'+\boldsymbol{q}''} \boldsymbol{X}'^{\alpha}_{\boldsymbol{q},\boldsymbol{b}} \boldsymbol{X}'^{\beta}_{\boldsymbol{q}',\boldsymbol{b}'} \boldsymbol{X}'^{\gamma}_{\boldsymbol{q}'',\boldsymbol{b}''} D^{\alpha,\beta,\gamma}_{\boldsymbol{q},\boldsymbol{b},\boldsymbol{q}',\boldsymbol{b}',\boldsymbol{q}'',\boldsymbol{b}''} \,. \tag{4.6}$$

In an annihilation and creation operator representation, the old unperturbed Hamiltonian was diagonal and of the form

$$\mathcal{H}^1 = \frac{1}{N^{1/2}} \sum_{\boldsymbol{q},p} (a^{\dagger}_{\boldsymbol{q},p} a_{\boldsymbol{q},p} + \tfrac{1}{2}) \hbar\omega_{\boldsymbol{q},p} \,. \tag{4.7}$$

The transformation that did this was (see Problem 2.22)

$$\boldsymbol{X}'_{\boldsymbol{q},\boldsymbol{b}} = -\mathrm{i} \sum_p \boldsymbol{e}^*_{\boldsymbol{q},\boldsymbol{b},p} \sqrt{\frac{\hbar}{2m_b\omega_{\boldsymbol{q},p}}} (a^{\dagger}_{\boldsymbol{q},p} - a_{-\boldsymbol{q},p}) \,. \tag{4.8}$$

Applying the same transformation on the perturbing part of Hamiltonian, we find

$$\begin{aligned} \mathcal{H}^1 = & \sum_{\boldsymbol{q},p,\boldsymbol{q}',p',\boldsymbol{q}'',p''} \delta^{\boldsymbol{G}_n}_{\boldsymbol{q}+\boldsymbol{q}'+\boldsymbol{q}''} (a^{\dagger}_{\boldsymbol{q},p} - a_{-\boldsymbol{q},p})(a^{\dagger}_{\boldsymbol{q}',p'} - a_{-\boldsymbol{q}',p'}) \\ & \times (a^{\dagger}_{\boldsymbol{q}',p'} - a_{-\boldsymbol{q}',p'}) M_{\boldsymbol{q},p,\boldsymbol{q}',p',\boldsymbol{q}'',p''} \,, \end{aligned} \tag{4.9}$$

where

$$M_{\boldsymbol{q},p,\boldsymbol{q}',p',\boldsymbol{q}'',p''} = f(D^{\alpha,\beta,\gamma}_{\boldsymbol{q},\boldsymbol{b},\boldsymbol{q}',\boldsymbol{b}',\boldsymbol{q}'',\boldsymbol{b}''}) \,, \tag{4.10}$$

i.e. it could be expressed in terms of the D if necessary.

4.2.2 Normal and Umklapp Processes (B)

Despite the apparent complexity of (4.9) and (4.10), they are in a transparent form. The essential thing is to find out what types of interaction processes are allowed by cubic anharmonic terms. Within the framework of first-order time-dependent perturbation theory (the Golden rule) this question can be answered.

In the first place, the only real (or direct) processes allowed are those that conserve energy:

$$E_{\text{initial}}^{\text{total}} = E_{\text{final}}^{\text{total}} . \tag{4.11}$$

In the second place, in order for the process to proceed, the Kronecker delta function in (4.9) says that there must be the following relation among wave vectors:

$$\boldsymbol{q} + \boldsymbol{q}' + \boldsymbol{q}'' = \boldsymbol{G}_n . \tag{4.12}$$

Within the limitations imposed by the constraints (4.11) and (4.12), the products of annihilation and creation operators that occur in (4.9) indicate the types of interactions that can take place. Of course, it is necessary to compute matrix elements (as required by the Golden rule) of (4.9) in order to assure oneself that the process is not only allowed by the conservation conditions, but is microscopically probable. In (4.9) a term of the form $a^{\dagger}_{q,p}a_{-q',p'}a_{-q'',p''}$ occurs. Let us assume all the p are the same and thus drop them as subscripts. This term corresponds to a process in which phonons in the modes $-\boldsymbol{q}'$ and $-\boldsymbol{q}''$ are destroyed, and a phonon in the mode $\boldsymbol{q}$ is created. This process can be diagrammatically presented as in Fig. 4.1. It is subject to the constraints

$$\boldsymbol{q} = -\boldsymbol{q}' + (-\boldsymbol{q}'') + \boldsymbol{G}_n \quad \text{and} \quad \hbar\omega_q = \hbar\omega_{-q'} + \hbar\omega_{-q''} .$$

If $\boldsymbol{G}_n = 0$, the vectors $\boldsymbol{q}$, $-\boldsymbol{q}'$, and $-\boldsymbol{q}''$ form a closed triangle and we have what is called a *normal* or *N-process*. If $\boldsymbol{G}_n \neq 0$, we have what is called a *U* or *umklapp* process.[2]

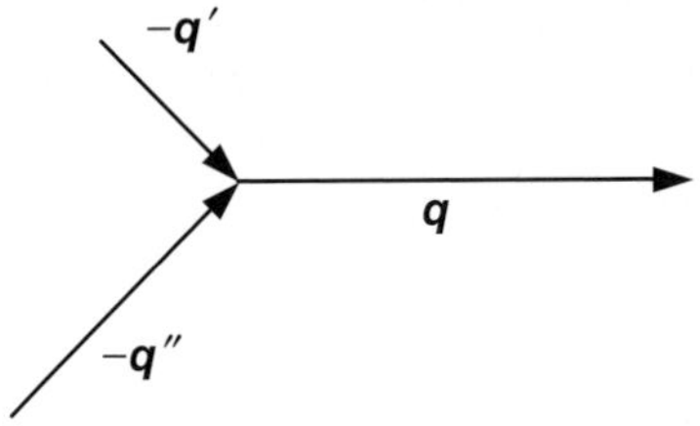

Fig. 4.1. Diagrammatic representation of a phonon–phonon interaction

Umklapp processes are very important in thermal conductivity as will be discussed later. It is possible to form a very simple picture of umklapp processes. Let us consider a two-dimensional reciprocal lattice as shown in Fig. 4.2. If $\boldsymbol{k}_1$ and $\boldsymbol{k}_2$ together add to a vector in reciprocal space that lies outside the first Brillouin zone, then a first Brillouin-zone description of $\boldsymbol{k}_1 + \boldsymbol{k}_2$, is $\boldsymbol{k}_3$, where $\boldsymbol{k}_1 + \boldsymbol{k}_2 = \boldsymbol{k}_3 - \boldsymbol{G}$. If $\boldsymbol{k}_1$ and $\boldsymbol{k}_2$ were the incident phonons and $\boldsymbol{k}_3$ the scattered phonon, we would call such a process a phonon–phonon umklapp process. From Fig. 4.2 we

---

2 Things may be a little more complicated, however, as the distinction between normal and umklapp may depend on the choice of primitive unit cell in $\boldsymbol{k}$ space [21, p. 502].

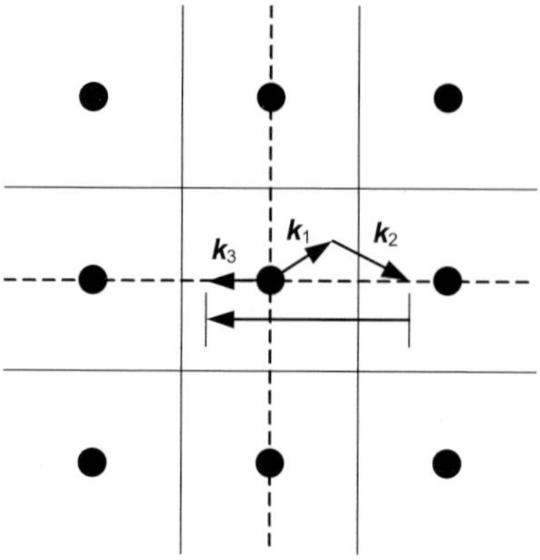

Fig. 4.2. Diagram for illustrating an umklapp process

see the reason for the name umklapp (which in German means "flop over"). We start out with two phonons going in one direction and end up with a phonon going in the opposite direction. This picture gives some intuitive understanding of how umklapp processes contribute to thermal resistance. Since high temperatures are needed to excite high-frequency (high-energy and thus probably large wave vector) phonons, we see that we should expect more umklapp processes as the temperature is raised. Thus we should expect the thermal conductivity of an insulator to drop with increase in temperature.

So far we have demonstrated that the cubic (and hence higher-order) terms in the potential cause the phonon–phonon interactions. There are several directly observable effects of cubic and higher-order terms in the potential. In an insulator in which the cubic and higher-order terms were absent, there would be no diffusion of heat. This is simply because the carriers of heat are the phonons. The phonons do not collide unless there are anharmonic terms, and hence the heat would be carried by "phonon radiation." In this case, the thermal conductivity would be infinite.

Without anharmonic terms, thermal expansion would not exist (see Sect. 2.3.4). Without anharmonic terms, the potential that each atom moved in would be symmetric, and so no matter what the amplitude of vibration of the atoms, the average position of the atoms would be constant and the lattice would not expand.

Anharmonic terms are responsible for small (linear in temperature) deviations from the classical specific heat at high temperature. We can qualitatively understand this by assuming that there is some energy involved in the interaction process. If this is so, then there are ways (in addition to the energy of the phonons) that energy can be carried, and so the specific heat is raised.

The spin–lattice interaction in solids depends on the anharmonic nature of the potential. Obviously, the way the location of a spin moves about in a solid will have a large effect on the total dynamics of the spin. The details of these interactions are not very easy to sort out.

More generally we have to consider that the anharmonic terms cause a temperature dependence of the phonon frequencies and also cause finite phonon lifetimes.

We can qualitatively understand the temperature dependence of the phonon frequencies from the fact that they depend on interatomic spacing that changes with temperature (thermal expansion). The finite phonon lifetimes obviously occur because the phonons scatter into different modes and hence no phonon lasts indefinitely in the same mode. For further details on phonon–phonon interactions see Ziman [99].

4.2.3 Comment on Thermal Conductivity (B)

In this Section a little more detail will be given to explain the way umklapp processes play a role in limiting the lattice thermal conductivity. The discussion in this Section involves only qualitative reasoning.

Let us define a phonon current density $\boldsymbol{J}$ by

$$\boldsymbol{J}_{\text{ph}} = \sum_{\boldsymbol{q}',p} \boldsymbol{q}' N_{\boldsymbol{q}'p} , \tag{4.13}$$

where $N_{\boldsymbol{q},p}$ is the number of phonons in mode $(\boldsymbol{q}, p)$. If this quantity is not equal to zero, then we have a phonon flux and hence heat transport by the phonons.

Now let us consider what the effect of phonon–phonon collisions on $\boldsymbol{J}_{\text{ph}}$ would be. If we have a phonon–phonon collision in which $\boldsymbol{q}_2$ and $\boldsymbol{q}_3$ disappear and $\boldsymbol{q}_1$ appears, then the new phonon flux becomes

$$\begin{aligned} \boldsymbol{J}'_{ph} &= \boldsymbol{q}_1(N_{\boldsymbol{q}_1 p} + 1) + \boldsymbol{q}_2(N_{\boldsymbol{q}_2 p} - 1) \\ &\quad + \boldsymbol{q}_3(N_{\boldsymbol{q}_3 p} - 1) + \sum_{\boldsymbol{q}(\neq \boldsymbol{q}_1, \boldsymbol{q}_2, \boldsymbol{q}_3), p} \boldsymbol{q} N_{\boldsymbol{q},p} . \end{aligned} \tag{4.14}$$

Thus

$$\boldsymbol{J}'_{\text{ph}} = \boldsymbol{q}_1 - \boldsymbol{q}_2 - \boldsymbol{q}_3 + \boldsymbol{J}_{\text{ph}} .$$

For phonon–phonon processes in which $\boldsymbol{q}_2$ and $\boldsymbol{q}_3$ disappear and $\boldsymbol{q}_1$ appears, we have that

$$\boldsymbol{q}_1 = \boldsymbol{q}_2 + \boldsymbol{q}_3 + \boldsymbol{G}_n ,$$

so that

$$\boldsymbol{J}'_{\text{ph}} = \boldsymbol{G}_n + \boldsymbol{J}_{\text{ph}} .$$

Therefore, if there were no umklapp processes the $\boldsymbol{G}_n$ would never appear and hence $\boldsymbol{J}'_{\text{ph}}$ would always equal $\boldsymbol{J}_{\text{ph}}$. This means that the phonon current density would not change; hence the heat flux would not change, and therefore the thermal conductivity would be infinite.

The contribution of umklapp processes to the thermal conductivity is important even at fairly low temperatures. To make a crude estimate, let us suppose that the temperature is much lower than the Debye temperature. This means that small $\boldsymbol{q}$ are important (in a first Brillouin-zone scheme for acoustic modes) because these are the $\boldsymbol{q}$ that are associated with small energy. Since for umklapp processes $\boldsymbol{q} + \boldsymbol{q}' + \boldsymbol{q}'' = \boldsymbol{G}_n$, we know that if most of the $\boldsymbol{q}$ are small, then one of the phonons involved in a phonon–phonon interaction must be of the order of $\boldsymbol{G}_n$, since the wave vectors in the interaction process must add up to $\boldsymbol{G}_n$.

By use of Bose statistics with $T << \theta_D$, we know that the mean number of phonons in mode $\boldsymbol{q}$ is given by

$$\overline{N}_{\boldsymbol{q}} = \frac{1}{\exp(\hbar\omega_{\boldsymbol{q}} / kT) - 1} \cong \exp(-\hbar\omega_{\boldsymbol{q}} / kT) . \qquad (4.15)$$

Let $\hbar\omega_q$ be the energy of the phonon with large $\boldsymbol{q}$, so that we have approximately

$$\hbar\omega_{\boldsymbol{q}} \cong k\theta_D , \qquad (4.16)$$

so that

$$\overline{N}_{\boldsymbol{q}} \cong \exp(-\theta_D / T) . \qquad (4.17)$$

The more $\overline{N}_q$s there are, the greater the possibility of an umklapp process, and since umklapp processes cause $\boldsymbol{J}_{\text{ph}}$ to change, they must cause a decrease in the thermal conductivity. Thus we would expect at least roughly

$$\overline{N}_{\boldsymbol{q}} \propto K^{-1} , \qquad (4.18)$$

where K is the thermal conductivity. Combining (4.17) and (4.18), we guess that the thermal conductivity of insulators at fairly low temperatures is given approximately by

$$K \propto \exp(\theta_D / T) . \qquad (4.19)$$

More accurate analysis suggests the form should be $T^n\exp(F\theta_D/T)$, where F is of order 1/2. At very low temperatures, other processes come into play and these will be discussed later. At high temperature, K (due to the umklapp) is proportional to T^{-1}. Expression (4.19) appears to predict this result, but since we assumed $T << \theta_D$ in deriving (4.19), we cannot necessarily believe (4.19) at high T.

It should be mentioned that there are many other types of phonon–phonon interactions besides the ones mentioned. We could have gone to higher-order terms in the Taylor expansion of the potential. A third-order expansion leads to three phonon (direct) processes. An Nth-order expansion leads to N phonon interactions. Higher-order perturbation theory allows additional processes. For example, it is possible to go indirectly from level i to level f via a virtual level k as is illustrated in Fig. 4.3.

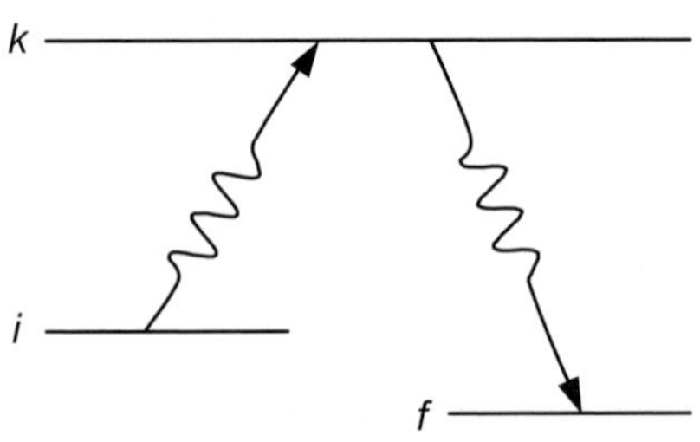

Fig. 4.3. Indirect $i \to f$ transitions via a virtual or short-lived level k

There are a great many more things that could be said about phonon–phonon interactions, but at least we should know what phonon–phonon interactions are by now.

The following statement is by way of summary: Without umklapp processes (and impurities and boundaries) there would be no resistance to the flow of phonon energy *at all temperatures* (in an insulator).

4.3 The Electron–Phonon Interaction

Physically it is easy to see why lattice vibrations scatter electrons. The lattice vibrations distort the lattice periodicity and hence the electrons cannot propagate through the lattice without being scattered.

The treatment of electron–phonon interactions that will be given is somewhat similar to the treatment of phonon–phonon interactions. Similar *selection rules* (or constraints) will be found. This is expected. The selection rules arise from conservation laws, and conservation laws arise from the fundamental symmetries of the physical system. The selection rules are: (1) energy is conserved, and (2) the total wave vector of the system before the scattering process can differ only by a reciprocal lattice vector from the total wave vector of the system after the scattering process. Again it is necessary to examine matrix elements in order to assure oneself that the process is microscopically probable as well as possible because it satisfies the selection rules.

The possibility of electron–phonon interactions has been introduced as if one should not be surprised by them. It is perhaps worth pointing out that electron–phonon interactions indicate a breakdown of the Born–Oppenheimer approximation. This is all right though. We assume that the Born–Oppenheimer approximation is the zeroth-order solution and that the corrections to it can be taken into account by first-order perturbation theory. It is almost impossible to rigorously justify this procedure. In order to treat the interactions adequately, we should go back and insert the terms that were dropped in deriving the Born–Oppenheimer approximation. It appears to be more practical to find a possible form for the interaction by phenomenological arguments. For further details on electron–phonon interactions than will be discussed in this book see Ziman [99].

4.3.1 Form of the Hamiltonian (B)

Whatever the form of the interaction, we know that it vanishes when there are no atomic displacements. For small displacements, the interaction should be linear in the displacements. Thus we write the phenomenological interaction part of the Hamiltonian as

$$\mathcal{H}_{\text{ep}} = \sum_{l,b} \boldsymbol{x}_{l,b} \cdot [\nabla_{\boldsymbol{x}_{l,b}} U(\boldsymbol{r}_{\text{e}})]_{\text{all } \boldsymbol{x}_{l,b}=0}\,, \tag{4.20}$$

where $\boldsymbol{r}_{\text{e}}$ represents the electronic coordinates.

As we will see later, the Boltzmann equation will require that we know the transition probability per unit time. The transition probability can be evaluated from the Golden rule of time-dependent first-order perturbation theory. Basically, the Golden rule requires that we evaluate $\langle f|\mathcal{H}_{\mathrm{ep}}|i\rangle$, where $|i\rangle$ and $\langle f|$ are formal ways of representing the initial and final states for both electron and phonon unperturbed states.

As usual it is convenient to write our expressions in terms of creation and destruction operators. The appropriate substitutions are the same as the ones that were previously used:

$$x_{l,b} = \frac{1}{\sqrt{N}} \sum\nolimits_{q} x'_{q,b} \mathrm{e}^{-\mathrm{i}q \cdot l} ,$$

$$x'_{q,b} = -\mathrm{i} \sum\nolimits_{p} e^{*}_{q,b,p} \sqrt{\frac{\hbar}{2m_b \omega_{q,p}}} (a^{\dagger}_{q,p} - a_{-q,p}) .$$

Combining these expressions, we find

$$x_{l,b} = -\mathrm{i} \sum\nolimits_{q,p} \sqrt{\frac{\hbar}{2Nm_b \omega_b}}\, \mathrm{e}^{-\mathrm{i}q \cdot l} e^{*}_{q,b,p} (a^{\dagger}_{q,p} - a_{-q,p}) . \tag{4.21}$$

If we assume that the electrons can be treated by a one-electron approximation, and that only harmonic terms are important for the lattice potential, a typical matrix element that will have to be evaluated is

$$T_{k,k'} \equiv \langle n_{q,p} \,|\, \int \psi^{*}_{k}(r) \mathcal{H}_{\mathrm{ep}} \psi_{k'}(r) \mathrm{d}r \,|\, n_{q,p} - 1 \rangle , \tag{4.22}$$

where $|n_{q,p}\rangle$ are phonon eigenkets and $\psi_k(r)$ are electron eigenfunctions. The phonon matrix elements can be evaluated by the usual rules (given below):

$$\langle n_{q,p} - 1 \,|\, a_{q',p'} \,|\, n_{q,p} \rangle = \sqrt{n_{q,p}}\, \delta^{q'}_{q} \delta^{p'}_{p} , \tag{4.23a}$$

and

$$\langle n_{q,p} + 1 \,|\, a^{\dagger}_{q',p'} \,|\, n_{q,p} \rangle = \sqrt{n_{q,p} + 1}\, \delta^{q'}_{q} \delta^{p'}_{p} . \tag{4.23b}$$

Combining (4.20), (4.21), (4.22), and (4.23), we find

$$T_{k,k'} = -\mathrm{i} \sum\nolimits_{l,b} \sqrt{\frac{\hbar n_{q,p}}{2Nm_b \omega_{q,b}}}\, \mathrm{e}^{-\mathrm{i}q \cdot l} \int_{\text{all space}} \psi^{*}_{k}(r) e^{*}_{q,b,p} \cdot [\nabla_{x_{l,b}} U(r)]_0 \psi_{k'}(r) \mathrm{d}^3 r . \tag{4.24}$$

Equation (4.24) can be simplified. In order to see how, let us consider a simple problem. Let

$$G = \sum\nolimits_{l} \mathrm{e}^{-\mathrm{i}ql} \int_{-L}^{L} f(x) U_l(x) \mathrm{d}x , \tag{4.25}$$

where

$$f(x + la) = e^{ikl} f(x), \tag{4.26}$$

l is an integer, and $U_l(x)$ is in general not a periodic function of x. In particular, let us suppose

$$U_l(x) \equiv \left(\frac{\partial U}{\partial x_l} \right)_{x_l = 0}, \tag{4.27}$$

where

$$U(x, x_l) = \sum_l \exp[-K(x - d_l)^2], \tag{4.28}$$

and

$$d_l = l + x_l . \tag{4.29}$$

$U(x, x_l)$ is periodic if $x_l = 0$. Combining (4.27) and (4.28), we have

$$\begin{aligned} U_l &= +2K \exp[-K(x - l)^2](x - l) \\ &\equiv F(x - l). \end{aligned} \tag{4.30}$$

Note that $U_l(x) = F(x - l)$ is a localized function.

Therefore we can write

$$G = \sum_l e^{-iql} \int_{-L}^{L} f(x) F(x - l) dx . \tag{4.31}$$

In (4.31), let us write $x' = x - l$ or $x = x' + l$. Then we must have

$$G = \sum_l e^{-iql} \int_{-L-l}^{L-l} f(x' + l) F(x') dx' . \tag{4.32}$$

Using (4.26), we can write (4.32) as

$$G = \sum_l e^{-i(q-k)l} \int_{-L-l}^{L-l} f(x') F(x') dx' . \tag{4.33}$$

If we are using periodic boundary conditions, then all of our functions must be periodic outside the basic interval $-L$ to $+L$. From this it follows that (4.33) can be written as

$$G = \sum_l e^{-i(q-k)l} \int_{-L}^{L} f(x') F(x') dx' . \tag{4.34}$$

The integral in (4.34) is independent of l. Also we shall suppose $F(x)$ is very small for x outside the basic one-dimensional unit cell Ω. From this it follows that we can write G as

$$G \cong \left(\int_{\Omega} f(x') F(x') dx' \right) \left(\sum_l e^{-i(q-k)l} \right). \tag{4.35}$$

A similar argument in three dimensions says that

$$\sum_{l,b} \mathrm{e}^{-\mathrm{i}\boldsymbol{q}\cdot\boldsymbol{l}} \int_{\text{all space}} \psi_{\boldsymbol{k}}^*(\boldsymbol{r}) \boldsymbol{e}_{\boldsymbol{q},b,p}^* [\nabla_{\boldsymbol{x}_{l,b}} U(\boldsymbol{r})]_0 \psi_{\boldsymbol{k}'}(\boldsymbol{r}) \mathrm{d}^3\boldsymbol{r}$$
$$\cong \sum_{l,b} \mathrm{e}^{-\mathrm{i}(\boldsymbol{k}'-\boldsymbol{k}-\boldsymbol{q})\cdot\boldsymbol{l}} \int_{\Omega} \psi_{\boldsymbol{k}}^*(\boldsymbol{r}) \boldsymbol{e}_{\boldsymbol{q},b,p}^* [\nabla_{\boldsymbol{x}_{l,b}} U(\boldsymbol{r})]_0 \psi_{\boldsymbol{k}'}(\boldsymbol{r}) \mathrm{d}^3\boldsymbol{r}.$$

Using the above, and the known delta function property of $\sum_l \mathrm{e}^{\mathrm{i}\boldsymbol{k}\cdot\boldsymbol{l}}$, we find that (4.24) becomes

$$T_{\boldsymbol{k},\boldsymbol{k}'} = -\mathrm{i}\sqrt{n_{\boldsymbol{q},p}} \sqrt{\frac{\hbar N}{2\omega_{\boldsymbol{q},b}}} \delta_{\boldsymbol{k}'-\boldsymbol{k}-\boldsymbol{q}}^{\boldsymbol{G}_n} \int_{\Omega} \psi_{\boldsymbol{k}}^* \sum_b \frac{1}{\sqrt{m_b}} \boldsymbol{e}_{\boldsymbol{q},b,p}^* \cdot [\nabla_{\boldsymbol{x}_{l,b}} U]_0 \psi_{\boldsymbol{k}'} \mathrm{d}^3 r \,. \quad (4.36)$$

Equation (4.36) gives us the usual but very important selection rule on the wave vector. The selection rule says that for all allowed electron–phonon processes; we must have

$$\boldsymbol{k}' - \boldsymbol{k} - \boldsymbol{q} = \boldsymbol{G}_n \,. \quad (4.37)$$

If $\boldsymbol{G}_n \neq 0$, then we have electron–phonon umklapp processes. Otherwise, we say we have normal processes. This distinction is not rigorous because it depends on whether or not the first Brillouin zone is consistently used.

The Golden rule also gives us a selection rule that represents energy conservation

$$E_{\boldsymbol{k}'} = E_{\boldsymbol{k}} + \hbar\omega_{\boldsymbol{q},p} \,. \quad (4.38)$$

Since typical phonon energies are much less than electron energies, it is usually acceptable to neglect $\hbar\omega_{\boldsymbol{q},p}$ in (4.38). Thus while technically speaking the electron scattering is inelastic, for practical purposes it is often elastic.[3] The matrix element considered was for the process of emission. A diagrammatic representation of this process is given in Fig. 4.4. There is a similar matrix element for phonon absorption, as represented in Fig. 4.5. One should remember that these processes came out of first-order perturbation theory. Higher-order perturbation theory would allow more complicated processes.

It is interesting that the selection rules for inelastic neutron scattering are the same as the rules for inelastic electron scattering. However, when thermal neutrons are scattered, $\hbar\omega_{\boldsymbol{q},p}$ is not negligible. The rules (4.37) and (4.38) are sufficient to map out the dispersion relations for lattice vibration. $E_{\boldsymbol{k}}$, $E_{\boldsymbol{k}'}$, $\boldsymbol{k}$, and $\boldsymbol{k}'$ are easily measured for the neutrons, and hence (4.37) and (4.38) determine $\omega_{\boldsymbol{q},p}$ versus q for phonons. In the hands of Brockhouse et al [4.5] this technique of slow *neutron diffraction* or inelastic neutron diffraction has developed into a very powerful modern research tool. It has also been used to determine dispersion relations for magnons. It is also of interest that tunneling experiments can sometimes be used to determine the phonon density of states.[4]

---

[3] This may not be true when electrons are scattered by polar optical modes.

[4] See McMillan and Rowell [4.29].

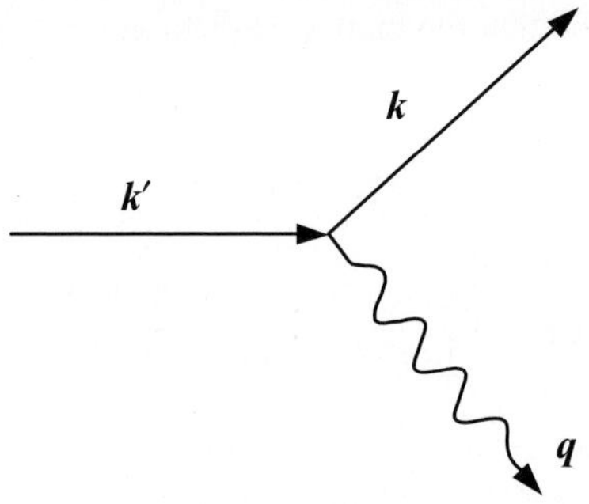

Fig. 4.4. Phonon emission in an electron–phonon interaction

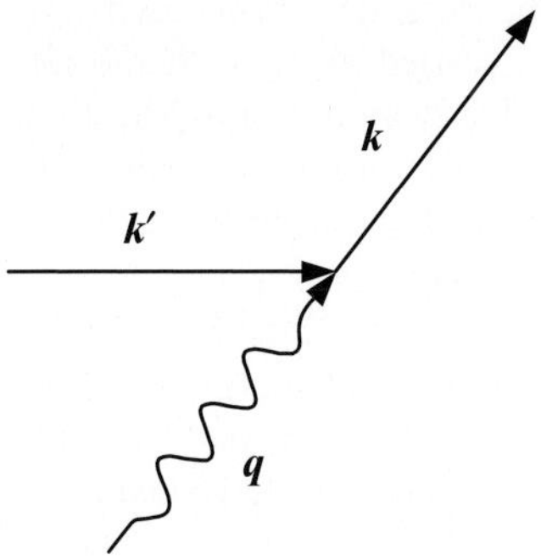

Fig. 4.5. Phonon absorption in an electron–phonon interaction

4.3.2 Rigid-Ion Approximation (B)

It is natural to wonder if all modes of lattice vibration are equally effective in the scattering of electrons. It is true that, in general, some modes are much more effective in scattering electrons than other modes. For example, it is usually possible to neglect optic mode scattering of electrons. This is because in optic modes the adjacent atoms tend to vibrate in opposite directions, and so the net effect of the vibrations tends to be very small due to cancellation. However, if the ions are charged, then the optic modes are polar modes and their effect on electron scattering is by no means negligible. In the discussion below, only one atom per unit cell is assumed. This assumption eliminates the possibility of optic modes. The polarization vectors are now real.

In what follows, an approximation called the *rigid-ion approximation* will be used to discuss differences in scattering between transverse and longitudinal acoustic modes. It appears that in some approximations, transverse phonons do not scatter electrons. However, this rule is only very approximate.

So far we have derived that the matrix element governing the scattering is

$$\left|T_{\boldsymbol{k},\boldsymbol{k}'}\right| = \sqrt{n_{\boldsymbol{q},p}}\sqrt{\frac{\hbar N}{2m\omega_{\boldsymbol{q},p}}}\,\delta^{\boldsymbol{G}_n}_{\boldsymbol{k}'-\boldsymbol{k}-\boldsymbol{q}}\left|H^{\boldsymbol{k},\boldsymbol{k}'}_{\boldsymbol{q},p}\right|, \tag{4.39}$$

where

$$\left|H^{\boldsymbol{k},\boldsymbol{k}'}_{\boldsymbol{q},p}\right| = \left|\int_\Omega \psi^*_{\boldsymbol{k}}\boldsymbol{e}_{\boldsymbol{q},p}\cdot(\nabla_{\boldsymbol{x}_{\boldsymbol{l},b}}U)_0\psi_{\boldsymbol{k}'}\mathrm{d}^3r\right|. \tag{4.40}$$

Equation (4.40) is not easily calculated, but it is the purpose of the rigid-ion approximation to make some comments about it anyway. The rigid-ion approximation assumes that the potential the electrons feel depends only on the vectors connecting the ions and the electron. We also assume that the total potential is the simple additive sum of the potentials from each ion. We thus assume that the potential from each ion is carried along with the ion and is undistorted by the motion of the ion. This is clearly an oversimplification, but it seems to have some degree of applicability, at least for simple metals. The rigid-ion approximation therefore says that the potential that the electron moves in is given by

$$U(\boldsymbol{r}) = \sum\nolimits_{\boldsymbol{l}'} v_a(\boldsymbol{r}-\boldsymbol{x}_{\boldsymbol{l}'}), \tag{4.41}$$

where $v_a(\boldsymbol{r}-\boldsymbol{x}_{\boldsymbol{l}'})$ refers to the potential energy of the electron in the field of the ion whose equilibrium position is at $\boldsymbol{l}'$. The v_a is the cell potential, which is used in the Wigner–Seitz approximation, so that we have inside a cell,

$$\left[-\frac{\hbar^2}{2m}\nabla^2 + v_a(\boldsymbol{r})\right]\psi_{\boldsymbol{k}'}(\boldsymbol{r}) = E_{\boldsymbol{k}'}\psi_{\boldsymbol{k}'}(\boldsymbol{r}). \tag{4.42}$$

The question is, how can we use these two results to evaluate the needed integrals in (4.40)? By (4.41) we see that

$$\nabla_{\boldsymbol{x}_{\boldsymbol{l}}}U = -\nabla_{\boldsymbol{r}}v_a \equiv -\nabla v_a. \tag{4.43}$$

What we need in (4.40) is thus an expression for ∇v_a. That is,

$$\left|H^{\boldsymbol{k},\boldsymbol{k}'}_{\boldsymbol{q},p}\right| = \left|\int_\Omega \psi^*_{\boldsymbol{k}}\boldsymbol{e}_{\boldsymbol{q},p}\cdot\nabla v_a\psi_{\boldsymbol{k}'}\mathrm{d}^3r\right|. \tag{4.44}$$

We can get an expression for the integrand in (4.44) by taking the gradient of (4.42) and multiplying by $\psi^*_{\boldsymbol{k}}$. We obtain

$$\psi^*_{\boldsymbol{k}}v_a\nabla\psi_{\boldsymbol{k}'} + \psi^*_{\boldsymbol{k}}(\nabla v_a)\psi_{\boldsymbol{k}'} = \psi^*_{\boldsymbol{k}}\frac{\hbar^2}{2m}\nabla^3\psi_{\boldsymbol{k}'} + E_{\boldsymbol{k}'}\psi^*_{\boldsymbol{k}}\nabla\psi_{\boldsymbol{k}'}. \tag{4.45}$$

Several transformations are needed before this gets us to a usable approximation: We can always use Bloch's theorem $\psi_{\boldsymbol{k}'} = \mathrm{e}^{i\boldsymbol{k}'\cdot\boldsymbol{r}}u_{\boldsymbol{k}'}(\boldsymbol{r})$ to replace $\nabla\psi_{\boldsymbol{k}'}$ by

$$\nabla\psi_{\boldsymbol{k}'} = \mathrm{e}^{\mathrm{i}\boldsymbol{k}'\cdot\boldsymbol{r}}\nabla u_{\boldsymbol{k}'}(\boldsymbol{r}) + \mathrm{i}\boldsymbol{k}'\psi_{\boldsymbol{k}'}. \tag{4.46}$$

We will also have in mind that any scattering caused by the motion of the rigid ions leads to only very small changes in the energy of the electrons, so that we will approximate E_k by $E_{k'}$ wherever needed. We therefore obtain from (4.45), (4.46), and (4.42)

$$\psi_{\boldsymbol{k}}^{*}(\nabla v_a)\psi_{\boldsymbol{k}'} = \psi_{\boldsymbol{k}}^{*}\frac{\hbar^2}{2m}\nabla^2(\mathrm{e}^{\mathrm{i}\boldsymbol{k}'\cdot\boldsymbol{r}}\nabla u_{\boldsymbol{k}'}) - \frac{\hbar^2}{2m}(\nabla^2\psi_{\boldsymbol{k}}^{*})\mathrm{e}^{\mathrm{i}\boldsymbol{k}'\cdot\boldsymbol{r}}\nabla u_{\boldsymbol{k}'} . \tag{4.47}$$

We can also write

$$\begin{aligned}
&\frac{\hbar^2}{2m}\int_{\text{surface S}}\{\psi_{\boldsymbol{k}}^{*}\nabla[\mathrm{e}^{\mathrm{i}\boldsymbol{k}'\cdot\boldsymbol{r}}(\nabla u_{\boldsymbol{k}'})_\alpha] - \mathrm{e}^{\mathrm{i}\boldsymbol{k}'\cdot\boldsymbol{r}}(\nabla u_{\boldsymbol{k}'})_\alpha\nabla\psi_{\boldsymbol{k}}^{*}\}\cdot\mathrm{d}\boldsymbol{S} \\
&\quad = \frac{\hbar^2}{2m}\int\nabla\cdot\{\psi_{\boldsymbol{k}}^{*}\nabla[\mathrm{e}^{\mathrm{i}\boldsymbol{k}'\cdot\boldsymbol{r}}(\nabla u_{\boldsymbol{k}'})_\alpha] - \mathrm{e}^{\mathrm{i}\boldsymbol{k}'\cdot\boldsymbol{r}}(\nabla u_{\boldsymbol{k}'})_\alpha\nabla\psi_{\boldsymbol{k}}^{*}\}\mathrm{d}\tau \\
&\quad = \frac{\hbar^2}{2m}\int\{\psi_{\boldsymbol{k}}^{*}\nabla^2[\mathrm{e}^{\mathrm{i}\boldsymbol{k}'\cdot\boldsymbol{r}}(\nabla u_{\boldsymbol{k}'})_\alpha] - \mathrm{e}^{\mathrm{i}\boldsymbol{k}'\cdot\boldsymbol{r}}(\nabla u_{\boldsymbol{k}'})_\alpha\nabla^2\psi_{\boldsymbol{k}}^{*}\}\mathrm{d}\tau ,
\end{aligned}$$

since we get a cancellation in going from the second step to the last step. This means by (4.44), (4.47), and the above that we can write

$$\left|H_{\boldsymbol{q},p}^{\boldsymbol{k},\boldsymbol{k}'}\right| = \left|\frac{\hbar^2}{2m}\int\{\psi_{\boldsymbol{k}}^{*}\nabla[\mathrm{e}^{\mathrm{i}\boldsymbol{k}'\cdot\boldsymbol{r}}(\boldsymbol{e}_{\boldsymbol{q},p}\cdot\nabla u_{\boldsymbol{k}'})] - \mathrm{e}^{\mathrm{i}\boldsymbol{k}'\cdot\boldsymbol{r}}\boldsymbol{e}_{\boldsymbol{q},p}\cdot(\nabla u_{\boldsymbol{k}'})\nabla\psi_{\boldsymbol{k}}^{*}\}\cdot\mathrm{d}\boldsymbol{S}\right| . \tag{4.48}$$

We will assume we are using a Wigner–Seitz approximation in which the Wigner–Seitz cells are spheres of radius r_0. The original integrals in $H_{\boldsymbol{q},p}^{\boldsymbol{k},\boldsymbol{k}'}$ involved only integrals over the Wigner–Seitz cell (because ∇v_a vanishes very far from the cell for v_a). Now $u_{k'} \cong \psi_{k'=0}$ in the Wigner–Seitz approximation, and also in this approximation we know $(\nabla\psi_{k'=0})_{r=r0} = 0$. Since $\nabla\psi_0 = \hat{\boldsymbol{r}}(\partial\psi_0/\partial r)$, by the above reasoning we can now write

$$\left|H_{\boldsymbol{q},p}^{\boldsymbol{k},\boldsymbol{k}'}\right| = \left|\int\psi_{\boldsymbol{k}}^{*}\mathrm{e}^{\mathrm{i}\boldsymbol{k}'\cdot\boldsymbol{r}}\frac{\hbar^2}{2m}\nabla^2\psi_0(\boldsymbol{e}_{\boldsymbol{k},p}\cdot\hat{\boldsymbol{r}})\mathrm{d}S\right| . \tag{4.49}$$

Consistent with the Wigner–Seitz approximation, we will further assume that v_a is spherically symmetric and that

$$\frac{\hbar^2}{2m}\nabla^2\psi_0 = [v_a(r_0) - E_0]\psi_0 ,$$

which means that

$$\begin{aligned}
\left|H_{\boldsymbol{q},p}^{\boldsymbol{k},\boldsymbol{k}'}\right| &= \left|[v_a(r_0) - E_0]\int\psi_{\boldsymbol{k}}^{*}\mathrm{e}^{\mathrm{i}\boldsymbol{k}'\cdot\boldsymbol{r}}\psi_0\boldsymbol{e}_{\boldsymbol{q},p}\cdot\hat{\boldsymbol{r}}\mathrm{d}S\right| \\
&\cong \left|[v_a(r_0) - E_0]\int\psi_{\boldsymbol{k}}^{*}\psi_{\boldsymbol{k}'}\boldsymbol{e}_{\boldsymbol{q},p}\cdot\hat{\boldsymbol{r}}\mathrm{d}S\right| \\
&\cong \left|[v_a(r_0) - E_0]\int_\Omega\boldsymbol{e}_{\boldsymbol{q},p}\cdot\nabla(\psi_{\boldsymbol{k}}^{*}\psi_{\boldsymbol{k}'})\mathrm{d}\tau\right| ,
\end{aligned} \tag{4.50}$$

where Ω is the volume of the Wigner–Seitz cell. We assume further that the main contribution to the gradient in (4.50) comes from the exponentials, which means that we can write

$$\nabla(\psi_{\boldsymbol{k}}^{*}\psi_{\boldsymbol{k}'}) \cong \mathrm{i}(\boldsymbol{k}'-\boldsymbol{k})\psi_{\boldsymbol{k}}^{*}\psi_{\boldsymbol{k}'} . \tag{4.51}$$

Finally, we obtain

$$\left|H_{\boldsymbol{q},p}^{\boldsymbol{k},\boldsymbol{k}'}\right| = \left|\boldsymbol{e}_{\boldsymbol{q},p}\cdot(\boldsymbol{k}'-\boldsymbol{k})[v_a(r_0)-E_0]\int\psi_{\boldsymbol{k}}^{*}\psi_{\boldsymbol{k}'}\mathrm{d}\tau\right| . \tag{4.52}$$

Neglecting umklapp processes, we have $\boldsymbol{k}' - \boldsymbol{k} = \boldsymbol{q}$ so

$$\left|H_{\boldsymbol{q},p}^{\boldsymbol{k},\boldsymbol{k}'}\right| \propto \boldsymbol{e}_{\boldsymbol{q},p}\cdot\boldsymbol{q} .$$

Since for transverse phonons, $\boldsymbol{e}_{q,p}$ is perpendicular to $\boldsymbol{q}$, $\boldsymbol{e}_{q,p}\cdot\boldsymbol{q} = 0$ and we get no scattering. We have the very approximate rule that transverse phonons do not scatter electrons. However, we should review all of the approximations that went into this result. By doing this, we can fully appreciate that the result is only very approximate [99].

4.3.3 The Polaron as a Prototype Quasiparticle (A)[5]

Introduction (A)

We look at a different kind of electron–phonon interaction in this section. Landau suggested that an F-center could be understood as a self-trapped electron in a polar crystal. Although this idea did not explain the F-center, it did give rise to the conception of polarons. Polarons occur when an electron polarizes the surrounding media, and this polarization reacts back on the electron and lowers the energy. The polarization field moves with the electron and the whole object is called a polaron, which will have an effective mass generally much greater than the electrons. Polarons also have different mobilities from electrons and this is one way to infer their existence. Much of the basic work on polarons has been done by Fröhlich. He approached polarons by considering electron–phonon coupling. His ideas about electron–phonon coupling also helped lead eventually to a theory of superconductivity, but he did not arrive at the correct treatment of the pairing interaction for superconductivity. Relatively simple perturbation theory does not work there.

There are large polarons (sometimes called Fröhlich polarons) where the lattice distortion is over many sites and small ones that are very localized (some people call these Holstein polarons). Polarons can occur in polar semiconductors or in polar insulators due to electrons in the conduction band or holes in the valence band. Only electrons will be considered here and the treatment will be limited to Fröhlich polarons. Then the polarization can be treated on a continuum basis.

---

[5] See, e.g., [4.26].

Once the effective Hamiltonian for electrons interact with the polarized lattice, perturbation theory can be used for the large-polaron case and one gets in a relatively simple manner the enhanced mass (beyond the Bloch effective mass) due to the polarization interaction with the electron. Apparently, the polaron was the first solid-state quasi particle treated by field theory, and its consideration has the advantage over relativistic field theories that there is no divergence for the self-energy. In fact, the polaron's main use may be as an academic example of a quasi particle that can be easily understood. From the field theoretic viewpoint, the polarization is viewed as a cloud of virtual phonons around the electron. The coupling constant is:

$$\alpha_c = \frac{1}{8\pi\varepsilon_0}\left(\frac{1}{K(\infty)} - \frac{1}{K(0)}\right)\frac{e^2}{\hbar\omega_L}\sqrt{\frac{2m\omega_L}{\hbar}} .$$

The $K(0)$ and $K(\infty)$ are the static and high-frequency dielectric constants, m is the Bloch effective mass of the electron, and ω_L is the long-wavelength longitudinal optic frequency. One can show that the total electron effective mass is the Bloch effective mass over the quantity $1 - \alpha_c/6$. The coupling constant α_c is analogous to the fine structure coupling constant $e^2/\hbar c$ used in a quantum-electrodynamics calculation of the electron–photon interaction.

The Polarization (A)

We first want to determine the electron–phonon interaction. The only coupling that we need to consider is for the longitudinal optical (LO) phonons, as they have a large electric field that interacts strongly with the electrons. We need to calculate the corresponding polarization of the unit cell due to the LO phonons. We will find this relates to the static and optical dielectric constants.

We consider a diatomic lattice of ions with charges $\pm e$. We examine the optical mode of vibrations with very long wavelengths so that the ions in neighboring unit cells vibrate in unison. Let the masses of the ions be $m_\pm$ and if k is the effective spring constant and E_f is the effective electric field acting on the ions we have $(e > 0)$

$$m_+\ddot{\boldsymbol{r}}_+ = -k(\boldsymbol{r}_+ - \boldsymbol{r}_-) + e\boldsymbol{E}_f , \tag{4.53a}$$

$$m_-\ddot{\boldsymbol{r}}_- = +k(\boldsymbol{r}_+ - \boldsymbol{r}_-) - e\boldsymbol{E}_f , \tag{4.53b}$$

where $\boldsymbol{r}_\pm$ is the displacement of the $\pm$ ions in the optic mode (related equations are more generally discussed in Sect. 10.10).

Subtracting, and defining the reduced mass in the usual way ($\mu^{-1} = m_+^{-1} + m_-^{-1}$), we have

$$\mu\ddot{\boldsymbol{r}} = -k\boldsymbol{r} + e\boldsymbol{E}_f , \tag{4.54a}$$

where

$$\boldsymbol{r} = \boldsymbol{r}_+ - \boldsymbol{r}_- . \tag{4.54b}$$

We assume E_{f} in the solid is given by the Lorentz field (derived in Chap. 9)

$$\boldsymbol{E}_{\mathrm{f}} = \boldsymbol{E} + \frac{\boldsymbol{P}}{3\varepsilon_0}, \tag{4.55}$$

where ε_0 is the permittivity of free space.

The polarization $\boldsymbol{P}$ is the dipole moment per unit volume. So if there are N unit cells in a volume V, and if the ± ions have polarizability of $\alpha_{\pm}$ so for both ions $\alpha = \alpha_+ + \alpha_-$, then

$$\boldsymbol{P} = \left(\frac{N}{V}\right)(e\boldsymbol{r} + \alpha\boldsymbol{E}_{\mathrm{f}}). \tag{4.56}$$

Inserting $\boldsymbol{E}_{\mathrm{f}}$ into this expression and solving for $\boldsymbol{P}$ we find:

$$\boldsymbol{P} = \left(\frac{N}{V}\right)\frac{e\boldsymbol{r} + \alpha\boldsymbol{E}}{1-(N\alpha/3V\varepsilon_0)}. \tag{4.57}$$

Putting $\boldsymbol{E}_{\mathrm{f}}$ into Eqs. (4.54a) and (4.56) and using (4.57) for $\boldsymbol{P}$, we find

$$\ddot{\boldsymbol{r}} = a\boldsymbol{r} + b\boldsymbol{E}, \tag{4.58a}$$

$$\boldsymbol{P} = c\boldsymbol{r} + d\boldsymbol{E}, \tag{4.58b}$$

where

$$b = \frac{e/\mu}{1-(N\alpha/3V\varepsilon_0)}, \tag{4.59a}$$

$$c = \left(\frac{N}{V}\right)\frac{e}{1-(N\alpha/3V\varepsilon_0)}, \tag{4.59b}$$

and a and d can be similarly evaluated if needed. Note that

$$b = \frac{V}{N\mu}c. \tag{4.60}$$

It is also convenient to relate these coefficients to the static and high-frequency dielectric constants $K(0)$ and $K(\infty)$. In general

$$\boldsymbol{D} = K\varepsilon_0\boldsymbol{E} = \varepsilon_0\boldsymbol{E} + \boldsymbol{P}, \tag{4.61}$$

so

$$\boldsymbol{P} = (K-1)\varepsilon_0\boldsymbol{E}. \tag{4.62}$$

For the static case $\ddot{\boldsymbol{r}} = 0$ and

$$\boldsymbol{r} = -\frac{b}{a}\boldsymbol{E}. \tag{4.63}$$

Thus

$$\boldsymbol{P} = [K(0)-1]\varepsilon_0 \boldsymbol{E} = \left(d - \frac{cb}{a}\right)\boldsymbol{E}\,. \tag{4.64}$$

For the high-frequency or optic case $\ddot{\boldsymbol{r}} \to \infty$, and $\boldsymbol{r} \to 0$ because the ions cannot follow the high-frequency fields so

$$\boldsymbol{P} = d\boldsymbol{E} = [K(\infty)-1]\varepsilon_0 \boldsymbol{E}\,. \tag{4.65}$$

From the above

$$d = [K(\infty)-1]\varepsilon_0\,, \tag{4.66}$$

$$d - \frac{bc}{a} = [K(0)-1]\varepsilon_0\,. \tag{4.67}$$

We can use the above to get an expression for the polarization, which in turn can be used to determine the electron–phonon interaction. First we need to evaluate $\boldsymbol{P}$.

We work out the polarization for the longitudinal optic mode, as that is all that is needed. Let

$$\boldsymbol{r} = \boldsymbol{r}_{\mathrm{T}} + \boldsymbol{r}_{\mathrm{L}}\,, \tag{4.68}$$

where T and L denote transverse and longitudinal. Since we assume

$$\boldsymbol{r}_{\mathrm{T}} = \boldsymbol{v}\exp[\mathrm{i}(\boldsymbol{q}\cdot\boldsymbol{r} + \omega t)]\,,\ \boldsymbol{v}\text{ a constant}\,, \tag{4.69a}$$

then

$$\nabla\cdot\boldsymbol{r}_{\mathrm{T}} = \mathrm{i}\boldsymbol{q}\cdot\boldsymbol{r}_{\mathrm{T}} = 0\,, \tag{4.69b}$$

by definition since $\boldsymbol{q}$ is the direction of motion of the vibrational wave and is perpendicular to $\boldsymbol{r}_{\mathrm{T}}$. There is no free charge to consider, so

$$\nabla\cdot\boldsymbol{D} = \nabla\cdot(\varepsilon_0\boldsymbol{E} + \boldsymbol{P}) = \nabla\cdot(\varepsilon_0\boldsymbol{E} + d\boldsymbol{E} + c\boldsymbol{r}) = 0$$

or

$$\nabla\cdot[(\varepsilon_0 + d)\boldsymbol{E} + c\boldsymbol{r}_{\mathrm{L}}] = 0\,, \tag{4.70}$$

using (4.69b). This gives as a solution for E

$$\boldsymbol{E} = \frac{-c}{\varepsilon_0 + d}\boldsymbol{r}_{\mathrm{L}}\,. \tag{4.71}$$

Therefore

$$\boldsymbol{P}_{\mathrm{L}} = c\boldsymbol{r}_{\mathrm{L}} + d\boldsymbol{E} = \frac{c\varepsilon_0}{\varepsilon_0 + d}\boldsymbol{r}_{\mathrm{L}}\,. \tag{4.72}$$

If

$$\boldsymbol{r}_{\mathrm{L}} = \boldsymbol{r}_{\mathrm{L}}(0)\exp(\mathrm{i}\omega_{\mathrm{L}}t)\,, \tag{4.73a}$$

and

$$\boldsymbol{r}_{\mathrm{T}} = \boldsymbol{r}_{\mathrm{T}}(0)\exp(\mathrm{i}\omega_{\mathrm{T}}t)\,, \tag{4.73b}$$

then

$$\ddot{\boldsymbol{r}}_{\mathrm{L}} = -\omega_{\mathrm{L}}^2\boldsymbol{r}_{\mathrm{L}}\,, \tag{4.74a}$$

and

$$\ddot{\boldsymbol{r}}_{\mathrm{T}} = -\omega_{\mathrm{T}}^2\boldsymbol{r}_{\mathrm{T}}\,. \tag{4.74b}$$

Thus by Eqs. (4.58a) and (4.71)

$$\ddot{\boldsymbol{r}}_{\mathrm{L}} = a\boldsymbol{r}_{\mathrm{L}} - \frac{cb}{\varepsilon_0 + d}\boldsymbol{r}_{\mathrm{L}}\,. \tag{4.75}$$

Also, using (4.71) and (4.58a)

$$\ddot{\boldsymbol{r}}_{\mathrm{T}} = a\boldsymbol{r}_{\mathrm{T}}\,, \tag{4.76}$$

so

$$a = -\omega_{\mathrm{T}}^2\,. \tag{4.77}$$

Using Eqs. (4.66) and (4.67)

$$a - \frac{bc}{\varepsilon_0 + d} = a\frac{K(0)}{K(\infty)}\,, \tag{4.78}$$

and so by (4.74a), (4.75) and (4.77)

$$\omega_{\mathrm{L}}^2 = -a\frac{K(0)}{K(\infty)} = \omega_{\mathrm{T}}^2\frac{K(0)}{K(\infty)}\,, \tag{4.79}$$

which is known as the LST (for Lyddane–Sachs–Teller) equation. See also Born and Huang [46 p. 87]. This will be further discussed in Chap. 9. Continuing, by (4.66),

$$\varepsilon_0 + d = K(\infty)\varepsilon_0\,, \tag{4.80}$$

and by (4.67)

$$d - [K(0) - 1]\varepsilon_0 = \frac{bc}{a}\,, \tag{4.81}$$

from which we determine by (4.60), (4.77), (4.78), (4.80), and (4.81)

$$c = \omega_{\mathrm{T}}\sqrt{\frac{N\mu}{V}}\sqrt{\varepsilon_0}\sqrt{K(0) - K(\infty)}\,. \tag{4.82}$$

Using (4.72) and the LST equation we find

$$\boldsymbol{P} = \omega_{\mathrm{L}} \sqrt{\varepsilon_0} \sqrt{\frac{N\mu}{V}} \sqrt{\frac{1}{K(0)K(\infty)}} \sqrt{K(0) - K(\infty)} \boldsymbol{r}_{\mathrm{L}} , \tag{4.83}$$

or if we define

$$\alpha_{\mathrm{c}} = \frac{e^2}{8\pi\varepsilon_0 \hbar\omega_{\mathrm{L}}} \frac{1}{r_0} \frac{1}{\overline{K}} , \tag{4.84}$$

with

$$\frac{1}{\overline{K}} = \frac{1}{K(\infty)} - \frac{1}{K(0)} , \tag{4.85}$$

and

$$r_0 = \sqrt{\frac{\hbar}{2m\omega_{\mathrm{L}}}} , \tag{4.86}$$

we can write a more convenient expression for P. Note we can think of $\overline{K}$ as the effective dielectric constant for the ion displacements. The quantity r_0 is called the radius of the polaron. A simple argument can be given to see why this is a good interpretation. The uncertainty in the energy of the electron due to emission or absorption of virtual phonons is

$$\Delta E = \hbar\omega_{\mathrm{L}} , \tag{4.87}$$

and if

$$\Delta E \cong \frac{\hbar^2}{2m} (\Delta k)^2 , \tag{4.88}$$

then

$$\frac{1}{\Delta k} \equiv r_0 = \sqrt{\frac{\hbar}{2m\omega_{\mathrm{L}}}} . \tag{4.89}$$

The quantity α_{c} is called the coupling constant and it can have values considerably less than 1 for for direct band gap semiconductors or greater than 1 for insulators. Using the above definitions:

$$\begin{aligned} \boldsymbol{P} &= \varepsilon_0 \omega_{\mathrm{L}} \sqrt{\frac{N\mu\alpha_c}{V} \frac{8\pi\hbar\omega_{\mathrm{L}}}{e^2} r_0} \, \boldsymbol{r}_{\mathrm{L}} \\ &\equiv A\boldsymbol{r}_{\mathrm{L}} . \end{aligned} \tag{4.90}$$

The Electron–Phonon Interaction due to the Polarization (A)

In the continuum approximation appropriate for large polarons, we can write the electron–phonon interaction as coming from dipole moments interacting with the gradient of the potential due to the electron (i.e. a dipole moment dotted with an electric field, $e > 0$) so

$$\mathcal{H}_{\text{ep}} = \frac{-e}{4\pi\varepsilon_0}\int \boldsymbol{P}(\boldsymbol{r})\nabla\frac{1}{|r - r_{\text{e}}|}\mathrm{d}\boldsymbol{r} = \frac{e}{4\pi\varepsilon_0}\int\frac{\boldsymbol{P}(\boldsymbol{r})\cdot(\boldsymbol{r}-\boldsymbol{r}_e)}{|\boldsymbol{r}-\boldsymbol{r}_{\text{e}}|^3}\mathrm{d}\boldsymbol{r}\,. \tag{4.91}$$

Since $\boldsymbol{P} = A\boldsymbol{r}_{\text{L}}$ and we have determined A, we need to write an expression for $\boldsymbol{r}_{\text{L}}$.

In the usual way we can express $\boldsymbol{r}_{\text{L}}$ at lattice position $\boldsymbol{R}_n$ in terms of an expansion in the normal modes for LO phonons (see Sect. 2.3.2):

$$\boldsymbol{r}_{\text{L}n} = \boldsymbol{r}_{n+} - \boldsymbol{r}_{n-} = \frac{1}{\sqrt{N}}\sum\nolimits_{\boldsymbol{q}} Q(\boldsymbol{q})\left[\frac{\boldsymbol{e}_+(\boldsymbol{q})}{\sqrt{m_+}} - \frac{\boldsymbol{e}_-(\boldsymbol{q})}{\sqrt{m_-}}\right]\exp(\mathrm{i}\boldsymbol{q}\cdot\boldsymbol{R}_n)\,. \tag{4.92}$$

The polarization vectors are normalized so

$$|\boldsymbol{e}_+|^2 + |\boldsymbol{e}_-|^2 = 1\,. \tag{4.93}$$

For long-wavelength LO modes

$$\boldsymbol{e}_+ = -\boldsymbol{e}_-\sqrt{\frac{m_-}{m_+}}\,. \tag{4.94}$$

Then we find a solution for the LO modes as

$$\boldsymbol{e}_+(\boldsymbol{q}) = \mathrm{i}\sqrt{\frac{\mu}{m_+}}\,\hat{\boldsymbol{e}}(\boldsymbol{q})\,, \tag{4.95a}$$

$$\boldsymbol{e}_-(\boldsymbol{q}) = -\mathrm{i}\sqrt{\frac{\mu}{m_-}}\,\hat{\boldsymbol{e}}(\boldsymbol{q})\,, \tag{4.95b}$$

where

$$\hat{\boldsymbol{e}}(\boldsymbol{q}) = \frac{\boldsymbol{q}}{q} \quad \text{as } q \to \infty\,.$$

Note the i allows us to satisfy

$$\boldsymbol{e}(\boldsymbol{q}) = \boldsymbol{e}^*(-\boldsymbol{q})\,, \tag{4.96}$$

as required. Thus

$$\boldsymbol{r}_{\text{L}n} = \frac{1}{\sqrt{N\mu}}\sum\nolimits_{\boldsymbol{q}} \mathrm{i}Q(\boldsymbol{q})\hat{\boldsymbol{e}}(\boldsymbol{q})\exp(\mathrm{i}\boldsymbol{q}\cdot\boldsymbol{R}_n)\,, \tag{4.97}$$

or in the continuum approximation

$$r_{Ln} = \frac{1}{\sqrt{N\mu}} \sum_{q} \mathrm{i}Q(q)\hat{e}(q)\exp(\mathrm{i}q \cdot r) . \tag{4.98}$$

Following the usual procedure:

$$Q(q) = \frac{1}{\mathrm{i}}\sqrt{\frac{\hbar}{2\omega_{\mathrm{L}}}}(a_{-q}^{+} - a_{q}) \tag{4.99}$$

(compare with Eqs. (2.140), (2.141)). Substituting and making a change in dummy summation variable:

$$r_{\mathrm{L}} = -\sqrt{\frac{\hbar}{2N\mu\omega_{\mathrm{L}}}} \sum_{q} (a_{q}^{+}\mathrm{e}^{-\mathrm{i}q\cdot r} + a_{q}\mathrm{e}^{\mathrm{i}q\cdot r})\frac{q}{q} . \tag{4.100}$$

Thus

$$\mathcal{H}_{\mathrm{ep}} = -\frac{\hbar\omega_{\mathrm{L}}}{4\pi}\sqrt{\frac{4\pi\alpha_{c}r_{0}}{V}} \int \mathrm{d}r \frac{r - r_{\mathrm{e}}}{|r - r_{\mathrm{e}}|^{3}} \frac{q}{q} \sum_{q} (a_{q}^{+}\mathrm{e}^{-\mathrm{i}q\cdot r} + a_{q}\mathrm{e}^{\mathrm{i}q\cdot r}) . \tag{4.101}$$

Using the identity from Madelung [4.26],

$$\int \exp[\pm \exp(\mathrm{i}q \cdot r)] \frac{(r - r_{e})}{|r - r_{e}|^{3}} \mathrm{d}r = \mp 4\pi i \frac{q}{q^{2}} \exp(\pm \mathrm{i}q \cdot r_{\mathrm{e}}) , \tag{4.102}$$

we find

$$\mathcal{H}_{\mathrm{ep}} = \mathrm{i}\hbar\omega_{\mathrm{L}}\sqrt{r_{0}}\sqrt{\frac{4\pi\alpha_{c}}{V}} \sum_{q} \frac{1}{q}[a_{q}\exp(\mathrm{i}q \cdot r_{\mathrm{e}}) - a_{q}^{+}\exp(-\mathrm{i}q \cdot r_{\mathrm{e}})] . \tag{4.103}$$

Energy and Effective Mass (A)

We consider only processes in which the polarizable medium is at absolute zero, and for which the electron does not have enough energy to create real optical phonons. We consider only the process described in Fig. 4.6. That is we consider the modification of self-energy of the electron due to virtual phonons. In perturbation theory we have as ground state $|k, 0_{q}\rangle$ with energy

$$E_{k} = \frac{\hbar^{2}k^{2}}{2m^{*}} \tag{4.104}$$

and no phonons. For the excited (virtual) state we have one phonon, $|k - q, 1_{q}\rangle$. By ordinary Rayleigh-Schrödinger perturbation theory, the perturbed energy of the ground state to second order is:

$$E_{k,0} = E_{k,0}^{(0)} + \langle k,0|\mathcal{H}_{\mathrm{ep}}|k,0\rangle + \sum_{q} \frac{|\langle k - q, 1|\mathcal{H}_{\mathrm{ep}}|k,0\rangle|^{2}}{E_{k,0}^{(0)} - E_{k-q,1}^{(0)}} . \tag{4.105}$$

But

$$E_{\boldsymbol{k},0}^{(0)} = \frac{\hbar^2 k^2}{2m^*},$$

$$\langle \boldsymbol{k},0|\mathcal{H}_{\text{ep}}|\boldsymbol{k},0\rangle = 0,$$

$$E_{\boldsymbol{k}-\boldsymbol{q},1}^{(0)} = \frac{\hbar^2}{2m^*}(\boldsymbol{k}-\boldsymbol{q})^2 + \hbar\omega_{\text{L}},$$

so

$$E_{\boldsymbol{k},0}^{(0)} - E_{\boldsymbol{k}-\boldsymbol{q},1}^{(0)} = \frac{\hbar^2}{2m^*}(2\boldsymbol{k}\cdot\boldsymbol{q} - q^2) - \hbar\omega_{\text{L}}, \tag{4.106}$$

and

$$\langle \boldsymbol{k}-\boldsymbol{q},1|\mathcal{H}_{\text{ep}}|\boldsymbol{k},0\rangle = -\text{i}\hbar\omega_{\text{L}}\sqrt{r_0}\sqrt{\frac{4\pi\alpha_{\text{c}}}{V}}\sum_{\boldsymbol{q}'}\frac{1}{q'}\langle \boldsymbol{k}-\boldsymbol{q},1|\text{e}^{(-\text{i}\boldsymbol{q}'\cdot\boldsymbol{r}_{\text{e}})}a_{\boldsymbol{q}'}^{+}|\boldsymbol{k},0\rangle. \tag{4.107}$$

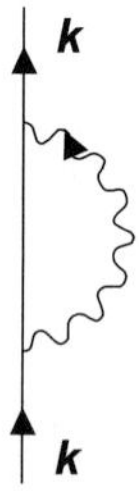

Fig. 4.6. Self-energy Feynman diagram (for interaction of electron and virtual phonon)

Since

$$\langle 1|a_{\boldsymbol{q}}^{+}|0\rangle = 1, \tag{4.108a}$$

$$\langle \boldsymbol{k}-\boldsymbol{q}|\exp(-\text{i}\boldsymbol{q}'\cdot\boldsymbol{r}_{\text{e}})|\boldsymbol{k}\rangle = \delta_{\boldsymbol{q},\boldsymbol{q}'}, \tag{4.108b}$$

we have

$$|\langle \boldsymbol{k}-\boldsymbol{q},1|\mathcal{H}_{\text{ep}}|\boldsymbol{k},0\rangle|^2 = (\hbar\omega_{\text{L}})^2 r_0 \frac{4\pi\alpha_c}{V}\frac{1}{q^2} \equiv \frac{C_{\text{H}}^2}{q^2}, \tag{4.109}$$

where

$$C_{\text{H}}^2 = (\hbar\omega_{\text{L}})^2 r_0 \frac{4\pi\alpha_{\text{c}}}{V}. \tag{4.110}$$

Replacing

$$\sum_{q} \text{ by } \frac{V}{(2\pi)^3}\int \mathrm{d}\boldsymbol{q}\,,$$

we have

$$E_{\boldsymbol{k},0} = \frac{\hbar^2 k^2}{2m^*} + \frac{VC_{\mathrm{H}}^2}{(2\pi)^3}\int \frac{1}{q^2} \frac{\mathrm{d}\boldsymbol{q}}{\left[\dfrac{\hbar^2 k^2}{2m^*}(2\boldsymbol{k}\cdot\boldsymbol{q} - q^2) - \hbar\omega_{\mathrm{L}}\right]}. \tag{4.111}$$

For small k we can show (see Problem 4.5)

$$E_{\boldsymbol{k},0} \cong -\alpha_{\mathrm{c}}\hbar\omega_{\mathrm{L}} + \frac{\hbar^2 k^2}{2m^{**}}, \tag{4.112}$$

where

$$m^{**} = \frac{m^*}{1-(\alpha_c/6)}. \tag{4.113}$$

Thus the self-energy is increased by the interaction of the cloud of virtual phonons surrounding the electrons.

Experiments and Numerical Results (A)

A discussion of experimental results for large polarons can be found in the paper by Appel [4.2, pp. 261-276]. Appel (pp. 366-391) also gives experimental results for small polarons. Polarons are real. However, there is not the kind of comprehensive comparisons of theory and experiment that one might desire. Cyclotron resonance and polaron mobility experiments are common experiments cited. Difficulties abound, however. For example, to determine m^{**} accurately, m^* is needed. Of course m^* depends on the band structure that then must be accurately known. Crystal purity is an important but limiting consideration in many experiments. The chapter by F. C. Brown in the book edited by Kuper and Whitfield [4.23] also reviews rather thoroughly the experimental situation. Some typical values for the coupling constant α_{c} (from Appel), are given below. Experimental estimates of α_{c} are also given by Mahan [4.27] on p. 508.

Table 4.4. Polaron coupling constant

Material	α_{c}
KBr	3.70
GaAs	0.031
InSb	0.015
CdS	0.65
CdTe	0.39

4.4 Brief Comments on Electron–Electron Interactions (B)

A few comments on electron–electron interactions have already been made in Chap. 3 (Sects. 3.1.4 and 3.2.2) and in the introduction to this chapter. Chapter 3 discussed in some detail the density functional technique (DFT), in which the density function plays a central role for accounting for effects of electron–electron interactions. Kohn [4.20] has given a nice summary of the limitation of this model. The DFT has become the traditional way nowadays for calculating the electronic structure of crystalline (and to some extent other types of) condensed matter. For actual electronic densities of interest in metals it has always been difficult to treat electron–electron interactions. We give below earlier results that have been obtained for high and low densities.

Results, which include correlations or the effect of electron–electron interactions, are available for a uniform electron gas with a uniform positive background (jellium). The results given below are in units of Rydberg (R_∞), see Appendix A. If ρ is the average electron density,

$$r_s = \left(\frac{3}{4\pi\rho} \right)^{1/3}$$

is the average distance between electrons. For high density ($r_s << 1$), the theory of Gell-mann and Bruckner gives for the energy per electron

$$\frac{E}{N} = \frac{2.21}{r_s^2} - \frac{0.916}{r_s} + 0.062 \ln r_s - 0.096 + (\text{higher order terms})(R_\infty) .$$

For low densities ($r_s >> 1$) the ideas of Wigner can be extended to give

$$\frac{E}{N} = -\frac{1.792}{r_s} + \frac{2.66}{r_s^{3/2}} + \text{higher order terms in } r_s^{-1/2} .$$

In the intermediate regime of metallic densities, the following expression is approximately true:

$$\frac{E}{N} = \frac{2.21}{r_s^2} - \frac{0.916}{r_s} + 0.031 \ln r_s - 0.115 \ (R_\infty) ,$$

for $1.8 \le r_s \le 5.5$. See Katsnelson et al [4.16]. This book is also excellent for DFT.

The best techniques for treating electrons in interaction that has been discussed in this book are the Hartree and Hartree–Fock approximation and especially the density functional method. As already mentioned, the Hartree–Fock method can give wrong results because it neglects the correlations between electrons with antiparallel spins. In fact, the correlation energy of a system is often defined as the difference between the exact energy (less the relativistic corrections if necessary) and the Hartree–Fock energy.

Even if we limit ourselves to techniques derivable from the variational principle, we can calculate the correlation energy at least in principle. All we have to do is to use a better trial wave function than a single Slater determinant. One way to do this is to use a linear combination of several Slater determinants (the method of superposition of configurations). The other method is to include interelectronic coordinates $r_{12} = |\boldsymbol{r}_1 - \boldsymbol{r}_2|$ in our trial wave function. In both methods there would be several independent functions weighted with coefficients to be determined by the variational principle. Both of these techniques are practical for atoms and molecules with a limited number of electrons. Both become much too complex when applied to solids. In solids, cleverer techniques have to be employed. Mattuck [4.28] will introduce you to some of these clever ideas and do it in a simple, understandable way, and density functional techniques (see Chap. 3) have become very useful, at least for ground-state properties.

It is well to keep in mind that most calculations of electronic properties in real solids have been done in some sort of one-electron approximation and they treat electron–electron interactions only approximately. There is no reason to suppose that electron correlations do not cause many types of new phenomena. For example, Mott has proposed that if we could bring metallic atoms slowly together to form a solid there would still be a *sudden* (so-called Mott) *transition* to the conducting or metallic state at a given distance between the atoms.[6] This *sudden transition* would be caused by electron–electron interactions and is to be contrasted with the older idea of conduction at all interatomic separations. The Mott view differs from the Bloch view that states that any material with well separated energy bands that are either filled or empty should be an insulator while any material with only partly filled bands (say about half-filled) should be a metal. Consider, for example, a hypothetical sodium lattice with N atoms in which the Na atoms are 1 meter apart. Let us consider the electrons that are in the outer unfilled shells. The Bloch theory says to put these electrons into the N lowest states in the conduction band. This leaves N higher states in the conduction band for conduction, and the lattice (even with the sodium atoms well separated) is a metal. This description allows two electrons with opposite spin to be on the same atom without taking into account the resulting increase in energy due to Coulomb repulsion. A better description would be to place just one electron on each atom. Now, the Coulomb potential energy is lower, but since we are using localized states, the kinetic energy is higher. For separations of 1 meter, the lowering of potential energy must dominate. In the better description as provided by the localized model, conduction takes place only by electrons hopping onto atoms that already have an outer electron. This requires considerable energy and so we expect the material to behave as an insulator at large atomic separations. Since the Bloch model so often works, we expect (usually) that the kinetic energy term dominates at actual interatomic spacing. Mott predicted that the transition to a metal from an insulator as the interatomic spacing is varied (in a situation such as we have

[6] See Mott [4.31].

described) should be a sudden transition. By now, many examples are known, NiO was one of the first examples of "Mott–Hubbard" insulators – following current usage. Anderson has predicted another kind of metal–insulator transition due to disorder.[6] Anderson's ideas are also discussed in Sect. 12.9.

Kohn has suggested another effect that may be due to electron–electron interactions. These interactions cause singularities in the dielectric constant (see, e.g., (9.167)) as a function of wave vector that can be picked up in the dispersion relation of lattice vibrations. This *Kohn* effect appears to offer a means of mapping out the *Fermi surface*.[7] Electron–electron interactions may also alter our views of impurity states.[8] We should continue to be hopeful about the possibility of finding new effects due to electron–electron interactions.[9]

4.5 The Boltzmann Equation and Electrical Conductivity

4.5.1 Derivation of the Boltzmann Differential Equation (B)

In this section, the Boltzmann equation for an electron gas will be derived. The principle lack of rigor will be our assumption that the electrons are described by wave packets made of one-electron Bloch wave packets (Bloch wave packets incorporate the effect of the fields due to the lattice ions which by definition change rapidly over inter ionic distances). We also assume these wave packets do not spread appreciably over times of interest. The external fields and temperatures will also be assumed to vary slowly over distances of the order of the lattice spacing.

Later, we will note that the Boltzmann equation is only relatively simple to solve in an iterated first order form when a relaxation time can be defined. The use of a relaxation time will further require that the collisions of the electrons with phonons (for example) do not appreciably alter their energies, that is that the relevant phonon energies are negligible compared to the electrons energies so that the scattering of the electrons may be regarded as elastic.

We start with the distribution function $f_{k\sigma}(\boldsymbol{r},t)$, where the normalization is such that

$$f_{\boldsymbol{k}\sigma}(\boldsymbol{r},t)\frac{\mathrm{d}\boldsymbol{k}\mathrm{d}\boldsymbol{r}}{(2\pi)^3}$$

is the number of electrons in $\mathrm{d}\boldsymbol{k}$ (= $\mathrm{d}k_x\mathrm{d}k_y\mathrm{d}k_z$) and $\mathrm{d}\boldsymbol{r}$ (= $\mathrm{d}x\mathrm{d}y\mathrm{d}z$) at time t with spin σ. In equilibrium, with a uniform distribution, $f_{k\sigma} \to f^0_{\boldsymbol{k}\sigma}$ becomes the Fermi–Dirac distribution.

[7] See [4.19]. See also Sect. 9.5.3.
[8] See Langer and Vosko [4.24].
[9] See also Sect. 12.8.3 where the half-integral quantum Hall effect is discussed.

If no collisions occurred, the r and k coordinates of every electron would evolve by the semiclassical equations of motion as will be shown (Sect. 6.1.2). That is:

$$v_{k\sigma} = \frac{1}{\hbar}\frac{\partial E_{k\sigma}}{\partial \boldsymbol{k}}, \tag{4.114}$$

and

$$\hbar\dot{\boldsymbol{k}} = \boldsymbol{F}_{\text{ext}}, \tag{4.115}$$

where $F = F_{\text{ext}}$ is the external force. Consider an electron having spin σ at $\boldsymbol{r}$ and $\boldsymbol{k}$ and time t started from $r - \boldsymbol{v}_{k\sigma}\mathrm{d}t$, $\boldsymbol{k} - \boldsymbol{F}\mathrm{d}t/\hbar$ at time $t - \mathrm{d}t$. Conservation of the number of electrons then gives us:

$$f_{k\sigma}(\boldsymbol{r},t)\mathrm{d}\boldsymbol{r}_t\mathrm{d}\boldsymbol{k}_t = f_{(\boldsymbol{k}-\boldsymbol{F}\mathrm{d}t/\hbar)\sigma}(\boldsymbol{r}-\boldsymbol{v}_{k\sigma}\mathrm{d}t, t-\mathrm{d}t)\mathrm{d}\boldsymbol{r}_{t-\mathrm{d}t}\mathrm{d}\boldsymbol{k}_{t-\mathrm{d}t}\,. \tag{4.116}$$

Liouville's theorem then says that the electrons, which move by their equation of motion, preserve phase space volume. Thus, if there were no collisions:

$$f_{k\sigma}(\boldsymbol{r},t) = f_{(\boldsymbol{k}-\boldsymbol{F}\mathrm{d}t/\hbar)\sigma}(\boldsymbol{r}-\boldsymbol{v}_{k\sigma}\mathrm{d}t, t-\mathrm{d}t)\,. \tag{4.117}$$

Scattering due to collisions must be considered, so let

$$Q(r,\boldsymbol{k},t) = \left.\frac{\partial f_{k\sigma}}{\partial t}\right)_{\text{collisions}} \tag{4.118}$$

be the net change, due to collisions, in the number of electrons (per $\mathrm{d}\boldsymbol{k}\mathrm{d}\boldsymbol{r}/(2\pi)^3$) that get to r, k at time t. By expanding to first order in infinitesimals,

$$f_{k\sigma}(\boldsymbol{r},t) = f_{k\sigma}(\boldsymbol{r},t) - \mathrm{d}t\left[\frac{\partial f_{k\sigma}}{\partial \boldsymbol{r}}\cdot\boldsymbol{v}_{k\sigma} + \frac{\partial f_{k\sigma}}{\partial \boldsymbol{k}}\cdot\frac{\boldsymbol{F}}{\hbar} + \frac{\partial f_{k\sigma}}{\partial t}\right] + Q(\boldsymbol{r},\boldsymbol{k},t)\mathrm{d}t\,, \tag{4.119}$$

so

$$Q(\boldsymbol{r},\boldsymbol{k},t) = \frac{\partial f_{k\sigma}}{\partial \boldsymbol{r}}\cdot\boldsymbol{v}_{k\sigma} + \frac{\partial f_{k\sigma}}{\partial \boldsymbol{k}}\cdot\frac{\boldsymbol{F}}{\hbar} + \frac{\partial f_{k\sigma}}{\partial t}\,. \tag{4.120}$$

If the steady state is assumed, then

$$\frac{\partial f_{k\sigma}}{\partial t} = 0\,. \tag{4.121}$$

Equation (4.120) may be the basic equation we need to solve, but it does us little good to write it down unless we can find useful expressions for Q. Evaluation of Q is by a detailed consideration of the scattering process. For many cases Q is determined by the scattering matrices as was discussed in Sects. 4.1 and 4.2. Even after Q is so determined, it is by no means a trivial problem to solve the Boltzmann integrodifferential (as it turns out to be) equation.

4.5.2 Motivation for Solving the Boltzmann Differential Equation (B)

Before we begin discussing the Q details, it is worthwhile to give a little motivation for solving the Boltzmann differential equation. We will show how two important quantities can be calculated once the solution to the Boltzmann equation is known. It is also very useful to approximate Q by a phenomenological argument and then obtain solutions to (4.120). Both of these points will be discussed before we get into the rather serious problems that arise when we try to calculate Q from first principles.

Solutions to (4.120) allow us, from $f_{k\sigma}$, to obtain the electric current density $\boldsymbol{J}$, and the electronic flux of heat energy $\boldsymbol{H}$. By definition of the distribution function, these two important quantities are given by

$$\boldsymbol{J} = \sum\nolimits_{\sigma} \int (-e) \boldsymbol{v}_{k\sigma} f_{k\sigma} \frac{\mathrm{d}\boldsymbol{k}}{(2\pi)^3} , \tag{4.122}$$

$$\boldsymbol{H} = \sum\nolimits_{\sigma} \int E_{k\sigma} \boldsymbol{v}_{k\sigma} f_{k\sigma} \frac{\mathrm{d}\boldsymbol{k}}{(2\pi)^3} . \tag{4.123}$$

Electrical conductivity σ and thermal conductivity κ [10] are defined by the relations

$$\boldsymbol{J} = \sigma \boldsymbol{E} , \tag{4.124}$$

$$\boldsymbol{H} = -\kappa \nabla T \tag{4.125}$$

(with a few additional restrictions as will be discussed, see, e.g., Sect. 4.6 and Table 4.5).

As long as we are this close, it is worthwhile to sketch the type of experimental results that are obtained for the transport coefficients κ and σ. In particular, it is useful to understand the particular form of the temperature dependences that are given in Fig. 4.7, Fig. 4.8, and Fig. 4.9. See Problems 4.2, 4.3, and 4.4.

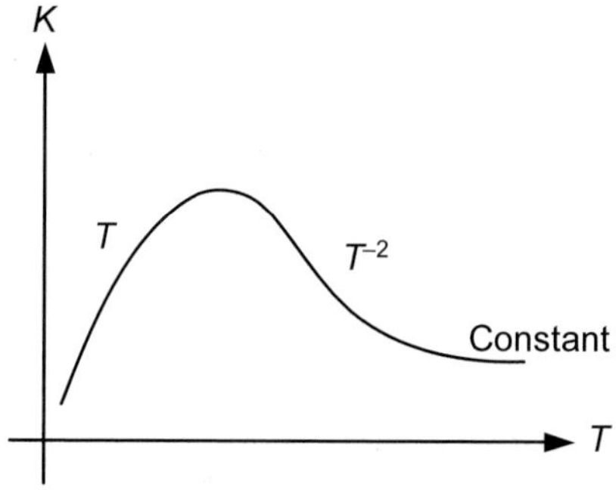

Fig. 4.7. The thermal conductivity of a good metal (e.g. Na as a function of temperature)

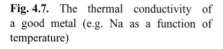

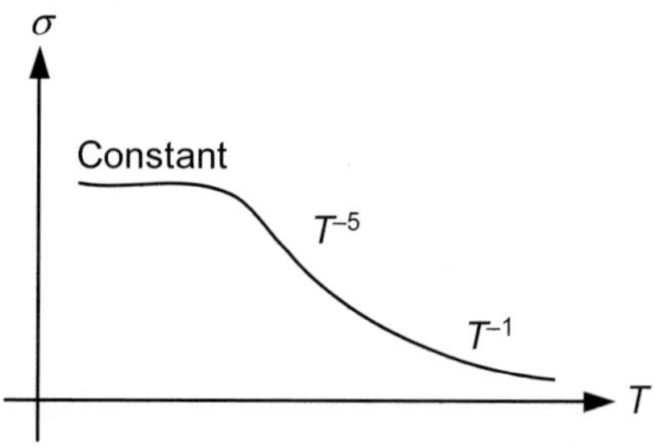

Fig. 4.8. The electrical conductivity of a good metal (e.g. Na as a function of temperature)

[10] See Table 4.5 for a more precise statement about what is held constant.

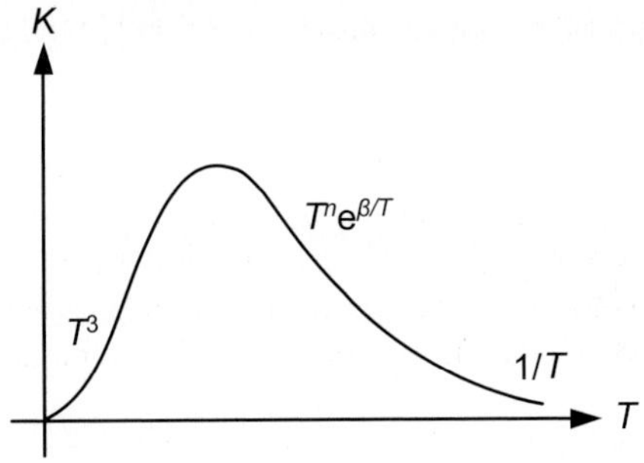

Fig. 4.9. The thermal conductivity of an insulator as a function of temperature, $\beta \cong \theta_D/2$

4.5.3 Scattering Processes and *Q* Details (B)

We now discuss the Q details. A typical situation in which we are interested is how to calculate the electron–phonon interaction and thus calculate the electrical resistivity. To begin with we consider how

$$\left.\frac{\partial f_{\boldsymbol{k}\sigma}}{\partial t}\right|_c = Q(\boldsymbol{r},\boldsymbol{k},t)$$

is determined by the interactions. Let $P_{\boldsymbol{k}\sigma,\boldsymbol{k}'\sigma'}$ be the probability per unit time to scatter from the state $\boldsymbol{k}'\sigma'$ to $\boldsymbol{k}\sigma$. This is typically evaluated from the Golden rule of time-dependent perturbation theory (see Appendix E):

$$P_{\boldsymbol{k}\sigma}^{\boldsymbol{k}'\sigma'} = \frac{2\pi}{\hbar}\left|\langle \boldsymbol{k}\sigma | V_{\text{int}} | \boldsymbol{k}'\sigma' \rangle\right|^2 \delta(E_{\boldsymbol{k}\sigma} - E_{\boldsymbol{k}'\sigma'})\,. \tag{4.126}$$

The probability that there is an electron at $\boldsymbol{r}$, $\boldsymbol{k}$, σ available to be scattered is $f_{\boldsymbol{k}\sigma}$ and $(1 - f_{\boldsymbol{k}'\sigma'})$ is the probability that $\boldsymbol{k}'\sigma'$ can accept an electron (because it is empty).

For scattering out of $\boldsymbol{k}\sigma$ we have

$$\left.\frac{\partial f_{\boldsymbol{k}\sigma}}{\partial t}\right)_{c,\text{out}} = -\sum\nolimits_{\boldsymbol{k}'\sigma'} P_{\boldsymbol{k}'\sigma',\boldsymbol{k}\sigma} f_{\boldsymbol{k}\sigma}(1 - f_{\boldsymbol{k}'\sigma'})\,. \tag{4.127}$$

By a similar argument for scattering into $\boldsymbol{k}\sigma$, we have

$$\left.\frac{\partial f_{\boldsymbol{k}\sigma}}{\partial t}\right)_{c,\text{in}} = +\sum\nolimits_{\boldsymbol{k}'\sigma'} P_{\boldsymbol{k}\sigma,\boldsymbol{k}'\sigma'} f_{\boldsymbol{k}'\sigma'}(1 - f_{\boldsymbol{k}\sigma})\,. \tag{4.128}$$

Combining these two we have an expression for Q:

$$\begin{aligned} Q(\boldsymbol{r},\boldsymbol{k},t) &= \left.\frac{\partial f_{\boldsymbol{k}\sigma}}{\partial t}\right)_c \\ &= \sum\nolimits_{\boldsymbol{k}'\sigma'}[P_{\boldsymbol{k}\sigma,\boldsymbol{k}'\sigma'} f_{\boldsymbol{k}'\sigma'}(1 - f_{\boldsymbol{k}\sigma}) - P_{\boldsymbol{k}'\sigma',\boldsymbol{k}\sigma} f_{\boldsymbol{k}\sigma}(1 - f_{\boldsymbol{k}'\sigma'})]\,. \end{aligned} \tag{4.129}$$

This rate equation for $f_{k\sigma}$ is a type of Master equation [11, p. 190]. At equilibrium, the above must yield zero and we have the principle of detailed balance.

$$P_{\boldsymbol{k}\sigma,\boldsymbol{k}'\sigma'} f^0_{\boldsymbol{k}'\sigma'}(1-f^0_{\boldsymbol{k}\sigma}) = P_{\boldsymbol{k}'\sigma',\boldsymbol{k}\sigma} f^0_{\boldsymbol{k}\sigma}(1-f^0_{\boldsymbol{k}'\sigma'}) \,. \tag{4.130}$$

Using the principle of detailed balance, we can write the rate equation as

$$\begin{aligned} Q(\boldsymbol{r},\boldsymbol{k},t) &= \left.\frac{\partial f_{\boldsymbol{k}\sigma}}{\partial t}\right)_c \\ &= \sum\nolimits_{\boldsymbol{k}'\sigma'} P_{\boldsymbol{k}'\sigma',\boldsymbol{k}\sigma} f^0_{\boldsymbol{k}\sigma}(1-f^0_{\boldsymbol{k}'\sigma'}) \left[\frac{f_{\boldsymbol{k}'\sigma'}(1-f_{\boldsymbol{k}\sigma})}{f^0_{\boldsymbol{k}'\sigma'}(1-f^0_{\boldsymbol{k}\sigma})} - \frac{f_{\boldsymbol{k}\sigma}(1-f_{\boldsymbol{k}'\sigma'})}{f^0_{\boldsymbol{k}\sigma}(1-f^0_{\boldsymbol{k}'\sigma'})}\right]. \end{aligned} \tag{4.131}$$

We now define a quantity $\varphi_{k\sigma}$ such that

$$f_{\boldsymbol{k}\sigma} = f^0_{\boldsymbol{k}\sigma} - \varphi_{\boldsymbol{k}\sigma}\frac{\partial f^0_{\boldsymbol{k}\sigma}}{\partial E_{\boldsymbol{k}\sigma}} \,, \tag{4.132}$$

where

$$f_{\boldsymbol{k}\sigma} = \frac{1}{\exp[\beta(E_{\boldsymbol{k}\sigma}-\mu)]+1} \,, \tag{4.133}$$

with $\beta = 1/k_{\mathrm{B}}T$ and $f^0_{k\sigma}$ is the Fermi function.

Noting that

$$\frac{\partial f^0_{\boldsymbol{k}\sigma}}{\partial E_{\boldsymbol{k}\sigma}} = -\beta f^0_{\boldsymbol{k}\sigma}(1-f^0_{\boldsymbol{k}\sigma}) \,, \tag{4.134}$$

we can show to linear order in $\varphi_{k\sigma}$ that

$$\beta(\varphi_{\boldsymbol{k}'\sigma'} - \varphi_{\boldsymbol{k}\sigma}) = \left[\frac{f_{\boldsymbol{k}'\sigma'}(1-f_{\boldsymbol{k}\sigma})}{f^0_{\boldsymbol{k}'\sigma'}(1-f^0_{\boldsymbol{k}\sigma})} - \frac{f_{\boldsymbol{k}\sigma}(1-f_{\boldsymbol{k}'\sigma'})}{f^0_{\boldsymbol{k}\sigma}(1-f^0_{\boldsymbol{k}'\sigma'})}\right]. \tag{4.135}$$

The Boltzmann transport equation can then be written in the form

$$\begin{aligned} &\frac{\partial f_{\boldsymbol{k}\sigma}}{\partial \boldsymbol{r}}\cdot \boldsymbol{v}_{\boldsymbol{k}\sigma} + \frac{\partial f_{\boldsymbol{k}\sigma}}{\partial \boldsymbol{k}}\cdot\frac{\boldsymbol{F}}{\hbar} + \frac{\partial f_{\boldsymbol{k}\sigma}}{\partial t} = \\ &\qquad \beta\sum\nolimits_{\boldsymbol{k}'\sigma'} P_{\boldsymbol{k}'\sigma',\boldsymbol{k}\sigma} f^0_{\boldsymbol{k}\sigma}(1-f^0_{\boldsymbol{k}'\sigma'})(\varphi_{\boldsymbol{k}'\sigma'} - \varphi_{\boldsymbol{k}\sigma}) \,. \end{aligned} \tag{4.136}$$

Since the sums over $\boldsymbol{k}'$ will be replaced by an integral, this is an integrodifferential equation.

Let us assume that in the Boltzmann equation, on the left-hand side, that there are small fields and temperature gradients so that $f_{k\sigma}$ can be replaced by its equilibrium value. Further, we will assume that $f^0_{k\sigma}$ characterizes local equilibrium in such a way that the spatial variation of $f^0_{k\sigma}$ arises from the temperature and chemical potential (μ). Thus

$$\frac{\partial f^0_{\boldsymbol{k}\sigma}}{\partial \boldsymbol{r}} = \frac{\partial f^0_{\boldsymbol{k}\sigma}}{\partial T}\nabla T + \frac{\partial f^0_{\boldsymbol{k}\sigma}}{\partial \mu}\nabla\mu = -\frac{(E_{\boldsymbol{k}\sigma}-\mu)}{T}\nabla T\frac{\partial f^0_{\boldsymbol{k}\sigma}}{\partial E_{\boldsymbol{k}\sigma}} - \frac{\partial f^0_{\boldsymbol{k}\sigma}}{\partial E_{\boldsymbol{k}\sigma}}\nabla\mu \,.$$

We also use

$$\frac{\partial f_{\boldsymbol{k}\sigma}}{\partial \boldsymbol{k}} = \hbar \boldsymbol{v}_{\boldsymbol{k}\sigma} \frac{\partial f^0_{\boldsymbol{k}\sigma}}{\partial E_{\boldsymbol{k}\sigma}}, \tag{4.137}$$

and assume an external electric field $\boldsymbol{E}$ so $\boldsymbol{F} = -e\boldsymbol{E}$. (The treatment of magnetic fields can be somewhat more complex, see, for example, Madelung [4.26, pp. 205 and following].)

We also replace the sums by integrals as follows:

$$\sum_{\boldsymbol{k}'\sigma'} \rightarrow \frac{V}{(2\pi)^3} \sum_{\sigma'} \int \mathrm{d}\boldsymbol{k}' .$$

We assume steady-state conditions so $\partial f_{k\sigma}/\partial t = 0$. We thus write for the Boltzmann integrodifferential equation:

$$\begin{aligned} &-\frac{(E_{\boldsymbol{k}\sigma} - \mu)}{T} \boldsymbol{v}_{\boldsymbol{k}\sigma} \cdot \nabla T \frac{\partial f^0_{\boldsymbol{k}\sigma}}{\partial E_{\boldsymbol{k}\sigma}} - e\left(\boldsymbol{E} + \frac{1}{e}\nabla\mu\right) \cdot \boldsymbol{v}_{\boldsymbol{k}\sigma} \frac{\partial f^0_{\boldsymbol{k}\sigma}}{\partial E_{\boldsymbol{k}\sigma}} \\ &= \frac{V}{(2\pi)^3 kT} \sum_{\sigma'} \int \mathrm{d}\boldsymbol{k}' P_{\boldsymbol{k}'\sigma',\boldsymbol{k}\sigma} f^0_{\boldsymbol{k}\sigma}(1 - f^0_{\boldsymbol{k}'\sigma'})(\varphi_{\boldsymbol{k}'\sigma'} - \varphi_{\boldsymbol{k}\sigma}) \\ &\equiv \left.\frac{\partial f_{\boldsymbol{k}\sigma}}{\partial t}\right)_c . \end{aligned} \tag{4.138}$$

We now want to see under what conditions we can have a relaxation time. To this end we now assume elastic scattering. This can be approximated by electrons scattering from phonons if the phonon energies are negligible. In this case we write:

$$-\frac{V}{(2\pi)^3} P_{\boldsymbol{k}'\sigma',\boldsymbol{k}\sigma} f^0_{\boldsymbol{k}\sigma}(1 - f^0_{\boldsymbol{k}'\sigma'}) = W(\boldsymbol{k}\sigma, \boldsymbol{k}'\sigma')\delta(E_{\boldsymbol{k}'\sigma'} - E_{\boldsymbol{k}\sigma}), \tag{4.139}$$

where the electron energies are given by $E_{k\sigma}$, so

$$\left.\frac{\partial f_{\boldsymbol{k}\sigma}}{\partial t}\right)_c = -\delta f_{\boldsymbol{k}\sigma} \sum_{\sigma'} \int \mathrm{d}\boldsymbol{k}' W(\boldsymbol{k}'\sigma', \boldsymbol{k}\sigma)\left(1 - \frac{\delta f_{\boldsymbol{k}'\sigma'}}{\delta f_{\boldsymbol{k}\sigma}}\right) \frac{1}{(\partial f^0_{\boldsymbol{k}\sigma} / \partial E_{\boldsymbol{k}\sigma})} \delta(E_{\boldsymbol{k}'\sigma'} - E_{\boldsymbol{k}\sigma}). \tag{4.140}$$

where $\delta f_{k\sigma} = f_{k\sigma} - f^0_{k\sigma}$ We will also assume that the effect of external fields in the steady state causes a displacement of the Fermi distribution in k space. If the energy surface is also assumed to be spherical so $E = E(k)$, with k equal to the magnitude of $\boldsymbol{k}$, (and $\boldsymbol{k}'$) we can write

$$f_{\boldsymbol{k}\sigma} = f^0_{\boldsymbol{k}\sigma} - \boldsymbol{k} \cdot \boldsymbol{c}(E) \frac{\partial f^0_{\boldsymbol{k}\sigma}}{\partial E_{\boldsymbol{k}\sigma}}, \tag{4.141}$$

where $\boldsymbol{c}$ is a constant vector in the direction that f is displaced in k space. Thus

$$\frac{\delta f_{\boldsymbol{k}\sigma}}{\partial f^0_{\boldsymbol{k}\sigma} / \partial E_{\boldsymbol{k}\sigma}} = -\boldsymbol{k} \cdot \boldsymbol{c}(E) , \tag{4.142}$$

and from Fig. 4.10, we see we can write:

$$\cos\Theta' = \frac{\boldsymbol{c}\cdot\boldsymbol{k}'}{ck} = \sin\theta \sin\Theta \cos\varphi' + \cos\Theta \cos\theta . \tag{4.143}$$

If we define a relaxation time by

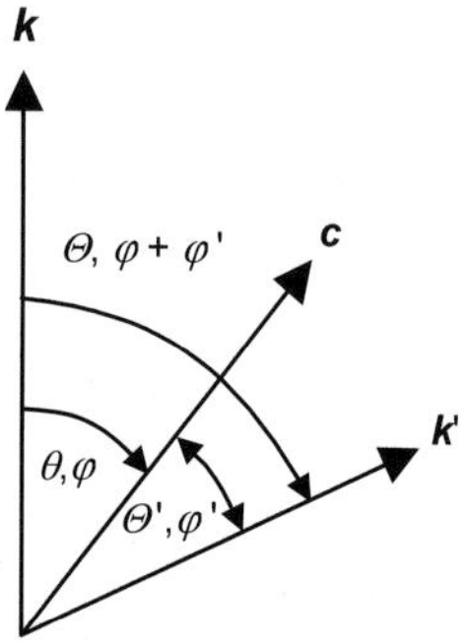

Fig. 4.10. Orientation of the constant $\boldsymbol{c}$ vector with respect to $\boldsymbol{k}$ and $\boldsymbol{k}'$ vectors

$$\left.\frac{\partial f_{\boldsymbol{k}\sigma}}{\partial t}\right)_{\mathrm{c}} = -\frac{\delta f_{\boldsymbol{k}\sigma}}{\tau(E)} , \tag{4.144}$$

then

$$\frac{1}{\tau(E)} = \sum_{\sigma'} \int \mathrm{d}\boldsymbol{k}' W(\boldsymbol{k}'\sigma', \boldsymbol{k}\sigma) \delta(E_{\boldsymbol{k}'\sigma'} - E_{\boldsymbol{k}\sigma}) \frac{(1-\cos\Theta)}{\partial f^0_{\boldsymbol{k}\sigma} / \partial E_{\boldsymbol{k}\sigma}} , \tag{4.145}$$

since the $\cos(\varphi')$ vanishes on integration.

Expressions for $\partial f_{k\sigma}/\partial t)_{\mathrm{c}}$ can be written down for various scattering processes. For example electron–phonon interactions can be sometimes evaluated as above using a relaxation-time approximation. Note if we were concerned with scattering of electrons from optical phonons, then in general their energies can not be neglected, and we would have neither an elastic scattering event, nor a relaxation-time approximation.[11] In any case, the evaluation of Q is complex and further approximations are typically made.

An assumption that is often made in deriving an expression for electrical conductivity, as controlled by the electron–phonon interaction, is called the *Bloch Ansatz*. The Bloch Ansatz is the assumption that the phonon distribution remains

[11] For a discussion of how to treat such cases, see, for example, Howarth and Sondheimer [4.13].

in equilibrium even though the phonons scatter electrons and vice versa. By carrying through an analysis of electron scattering by phonons, using the approximations equivalent to the relaxation-time approximation (above), neglecting umklapp processes, and also making the Debye approximation for the phonons, Bloch evaluated the equilibrium resistivity of electrons as a function of temperature. He found that the electrical resistivity is approximated by

$$\frac{1}{\sigma} \propto \left(\frac{T}{\theta_{\mathrm{D}}}\right)^5 \int_0^{\theta_{\mathrm{D}}/T} \frac{x^5 \mathrm{d}x}{(\mathrm{e}^x - 1)(1 - \mathrm{e}^{-x})}. \tag{4.146}$$

This is called the Bloch–Gruneisen relation. In (4.146), θ_{D} is the Debye temperature. Note that (4.146) predicts the resistivity curve goes as T^5 at low temperatures, and as T at higher temperatures.[12] In (4.146), $1/\sigma$ is the resistivity ρ, and for real materials one should include a residual resistivity ρ_0 as a further additive factor. The purity of the sample determines ρ_0.

4.5.4 The Relaxation-Time Approximate Solution of the Boltzmann Equation for Metals (B)

A phenomenological form of

$$Q = \left(\frac{\partial f}{\partial t}\right)_{\mathrm{scatt}}$$

will be stated. We assume that $(\partial f/\partial t)_{\mathrm{scatt}}$ $(= \partial f/\partial t)_{\mathrm{c}})$ is proportional to the difference of f from its equilibrium f_0 and is also proportional to the probability of a collision $1/\tau$, where τ is the relaxation time, as in (4.144) and (4.145). Then

$$\left(\frac{\partial f}{\partial t}\right)_{\mathrm{scatt}} = -\frac{f - f_0}{\tau}. \tag{4.147}$$

Integrating (4.147) gives

$$f - f_0 = A\mathrm{e}^{-t/\tau}, \tag{4.148}$$

which simply says that in the absence of external perturbations, any system will reach its equilibrium value when t becomes infinite. Equation (4.148) assumes that collisions will bring the system to equilibrium. This may be hard to prove, but it is physically very reasonable. There may be only a few cases where the assumption of a relaxation time is fully justified. To say more about this point requires a discussion of the Q details of the system. In (4.131), τ will be assumed to be a function of E_k only. A more drastic assumption would be that τ is a constant, and a less drastic assumption would be that τ is a function of k.

---

[12] As emphasized by Arajs [4.3], (4.146) should not be applied blindly with the expectation of good results in all metals (particularly for low temperature).

With all of the above assumptions and assuming steady state, the Boltzmann differential equation is[13]

$$\boldsymbol{v}_{\boldsymbol{k}} \cdot \nabla T \frac{\partial f_{\boldsymbol{k}}}{\partial T} - e(\boldsymbol{E} + \boldsymbol{v}_{\boldsymbol{k}} \times \boldsymbol{B}) \cdot \boldsymbol{v}_{\boldsymbol{k}} \frac{\partial f_{\boldsymbol{k}}}{\partial E_{\boldsymbol{k}}} = -\frac{f_{\boldsymbol{k}} - f_{\boldsymbol{k}}^0}{\tau(E_{\boldsymbol{k}})} . \tag{4.149}$$

Since electrons are being considered, if we ignore the possibility of electron correlations, then $f_{\boldsymbol{k}}^0$ is the Fermi–Dirac distribution function. (as in (4.154)).

In order to show the utility of (4.149), a calculation of the electrical conductivity using (4.149) will be made. We assume $\nabla T = 0$, $\boldsymbol{B} = 0$, and $E = E\hat{z}$. Then (4.149) reduces to

$$f_{\boldsymbol{k}} = f_{\boldsymbol{k}}^0 + e\tau E v_{\boldsymbol{k}}^z \frac{\partial f_{\boldsymbol{k}}}{\partial E_{\boldsymbol{k}}} . \tag{4.150}$$

If we assume that there is only a small deviation from *equilibrium*, a first iteration yields

$$f_{\boldsymbol{k}} = f_{\boldsymbol{k}}^0 - e\tau E v_{\boldsymbol{k}}^z \frac{\partial f_{\boldsymbol{k}}^0}{\partial E_{\boldsymbol{k}}} . \tag{4.151}$$

Since there is no electrical current in equilibrium, substitution of (4.151) into (4.122) gives

$$J_z = -\frac{e^2}{4\pi^2} \int (v_{\boldsymbol{k}}^z)^2 \tau \frac{\partial f_{\boldsymbol{k}}^0}{\partial E_{\boldsymbol{k}}} E \mathrm{d}^3 k . \tag{4.152}$$

If we have spherical symmetry in k space,

$$\boldsymbol{J} = -\frac{1}{3} \frac{e^2}{4\pi^2} E \int v_{\boldsymbol{k}}^2 \tau \frac{\partial f_{\boldsymbol{k}}^0}{\partial E_{\boldsymbol{k}}} \mathrm{d}^3 k . \tag{4.153}$$

Since $f_{\boldsymbol{k}}^0$ represents the value of the number of electrons, by our normalization (4.5.1)

$$f_{\boldsymbol{k}}^0 = F \quad \text{the Fermi function.} \tag{4.154}$$

At temperatures lower than several thousand degrees $F \cong 1$ for $E_k < E_F$ and $F \cong 0$ for $E_k > E_F$, and so

$$\frac{\partial F}{\partial E_{\boldsymbol{k}}} \cong -\delta(E_{\boldsymbol{k}} - E_F) , \tag{4.155}$$

[13] Equation (4.149) is the same as (4.138) and (4.145) with $\nabla\mu = 0$ and $\boldsymbol{B} = 0$. These are typical conditions for metals, although not necessarily for semiconductors.

where δ is the Dirac delta function and E_F is the Fermi energy. Now since a volume in $\boldsymbol{k}$-space may be written as

$$\mathrm{d}^3k = \frac{\mathrm{d}S\mathrm{d}E}{|\nabla_{\boldsymbol{k}}E|} = \frac{\mathrm{d}S\mathrm{d}E}{\hbar v_{\boldsymbol{k}}}, \tag{4.156}$$

where S is a surface of constant energy, (4.153), (4.154), (4.155), and (4.156) imply

$$J = \frac{e^2 E}{12\pi^3\hbar}\int\left[\int v_{\boldsymbol{k}}\tau\delta(E_{\boldsymbol{k}} - E_F)\mathrm{d}E\right]\mathrm{d}S\,. \tag{4.157}$$

Using $E_{\boldsymbol{k}} = \hbar^2\boldsymbol{k}^2/2m$, (4.157) becomes

$$J = \frac{e^2 E}{12\pi^3\hbar}(v_{\boldsymbol{k}}^F)(\tau_F)4\pi k_F^2\,, \tag{4.158}$$

where the subscript F means that the function is to be evaluated at the Fermi energy. If n is the number of conduction electrons per unit volume, then

$$n = \frac{1}{4\pi^3}\int F\mathrm{d}^3k = \frac{4\pi}{3}k_F^3\frac{1}{4\pi^3}\,. \tag{4.159}$$

Combining (4.158) and (4.159), we find that

$$J = \frac{ne^2 E\tau_F}{m} = \sigma E \quad \text{or} \quad \sigma = \frac{ne^2\tau_F}{m}\,. \tag{4.160}$$

This is (3.220) that was derived earlier. Now it is clear that all pertinent quantities are to be evaluated at the Fermi energy. There are several general techniques for solving the Boltzmann equation, for example the variation principle. The book by Ziman can be consulted [99, p275ff].

4.6 Transport Coefficients

As mentioned, if we have no magnetic field (in the presence of a magnetic field, several other characteristic effects besides those mentioned below are of importance [4.26, p 205] and [73]), then the approximate Boltzmann differential equation is (in the relaxation-time approximation)

$$v_{\boldsymbol{k}}\cdot\left(-\nabla T\frac{\partial f_{\boldsymbol{k}}^0}{\partial T} + e\boldsymbol{E}\frac{\partial f_{\boldsymbol{k}}^0}{\partial E_{\boldsymbol{k}}}\right) = \frac{f_{\boldsymbol{k}} - f_{\boldsymbol{k}}^0}{\tau}\,. \tag{4.161}$$

Using the definitions of $\boldsymbol{J}$ and $\boldsymbol{H}$ in terms of the distribution function ((4.122) and (4.123)), and using (4.161), we have

$$\boldsymbol{J} = a\boldsymbol{E} + b\nabla T\,, \tag{4.162}$$

$$\boldsymbol{H} = c\boldsymbol{E} + d\nabla T\,. \tag{4.163}$$

For cubic crystals *a*, *b*, *c*, and *d* are scalars. Equations (4.162) and (4.163) are more general than their derivation based on (4.161) might suggest. The equations must be valid for sufficiently small $\boldsymbol{E}$ and $\nabla \boldsymbol{T}$. This is seen by a Taylor series expansion and by the fact that $\boldsymbol{J}$ and $\boldsymbol{H}$ must vanish when $\boldsymbol{E}$ and $\nabla \boldsymbol{T}$ vanish. The point of this Section will be to show how experiments determine *a*, *b*, *c*, and *d* for materials in which electrons carry both heat and electricity.

4.6.1 The Electrical Conductivity (B)

The electrical conductivity measurement is the simplest of all. We simply set $\nabla \boldsymbol{T} = 0$ and measure the electrical current. Equation (4.162) becomes $\boldsymbol{J} = a\boldsymbol{E}$, and so we obtain $a = \sigma$.

4.6.2 The Peltier Coefficient (B)

This is also an easy measurement to describe. We use the same experimental setup as for electrical conductivity, but now we measure the heat current. Equation (4.163) becomes

$$\boldsymbol{H} = c\boldsymbol{E} = c\frac{\boldsymbol{J}}{\sigma} = \frac{c}{a}\boldsymbol{J}. \tag{4.164}$$

The Peltier coefficient is the heat current per unit electrical current and so it is given by $\Pi = c/a$.

4.6.3 The Thermal Conductivity (B)

This is just a little more complicated than the above, because we usually do the thermal conductivity measurements with no electrical current rather than no electrical field. By the definition of thermal conductivity and (4.163), we obtain

$$\boldsymbol{K} = -\frac{|\boldsymbol{H}|}{|\nabla \boldsymbol{T}|} = -\frac{|c\boldsymbol{E} + d\nabla \boldsymbol{T}|}{|\nabla \boldsymbol{T}|}. \tag{4.165}$$

Using (4.162) with no electrical current, we have

$$\boldsymbol{E} = -\frac{b}{a}\nabla T. \tag{4.166}$$

The thermal conductivity is then given by

$$\boldsymbol{K} = -d + \frac{cb}{a}. \tag{4.167}$$

We might expect the thermal conductivity to be $-d$, but we must remember that we required there to be no electrical current. This causes an electric field to appear, which tends to reduce the heat current.

4.6.4 The Thermoelectric Power (B)

We use the same experimental setup as for thermal conductivity but now we measure the electric field. The absolute thermoelectric power Q is defined as the proportionality constant between electric field and temperature gradient. Thus

$$\boldsymbol{E} = Q\nabla T \,. \tag{4.168}$$

Comparing with (4.166) gives

$$Q = -\frac{b}{a} \,. \tag{4.169}$$

We generally measure the difference of two thermoelectric powers rather than the absolute thermoelectric power. We put two unlike metals together in a loop and make a break somewhere in the loop as shown in Fig. 4.11. If V_{AB} is the voltage across the break in the loop, an elementary calculation shows

$$|Q_2 - Q_1| \cong \frac{|V_{\mathrm{AB}}|}{|T_2 - T_1|} \,. \tag{4.170}$$

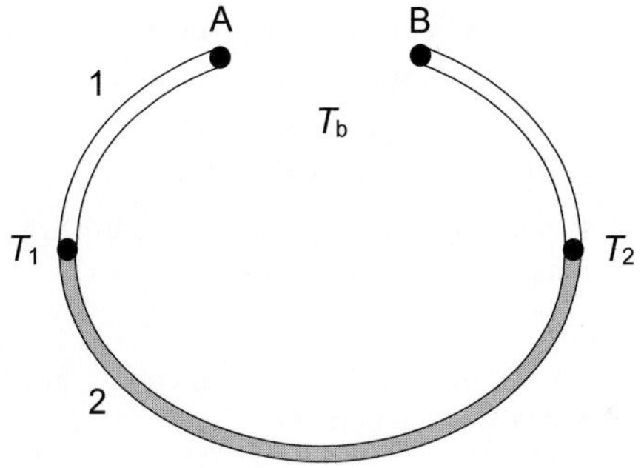

Fig. 4.11. Circuit for measuring the thermoelectric power. The junctions of the two metals are at temperature T_1 and T_2

4.6.5 Kelvin's Theorem (B)

A general theorem originally stated by Lord Kelvin, which can be derived from the thermodynamics of irreversible process, states that [99]

$$\Pi = QT \,. \tag{4.171}$$

Summarizing, by using (4.162), (4.163), $\sigma = a$, (4.165), (4.167), (4.164), and (4.171), we can write

$$\boldsymbol{J} = \sigma\boldsymbol{E} - \frac{\sigma\Pi}{T}\nabla T \,, \tag{4.172}$$

$$\boldsymbol{H} = \sigma\Pi\boldsymbol{E} - \left(K + \sigma\frac{\Pi^2}{T}\right)\nabla T \,. \tag{4.173}$$

If, in addition, we assume that the Wiedemann–Franz law holds, then $K = CT\sigma$, where $C = (\pi^2/3)(k/e)^2$, and we obtain

$$\boldsymbol{J} = \sigma\boldsymbol{E} - \frac{\sigma\Pi}{T}\nabla T , \tag{4.174}$$

$$\boldsymbol{H} = \sigma\Pi\boldsymbol{E} - \sigma\left(CT + \frac{\Pi^2}{T}\right)\nabla T . \tag{4.175}$$

We summarize these results in Table 4.5. As noted in the references there are several other transport coefficients including magnetoresistance, Rigli–Leduc, Ettinghausen, Nernst, and Thompson.

Table 4.5. Transport coefficients

Quantity	Definition	Comment
Electrical conductivity	Electric current density at unit electric field (no magnetic (B) field, no temperature gradient).	See Sect. 4.5.4 and 4.6.1
Thermal conductivity	Heat flux per unit temp. gradient (no electric current).	See Sect. 4.6.3
Peltier coefficient	Heat exchanged at junction per electric current density.	See Sect. 4.6.2
Thermoelectric power (related to Seebeck effect)	Electric field per temperature gradient (no electric current).	See Sect. 4.6.4
Kelvin relations	Relates thermopower, Peltier coefficient and temperature.	See Sect. 4.6.5

References:
[4.1, 4.32, 4.39]

4.6.6 Transport and Material Properties in Composites (MET, MS)

Introduction (MET, MS)

Sometimes the term composite is used in a very restrictive sense to mean fibrous structures that are used, for example, in the aircraft industry. The term composite is used much more generally here as any material composed of constituents that themselves are well defined. A rock composed of minerals, is thus a composite using this definition. In general, composite materials have become very important not only in the aircraft industry, but in the manufacturing of cars, in many kinds of building materials, and in other areas.

A typical problem is to find the effective dielectric constant of a composite media. As we will show below, if we can find the potential as a function of position, we can evaluate the effective dielectric constant. First, we want to illustrate that this is also the same problem as the effective thermal conductivity, the effective electrical conductivity, or the effective magnetic permeability of a composite. For in each case, we end up solving the same differential equation as shown in Table 4.6.

Table 4.6. Equivalent problems

Dielectric constant	*Magnetic permeability*
$\boldsymbol{D} = \varepsilon \boldsymbol{E}$ ε is dielectric constant $\boldsymbol{E}$ is electric field $\boldsymbol{D}$ is electric displacement vector	$\boldsymbol{B} = \mu \boldsymbol{H}$ μ is magnetic permeability $\boldsymbol{H}$ is magnetic field intensity $\boldsymbol{B}$ is magnetic flux density
$\nabla \times \boldsymbol{E} = 0$ (no changing $\boldsymbol{B}$) $\boldsymbol{E} = -\nabla(\phi)$ $\nabla \cdot \boldsymbol{D} = 0$ (no free charge) $\nabla \cdot (\varepsilon \nabla(\phi)) = 0$	$\nabla \times \boldsymbol{B} = 0$ (no current, no changing $\boldsymbol{E}$) $\boldsymbol{H} = -\nabla(\Phi)$ $\nabla \cdot \boldsymbol{B} = 0$ (Maxwell equation) $\nabla \cdot (\mu \nabla(\Phi)) = 0$
B.C. ϕ constant at top and bottom $\nabla(\phi) = 0$ on side surfaces	analogous B.C.
Electrical conductivity	*Thermal conductivity*
$\boldsymbol{J} = \sigma \boldsymbol{E}$ and only driven by $\boldsymbol{E}$ σ is electrical conductivity E is electric field J is electrical current density	$J = -K\nabla(T)$ and only driven by ∇T K is the thermal conductivity T is the temperature J is the heat flux
$\nabla \times \boldsymbol{E} = 0$ (no changing B) $\boldsymbol{E} = -\nabla(\phi)$ $\nabla \cdot \boldsymbol{J} = 0$ (cont. equation, steady state) $\nabla \cdot (s\nabla(\phi)) = 0$ analogous B.C.	$\nabla \times \nabla(T) = 0$, an identity grad dot $J = 0$ (cont. equation, steady state) $\nabla \cdot K(\nabla(T)) = 0$ analogous B.C.

To begin with we must define the desired property for the composite. Consider the case of the dielectric constant. Once the overall potential is known (and it will depend on boundary conditions in general as well as the appropriate differential equation), the effective dielectric constant may ε_c be defined such that it would lead to the same over all energy. In other words

$$\varepsilon_c E_0^2 = \frac{1}{V}\int \varepsilon(\boldsymbol{r})E^2(\boldsymbol{r})\mathrm{d}V\,, \tag{4.176}$$

where

$$E_0 = \frac{1}{V}\int E(\boldsymbol{r})\mathrm{d}V\,, \tag{4.177}$$

where V is the volume of the composite, and the electric field $E(\boldsymbol{r})$ is known from solving for the potential. The spatial dependence of the dielectric constant, $\varepsilon(\boldsymbol{r})$, is known from the way the materials are placed in the composite.

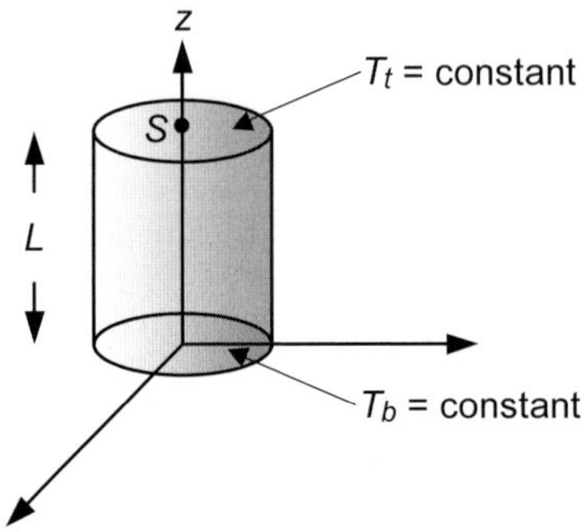

Fig. 4.12. The right-circular cylinder shown is assumed to have sides insulated and it has volume $V = LS$

One may similarly define the effective thermal conductivity. Let $\boldsymbol{b} = -\nabla T$, where T is the temperature, and $\boldsymbol{h} = -K\nabla T$, where K is the thermal conductivity. The equivalent definition for the thermal conductivity of a composite is

$$K_c = \frac{V\int \boldsymbol{h}\cdot\boldsymbol{b}\mathrm{d}V}{\left(\int \boldsymbol{b}\mathrm{d}V\right)^2}\,. \tag{4.178}$$

For the geometry and boundary conditions shown in Fig. 4.12, we show this expression reduces to the usual definition of thermal conductivity.

Note since $\nabla\cdot\boldsymbol{h} = 0$ in the steady state that $-\nabla\cdot(T\boldsymbol{h}) = \boldsymbol{h}\cdot\boldsymbol{b}$, and so $\int\boldsymbol{h}\cdot\boldsymbol{b}\mathrm{d}V = -(T_t - T_b)\int h_z\mathrm{d}S_z$, where the law of Gauss has been used, and the integral is over the top of the cylinder. Also note, by the Gauss law $\hat{z}\cdot\int\boldsymbol{b}\mathrm{d}V = (T_t - T_b)S$, where S is the top or bottom area. We assume either parallel slabs, or macroscopically dilute solutions of ellipsoidally shaped particles so that the average temperature gradient will be along the z-axis, then

$$-K_c S(T_t - T_b)/L = \int_{\text{top}} h_z \mathrm{d}S_z\,, \tag{4.179}$$

as required by the usual definition of thermal conductivity.

It is an elementary exercise to compute the effective material property for the series and parallel cases. For example, consider the thermal conductivity. If one has a two-component system with volume fractions φ_1 and φ_2, then for the series case one obtains for the effective thermal conductivity K_c of the composite:

$$\frac{1}{K_c} = \frac{\varphi_1}{K_1} + \frac{\varphi_2}{K_2} . \tag{4.180}$$

This is easily shown as follows. Suppose we have a rod of total length $L = (l_1 + l_2)$ and uniform cross-sectional area composed of a smaller length l_1 with thermal conductivity K_1 and an upper length l_2 with K_2. The sides of the rod are assumed to be insulated and we maintain the bottom temperature at T_0, the interface at T_1, and the top at T_2. Then since $\Delta T_1 = T_0 - T_1$ and $\Delta T_2 = T_1 - T_2$ we have $\Delta T = \Delta T_1 + \Delta T_2$ and since the temperature changes linearly along the length of each rod:

$$K_1 \frac{\Delta T_1}{l_1} = K_2 \frac{\Delta T_2}{l_2} = K_c \frac{\Delta T}{L} , \tag{4.181}$$

where K_c is the effective thermal conductivity of the rod. We can thus write:

$$\Delta T_1 = \frac{K}{K_1} l_1 \frac{\Delta T}{L} , \quad \Delta T_2 = \frac{K}{K_2} l_2 \frac{\Delta T}{L} , \tag{4.182}$$

and so

$$\Delta T = \Delta T_1 + \Delta T_2 = \left(\frac{K}{K_1} \frac{l_1}{L} + \frac{K}{K_2} \frac{l_2}{L} \right) \Delta T , \tag{4.183}$$

and since the volume fractions are given by $\varphi_1 = (Al_1/AL) = l_1/L$ and $\varphi_2 = l_2/L$, this yields the desired result.

Similarly for the parallel case, one can show:

$$K_c = \varphi_1 K_1 + \varphi_2 K_2 . \tag{4.184}$$

Consider two equal length slabs of length L and areas A_1 and A_2. These are placed parallel to each other with the sides insulated and the tops and bottoms maintained at T_0 and T_2. Then if $\Delta T = T_0 - T_2$, the effective thermal conductivity can be defined by

$$K(A_1 + A_2) \frac{\Delta T}{L} = K_1 A_1 \frac{\Delta T}{L} + K_2 A_2 \frac{\Delta T}{L} , \tag{4.185}$$

where we have used that the temperature changes linearly along the slabs. Solving for K yields the desired relation, with the volume fractions defined by $\varphi_1 = A_1/(A_1{+}A_2)$ and $\varphi_2 = A_2/(A_1{+}A_2)$.

General Theory (MET, MS)[14]

Let

$$\boldsymbol{u} = \frac{\int \boldsymbol{b}\mathrm{d}V}{\left|\int \boldsymbol{b}\mathrm{d}V\right|}, \tag{4.186}$$

and with the boundary conditions and material assumptions we have made, $\boldsymbol{u} = \hat{z}$. Define the following averages:

$$\bar{h} = \frac{1}{V}\int_V \boldsymbol{u}\cdot\boldsymbol{h}\mathrm{d}V, \tag{4.187}$$

$$\bar{b} = \frac{1}{V}\int_V \boldsymbol{u}\cdot\boldsymbol{b}\mathrm{d}V, \tag{4.188}$$

$$\bar{h}_i = \frac{1}{V_i}\int_{V_i} \boldsymbol{u}\cdot\boldsymbol{h}\mathrm{d}V_i, \tag{4.189}$$

$$\bar{b}_i = \frac{1}{V_i}\int_{V_i} \boldsymbol{u}\cdot\boldsymbol{b}\mathrm{d}V_i, \tag{4.190}$$

where V is the overall volume, and V_i is the volume of each constituent so $V = \sum V_i$. From this we can show (using Gauss-law manipulations similar to that already given) that

$$K_c = \frac{\bar{h}}{\bar{b}} \tag{4.191}$$

will give the same value for the effective thermal conductivity as the original definition. Letting $\varphi_i = V_i/V$ be the volume fractions and $f_i = \bar{b}_i/\bar{b}$ be the "field ratios" we have

$$K_i f_i = \frac{\bar{h}_i}{\bar{b}}, \tag{4.192}$$

and

$$\sum \bar{h}_i \varphi_i = \bar{h}, \tag{4.193}$$

so

$$K = \sum K_i f_i \varphi_i . \tag{4.194}$$

[14] This is basically Maxwell–Garnett theory. See Garnett [4.9]. See also Reynolds and Hough [4.36].

Also

$$\sum f_i \varphi_i = 1\,, \tag{4.195}$$

and

$$\sum \varphi_i = 1\,. \tag{4.196}$$

The field ratios f_i, the volume fractions φ_i, and the thermal conductivities K_i of the constituents determine the overall thermal conductivity. The f_i will depend on the K_i and the geometry. They are only known for the case of parallel slabs or very dilute solutions of ellipsoidally shaped particles. We have already assumed this, and we will only treat these cases. We also only consider the case of two phases, although it is relatively easy to generalize to several phases.

The field ratios can be evaluated from the equivalent electrostatic problem. The b inside an ellipsoid b_i are given in terms of the externally applied $\boldsymbol{b}(\boldsymbol{b}_0)$ by[15]

$$b_i = g_i b_{0i}\,, \tag{4.197}$$

where the i refer to the principle axis of the ellipsoid. With the ellipsoid having thermal conductivity K_j and its surrounding K^* the g_i are

$$g_i = \frac{1}{1 + N_i[(K_j / K^*) - 1]}\,, \tag{4.198}$$

where the N_i are the depolarization factors. As usual,

$$\sum_{i=1}^{3} N_i = 1\,.$$

Redefine (equivalently, e.g. using our conventions, we would apply an external thermal gradient along the z-axis)

$$\boldsymbol{u} = \frac{\boldsymbol{b}_0}{b_0}\,,$$

and let θ_i be the angle between the principle axes of the ellipsoid and $\boldsymbol{u}$. Then

$$\boldsymbol{u} \cdot \boldsymbol{b} = \sum_{i=1}^{3} g_i b_0 \cos^2\theta_i\,, \tag{4.199}$$

so

$$f_j = \sum_i g_i \cos^2\theta_i\,, \tag{4.200}$$

[15] See Stratton [4.38].

where the sum over i is over the principle axis directions and j refers to the constituents. *Conditions that insure* that $\overline{b} = b_0$ have already been assumed. We have

$$f_j = \sum_{i=1}^{3} \frac{\cos^2 \theta_i}{1 + N_i[(K_j / K^*) - 1]}, \tag{4.201}$$

K_j is the thermal conductivity of the ellipsoid surrounded by K^*.

Case 1: Thin slab parallel to $\boldsymbol{b}_0$, with $K^* = K_2$. Assuming an ellipsoid of revolution,

$$\begin{aligned} &N = 0, \quad \text{(depolarization factor along } b_0) \\ &f_1 = 1, \\ &f_2 = 1. \end{aligned}$$

Using

$$K = \sum K_i f_i \varphi_i ,$$

we get

$$K = K_1 \varphi_1 + K_2 \varphi_2 . \tag{4.202}$$

We have already seen this is appropriate for the parallel case.

Case 2: Thin slab with plane normal to $\boldsymbol{b}_0$, $K^* = K_2$.

$$N = 1, \quad f_1 = \frac{1}{1 + (K_1 / K_2) - 1} = \frac{K_2}{K_1}, \quad f_2 = 1,$$

so we get

$$\frac{1}{K} = \frac{\varphi_1}{K_1} + \frac{\varphi_2}{K_2}. \tag{4.203}$$

Again as before.

Case 3: Spheres with $K^* = K_2$ (where by (4.195), the denominator in (4.204) is 1)

$$N = \frac{1}{3}, \quad f_1 = \frac{1}{2 + (K_1 / K_2)}, \quad f_2 = 1$$

$$K = \frac{K_2 \varphi_2 + K_1 \varphi_1 \dfrac{3}{2 + (K_1 / K_2)}}{\varphi_2 + \varphi_1 \dfrac{3}{2 + (K_1 / K_2)}}. \tag{4.204}$$

These are called the Maxwell (composite) equations (interchanging 1 and 2 gives the second one).

The parallel and series combinations can be shown to provide absolute upper and lower bounds on the thermal conductivity of the composite.[16] The Maxwell

[16] See Bergmann [4.4].

equations provide bounds if the material is microscopically isotropic and homogenous.[16] If $K_2 > K_1$ then the Maxwell equation written out above is a lower bound.

As we have mentioned, generalizations to more than two components is relatively straightforward.

The empirical equation

$$K = K_1^{\varphi_1} K_2^{\varphi_2} \tag{4.205}$$

is known as Lictenecker's equation and is commonly used when K_1 and K_2 are not too drastically different.[17]

Problems

4.1 According to the equation

$$K = \tfrac{1}{3}\sum_m C_m \overline{V}_m \lambda_m ,$$

the specific heat C_m can play an important role in determining the thermal conductivity K. (The sum over m means a sum over the modes m carrying the energy.) The total specific heat of a metal at low temperature can be represented by the equation

$$C_v = AT^3 + BT ,$$

where A and B are constants. Explain where the two terms come from.

4.2 Look at Fig. 4.7 and Fig. 4.9 for the thermal conductivity of metals and insulators. Match the temperature dependences with the "explanations." For (3) and (6) you will have to decide which figure works for an explanation.

(1) T	(a) Boundary scattering of phonons K $= C\overline{V}\lambda/3$, and $\overline{V}$, λ approximately constant.
(2) T^2	(b) Electron–phonon interactions at low temperature changes cold to hot electrons and vice versa.
(3) constant	(c) $C_v \propto T$.
(4) T^3	(d) $T > \theta_D$, you know ρ from Bloch (see Problem 4.4), and use the Wiedeman–Franz law.
(5) $T^n e^{\beta/T}$	(e) C and $\overline{V} \cong$ constant. The mean squared displacement of the ions is proportional to T and is also inversely proportional to the mean free path of phonons. This is high-temperature umklapp.
(6) T^1	(f) Umklapp processes at not too high temperatures.

[17] Also of some interest is the variation in K due to inaccuracies in the input parameters (such as K_1, K_2) for different models used for calculating K for a composite. See, e.g., Patterson [4.34].

4.3 Calculate the thermal conductivity of a good metal at high temperature using the Boltzmann equation and the relaxation-time approximation. Combine your result with (4.160) to derive the law of Wiedeman and Franz.

4.4 From Bloch's result (4.146) show that σ is proportional to T^{-1} at high temperatures and that σ is proportional to T^{-5} at low temperatures. Many solids show a constant residual resistivity at low temperatures (Matthiessen's rule). Can you suggest a reason for this?

4.5 Feynman [4.7, p. 226], while discussing the polaron, evaluates the integral

$$I = \int \frac{\mathrm{d}\boldsymbol{q}}{q^2 f(\boldsymbol{q})} ,$$

(compare (4.112)) where

$$\mathrm{d}\boldsymbol{q} = \mathrm{d}q_x \, \mathrm{d}q_y \, \mathrm{d}q_z ,$$

and

$$f(\boldsymbol{q}) = \frac{\hbar^2}{2m}(2\boldsymbol{k} \cdot \boldsymbol{q} - q^2) - \hbar\omega_{\mathrm{L}} ,$$

by using the identity:

$$\frac{1}{K_1 K_2} = \int_0^1 \frac{\mathrm{d}x}{[K_1 a + K_2(1-x)]^2} .$$

a. Prove this identity

b. Then show the integral is proportional to

$$\frac{1}{k} \sin^{-1} \frac{K_3 k}{\sqrt{2}} ,$$

and evaluate K_3.

c. Finally, show the desired result:

$$E_{\boldsymbol{k},0} = -\alpha_{\mathrm{c}} \hbar\omega_{\mathrm{L}} + \frac{\hbar^2 k^2}{2m^{**}} ,$$

where

$$m^{**} = \frac{m^*}{1 - \dfrac{\alpha_c}{6}} ,$$

and m^* is the ordinary effective mass.

5 Metals, Alloys, and the Fermi Surface

Metals are one of our most important sets of materials. The study of bronzes (alloys of copper and tin) dates back thousands of years. Metals are characterized by high electrical and thermal conductivity and by electrical resistivity (the inverse of conductivity) increasing with temperature. Typically, metals at high temperature obey the Wiedeman–Franz law (Sect. 3.2.2). They are ductile and deform plastically instead of fracturing. They are also opaque to light for frequencies below the plasma frequency (or the plasma edge as discussed in the chapter on optical properties). Many of the properties of metals can be understood, at least partly, by considering metals as a collection of positive ions in a sea of electrons (the jellium model). The metallic bond, as discussed in Chap. 1, can also be explained to some extent with this model.

Metals are very important but this chapter is relatively short. The reason for this is that various properties of metals are discussed in other chapters. For example in Chap. 3 the free-electron model, the pseudopotential, and band structure were discussed, as well as some aspects of electron correlations. Electron correlations were also mentioned in Chap. 4 along with the electrical and thermal conductivity of solids including metals. Metals are also important for the study of magnetism (Chap. 7) and superconductors (Chap. 8). The effect of electron screening is discussed in Chap. 9 and free-carrier absorption by electrons in Chap. 10.

Metals occur whenever one has partially filled bands because of electron concentration and/or band overlapping. Many elements and alloys form metals (see Sect. 5.10). The elemental metals include alkali metals (e.g. Na), noble metals (Cu and Ag are examples), polyvalent metals (e.g. Al), transition metals with incomplete d shells, rare earths with incomplete f shells, lanthanides, and actinides. Even nonmetallic materials such as iodine may become metallic under very high pressure.

Also, in this chapter we will include some relatively new and novel ideas such as heavy electron systems, and so-called linear metals.

We start by discussing one of the most important properties of metals—the Fermi surface, and show how one can use simple free-electron ideas along with the Brillouin zone to get a first orientation.

5.1 Fermi Surface (B)

Mackintosh has defined a metal as a solid with a Fermi-Surface [5.19]. This tacitly assumes that the highest occupied band is only partly filled. At absolute zero, the Fermi surface is the highest filled energy surface in $\boldsymbol{k}$ or wave vector space.

When one has a constant potential, the metal has free-electron spherical energy surfaces, but a periodic potential can cause many energy surface shapes. Although the electrons populate the energy surfaces according to Fermi–Dirac statistics, the transition from fully populated to unpopulated energy surfaces is relatively sharp at room temperature. The Fermi surface at room temperature is typically as well defined as is the surface of a peach, i.e. the surface has a little "fuzz", but the overall shape is well defined.

For many electrical properties, only the electrons near the Fermi surface are active. Therefore, the nature of the Fermi surface is very important. Many Fermi surfaces can be explained by starting with a free-electron Fermi surface in the extended-zone scheme and, then, mapping surface segments into the reduced-zone scheme. Such an approach is said to be an empty-lattice approach. We are not considering interactions but we have already noted that the calculations of Luttinger and others (see Sect. 3.1.4) indicate that the concept of a Fermi surface should have meaning, even when electron–electron interactions are included. Experiments, of course, confirm this point of view (the Luttinger theorem states that the volume of the Fermi surface is unchanged by interactions).

When Fermi surfaces intersect Brillouin zone boundaries, useful Fermi surfaces can often be constructed by using an extended or repeated-zone scheme. Then constant-energy surfaces can be mapped in such a way that electrons on the surface can travel in a closed loop (i.e. without "Bragg scattering"). See, e.g. [5.36, p. 66].

Going beyond the empty-lattice approach, we can use the results of calculations based on the one-electron theory to construct the Fermi surface. We first solve the Schrödinger equation for the crystal to determine $E_b(\boldsymbol{k})$ for the electrons (b labels the different bands). We assume the temperature is zero and we find the highest occupied band $E_{b'}(\boldsymbol{k})$. For this band, we construct constant-energy surfaces in the first Brillouin zone in $\boldsymbol{k}$-space. The highest occupied surface is the Fermi surface. The effects of nonvanishing temperatures and of overlapping bands may make the situation more complicated. As mentioned, finite temperatures only smear out the surface a little. The highest occupied energy surface(s) at absolute zero is (are) still the Fermi surface(s), even with overlapping bands. It is possible to generalize somewhat. One can plot the surface in other zones besides the first zone. It is possible to imagine a Fermi surface for holes as well as electrons, where appropriate.

However, this approach is often complex so we start with the empty-lattice approach. Later we will give an example of the results of a band-structure calculation (Fig. 5.2). We then discuss (Sects. 5.3 and 5.4) how experiments can be used to elucidate the Fermi surface.

5.1.1 Empty Lattice (B)

Suppose the electrons are characterized by free electrons with effective mass m^* and let E_F be the Fermi energy. Then we can say:

a) $$E = \frac{\hbar^2 k^2}{2m^*},$$

b) $k_F = \sqrt{\frac{2m^* E_F}{\hbar^2}}$ is the Fermi radius,

c) $n = \frac{1}{3\pi^2} k_F^3$ is the number of electrons per unit volume,

$$\left[n = \frac{N}{V} = \left(\frac{2}{8\pi^3} \right) \left(\frac{4}{3} \pi k_F^3 \right) \right],$$

d) in a volume Δk_V of k-space, there are

$$\Delta n = \frac{1}{4\pi^3} \Delta k_V$$

electrons per unit volume of real space, and finally

e) the density of states per unit volume is

$$\mathrm{d}n = \frac{1}{2\pi^2} \left(\frac{2m^*}{\hbar^2} \right)^{3/2} \sqrt{E}\, \mathrm{d}E .$$

We consider that each band is formed from an atomic orbital with two spin states. There are, thus, $2N$ states per band if there are N atoms associated with N lattice points. If each atom contributes one electron, then the band is half-full, and one has a metal, of course. The total volume enclosed by the Fermi surface is determined by the electron concentration.

5.1.2 Exercises (B)

In 2D, find the reciprocal lattice for the lattice defined by the unit cell, given next.

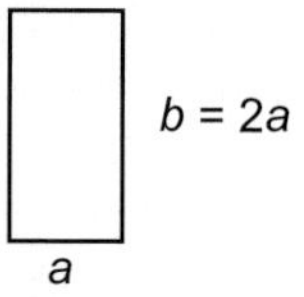

The direct lattice is defined by

$$\boldsymbol{a} = a\boldsymbol{i} \quad \text{and} \quad \boldsymbol{b} = b\boldsymbol{j} = 2a\boldsymbol{j} . \tag{5.1}$$

The reciprocal lattice is defined by vectors

$$\boldsymbol{A} = A_x\boldsymbol{i} + A_y\boldsymbol{j} \quad \text{and} \quad \boldsymbol{B} = B_x\boldsymbol{i} + B_y\boldsymbol{j} , \tag{5.2}$$

with

$$\boldsymbol{A}\cdot\boldsymbol{a}=\boldsymbol{B}\cdot\boldsymbol{b}=2\pi \quad \text{and} \quad \boldsymbol{A}\cdot\boldsymbol{b}=\boldsymbol{B}\cdot\boldsymbol{a}=0\,.$$

Thus

$$\boldsymbol{A}=\frac{2\pi}{a}\boldsymbol{i}\,, \tag{5.3}$$

$$\boldsymbol{B}=\frac{2\pi}{b}\boldsymbol{j}=\frac{\pi}{a}\boldsymbol{j}\,, \tag{5.4}$$

where the 2π now inserted in an alternative convention for reciprocal-lattice vectors. The unit cell of the reciprocal lattice looks like:

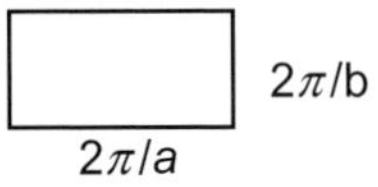

Now we suppose there is one electron per atom and one atom per unit cell. We want to calculate (a) the radius of the Fermi surface and (b) the radius of an energy surface that just manages to touch the first Brillouin zone boundary. The area of the first Brillouin zone is

$$A_{\mathrm{BZ}}=\frac{(2\pi)^2}{ab}=\frac{2\pi^2}{a^2}\,. \tag{5.5}$$

The radius of the Fermi surface is determined by the fact that its area is just 1/2 of the full Brillouin zone area

$$\pi k_F^2=\frac{1}{2}A_{\mathrm{BZ}} \quad \text{or} \quad k_F=\frac{\sqrt{\pi}}{a}\,. \tag{5.6}$$

The radius to touch the Brillouin zone boundary is

$$k_T=\frac{1}{2}\cdot\frac{2\pi}{b}=\frac{\pi}{2a}\,. \tag{5.7}$$

Thus,

$$\frac{k_T}{k_F}=\frac{\sqrt{\pi}}{2}=0.89\,,$$

and the circular Fermi surface extends into the second Brillouin zone. The first two zones are sketched in Fig. 5.1.

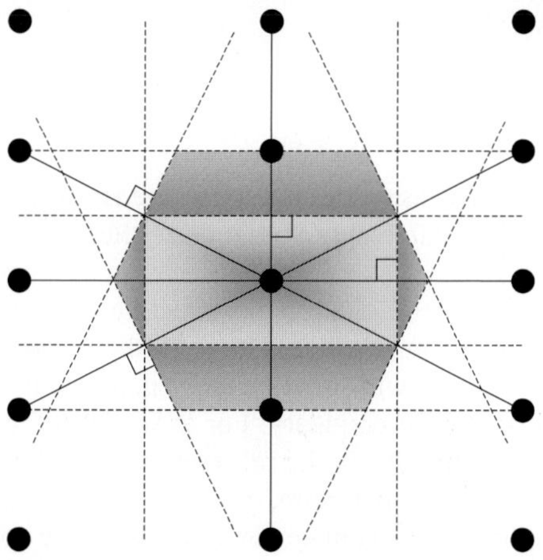

Fig. 5.1. First (light-shaded area) and second (dark-shaded area) Brillouin zones

As another example, let us consider a body-centered cubic lattice (bcc) with a standard, nonprimitive, cubic unit cell containing two atoms. The reciprocal lattice is fcc. Starting from a set of primitive vectors, one can show that the first Brillouin zone is a dodecahedron with twelve faces that are bounded by planes with perpendicular vector from the origin at

$$\frac{\pi}{a}\{(\pm 1,\pm 1,0),(\pm 1,0,\pm 1),(0,\pm 1,\pm 1)\}\,.$$

Since there are two atoms per unit cell, the volume of a primitive unit cell in the bcc lattice is

$$V_C = \frac{a^3}{2}\,. \tag{5.8}$$

The Brillouin zone, therefore, has volume

$$V_{BZ} = \frac{(2\pi)^3}{V_C} = \frac{16\pi^3}{a^3}\,. \tag{5.9}$$

Let us assume we have one atom per primitive lattice point and each atom contributes one electron to the band. Then, since the Brillouin zone is half-filled, if we assume a spherical energy surface, the radius is determined by

$$\frac{4\pi k_F^3}{3} = \frac{1}{2}\cdot\frac{16\pi^3}{a^3} \quad \text{or} \quad k_F = \frac{\sqrt[3]{6\pi^2}}{a}\,. \tag{5.10}$$

From (5.11), a sphere of maximum radius k_{T}, as given below, can just be inscribed within the first Brillouin zone

$$k_T = \frac{\pi}{a}\sqrt{2}\,. \tag{5.11}$$

Direct computation yields

$$\frac{k_T}{k_F} = 1.14\,,$$

so the Fermi surface in this case, does not touch the Brillouin zone. We might expect, therefore, that a reasonable approximation to the shape of the Fermi surface would be spherical.

By alloying, it is possible to change the effective electron concentration and, hence, the radius of the Fermi surface. Hume-Rothery has predicted that phase changes to a crystal structure with lower energy may occur when the Fermi surface touches the Brillouin zone boundary. For example in the AB alloy $Cu_{1-x}Zn_x$, Cu has one electron to contribute to the relevant band, and Zn has two. Thus, the number of electrons on average per atom, α, varies from 1 to 2.

For another example, let us estimate for a fcc structure (bcc in reciprocal lattice) at what $\alpha = \alpha_T$ the Brillouin zone touches the Fermi surface. Let k_T be the radius that just touches the Brillouin zone. Since the number of states per unit volume of reciprocal space is a constant,

$$\frac{\alpha_T N}{4\pi k_T^3/3} = \frac{2N}{V_{BZ}}\,, \tag{5.12}$$

where N is the number of atoms. In a fcc lattice, there are 4 atoms per nonprimitive unit cell. If V_C is the volume of a primitive cell, then

$$V_{BZ} = \frac{(2\pi)^3}{V_C} = \frac{4}{a^3}(2\pi)^3\,. \tag{5.13}$$

The primitive translation vectors for a bcc unit cell are

$$\boldsymbol{A} = \frac{2\pi}{a}(\boldsymbol{i} + \boldsymbol{j} - \boldsymbol{k})\,, \tag{5.14}$$

$$\boldsymbol{B} = \frac{2\pi}{a}(-\boldsymbol{i} + \boldsymbol{j} + \boldsymbol{k})\,, \tag{5.15}$$

$$\boldsymbol{C} = \frac{2\pi}{a}(\boldsymbol{i} + \boldsymbol{j} + \boldsymbol{k})\,. \tag{5.16}$$

From this we easily conclude

$$k_T = \left(\frac{2\pi}{a}\right)\left(\frac{1}{2}\right)\sqrt{3}\,.$$

So we find

$$\alpha_T = 2\left(\frac{a^3}{4}\right)\frac{1}{8\pi^3}\cdot\frac{4}{3}\pi\left[\frac{(2\pi)^3}{a^3}\right]\frac{1}{8}3^{3/2} \quad \text{or} \quad \alpha_T = 1.36\,.$$

5.2 The Fermi Surface in Real Metals (B)

5.2.1 The Alkali Metals (B)

For many purposes, the Fermi surface of the alkali metals (e.g. Li) can be considered to be spherical. These simple metals have one valence electron per atom. The conduction band is only half-full, and this means that the Fermi surface will not touch the Brillouin zone boundary (includes Li, Na, K, Rb, Cs, and Fr).

5.2.2 Hydrogen Metal (B)

At a high enough pressure, solid molecular hydrogen presumably becomes a metal with high conductivity due to relatively free electrons.[1] So far, this high pressure (about two million atmospheres at about 4400 K) has only been obtained explosively in the laboratory. The metallic hydrogen produced was a fluid. There may be metallic hydrogen on Jupiter (which is 75% hydrogen). It is premature, however, to give the phenomenon extended discussion, or to say much about its Fermi surface.

5.2.3 The Alkaline Earth Metals (B)

These are much more complicated than the alkali metals. They have two valence electrons per atom, but band overlapping causes the alkaline earths to form metals rather than insulators. Fig. 5.2 shows the Fermi surfaces for Mg. The case for second-zone holes has been called "Falicov's Monster". Examples of the alkaline earth metals include Be, Mg, Ca, Sr, and Ra. A nice discussion of this as well as other Fermi surfaces is given by Harrison [56, Chap. 3].

5.2.4 The Noble Metals (B)

The Fermi surface for the noble metals is typically more complicated than for the alkali metals. The Fermi surface of Cu is shown in Fig. 5.3. Other examples are Zn, Ag, and Au. Further information about Fermi surfaces is given in Table 5.1.

---

[1] See Wigner and Huntington [5.32].

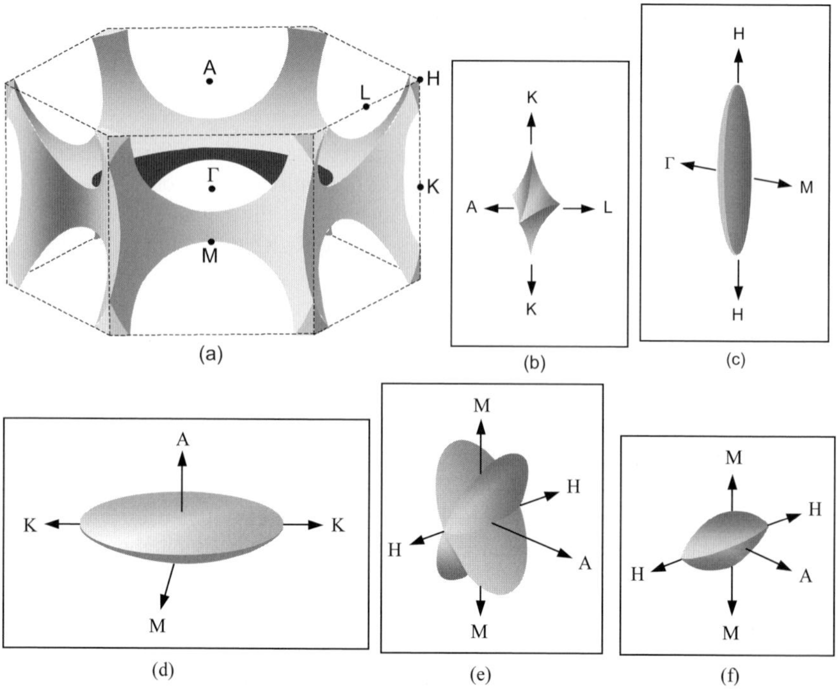

Fig. 5.2. Fermi surfaces in magnesium based on the single OPW model: (**a**) second-zone holes, (**b**) first-zone holes, (**c**) third-zone electrons, (**d**) third-zone electrons, (**e**) third-zone electrons, (**f**) fourth-zone electrons. [Reprinted with permission from Ketterson JB and Stark RW, *Physical Review*, **156**(3), 748 (1967). Copyright 1967 by the American Physical Society.]

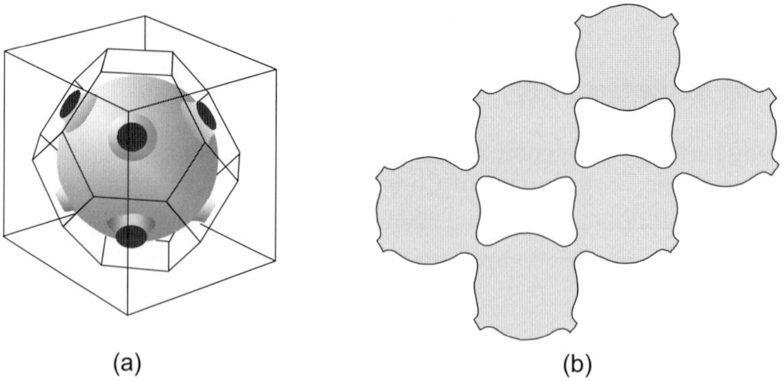

Fig. 5.3. Sketch of the Fermi surface of Cu (**a**) in the first Brillouin zone, (**b**) in a cross Section of an extended zone representation

Table 5.1. Summary of metals and Fermi surface

The Fermi energy E_F is the highest filled electron energy at absolute zero. The Fermi surface is the locus of points in k space such that $E(\boldsymbol{k}) = E_F$.

Type of metal	Fermi surface	Comment
Free-electron gas	Sphere	
Alkali (bcc) (monovalent, Na, K, Rb, Cs)	Nearly spherical	Specimens hard to work with
Alkaline earth (fcc) (Divalent, Be, Mg, Ca, Sr, Ba)	See Fig. 5.2.	Can be complex
Noble (monovalent, Cu Ag, Au)	Distorted sphere makes contact with hexagonal faces – complex in repeated zone scheme. See Fig. 5.3.	Specimens need to be pure and single crystal

Many more complex examples are discussed in Ashcroft and Mermin [21 Chap. 15]. Examples include tri (e.g. Al and tetravalent (e.g. Pb) metals, transition metals, rare earth metals, and semimetals (e.g. graphite).

There were many productive scientists connected with the study of Fermi surfaces, we mention only: A. B. Pippard, D. Schoenberg, A. V. Gold, and A. R. Mackintosh.

Experimental methods for studying the Fermi surface include the de Haas–van Alphen effect, the magnetoacoustic effect, ultrasonic attenuation, magnetoresistance, anomalous skin effect, cyclotron resonance, and size effects (see Ashcroft and Mermin [21 Chap. 14]. See also Pippard [5.24]. We briefly discuss some of these in Sect. 5.3.

5.3 Experiments Related to the Fermi Surface (B)

We will describe the *de Haas–van Alphen effect* in more detail in the next section. Under suitable conditions, if we measure the magnetic susceptibility of a metal as a function of external magnetic field, we find oscillations. Extreme cross-sections of the Fermi surface normal to the direction of the magnetic field are determined by the change of magnetic field that produces one oscillation. For similar physics reasons, we may also observe oscillations in the Hall effect, and thermal conductivity, among others.

We can also measure the dc electrical conductivity as a function of applied magnetic field as in *magnetoresistance experiments*. Under appropriate conditions, we may see an oscillatory change with the magnetic field as in the *de Haas–-Schubnikov effect*. Under other conditions, we may see a steady change of the conductivity with magnetic field. The interpretation of these experiments may be somewhat complex.

In Chap. 6, we will discuss *cyclotron resonance* in semiconductors. As we will see then, cyclotron resonance involves absorption of energy from an alternating electric field by an electron that is circling about a magnetic field. In metals, due to skin-depth problems, we need to use the *Azbel–Kaner* geometry that places both the electric and magnetic fields parallel to the metallic surface. Cyclotron resonance provides a way of finding the effective mass m^* appropriate to extremal sections of the Fermi surface. This can be used to extrapolate $E(k)$ away from the Fermi surface.

Magnetoacoustic experiments can determine extremal dimensions of the Fermi surface normal to the plane formed by the ultrasonic wave and perpendicular magnetic field. It turns out that as we vary the magnetic field we find oscillations in the ultrasonic absorption. The oscillations depend on the wavelength of the ultrasonic waves. Proper interpretation gives the information indicated. Another technique for learning about the Fermi surface is the *anomalous skin effect*. We shall not discuss this technique here.

5.4 The de Haas–van Alphen effect (B)

The de Haas–van Alphen effect will be studied as an example of how experiments can be used to determine the Fermi surface and as an example of the wave-packet description of electrons. The most important factor in the de Haas–van Alphen effect involves the quantization of electron orbits in a constant magnetic field. Classically, the electrons revolve around the magnetic field with the cyclotron frequency

$$\omega_c = \frac{eB}{m}. \tag{5.17}$$

There may also be a translational motion along the direction of the field. Let τ be the mean time between collisions for the electrons, T be the temperature, and k be the Boltzmann constant.

In order for the de Haas–van Alphen effect to be detected, two conditions must be satisfied. First, despite scattering, the orbits must be well defined, or

$$\omega_c \tau > 2\pi . \tag{5.18}$$

Second, the quantization of levels should not be smeared out by the thermal motion so

$$\hbar\omega_c > kT . \tag{5.19}$$

The energy difference between the quantized orbits is $\hbar\omega_c$, and kT is the average energy of thermal motion. To satisfy these conditions, we need large τ and large ω_c, or high purity, low temperatures, and high magnetic fields.

We now consider the motions of the electrons in a magnetic field. For electrons in a magnetic field B, we can write ($e > 0$, see Sect. 6.1.2)

$$\boldsymbol{F} = \hbar\dot{\boldsymbol{k}} = -e(\boldsymbol{v} \times \boldsymbol{B}) , \tag{5.20}$$

and taking magnitudes

$$dk = \frac{eB}{\hbar} v_{\perp}^{1} dt , \tag{5.21}$$

where $v_{\perp}^{1}$ is the component of velocity perpendicular to $\boldsymbol{B}$ and $\boldsymbol{F}$.

It will take an electron the same length of time to complete a cycle of motion in real space as in $\boldsymbol{k}$-space. Therefore, for the period of the orbit, we can write

$$T = \frac{2\pi}{\omega_c} = \oint \mathrm{d}t = \frac{\hbar}{eB} \oint \frac{\mathrm{d}k}{v_{\perp}^{1}} . \tag{5.22}$$

Since the force is perpendicular to the velocity of the electron, the constant magnetic field cannot change the energy of the electron. Therefore, in $\boldsymbol{k}$-space, the electron must stay on the same constant energy surface. Only electrons near the Fermi surface will be important for most effects, so let us limit our discussion to these. That the motion must be along the Fermi surface follows not only from the fact that the motion must be at constant energy, but that d$\boldsymbol{k}$ is perpendicular to

$$\boldsymbol{v} = \left(\frac{1}{\hbar}\right) \nabla_{\boldsymbol{k}} E(\boldsymbol{k}) , \tag{5.23}$$

because $\nabla_k E(\boldsymbol{k})$ is perpendicular to constant-energy surfaces. Equation (5.23) is derived in Sect. 6.1.2. The orbit in $\boldsymbol{k}$-space is confined to the intersection of the Fermi surface and a plane perpendicular to the magnetic field.

In order to consider the de Haas–van Alphen effect, we need to relate the energy of the electron to the area of its orbit in k-space. We do this by considering two orbits in k-space, which differ in energy by the small amount ΔE.

$$v_{\perp} = \frac{1}{\hbar} \cdot \frac{\Delta E}{\Delta k_{\perp}} , \tag{5.24}$$

where $v_{\perp}$ is the component of electron velocity perpendicular to the energy surface. From Fig. 5.4, note

$$v_{\perp}^{1} = v_{\perp} \sin\theta = \frac{1}{\hbar} \cdot \frac{\Delta E}{\Delta k_{\perp}} \sin\theta = \frac{1}{\hbar} \cdot \frac{\Delta E}{\Delta k_{\perp} / \sin\theta} = \frac{1}{\hbar} \cdot \frac{\Delta E}{\Delta k_{\perp}^{1}} . \tag{5.25}$$

Therefore,

$$\frac{2\pi}{\omega_c} = \frac{\hbar}{eB} \oint \frac{\mathrm{d}k}{\frac{1}{\hbar} \cdot \Delta E / \Delta k_{\perp}^{1}} = \frac{\hbar^2}{eB} \cdot \frac{1}{\Delta E} \oint \Delta k_{\perp}^{1} \, \mathrm{d}k , \tag{5.26}$$

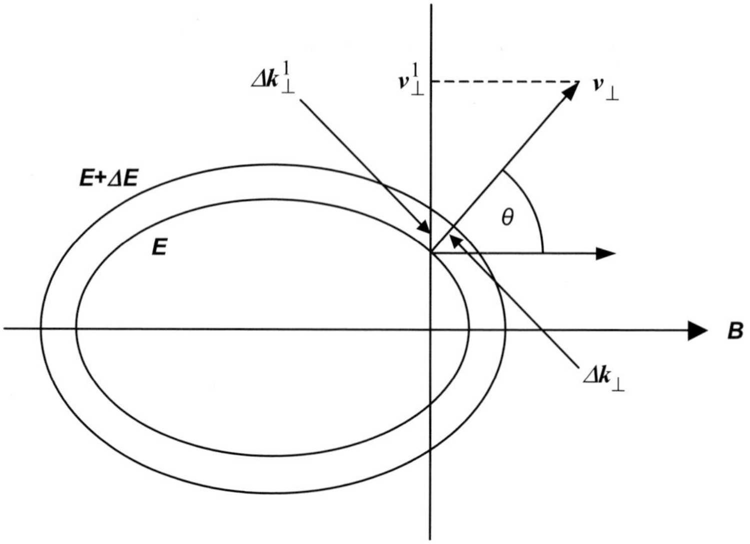

Fig. 5.4. Constant-energy surfaces for the de Haas–van Alphen effect

and

$$\frac{2\pi}{\omega_c} = \frac{\hbar^2}{eB} \cdot \frac{\Delta A}{\Delta E}, \tag{5.27}$$

where ΔA is the area between the two Fermi surfaces in the plane perpendicular to $\boldsymbol{B}$. This result was first obtained by Onsager in 1952 [5.20].

Recall that we have already found that the energy levels of an electron in a magnetic field (in the z direction) are given by (3.201)

$$E_{n,k_z} = \hbar\omega_c\left(n + \frac{1}{2}\right) + \frac{\hbar^2 k_z^2}{2m}. \tag{5.28}$$

This equation tells us that the difference in energy between different orbits with the same k_z is $\hbar\omega_c$. Let us identify the ΔE in the equations of the preceding figure with the energy differences of $\hbar\omega_c$. This tells us that the area (perpendicular to B) between adjacent quantized orbits in k-space is given by

$$\Delta A = \frac{eB}{\hbar^2} \cdot \frac{2\pi}{\omega_c} \hbar\omega_c = \frac{2\pi eB}{\hbar}. \tag{5.29}$$

The above may be interesting, but it is not yet clear what it has to do with the Fermi surface or with the de Haas–van Alphen effect. The effect of the magnetic field along the z-axis is to cause the quantization in k-space to be along energy tubes (with axis along the z-axis perpendicular to the cross-sectional area). Each tube has a different quantum number with corresponding energy

$$\hbar\omega_c \cdot \left(n + \frac{1}{2}\right) + \frac{\hbar^2 k_z^2}{2m}.$$

We think of these tubes existing only when the magnetic field along the z-axis is turned on. When it is turned on, the tubes furnish the only available states for the electrons. If the magnetic field is not too strong, this shifting of states onto the tube does not change the overall energy very much. We want to consider what happens as we increase the magnetic field. This increases the area of each tube of fixed n. It is convenient to think of each tube with only small extension in the k_z direction, Ziman makes this clear [5.35, Fig. 140, 1st edn.]. For some value of B, the tube of fixed n will break away from that part of the Fermi surface (*with maximum cross-sectional area*, see comment after (5.31)). As the tube breaks away, it pulls the allowed states (and, hence, electrons) at the Fermi surface with it. This causes an increase in energy. This increase continues until the next tube approaches from below. The electrons with energy just above the Fermi energy then hop down to this new tube. This results in a decrease in energy. Thus, the energy undergoes oscillations as the magnetic field is increased. These oscillations in energy can be detected as an oscillation in the magnetic susceptibility, and this is the de Haas–van Alphen effect. The oscillations look somewhat as sketched in Fig. 5.5. Such oscillations have now been seen in many metals.

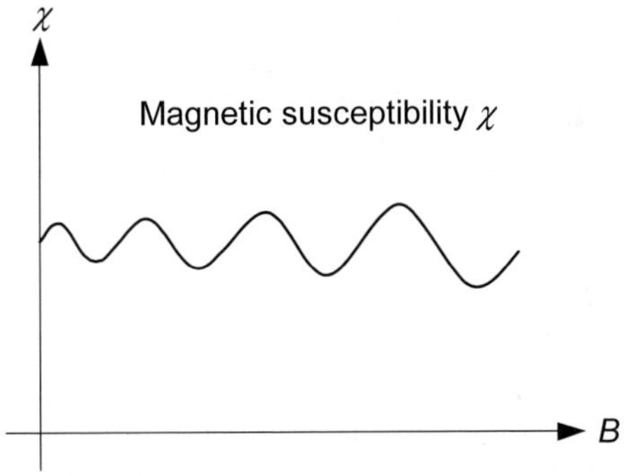

Fig. 5.5. Sketch of de Haas-Van Alphen oscillations in Cu

One might still ask why the electrons hop down to the lower tube. That is, why do states become available on the lower tube? The states become available because the number of states on each tube increases with the increase in magnetic field (the density of states per unit area is eB/h, see Sect. 12.7.3). This fact also explains why the total number of states inside the Fermi surface is conserved (on average) even though tubes containing states keep moving out of the Fermi surface with increasing magnetic field.

The difference in area between the $n = 0$ tube and the $n = n$ tube is

$$\Delta A_{0n} = \frac{2\pi eB}{\hbar} \cdot n \,. \tag{5.30}$$

Thus, the area of the tube n is

$$A_n = \frac{2\pi eB}{\hbar}(n + \text{constant}) \,. \tag{5.31}$$

If A_0 is the area of an extremal (where one gets the dominant response, see Ziman [5.35, p 322]) cross-sectional area (perpendicular to B) of the Fermi surface and if B_1 and B_2 are the two magnetic fields that make adjacent tubes equal in area to A_0, then

$$\frac{1}{B_2} = \frac{2\pi e}{\hbar A_0}[(n+1) + \text{constant}] \,, \tag{5.32}$$

and

$$\frac{1}{B_1} = \frac{2\pi e}{\hbar A_0}(n + \text{constant}) \,, \tag{5.33}$$

and so, by subtraction

$$\Delta\left(\frac{1}{B}\right) = \frac{2\pi e}{\hbar A_0} \,. \tag{5.34}$$

$\Delta(1/B)$ is the change in the reciprocal of the magnetic field necessary to induce one fluctuation of the magnetic susceptibility. Thus, experiments combined with the above equation determine A_0. For various directions of B, A_0 gives considerable information about the Fermi surface.

5.5 Eutectics (MS, ME)

In metals, the study of alloys is very important, and one often encounters phase diagrams as in Fig. 5.6. This is a particularly important technical example as discussed below. The subject of binary mixtures, phase diagrams, and eutectics is well treated in Kittel and Kroemer [5.15].

Alloys that are mixtures of two or more substances with two liquidus branches, as shown in Fig. 5.6, are especially interesting. They are called eutectics and the eutectic mixture is the composition that has the lowest freezing point, which is called the eutectic point (0.3 in Fig. 5.6). At the eutectic, the mixture freezes relatively uniformly (on the large scale) but consists of two separate intermixed phases. In solid-state physics, an important eutectic mixture occurs in the $Au_{1-x}Si_x$ system. This system occurs when gold contacts are made on Si devices. The resulting freezing point temperature is lowered, as seen in Fig. 5.6.

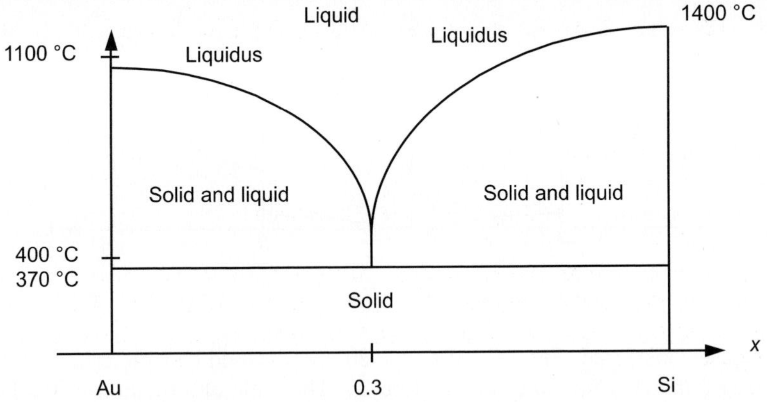

Fig. 5.6. Sketch of eutectic for $Au_{1-x}Si_x$ *Adapted from Kittel and Kroemer (op. cit.)*

5.6 Peierls Instability of Linear Metals (B)

The Peierls transition [75 pp 108-112, 23 p 203] is an example of a broken symmetry (see Sect. 7.2.6) in which the ground state has a lower symmetry than the Hamiltonian. It is a sort of metal–insulator phase transition that happens because a bandgap can occur at the Fermi surface, which results in an overall lowering of energy. One thinks of there being displacements in the regular array of lattice ions, induced by a strong electron–phonon interaction, that decreases the electronic energy without a larger increase in lattice elastic energy. The charge density then is nonuniform but has a periodic spatial variation.

We will only consider one dimension in this section. However, Peierls transitions have been discovered in (very special kinds of) real three-dimensional solids with weakly coupled molecular chains.

As Fig. 5.7 shows, a linear metal (in which the nearly free-electron model is appropriate) could lower its total electron energy by spontaneously distorting, that is reducing its symmetry, with a wave vector equal to twice the Fermi wave vector. From Fig. 5.7 we see that the states that increase in energy are empty, while those that decrease in energy are full. This implies an additional periodicity due to the distortion of

$$p = \frac{2\pi}{2k_F} = \frac{\pi}{k_F},$$

or a corresponding reciprocal lattice vector of

$$\frac{2\pi}{p} = 2k_F .$$

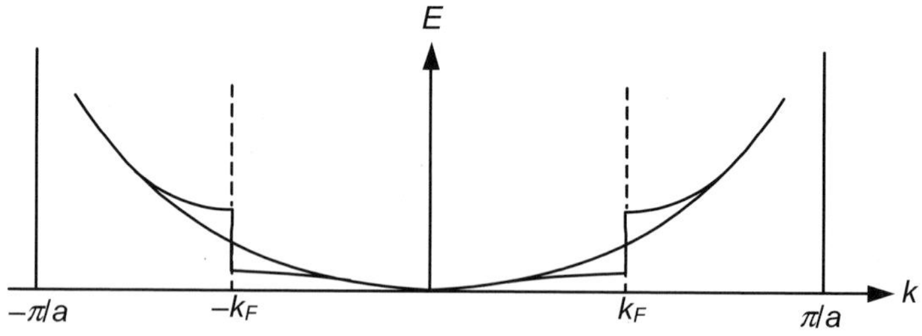

Fig. 5.7. Splitting of energy bands at Fermi wave vector due to distortion

In the case considered (Fig. 5.7), if $k_F = \pi/2a$, there would be a dimerization of the lattice and the new periodicity would be $2a$. Thus, the deformation in the lattice can be approximated by

$$d = c \cdot \cos(2k_F z)\,, \tag{5.35}$$

which is periodic with period π/k_F as desired, and c is a constant. As Fig. 5.7 shows, the creation of an energy gap at the Fermi surface leads to a lowering of the electronic energy, but there still is a question as to what electron–lattice interaction drives the distortion. A clue to the answer is obtained from the consideration of screening of charges by free electrons. As (9.167) shows, there is a singularity in the dielectric function at $2k_F$ that causes a long-range screened potential proportional to $r^{-3}\cos(2k_F r)$, in 3D. This can relate to the distortion with period $2\pi/2k_F$. Of course, the deformation also leads to an increase in the elastic energy, and it is the sum of the elastic and electronic energies that must be minimized.

For the case where k and k' are near the Brillouin zone boundary at $k_F = K'/2$, we assume, with c_1 a constant, that the potential energy due to the distortion is proportional to the distortion, so[2]

$$V(z) = c_1 d = c_1 c \cdot \cos(2k_F z)\,. \tag{5.36}$$

So $2V(K') \equiv 2V(2k_F) = c_1 c$, and in the nearly free-electron model we have shown (by (3.231) to (3.233))

$$E_k = \frac{1}{2}(E_k^0 + E_{k'}^0) \pm \frac{1}{2}\{4[V(K')]^2 + (E_k^0 - E_{k'}^0)^2\}^{1/2}\,,$$

where

$$E_k^0 = V(0) + \frac{\hbar^2 k^2}{2m}\,,$$

and

$$E_{k'}^0 = V(0) + \frac{\hbar^2}{2m}|k + K'|^2\,.$$

[2] See e.g. Marder [3.34, p. 277]

Let $k = \Delta - K'/2$, so

$$k^2 - (k + K')^2 = -K'(2\Delta),$$

$$\frac{1}{2}(k^2 + |k + K'|^2) = \Delta^2 + k_F^2 .$$

For the lower branch, we find:

$$E_k = V(0) + \frac{\hbar^2}{2m}(\Delta^2 + k_F^2) - \left[\frac{1}{4}c_1^2 c^2 + 4k_F^2\Delta^2\left(\frac{\hbar^2}{2m}\right)^2\right]^{1/2} . \tag{5.37}$$

We compute an expression relating to the lowering of electron energy due to the gap caused by shifting of lattice ion positions. If we define

$$y_F = \frac{\hbar^2 k_F^2}{2m} \quad \text{and} \quad y = \frac{\hbar^2 \Delta k_F}{2m}, \tag{5.38}$$

we can write[3]

$$\begin{aligned}\frac{\mathrm{d}E_{\mathrm{el}}}{\mathrm{d}c} &= \frac{2}{\pi}\int_0^{k_F} \mathrm{d}\Delta \frac{\mathrm{d}E_k}{\mathrm{d}c} \\ &= -\frac{c_1^2 c}{2\pi}\left(\frac{k_F}{y_F}\right)\int_0^{2y_F}\left(4y^2 + \frac{c_1^2 c^2}{4}\right)^{-1/2} \mathrm{d}y \\ &= -\frac{c_1^2 c k_F}{4\pi y_F}\cdot \ln\left(\frac{8y_F}{cc_1}\right), \quad \text{if } \frac{8y_F}{cc_1} >> 1.\end{aligned} \tag{5.39}$$

As noted by R. Peierls in [5.23], this logarithmic dependence on displacement is important so that this instability not be swamped other effects. If we assume the average elastic energy per unit length is

$$E_{\mathrm{elastic}} = \frac{1}{4}c_{\mathrm{el}}c^2; \propto d^2, \tag{5.40}$$

we find the minimum (total $E_{\mathrm{el}} + E_{\mathrm{elastic}}$) energy occurs at

$$\frac{c_1 c}{2} \cong \frac{2\hbar^2 k_F^2}{m}\exp\left(-\frac{\hbar^2 k_F \pi c_{\mathrm{el}}}{m c_1^2}\right). \tag{5.41}$$

[3] The number of states per unit length with both spins is $2\mathrm{d}k/2\pi$ and we double as we only integrate from $\Delta = 0$ to k_F or $-k_F$ to 0. We compute the derivative, as this is all we need in requiring the total energy to be a minimum.

The lattice distorts if the quasifree-electron energy is lowered more by the distortions than the elastic energy increases. Now, as defined above,

$$y_F = \frac{\hbar^2 k_F{}^2}{2m} \tag{5.42}$$

is the free-electron bandwidth, and

$$\left.\frac{1}{\pi}\cdot\frac{\mathrm{d}k}{\mathrm{d}E}\right)_{k=k_F} = N(E_F) = \frac{1}{\pi}\cdot\frac{m}{\hbar^2 k_F} \tag{5.43}$$

equals the density (per unit length) of orbitals at the Fermi energy (for free electrons), and we define

$$V_1 = \frac{c_1^2}{c_{\mathrm{el}}} \tag{5.44}$$

as an effective interaction energy. Therefore, the distortion amplitude c is proportional to y_F times an exponential;

$$c \propto y_F \exp\left(-\frac{1}{N(E_F)V_1}\right). \tag{5.45}$$

Our calculation is of course done at absolute zero, but this equation has a formal similarity to the equation for the transition temperature or energy gap as in the superconductivity case. See, e.g., Kittel [23, p 300], and (8.215). Comparison can be made to the Kondo effect (Sect. 7.5.2) where the Kondo temperature is also given by an exponential.

5.6.1 Relation to Charge Density Waves (A)

The Peierls instability in one dimension is related to a mechanism by which charge density waves (CDW) may form in three dimensions. A charge density wave is the modulation of the electron density with an associated modulation of the location of the lattice ions. These are observed in materials that conduct primarily in one (e.g. $NbSe_3$, $TaSe_3$) or two (e.g. $NbSe_2$, $TaSe_2$) dimensions. Limited dimensionality of conduction is due to weak coupling. For example, in one direction the material is composed of weakly coupled chains. The Peierls transitions cause a modulation in the periodicity of the ionic lattice that leads to lowering of the energy. The total effect is of course rather complex. The effect is temperature dependent, and the CDW forms below a transition temperature with the strength p (see as in (5.46)) growing as the temperature is lowered.

The charge density assumes the form

$$\rho(\boldsymbol{r}) = \rho_0(\boldsymbol{r})[1 + p\cos(\boldsymbol{k}\cdot\boldsymbol{r} + \phi)], \tag{5.46}$$

where ϕ is the phase, and the length of the CDW determined by $\boldsymbol{k}$ is, in general, not commensurate with the lattice. $\boldsymbol{k}$ is given by $2k_F$ where k_F is the Fermi wave vector. CDWs can be detected as satellites to Bragg peaks in X-ray diffraction. See, e.g., Overhauser [5.21]. See also Thorne [5.31].

CDW's have a long history. Peierls considered related mechanisms in the 1930s. Frolich and Peierls discussed CDWs in the 1950s. Bardeen and Frolich actually considered them as a model for superconductivity. It is true that some CDW systems show collective transport by sliding in an electric field but the transport is damped. It also turns out that the total electron conduction charge density is involved in the conduction.

It is well to point out that CDWs have three properties (see, e.g., Thorne op cit)

a. An instability associated with the Fermi surface caused by electron–phonon and electron–electron interactions.

b. An opening of an energy gap at the Fermi surface.

c. The wavelength of the CDW is π/k_F.

5.6.2 Spin Density Waves (A)

Spin density waves (SDW) are much less common than CDW. One thinks here of a "spin Peierls" transition. SDWs have been found in chromium. The charge density of a SDW with up (↑ or +) and down (↓ or −) spins looks like

$$\rho_\pm(\boldsymbol{r}) = \frac{1}{2}\rho_0(\boldsymbol{r})[1 \pm p\cos(\boldsymbol{k}\cdot\boldsymbol{r} + \phi)]. \tag{5.47}$$

So, there is no change in charge density $[\rho_+ + \rho_- = \rho_0(\boldsymbol{r})]$ except for that due to lattice periodicity. The spin density, however, looks like

$$\rho_S(\boldsymbol{r}) = \hat{\varepsilon}\rho_0(\boldsymbol{r})\cos(\boldsymbol{k}\cdot\boldsymbol{r} + \phi), \tag{5.48}$$

where $\hat{\varepsilon}$ defines the quantization axis for spin. In general, the SDW is not commensurate with the lattice. SDWs can be observed by magnetic satellites in neutron diffraction. See, e.g., Overhauser [5.21]. Overhauser first discussed the possibility of SDWs in 1962. See also Harrison [5.10].

5.7 Heavy Fermion Systems (A)

This has opened a new branch of metal physics. Certain materials exhibit huge ($\sim 1000m_e$) electron effective masses at very low temperatures. Examples are

$CeCu_2Si_2$, UBe_{13}, UPt_3, $CeAl_3$, UAl_2, and $CeAl_2$. In particular, they may show large, low-T electronic specific heat. Some materials show f-band superconductivity—perhaps the so-called "triplet superconductivity" where spins do not pair. The novel results are interpreted in terms of quasiparticle interactions and incompletely filled shells. The heavy fermions represent low-energy excitations in a strongly correlated, many-body state. See Stewart [5.30], Radousky [5.25]. See also Fisk et al [5.8].

5.8 Electromigration (EE, MS)

Electromigration is of great interest because it is an important failure mechanism as aluminum interconnects in integrated circuits are becoming smaller and smaller in very large scale integrated (VLSI) circuits. Simply speaking, if the direct current in the interconnect is large, it can start some ions moving. The motion continues under the "push" of the moving electrons.

More precisely, electromigration is the motion of ions in a conductor due to momentum exchange with flowing electrons and also due to the Coulomb force from the electric field.[4] The momentum exchange is dubbed the electron wind and we will assume it is the dominant mechanism for electromigration. Thus, electromigration is diffusion with a driving force that increases with electric current density. It increases with decreasing cross section. The resistance is increased and the heating is larger as are the lattice vibration amplitudes. We will model the inelastic interaction of the electrons with the ion by assuming the ion is in a potential hole, and later simplify even that assumption.

Damage due to electromigration can occur when there is a divergence in the flux of aluminum ions. This can cause the appearance of a void and hence a break in the circuit or a hillock can appear that causes a short circuit. Aluminum is cheaper than gold, but gold has much less electromigration-induced failures when used in interconnects. This is because the ions are much more massive and hence harder to move.

Electromigration is a very complex process and we follow Fermi's purported advice to use simpler models for complex situations. We do a one-dimensional classical calculation to illustrate how the electron wind force can assist in breaking atoms loose and how it contributes to the steady flow of ions. We let p and P be

[4] To be even more precise the phenomena and technical importance of electromigration is certainly real. The explanations have tended to be controversial. Our explanation is the simplest and probably has at least some of the truth (See, e.g., Borg and Dienes [5.3].) The basic physics involving momentum transfer was discussed early on by Fiks [5.7] and Huntington and Grove [5.13]. Modern work is discussed by R. S Sorbello as referred to at the end of this section.

the momentum of the electron before and after collision, and p_a and P_a be the momentum of the ion before and after. By momentum and energy conservation we have:

$$p + p_a = P + P_a, \tag{5.49}$$

$$\frac{p^2}{2m} + \frac{p_a^2}{2m_a} = \frac{P^2}{2m} + \frac{P_a^2}{2m_a} + V_0, \tag{5.50}$$

where V_0 is the magnitude of the potential hole the ion is in before collision, and m and m_a are the masses of the electron and the ion, respectively. Solving for P_a and P in terms of p_a and p, retaining only the physically significant roots and assuming $m << m_a$:

$$P_a = (p + p_a) + \sqrt{p^2 - 2mV_0}, \tag{5.51}$$

$$P = -\sqrt{p^2 - 2mV_0}. \tag{5.52}$$

In order to move the ion, the electron's kinetic energy must be greater than V_0 as perhaps is obvious. However, the process by which ions are started in motion is surely more complicated than this description, and other phenomena, such as the presence of vacancies are involved. Indeed, electromigration is often thought to occur along grain boundaries.

For the simplest model, we may as well start by setting V_0 equal to zero. This makes the collisions elastic. We will assume that the ions are pushed along by the electron wind, but there are other forces that cancel out the wind force, so that the flow is in steady state. The relevant conservation equations become:

$$P_a = p_a + 2p, \quad P = -p.$$

We will consider motion in one dimension only. The ions drift along with a momentum p_a. The electrons move back and forth between the drifting ions with momentum p. We assume the electron's velocity is so great that the ions are stationary in comparison. Assume the electric field points along the $-x$-axis. Electrons moving to the right collide and increase the momentum of the ions, and those moving to the left decrease their momentum. Because of the action of the electric field, electrons moving to the right have more momentum so the net effect is a small increase in the momentum of the ions (which, as mentioned, is removed by other effects to produce a steady-state drift). If E is the electric field, then in time τ, (the time taken for electrons to move between ions), an electron of charge $-e$ gains momentum

$$\Delta = eE\tau, \tag{5.53}$$

if it moves against the field, and it loses a similar amount of momentum if it goes in the opposite direction. Assume the electrons have momentum p when they are halfway between ions. The net effect of collisions to the left and to the right of the ion is to transfer an amount of momentum of

$$\Delta = 2eE\tau\,. \tag{5.54}$$

This amount of momentum is gained per pair of collisions. Each ion experiences such pair collisions every 2τ. Thus, each ion gains on average an amount of momentum $eE\tau$ in time τ. If n is the electron density, v the average velocity of electrons and σ the cross section, then the number of collisions per unit time is $nv\sigma$, and the net force is this times the momentum transferred per collision. Since the mean free path is $\lambda = v\tau$, we find for the magnitude of the wind force

$$F_W = eE\tau n(\lambda / \tau)\sigma = eEn\lambda\sigma\,. \tag{5.55}$$

If Ze is the charge of the ion, then the net force on the ion, including the electron wind and direct Coulomb force can be written

$$F = -Z^{*}eE\,, \tag{5.56}$$

where the effective charge of the ion is

$$Z^{*} = n\lambda\sigma - Z\,, \tag{5.57}$$

and the sign has been chosen so a positive electric field gives a negative wind force (see Borg and Dienes, op cit). The subject is of course much more complicated that this. Note also, if the mobility of the ions is μ, then the ion flux under the wind force has magnitude $Z^{*}n_a\mu E$, where n_a is the concentration of the ions. For further details, see, e.g., Lloyd [5.18]. See also Sorbello [5.28]. Sorbello summarizes several different approaches. Our approach could be called a rudimentary ballistic method.

5.9 White Dwarfs and Chandrasekhar's Limit (A)

This Section is a bit of an excursion. However, metals have electrons that are degenerate as do white dwarfs, except the electrons here are at a much higher degeneracy. White dwarfs evolve from hydrogen-burning stars such as the sun unless, as we shall see, they are much more massive than the sun. In such stars, before white-dwarf formation, the inward pressure due to gravitation is balanced by the outward pressure caused by the "burning" of nuclear fuel.

Eventually the star runs out of nuclear fuel and one is left with a collection of electrons and ions. This collection then collapses under gravitational pressure. the electron gas becomes degenerate when the de Broglie wavelength of the electrons becomes comparable with their average separation. Ions are much more massive. Their de Broglie wavelength is much shorter and they do not become degenerate. The outward pressure of the electrons, which arises because of the Pauli principle and the electron degeneracy, balances the inward pull of gravity and eventually the star reaches stability. However, by then it is typically about the size of the earth and is called a white dwarf.

A white dwarf is a mass of atoms with major composition of C^{12} and O^{16}. We assume the gravitational pressure is so high that the atoms are completely ionized, so the white dwarf is a compound of ions and degenerate electrons.

For typical conditions, the actual temperature of the star is much less than the Fermi temperature of the electrons. Therefore, the star's electron gas can be regarded as an ideal Fermi gas in the ground state with an effective temperature of absolute zero.

In white dwarfs, it is very important to note that the density of electrons is such as to require a relativistic treatment. A nonrelativistic limit does not put a mass limit on the white dwarf star.

Some reminders of results from special relativity: The momentum p is given by

$$p = mv = m_0 \gamma v , \tag{5.58}$$

where m_0 is the rest mass.

$$\beta = \frac{v}{c} \tag{5.59}$$

$$\gamma = (1-\beta^2)^{-1/2} \tag{5.60}$$

$$\begin{aligned} E &= K + m_0 c^2 = \text{kinetic energy plus rest energy} \\ &= \gamma m_0 c^2 \\ &= mc^2 = \sqrt{p^2c^2 + m_0^2 c^4} . \end{aligned} \tag{5.61}$$

5.9.1 Gravitational Self-Energy (A)

If G is the gravitational constant, the gravitational self-energy of a mass M with radius R is

$$U = -G\alpha\left(\frac{M^2}{R}\right). \tag{5.62}$$

For uniform density, $\alpha = 3/5$, which is an oversimplification. We simply assume $\alpha = 1$ for stars.

5.9.2 Idealized Model of a White Dwarf (A)[5]

We will simply assume that we have N electrons in their lowest energy state, which is of such high density that we are forced to use relativistic dynamics. This

5 See e.g. Huang [5.12]. See also Shapiro and Teukolsky [5.26].

leads to less degeneracy pressure than in the nonrelativistic case and hence collapse. The nuclei will be assumed motionless, but they will provide the gravitational force holding the white dwarf together. The essential features of the model are the Pauli principle, relativistic dynamics, and gravity.

We first need to calculate the relativistic pressure exerted by the Fermi gas of electrons in their ground state. The combined first and second laws of thermodynamics for open systems states:

$$dU = TdS - pdV + \mu dN\,. \tag{5.63}$$

As $T \to 0$, $U \to E_0$, so

$$p = -\left.\frac{\partial E_0}{\partial V}\right)_{N,T=0}. \tag{5.64}$$

For either up or down spin, the electron energy is given by

$$\varepsilon_p = \sqrt{(pc)^2 + (m_e c^2)^2}\,, \tag{5.65}$$

where m_e is the rest mass of the electrons. Including spin, the ground-state energy of the Fermi gas is given by (with $p = \hbar k$)

$$E_0 = 2\sum_{k<k_F}\sqrt{(\hbar kc)^2 + (m_e c^2)^2} = \frac{V}{\pi^2}\int_0^{k_F} k^2\sqrt{(\hbar kc)^2 + (m_e c^2)^2}\,\mathrm{d}k\,. \tag{5.66}$$

The Fermi momentum k_F is determined from

$$\frac{k_F^3 V}{3\pi^3} = N\,, \tag{5.67}$$

where N is the number of electrons, or

$$k_F = \left(\frac{3\pi^2 N}{V}\right)^{1/3}. \tag{5.68}$$

From the above we have

$$\frac{E_0}{N} \propto \int_0^{\hbar k_F/m_e c} x^2\sqrt{1+x^2}\,\mathrm{d}x\,, \tag{5.69}$$

where $x = \hbar k/m_e c$. The volume of the star is related to the radius by

$$V = \frac{4}{3}\pi R^3 \tag{5.70}$$

and the mass of the star is, neglecting electron mass and assuming the neutron mass equals the proton mass (m_p) and that there are the same number of each

$$M = 2m_{\rm p}N \,. \tag{5.71}$$

Using (5.64) we can then show for highly relativistic conditions ($x_F >> 1$) that

$$p_0 \propto \beta'^2 - \beta' \,, \tag{5.72}$$

where

$$\beta' \propto \frac{M^{2/3}}{R^2} \,. \tag{5.73}$$

We now want to work out the conditions for equilibrium. Without gravity, the work to compress the electrons is

$$-\int_{\infty}^{R} p_0 4\pi r^2 \cdot {\rm d}r \,. \tag{5.74}$$

Gravitational energy is approximately (with $\alpha = 1$)

$$-\frac{GM^2}{R} \,. \tag{5.75}$$

If R is the equilibrium radius of the star, since gravitational self-energy plus work to compress = 0, we have

$$\int_{\infty}^{R} p_0 4\pi r^2 \cdot {\rm d}r + \frac{GM^2}{R} = 0 \,. \tag{5.76}$$

Differentiating, we get the condition for equilibrium

$$p_0 \propto \frac{M^2}{R} \,. \tag{5.77}$$

Using the expression for p_0 (5.72) with $x_{\rm F} >> 1$, we find

$$R \propto M^{1/3}\sqrt{1-\left(\frac{M}{M_0}\right)^2} \,, \tag{5.78}$$

where

$$M_0 \cong M_{\rm sun} \,, \tag{5.79}$$

and this result is good for small R (and large x_F). A more precise derivation predicts $M_0 \cong 1.4M_{\rm sun}$. Thus, there is *no* white dwarf star with mass $M \geq M_0 \cong M_{\rm sun}$. See Fig. 5.8. M_0 is known as the mass for the Chandrasekhar limit. When the mass is greater than M_0, the Pauli principle is not sufficient to support the star against gravitational collapse. It may then become a neutron star or even a black hole, depending upon the mass.

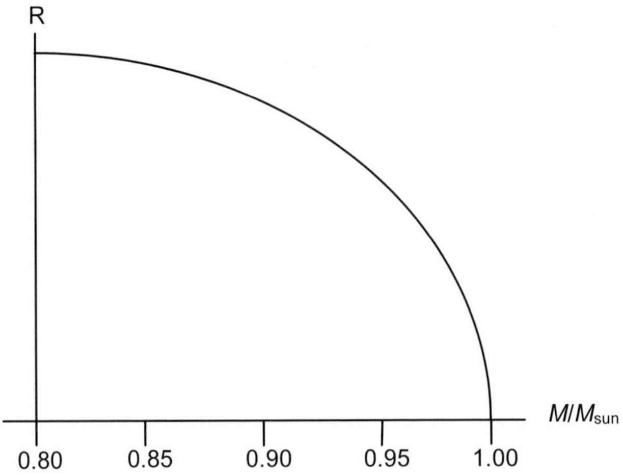

Fig. 5.8. The Chandrasekhar limit

5.10 Some Famous Metals and Alloys (B, MET)[6]

We finish the chapter on a much less abstract note. Many of us became familiar with the solid-state by encountering these metals.

Iron This, of course, is the most important metal. Alloying with carbon, steel of much greater strength is produced.

Aluminum The second most important metal. It is used everywhere from aluminum foil to alloys for aircraft.

Copper Another very important metal used for wires because of its high conductivity. It is also very important in brasses (copper-zinc alloys).

Zinc Zinc is widely used in making brass and for inhibiting rust in steel (galvanization).

Lead Used in sheathing of underground cables, making pipes, and for the absorption of radiation.

Tin Well known for its use as tin plate in making tin cans. Originally, the word "bronze" was meant to include copper-tin alloys, but its use has been generalized to include other materials.

Nickel Used for electroplating. Nickel steels are known to be corrosion resistant. Also used in low-expansion "Invar" alloys (36% Ni-Fe alloy).

[6] See Alexander and Street [5.1].

Chromium Chrome plated over nickel to produce an attractive finish is a major use. It is also used in alloy steels to increase hardness.

Gold Along with silver and platinum, gold is one of the precious metals. Its use as a semiconductor connection in silicon is important.

Titanium Much used in the aircraft industry because of the strength and lightness of its alloys.

Tungsten Has the highest melting point of any metal and is used in steels, as filaments in light bulbs and in tungsten carbide. The hardest known metal.

Problems

5.1 For the Hall effect (metals-electrons only), find the Hall coefficient, the effective conductance j_x/E_x, and σ_{yx}. For high magnetic fields, relate σ_{yx} to the Hall coefficient. Assume the following geometry:

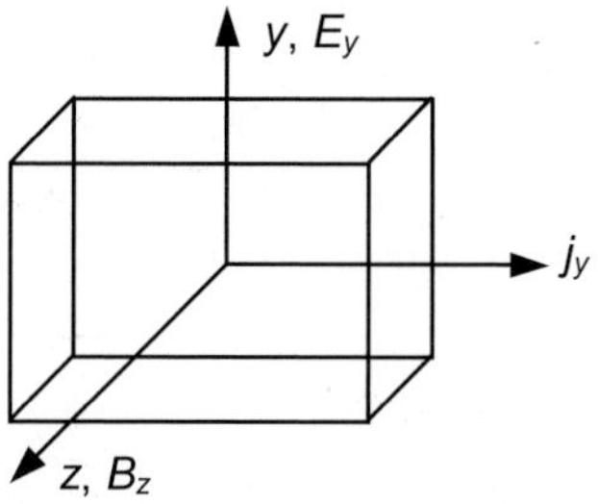

Reference can be made to Sect. 6.1.5 for the definition of the Hall effect.

5.2 (a) A twodimensional metal has one atom of valence one in a simple rectangular primitive cell $a = 2$, $b = 4$ (units of angstroms). Draw the First Brillouin zone and give dimensions in cm^{-1}.

(b) Calculate the areal density of electrons for which the free electron Fermi surface first touches the Brillouin zone boundary.

5.3 For highly relativistic conditions within a white dwarf star, derive the relationship for pressure p_0 as a function of mass $\bar{M}$ and radius $\bar{R}$ using $p_0 = -\partial E_0/\partial V$.

5.4 Consider the current due to metal-insulator-metal tunneling. Set up an expression for calculating this current. Do not necessarily assume zero temperature. See, e.g., Duke [5.6].

5.5 Derive (5.37).

6 Semiconductors

Starting with the development of the transistor by Bardeen, Brattain, and Shockley in 1947, the technology of semiconductors has exploded. With the creation of integrated circuits and chips, semiconductor devices have penetrated into large parts of our lives. The modern desktop or laptop computer would be unthinkable without microelectronic semiconductor devices, and so would a myriad of other devices.

Recalling the band theory of Chap. 3, one could call a semiconductor a narrow-gap insulator in the sense that its energy gap between the highest filled band (the valence band) and the lowest unfilled band (the conduction band) is typically of the order of one electron volt. The electrical conductivity of a semiconductor is consequently typically much less than that of a metal.

The purity of a semiconductor is very important and controlled doping is used to vary the electrical properties. As we will discuss, *donor* impurities are added to increase the number of electrons and *acceptors* are added to increase the number of holes (which are caused by the absence of electrons in states normally electron occupied – and as discussed later in the chapter, holes act as positive charges). Donors are impurities that become positively ionized by contributing an electron to the conduction band, while acceptors become negatively ionized by accepting electrons from the valence band. The electrons and holes are thermally activated and in a temperature range in which the charged carriers contributed by the impurities dominate, the semiconductor is said to be in the extrinsic temperature range, otherwise it is said to be intrinsic. Over a certain temperature range, donors can add electrons to the conduction band (and acceptors can add holes to the valence band) as temperature is increased. This can cause the electrical resistivity to decrease with increasing temperature giving a negative coefficient of resistance. This is to be contrasted with the opposite behavior in metals. For group IV semiconductors (Si, Ge) typical donors come from column V of the periodic table (P, As, Sb) and typical acceptors from column III (B, Al, Ga, In).

Semiconductors tend to be bonded tetrahedrally and covalently, although binary semiconductors may have polar, as well as covalent character. The simplest semiconductors are the nonpolar semiconductors from column 4 of the Periodic Table: Si and Ge. Compound III-V semiconductors are represented by, e.g., InSb and GaAs while II-VI semiconductors are represented by, e.g., CdS and CdSe. The pseudobinary compound $Hg_{(1-x)}Cd_{(x)}Te$ is an important narrow gap semiconductor whose gap can be varied with concentration x and it is used as an infrared detector. There are several other pseudobinary alloys of technical importance as well.

As already alluded to, there are many applications of semiconductors, see for example Sze [6.42]. Examples include diodes, transistors, solar cells, microwave generators, light-emitting diodes, lasers, charge-coupled devices, thermistors, strain gauges, and photoconductors. Semiconductor devices have been found to be highly economical because of their miniaturization and reliability. We will discuss several of these applications.

The technology of semiconductors is highly developed, but cannot be discussed in this book. The book by Fraser [6.14] is a good starting point for a physics oriented discussion of such topics as planar technology, information technology, computer memories, etc.

Table 6.1 and Table 6.2 summarize several semiconducting properties that will be used throughout this chapter. Many of the concepts within these tables will become clearer as we go along. However, it is convenient to collect several values all in one place for these properties. Nevertheless, we need here to make a few introductory comments about the quantities given in Table 6.1 and Table 6.2.

In Table 6.1 we mention bandgaps, which as already stated, express the energy between the top of the valence band and the bottom of the conduction band. Note that the bandgap depends on the temperature and may slowly and linearly decrease with temperature, at least over a limited range.

In Table 6.1 we also talk about direct (D) and indirect (I) semiconductors. If the conduction-band minimum (in energy) and the valence-band maximum occur at the same $\boldsymbol{k}$ (wave vector) value one has a direct (D) semiconductor, otherwise the semiconductor is indirect (I). Indirect and direct transitions are also discussed in Chap. 10, where we discuss optical measurement of the bandgap.

In Table 6.2 we mention several kinds of effective mass. Effective masses are used to take into account interactions with the periodic lattice as well as other interactions (when appropriate). Effective masses were defined earlier in Sect. 3.2.1 (see (3.163)) and discussed in Sect. 3.2.2 as well as Sect. 4.3.3. They will be further discussed in this chapter as well as in Sect. 11.3. Hole effective masses are defined by (6.65).

When, as in Sect. 6.1.6 on cyclotron resonance, electron-energy surfaces are represented as ellipsoids of revolution, we will see that we may want to represent them with longitudinal and transverse effective masses as in (6.103). The relation of these to the so-called 'density of states effective mass' is given in Sect. 6.1.6 under "Density of States Effective Electron Masses for Si." Also, with certain kinds of band structure there may be, for example, two different $E(k)$ relations for holes as in (6.144) and (6.145). One may then talk of light and heavy holes as in Sect. 6.2.1.

Finally, mobility, which is drift velocity per unit electric field, is discussed in Sect. 6.1.4 and the relative static dielectric constant is the permittivity over the permittivity of the vacuum.

The main objective of this chapter is to discuss the basic physics of semiconductors, including the physics necessary for understanding semiconductor devices. We start by discussing electrons and holes—their concentration and motion.

Table 6.1. Important properties of representative semiconductors (A)

Semiconductor	Direct/indirect, crystal struct.	Lattice constant	Bandgap (eV)	
	D/I	300 K (Å)*	0 K	300 K
Si	I, diamond	5.43	1.17	1.124
Ge	I, diamond	5.66	0.78	0.66
InSb	D, zincblende	6.48	0.23	0.17
GaAs	D, zincblende	5.65	1.519	1.424
CdSe	D, zincblende	6.05	1.85	1.70
GaN	D, wurtzite	$a = 3.16$, $c = 5.12$	3.5	3.44

* Adapted from Sze SM (ed), *Modern Semiconductor Device Physics*, Copyright © 1998, John Wiley & Sons, Inc, New York, pp. 537-540. This material is used by permission of John Wiley & Sons, Inc.

Table 6.2. Important properties of representative semiconductors (B)

Semiconductor	Effective masses (units of free electron mass)		Mobility (300 K) (cm^2/Vs)		Relative static dielectric constant
	Electron*	Hole**	Electron	Hole	
Si	$m_l = 0.92$ $m_t = 0.19$	$m_{lh} = 0.15$ $m_{hh} = 0.54$	1450	505	11.9
Ge	$m_l = 1.57$ $m_t = 0.082$	$m_{lh} = 0.04$ $m_{hh} = 0.28$	3900	1800	16.2
InSb	0.0136	$m_{lh} = 0.0158$ $m_{hh} = 0.34$	77 000	850	16.8
GaAs	0.063	$m_{lh} = 0.076$ $m_{hh} = 0.50$	9200	320	12.4
CdSe	0.13	0.45	800	—	10
GaN	0.22	0.96	440	130	10.4

* m_l is longitudinal, m_t is transverse.
** m_{lh} is light hole, m_{hh} is heavy hole.
Adapted from Sze SM (ed), *Modern Semiconductor Device Physics*, Copyright © 1998, John Wiley & Sons, Inc, New York, pp. 537-540. This material is used by permission of John Wiley & Sons, Inc.

6.1 Electron Motion

6.1.1 Calculation of Electron and Hole Concentration (B)

Here we give the standard calculation of carrier concentration based on (a) excitation of electrons from the valence to the conduction band leaving holes in the valence band, (b) the presence of impurity donors and acceptors (of electrons) and (c) charge neutrality. This discussion is important for electrical conductivity among other properties.

We start with a simple picture assuming a parabolic band structure of semiconductors involving conduction and valence bands as shown in Fig. 6.1. We will later find our results can be generalized using a suitable effective mass (Sect.6.1.6). Here when we talk about donor and acceptor impurities we are talking about shallow defects only (where the energy levels of the donors are just below the conduction band minimum and of acceptors just above the valence-band maximum). Shallow defects are further discussed in Sect. 11.2. Deep defects are discussed and compared to shallow defects in Sect. 11.3 and Table 11.1. We limit ourselves in this chapter to impurities that are sufficiently dilute that they form localized and discrete levels. Impurity bands can form where $4\pi a^3 n/3 \cong 1$ where a is the lattice constant and n is the volume density of impurity atoms of a given type.

The charge-carrier population of the levels is governed by the Fermi function f. The Fermi function evaluated at the Fermi energy $E = \mu$ is 1/2. We have assumed μ is near the middle of the band. The Fermi function is given by

$$f(E) = \frac{1}{\exp\left(\frac{E-\mu}{kT}\right)+1} . \tag{6.1}$$

In Fig. 6.1 E_C is the energy of the bottom of the conduction band. E_V is the energy of the top of the valence band. E_D is the donor state energy (energy with one electron and in which case the donor is assumed to be neutral). E_A is the acceptor state energy (which when it has two electrons and no holes is singly charged). For more on this model see Table 6.3 and Table 6.4. Some typical donor and acceptor energies for column IV semiconductors are 44 and 39 meV for P and Sb in Si, 46 and 160 meV for B and In in Si.[1]

We now evaluate expressions for the electron concentration in the conduction band and the hole concentration in the valence band. We assume the nondegenerate case when E in the conduction band implies $(E - \mu) >> kT$, so

$$f(E) \cong \exp\left(-\frac{E-\mu}{kT}\right) . \tag{6.2}$$

[1] [6.2, p. 580]

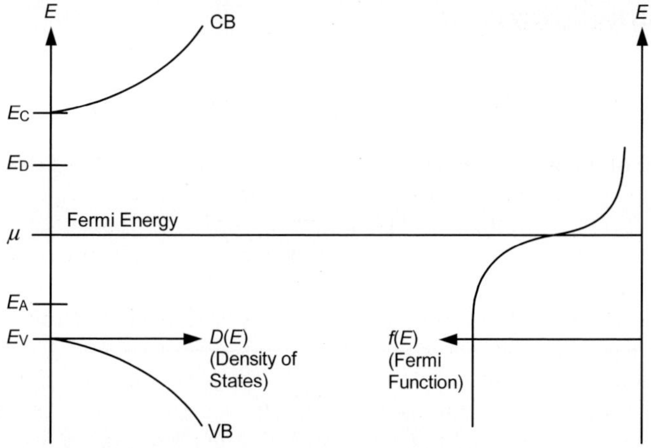

Fig. 6.1. Energy gaps, Fermi function, and defect levels (sketch). Direction of increase of $D(E)$, $f(E)$is indicated by arrows

We further assume a parabolic band, so

$$E = \frac{\hbar^2 k^2}{2m_{\mathrm{e}}^*} + E_{\mathrm{C}}\,, \tag{6.3}$$

where m_e^* is a constant. For such a case we have shown (in Chap. 3) the density of states is given by

$$D(E) = \frac{1}{2\pi^2}\left(\frac{2m_{\mathrm{e}}^*}{\hbar^2}\right)^{3/2}\sqrt{E - E_{\mathrm{C}}}\,. \tag{6.4}$$

The number of electrons per unit volume in the conduction band is given by:

$$n = \int_{E_{\mathrm{C}}}^{\infty} D(E) f(E) \mathrm{d}E\,. \tag{6.5}$$

Evaluating the integral, we find

$$n = 2\left(\frac{m_{\mathrm{e}}^* kT}{2\pi\hbar^2}\right)^{3/2}\exp\left(\frac{\mu - E_{\mathrm{C}}}{kT}\right). \tag{6.6}$$

For holes, we assume, following (6.3),

$$E = E_{\mathrm{V}} - \frac{\hbar^2 k^2}{2m_{\mathrm{h}}^*}, \tag{6.7}$$

which yields the density of states

$$D_{\rm h}(E) = \frac{1}{2\pi^2}\left(\frac{2m_{\rm n}^*}{\hbar^2}\right)^{3/2}\sqrt{E_{\rm V} - E}\,. \tag{6.8}$$

The number of holes per state is

$$f_{\rm h} = 1 - f(E) = \frac{1}{\exp\left(\frac{\mu - E}{kT}\right) + 1}\,. \tag{6.9}$$

Again, we make a nondegeneracy assumption and assume $(\mu - E) >> kT$ for E in the valence band, so

$$f_{\rm h} \cong \exp\left(\frac{E - \mu}{kT}\right). \tag{6.10}$$

The number of holes/volume in the valence band is then given by

$$p = \int_{-\infty}^{E_V} D_{\rm h}(E) f_{\rm h}(E) \mathrm{d}E\,, \tag{6.11}$$

from which we find

$$p = 2\left(\frac{m_{\rm h}^* kT}{2\pi\hbar^2}\right)^{3/2} \exp\left(\frac{E_{\rm V} - \mu}{kT}\right). \tag{6.12}$$

Since the density of states in the valence and conduction bands is essentially unmodified by the presence or absence of donors and acceptors, the equations for n and p are valid with or without donors or acceptors. (Donors or acceptors, as we will see, modify the value of the chemical potential, μ.) Multiplying n and p, we find

$$np = n_i^2\,, \tag{6.13}$$

where

$$n_i = 2\left(\frac{kT}{2\pi\hbar^2}\right)^{3/2} (m_{\rm e}^* m_{\rm h}^*)^{3/4} \exp\left(-\frac{E_{\rm g}}{2kT}\right), \tag{6.14}$$

where $E_{\rm g} = E_{\rm C} - E_{\rm V}$ is the bandgap and n_i is the intrinsic (without donors or acceptors) electron concentration. Equation (6.13) is sometimes called the Law of Mass Action and is generally true since it is independent of μ.

We now turn to the question of calculating the number of electrons on donors and holes on acceptors. We use the basic theorem for a grand canonical ensemble (see, e.g., Ashcroft and Mermin, [6.2, p 581])

$$\langle n \rangle = \frac{\sum_j \mathrm{N}_j \exp[-\beta(E_j - \mu N_j)]}{\sum_j \exp[-\beta(E_j - \mu N_j)]}\,, \tag{6.15}$$

where $\beta = 1/kT$ and $\langle n \rangle$ = mean number of electrons in a system with states j, with energy E_j, and number of electrons N_j.

Table 6.3. Model for energy and degeneracy of donors

Number of electrons	Energy	Degeneracy of state
$N_j = 0$	0	1
$N_j = 1$	E_d	2
$N_j = 2$	$\to \infty$	neglect as too improbable

We are considering a model of a donor level that is doubly degenerate (in a single-particle model). Note that it is possible to have other models for donors and acceptors. There are basically three cases to look at, as shown in Table 6.3. Noting that when we sum over states, we must include the degeneracy factors. For the mean number of electrons on a state j as defined in Table 6.3

$$\langle n \rangle = \frac{(1)(2)\exp[-\beta(E_d - \mu)]}{1 + 2\exp[-\beta(E_d - \mu)]}, \tag{6.16}$$

or

$$\langle n \rangle = \frac{1}{\frac{1}{2}\exp[-\beta(E_d - \mu)] + 1} = \frac{n_d}{N_d}, \tag{6.17}$$

where n_d is the number of electrons/volume on donor atoms and N_d is the number of donor atoms/volume. For the acceptor case, our model is given by Table 6.4.

Table 6.4. Model for energy and degeneracy of acceptors

Number of electrons	Number of holes	Energy	Degeneracy
0	2	very large	neglect
1	1	0	2
2	0	E_A	1

The number of electrons per acceptor level of the type defined in Table 6.4 is

$$\langle n \rangle = \frac{(1)(2)\exp[-\beta(-\mu)] + 2(1)\exp[-\beta(E_a - 2\mu)]}{2\exp[\beta\mu] + \exp[-\beta(E_a - 2\mu)]}, \tag{6.18}$$

which can be written

$$\langle n \rangle = \frac{\exp[\beta(\mu - E_a)] + 1}{\frac{1}{2}\exp[\beta(\mu - E_a)] + 1}. \tag{6.19}$$

Now, the average number of electrons plus the average number of holes associated with the acceptor level is 2. So, $\langle n\rangle + \langle p\rangle = 2$. We thus find

$$\langle p\rangle = \frac{p_{\mathrm{a}}}{N_{\mathrm{a}}} = \frac{1}{\frac{1}{2}\exp[\beta(\mu - E_{\mathrm{a}})] + 1}, \tag{6.20}$$

where p_{a} is the number of holes/volume on acceptor atoms. N_{a} is the number of acceptor atoms/volume.

So far, we have four equations for the five unknowns n, p, n_{d}, p_{a}, and μ. A fifth equation, determining μ can be found from the condition of electrical neutrality. Note:

$$N_{\mathrm{d}} - n_{\mathrm{d}} \equiv \text{number of ionized and, hence, positive donors} \equiv N_{\mathrm{d}}^{+},$$

$$N_{\mathrm{a}} - p_{\mathrm{a}} \equiv \text{number of negative acceptors} = N_{\mathrm{a}}^{-}.$$

Charge neutrality then says,

$$p + N_{\mathrm{d}}^{+} = n + N_{\mathrm{a}}^{-}, \tag{6.21}$$

or

$$n + N_{\mathrm{a}} + n_{\mathrm{d}} = p + N_{\mathrm{d}} + p_{\mathrm{a}}. \tag{6.22}$$

We start by discussing an example of the exhaustion region where all the donors are ionized. We assume $N_{\mathrm{a}} = 0$, so also $p_{\mathrm{a}} = 0$. We assume $kT << E_{\mathrm{g}}$, so also $p = 0$. Thus, the electrical neutrality condition reduces to

$$n + n_{\mathrm{d}} = N_{\mathrm{d}}. \tag{6.23}$$

We also assume a temperature that is high enough that all donors are ionized. This requires $kT >> E_{\mathrm{c}} - E_{\mathrm{d}}$. This basically means that the probability that states in the donor are occupied is the same as the probability that states in the conduction band are occupied. But, there are many more states in the conduction band compared to donor states, so there are many more electrons in the conduction band. Therefore $n_{\mathrm{d}} << N_{\mathrm{d}}$ or $n \cong N_{\mathrm{d}}$. This is called the exhaustion region of donors.

As a second example, we consider the same situation, but now the temperature is not high enough that all donors are ionized. Using

$$n_{\mathrm{d}} = \frac{N_{\mathrm{d}}}{1 + a\exp[\beta(E_{\mathrm{d}} - \mu)]}. \tag{6.24}$$

In our model $a = 1/2$, but different models could yield different a. Also

$$n = N_{\mathrm{c}} \exp[-\beta(E_{\mathrm{c}} - \mu)], \tag{6.25}$$

where

$$N_{\mathrm{c}} = 2\left(\frac{m_{\mathrm{e}}^{*}kT}{2\pi\hbar^{2}}\right)^{3/2}. \tag{6.26}$$

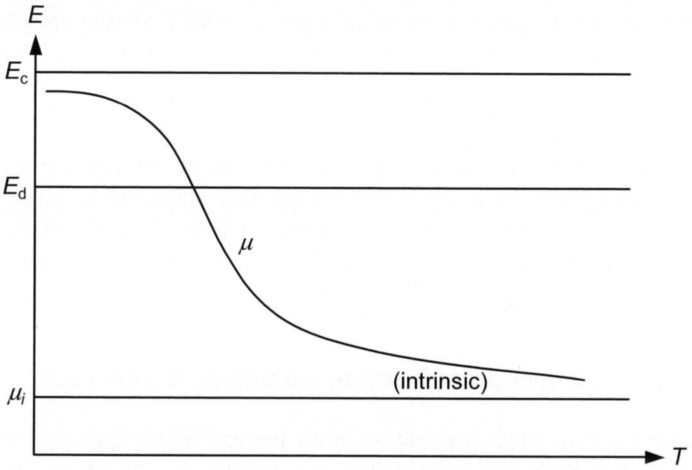

Fig. 6.2. Sketch of variation of Fermi energy or chemical potential μ, with temperature for $N_a = 0$ and $N_d > 0$

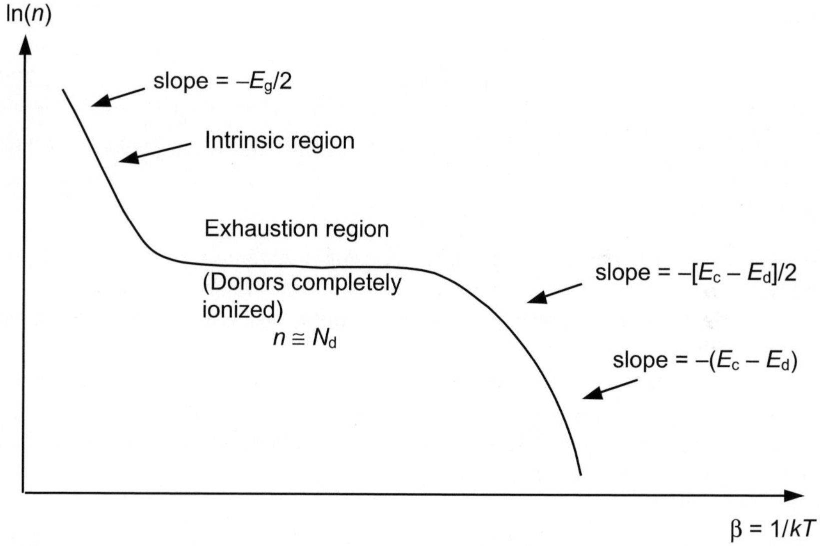

Fig. 6.3. Energy gaps, Fermi function, and defect levels (sketch)

The neutrality condition then gives

$$N_c \exp[-\beta(E_c - \mu)] + \frac{N_d}{1 + a\exp[\beta(E_d - \mu)]} = N_d \, . \tag{6.27}$$

Defining $x = e^{\beta\mu}$, the above gives a quadratic equation for x. Finding the physically realistic solution for low temperatures, $kT << (E_c - E_d)$, we find x and, hence,

$$n = \sqrt{a}\sqrt{N_c N_d}\exp[-\beta(E_c - E_d)/2]\,. \tag{6.28}$$

This result is valid only in the case that acceptors can be neglected, but in actual impure semiconductors this is not true in the low-temperature limit. More detailed considerations give the variation of Fermi energy with temperature for $N_a = 0$ and $N_d > 0$ as sketched in Fig. 6.2. For the variation of the majority carrier density for $N_d > N_a \neq 0$, we find something like Fig. 6.3.

6.1.2 Equation of Motion of Electrons in Energy Bands (B)

We start by discussing the dynamics of wave packets describing electrons [6.33, p23]. We need to do this in order to discuss properties of semiconductors such as the Hall effect, electrical conductivity, cyclotron resonance, and others. In order to think of the motion of charge, we need to think of the charge being transported by the wave packets.[2] The three-dimensional result using free-electron wave packets can be written as

$$\boldsymbol{v} = \frac{1}{\hbar}\nabla_{\boldsymbol{k}} E(\boldsymbol{k})\,. \tag{6.29}$$

This result, as we now discuss, is appropriate even if the wave packets are built out of Bloch waves.

Let a Bloch state be represented by

$$\psi_{n\boldsymbol{k}} = u_{n\boldsymbol{k}}(\boldsymbol{r})e^{i\boldsymbol{k}\cdot\boldsymbol{r}}\,, \tag{6.30}$$

where n is the band index and $u_{n\boldsymbol{k}}(\boldsymbol{r})$ is periodic in the space lattice. With the Hamiltonian

$$\mathcal{H} = \frac{1}{2m}\left(\frac{\hbar}{i}\nabla\right)^2 + V(\boldsymbol{r})\,, \tag{6.31}$$

where $V(\boldsymbol{r})$ is periodic,

$$\mathcal{H}\psi_{n\boldsymbol{k}} = E_{n\boldsymbol{k}}\psi_{n\boldsymbol{k}}\,, \tag{6.32}$$

and we can show

$$\mathcal{H}_{k}u_{n\boldsymbol{k}} = E_{n\boldsymbol{k}}u_{n\boldsymbol{k}}\,, \tag{6.33}$$

---

2 The standard derivation using wave packets is given by, e.g., Merzbacher [6.24]. In Merzbacher's derivation, the peak of the wave packet moves with the group velocity.

where

$$\mathcal{H}_k = \frac{\hbar^2}{2m}\left(\frac{1}{\mathrm{i}}\nabla + \boldsymbol{k}\right)^2 + V(\boldsymbol{r})\,. \tag{6.34}$$

Note

$$\mathcal{H}_{k+q} u_{nk+q} = E_{nk+q} u_{nk+q}\,, \tag{6.35}$$

and to first order in $\boldsymbol{q}$:

$$\mathcal{H}_{k+q} = \mathcal{H}_k + \frac{\hbar^2}{m}\boldsymbol{q}\cdot\left(\frac{1}{\mathrm{i}}\nabla + \boldsymbol{k}\right). \tag{6.36}$$

To first order

$$E_n(\boldsymbol{k}+\boldsymbol{q}) = E_n(\boldsymbol{k}) + \boldsymbol{q}\cdot\nabla_k E_{nk}\,. \tag{6.37}$$

Also by first-order perturbation theory

$$E_n(\boldsymbol{k}+\boldsymbol{q}) = E_n(\boldsymbol{k}) + \int u_{nk}\frac{\hbar^2}{m}\boldsymbol{q}\cdot\left(\frac{1}{\mathrm{i}}\nabla + \boldsymbol{k}\right)u_{nk}\mathrm{d}V\,. \tag{6.38}$$

From this we conclude

$$\begin{aligned}\nabla_k E_{nk} &= \int u_{nk}\frac{\hbar^2}{m}\left(\frac{1}{\mathrm{i}}\nabla + \boldsymbol{k}\right)u_{nk}\mathrm{d}V \\ &= \hbar\int \psi_{nk}\frac{\hbar}{m\mathrm{i}}\nabla\psi_{nk}\mathrm{d}V \\ &= \hbar\left\langle\psi_{nk}\left|\frac{p}{m}\right|\psi_{nk}\right\rangle.\end{aligned} \tag{6.39}$$

Thus if we define

$$\boldsymbol{v} = \left\langle\psi_{nk}\left|\frac{p}{m}\right|\psi_{nk}\right\rangle, \tag{6.40}$$

then $\boldsymbol{v}$ equals the average velocity of the electron in the Bloch state $\boldsymbol{nk}$. So we find

$$\boldsymbol{v} = \frac{1}{\hbar}\nabla_k E_{nk}\,.$$

Note that $\boldsymbol{v}$ is a constant velocity (for a given $\boldsymbol{k}$). We interpret this as meaning that a Bloch electron in a periodic crystal is not scattered.

Note also that we should use a packet of Bloch waves to describe the motion of electrons. Thus we should average this result over a set of states peaked at $\boldsymbol{k}$. It can also be shown following standard arguments (Smith [6.38], Sect. 4.6) that (6.29) is the appropriate velocity of such a packet of waves.

We now apply external fields and ask what is the effect of these external fields on the electrons. In particular, what is the effect on the electrons if they are already in a periodic potential? If an external force F_{ext} acts on an electron during a time interval δt, it produces a change in energy given by

$$\delta E = F_{\text{ext}} \delta x = F v_{\text{g}} \delta t \,. \tag{6.41}$$

Substituting for v_{g},

$$\delta E = F_{\text{ext}} \frac{1}{\hbar} \frac{\delta E}{\delta k} \delta t \,. \tag{6.42}$$

Canceling out δE, we find

$$F_{\text{ext}} = \hbar \frac{\delta k}{\delta t} \,. \tag{6.43}$$

The three-dimensional result may formally be obtained by analogy to the above:

$$\boldsymbol{F}_{\text{ext}} = \hbar \frac{\mathrm{d}\boldsymbol{k}}{\mathrm{d}t} \,. \tag{6.44}$$

In general, $\boldsymbol{F}$ is the external force, so if $\boldsymbol{E}$ and $\boldsymbol{B}$ are electric and magnetic fields, then

$$\hbar \frac{\mathrm{d}\boldsymbol{k}}{\mathrm{d}t} = -e(\boldsymbol{E} + \boldsymbol{v} \times \boldsymbol{B}) \tag{6.45}$$

for an electron with charge $-e$. See Problem 6.3 for a more detailed derivation. This result is often called the acceleration theorem in $\boldsymbol{k}$-space.

We next introduce the concept of effective mass. In one dimension, by taking the time derivative of the group velocity we have

$$\frac{\mathrm{d}v}{\mathrm{d}t} = \frac{1}{\hbar} \frac{\mathrm{d}^2 E}{\mathrm{d}k^2} \frac{\mathrm{d}k}{\mathrm{d}t} = \frac{1}{\hbar^2} \frac{\mathrm{d}^2 E}{\mathrm{d}k^2} F_{\text{ext}} \,. \tag{6.46}$$

Defining the effective mass so

$$F_{\text{ext}} = m^* \frac{\mathrm{d}v}{\mathrm{d}t} \,, \tag{6.47}$$

we have

$$m^* = \frac{\hbar^2}{\mathrm{d}^2 E / \mathrm{d}k^2} \,. \tag{6.48}$$

In three dimensions:

$$\left(\frac{1}{m^*} \right)_{\alpha\beta} = \frac{1}{\hbar^2} \frac{\partial^2 E}{\partial k_\alpha \partial k_\beta} \,. \tag{6.49}$$

Notice in the free-electron case when $E = \hbar^2 k^2/2m$,

$$\left(\frac{1}{m^*}\right)_{\alpha\beta} = \frac{\delta_{\alpha\beta}}{m}. \tag{6.50}$$

6.1.3 Concept of Hole Conduction (B)

The totality of the electrons in a band determines the conduction properties of that band. But, when a band is nearly full it is usually easier to consider holes that represent the absent electrons. There will be far fewer holes than electrons and this in itself is a huge simplification.

It is fairly easy to see why an absent electron in the valence band acts as a positive electron. See also Kittel [6.17, p206ff]. Let f label filled electron states, and g label the states that will later be emptied. For a full band in a crystal, with volume V, for conduction in the x direction,

$$j_x = -\frac{e}{V}\sum\nolimits_f v_x^f - \frac{e}{V}\sum\nolimits_g v_x^g = 0, \tag{6.51}$$

so that

$$\sum\nolimits_f v_x^f = -\sum\nolimits_g v_x^g . \tag{6.52}$$

If g states of the band are now emptied, then the current is given by

$$j_x = -\frac{e}{V}\sum\nolimits_f v_x^f = \frac{e}{V}\sum\nolimits_g v_x^g . \tag{6.53}$$

Notice this argument means that the current in a partially empty band can be considered as due to holes of charge $+e$, which move with the velocities of the states that are missing electrons. In other words, $q_{\mathrm{h}} = +e$ and $v_{\mathrm{h}} = v_{\mathrm{e}}$.

Now, let us talk about the energy of the holes. Consider a full band with one missing electron. Let the wave vector of the missing electron be k_{e} and the corresponding energy $E_{\mathrm{e}}(k_{\mathrm{e}})$:

$$E_{\text{solid, full band}} = E_{\text{solid, one missing electron}} + E_{\mathrm{e}}(k_{\mathrm{e}}) . \tag{6.54}$$

Since the hole energy is the energy it takes to remove the electron, we have

$$\text{Hole energy} = E_{\text{solid, one missing electron}} - E_{\text{solid, full band}} = -E_{\mathrm{e}}(k_{\mathrm{e}}) \tag{6.55}$$

by using the above. Now in a full band the sum of the $\boldsymbol{k}$ is zero. Since we identify the hole wave vector as the totality of the filled electronic states

$$k_{\mathrm{e}} + \sum{}' \boldsymbol{k} = 0, \tag{6.56}$$

$$k_{\mathrm{h}} = \sum{}' \boldsymbol{k} = -k_{\mathrm{e}}, \tag{6.57}$$

where $\Sigma' \boldsymbol{k}$ means the sum over $\boldsymbol{k}$ omitting k_e. Thus, we have, assuming symmetric bands with $E_e(k_e) = E_e(-k_e)$:

$$E_h(k_h) = -E_e(-k_e) , \tag{6.58}$$

or

$$E_h(k_h) = -E_e(k_e) . \tag{6.59}$$

Notice also, since

$$\hbar \frac{d\boldsymbol{k}_e}{dt} = -e(E + \boldsymbol{v}_e \times \boldsymbol{B}) , \tag{6.60}$$

with $q_h = +e$, $k_h = -k_e$ and $v_e = v_h$, we have

$$\hbar \frac{d\boldsymbol{k}_h}{dt} = +e(E + \boldsymbol{v}_h \times \boldsymbol{B}) , \tag{6.61}$$

as expected. Now, since

$$v_e = \frac{1}{\hbar}\frac{\partial E_e(k_e)}{\partial (k_e)} = \frac{1}{\hbar}\frac{\partial(-E_h(k_h))}{\partial(-k_h)} = \frac{1}{\hbar}\frac{\partial E_h}{\partial k_h} , \tag{6.62}$$

and since $v_e = v_h$, then

$$v_h = \frac{1}{\hbar}\frac{\partial E_h}{\partial k_h} . \tag{6.63}$$

Now,

$$\frac{dv_h}{dt} = \frac{1}{\hbar}\frac{\partial^2 E_h}{\partial k_h^2}\frac{dk_h}{dt} = \frac{1}{\hbar^2}\frac{\partial^2 E_h}{\partial k_h^2} F_h . \tag{6.64}$$

Defining the hole effective mass as

$$\frac{1}{m_h^*} = \frac{1}{\hbar^2}\frac{\partial^2 E_h}{\partial k_h^2} , \tag{6.65}$$

we see

$$\frac{1}{m_h^*} = -\frac{1}{\hbar^2}\frac{\partial^2 E_e}{\partial(-k_e)^2} = -\frac{1}{m_e^*} , \tag{6.66}$$

or

$$m_e^* = -m_h^* . \tag{6.67}$$

Notice that if $E_e = Ak^2$, where A is constant then $m_e^* > 0$, whereas if $E_e = -Ak^2$, then $m_h^* = -m_e^* > 0$, and concave down bands have negative electron masses but positive hole masses. Later we note that electrons and holes may interact so as to form excitons (Sect. 10.7, Exciton Absorption).

6.1.4 Conductivity and Mobility in Semiconductors (B)

Current can be produced in semiconductors by, e.g., potential gradients (electric fields) or concentration gradients. We now discuss this.

We assume, as is usually the case, that the lifetime of the carriers is very long compared to the mean time between collisions. We also assume a Drude model with a unique collision or relaxation time τ. A more rigorous presentation can be made by using the Boltzmann equation where in effect we assume $\tau = \tau(E)$. A consequence of doing this is mentioned in (6.102).

We are actually using a semiclassical Drude model where the effect of the lattice is taken into account by using an effective mass, derived from the band structure, and we treat the carriers classically except perhaps when we try to estimate their scattering. As already mentioned, to regard the carriers classically we must think of packets of Bloch waves representing them. These wave packets are large compared to the size of a unit cell and thus the field we consider must vary slowly in space. An applied field also must have a frequency much less than the bandgap over $\hbar$ in order to avoid band transitions.

We consider current due to drift in an electric field. Let v be the drift velocity of electrons, m^* be their effective mass, and τ be a relaxation time that characterizes the friction drag on the electrons. In an electric field E, we can write (for $e > 0$)

$$m^* \frac{\mathrm{d}v}{\mathrm{d}t} = -\frac{m^* v}{\tau} - eE \,. \tag{6.68}$$

Thus in the steady state

$$v = -\frac{e\tau E}{m^*} \,. \tag{6.69}$$

If n is the number of electrons per unit volume with drift velocity v, then the current density is

$$j = -nev \,. \tag{6.70}$$

Combining the last two equations gives

$$j = \frac{ne^2 \tau E}{m^*} \,. \tag{6.71}$$

Thus, the electrical conductivity σ, defined by j/E, is given by

$$\sigma = \frac{ne^2 \tau}{m^*} \,. \tag{6.72}$$

[3] The electrical mobility is the magnitude of the drift velocity per unit electric field $|v/E|$, so

$$\mu = \frac{e\tau}{m^*} \,. \tag{6.73}$$

[3] We have already derived this, see, e.g., (3.214) where effective mass was not used and in (4.160) where again the m used should be effective mass and τ is more precisely evaluated at the Fermi energy.

Notice that the mobility measures the scattering, while the electrical conductivity measures both the scattering and the electron concentration. Combining the last two equations, we can write

$$\sigma = ne\mu \,. \tag{6.74}$$

If we have both electrons (e) and holes (h) with concentration n and p, then

$$\sigma = ne\mu_{\mathrm{e}} + pe\mu_{\mathrm{h}} \,, \tag{6.75}$$

where

$$\mu_{\mathrm{e}} = \frac{e\tau_{\mathrm{e}}}{m_{\mathrm{e}}^{*}} \,, \tag{6.76}$$

and

$$\mu_{\mathrm{h}} = \frac{e\tau_{\mathrm{h}}}{m_{\mathrm{h}}^{*}} \,. \tag{6.77}$$

The drift current density J_{d} can be written either as

$$J_{\mathrm{d}} = -nev_{\mathrm{e}} + pev_{\mathrm{h}} \,, \tag{6.78}$$

or

$$J_{\mathrm{d}} = [(ne\mu_{\mathrm{e}}) + (pe\mu_{\mathrm{h}})]E \,. \tag{6.79}$$

As mentioned, in semiconductors we can also have current due to concentration gradients. By Fick's Law, the diffusion number current is negatively proportional to the concentration gradient with the proportionality constant equal to the diffusion constant. Multiplying by the charge gives the electrical current density. Thus,

$$J_{\mathrm{e,\,diffusion}} = eD_{\mathrm{e}} \frac{\mathrm{d}n}{\mathrm{d}x} \tag{6.80}$$

$$J_{\mathrm{h,\,diffusion}} = -eD_{\mathrm{h}} \frac{\mathrm{d}p}{\mathrm{d}x} \,. \tag{6.81}$$

For both drift and diffusion currents, the electronic current density is

$$J_{\mathrm{e}} = \mu_{\mathrm{e}} enE + eD_{\mathrm{e}} \frac{\mathrm{d}n}{\mathrm{d}x} \,, \tag{6.82}$$

and the hole current density is

$$J_{\mathrm{h}} = \mu_{\mathrm{h}} epE - eD_{\mathrm{h}} \frac{\mathrm{d}p}{\mathrm{d}x} \,. \tag{6.83}$$

In both cases, the diffusion constant can be related to the mobility by the Einstein relationship (valid for both Drude and Boltzmann models)

$$eD_e = \mu_e kT \,, \tag{6.84}$$

$$eD_h = \mu_h kT \,. \tag{6.85}$$

6.1.5 Drift of Carriers in Electric and Magnetic Fields: The Hall Effect (B)

The Hall effect is the production of a transverse voltage (a voltage change along the "y direction") due to a transverse B-field (in the "z direction") with current flowing in the "x direction." It is useful for determining information on the sign and concentration of carriers. See Fig. 6.4.

If the collisional force is described by a relaxation time τ,

$$m_e \frac{d\boldsymbol{v}}{dt} = -e(\boldsymbol{E} + \boldsymbol{v} \times \boldsymbol{B}) - m_e \frac{\boldsymbol{v}}{\tau_e}\,, \tag{6.86}$$

where $\boldsymbol{v}$ is the drift velocity. We treat the steady state with $d\boldsymbol{v}/dt = 0$. The magnetic field is assumed to be in the z direction and we define

$$\omega_e = \frac{eB}{m_e}\,, \text{ the cyclotron frequency,} \tag{6.87}$$

and

$$\mu_e = \frac{e\tau_e}{m_e}\,, \text{ the mobility.} \tag{6.88}$$

For electrons, from (6.86) we can write the components of drift velocity as (steady state)

$$v_x^e = -\mu_e E_x - \omega_e \tau_e v_y^e\,, \tag{6.89}$$

$$v_y^e = -\mu_e E_y + \omega_e \tau_e v_x^e\,, \tag{6.90}$$

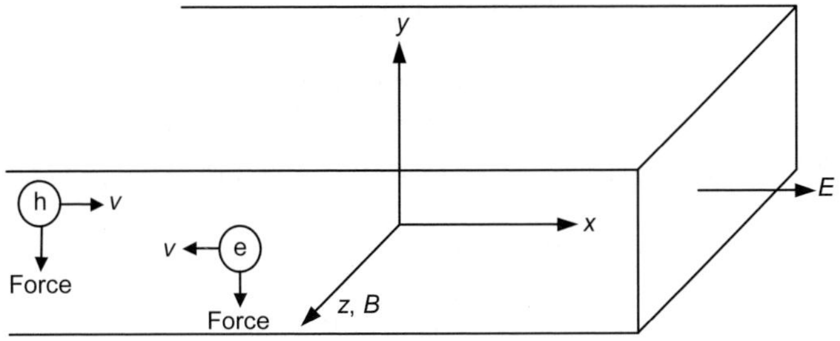

Fig. 6.4. Geometry for the Hall effect

where $v_z^{\mathrm{e}} = 0$, since $E_z = 0$. With similar definitions, the equations for holes become

$$v_x^{\mathrm{h}} = +\mu_{\mathrm{h}} E_x + \omega_{\mathrm{h}} \tau_{\mathrm{h}} v_y^{\mathrm{h}}, \tag{6.91}$$

$$v_y^{\mathrm{h}} = +\mu_{\mathrm{h}} E_y - \omega_{\mathrm{h}} \tau_{\mathrm{h}} v_x^{\mathrm{h}}. \tag{6.92}$$

Due to the electric field in the x direction, the current is

$$\boldsymbol{j}_x = -nev_x^{\mathrm{e}} + pev_x^{\mathrm{h}}. \tag{6.93}$$

Because of the magnetic field in the z direction, there are forces also in the y direction, which end up creating an electric field E_y in that direction. The Hall coefficient is defined as

$$R_{\mathrm{H}} = \frac{E_y}{j_x B}. \tag{6.94}$$

Equations (6.89) and (6.90) can be solved for the electrons drift velocity and (6.91) and (6.92) for the hole's drift velocity. We assume weak magnetic fields and neglect terms of order ω_{e}^2 and ω_{h}^2, since ω_{e} and ω_{h} are proportional to the magnetic field. This is equivalent to neglecting magnetoresistance, i.e. the variation with resistance in a magnetic field. It can be shown that for carriers of two types if we retain terms of second order then we have a magnetoresistance. So far we have not considered a distribution of velocities as in the Boltzmann approach. Combining these assumptions, we get

$$v_x^{\mathrm{e}} = -\mu_{\mathrm{e}} E_x + \mu_{\mathrm{e}} \omega_{\mathrm{e}} \tau_{\mathrm{e}} E_y, \tag{6.95}$$

$$v_x^{\mathrm{h}} = +\mu_{\mathrm{h}} E_x + \mu_{\mathrm{h}} \omega_{\mathrm{h}} \tau_{\mathrm{h}} E_y, \tag{6.96}$$

$$v_y^{\mathrm{e}} = -\mu_{\mathrm{e}} E_y - \mu_{\mathrm{e}} \omega_{\mathrm{e}} \tau_{\mathrm{e}} E_x, \tag{6.97}$$

$$v_y^{\mathrm{h}} = +\mu_{\mathrm{h}} E_y - \mu_{\mathrm{h}} \omega_{\mathrm{h}} \tau_{\mathrm{h}} E_x. \tag{6.98}$$

Since there is no net current in the y direction,

$$\boldsymbol{j}_y = -nev_y^{\mathrm{e}} + pev_y^{\mathrm{h}} = 0. \tag{6.99}$$

Substituting (6.97) and (6.98) into (6.99) gives

$$E_x = -E_y \frac{n\mu_{\mathrm{e}} + p\mu_{\mathrm{h}}}{n\mu_{\mathrm{e}}\omega_{\mathrm{e}}\tau_{\mathrm{e}} - p\mu_{\mathrm{h}}\omega_{\mathrm{h}}\tau_{\mathrm{h}}}. \tag{6.100}$$

Putting (6.95) and (6.96) into $\boldsymbol{j}_x$, using (6.100) and putting the results into R_{H}, we find

$$R_{\mathrm{H}} = \frac{1}{e} \frac{p - nb^2}{(p + nb)^2}, \tag{6.101}$$

where $b = \mu_e/\mu_h$. Note if $p = 0$, $R_H = -1/ne$ and if $n = 0$, $R_H = +1/pe$. Both the sign and concentration of carriers are included in the Hall coefficient. As noted, this development did not take into account that the carrier would have a velocity distribution. If a Boltzmann distribution is assumed,

$$R_H = r\left(\frac{1}{e}\right)\frac{p - nb^2}{(p + nb)^2}, \tag{6.102}$$

where r depends on the way the electrons are scattered (different scattering mechanisms give different r).

The Hall effect is further discussed in Sects. 12.6 and 12.7, where peculiar effects involved in the quantum Hall effect are dealt with. The Hall effect can be used as a sensor of magnetic fields since it is proportional to the magnetic field for fixed currents.

6.1.6 Cyclotron Resonance (A)

Cyclotron resonance is the absorption of electromagnetic energy by electrons in a magnetic field at multiples of the cyclotron frequency. It was predicted by Dorfmann and Dingel and experimentally demonstrated by Kittel all in the early 1950s.

In this section, we discuss cyclotron resonance only in semiconductors. As we will see, this is a good way to determine effective masses but few carriers are naturally excited so external illumination may be needed to enhance carrier concentration (see further comments at the end of this section). Metals have plenty of carriers but skin-depth effects limit cyclotron resonance to those electrons near the surface (as discussed in Sect. 5.4).

We work on the case for Si. See also, e.g. [6.33, pp. 78-83]. We impose a magnetic field and seek the natural frequencies of oscillatory motion. Cyclotron resonance absorption will occur when an electric field with polarization in the plane of motion has a frequency equal to the frequency of oscillatory motion due to the magnetic field. We first look at motion for the energy lobes along the k_z-axis (see Si in Fig. 6.6). The energy ellipsoids are not centered at the origin. Thus, the two constant energy ellipsoids along the k_z-axis can be written

$$E = \frac{\hbar^2}{2}\left[\frac{k_x^2 + k_y^2}{m_T} + \frac{(k_z - k_0)^2}{m_L}\right]. \tag{6.103}$$

The shape of the ellipsoid determines the effective mass (T for transverse, L for longitudinal) in (6.103). The star on the effective mass is eliminated for simplicity. The velocity is given by

$$\boldsymbol{v} = \frac{1}{\hbar}\nabla_{\boldsymbol{k}} E_{\boldsymbol{k}}, \tag{6.104}$$

so

$$v_x = \frac{\hbar k_x}{m_T} \tag{6.105}$$

$$v_y = \frac{\hbar k_y}{m_T} \tag{6.106}$$

$$v_z = \frac{\hbar(k_z - k_0)}{m_L}. \tag{6.107}$$

The equation of motion for charge q is

$$\hbar\frac{\mathrm{d}\boldsymbol{k}}{\mathrm{d}t} = q\boldsymbol{v}\times\boldsymbol{B}. \tag{6.108}$$

Writing out the three components of this equation, and substituting the equations for the velocity, we find with (see Fig. 6.5)

$$B_x = B\sin\theta\cos\phi, \tag{6.109}$$

$$B_y = B\sin\theta\sin\phi, \tag{6.110}$$

$$B_z = B\cos\theta, \tag{6.111}$$

$$\frac{\mathrm{d}k_x}{\mathrm{d}t} = qB\left[\frac{k_y\cos\theta}{m_T} - \frac{(k_z - k_0)}{m_L}\sin\theta\sin\phi\right], \tag{6.112}$$

$$\frac{\mathrm{d}k_y}{\mathrm{d}t} = qB\left[\frac{(k_z - k_0)}{m_L}\sin\theta\cos\phi - \frac{k_x}{m_T}\cos\theta\right], \tag{6.113}$$

$$\frac{\mathrm{d}k_z}{\mathrm{d}t} = qB\left[\frac{k_x}{m_T}\sin\theta\sin\phi - \frac{k_y}{m_T}\sin\theta\cos\phi\right]. \tag{6.114}$$

Seeking solutions of the form

$$k_x = A_1\exp(\mathrm{i}\omega t), \tag{6.115}$$

$$k_y = A_2\exp(\mathrm{i}\omega t), \tag{6.116}$$

$$(k_z - k_0) = A_3\exp(\mathrm{i}\omega t), \tag{6.117}$$

and defining a, b, c, and γ for convenience,

$$a = \frac{qB\cos\theta}{m_T}, \tag{6.118}$$

$$b = \frac{qB}{m_L}\sin\theta\sin\phi, \tag{6.119}$$

$$c = \frac{qB}{m_{\mathrm{L}}} \sin\theta \cos\phi\,, \tag{6.120}$$

$$\gamma = \frac{m_{\mathrm{L}}}{m_{\mathrm{T}}}\,, \tag{6.121}$$

we can express (6.112), (6.113), and (6.114) in the matrix form

$$\begin{bmatrix} \mathrm{i}\omega & -a & b \\ a & \mathrm{i}\omega & -c \\ -b\gamma & \gamma c & \mathrm{i}\omega \end{bmatrix} \begin{bmatrix} a \\ b \\ c \end{bmatrix} = 0\,. \tag{6.122}$$

Setting the determinant of the coefficient matrix equal to zero gives three solutions for ω,

$$\omega = 0\,, \tag{6.123}$$

and

$$\omega^2 = a^2 + \gamma(b^2 + c^2)\,. \tag{6.124}$$

After simplification, the nonzero frequency solution (6.124) can be written:

$$\omega^2 = (qB)^2 \left[\frac{\cos^2\theta}{m_{\mathrm{T}}^2} + \frac{\sin^2\theta}{m_{\mathrm{L}} m_{\mathrm{T}}} \right]. \tag{6.125}$$

Since we have two other sets of lobes in the electronic wave function in Si (along the x-axis and along the y-axis), we have two other sets of frequencies that can be obtained by substituting θ_x and θ_y for θ (Fig. 6.5 and Fig. 6.6).

Note from Fig. 6.5

$$\cos\theta_x = \frac{\boldsymbol{B}\cdot\boldsymbol{i}}{B} = \sin\theta\cos\phi \tag{6.126}$$

$$\cos\theta_y = \frac{\boldsymbol{B}\cdot\boldsymbol{j}}{B} = \sin\theta\sin\phi\,. \tag{6.127}$$

Thus, the three resonance frequencies can be determined. For the (energy) lobes along the z-axis, we have found

$$\omega_z^2 = (qB)^2 \left[\frac{\cos^2\theta}{m_{\mathrm{T}}^2} + \frac{\sin^2\theta}{m_{\mathrm{L}} m_{\mathrm{T}}} \right]. \tag{6.128}$$

For the lobes along the x-axis, replace θ with θ_x and get

$$\omega_x^2 = (qB)^2 \left[\frac{\sin^2\theta\cos^2\phi}{m_{\mathrm{T}}^2} + \frac{1-\sin^2\theta\cos^2\phi}{m_{\mathrm{L}} m_{\mathrm{T}}} \right], \tag{6.129}$$

and for the lobes along the y-axis, replace θ with θ_y and get

$$\omega_y^2 = (qB)^2 \left[\frac{\sin^2\theta\sin^2\phi}{m_{\mathrm{T}}^2} + \frac{1-\sin^2\theta\sin^2\phi}{m_{\mathrm{L}} m_{\mathrm{T}}} \right]. \tag{6.130}$$

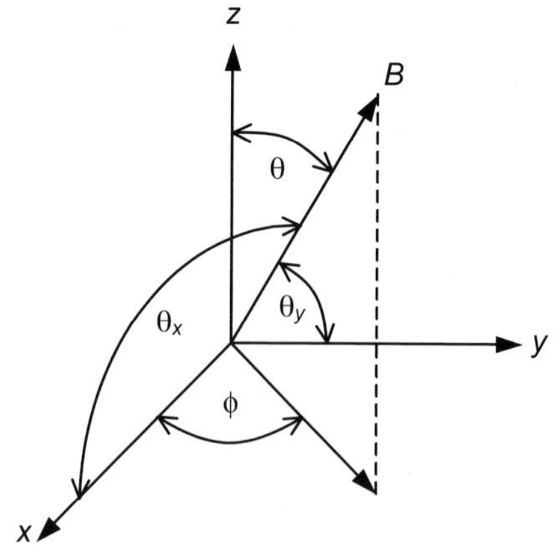

Fig. 6.5. Definition of angles used for cyclotron-resonance discussion

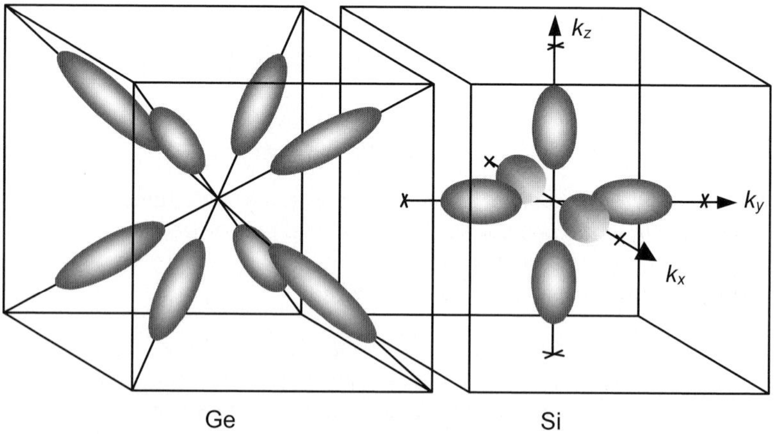

Fig. 6.6. Constant energy ellipsoids in Ge and Si. From Ziman JM, *Electrons and Phonons: The Theory of Transport Phenomena in Solids*, Clarendon Press, Oxford (1960). By permission of Oxford University Press

In general, then we get three resonance frequencies. Obviously, for certain directions of $\boldsymbol{B}$, some or all of these frequencies may become degenerate.

Several comments:

1. When $m_L = m_T$, these frequencies reduce to the cyclotron frequency $\omega_c = qB/m$.
2. In general, one will have to illuminate the sample to produce enough electrons and holes to detect the absorption, as with laser illumination.
3. In order to see the absorption, one wants collisions to be rare. If τ is the mean time between collisions, we then require $\omega_c\tau > 1$ or low temperatures, high purity, and high magnetic fields are required.

4. The resonant frequencies can be used to determine the longitudinal and transverse effective mass m_L, m_T.
5. Extremal orbits, with high density of states, are most important for effective absorption.

Some classic cyclotron resonance results obtained at Berkeley in 1955 by Dresselhaus, Kip, and Kittel are sketched in Fig. 6.7. See also the Section below "Power Absorption in Cyclotron Resonance."

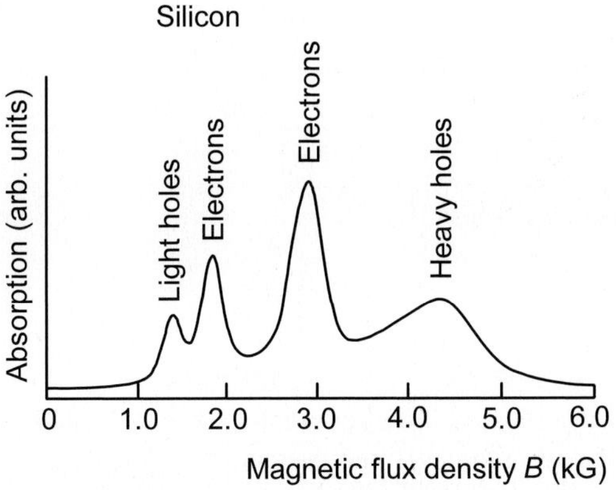

Fig. 6.7. Sketch of cyclotron resonance for silicon. (near 24×10^3 Mc/s and 4 K, B at 30° with [100] and in (110) plane). Adapted from Dresselhaus, Kip and Kittel [6.11]

Density of States Effective Electron Masses for Si (A)

We can now generalize the concept of density of states effective mass so as to extend the use of equations like (6.4). For Si, we relate the transverse and longitudinal effective masses to the density of states effective mass. See "Density of States for Effective Hole Masses" in Sect. 6.2.1 for light and heavy hole effective masses. For electrons in the conduction band we have used the density of states.

$$D(E) = \frac{1}{2\pi^2}\left(\frac{2m_e^*}{\hbar^2}\right)^{3/2}\sqrt{E}\,. \tag{6.131}$$

This can be derived from

$$D(E) = \frac{dn(E)}{dE} = \frac{dn(E)}{dV_k}\frac{dV_k}{dE}\,,$$

where $n(E)$ is the number of states per unit volume of real space with energy E and $\mathrm{d}V_k$ is the volume of k-space with energy between E and $E + \mathrm{d}E$. Since we have derived (see Sect. 3.2.3)

$$\mathrm{d}n(E) = \frac{2}{(2\pi)^3}\mathrm{d}V_k\,,$$

$$D(E) = \frac{1}{4\pi^3}\frac{\mathrm{d}V_k}{\mathrm{d}E}\,,$$

for

$$E = \frac{\hbar^2}{2m_\mathrm{e}^*}k^2\,,$$

with a spherical energy surface,

$$V_k = \frac{4}{3}\pi k^3\,,$$

so we get (6.131).

We know that an ellipsoid with semimajor axes a, b, and c has volume $V = 4\pi abc/3$. So for Si with an energy represented by ((6.110) with origin shifted so $k_0 = 0$)

$$E = \frac{1}{2}\left(\frac{k_x^2 + k_y^2}{m_\mathrm{T}} + \frac{k_z^2}{m_\mathrm{L}}\right),$$

the volume in k-space with energy E is

$$V = \frac{4}{3}\pi\left(\frac{2m_\mathrm{T}^{2/3}m_\mathrm{L}^{1/3}}{\hbar^2}\right)^{3/2}E^{3/2}\,. \tag{6.132}$$

So

$$D(E) = \frac{1}{2\pi^2}\left(\frac{2(m_\mathrm{T}^2 m_\mathrm{L})^{1/3}}{\hbar^2}\right)^{3/2}\sqrt{E}\,. \tag{6.133}$$

Since we have six ellipsoids like this, we must replace in (6.131)

$$(m_\mathrm{e}^*)^{3/2} \quad \text{by} \quad 6(m_\mathrm{L}m_\mathrm{T}^2)^{1/2}\,,$$

or

$$m_\mathrm{e}^* \quad \text{by} \quad 6^{2/3}(m_\mathrm{L}m_\mathrm{T}^2)^{1/3}$$

for the electron density of states effective mass.

Power Absorption in Cyclotron Resonance (A)

Here we show how a resonant frequency gives a maximum in the power absorption versus field, as for example in Fig. 6.7. We will calculate the power absorption by evaluating the complex conductivity. We use (6.86) with v being the drift velocity of the appropriate charge carrier with effective mass m^* and charge $q = -e$. This equation neglects interactions between charge carriers in semiconductors since the carrier density is low and they can stay out of each others way. In (6.86), τ is the relaxation time and the $1/\tau$ terms take care of the damping effect of collisions. As usual the carriers will be assumed to be quasifree (free electrons with an effective mass to include lattice effects) and we assume that the wave packets describing the carriers spread little so the carriers can be treated classically.

Let the B field be a static field along the z-axis and let $\boldsymbol{E} = E_x e^{i\omega t}\boldsymbol{i}$ be the plane-polarized electric field. Solutions of the form

$$\boldsymbol{v}(t) = \boldsymbol{v}e^{i\omega t}, \tag{6.134}$$

will be sought. Then (6.86) may be written in component form as

$$m^*(i\omega)v_x = qE_x + qv_yB - \frac{m^*}{\tau}v_x, \tag{6.135}$$

$$m^*(i\omega)v_y = -qv_xB - \frac{m^*}{\tau}v_y. \tag{6.136}$$

If we assume the carriers are electrons then $j = n_e v_x(-e) = \sigma E_x$ so the complex conductivity is

$$\sigma = -\frac{e n_e v_x}{E_x}, \tag{6.137}$$

where n_e is the concentration of electrons. By solving (6.136) and (6.137) we find

$$\sigma = \sigma_0 \frac{[1+(\omega_c^2-\omega^2)\tau^2]+2\omega^2\tau^2}{[1+(\omega_c^2-\omega^2)\tau^2]^2+4\omega^2\tau^2} + i\sigma_0 \frac{\omega\tau[1+(\omega_c^2-\omega^2)\tau^2-2]}{[1+(\omega_c^2-\omega^2)\tau^2]^2+4\omega^2\tau^2}, \tag{6.138}$$

where $\sigma_0 = n_e e^2\tau/m^*$ is the dc conductivity and $\omega_c = eB/m^*$.

The rate at which energy is lost (per unit volume) due to Joule heating is $\boldsymbol{j}\cdot\boldsymbol{E} = j_x E_x$. But

$$\begin{aligned}\mathrm{Re}(j_x) &= \mathrm{Re}(\sigma E_x) \\ &= \mathrm{Re}[(\sigma_r + i\sigma_i)(E_x\cos\omega t + iE_x\sin\omega t)] \\ &= \sigma_r E_x\cos\omega t - \sigma_i E_x \sin\omega t .\end{aligned} \tag{6.139}$$

So

$$\mathrm{Re}(j_x)\mathrm{Re}(E_c) = E_x^2(\sigma_r\cos^2\omega t - \sigma_i\cos\omega t\sin\omega t). \tag{6.140}$$

The average energy (over a cycle) dissipated per unit volume is thus

$$\overline{P} = \overline{\mathrm{Re}(j_x)\mathrm{Re}(E_c)} = \frac{1}{2}\sigma_{\mathrm{r}}\,|E|^2, \tag{6.141}$$

where $|E| \equiv E_x$. Thus

$$\overline{P} \propto \mathrm{Re}\left(\frac{\sigma}{\sigma_0}\right) \propto \frac{1+g_{\mathrm{c}}^2+g^2}{(1+g_{\mathrm{c}}^2-g^2)^2+4g^2},$$

where $g = \omega\tau$ and $g_{\mathrm{c}} = \omega_{\mathrm{c}}\tau$. We get a peak when $g = g_{\mathrm{c}}$. If there is more than one resonance there is more than one maximum as we have already noted. See Fig. 6.7.

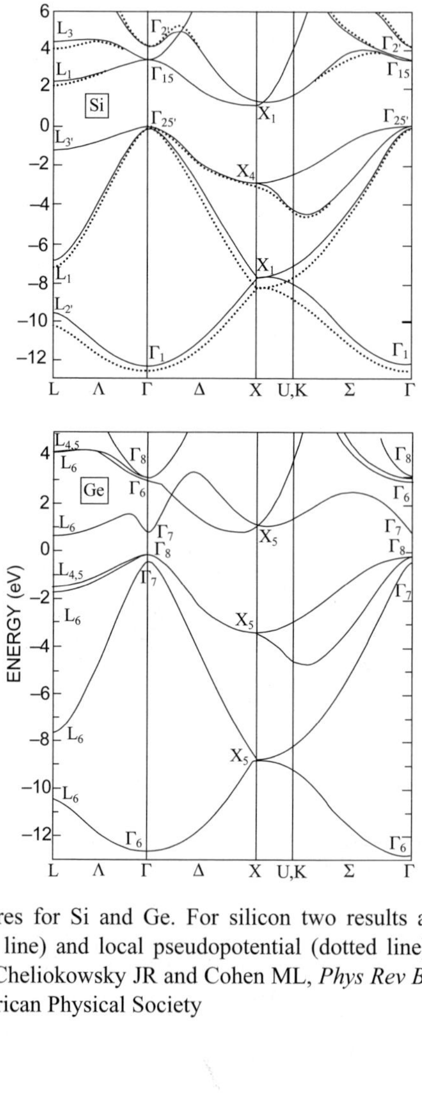

Fig. 6.8. Band structures for Si and Ge. For silicon two results are presented: nonlocal pseudopotential (solid line) and local pseudopotential (dotted line). Adaptation reprinted with permission from Cheliokowsky JR and Cohen ML, *Phys Rev B* **14**, 556 (1976). Copyright 1976 by the American Physical Society

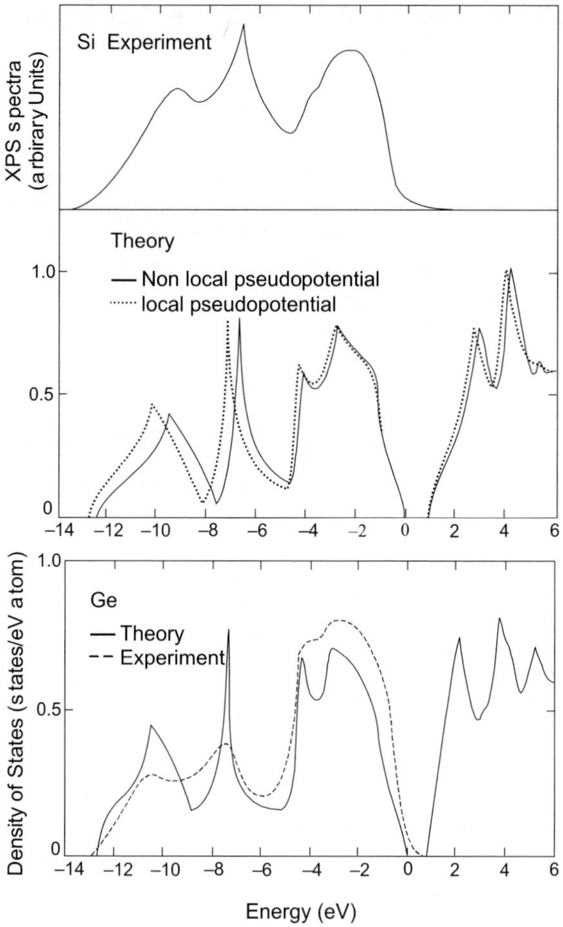

Fig. 6.9. Theoretical pseudopotential electronic valence densities of states compared with experiment for Si and Ge. Adaptation reprinted with permission from Cheliokowsky JR and Cohen ML, *Phys Rev B* **14**, 556 (1976). Copyright 1976 by the American Physical Society

6.2 Examples of Semiconductors

6.2.1 Models of Band Structure for Si, Ge and II-VI and III-V Materials (A)

First let us give some band structure and density of states for Si and Ge. See Fig. 6.8 and Fig. 6.9. The figures illustrate two points. First, that model calculation tools using the pseudopotential (see "The Pseudopotential Method" under Sect. 3.2.3) have been able to realistically model actual semiconductors. Second, that the models we often use (such as the simplified pseudopotential) are oversimplified but still useful in getting an idea about the complexities involved. As discussed by Cohen and Chelikowsky [6.8], optical properties have been very useful in obtaining experimental results about actual band structures.

For very complicated cases, models are still useful. A model by Kane has been found useful for many II-VI and III-V semiconductors [6.16]. It yields a conduction band that is not parabolic, as well as having both heavy and light holes and a split-off band as shown in Fig. 6.10. It even applies to pseudobinary alloys such as mercury cadmium telluride (MCT) provided one uses a virtual crystal approximation (VCA), in which alloy disorder later can be put in as a perturbation, e.g. to discuss mobility. In the VCA, $Hg_{1-x}Cd_xTe$ is replaced by ATe, where A is some "average" atom representing the Hg and Cd.

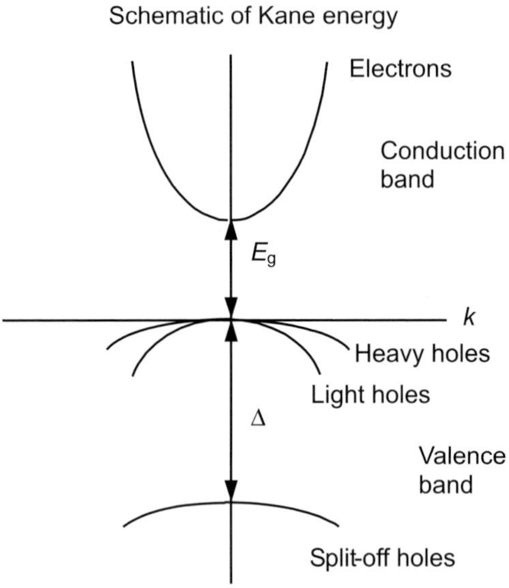

Fig. 6.10. Energy bands for zincblende lattice structure

If one solves the secular equation of the Kane [6.16] model, one finds the following equation for the conduction, light holes, and split-off band:

$$E^3 + (\Delta - E_g)E^2 - (E_g\Delta + P^2k^2)E - \frac{2}{3}\Delta P^2k^2 = 0, \tag{6.142}$$

where Δ is a constant representing the spin-orbit splitting, E_g is the bandgap, and P is a constant representing a momentum matrix element. With the energy origin chosen to be at the top of the valence band, if $\Delta >> E_g$ and Pk, and including heavy holes, one can show:

$$E = E_g + \frac{\hbar^2 k^2}{2m} + \frac{1}{2}\left(\sqrt{E_g^2 + \frac{8P^2k^2}{3}} - E_g\right) \text{ for the conduction band,} \quad (6.143)$$

$$E = -\frac{\hbar^2 k^2}{2m_{hh}}, \text{ for the heavy holes,} \quad (6.144)$$

$$E = -\frac{\hbar^2 k^2}{2m} - \frac{1}{2}\left(\sqrt{E_g^2 + \frac{8P^2k^2}{3}} - E_g\right) \text{ for the light holes, and} \quad (6.145)$$

$$E = -\Delta - \frac{\hbar^2 k^2}{2m} - \frac{P^2k^2}{3E_g + 3\Delta} \text{ for the split-off band.} \quad (6.146)$$

In the above, m is the mass of a free electron (Kane [6.16]).

Knowing the E vs. k relation, as long as E depends only on $|\boldsymbol{k}|$, the density of states per unit volume is given by

$$D(E)\mathrm{d}E = 2 \times \frac{4\pi k^2 \mathrm{d}k}{(2\pi)^3}, \quad (6.147)$$

or

$$D(E) = \frac{h^2 \mathrm{d}k}{\pi^2 \mathrm{d}E}. \quad (6.148)$$

Finally, for the conduction band, if $\hbar^2k^2/2m$ is negligible compared to the other terms, we can show for the conduction band that

$$E\left(\frac{E - E_g}{E_g}\right) = \frac{\hbar^2 k^2}{2m_1}, \quad (6.149)$$

where

$$m_1 = \frac{3\hbar^2}{4P^2} E_g. \quad (6.150)$$

This clearly leads to changes in effective mass from the parabolic case ($E \propto k^2$).

Brief properties of MCT, as an example of a II-VI alloy, [6.5, 6.7] showing its importance:

1. A pseudobinary II-VI compound with structure isomorphic to zincblende.
2. $Hg_{1-x}Cd_xTe$ forms a continuous range of solid solutions between the semimetals HgTe and CdTe. The bandgap is tunable from 0 to about 1.6 eV as x varies from about 0.15 (at low temperature) to 1.0. The bandgap also depends on temperature, increasing (approximately) linearly with temperature for a fixed value of x.
3. Useful as an infrared detector at liquid nitrogen temperature in the wavelength 8–12 micrometers, which is an atmospheric window. A higher operating temperature than alternative materials and MCT has high detectivity, fast response, high sensitivity, IC compatible and low power.
4. The band structure involves mixing of unperturbed valence and conduction band wave function, as derived by the Kane theory. They have nonparabolic bands, which makes their analysis more difficult.
5. Typical carriers have small effective mass (about 10^{-2} free-electron mass), which implies large mobility and enhances their value as IR detectors.
6. At higher temperatures (well above 77 K) the main electron scattering mechanism is the scattering by longitudinal optic modes. These modes are polar modes as discussed in Sect. 10.10. This scattering process is inelastic, and it makes the calculation of electron mobility by the Boltzmann equation more difficult (noniterated techniques for solving this equation do not work). At low temperatures the scattering may be dominated by charged impurities. See Yu and Cardona [6.44, p. 207]. See also Problem 6.7.
7. The small bandgap and relatively high concentration of carriers make it necessary to include screening in the calculation of the scattering of carriers by several interactions.
8. It is a candidate for growth in microgravity in order to make a more perfect crystal.

The figures below may further illustrate II-VI and III-V semiconductors, which have a zincblende structure. Figure 6.11 shows two interpenetrating lattices in the zincblende structure. Figure 6.12 shows the first Brillouin zone. Figure 6.13 sketches results for GaAs (which is zincblende in structure) which can be compared to Si and Ge (see Fig. 6.8). The study of complex compound semiconductors is far from complete.[4]

[4] See, e.g., Patterson [6.30].

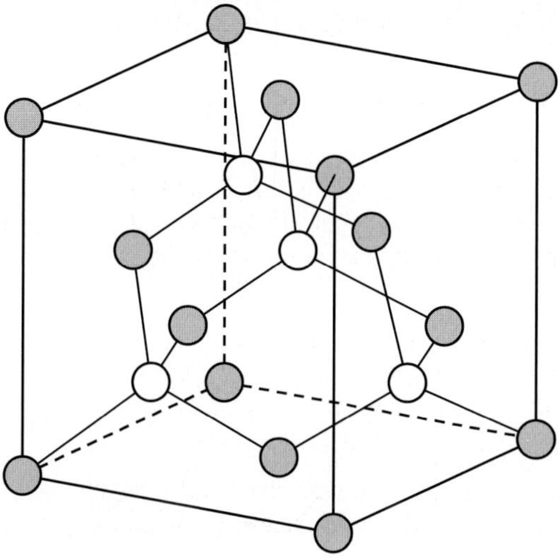

Fig. 6.11. Zincblende lattice structure. The shaded sites are occupied by one type of ion, the unshaded by another type

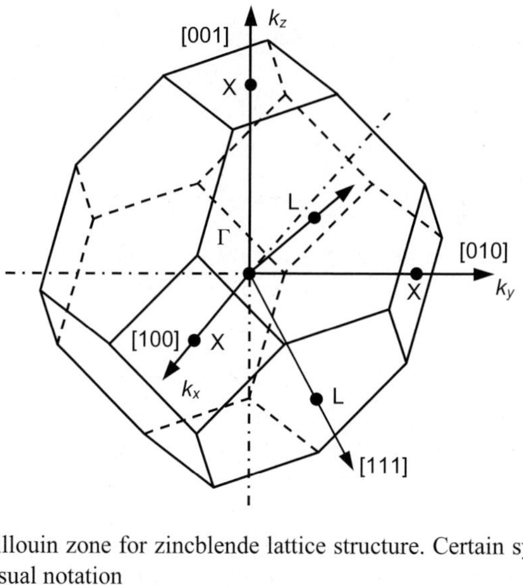

Fig. 6.12. First Brillouin zone for zincblende lattice structure. Certain symmetry points are denoted with the usual notation

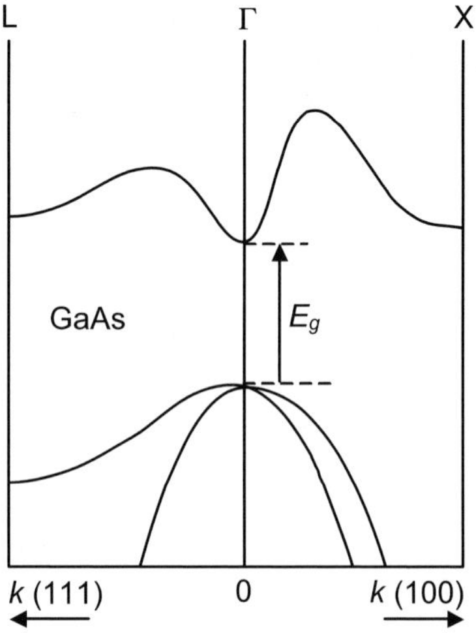

Fig. 6.13. Sketch of the band structure of GaAs in two important directions. Note that in the valence bands there are both light and heavy holes. For more details see Cohen and Chelikowsky [6.8]

Density of States for Effective Hole Masses (A)

If we have light and heavy holes with energies

$$\left|E_{l,h}\right| = \frac{\hbar^2 k^2}{2m_{lh}},$$

$$\left|E_{h,h}\right| = \frac{\hbar^2 k^2}{2m_{hh}},$$

each will give a density of states and these density of states will add so we must replace in an equation analogous to (6.131),

$$(m_h^*)^{3/2} \quad by \quad m_{lh}^{3/2} + m_{hh}^{3/2}.$$

Alternatively, the effective hole mass for density of states is given by the replacement of

$$m_h \quad \text{by} \quad (m_{lh}^{3/2} + m_{hh}^{3/2})^{2/3}.$$

6.2.2 Comments about GaN (A)

GaN is a III-V material that has been of much interest lately. It is a direct wide bandgap semiconductor (3.44 electron volts at 300 K). It has applications in blue and UV light emitters (LEDs) and detectors. It forms a heterostructure (see

Sect. 12.4) with AlGaN and thus HFETs (heterostructure field effect transistors) have been made. Transistors of both high power and high frequency have been produced with GaN. It also has good mechanical properties, and can work at higher temperature as well as having good thermal conductivity and a high breakdown field. GaN has become very important for recent advances in solid-state lighting. Studies of dopants, impurities, and defects are important for improving the light-emitting efficiency.

GaN is famous for its use in making blue lasers. See Nakamura et al [6.26], Pankove and Moustaka (eds) [6.28], and Willardson and Weber [6.43].

6.3 Semiconductor Device Physics

This Section will give only some of the flavor and some of the approximate device equations relevant to semiconductor applications. The book by Dalven [6.10] is an excellent introduction to this subject. So is the book by Fraser [6.14]. The most complete book is by Sze [6.41]. In recent years layered structures with quantum wells and other new effects are being used for semiconductor devices. See Chap. 12 and references [6.1, 6.19]

6.3.1 Crystal Growth of Semiconductors (EE, MET, MS)

The engineering of semiconductors has been as important as the science. By engineering we mean growth, purification, and controlled doping. In Chap. 12 we go a little further and talk of the band engineering of semiconductors. Here we wish to consider growth and related matters. For further details, see Streetman [6.40, p12ff]. Without the ability to grow extremely pure single crystal Si, the semiconductor industry as we know it would not have arisen. With relatively few electrons and holes, semiconductors are just too sensitive to impurities.

To obtain the desired pure crystal semiconductor, elemental Si, for example, is chemically deposited from compounds. Ingots are then poured that become polycrystalline on cooling.

Single crystals can be grown by starting with a seed crystal at one end and passing a molten zone down a "boat" containing the seed crystal (the molten zone technique), see Fig. 6.14.

Since the boat can introduce stresses (as well as impurities) an alternative method is to grow the crystal from the melt by pulling a rotating seed from it (the *Czochralski technique*), see Fig. 6.14b.

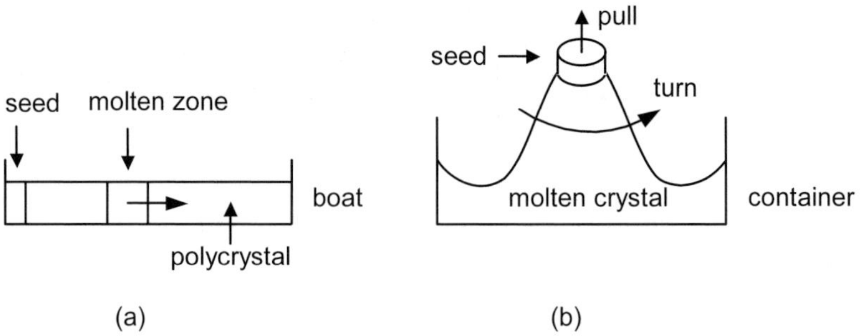

Fig. 6.14. (a) The molten zone technique for crystal growth and (b) the Czochralski Technique for crystal growth

Purification can be achieved by passing a molten zone through the crystal. This is called *zone refining*. The impurities tend to concentrate in the molten zone, and more than one pass is often useful. A variation is the *floating zone* technique where the crystal is held vertically and no walls are used.

There are other crystal growth techniques. *Liquid phase epitaxy* and *vapor phase epitaxy*, where crystals are grown below their melting point, are discussed by Streetman (see reference above). We discuss *molecular beam epitaxy*, important in molecular engineering, in Chap. 12.

In order to make a semiconductor device, initial purity and controlled introduction of impurities is necessary. Diffusion at high temperatures is often used to dope or introduce impurities. An alternative process is ion implantation that can be done at low temperature, producing well-defined doping layers. However, lattice damage may result, see Streetman [6.40, p128ff], but this can often be removed by annealing.

6.3.2 Gunn Effect (EE)

The Gunn effect is the generation of microwave oscillations in a semiconductor like GaAs or InP (or other III-V materials) due to a high (of order several thousand V/cm) electric field. The effect arises due to the energy band structure sketched in Fig. 6.15.

Since $m^* \propto (\mathrm{d}^2E/\mathrm{d}k^2)^{-1}$, we see $m_2^* > m_1^*$, or m_2 is heavy compared to m_1. The applied electric field can supply energy to the electrons and raise them from the m_1^* (where they would tend to be) part of the band to the m_2^* part. With their gain in mass, it is possible for the electrons to experience a drop in drift velocity (mobility = $v/E \propto 1/m^*$).

If we make a plot of drift velocity *versus* electric field, we get something like Fig. 6.16. The differential conductivity is

$$\sigma_d = \frac{\mathrm{d}J}{\mathrm{d}E}, \tag{6.151}$$

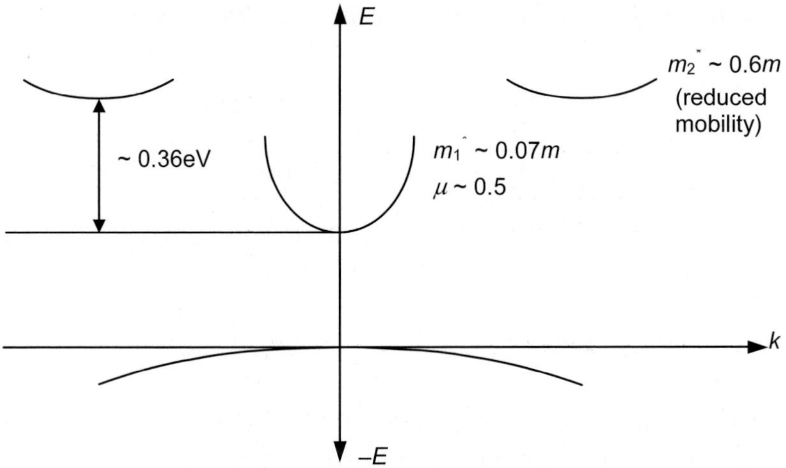

Fig. 6.15. Schematic of energy band structure for GaAs used for Gunn effect

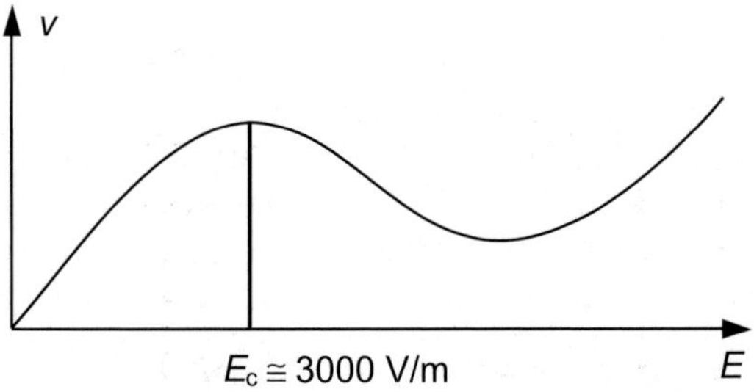

Fig. 6.16. Schematic of electron drift velocity vs. electric field in GaAs

where J is the electrical current density that for electrons we can write as $J = nev$, where $v = |\mathbf{v}|$, $e > 0$. Thus,

$$\sigma_d = ne\frac{\mathrm{d}v}{\mathrm{d}E} < 0, \tag{6.152}$$

when $E > E_c$ and is not too large. This is the region of bulk negative conductivity (BNC), and it is unstable and leads to the Gunn effect. The generation of Gunn microwave oscillations may be summarized by the following three statements:

1. Because the electrons gain energy from the electric field, they transfer to a region of E(**k**) space where they have higher masses. There, they slow down, "pile up", and form space-charge domains that move with an overall drift velocity v.

2. We assume the length of the sample is l. A current pulse is delivered for every domain transit.

3. Because of reduction of the electric field external to the domain, once a domain is formed, another is not formed until the first domain drifts across.

The frequency of the oscillation is approximately

$$f = \frac{v}{l} \approx \frac{10^7\,\text{m/s}}{10^{-3}\,\text{m}} \approx 10\text{GHz}\,. \tag{6.153}$$

The instability with respect to charge domain-foundation can be simply argued. In one dimension from the continuity equation and Gauss' law, we have

$$\frac{\partial J}{\partial x} + \frac{\partial \rho}{\partial t} = 0\,, \tag{6.154}$$

$$\frac{\partial E}{\partial x} = \frac{\rho}{\varepsilon}\,, \tag{6.155}$$

$$\frac{\partial J}{\partial x} = \frac{\partial J}{\partial E} \cdot \frac{\partial E}{\partial x} = \sigma_d \frac{\rho}{\varepsilon}\,. \tag{6.156}$$

So,

$$\frac{\partial \rho}{\partial \tau} = -\frac{\partial J}{\partial x} = -\sigma_d \frac{\rho}{\varepsilon}\,, \tag{6.157}$$

or

$$\rho = \rho(0) \exp\left(-\frac{\sigma_d}{\varepsilon} t \right). \tag{6.158}$$

If $\sigma_d < 0$, and there is a random charge fluctuation, then ρ is unstable with respect to growth. A major application of Gunn oscillations is in RADAR.

We should mention that GaN (see Sect. 6.2.2) is being developed for high-power and high-frequency (~ 750 GHz) Gunn diodes.

6.3.3 *pn*-Junctions (EE)

The *pn* junction is fundamental for constructing transistors and many other important applications. We assume a linear junction, which is abrupt, with acceptor doping for $x < 0$ and donor doping for $x > 0$ as in Fig. 6.17. Of course, this is an approximation. No doping profile is absolutely sharp. In some cases a graded junction (discussed later) may be a better approximation. We now develop approximately valid results concerning the *pn* junction. We use simple principles and develop what we call device equations.

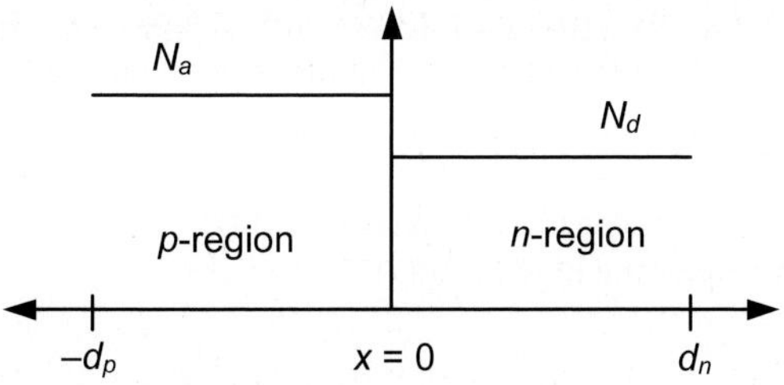

Fig. 6.17. Model of doping profile of abrupt *pn* junction

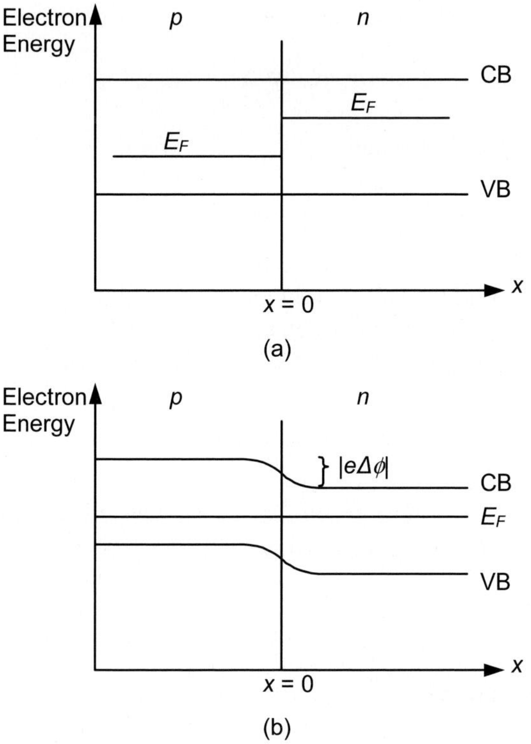

Fig. 6.18. The *pn* junction: (a) Hypothetical junction just after doping but before equilibrium (i.e. before electrons and holes are transferred). (b) *pn* junction in equilibrium. CB = conduction band, VB = valence band

For $x < -d_p$ we assume $p = N_a$ and for $x > +d_n$ we assume $p = N_d$, i.e. exhaustion in both cases. Near the junction at $x = 0$, holes will tend to diffuse into the $x > 0$ region and electrons will tend to diffuse into the $x < 0$ region. This will cause a built-in potential that will be higher on the n-side ($x > 0$) than the p-side ($x < 0$). The potential will increase until it is of sufficient size to stop the net diffusion of electrons to the p-side and holes to the n-side. See Fig. 6.18. The region between

$-d_p$ and d_n is called the depletion region. We further make the depletion layer approximation that assumes there are negligible free carriers in this depletion region. We assume this occurs because the large electric field in the region quickly sweeps any free carriers across it. It is fairly easy to calculate the built-in potential from the fact that the net hole (or electron) current is zero.

Consider, for example, the hole current:

$$J_p = e\left(p\mu_p E - D_p \frac{\mathrm{d}p}{\mathrm{d}x} \right) = 0\,. \tag{6.159}$$

The electric field is related to the potential by $E = -\mathrm{d}\varphi/\mathrm{d}x$, and using the Einstein relation, $D_p = \mu_p kT/e$, we find

$$-\frac{e}{kT}\mathrm{d}\varphi = \frac{\mathrm{d}p}{p}\,. \tag{6.160}$$

Integrating from $-d_p$ to d_n, we find

$$\frac{p_{p0}}{p_{n0}} = \exp\left(\frac{e}{kT}(\varphi_n - \varphi_p) \right), \tag{6.161}$$

where p_{p0} and p_{n0} mean the hole concentrations located in the homogeneous part of the semiconductor beyond the depletion region. The Law of Mass Action tells us that $np = n_i^2$, and we know that $p_{p0} = N_a$, $n_{n0} = N_d$, and $n_{n0}p_{n0} = n_i^2$; so

$$p_{n_0} = n_i^2 / N_d\,. \tag{6.162}$$

Thus, we find

$$e(\varphi_n - \varphi_p) = kT\ln\left(\frac{N_a N_d}{n_i^2} \right), \tag{6.163}$$

for the built-in potential. The same built-in potential results from the constancy of the chemical potential. We will leave this as a problem.

We obtain the width of the depletion region by solving Gauss's law for this region. We have assumed negligible carriers in the depletion region $-d_p$ to d_n:

$$\frac{\mathrm{d}E}{\mathrm{d}x} = -\frac{eN_a}{\varepsilon} \quad \text{for} \quad -d_p \le x \le 0\,, \tag{6.164}$$

and

$$\frac{\mathrm{d}E}{\mathrm{d}x} = +\frac{eN_d}{\varepsilon} \quad \text{for} \quad 0 \le x \le d_n\,. \tag{6.165}$$

Integrating and using $E = 0$ at both edges of the depletion region

$$E = -\frac{eN_a}{\varepsilon}(x + d_p) \quad \text{for} \quad -d_p \le x \le 0\,, \tag{6.166}$$

$$E = +\frac{eN_d}{\varepsilon}(x - d_n) \quad \text{for} \quad 0 \le x \le d_n\,. \tag{6.167}$$

Since E must be continuous at $x = 0$, we find

$$N_a d_p = N_d d_n\,, \tag{6.168}$$

which is just an expression of charge neutrality. Using $E = -\mathrm{d}\varphi/\mathrm{d}x$, integrating these equations one more time, and using the fact that φ is continuous at $x = 0$, we find

$$\Delta\varphi = \varphi(d_n) - \varphi(-d_p) = \frac{e}{2\varepsilon}\left[N_d d_n^2 + N_a d_p^2\right]. \tag{6.169}$$

Using the electrical neutrality condition, $N_a d_p = N_d d_n$, we find

$$d_p = \sqrt{\Delta\varphi\left(\frac{2\varepsilon}{eN_a}\right)\left(\frac{N_d}{N_a + N_d}\right)}\,, \tag{6.170}$$

$$d_n = \sqrt{\Delta\varphi\left(\frac{2\varepsilon}{eN_d}\right)\left(\frac{N_a}{N_d + N_a}\right)}\,, \tag{6.171}$$

and the width of the depletion region is $W = d_p + d_n$. Notice d_p increases as N_a decreases, as would be expected from electrical neutrality. Similar comments about d_n and N_d may be made.

6.3.4 Depletion Width, Varactors, and Graded Junctions (EE)

From the previous results, we can show for the depletion width at an abrupt *pn*-junction

$$W = \sqrt{\frac{2\varepsilon\Delta\varphi}{e}\left(\frac{N_a + N_d}{N_a N_d}\right)}\,. \tag{6.172}$$

Also,

$$d_n = \left(\frac{N_a}{N_d + N_a} \right) W \,, \tag{6.173}$$

$$d_p = \left(\frac{N_d}{N_d + N_a} \right) W \,. \tag{6.174}$$

If we add a bias voltage φ_b selected so $\varphi_b > 0$ when a positive bias is applied on the p-side, then

$$W = \sqrt{\frac{2\varepsilon(\Delta\varphi - \varphi_b)}{e} \left(\frac{N_a + N_d}{N_a N_d} \right)} \,. \tag{6.175}$$

For noninfinite current, $\Delta\varphi > \varphi_b$.

The charge associated with the space charge on the p-side is $Q = eAd_pN_a$, where A is the cross-sectional area of the pn-junction. We find

$$Q = A\sqrt{2e\varepsilon(\Delta p - \varphi_b)\frac{N_a N_d}{N_a + N_d}} \,. \tag{6.176}$$

The junction capacitance is then defined as

$$C_J = \left| \frac{\mathrm{d}Q}{\mathrm{d}\varphi_b} \right| , \tag{6.177}$$

which, perhaps, not surprisingly comes out

$$C_J = \frac{\varepsilon A}{W} , \tag{6.178}$$

just like a parallel-plate capacitor. Note that C_J depends on the voltage through W. When the pn-junction is used in such a way as to make use of the voltage dependence of C_J, the resulting device is called a *varactor*. A varactor is useful when it is desired to vary the capacitance electronically rather than mechanically.

To introduce another kind of pn-junction, and to see how this affects the concept of a varactor, let us consider the graded junction. Any simple model of a junction only approximately describes reality. This is true for both abrupt and graded junctions. The abrupt model may approximate an alloyed junction. When the junction is formed by diffusion, it may be better described by a graded junction. For a graded junction, we assume

$$N_d - N_a = Gx \,, \tag{6.179}$$

which is p-type for $x < 0$ and n-type for $x > 0$. Note the variation is now smooth rather than abrupt. We assume, as before, that within the transition region we have complete ionization of impurities and that carriers there can be neglected in terms of their effect on net charge. Gauss' law becomes

$$\frac{\mathrm{d}E}{\mathrm{d}z} = \frac{e}{\varepsilon}(N_d - N_a) = \frac{eGx}{\varepsilon}. \tag{6.180}$$

Integrating

$$E = \frac{eG}{2\varepsilon}x^2 + k. \tag{6.181}$$

The doping is symmetrical, so the electric field should vanish at the same distance on either side from $x = 0$. Therefore,

$$d_p = d_n = \frac{W}{2}, \tag{6.182}$$

and

$$E = \frac{eG}{2\varepsilon}\left[x^2 - \left(\frac{W}{2}\right)^2\right]. \tag{6.183}$$

Integrating

$$\varphi(z) = -\frac{eG}{2\varepsilon}\left[\frac{x^3}{3} - \left(\frac{W}{2}\right)^2 x\right] + k_2. \tag{6.184}$$

Thus,

$$\Delta\varphi = \varphi\left(\frac{W}{2}\right) - \varphi\left(\frac{-W}{2}\right) = \frac{W^3}{12}\left(\frac{eG}{\varepsilon}\right), \tag{6.185}$$

or

$$W = \left(\frac{12\varepsilon}{eG}\Delta\varphi\right)^{1/3}. \tag{6.186}$$

With an applied voltage, this becomes

$$W = \left[\frac{12\varepsilon}{eG}(\Delta\varphi - \varphi_b)\right]^{1/3}. \tag{6.187}$$

The charge associated with the right dipole layer is

$$Q = \int_0^{W/2} eGxA\mathrm{d}x = \frac{eGW^2}{8}A. \tag{6.188}$$

The junction capacitance therefore is

$$C_J = \left|\frac{\mathrm{d}Q}{\mathrm{d}\varphi_b}\right| = \left|\frac{\mathrm{d}Q}{\mathrm{d}W}\right|\left|\frac{\mathrm{d}W}{\mathrm{d}\varphi_b}\right|, \tag{6.189}$$

which, finally, gives again

$$C_J = \frac{A\varepsilon}{W}.$$

But, now W depends on φ_b in a 1/3 power way rather than a 1/2 power. Different approximate models lead to different approximate device equations.

6.3.5 Metal Semiconductor Junctions — the Schottky Barrier (EE)

We consider the situation shown in Fig. 6.19 where an n-type semiconductor is in contact with the metal. Before contact we assume the Fermi level of the semiconductor is above the Fermi level of the metal. After contact electrons flow from the semiconductor to the metal and the Fermi levels equalize. The work functions Φ_m, Φ_s are defined in Fig. 6.19. We assume $\Phi_m > \Phi_s$. If $\Phi_m < \Phi_s$ an ohmic contact with a much smaller barrier is formed (Streetman [6.40, p185ff]). The internal electric fields cause a varying potential and hence band bending as shown. The concept of band bending requires the semiclassical approximation (Sect. 6.1.4). Let us analyze this in a bit more detail. Choose $x > 0$ in the semiconductor and $x < 0$ in the metal. We assume the depletion layer has width x_b. For $x_b > x > 0$, Gauss' equation is

$$\frac{\mathrm{d}E}{\mathrm{d}x} = \frac{N_d e}{\varepsilon}. \tag{6.190}$$

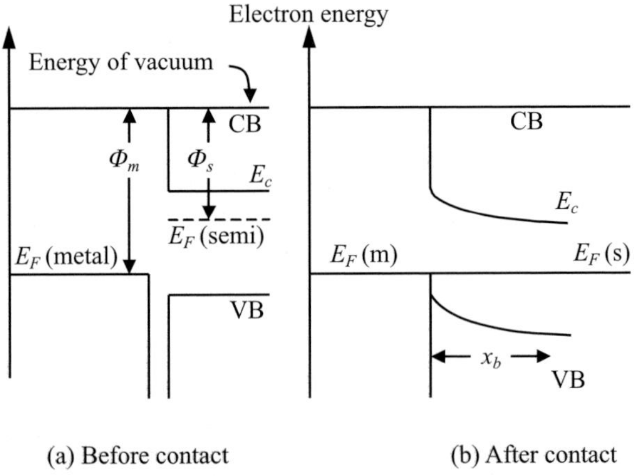

Fig. 6.19. Schottky barrier formation (sketch)

Using $E = -\mathrm{d}\varphi/\mathrm{d}x$, setting the potential at 0 and x_b equal to φ_0 and φ_{xb}, and requiring the electric field to vanish at $x = x_b$, by integrating the above for φ we find

$$\varphi_0 - \varphi_{x_b} = -\frac{N_d e x_b^2}{2\varepsilon} . \tag{6.191}$$

If the potential energy difference for electrons across the barrier is

$$\Delta V = -e(\varphi_0 - \varphi_{z_b}) ,$$

we know

$$\Delta V = +E_F(s) - E_F(m) \quad \text{(before contact).} \tag{6.192}$$

Solving the above for x_b gives the width of the depletion layer as

$$x_b = \sqrt{\frac{2\varepsilon\Delta V}{N_d e^2}} . \tag{6.193}$$

Shottky barrier diodes have been used as high-voltage rectifiers. The behavior of these diodes can be complicated by "dangling bonds" where the rough semiconductor surface joins the metal. See Bardeen [6.3].

6.3.6 Semiconductor Surface States and Passivation (EE)

The subject of passivation is complex, and we will only make brief comments. The most familiar passivation layer is SiO_2 over Si, which reduces the number of surface states. A mixed layer of GaAs-AlAs on GaAs is also a passivating layer that reduces the number of surface states. The ease of passivation of the Si surface by oxygen is a major reason it is the dominant semiconductor for device usage.

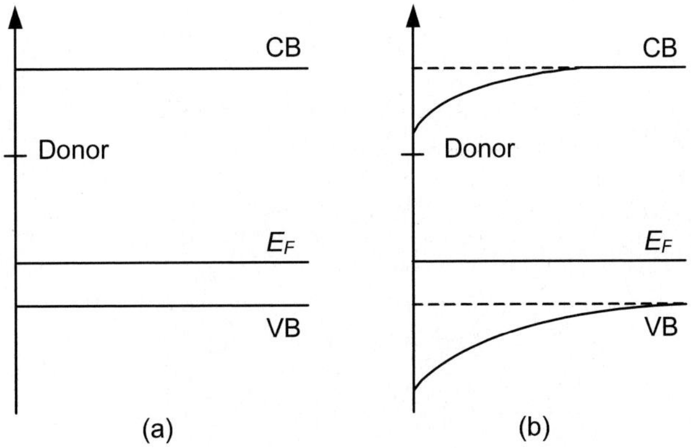

Fig. 6.20. *p*-type semiconductor with donor surface states (a) before equilibrium, (b) after equilibrium ($T = 0$). In both (a) and (b) only relative energies are sketched

What are surface states? A solid surface is a solid terminated at a two-dimensional surface. The effect on charge carriers is modeled by using a surface potential barrier. This can cause surface states with energy levels in the forbidden gap. The name "surface states" is used because the corresponding wave function is localized near the surface. Further comments about surface states are found in Chap. 11.

Surface states can have interesting effects, which we will illustrate with an example. Let us consider a p-type semiconductor (bulk) with surface states that are donors. The situation before and after equilibrium is shown in Fig. 6.20. For the equilibrium case (b), we assume that all donor states have given up their electrons, and hence, are positively charged. Thus, the Fermi energy is less than the donor-level energy. A particularly interesting case occurs when the Fermi level is *pinned* at the surface donor level. This occurs when there are so many donor states on the surface that not all of them can be ionized. In that case (b), the Fermi level would be drawn on the same level as the donor level.

One can calculate the amount of band bending by a straightforward calculation. The band bending is caused by the electrons flowing from the donor states at the surface to the acceptor states in the bulk. For the depletion region, we assume,

$$\rho(x) = -eN_a \tag{6.194}$$

$$\frac{\mathrm{d}E}{\mathrm{d}x} = \frac{-eN_a}{\varepsilon} . \tag{6.195}$$

So,

$$\frac{\mathrm{d}^2V}{\mathrm{d}x^2} = \frac{eN_a}{\varepsilon} . \tag{6.196}$$

If n_d is the number of donors per unit area, the surface charge density is $\sigma = en_d$. The boundary condition at the surface is then

$$E_{\text{surface}} = -\left.\frac{\mathrm{d}V}{\mathrm{d}x}\right|_{x=0} = \frac{en_d}{\varepsilon} . \tag{6.197}$$

If the width of the depletion layer is d, then

$$E(x = d) = 0 . \tag{6.198}$$

Integrating (6.196) with boundary condition (6.198) gives

$$E = \frac{eN_a}{\varepsilon}(d - x) . \tag{6.199}$$

Using the boundary condition (6.197), we find

$$d = \frac{n_d}{N_a} . \tag{6.200}$$

Integrating a second time, we find

$$V = \frac{eN_a}{2\varepsilon}x^2 - \frac{eN_a d}{\varepsilon}x + \text{constant}\,. \tag{6.201}$$

Therefore, the total amount of band bending is

$$e[V(0) - V(d)] = \frac{e^2 N_a d^2}{2\varepsilon} = \frac{e^2 n_d^2}{2\varepsilon N_a}\,. \tag{6.202}$$

This band bending is caused entirely by the assumed ionized donor surface states. We have already mentioned that surface states can complicate the analysis of metal-semiconductor junctions.

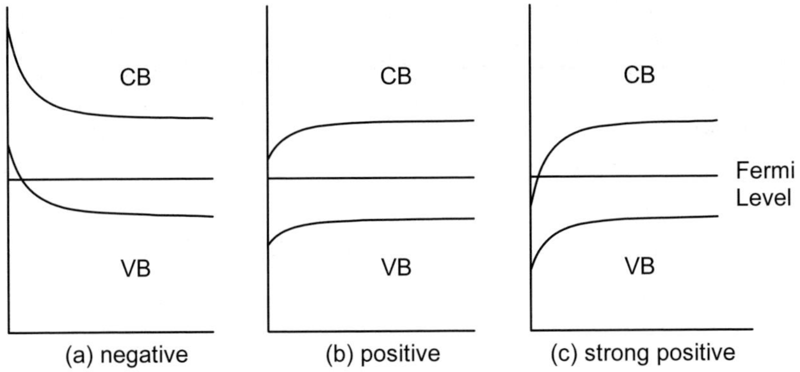

Fig. 6.21. *p*-type semiconductor under bias voltage (energies in each figure are relative)

6.3.7 Surfaces Under Bias Voltage (EE)

Let us consider a *p*-type surface under three kinds of voltage shown in Fig. 6.21: (a) a negative bias voltage, (b) a positive bias voltage, and then (c) a very strong, positive bias voltage.

In case (a), the bands bend upward, holes are attracted to the surface, and thus, an accumulation layer of holes is founded. In (b), holes are repelled from the surface forming the depletion layer. In (c) the bands are bent sufficiently such that the conduction band bottom is below the Fermi energy and the semiconductor becomes *n*-type, forming an inversion region. In all these cases, we are essentially considering a capacitor with the semiconductor forming one plate. These ideas have been further developed into the MOSFET (metal-oxide semiconductor field-effect transistor, see Sect. 6.3.10).

6.3.8 Inhomogeneous Semiconductors Not in Equilibrium (EE)

Here we will discuss *pn*-junctions under bias and how this leads to electron and hole injection. We will start with a qualitative treatment and then do a more quantitative analysis. The study of *pn*-junctions is fundamental for the study of transistors.

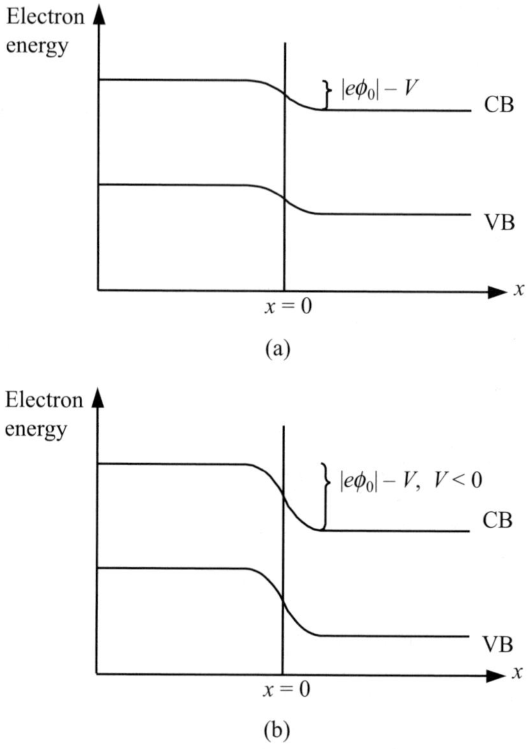

Fig. 6.22. The *pn*-junction under bias V: (a) Forward bias, (b) Reverse bias. (Only relative shift is shown)

We start by looking at a *pn*-junction in equilibrium where there are two types of electron flow that balance in equilibrium (as well as two types of hole flow which also balance in equilibrium). See also, e.g., Kittel [6.17, p. 572] or Ashcroft and Mermin [6.2, p. 600].

From the *n*-side to the *p*-side, there is an electron recombination (r) or diffusion current (J_{nr}) where n denotes electrons. This is due to the majority carrier electrons, which have enough energy to surmount the potential barrier. This current is very sensitive to a bias field that would change the potential barrier. On the *p*-side, there are thermally generated electrons, which in the space-charge region may be swiftly swept downhill into the *n*-region. This causes the thermal generation (g) or drift current (J_{ng}). Electrons produced farther than a *diffusion length* (to be defined) recombine before being swept across. As mentioned, in the absence of potential, the electron currents balance and we have

$$J_{nr}(0) + J_{ng}(0) = 0, \tag{6.203}$$

where the 0 in $J_{nr}(0)$, etc. means zero bias voltage. Similarly, for holes, denoted by p,

$$J_{pr}(0) + J_{pg}(0) = 0. \tag{6.204}$$

We set the notation that forward bias ($V > 0$) is when the p-side is higher in potential than the n-side. See Fig. 6.22. Since the barrier responds exponentially to the bias voltage, we might expect the electron *injection current*, from n to p, to be given by

$$J_{nr}(V) = J_{nr}(0)\exp\left(\frac{eV}{kT}\right). \tag{6.205}$$

The thermal generation current is essentially independent of voltage so

$$J_{ng}(V) = J_{ng}(0) = -J_{nr}(0). \tag{6.206}$$

Similarly, for injection of holes from p to n, we expect

$$J_{pr}(V) = J_{pr}(0)\exp\left(\frac{eV}{kT}\right), \tag{6.207}$$

and similarly for the generation current,

$$J_{pg}(V) = J_{pg}(0) = -J_{pr}(0). \tag{6.208}$$

Adding everything up, we get the Shockley diode equation for a pn-junction under bias

$$\begin{aligned} J &= J_{nr}(V) + J_{ng}(V) + J_{pr}(V) + J_{pg}(V) \\ &= J_0[\exp(eV/kT) - 1] \end{aligned} \tag{6.209}$$

where $J_0 = J_{nr}(0) + J_{pr}(0)$.

We now give a more detailed derivation, in which the exponential term is more carefully argued, and J_0 is calculated. We assume that both electrons and holes recombine (due to various processes) with characteristic recombination times τ_n and τ_p. The usual assumption is, that as far as net recombination goes *with no flow*,

$$\left.\frac{\partial p}{\partial \tau}\right)_r = -\frac{p - p_0}{\tau_p}, \tag{6.210}$$

and

$$\left.\frac{\partial n}{\partial \tau}\right)_r = -\frac{n - n_0}{\tau_n}, \tag{6.211}$$

where r denotes recombination. Assuming no external generation of electrons or holes, the continuity equation with *flow and recombination* can be written (in one dimension):

$$\frac{\partial J_p}{\partial x} + e\frac{\partial p}{\partial \tau} = -e\left(\frac{p - p_0}{\tau_p}\right), \tag{6.212}$$

$$\frac{\partial J_n}{\partial x} - e\frac{\partial n}{\partial \tau} = +e\left(\frac{n - n_0}{\tau_n}\right). \tag{6.213}$$

The electron and hole current densities are given by

$$J_p = -eD_p\frac{\partial p}{\partial x} + ep\mu_p E, \tag{6.214}$$

$$J_n = eD_n\frac{\partial n}{\partial x} + en\mu_n E. \tag{6.215}$$

And, as always, we assume Gauss' law, where ρ is the total charge density

$$\frac{\partial E}{\partial x} = \frac{\rho}{\varepsilon}. \tag{6.216}$$

We will also assume a steady state, so

$$\frac{\partial p}{\partial t} = \frac{\partial n}{\partial t} = 0. \tag{6.217}$$

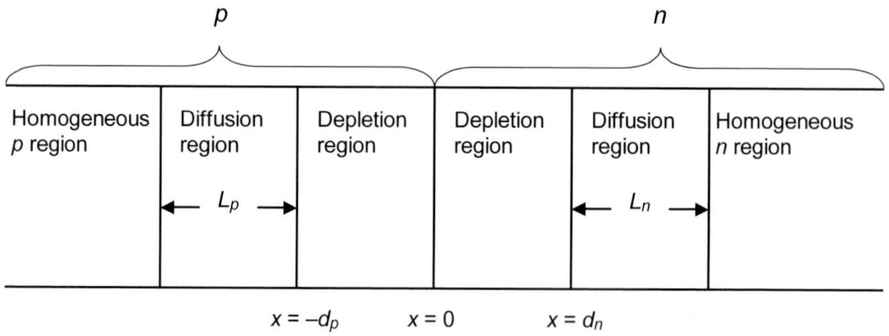

Fig. 6.23. Schematic of *pn*-junction (p region for $x < 0$ and n region for $x > 0$). L_n and L_p are n and p diffusion lengths

An explicit solution is fairly easy to obtain if we make three further assumptions (See Fig. 6.23):

(a) The electric field is very small outside the depletion region, so whatever drop in potential there is occurs across the depletion region.

(b) The concentrations of injected minority carriers in the region outside the depletion region is negligible compared to the majority carrier concentration. Also, the majority carrier concentration is essentially constant beyond the depletion and diffusion regions.

(c) Finally, we assume negligible generation or recombination of carriers in the depletion region. We can argue that this ought to be a good approximation if the depletion layer is sufficiently thin. Under this approximation, the electron and hole currents are constant across the depletion region.

A few further comments are necessary before we analyze the pn-junction. In the depletion region there are both drift and diffusion currents that are large. In the nonequilibrium case they do not quite cancel. Consistent with this the electric fields, gradient of carrier densities and space charge are all large. Electric fields can be so large here as to lead to the validity of the semiclassical model being open to question. However, we are only trying to develop approximate device equations so our approximations are probably OK.

The diffusion region only exists under applied voltage. The minority drift current is negligible here but the gradient of carrier densities can still be appreciable as can the drift current even though electric fields and space charges are small. The majority drift current is not small as the majority density is large.

In the homogeneous region the whole current is carried by drift and both diffusion currents are negligible. The carrier densities are nearly the same as in equilibrium, but the electric field, space charge, and gradient of carrier densities are all small.

For any x (the direction along the *pn*-junction, see Fig. 6.23), the total current should be given by

$$J_{\text{total}} = J_n(x) + J_p(x) \,. \tag{6.218}$$

Since by (c) both J_n and J_p are independent of x in the depletion region, we can evaluate them for the x that is most convenient, see Fig. 6.23,

$$J_{\text{total}} = J_n(-d_p) + J_p(d_n) \,. \tag{6.219}$$

That is, we need to evaluate only minority current densities. Also, since by (a) and (b), the minority current drift densities are negligible, we can write

$$J_{\text{total}} = eD_n \left.\frac{\partial n}{\partial x}\right|_{x=-d_p} - eD_p \left.\frac{\partial p}{\partial x}\right|_{x=-d_n} , \tag{6.220}$$

which means we only need to find the minority carrier concentrations. In the steady state, neglecting carrier drift currents, we have

$$\frac{\mathrm{d}^2 p_n}{\mathrm{d}x^2} - \frac{p_n - p_{n0}}{L_p^2} = 0\,,\ \text{for } x \geq d_n\,, \tag{6.221}$$

and

$$\frac{\mathrm{d}^2 n_p}{\mathrm{d}x^2} - \frac{n_p - n_{p0}}{L_n^2} = 0\,,\ \text{for } x \leq -d_p\,, \tag{6.222}$$

where the diffusion lengths are defined by

$$L_p^2 = D_p \tau_p\,, \tag{6.223}$$

and

$$L_n^2 = D_n \tau_n\,. \tag{6.224}$$

Diffusion lengths measure the distance a carrier goes before recombining. The solutions obeying appropriate boundary conditions can be written

$$p_n(x) - p_{n_0} = [p_n(d_n) - p_{n_0}]\exp\left(-\frac{(x-d_n)}{L_p}\right), \tag{6.225}$$

and

$$n_p(x) - n_{p_0} = [n_p(-dp) - n_{p_0}]\exp\left(+\frac{(x+d_p)}{L_n}\right). \tag{6.226}$$

Thus,

$$-\left.\frac{\partial p_n}{\partial x}\right|_{x=d_n} = \frac{[p_n(d_n) - p_{n_0}]}{L_p}, \tag{6.227}$$

and

$$+\left.\frac{\partial n_p}{\partial x}\right|_{x=-d_p} = \frac{[n_p(-d_p) - n_{p_0}]}{L_n}. \tag{6.228}$$

Thus,

$$J_{\text{total}} = \left(\frac{eD_n}{L_n}\right)[n_p(-d_p) - n_{p_0}] + \left(\frac{eD_p}{L_p}\right)[p_n(d_n) - p_{n_0}]\,. \tag{6.229}$$

To finish the calculation, we need expressions for $n_p(-d_p) - n_{p0}$ and $p_n(-d_n) - p_{n0}$, which are determined by the injected minority carrier densities.

Across the depletion region, even with applied bias, J_n and J_p are very small compared to individual drift and diffusion currents of electrons and holes (which nearly cancel). Therefore, we can assume $J_n \cong 0$ and $J_p \cong 0$ across the depletion regions. Using the Einstein relations, as well as the definition of drift and diffusion currents, we have

$$kT \frac{\partial n}{\partial x} = en \frac{\partial \varphi}{\partial x}, \tag{6.230}$$

and

$$kT \frac{\partial p}{\partial x} = -ep \frac{\partial \varphi}{\partial x}. \tag{6.231}$$

Integrating across the depletion region

$$\frac{n(d_n)}{n(-d_p)} = \exp\left(+ \frac{e}{kT} [\varphi(d_n) - \varphi(-d_p)] \right), \tag{6.232}$$

and

$$\frac{p(d_n)}{p(-d_p)} = \exp\left(- \frac{e}{kT} [\varphi(d_n) - \varphi(-d_p)] \right). \tag{6.233}$$

If $\Delta\varphi$ is the built-in potential and φ_b is the bias voltage with the conventional sign

$$\varphi(d_n) - \varphi(-d_p) = \Delta\varphi - \varphi_b. \tag{6.234}$$

Thus,

$$\frac{n(d_n)}{n(-d_p)} = \exp\left(\frac{e\Delta\varphi}{kT} \right) \exp\left(- \frac{e\varphi_b}{kT} \right) = \frac{n_{n0}}{n_{p0}} \exp\left(- \frac{e\varphi_b}{kT} \right), \tag{6.235}$$

and

$$\frac{p(d_n)}{p(-d_p)} = \exp\left(- \frac{e\Delta\varphi}{kT} \right) \exp\left(\frac{e\varphi_b}{kT} \right) = \frac{p_{n0}}{p_{p0}} \exp\left(- \frac{e\varphi_b}{kT} \right). \tag{6.236}$$

By assumption (b)

$$n(d_n) \cong n_{n0}, \tag{6.237}$$

and

$$p(-d_p) \cong p_{p0}. \tag{6.238}$$

So, we find

$$n_p(-d_p) = n_{p_0} \exp\left(\frac{e\varphi_b}{kT}\right), \tag{6.239}$$

and

$$p_n(d_n) = p_{n_0} \exp\left(\frac{e\varphi_b}{kT}\right). \tag{6.240}$$

Substituting, we can find the total current, as given by the Shockley diode equation

$$J_{\text{total}} = e\left(\frac{D_n}{L_n} n_{p_0} + \frac{D_p}{L_p} p_{n_0}\right)\left[\exp\left(\frac{e\varphi_b}{kT}\right) - 1\right]. \tag{6.241}$$

Reverse Bias Breakdown (EE)

The Shockley diode equation indicates that the current attains a constant value of $-J_0$ when the reverse bias is sufficiently strong. Actually, under large reverse bias, the Shockley diode equation is no longer valid and the current becomes arbitrarily large and negative. There are two mechanisms for this reverse current breakdown, as we discuss below (which may or may not destroy the device).

One is called the Zener breakdown. This is due to quantum-mechanical interband tunneling and involves a breakdown of the quasiclassical approximation. It can occur at lower voltages in narrow junctions with high doping. At higher voltages, another mechanism for reverse bias breakdown is dominant. This is the avalanche mechanism. The electric field in the junction accelerates electrons in the electric field. When the electron gains kinetic energy equal to the gap energy, then the electron can create an electron–hole pair ($e^- \to e^- + e^- + h$). If the sample is wide enough to allow further accelerations and/or if the electrons themselves retain sufficient energy, then further electron–hole pairs can form, etc. Since a very narrow junction is required for tunneling, avalanching is usually the mode by which reverse bias breakdown occurs.

6.3.9 Solar Cells (EE)

One of the most important applications of *pn*-junctions is for obtaining energy of the sun. Compare, e.g., Sze, [6.42, p. 473]. The photovoltaic effect is the appearance of a forward voltage across an illuminated junction. By use of the photovoltaic effect, the energy of the sun, as received at the earth, can be converted directly into electrical power. When the light is absorbed, mobile electron–hole pairs are created, and they may diffuse to the *pn*-junction region if they are created nearby (within a diffusion length). Once in this region, the large built-in electric field acts on electrons on the *p*-side, and holes on the *n*-side to produce a voltage that drives a current in the external circuit.

The first practical solar cell was developed at Bell Labs in 1954 (by Daryl M. Chapin, Calvin S. Fuller, and Gerald L. Pearson). A photovoltaic cell converts sunlight directly into electrical energy. An antireflective coating is used to maximize energy transfer. The surface of the earth receives about 1000 W/m^2 from the sun. More specifically, AM0 (air mass zero) has 1367 W/m^2, while AM1 (directly overhead through atmosphere without clouds) is 1000 W/m^2. Solar cells are used in spacecraft as well as in certain remote terrestrial regions where an economical power grid is not available.

If $P_{\rm M}$ is the maximum power produced by the solar cell and $P_{\rm I}$ is the incident solar power, the efficiency is

$$E = 100 \frac{P_{\rm M}}{P_{\rm I}} \% . \tag{6.242}$$

A typical efficiency is of order 10%. Efficiencies are limited because photons with energy less than the bandgap energy do not create electron–hole pairs and so, cannot contribute to the output power. On the other hand, photons with energy much greater than the bandgap energy tend to produce carriers that dissipate much of their energy by heat generation. For maximum efficiency, the bandgap energy needs to be just less than the energy of the peak of the solar energy distribution. It turns out that GaAs with $E \cong 1.4$ eV tends to fit the bill fairly well. In principle, GaAs can produce an efficiency of 20% or so.

The GaAs cell is covered by a thin epitaxial layer of mixed GaAs-AlAs that has a good lattice match with the GaAs and that has a large energy gap thus being transparent to sunlight. The purpose of this over-layer is to reduce the number of surface states (and, hence, the surface recombination velocity) at the GaAs surface. Since GaAs is expensive, focused light can be used effectively. Less expensive Si is often used as a solar cell material.

Single-crystal Si *pn*-junctions still have the disadvantage of relatively high cost. Amorphous Si is much cheaper, but one cannot make a solar cell with it unless it is treated with hydrogen. Hydrogenated amorphous Si can be used since the hydrogen apparently saturates some dangling or broken bonds and allows *pn*-junction solar cells to be built. We should mention also that new materials for photovoltaic solar cells are constantly under development. For example, copper indium gallium selenide (CIGS) thin films are being considered as a low-cost alternative.

Let us start with a one-dimensional model. The dark current, neglecting the series resistance of the diode can be written

$$I = I_0 \left[\exp\left(\frac{eV}{kT} \right) - 1 \right] . \tag{6.243}$$

The illuminated current is

$$I = I_0 \left[\exp\left(\frac{eV}{kT} \right) - 1 \right] - I_S , \tag{6.244}$$

where

$$I_S = \eta e p \tag{6.245}$$

(p = photons/s, η = quantum efficiency). Solving for the voltage, we find

$$V = \frac{kT}{e}\ln\left(\frac{I + I_0 + I_S}{I_0}\right). \tag{6.246}$$

The open-circuit voltage is

$$V_{\mathrm{OC}} = \frac{kT}{e}\ln\left(\frac{I_S + I_0}{I_0}\right), \tag{6.247}$$

because the dark current $I = 0$ in an open circuit. The short circuit current (with $V = 0$) is

$$I_{\mathrm{SC}} = -I_S. \tag{6.248}$$

The power is given by

$$P = VI = V\left[I_0\left(\exp\left(\frac{eV}{kT}\right) - 1\right) - I_S\right]. \tag{6.249}$$

The voltage V_{M} and current I_{M} for maximum power can be obtained by solving $\mathrm{d}P/\mathrm{d}V = 0$. Since $P = IV$, this means that $\mathrm{d}I/\mathrm{d}V = -I/V$. Figure 6.24 helps to show this. If P is the point of maximum power, then at P,

$$\frac{\mathrm{d}V}{\mathrm{d}I} = -\frac{V_M}{I_M} > 0 \quad \text{since} \quad I_M < 0. \tag{6.250}$$

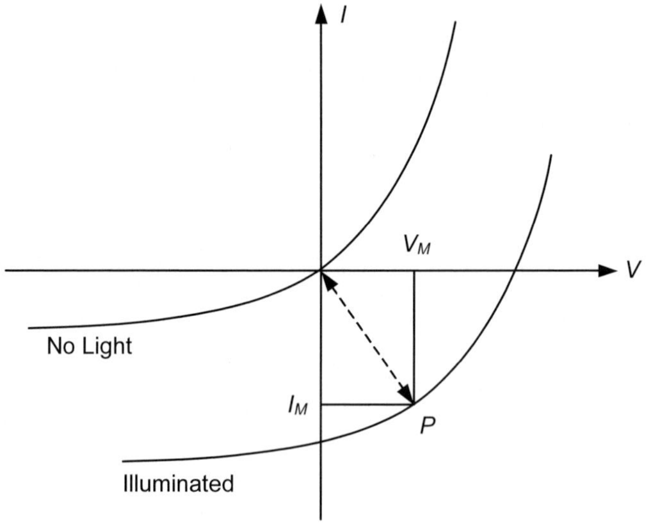

Fig. 6.24. Current–voltage relation for a solar cell

No current or voltage can be measured across the *pn*-junction unless light shines on it. In a complete circuit, the contact voltages of metallic leads will always be what is needed to cancel out the built-in voltage at the *pn*-junction. Otherwise, energy would not be conserved.

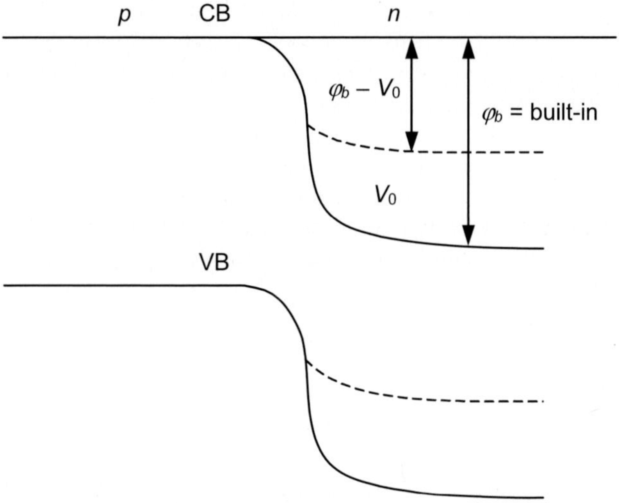

Fig. 6.25. The photoelectric effect for a *pn*-junction before and after illumination. The "before" are the solid lines and the "after" are the dashed lines. φ_b is the built-in potential and V_0 is the potential produced by the cell

To understand physically the photovoltaic effect, consider Fig. 6.25. When light shines on the cell, electron–hole pairs are produced. Electrons produced in the *p*-region (within a diffusion length of the *pn*-junction) will tend to be swept over to the *n*-side and similarly for holes on the *n*-side. This reduces the voltage across the *pn*-junction from φ_b to $\varphi_b - V_0$, say, and thus, produces a measurable forward voltage of V_0. The maximum value of the output potential V_0 from the solar cell is limited by the built-in potential φ_b.

$$V_0 \leq \varphi_b , \tag{6.251}$$

for if $V_0 = \varphi_b$, then the built-in potential has been canceled and there is no potential left to separate electron–hole pairs.

In nondegenerate semiconductors suppose, before the *p*- and *n*- sides were "joined," we let the Fermi levels be $E_F(p)$ and $E_F(n)$. When they are joined, equilibrium is established by electron–hole flow, which equalizes the Fermi energies. Thus, the built-in potential simply equals the original difference of Fermi energies

$$e\varphi_b = E_F(n) - E_F(p) . \tag{6.252}$$

But, for the nondegenerate case

$$E_F(n) - E_F(p) \leq E_C - E_V = E_g . \quad (6.253)$$

Therefore,

$$eV_0 \leq E_g . \quad (6.254)$$

Smaller E_g means smaller photovoltages and, hence, less efficiency. By connecting several solar cells together in series, we can build a significant potential with arrays of *pn*-junctions. These connected cells power space satellites.

We give, now, an introduction to a more quantitative calculation of the behavior of a solar cell. Just as in our discussion of *pn*-junctions, we can find the total current by finding the minority current injected on each side. The only difference is that the external photons of light create electron–hole pairs. We assume the flux of photons is given by (see Fig. 6.26)

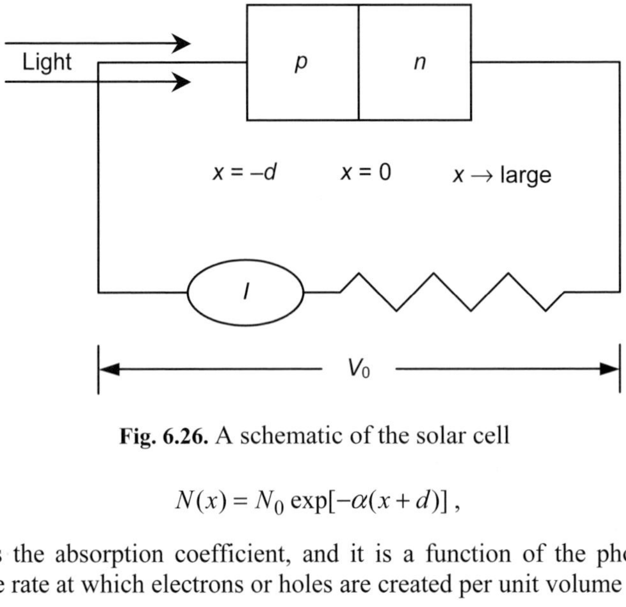

Fig. 6.26. A schematic of the solar cell

$$N(x) = N_0 \exp[-\alpha(x+d)] , \quad (6.255)$$

where α is the absorption coefficient, and it is a function of the photon wavelength. The rate at which electrons or holes are created per unit volume is

$$-\frac{\mathrm{d}N}{\mathrm{d}x} = \alpha N_0 \exp[-\alpha(x+d)] . \quad (6.256)$$

The equations for the minority carrier concentrations are just like those used for the *pn*-junction in (6.221) and (6.222), except now we must take into account the creation of electrons and holes by light from (6.256). We have

$$\frac{\mathrm{d}^2(n_p - n_{p0})}{\mathrm{d}x^2} - \frac{n_p - n_{p0}}{L_n^2} = -\frac{\alpha N_0}{D_n} \exp[-\alpha(x+d)] , \quad x < 0 , \quad (6.257)$$

and

$$\frac{d^2(p_n - p_{n0})}{dx^2} - \frac{p_n - p_{n0}}{L_p^2} = -\frac{\alpha N_0}{D_p}\exp[-\alpha(x+d)], \quad x>0. \tag{6.258}$$

Both equations apply outside the depletion region when drift currents are negligible. The depletion region is so thin it is assumed to be treatable as being located in the plane $x = 0$.

By adding a particular solution of the inhomogeneous equation to a general solution of the homogeneous equation, we find

$$n_p(x) - n_{p0} = a\cosh\left(\frac{x}{L_n}\right) + b\sinh\left(\frac{x}{L_n}\right) + \frac{\alpha N_0 \tau_n}{1-\alpha^2 L_n^2}\exp[-\alpha(x+d)], \tag{6.259}$$

and

$$p_n(x) - p_{n0} = d\exp\left(-\frac{x}{L_p}\right) + \frac{\alpha N_0 \tau_p}{1-\alpha^2 L_p^2}\exp[-\alpha(x+d)], \tag{6.260}$$

where it has been assumed that p_n approaches a finite value for large x. We now have three constants to evaluate (a), (b), and (d). We can use the following boundary conditions:

$$\frac{n_p(0)}{n_{p0}} = \exp\left(\frac{eV_0}{kT}\right), \tag{6.261}$$

$$\frac{p_n(0)}{p_{n0}} = \exp\left(\frac{eV_0}{kT}\right), \tag{6.262}$$

and

$$D_n\left[\frac{d}{dx}(n_p - n_{p0})\right]_{x=-d} = S_p\left[n_p(-d) - n_{p0}\right]. \tag{6.263}$$

This is a standard assumption that introduces a surface recombination velocity S_p. The total current as a function of V_0 can be evaluated from

$$I = eA[J_p(0) - J_n(0)], \tag{6.264}$$

where A is the cross-sectional area of the p-n junction. V_0 is now the bias voltage across the pn-junction. The current can be evaluated from (with a negligibly thick depletion region)

$$J_{\text{Total}} = qD_n \left.\frac{dn_p}{dx}\right|_{\substack{x<0 \\ x\to 0}} - qD_p \left.\frac{dp_n}{dx}\right|_{\substack{x>0 \\ x\to 0}}. \tag{6.265}$$

For a modern update, see Martin Green, "Solar Cells" (Chap. 8 in Sze, [6.42]).

6.3.10 Transistors (EE)

A power-amplifying structure made with *pn*-junctions is called a transistor. There are two main types of transistors: bipolar junction transistors (BJTs) and metal-oxide semiconductor field effect transistors (MOSFETs). MOSFETs are unipolar (electrons or holes are the carriers) and are the most rapidly developing type partly because they are easier to manufacture. However, MOSFETs have large gate capacitors and are slower. The huge increase in the application of microelectronics is due to integrated circuits and planar manufacturing techniques (Sapoval and Hermann, [6.33, p 258]; Fraser, [6.14, Chap. 6]). MOSFETs may have smaller transistors and can thus be used for higher integration. A serious discussion of the technology of these devices would take us too far aside, but the student should certainly read about it. Three excellent references for this purpose are Streetman [6.40] and Sze [6.41, 6.42].

Although J. E. Lilienfied was issued a patent for a field effect device in 1935, no practical commercial device was developed at that time because of the poor understanding of surfaces and surface states. In 1947, Shockley, Bardeen, and Brattrain developed the point constant transistor and won a Nobel Prize for that work. Shockley invented the bipolar junction transistor in 1948. This work had been stimulated by earlier work of Schottky on rectification at a metal-semiconductor interface. A field effect transistor was developed in 1953, and the more modern MOS transistors were invented in the 1960s.

6.3.11 Charge-Coupled Devices (CCD) (EE)

Charge-coupled devices (CCDs) were developed at Bell Labs in the 1970s and are now used extensively by astronomers for imaging purposes, and in digital cameras.

CCDs are based on ideas similar to those in metal-insulator-semiconductor structures that we just discussed. These devices are also called charge-transfer devices. The basic concept is shown in Fig. 6.27. Potential wells can be created under each electrode by applying the proper bias voltage.

$$V_1, V_2, V_3 < 0 \quad \text{and} \quad |V_2| > |V_1| \,\text{or}\, |V_3| \,.$$

By making V_2 more negative than V_1, or V_3, one can create a hole inversion layer under V_2. Generally, the biasing is changed frequently enough that holes under V_2 only come by transfer and not thermal excitation. For example, if we have holes under V_2, simply by exchanging the voltages on V_2 and V_3 we can move the hole to under V_3.

Since the presence or absence of charge is information in binary form, we have a way of steering or transferring information. CCDs have also been used to temporarily store an image. If we had large negative potentials at each V_i, then only those V_is, where light was strong enough to create electron–hole pairs, would have holes underneath them. The image is digitized and can be stored on a disk, which later can be used to view the image through a monitor.

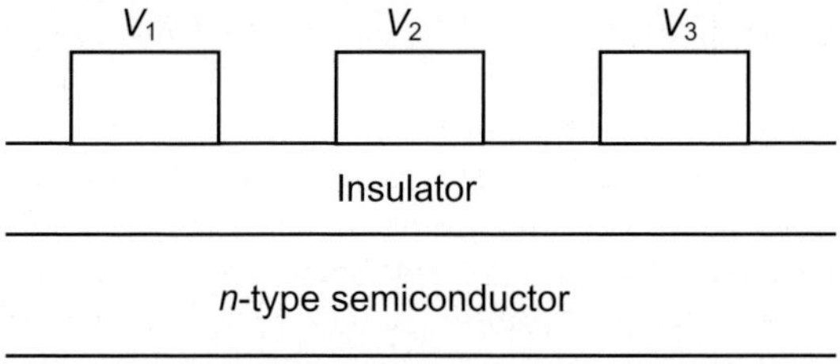

Fig. 6.27. Schematic for a charge-coupled device

Problems

6.1 For the nondegenerate case where $E - \mu >> kT$, calculate the number of electrons per unit volume in the conduction band from the integral

$$n = \int_{E_c}^{\infty} D(E) f(E) \mathrm{d}E .$$

$D(E)$ is the density of states, $f(E)$ is the Fermi function.

6.2 Given the neutrality condition

$$N_c \exp[-\beta(E_c - \mu)] + \frac{N_d}{1 + a\exp[\beta(E_d - \mu)]} = N_d ,$$

and the definition $x = \exp(\beta\mu)$, solve the condition for x. Then solve for n in the region $kT << E_c - E_d$, where $n = N_c \exp[-\beta(E_c - \mu)]$.

6.3 Derive (6.45). Hint – look at Sect. 8.8 and Appendix 1 of Smith [6.38].

6.4 Discuss in some detail the variation with temperature of the position of the Fermi energy in a fairly highly donor doped *n*-type semiconductor.

6.5 Explain how the junction between two dissimilar metals can act as a rectifier.

6.6 Discuss the mobility due to the lattice scattering of electrons in silicon or germanium. See, for example, Seitz [6.35].

6.7 Discuss the scattering of charge carriers in a semiconductor by ionized donors or acceptors. See, for example, Conwell and Weisskopf [6.9].

6.8 A sample of Si contains 10^{-4} atomic per cent of phosphorous donors that are all singly ionized at room temperature. The electron mobility is 0.15 $m^2V^{-1}s^{-1}$. Calculate the extrinsic resistivity of the sample (for Si, atomic weight = 28, density = 2300 kg/m^3).

6.9 Derive (6.163) by use of the spatial constancy of the chemical potential.

7 Magnetism, Magnons, and Magnetic Resonance

The first chapter was devoted to the solid-state medium (i.e. its crystal structure and binding). The next two chapters concerned the two most important types of energy excitations in a solid (the electronic excitations and the phonons). *Magnons* are another important type of energy excitation and they occur in magnetically ordered solids. However, it is not possible to discuss magnons without laying some groundwork for them by discussing the more elementary parts of magnetic phenomena. Also, there are many magnetic properties that cannot be discussed by using the concept of magnons. In fact, the study of magnetism is probably the first solid-state property that was seriously studied, relating as it does to lodestone and compass needles.

Nearly all the magnetic effects in solids arise from electronic phenomena, and so it might be thought that we have already covered at least the fundamental principles of magnetism. However, we have not yet discussed in detail the electron's spin degree of freedom, and it is this, as well as the orbital angular moment that together produce magnetic moments and thus are responsible for most magnetic effects in solids. When all is said and done, because of the richness of this subject, we will end up with a rather large chapter devoted to magnetism.

We will begin by briefly surveying some of the larger-scale phenomena associated with magnetism (diamagnetism, paramagnetism, ferromagnetism, and allied topics). These are of great technical importance. We will then show how to understand the origin of ordered magnetic structures from a quantum-mechanical viewpoint (in fact, strictly speaking this is the only way to understand it). This will lead to a discussion of the Heisenberg Hamiltonian, mean field theory, spin waves and magnons (the quanta of spin waves). We will also discuss the behavior of ordered magnetic systems near their critical temperature, which turns out also to be incredibly rich in ideas.

Following this we will discuss magnetic domains and related topics. This is of great practical importance.

Some of the simpler aspects of magnetic resonance will then be discussed as it not only has important applications, but magnetic resonance experiments provide direct measurements of the very small energy differences between magnetic sublevels in solids, and so they can be very sensitive probes into the inner details of magnetic solids.

We will end the chapter with some brief discussion of recent topics: the Kondo effect, spin glasses, magnetoelectronics, and solitons.

7.1 Types of Magnetism

7.1.1 Diamagnetism of the Core Electrons (B)

All matter shows diamagnetic effects, although these effects are often obscured by other stronger types of magnetism. In a solid in which the diamagnetic effect predominates, the solid has an induced magnetic moment that is in the opposite direction to an external applied magnetic field.

Since the diamagnetism of conduction electrons (Landau diamagnetism) has already been discussed (Sect. 3.2.2), this Section will concern itself only with the diamagnetism of the core electrons.

For an external magnetic field H in the z direction, the Hamiltonian (SI, $e > 0$) is given by

$$\mathcal{H} = \frac{p^2}{2m} + V(r) + \frac{e\hbar\mu_0 H}{2mi}\left(x\frac{\partial}{\partial y} - y\frac{\partial}{\partial x} \right) + \frac{e^2\mu_0^2 H^2}{8m}(x^2 + y^2) .$$

For purely diamagnetic atoms with zero total angular momentum, the term involving first derivatives has zero matrix elements and so will be neglected. Thus, with a spherically symmetric potential $V(r)$, the one-electron Hamiltonian is

$$\mathcal{H} = \frac{p^2}{2m} + V(r) + \frac{e^2\mu_0^2 H^2}{8m}(x^2 + y^2) . \tag{7.1}$$

Let us evaluate the susceptibility of such a diamagnetic substance. It will be assumed that the eigenvalues of (7.1) (with $H = 0$) and the eigenkets $|n\rangle$ are precisely known. Then by first-order perturbation theory, the energy change in state n due to the external magnetic field is

$$E' = \frac{e^2\mu_0^2 H^2}{8m}\langle n | x^2 + y^2 | n \rangle . \tag{7.2}$$

For simplicity, it will be assumed that $|n\rangle$ is spherically symmetric. In this case

$$\langle n | x^2 + y^2 | n \rangle = \tfrac{2}{3}\langle n | r^2 | n \rangle . \tag{7.3}$$

The induced magnetic moment μ can now be readily evaluated:

$$\mu = -\frac{\partial E'}{\partial(\mu_0 H)} = -\frac{e^2\mu_0 H}{6m}\langle n | r^2 | n \rangle . \tag{7.4}$$

If N is the number of atoms per unit volume, and Z is the number of core electrons, then the magnetization M is $ZN\mu$, and the magnetic susceptibility χ is

$$\chi = \frac{\partial M}{\partial H} = -\frac{ZNe^2\mu_0}{6m}\langle n | r^2 | n \rangle . \tag{7.5}$$

If we make an obvious reinterpretation of $\langle n|r^2|n\rangle$, then this result agrees with the classical result [7.39 p. 418]. The derivation of (7.5) assumes that the core electrons do not interact and that they are all in the same state $|n\rangle$. For core electrons on different atoms noninteraction would appear to be reasonable. However, it is not clear that this would lead to reasonable results for core electrons on the same atom. A generalization to core atoms in different states is fairly obvious.

A measurement of the diamagnetic susceptibility, when combined with theory (similar to the above), can sometimes provide a good test for any proposed forms for the core wave functions. However, if paramagnetic or other effects are present, they must first be subtracted out, and this procedure can lead to uncertainty in interpretation.

In summary, we can make the following statements about diamagnetism:

1. Every solid has diamagnetism although it may be masked by other magnetic effects.
2. The diamagnetic susceptibility (which is negative) is temperature independent (assuming we can regard $\langle n|r^2|n\rangle$ as temperature independent).

7.1.2 Paramagnetism of Valence Electrons (B)

This Section is begun by making several comments about paramagnetism:

1. One form of paramagnetism has already been studied. This is the Pauli paramagnetism of the free electrons (Sect. 3.2.2).
2. When discussing paramagnetic effects, in general both the orbital and intrinsic spin properties of the electrons must be considered.
3. A paramagnetic substance has an induced magnetic moment in the same direction as the applied magnetic field.
4. When paramagnetic effects are present, they generally are much larger than the diamagnetic effects.
5. At high enough temperatures, all substances appear to behave in either a paramagnetic fashion or a diamagnetic fashion (even ferromagnetic solids, as we will discuss, become paramagnetic above a certain temperature).
6. The calculation of the paramagnetic susceptibility is a statistical problem, but the general reason for paramagnetism is unpaired electrons in unfilled shells of electrons.
7. The study of paramagnetism provides a natural first step for understanding ferromagnetism.

The calculation of a paramagnetic susceptibility will only be outlined. The perturbing part of the Hamiltonian is of the form [94], $e > 0$,

$$\mathcal{H}' = \frac{e\mu_0 \boldsymbol{H}}{2m} \cdot (\boldsymbol{L} + 2\boldsymbol{S}), \tag{7.6}$$

where $\boldsymbol{L}$ is the total orbital angular momentum operator, and $\boldsymbol{S}$ is the total spin operator. Using a canonical ensemble, we find the magnetization of a sample to be given by

$$\langle \boldsymbol{M} \rangle = N\mathrm{Tr}\left[\boldsymbol{\mu} \exp\left(\frac{F - \mathcal{H}'}{kT} \right) \right], \tag{7.7}$$

where N is the number of atoms per unit volume, $\boldsymbol{\mu}$ is the magnetic moment operator proportional to $(\boldsymbol{L} + 2\boldsymbol{S})$, and F is the Helmholtz free energy.

Once (7.7) has been computed, the magnetic susceptibility is easily evaluated by means of

$$\chi \equiv \frac{\partial \langle M \rangle}{\partial H}. \tag{7.8}$$

Equations (7.7) and (7.8) are always appropriate for evaluating χ, but the form of the Hamiltonian is modified if one wants to include complicated interaction effects.

At lower temperatures we expect that interactions such as crystal-field effects will become important. Properly including these effects for a specific problem is usually a research problem. The effects of crystal fields will be discussed later in the chapter.

Let us consider a particularly simple case of paramagnetism. This is the case of a particle with spin S (and no other angular momentum). For a magnetic field in the z-direction we can write the Hamiltonian as (charge on electron is $e > 0$)

$$\mathcal{H}' = \frac{e\mu_0 H}{2m} \cdot 2S_z. \tag{7.9}$$

Let us define $g\mu_B$ in such a way that the eigenvalues of (7.9) are

$$E = g\mu_B \mu_0 H M_S, \tag{7.10}$$

where $\mu_B = e\hbar/2m$ is the Bohr magneton, and g is sometimes called simply the g-factor. The use of a g-factor allows our formalism to include orbital effects if necessary. In (7.10) $g = 2$ (spin only).

If N is the number of particles per unit volume, then the average magnetization can be written as[1]

$$\langle M \rangle = N \frac{\sum_{M_S=-S}^{S} M_S g\mu_B \exp(M_S g\mu_B \mu_0 H / kT)}{\sum_{M_S=-S}^{S} \exp(M_S g\mu_B \mu_0 H / kT)}. \tag{7.11}$$

[1] Note that μ_B has absorbed the $\hbar$ so M_S and S are either integers or half-integers. Also note (7.11) is invariant to a change of the dummy summation variable from M_S to $-M_S$.

For high temperatures (and/or weak magnetic fields, so only the first two terms of the expansion of the exponential need be retained) we can write

$$\langle M \rangle \cong Ng\mu_B \frac{\sum_{M_S=-S}^{S} M_S (1 + M_S g\mu_B \mu_0 H / kT)}{\sum_{M_S=-S}^{S} (1 + M_S g\mu_B \mu_0 H / kT)},$$

which, after some manipulation, becomes to order H

$$\langle M \rangle = g^2 S(S+1) \frac{N\mu_B^2 \mu_0 H}{3kT},$$

or

$$\chi \equiv \frac{\partial \langle M \rangle}{\partial H} = \mu_0 \frac{N p_{\text{eff}}^2 \mu_B^2}{3kT}, \tag{7.12}$$

[2] where $p_{\text{eff}} = g[S(S+1)]^{1/2}$ is called the *effective magneton number.* Equation (7.12) is the *Curie law.* It expresses the $(1/T)$ dependence of the magnetic susceptibility at high temperature. Note that when $H \to 0$, (7.12) is an exact consequence of (7.11).

It is convenient to have an expression for the magnetization of paramagnets that is valid at all temperatures and magnetic fields.

If we define

$$X = \frac{g\mu_B \mu_0 H}{kT}, \tag{7.13}$$

then

$$\langle M \rangle = Ng\mu_B \frac{\sum_{M_S=-S}^{S} M_S e^{M_S X}}{\sum_{M_S=-S}^{S} e^{M_S X}}. \tag{7.14}$$

With a little elementary manipulation, it is possible to perform the sums indicated in (7.14):

$$\langle M \rangle = Ng\mu_B \frac{\mathrm{d}}{\mathrm{d}X}\left[\ln\left(\frac{\sinh[(S+\frac{1}{2})X]}{\sinh(X/2)}\right)\right],$$

or

$$\langle M \rangle = Ng\mu_B S\left[\frac{2S+1}{2S}\coth\left(\frac{2S+1}{2S}SX\right) - \frac{1}{2S}\coth\left(\frac{SX}{2S}\right)\right]. \tag{7.15}$$

[2] A temperature-independent contribution known as van Vleck paramagnetism may also be important for some materials at low temperature. It may occur due to the effect of excited states that can be treated by second-order perturbation theory. It is commonly important when first-order terms vanish. See Ashcroft and Mermin [7.2 p. 653].

Defining the Brillouin function $B_J(y)$ as

$$B_J(y) = \frac{2J+1}{2J}\coth\left(\frac{2J+1}{2J}y\right) - \frac{1}{2J}\coth\frac{y}{2J}, \tag{7.16}$$

we can write the magnetization ⟨M⟩ as

$$\langle M \rangle = NgS\mu_B B_S(SX). \tag{7.17}$$

It is easy to recover the high-temperature results (7.12) from (7.17). All we have to do is use

$$B_J(y) = \frac{J+1}{3J}y \quad \text{if} \quad y << 1. \tag{7.18}$$

Then

$$\langle M \rangle \to NgS\mu_B \frac{S(S+1)}{3S} SX,$$

or using (7.13),

$$\langle M \rangle = \frac{Ng^2\mu_B^2 S(S+1)\mu_0 H}{3kT}.$$

7.1.3 Ordered Magnetic Systems (B)

Ferromagnetism and the Weiss Mean Field Theory (B)

Ferromagnetism refers to solids that are magnetized without an applied magnetic field. These solids are said to be spontaneously magnetized. Ferromagnetism occurs when paramagnetic ions in a solid "lock" together in such a way that their magnetic moments all point (on the average) in the same direction. At high enough temperatures, this "locking" breaks down and ferromagnetic materials become paramagnetic. The temperature at which this transition occurs is called the *Curie temperature.*

There are two aspects of ferromagnetism. One of these is the description of what goes on inside a single magnetized *domain* (where the magnetic moments are all aligned). The other is the description of how domains interact to produce the observed magnetic effects such as hysteresis. Domains will be briefly discussed later (Sect. 7.3).

We start by considering various magnetic structures without the complication of domains. Ferromagnetism, especially ferromagnetism in metals, is still not quantitatively and completely understood in all magnetic materials. We will turn to a more detailed study of the fundamental origin of ferromagnetism in Sect. 7.2. Our aim in this Section is to give a brief survey of the phenomena and of some phenomenological ideas.

In the ferromagnetic state at low temperatures, the spins on the various atoms are aligned parallel. There are several other types of ordered magnetic structures. These structures order for the same physical reason that ferromagnetic structures do (i.e. because of exchange coupling between the spins as we will discuss in Sect. 7.2). They also have more complex domain effects that will not be discussed.

Examples of elements that show spontaneous magnetism or ferromagnetism are (1) transition or iron group elements (e.g. Fe, Ni, Co), (2) rare earth group elements (e.g. Gd or Dy), and (3) many compounds and alloys. Further examples are given in Sect. 7.3.2.

The Weiss theory is a mean field theory and is perhaps the simplest way of discussing the appearance of the ferromagnetic state. First, what is mean field theory? Basically, mean field theory is a linearized theory in which the Hamiltonian products of operators representing dynamical observables are approximated by replacing these products by a dynamical observable times the mean or average value of a dynamic observable. The average value is then calculated self-consistently from this approximated Hamiltonian. The nature of this approximation is such that thermodynamic fluctuations are ignored. Mean field theory is often used to get an idea as to what structures or phases are present as the temperature and other parameters are varied. It is almost universally used as a first approximation, although, as discussed below, it can even be qualitatively wrong (in, for example, predicting a phase transition where there is none).

The Weiss mean field theory does the main thing that we want a theory of the magnetic state to do. It predicts a phase transition. Unfortunately, the quantitative details of real phase transitions are typically not what the Weiss theory says they should be. Still, it has several advantages:

1. It provides a comprehensive if at times only qualitative description of most magnetic materials. The Weiss theory (augmented with the concept of domains) is still the most important theory for a practical discussion of many types of magnetic behavior. Many experimental results are still presented within the context of this theory, and so in order to read the experimental papers it is necessary to understand Weiss theory.

2. It is rigorous for infinite-range interactions between spins (which never occur in practice).

3. The Weiss theory originally postulated a mysterious molecular field that was the "real" cause of the ordered magnetic state. This molecular field was later given an explanation based on the exchange effects described by the Heisenberg Hamiltonian (see Sect. 7.2). The Weiss theory gives a very simple way of relating the occurrence of a phase transition to the description of a magnetic system by the Heisenberg Hamiltonian. Of course, the way it relates these two is only qualitatively correct. However, it is a good starting place for more general theories that come closer to describing the behavior of the actual magnetic systems.[3]

---

[3] Perhaps the best simple discussion of the Weiss and related theories is contained in the book by J. S. Smart [92], which can be consulted for further details. By using two sublattices, it is possible to give a similar (to that below) description of antiferromagnetism. See Sect. 7.1.3.

For the case of a simple paramagnet, we have already derived that (see Sect. 7.1.2)

$$M = NgS\mu_B B_S(a)\,, \tag{7.19}$$

[4] where B_S is defined by (7.16) and

$$a \equiv \frac{Sg\mu_B\mu_0 H}{kT}\,. \tag{7.20}$$

Recall also high-temperature (7.18) for $B_S(a)$ can be used.

Following a modern version of the original Weiss theory, we will give a qualitative description of the occurrence of spontaneous magnetization. Based on the concept of the mean or molecular field the spontaneous magnetization must be caused by some sort of atomic interaction. Whatever the physical origin of this interaction, it tends to bring about an ordering of the spins. Weiss did not attempt to derive the origin of this interaction. In fact, all he did was to postulate the existence of a molecular field that would tend to align the spins. His basic assumption was that the interaction would be taken account of if H (the applied magnetic field) were replaced by $H + \gamma M$, where γM is the molecular field. (γ is called the molecular field constant, sometimes the Weiss constant, and has nothing to do with the gyromagnetic ratio γ that will be discussed later.)

Thus the basic equation for ferromagnetic materials is

$$M = Ng\mu_B S B_S(a')\,, \tag{7.21}$$

where

$$a' = \frac{\mu_0 Sg\mu_B}{kT}(H + \gamma M)\,. \tag{7.22}$$

That is, the basic equations of the molecular field theory are the same as the paramagnetic case plus the $H \to H + \gamma M$ replacement. Equations (7.21) and (7.22) are really all there is to the molecular field model. We shall derive other results from these equations, but already the basic ideas of the theory have been covered.

Let us now indicate how this predicts a phase transition. By a phase transition, we mean that spontaneous magnetization ($M \neq 0$ with $H = 0$) will occur for all temperatures below a certain temperature T_c called the *ferromagnetic Curie temperature.*

At the Curie temperature, for a consistent solution of (7.21) and (7.22) we require that the following two equations shall be identical as $a' \to 0$ and $H = 0$:

$$M_1 = Ng\mu_B S B_S(a')\,,\ \text{((7.21) again)}$$

$$M_2 = \frac{kTa'}{Sg\mu_B\gamma\mu_0}\,,\ \text{((7.22) with } H \to 0\text{)}.$$

[4] Here e can be treated as $|e|$ and so as usual, $\mu_B = |e|\hbar/2m$.

If these equations are identical, then they must have the same slope as $a' \to 0$. That is, we require

$$\left(\frac{\mathrm{d}M_1}{\mathrm{d}a'}\right)_{a'\to 0} = \left(\frac{\mathrm{d}M_2}{\mathrm{d}a'}\right)_{a'\to 0} . \tag{7.23}$$

Using the known behavior of $B_S(a')$ as $a' \to 0$, we find that condition (7.23) gives

$$T_c = \frac{\mu_0 N g^2 S(S+1)\mu_B^2}{3k}\gamma . \tag{7.24}$$

Equation (7.24) provides the relationship between the Curie constant T_c and the Weiss molecular field constant γ. Note that, as expected, if $\gamma = 0$, then $T_c = 0$ (i.e. if $\gamma \to 0$, there is no phase transition). Further, numerical evaluation shows that if $T > T_c$, (7.21) and (7.22) with $H = 0$ have a common solution for M only if $M = 0$. However, for $T < T_c$, numerical evaluation shows that they have a common solution $M \neq 0$, corresponding to the spontaneous magnetization that occurs when the molecular field overwhelms thermal effects.

There is another Curie temperature besides T_c. This is the so-called *paramagnetic Curie temperature* θ that enters into the equation for the high-temperature behavior of the magnetic susceptibility. Within the context of the Weiss theory, these two temperatures turn out to be the same. However, if one makes an experimental determination of T_c (from the transition temperature) and of θ from the high-temperature magnetic susceptibility, θ and T_c do not necessarily turn out to be identical (See Fig. 7.1). We obtain an explicit expression for θ below.

For $\mu_0 HSg\mu_B/kT << 1$ we have (by (7.17) and (7.18))

$$M = \frac{\mu_0 N g^2 \mu_B^2 S(S+1)}{3kT}h = C'h . \tag{7.25}$$

For ferromagnetic materials we need to make the replacement $H \to H + \gamma M$ so that $M = C'H + C'\gamma M$ or

$$M = \frac{C'H}{1 - C'\gamma} . \tag{7.26}$$

Substituting the definition of C', we find that (7.26) gives for the susceptibility

$$\chi = \frac{M}{H} = \frac{C}{T - \theta} , \tag{7.27}$$

where

$$C \equiv \text{the Curie–Weiss} = \frac{\mu_0 N g^2 \mu_B^2 S(S+1)}{3k} ,$$

$$\theta \equiv \text{the paramagnetic Curie temperature} = \frac{\mu_0 N g^2 S(S+1)}{3k}\mu_B^2 \gamma .$$

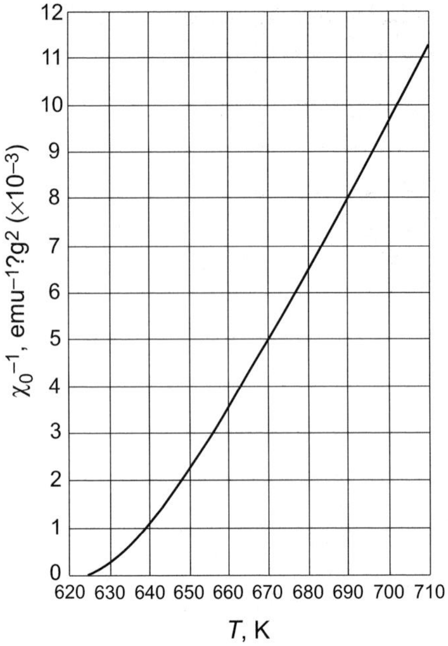

Fig. 7.1. Inverse susceptibility χ_0^{-1} of Ni. [Reprinted with permission from Kouvel JS and Fisher ME, *Phys Rev* **136**, A1626 (1964). Copyright 1964 by the American Physical Society. Original data from Weiss P and Forrer R, *Annales de Physique* (*Paris*), **5**, 153 (1926).]

The Weiss theory gives the same result:

$$C\gamma = \theta = T_c = \frac{N\mu_B^2}{3k}(p_{\text{eff}})^2 \mu_0 \gamma \,, \tag{7.28}$$

where $p_{\text{eff}} = g[S(S+1)]^{1/2}$ is the effective magnetic moment in units of the Bohr magneton. Equation (7.27) is valid experimentally only if $T >> \theta$. See Fig. 7.1.

It may not be apparent that the above discussion has limited validity. We have predicted a phase transition, and of course γ can be chosen so that the predicted T_c is exactly the experimental T_c. The Weiss prediction of the $(T - \theta)^{-1}$ behavior for χ also fits experiment at high enough temperatures.

However, we shall see that when we begin to look at further details, the Weiss theory begins to break down. In order to keep the algebra fairly simple it is convenient to absorb some of the constants into the variables and thus define new variables. Let us define

$$b \equiv \frac{\mu_0 g \mu_B}{kT}(H + \gamma M) \,, \tag{7.29}$$

and

$$m \equiv \frac{M}{Ng\mu_B S} \equiv B_S(bS) \,, \tag{7.30}$$

which should not be confused with the magnetic moment.

It is also convenient to define a quantity J_{ex} by

$$\gamma = \frac{2ZJ_{ex}}{\mu_0 N g^2 \mu_B^2} \hbar^2 , \tag{7.31}$$

where Z is the number of nearest neighbors in the lattice of interest, and J_{ex} is the *exchange integral.* Compare this to (7.95), which is the same. That is, we will see that (7.31) makes sense from the discussion of the physical origin of the molecular field.

Finally, let us define

$$b_0 = \frac{g\mu_B}{kT} \mu_0 H , \tag{7.32}$$

and

$$\tau = T / T_c .$$

With these definitions, a little manipulation shows that (7.29) is

$$bS = b_0 S + \frac{3S}{S+1} \frac{m}{\tau} . \tag{7.33}$$

Equations (7.30) and (7.33) can be solved simultaneously for m (which is proportional to the magnetization). With b_0 equal to zero (i.e. $H = 0$) we combine (7.30) and (7.33) to give a single equation that determines the spontaneous magnetization:

$$m = B_S \left(\frac{3S}{S+1} \frac{m}{\tau} \right) . \tag{7.34}$$

A plot similar to that yielded by (7.34) is shown in Fig. 7.16 ($H = 0$). The fit to experiment of the molecular field model is at least qualitative. Some classic results for Ni by Weiss and Forrer as quoted by Kittel [7.39 p. 448] yield a reasonably good fit.

We have reached the point where we can look at sufficiently fine details to see how the molecular field theory gives predictions that do not agree with experiment. We can see this by looking at the solutions of (7.34) as $\tau \to 0$ (i.e. $T << T_c$) and as $\tau \to 1$ (i.e. $T \to T_c$).

We know that for any y that $B_S(y)$ is given by (7.16). We also know that

$$\coth X = \frac{1+e^{-2X}}{1-e^{-2X}} . \tag{7.35}$$

Since for large X

$$\coth X \cong 1 + 2e^{-2X} ,$$

we can say that for large y

$$B_S(y) \cong 1 + \frac{2S+1}{S}\exp\left(-\frac{2S+1}{S}y\right) - \frac{1}{S}\exp\left(-\frac{y}{S}\right). \tag{7.36}$$

Therefore by (7.34), m can be written for $T \to 0$ as

$$m \cong 1 + \left(\frac{2S+1}{S}\right)\exp\left[-\frac{3(2S+1)m}{(S+1)\tau}\right] - \frac{1}{S}\exp\left[-\frac{3m}{(S+1)\tau}\right]. \tag{7.37}$$

By iteration, it is clear that $m = 1$ can be used in the exponentials. Further,

$$\exp\left[-2\frac{3}{(S+1)\tau}\right] << \exp\left[-\frac{3}{(S+1)\tau}\right],$$

so that the second term can be neglected for all $S \neq 0$ (for $S = 0$ we do not have ferromagnetism anyway). Thus at lower temperature, we finally find

$$m \cong 1 - \frac{1}{S}\exp\left(-\frac{3}{S+1}\frac{T_c}{T}\right). \tag{7.38}$$

Experiment does not agree well with (7.38). For many materials, experiment agrees with

$$m \cong 1 - CT^{3/2}, \tag{7.39}$$

where C is a constant. As we will see in Sect. 7.2, (7.39) is correctly predicted by spin wave theory.

It also turns out that the Weiss molecular field theory disagrees with experiment at temperatures just below the Curie temperature. By making a Taylor series expansion, one can show that for $y << 1$,

$$B_S(y) \cong \frac{(2S+1)^2-1}{(2S)^2}\cdot\frac{y}{3} - \frac{(2S+1)^4-1}{(2S)^4}\cdot\frac{y^3}{45}. \tag{7.40}$$

Combining (7.40) with (7.34), we find that

$$m = K(T_c - T)^{1/2}, \tag{7.41}$$

and

$$\frac{dm^2}{dT} = -K^2 \quad \text{as} \quad T \to T_c^-. \tag{7.42}$$

Equations (7.41) and (7.42) agree only qualitatively with experiment. For many materials, experiment predicts that just below the Curie temperature

$$m \cong A(T_c - T)^{1/3}. \tag{7.43}$$

Perhaps the most dramatic failure of the Weiss molecular field theory occurs when we consider the specific heat. As we will see, the Weiss theory flatly predicts that the specific heat (with no external field) should vanish for temperatures above the Curie temperature. Experiment, however, says nothing of the sort. There is a small residual specific heat above the Curie temperature. This specific heat drops off with temperature. The reason for this failure of the Weiss theory is the neglect of short-range ordering above the Curie temperature.

Let us now look at the behavior of the Weiss predictions for the magnetic specific heat in a little more detail. The energy of a spin in a γM field in the z direction due to the molecular field is

$$E_i = \frac{\mu_0 g \mu_B}{\hbar} S_{iz} \gamma M \,. \tag{7.44}$$

Thus the internal energy U obtained by averaging E_i for N spins is,

$$U = \mu_0 \frac{N}{2} \frac{g\mu_B}{\hbar} \gamma M \langle S_{iz} \rangle = -\tfrac{1}{2} \mu_0 \gamma M^2 \,, \tag{7.45}$$

where the factor 1/2 comes from the fact that we do not want to count bonds twice, and $M = -Ng\mu_B \langle S_{iz} \rangle / \hbar$ has been used.

The specific heat in zero magnetic field is then given by

$$C_0 = \frac{\partial U}{\partial T} = -\tfrac{1}{2} \mu_0 \gamma \frac{\mathrm{d}M^2}{\mathrm{d}T} \,. \tag{7.46}$$

For $T > T_c$, $M = 0$ (with no external magnetic field) and so the specific heat vanishes, which contradicts experiment.

The precise behavior of the magnetic specific heat just above the Curie temperature is of more than passing interest. Experimental results suggest that the specific heat should exhibit a logarithmic singularity or near logarithmic singularity as $T \to T_c$. The Weiss theory is inadequate even to begin attacking this problem.

Antiferromagnetism, Ferrimagnetism, and Other Types of Magnetic Order (B)

Antiferromagnetism is similar to ferromagnetism except that the lowest-energy state involves adjacent spins that are antiparallel rather than parallel (but see the end of this section). As we will see, the reason for this is a change in sign (compared to ferromagnetism) for the coupling parameter or *exchange integral.*

Ferrimagnetism is similar to antiferromagnetism except that the paired spins do not cancel and thus the lowest-energy state has a net spin.

Examples of antiferromagnetic substances are FeO and MnO. Further examples are given in Sect. 7.3.2. The temperature at which an antiferromagnetic substance becomes paramagnetic is known as the *Néel* temperature.

Examples of ferrimagnetism are $MnFe_2O_4$ and $NiFe_2O_7$. Further examples are also given in Sect. 7.3.2.

We now discuss these in more detail by use of mean field theory.[5] We assume near-neighbor and next-nearest-neighbor coupling as shown schematically in Fig. 7.2. The figure is drawn for an assumed ferrimagnetic order below the transition temperature. A and B represent two sublattices with spins S_A and S_B. The coupling is represented by the exchange integrals J (we assume $J_{BA} = J_{AB} < 0$ and these J dominate J_{AA} and $J_{BB} > 0$). Thus we assume the effective field between A and B has a negative sign. For the effective field we write:

$$B_A = -\omega\mu_0 M_B + \alpha_A \mu_0 M_A + B\,, \tag{7.47}$$

$$B_B = -\omega\mu_0 M_A + \beta_B \mu_0 M_B + B\,, \tag{7.48}$$

where $\omega > 0$ is a constant proportional to $|J_{AB}| = |J_{BA}|$, while α_A and β_B are constants proportional to J_{AA} and J_{BB}. The M represent magnetization and B is the external field (that is the magnetic induction $B = \mu_0 H_{\text{external}}$).

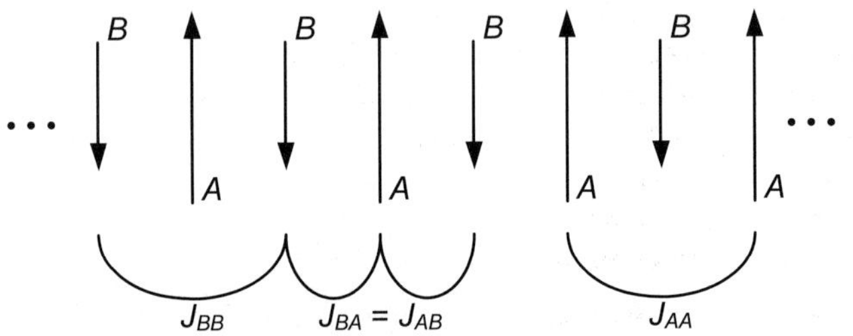

Fig. 7.2. Schematic to represent ferrimagnets

By the mean field approximation with B_{SA} and B_{SB} being the appropriate Brillouin functions (defined by (7.16)):

$$M_A = N_A g_A S_A \mu_B B_{S_A}(\beta g_A \mu_B S_A B_A)\,, \tag{7.49}$$

$$M_B = N_B g_B S_B \mu_B B_{S_B}(\beta g_B \mu_B S_B B_B)\,. \tag{7.50}$$

The S_A, S_B are quantum numbers (e.g. 1, 3/2, etc., labeling the spin). We also will use the result (7.40) for $B_S(x)$ with $x << 1$. In the above, N_i is the number of ions of type i per unit volume, g_A and g_B are the *Lande g-factors* (note we are using B not $\mu_0 H$), μ_B is the Bohr magneton and $\beta = 1/(k_B T)$.

Defining the Curie constants

$$C_A = \frac{N_A S_A (S_A + 1) g_A^2 \mu_B^2}{3k}\,, \tag{7.51}$$

$$C_B = \frac{N_B S_B (S_B + 1) g_B^2 \mu_B^2}{3k}\,, \tag{7.52}$$

[5] See also, e.g., Kittel [7.39 p458ff].

we have if B_A/T and B_B/T are small:

$$M_A = \frac{C_A B_A}{T}, \tag{7.53}$$

$$M_B = \frac{C_B B_B}{T}. \tag{7.54}$$

This holds above the ordering temperature when $B \to 0$ and even just below the ordering temperature provided $B \to 0$ and M_A, M_B are very small. Thus the equations determining the magnetization become:

$$(T - \alpha_A \mu_0 C_A) M_A + \omega \mu_0 C_A M_B = C_A B, \tag{7.55}$$

$$\omega \mu_0 C_B M_A + (T - \beta_B \mu_0 C_B) M_B = C_B B. \tag{7.56}$$

If the external field $B \to 0$, we can have nonzero (but very small) solutions for M_A, M_B provided

$$(T - \alpha_A \mu_0 C_A)(T - \beta_B \mu_0 C_B) = \omega^2 \mu_0^2 C_A C_B. \tag{7.57}$$

So

$$T_c^{\pm} = \frac{\mu_0}{2}\left(\alpha_A C_A + \beta_B C_B \pm \sqrt{4\omega^2 C_A C_B + (\alpha_A C_A - \beta_B C_B)^2} \right). \tag{7.58}$$

The critical temperature is chosen so $T_c = \omega \mu_0 (C_A C_B)^{1/2}$ when $\alpha_A \to \beta_B \to 0$, and so $T_c = T_c^{+}$. Above T_c for $B \neq 0$ (and small) with

$$D \equiv (T - T_c^{+})(T - T_c^{-}),$$

$$M_A = D^{-1}[(T - \beta_B \mu_0 C_B) C_A - \omega \mu_0 C_A C_B] B,$$

$$M_B = D^{-1}[(T - \alpha_A \mu_0 C_A) C_B - \omega \mu_0 C_A C_B] B.$$

The reciprocal magnetic susceptibility is then given by

$$\frac{1}{\chi} = \frac{B}{\mu_0 (M_A + M_B)} = \frac{D}{\mu_0 \{ T(C_A + C_B) - [(\alpha_A + \beta_B) + 2\omega] \mu_0 C_A C_B \}}. \tag{7.59}$$

Since D is quadratic in T, $1/\chi$ is linear in T only at high temperatures (ferrimagnetism). Also note

$$\frac{1}{\chi} = 0 \quad \text{at} \quad T = T_c^{+} = T_c.$$

In the special case where two sublattices are identical (and $\omega > 0$), since $C_A = C_B \equiv C_1$ and $\alpha_A = \beta_B \equiv \alpha_1$,

$$T_c^{+} = (\alpha_1 + \omega) C_1 \mu_0, \tag{7.60}$$

and after canceling,

$$\chi^{-1} = \frac{[T - C_1\mu_0(\alpha_1 - \omega)]}{2C_1\mu_0}, \tag{7.61}$$

which is linear in T (antiferromagnetism).

This equation is valid for $T > T_c^+ = \mu_0(\alpha_1+\omega)C_1 \equiv T_N$, the Néel temperature. Thus, if we define

$$\theta \equiv C_1(\omega - \alpha_1)\mu_0,$$

$$\chi_{AF} = \frac{2\mu_0 C_1}{T + \theta}. \tag{7.62}$$

Note:

$$\frac{\theta}{T_N} = \frac{\omega - \alpha_1}{\omega + \alpha_1}.$$

We can also easily derive results for the ferromagnetic case. We choose to drop out one sublattice and in effect double the effect of the other to be consistent with previous work.

$$C_A = C_A^F \equiv 2C_1, \; \beta_B = 0, \; C_B = 0,$$

so

$$T_c = \mu_0 \alpha_A^F C_A^F = 2C_1\mu_0\alpha_1 \quad (\text{if } \alpha_1 \equiv \alpha_A^F).$$

Then,[6]

$$\chi = \frac{\mu_0 M_A}{B} = \frac{\mu_0 T(2C_1)}{T(T - 2C_1\mu_0\alpha_1)} = \frac{2C_1\mu_0}{T - 2C_1\mu_0\alpha_1}. \tag{7.63}$$

The paramagnetic case is obtained from neglecting the coupling so

$$\chi = \frac{2C_1\mu_0}{T}. \tag{7.64}$$

The reality of antiferromagnetism has been absolutely determined by neutron diffraction that shows the appearance of magnetic order below the critical temperature. See Fig. 7.3 and Fig. 7.4. Figure 7.5 summarizes our results.

[6] $2C_1\mu_0 = C$ of (7.27).

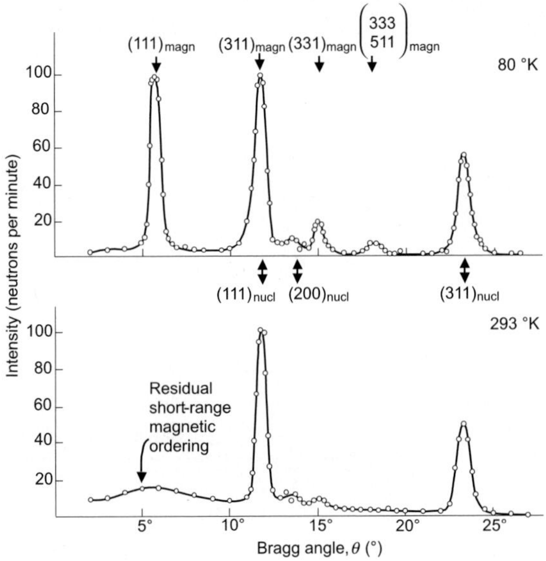

Fig. 7.3. Neutron-diffraction patterns of MnO at 80 K and 293 K. The Curie temperature is 120 K. The low-temperature pattern has extra antiferromagnetic reflections for a magnetic unit twice that of the chemical unit cell. From Bacon GE, *Neutron Diffraction*, Oxford at the Clarendon Press, London, 1962 2nd edn, Fig. 142 p.297. By permission of Oxford University Press. Original data from Shull CG and Smart JS, *Phys Rev*, **76**, 1256 (1949)

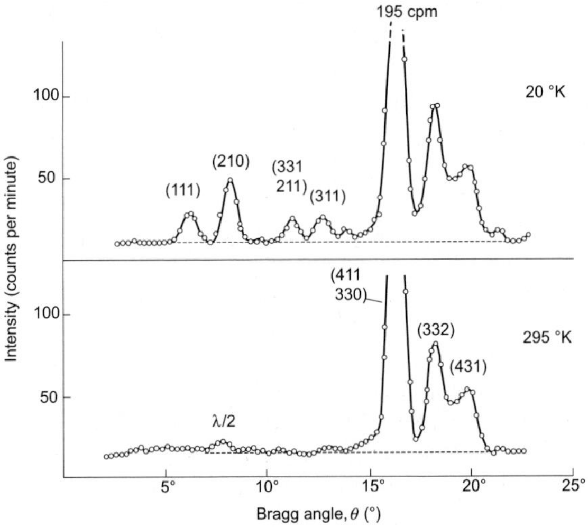

Fig. 7.4. Neutron-diffraction patterns for α-manganese at 20 K and 295 K. Note the antiferromagnetic reflections at the lower temperature. From Bacon GE, *Neutron Diffraction*, Oxford at the Clarendon Press, London, 1962 2nd edn, Fig. 129 p.277. By permission of Oxford University Press. Original data from Shull CG and Wilkinson MK, *Rev Mod Phys*, **25**, 100 (1953)

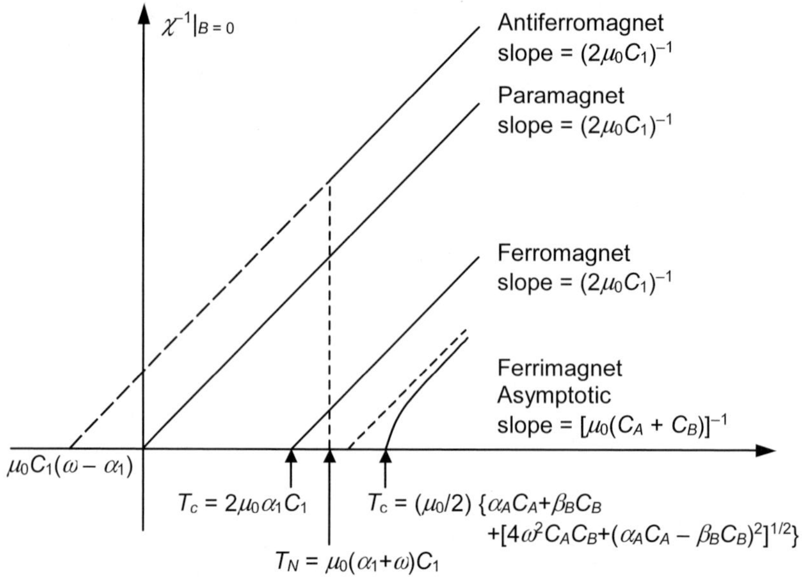

Fig. 7.5. Schematic plot of reciprocal magnetic susceptibility. Note the constants for the various cases can vary. For example α_1 could be negative for the antiferromagnetic case and α_A, β_B could be negative for the ferrimagnetic case. This would shift the zero of χ^{-1}

The above definitions of antiferromagnetism and ferrimagnetism are the old definitions (due to Néel). In recent years it has been found useful to generalize these definitions somewhat. Antiferromagnetism has been generalized to include solids with more than two sublattices and to include materials that have triangular, helical or spiral, or canted spin ordering (which may not quite have a net zero magnetic moment). Similarly, ferrimagnetism has been generalized to include solids with more than two sublattices and with spin ordering that may be, for example, triangular or helical or spiral. For ferrimagnetism, however, we are definitely concerned with the case of nonvanishing magnetic moment.

It is also interesting to mention a remarkable theorem of Bohr and Van Leeuwen [94]. This theorem states that for classical, nonrelativistic electrons for all finite temperatures and applied electric and magnetic fields, the net magnetization of a collection of electrons in thermal equilibrium vanishes. This is basically due to the fact that the paramagnetic and diamagnetic terms exactly cancel one another on a classical and statistical basis. Of course, if one cleverly makes omissions, one can discuss magnetism on a classical basis. The theorem does tell us that if we really want to understand magnetism, then we had better learn quantum mechanics. See Problem 7.17.

It might be well to learn relativity also. Relativity tells us that the distinction between electric and magnetic fields is just a distinction between reference frames.

7.2 Origin and Consequences of Magnetic Order

7.2.1 Heisenberg Hamiltonian

The Heitler–London Method (B)

In this Section we develop the Heisenberg Hamiltonian and then relate our results to various aspects of the magnetic state. The first method that will be discussed is the Heitler–London method. This discussion will have at least two applications. First, it helps us to understand the covalent bond, and so relates to our previous discussion of valence crystals. Second, the discussion gives us a qualitative understanding of the Heisenberg Hamiltonian. This Hamiltonian is often used to explain the properties of coupled spin systems. The Heisenberg Hamiltonian will be used in the discussion of magnons. Finally, as we will show, the Heisenberg Hamiltonian is useful in showing how an electrostatic exchange interaction approximately predicts the existence of a molecular field and hence gives a fundamental qualitative explanation of the existence of ferromagnetism.

Let a and b label two hydrogen atoms separated by R (see Fig. 7.6). Let the separated ($R \to \infty$) hydrogen atoms be described by the Hamiltonians

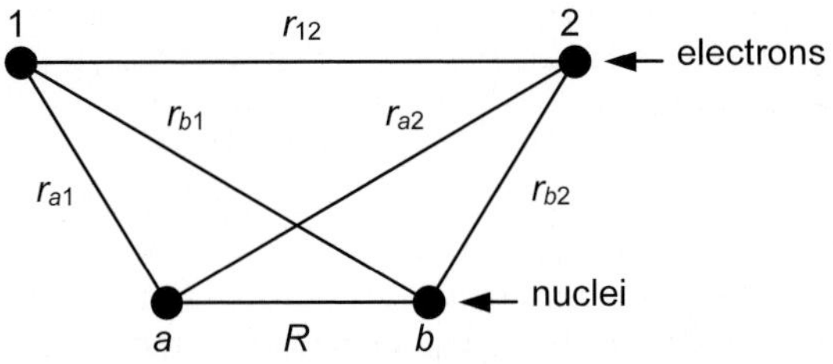

Fig. 7.6. Model for two hydrogen atoms

$$\mathcal{H}_0^a(1) = -\frac{\hbar^2}{2m}\nabla_1^2 - \frac{e^2}{4\pi\varepsilon_0 r_{a1}}, \tag{7.65}$$

and

$$\mathcal{H}_0^b(2) = -\frac{\hbar^2}{2m}\nabla_2^2 - \frac{e^2}{4\pi\varepsilon_0 r_{b2}}. \tag{7.66}$$

Let $\psi_a(1)$ and $\psi_b(2)$ be the spatial ground-state wave functions, that is

$$\mathcal{H}_0^a \psi_a(1) = E_0 \psi_a(1), \tag{7.67}$$

or

$$\mathcal{H}_0^b \psi_b(2) = E_0 \psi_b(2),$$

where E_0 is the ground-state energy of the hydrogen atom. The zeroth-order hydrogen molecular wave functions may be written

$$\psi_{\pm} = \psi_a(1)\psi_b(2) \pm \psi_a(2)\psi_b(1)\,. \tag{7.68}$$

In the Heitler–London approximation for un-normalized wave functions

$$E_{\pm} \cong \frac{\int \psi_{\pm} \mathcal{H} \psi_{\pm} \mathrm{d}\tau_1 \mathrm{d}\tau_2}{\int \psi_{\pm}^2 \mathrm{d}\tau_1 \mathrm{d}\tau_2}\,, \tag{7.69}$$

where $\mathrm{d}\tau_i = \mathrm{d}x_i \mathrm{d}y_i \mathrm{d}z_i$ and we have used that wave functions for stationary states can be chosen to be real. In (7.69),

$$\mathcal{H} = \mathcal{H}_0^a(1) + \mathcal{H}_0^b(2) - \frac{e^2}{4\pi\varepsilon_0}\left(\frac{1}{r_{a2}} + \frac{1}{r_{b1}} - \frac{1}{r_{12}} - \frac{1}{R}\right). \tag{7.70}$$

Working out the details when (7.68) is put into (7.69) and assuming $\psi_a(1)$ and $\psi_b(2)$ are normalized we find

$$E_{\pm} = 2E_0 + \frac{e^2}{4\pi\varepsilon_0 R} + \frac{K \pm J_E}{1 \pm S}\,, \tag{7.71}$$

where

$$S = \int \psi_a(1)\psi_b(1)\psi_a(2)\psi_b(2)\mathrm{d}\tau_1 \mathrm{d}\tau_2 \tag{7.72}$$

is the overlap integral,

$$K = \frac{e^2}{4\pi\varepsilon_0} \int \psi_a^2(1)\psi_b^2(2) V(1,2)\mathrm{d}\tau_1 \mathrm{d}\tau_2 \tag{7.73}$$

is the Coulomb energy of interaction, and

$$J_E = \frac{e^2}{4\pi\varepsilon_0} \int \psi_a(1)\psi_a(2)\psi_b(1)\psi_b(2) V(1,2)\mathrm{d}\tau_1 \mathrm{d}\tau_2 \tag{7.74}$$

is the exchange energy. In (7.73) and (7.74),

$$V(1,2) = \frac{e^2}{4\pi\varepsilon_0}\left(\frac{1}{r_{12}} - \frac{1}{r_{a2}} - \frac{1}{r_{b1}}\right). \tag{7.75}$$

The corresponding normalized eigenvectors are

$$\psi^{\pm}(1,2) = \frac{1}{\sqrt{2(1 \pm S)}}[\psi_1(1,2) \pm \psi_2(1,2)]\,, \tag{7.76}$$

where

$$\psi_1(1,2) = \psi_a(1)\psi_b(2)\,, \tag{7.77}$$

$$\psi_2(1,2) = \psi_a(2)\psi_b(1)\,. \tag{7.78}$$

So far there has been no need to discuss spin, as the Hamiltonian did not explicitly involve it. However, it is easy to see how spin enters. ψ^+ is a symmetric function in the interchange of coordinates 1 and 2, and ψ^- is an antisymmetric function in the interchange of coordinates 1 and 2. The total wave function that includes both space and spin coordinates must be antisymmetric in the interchange of all coordinates. Thus in the total wave function, an antisymmetric function of spin must multiply ψ^+, and a symmetric function of spin must multiply ψ^-. If we denote $\alpha(i)$ as the "spin-up" wave function of electron i and $\beta(j)$ as the "spin-down" wave function of electron j, then the total wave functions can be written as

$$\psi_T^+ = \frac{1}{\sqrt{2(1+S)}}(\psi_1 + \psi_2)\frac{1}{\sqrt{2}}[\alpha(1)\beta(2) - \alpha(2)\beta(1)]\,, \tag{7.79}$$

$$\psi_T^- = \frac{1}{\sqrt{2(1-S)}}(\psi_1 - \psi_2)\begin{cases} \alpha(1)\alpha(2), \\ \frac{1}{\sqrt{2}}[\alpha(1)\beta(2) + \alpha(2)\beta(1)], \\ \beta(1)\beta(2). \end{cases} \tag{7.80}$$

Equation (7.79) has total spin equal to zero, and is said to be a *singlet* state. It corresponds to antiparallel spins. Equation (7.80) has total spin equal to one (with three projections of +1, 0, −1) and is said to describe a *triplet* state. This corresponds to parallel spins. For hydrogen atoms, J in (7.74) is called the *exchange integral* and is negative. Thus E_+ (corresponding to ψ_T^+) is lower in energy than E_- (corresponding to ψ_T^-), and hence the singlet state is lowest in energy. A calculation of $E_\pm - E_0$ for E_0 labeling the ground state of hydrogen is sketched in Fig. 7.7. Let us now pursue this two-spin case in order to write an effective spin Hamiltonian that describes the situation. Let $\boldsymbol{S}_1$ and $\boldsymbol{S}_2$ be the spin operators for particles 1 and 2. Then

$$(\boldsymbol{S}_1 + \boldsymbol{S}_2)^2 = \boldsymbol{S}_1^2 + \boldsymbol{S}_2^2 + 2\boldsymbol{S}_1 \cdot \boldsymbol{S}_2\,. \tag{7.81}$$

Since the eigenvalues of $\boldsymbol{S}_2^1$ and $\boldsymbol{S}_2^2$ are $3\hbar^2/4$ we can write for appropriate ϕ in the space of interest

$$\boldsymbol{S}_1 \cdot \boldsymbol{S}_2\phi = \tfrac{1}{2}[(\boldsymbol{S}_1 + \boldsymbol{S}_2)^2 - \tfrac{3}{2}\hbar^2]\phi\,. \tag{7.82}$$

In the triplet (or parallel spin) state, the eigenvalue of $(\boldsymbol{S}_1 + \boldsymbol{S}_2)^2$ is $2\hbar^2$, so

$$\boldsymbol{S}_1 \cdot \boldsymbol{S}_2\phi_{\text{triplet}} = \tfrac{1}{4}\hbar^2\phi_{\text{triplet}}\,. \tag{7.83}$$

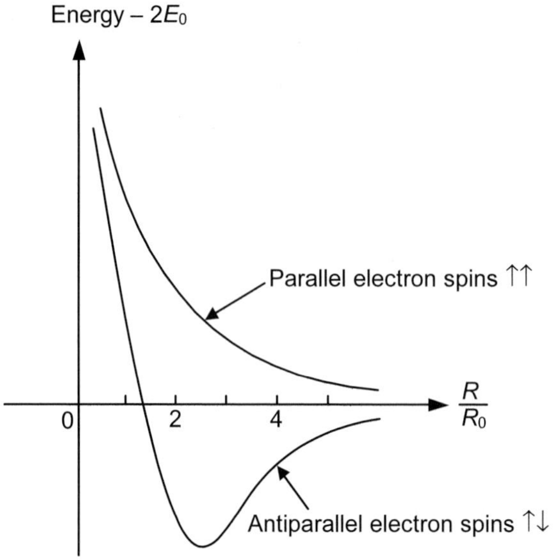

Fig. 7.7. Sketch of results of the Heitler–London theory applied to two hydrogen atoms (R/R_0 is the distance between the two atoms in Bohr radii). See also, e.g., Heitler [7.26].

In the singlet (or antiparallel spin) state, the eigenvalue of $(\boldsymbol{S}_1 + \boldsymbol{S}_2)^2$ is 0, so

$$\boldsymbol{S}_1 \cdot \boldsymbol{S}_2 \phi_{\text{singlet}} = -\tfrac{3}{4}\hbar^2 \phi_{\text{singlet}} \,. \tag{7.84}$$

Comparing these results to Fig. 7.7, we see we can formally write an effective spin Hamiltonian for the two electrons on the two different atoms:

$$\mathcal{H} = -2J\boldsymbol{S}_1 \cdot \boldsymbol{S}_2 \,, \tag{7.85}$$

where J is often simply called the *exchange constant* and $J = J(R)$, i.e. it depends on the separation R between atoms. By suitable choice of $J(R)$, the eigenvalues of $\mathcal{H} - 2E_0$ can reproduce the curves of Fig. 7.7. Note that $J > 0$ gives the parallel-spin case the lowest energy (ferromagnetism) and $J < 0$ (the two-hydrogen-atom case – this does not *always* happen, especially in a solid) gives the antiparallel-spin case the lowest energy (antiferromagnetism). If we have many atoms on a lattice, and if there is an exchange coupling between the spins of the atoms, we assume that we can write a Hamiltonian:

$$\mathcal{H} = -\sum_{\substack{\alpha,\beta \\ \text{(electrons)}}}{}' J_{\alpha,\beta} \boldsymbol{S}_\alpha \cdot \boldsymbol{S}_\beta \,. \tag{7.86}$$

If there are several electrons on the same atom and if J is constant for all electrons on the same atom, then we assume we can write

$$\begin{aligned}\sum J_{\alpha,\beta}\boldsymbol{S}_\alpha\cdot\boldsymbol{S}_\beta &\cong \underset{\text{(atoms)}}{\sum\nolimits_{k,l}} J_{k,l} \underset{\text{(electrons on } k,l \text{ atoms)}}{\sum\nolimits_{i,j}} \boldsymbol{S}_{ki}\cdot\boldsymbol{S}_{lj}\\ &= \sum\nolimits_{k,l} J_{k,l}(\sum\nolimits_i \boldsymbol{S}_{ki})(\sum\nolimits_j \boldsymbol{S}_{lj})\\ &= \sum\nolimits_{k,l} J_{k,l}\boldsymbol{S}_k^T\cdot\boldsymbol{S}_l^T,\end{aligned} \tag{7.87}$$

where $\boldsymbol{S}_k^T$ and $\boldsymbol{S}_l^T$ refer to the spin operators associated with atoms k and l. Since $\sum' \boldsymbol{S}_\alpha\cdot\boldsymbol{S}_\beta J_{\alpha\beta}$ differs from $\sum \boldsymbol{S}_\alpha\cdot\boldsymbol{S}_\beta J_{\alpha\beta}$ by only a constant and $\sum'_{k,l} \mathrm{J}_{kl}\boldsymbol{S}_k^T\boldsymbol{S}_l^T$ differs from $\sum_{k,l} \mathrm{J}_{kl}\boldsymbol{S}_k^T\boldsymbol{S}_l^T$ by only a constant, we can write the effective spin Hamiltonian as

$$\mathcal{H} = -\sum\nolimits_{k,l}' J_{k,l}\boldsymbol{S}_k^T\cdot\boldsymbol{S}_l^T, \tag{7.88}$$

here unimportant constants have not been retained. This last expression is called the *Heisenberg Hamiltonian* for a system of interacting spins in the absence of an external field.

This form of the Heisenberg Hamiltonian already tells us two important things:

1. It is applicable to atoms with arbitrary spin.
2. Closed shells contribute nothing to the Heisenberg Hamiltonian because the spin is zero for a closed shell.

Our development of the Heisenberg Hamiltonian has glossed over the approximations that were made. Let us now return to them. The first obvious approximation was made in going from the two-spin case to the N-spin case. The presence of a third atom can and does affect the interaction between the original pair. In addition, we assumed that the exchange interaction between all electrons on the same atom was a constant.

Another difficulty with the extension of the Heitler–London method to the n-electron problem is the so-called "overlap catastrophe." This will not be discussed here as we apparently do not have to worry about it when using the simple Heisenberg theory for insulators.[7] There are also no provisions in the Heisenberg Hamiltonian for crystalline anisotropy, which must be present in any real crystal. We will discuss this concept in Sects. 7.2.2 and 7.3.1. However, so far as energy goes, the Heisenberg model does seem to contain the main contributions.

But there are also several approximations made in the Heitler–London theory itself. The first of these assumptions is that the wave functions associated with the electrons of interest are well-localized wave functions. Thus we expect the Heisenberg Hamiltonian to be more nearly valid in insulators than in metals. The assumption is necessary in order that the perturbation approach used in the Heitler–London method will be valid. It is also assumed that the electrons are in non-degenerate orbital states and that the excited states can be neglected. This makes it

[7] For a discussion of this point see the article by Keffer, [7.37].

harder to see what to do in states that are not "spin only" states, i.e. in states in which the total orbital angular momentum L is not zero or is not quenched. Quenching of angular momentum means that the expectation value of $\boldsymbol{L}$ (but not L^2) for electrons of interest is zero when the atom is in the solid. For the nonspin only case, we have orbital degeneracy (plus the effects of crystal fields) and thus the basic assumptions of the *simple* Heitler–London method are not met.

The Heitler–London theory does, however, indicate one useful approximation: that $J\hbar^2$ is of the same order of magnitude as the electrostatic interaction energy between two atoms and that this interaction depends on the overlap of the wave functions of the atoms. Since the overlap seems to die out exponentially, we expect the *direct* exchange interaction between any two atoms to be of rather short range. (Certain indirect exchange effects due to the presence of a third atom may extend the range somewhat and in practice these indirect exchange effects may be very important. Indirect exchange can also occur by means of the conduction electrons in metals, as discussed later.)

Before discussing further the question of the applicability of the Heisenberg model, it is useful to get a physical picture of why we expect the spin-dependent energy that it predicts. In considering the case of two interacting hydrogen atoms, we found that we had a parallel spin case and an antiparallel spin case. By the Pauli principle, the parallel spin case requires an antisymmetric spatial wave function, whereas the antiparallel case requires a symmetric spatial wave function. The antisymmetric case concentrates less charge in the region between atoms and hence the electrostatic potential energy of the electrons ($e^2/4\pi\varepsilon_0 r$) is smaller. However, the antisymmetric case causes the electronic wave function to "wiggle" more and hence raises the kinetic energy T ($T_{\text{op}} \propto \nabla^2$). In the usual situation (in the two-hydrogen-atom case and in the much more complicated case of many insulating solids) the kinetic energy increase dominates the potential energy decrease; hence the antiparallel spin case has the lowest energy and we have antiferromagnetism ($J < 0$). In exceptional cases, the potential energy decrease can dominate the kinetic energy increases, and hence the parallel spin case has the least energy and we have ferromagnetism ($J > 0$). In fact, most insulators that have an ordered magnetic state become antiferromagnets at low enough temperature.

Few rigorous results exist that would tend either to prove or disprove the validity of the Heisenberg Hamiltonian for an actual physical situation. This is one reason for doing calculations based on the Heisenberg model that are of sufficient accuracy to yield results that can usefully be compared to experiment. Dirac[8] has given an explicit proof of the Heisenberg model in a situation that is oversimplified to the point of not being physical. Dirac assumes that each of the electrons is confined to a different specified orthogonal orbital. He also assumes that these orbitals can be thought of as being localizable. It is clear that this is never the situation in a real solid. Despite the lack of rigor, the Heisenberg Hamiltonian appears to be a good starting place for any theory that is to be used to explain experimental magnetic phenomena in insulators. The situation in metals is more complex.

[8] See, for example, Anderson [7.1].

Another side issue is whether the exchange "constants" that work well above the Curie temperature also work well below the Curie temperature. Since the development of the Heisenberg Hamiltonian was only phenomenological, this is a sensible question to ask. It is particularly sensible since J depends on R and R increases as the temperature is increased (by thermal expansion). Charap and Boyd[9] and Wojtowicz[10] have shown for EuS (which is one of the few "ideal" Heisenberg ferromagnets) that the same set of J will fit both the low-temperature specific heat and magnetization and the high-temperature specific heat.

We have made many approximations in developing the Heisenberg Hamiltonian. The use of the Heitler–London method is itself an approximation. But there are other ways of understanding the binding of the hydrogen atoms and hence of developing the Heisenberg Hamiltonian. The Hund–Mulliken[11] method is one of these techniques. The Hund–Mulliken method should work for smaller R, whereas the Heitler–London works for larger R. However, they both qualitatively lead to a Heisenberg Hamiltonian.

We should also mention the Ising model, where $\mathcal{H} = -\sum J_{ij}\sigma_{iz}\sigma_{jz}$, and the σ are the Pauli spin matrices. Only nearest-neighbor coupling is commonly used. This model has been solved exactly in two dimensions (see Huang [7.32 p341ff]). The Ising model has spawned a huge number of calculations.

The Heisenberg Hamiltonian and its Relationship to the Weiss Mean Field Theory (B)

We now show how the mean molecular field arises from the Heisenberg Hamiltonian. If we assume a mean field $\gamma\boldsymbol{M}$ then the interaction energy of moment $\boldsymbol{\mu}_k$ with this field is

$$E_k = -\mu_0\gamma\boldsymbol{M}\cdot\boldsymbol{\mu}_k\,. \tag{7.89}$$

Also from the Heisenberg Hamiltonian

$$E_k = -{\sum_i}' J_{ik}\boldsymbol{S}_i\cdot\boldsymbol{S}_k - {\sum_j}' J_{kj}\boldsymbol{S}_k\cdot\boldsymbol{S}_j\,,$$

and since $J_{ij} = J_{ji}$, and noting that j is a dummy summation variable

$$E_k = -2{\sum_i}' J_{ik}\boldsymbol{S}_i\cdot\boldsymbol{S}_k\,. \tag{7.90}$$

---

9 See [7.10].

10 See Wojtowicz [7.70].

11 See Patterson [7.53 p176ff].

In the spirit of the mean-field approximation we replace $\boldsymbol{S}_i$ by its average $\overline{\boldsymbol{S}}_i = \boldsymbol{S}$ since the average of each site is the same. Further, we assume only nearest-neighbor interactions so $J_{ik} = J$ for each of the Z nearest neighbors. So

$$E_k \cong -2ZJ\boldsymbol{S} \cdot \boldsymbol{S}_k \, . \tag{7.91}$$

But

$$\mu_k \cong -\frac{g\mu_B \boldsymbol{S}_k}{\hbar} \tag{7.92}$$

(with $\mu_B = |e|\hbar/2m$), and the magnetization $\boldsymbol{M}$ is

$$\boldsymbol{M} \cong -\frac{Ng\mu_B \boldsymbol{S}}{\hbar} , \tag{7.93}$$

where N is the number of atomic moments per unit volume ($\equiv 1/\Omega$, where Ω is the atomic volume). Thus we can also write

$$E_k \cong -2ZJ \frac{\Omega \boldsymbol{M} \cdot \boldsymbol{\mu}_k}{(g\mu_B)^2} \hbar^2 \, . \tag{7.94}$$

Comparing (7.89) and (7.94)

$$J = \frac{\mu_0 \gamma (g\mu_B)^2}{2Z\Omega\hbar^2} . \tag{7.95}$$

This not only shows how Heisenberg's theory "explains" the Weiss mean molecular field, but also gives an approximate way of evaluating the parameter J. Slight modifications in (7.95) result for other than nearest-neighbor interactions.

RKKY Interaction[12] *(A)*

The Ruderman, Kittel, Kasuya, Yosida, (RKKY) interaction is important for rare earths. It is an interaction between the conduction electrons with the localized moments associated with the 4f electrons. Since the spins cause the localized moments, the conduction electrons can mediate an indirect exchange interaction between the spins. This interaction is called RKKY interaction.

We assume, following previous work, that the total exchange interaction is of the form

$$\mathcal{H}_{\text{ex}}^{\text{Total}} = -\sum\nolimits_{i,\alpha} J_x(\boldsymbol{r}_i - \boldsymbol{R}_\alpha) \boldsymbol{S}_\alpha \cdot \boldsymbol{S}_i \, , \tag{7.96}$$

---

[12] Kittel [60, pp 360-366] and White [7.68 pp 197-200].

where S_α is an ion spin and S_i is the conduction spin. For convenience we assume the $\boldsymbol{S}$ are dimensionless with $\hbar$ absorbed in the J. We assume $J_x(\boldsymbol{r}_i - \boldsymbol{R}_\alpha)$ is short range (the size of 4f orbitals) and define

$$J = \int J_x(\boldsymbol{r} - \boldsymbol{R}_\alpha)\mathrm{d}\boldsymbol{r} \,. \tag{7.97}$$

Consistent with (7.97), we assume

$$J_x(\boldsymbol{r}_i - \boldsymbol{R}_\alpha) = J\delta(\boldsymbol{r}) \,, \tag{7.98}$$

where $\boldsymbol{r} = \boldsymbol{r}_i - \boldsymbol{R}_\alpha$ and write

$$\mathcal{H}_{ex} = -J\boldsymbol{S}_\alpha \cdot \boldsymbol{S}_i \delta(\boldsymbol{r})$$

for the exchange interaction between the ion α and the conduction electron. This is the same form as the Fermi contact term, but the physical basis is different. We can regard $S_i\delta(\boldsymbol{r}) = S_i(\boldsymbol{r})$ as the electronic conduction spin density. Now, the interaction between the ion spin $\boldsymbol{S}_\alpha$ and the conduction spin $\boldsymbol{S}_i$ can be written (gaussian units, $\mu_0 = 1$)

$$-J\boldsymbol{S}_\alpha \cdot \boldsymbol{S}_i \delta(\boldsymbol{r}) = -(-g\mu_B \boldsymbol{S}_i) \cdot H_{\mathrm{eff}}(\boldsymbol{r}) \,,$$

so this defines an effective field

$$H_{\mathrm{eff}} = -\frac{J\boldsymbol{S}_\alpha}{g\mu_B}\delta(\boldsymbol{r}) \,. \tag{7.99}$$

The Fourier component of the effective field can be written

$$H_{\mathrm{eff}}(\boldsymbol{q}) = \int H_{\mathrm{eff}}(\boldsymbol{r})\mathrm{e}^{-\mathrm{i}\boldsymbol{q}\cdot\boldsymbol{r}}\mathrm{d}\boldsymbol{r} = -\frac{J}{g\mu_B}\boldsymbol{S}_\alpha \,. \tag{7.100}$$

We can now determine the magnetization induced by the effective field by use of the magnetic susceptibility. In Fourier space

$$\chi(q) = \frac{M(q)}{H(q)} \,. \tag{7.101}$$

This gives us the response in magnetization of a free-electron gas to a magnetic field. It turns out that this response (at $T = 0$) is functionally just like the response to an electric field (see Sect. 9.5.3 where Friedel oscillation in the screening of a point charge is discussed).

We find

$$\chi(\boldsymbol{q}) = \frac{3g^2\mu_B^2}{8E_F}\frac{N}{V}A(q/2k_F) \,, \tag{7.102}$$

where N/V is the number of electrons per unit volume and

$$A(q/2k_F) = \frac{1}{2} + \frac{k_F}{2q}\left(1 - \frac{q^2}{4k_F^2}\right)\ln\left|\frac{2k_F + q}{2k_F - q}\right|. \tag{7.103}$$

The magnetization $M(\boldsymbol{r})$ of the conduction electrons can now be calculated from (7.101), (7.102), and (7.103).

$$\begin{aligned} M(\boldsymbol{r}) &= \frac{1}{V}\sum\nolimits_{\boldsymbol{q}} M(\boldsymbol{q})\mathrm{e}^{\mathrm{i}\boldsymbol{q}\cdot\boldsymbol{r}} \\ &= \frac{1}{V}\sum\nolimits_{q} \chi(q) H_{\mathrm{eff}}(q)\mathrm{e}^{\mathrm{i}\boldsymbol{q}\cdot\boldsymbol{r}} \\ &= -\frac{J}{g\mu_B V}\boldsymbol{S}_\alpha \sum\nolimits_{q} \chi(q)\mathrm{e}^{\mathrm{i}\boldsymbol{q}\cdot\boldsymbol{r}}. \end{aligned} \tag{7.104}$$

With the aid of (7.102) and (7.103), we can evaluate (7.104) to find

$$M(\boldsymbol{r}) = -\frac{J}{g\mu_B} K G(r)\boldsymbol{S}_\alpha, \tag{7.105}$$

where

$$K = \frac{3g^2\mu_B^2}{8E_F}\frac{N}{V}\frac{k_F^3}{16\pi}, \tag{7.106}$$

and

$$G(r) = \frac{\sin(2k_F r) - 2k_F r\cos(2k_F r)}{(k_F r)^4}. \tag{7.107}$$

The localized moment $\boldsymbol{S}_\alpha$ causes conduction spins to develop an oscillating polarization in the vicinity of it. The spin-density oscillations have the same form as the charge-density oscillations that result when an electron gas screens a charged impurity.[13]

Let us define

$$F(x) = \frac{\sin x - x\cos x}{x^4},$$

so

$$G(r) = 2^4 F(2k_F r).$$

$F(x)$ is the basic function that describes spatial oscillating polarization induced by a localized moment in its vicinity. It is sketched in Fig. 7.8. Note as $x \to \infty$, $F(x) \to -\cos(x)/x^3$ and as $x \to 0$, $F(x) \to 1/(3x)$.

[13] See Langer and Vosko [7.42].

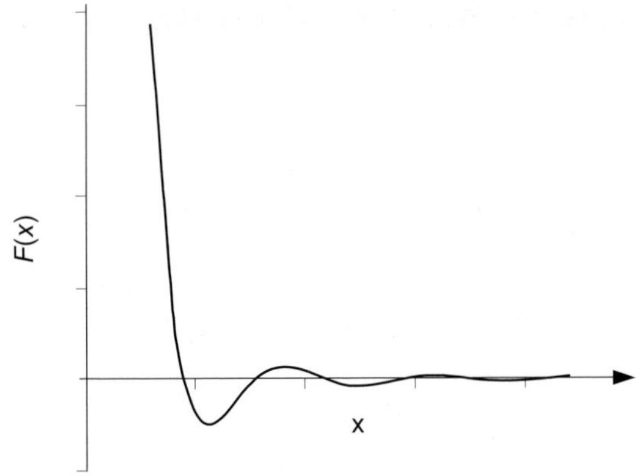

Fig. 7.8. Sketch of $F(x) = [\sin(x) - x\cos(x)]/x^4$, which describes the RKKY exchange interaction

Using (7.105), if $S(\boldsymbol{r})$ is the spin density,

$$S(\boldsymbol{r}) = \frac{M(\boldsymbol{r})}{(-g\mu_B)} = \frac{J}{(g\mu_B)^2} KG\boldsymbol{S}_\alpha . \tag{7.108}$$

Another localized ionic spin at $\mathbf{S}_\beta$ interacts with $\mathrm{S}(\mathbf{r})$

$$\mathcal{H}_{\alpha \text{ and } \beta}^{\text{indirect}} = -J\boldsymbol{S}_\beta \cdot S(\boldsymbol{r}_\alpha - \boldsymbol{r}_\beta) .$$

Now, summing over all α, β interactions and being careful to avoid double counting spins, we have

$$\mathcal{H}_{RKKY} = -\frac{1}{2}\sum\nolimits_{\alpha,\beta} J_{\alpha\beta}\boldsymbol{S}_\alpha \cdot \boldsymbol{S}_\beta , \tag{7.109}$$

where

$$J_{\alpha\beta} = \frac{J^2}{(g\mu_B)^2} KG(r = r_{\alpha\beta}) . \tag{7.110}$$

For strong spin-orbit coupling, it would be more natural to express the Hamiltonian in terms of $\boldsymbol{J}$ (the total angular momentum) rather than $\boldsymbol{S}$. $\boldsymbol{J} = \boldsymbol{L} + \boldsymbol{S}$ and within the set of states of constant J, g_J is defined so

$$g_J\mu_B\boldsymbol{J} = \mu_B(\boldsymbol{L} + 2\boldsymbol{S}) = \mu_B(\boldsymbol{J} + \boldsymbol{S}) ,$$

where remember the g factor for $\boldsymbol{L}$ is 1, while for spin $\boldsymbol{S}$ it is 2. Thus, we write

$$(g_J - 1)\boldsymbol{J} = \boldsymbol{S} .$$

If $\boldsymbol{J}_\alpha$ is the total angular momentum associated with site α, by substitution

$$\mathcal{H}_{RKKY} = -\frac{1}{2}(g_J - 1)^2 \sum\nolimits_{\alpha,\beta} J_{\alpha\beta}\boldsymbol{J}_\alpha \cdot \boldsymbol{J}_\beta , \tag{7.111}$$

where $(g_J - 1)^2$ is called the deGennes factor.

Magnetic Structure and Mean Field Theory (A)

We assume the Heisenberg Hamiltonian where the lattice is assumed to have transitional symmetry, $\boldsymbol{R}$ labels the lattice sites, $J(0) = 0$, $J(\boldsymbol{R} - \boldsymbol{R}') = J(\boldsymbol{R}' - \boldsymbol{R})$. We wish to investigate the ground state of a Heisenberg-coupled classical spin system, and for simplicity, we will assume:

a. $T = 0$ K
b. The spins can be treated classically
c. A one-dimensional structure (say in the z direction), and
d. The $\boldsymbol{S}_R$ are confined to the (x,y)-plane

$$S_{R_x} = S\cos\varphi_R\,,\ S_{R_y} = S\sin\varphi_R\,.$$

Thus, the Heisenberg Hamiltonian can be written:

$$\mathcal{H} = -\frac{1}{2}\sum\nolimits_{R,R'} S^2 J(\boldsymbol{R} - \boldsymbol{R}')\cos(\varphi_{\boldsymbol{R}} - \varphi_{\boldsymbol{R}'})\,.$$

e. We are going to further consider the possibility that the spins will have a constant turn angle of qa (between each spin), so $\varphi_R = qR$, and for adjacent spins $\Delta\varphi_R = q\Delta R = qa$.

Substituting (in the Hamiltonian above), we find

$$\mathcal{H} = -\frac{NS^2}{2}J(q)\,, \tag{7.112}$$

where

$$J(q) = \sum\nolimits_R J(R)\mathrm{e}^{\mathrm{i}qR} \tag{7.113}$$

and $J(q) = J(-q)$. Thus, the problem of finding $\mathcal{H}_{\min}$ reduces to the problem of finding $J(q)_{\max}$.

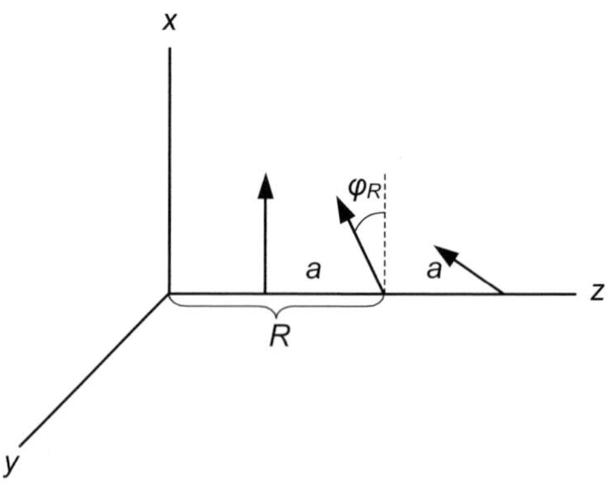

Fig. 7.9. Graphical depiction of the classical spin system assumptions

Note if $J(q) \to$ max for $\begin{cases} q = 0, & \text{get ferromagnetism,} \\ q = \pi/a, & \text{get antiferromagnetism,} \\ qa \neq 0 \text{ or } \pi, & \text{get heliomagnetism with } qa \\ & \text{defining the turn angles.} \end{cases}$

It may be best to give an example. We suppose that $J(a) = J_1$, $J(2a) = J_2$ and the rest are zero. Using (7.113) we find:

$$J(q) = 2J_1 \cos(qa) + 2J_2 \cos(2qa) . \tag{7.114}$$

For a minimum of energy [maximum $J(q)$] we require

$$\frac{\partial J}{\partial q} = 0 \to J_1 = -4J_2 \cos(qa) \quad \text{or} \quad q = 0 \text{ or } \frac{\pi}{a} ,$$

and

$$\frac{\partial^2 J}{\partial q^2} < 0 \quad \text{or} \quad J_1 \cos(qa) > -4J_2 \cos(2qa) .$$

The three cases give:

$q = 0$	$q = \pi/a$	$q \neq 0, \pi/a$
$J_1 > -4J_2$ Ferromagnetism e.g. $J_1 > 0, J_2 = 0$	$J_1 < 4J_2$ Antiferromagnetism e.g. $J_1 < 0, J_2 = 0$	Turn angle qa defined by $\cos(qa) = -J_1/4J_2$ and $J_1\cos(qa) > -4J_2\cos(2qa)$

7.2.2 Magnetic Anisotropy and Magnetostatic Interactions (A)

Anisotropy

Exchange interactions drive the spins to lock together at low temperature into an ordered state, but often the exchange interaction is isotropic. So, the question arises as to why the solid magnetizes in a particular direction. The answer is that other interactions are active that lock in the magnetization direction. These interactions cause magnetic anisotropy.

Anisotropy can be caused by different mechanisms. In rare earths, because of the strong-spin orbit coupling, magnetic moments arise from both spin and orbital motion of electrons. Anisotropy, then, can be caused by direct coupling between the orbit and lattice.

There is a different situation in the iron group magnetic materials. Here we think of the spins of the 3d electrons as causing ferromagnetism. However, the spins are not directly coupled to the lattice. Anisotropy arises because the orbit "feels" the lattice, and the spins are coupled to the orbit by the spin-orbit coupling.

Let us first discuss the rare earths, which are perhaps the easier of the two to understand. As mentioned, the anisotropy comes from a direct coupling between the crystalline field and the electrons. In this connection, it is useful to consider the classical multipole expansion for the energy of a charge distribution in a potential Φ. The first three terms are given below:

$$u = q\Phi(0) - \mathbf{p}\cdot\mathbf{E}(0) - \frac{1}{6}\sum\nolimits_{i,j} Q_{ij}\left(\frac{\partial E_j}{\partial x_i}\right)_0 + \text{higher-order terms.} \tag{7.115}$$

Here, q is the total charge, $\boldsymbol{p}$ is the dipole moment, Q_{ij} is the quadrupole moment, and the electric field is $\mathrm{E} = -\nabla\Phi$. For charge distributions arising from states with definite parity, $\boldsymbol{p} = 0$. (We assume this, or equivalently we assume the parity operator commutes with the Hamiltonian.) Since the term $q\Phi(0)$ is an additive constant, and since $\boldsymbol{p} = 0$, the first term that merits consideration is the quadrupole term. The quadrupole term describes the interaction of the quadrupole moment with the gradient of the electric field. Generally, the quadrupole moments will vary with $|J, M\rangle$ (J = total angular momentum quantum number and M refers to the z component), which will enable us to construct an effective Hamiltonian. This Hamiltonian will include the anisotropy in which different states within a manifold of constant J will have different energies, hence anisotropy. We now develop this idea in quantum mechanics below.

We suppose the crystal field is caused by an array of charges described by $\rho(\boldsymbol{R})$. Then, the potential energy of $-e$ at the point $\boldsymbol{r}_i$ is given by

$$V(\boldsymbol{r}_i) = -\int \frac{e\rho(\boldsymbol{R})\mathrm{d}\boldsymbol{R}}{4\pi\varepsilon_0|\boldsymbol{r}_i - \boldsymbol{R}|}\,. \tag{7.116}$$

If we further suppose $\rho(\boldsymbol{R})$ is outside the ion in question, then in the region of the ion, $V(\boldsymbol{r})$ is a solution of the Laplace equation, and we can expand it as a solution of this equation:

$$V(\boldsymbol{r}_i) = \sum\nolimits_{l,m} B_l^m r^l Y_l^m(\theta,\phi)\,, \tag{7.117}$$

where the constants B_l^m can be computed from $\rho(\boldsymbol{R})$. For rare earths, the effects of the crystal field, typically, can be adequately calculated in first-order perturbation theory. Let $|v\rangle$ be all states $|J, M\rangle$, which are formed of fixed J manifolds from $|l, m\rangle$, and $|s, m_s\rangle$ where $l = 3$ for $4f$ electrons. The type of matrix element that we need to evaluate can be written:

$$\langle v|\sum\nolimits_i V(\boldsymbol{r}_i)|v'\rangle\,, \tag{7.118}$$

summing over the 4f electrons. By (7.117), this eventually means we will have to evaluate matrix elements of the form

$$\langle lm_i|Y_{l'}^{m'}|lm_i'\rangle\,, \tag{7.119}$$

and since $l = 3$ for 4f electrons, this must vanish if $l' > 6$. Also, the parity of the functions in (7.119) is $(-)^{2l+l'}$ the matrix element must vanish if l' is odd since

$2l = 6$, and the integral over all space is of an odd parity function is zero. For 4f electrons, we can write

$$V(\boldsymbol{r}_i) = \sum_{\substack{l'=0 \\ (even)}}^{6} \sum_{m'} B_{l'}^{m'} r^{l'} Y_{l'}^{m'}(\theta,\phi) . \tag{7.120}$$

We define the effective Hamiltonian as

$$\mathcal{H}_A = \sum_i \langle V(\boldsymbol{r}_i) \rangle_{\text{doing radial integrals only}} .$$

If we then apply the Wigner–Eckhart theorem [7.68 p33], in which one replaces (x'/r), etc. by their operator equivalents J_x, etc., we find for hexagonal symmetry

$$\mathcal{H}_A = K_1 J_z^2 + K_2 J_z^4 + K_3 J_z^6 + K_4 (J_+^6 + J_-^6), \quad (J_\pm = J_x \pm iJ_y) . \tag{7.121}$$

We now discuss the anisotropy that is appropriate to the iron group [7.68 p57]. This is called single-ion anisotropy. Under the action of a crystalline field we will assume the relevant atomic states include a ground state (G) of energy ε_0 and appropriate excited (E) states of energy $\varepsilon_0 + \Delta$. We will consider only one excited state, although in reality there would be several. We assume $|G\rangle$ and $|E\rangle$ are separated by energy Δ.

The states $|G\rangle$ and $|E\rangle$ are assumed to be spatial functions only and not spin functions. In our argument, we will carry the spin $\boldsymbol{S}$ along as a classical vector. The argument we will give is equivalent to perturbation theory.

We assume a spin-orbit interaction of the form $V = \lambda \boldsymbol{L}\cdot\boldsymbol{S}$, which mixes some of the excited state into the ground state to produce a new ground state.

$$|G\rangle \to |G_T\rangle = |G\rangle + a|E\rangle , \tag{7.122}$$

where a is in general complex. We further assume $\langle G|G\rangle = \langle E|E\rangle = 1$ and $\langle E|G\rangle = 0$ so $\langle G_T|G_T\rangle = 1$ to $O(a)$. Also note the probability that $|E\rangle$ is contained in $|G_T\rangle$ is $|a|^2$. The increase in energy due to the mixture of the excited state is (after some algebra)

$$\varepsilon_1 = \frac{\langle G_T|H|G_T\rangle}{\langle G_T|G_T\rangle} - \varepsilon_0 = \frac{\langle aE + G|H|aE + G\rangle}{1 + |a|^2} - \varepsilon_0 ,$$

or

$$\varepsilon_1 = |a|^2 \Delta . \tag{7.123}$$

In addition, due to first-order perturbation theory, the spin-orbit interaction will cause a change in energy given by

$$\varepsilon_2 = \lambda \langle G_T|\boldsymbol{L}|G_T\rangle \cdot \boldsymbol{S} . \tag{7.124}$$

We assume the angular momentum $\boldsymbol{L}$ is quenched in the original ground state so by definition $\langle G|\boldsymbol{L}|G\rangle = 0$. (See also White, [7.68 p43]. White explains that if a crystal field removes the orbital degeneracy, then the matrix element of $\boldsymbol{L}$ must be zero. This does not mean the matrix element of L^2 in the same state is zero.) Thus to first order in a,

$$\varepsilon_2 = \lambda a^* \langle E|\boldsymbol{L}|G\rangle \cdot \boldsymbol{S} + \lambda a \langle G|\boldsymbol{L}|E\rangle \cdot \boldsymbol{S} . \tag{7.125}$$

The total change in energy given by (7.123) and (7.125) is $\varepsilon = \varepsilon_1 + \varepsilon_2$. Since a and a^* are complex with two components we can treat them as linearly independent, so $\partial\varepsilon/\partial a^* = 0$, which gives

$$a = \frac{-\langle E|\lambda \boldsymbol{L}|G\rangle \cdot \boldsymbol{S}}{\Delta} .$$

Therefore, after some algebra $\varepsilon = \varepsilon_1 + \varepsilon_2$ becomes

$$\varepsilon = -|a|^2 \Delta = \frac{-|\langle E|\lambda \boldsymbol{L}|G\rangle \cdot \boldsymbol{S}|^2}{\Delta} < 0 ,$$

a decrease in energy. If we let

$$\boldsymbol{A} = \frac{\langle E|\lambda \boldsymbol{L}|G\rangle}{\sqrt{\Delta}} ,$$

then

$$\varepsilon = -\boldsymbol{A} \cdot \boldsymbol{S}\boldsymbol{A}^* \boldsymbol{S} = -\sum\nolimits_{\mu,\nu} S_\mu B_{\mu\nu} S_\nu ,$$

where $B_{\mu\nu} = A_\mu A_\nu^*$. If we let $\boldsymbol{S}$ become a spin operator, we get the following Hamiltonian for single-ion anisotropy:

$$\mathcal{H}_{spin} = -\sum\nolimits_{\mu,\nu} S_\mu B_{\mu\nu} S_\nu . \tag{7.126}$$

When we have axial symmetry, this simplifies to

$$\mathcal{H}_{spin} = -DS_z^2 .$$

For cubic crystal fields, the quadratic (in S) terms go to a constant and can be neglected. In that case, we have to go to a higher order. Things are also more complicated if the ground state has orbital degeneracy. Finally, it is also possible to have anisotropic exchange. Also, as we show below, the shape of the sample can generate anisotropy.

Magnetostatics (B)

The magnetostatic energy can be regarded as the quantity whose reduction causes domains to form. The other interactions then, in a sense, control the details of how the domains form. Domain formation will be considered in Sect. 7.3. Here we will show how the domain magnetostatic interaction can cause shape anisotropy.

Consider a magnetized material in which there is no real or displacement current. The two relevant Maxwell equations can be written in the absence of external currents and in the static situation

$$\nabla \times \boldsymbol{H} = 0, \tag{7.127}$$

$$\nabla \cdot \boldsymbol{B} = 0. \tag{7.128}$$

Equation (7.127) implies there is a potential Φ from which the magnetic field $\boldsymbol{H}$ can be derived:

$$\boldsymbol{H} = -\nabla \Phi. \tag{7.129}$$

We assume a constitutive equation linking the magnetic induction $\boldsymbol{B}$, the magnetization $\boldsymbol{M}$ and $\boldsymbol{H}$;

$$\boldsymbol{B} = \mu_0(\boldsymbol{H} + \boldsymbol{M}), \tag{7.130}$$

where μ_0 is called the permeability of free space. Equations (7.128) and (7.130) become

$$\nabla \cdot \boldsymbol{H} = -\nabla \cdot \boldsymbol{M}. \tag{7.131}$$

In terms of the magnetic potential Φ,

$$\nabla^2 \Phi = \nabla \cdot \boldsymbol{M}. \tag{7.132}$$

This is analogous to Poisson's equation of electrostatics with $\rho_M = -\nabla \cdot \boldsymbol{M}$ playing the role of a magnetic source density.

By analogy to electrostatics, and in terms of equivalent surface and volume pole densities, we have

$$\Phi = \frac{1}{4\pi}\left[\int_S \frac{\boldsymbol{M} \cdot \mathrm{d}\boldsymbol{S}}{r} - \int_V \frac{\nabla \cdot \boldsymbol{M}}{r} \mathrm{d}V\right], \tag{7.133}$$

where S and V refer to the surface and volume of the magnetized body. By analogy to electrostatics the magnetostatic self-energy is

$$U_M = \frac{\mu_0}{2} \int \rho_M \Phi \mathrm{d}V = -\frac{\mu_0}{2} \int \nabla \cdot \boldsymbol{M} \Phi \mathrm{d}V = -\frac{\mu_0}{2} \int \boldsymbol{M} \cdot \boldsymbol{H} \mathrm{d}V \quad \left(\text{since } \int_{\text{all space}} \nabla \cdot (\boldsymbol{M}\Phi) \mathrm{d}V = 0\right), \tag{7.134}$$

which also would follow directly from the energy of a dipole $\boldsymbol{\mu}$ in a magnetic field ($-\boldsymbol{\mu} \cdot \boldsymbol{B}$), with a 1/2 inserted to eliminate double counting. Using $\nabla \cdot \boldsymbol{M} = -\nabla \cdot \boldsymbol{H}$ and $\int_{\text{all space}} \nabla \cdot (\boldsymbol{H}\Phi) \mathrm{d}V = 0$, we get

$$U_M = \frac{\mu_0}{2} \int H^2 \mathrm{d}V. \tag{7.135}$$

For ellipsoidal specimens the magnetization is uniform and

$$\boldsymbol{H}_D = -D\boldsymbol{M} , \tag{7.136}$$

where $\boldsymbol{H}_D$ is the demagnetization field, D is the demagnetization factor that depends on the shape of the sample and the direction of magnetization and hence one has shape isotropy, since (7.135) would have different values for $\boldsymbol{M}$ in different directions. For ellipsoidal magnets, the demagnetization energy per unit volume is then

$$u_M = \frac{\mu_0}{2} D^2 M^2 . \tag{7.137}$$

7.2.3 Spin Waves and Magnons (B)

If there is an external magnetic field $B = \mu_0 H\hat{z}$, and if the magnetic moment of each atom is $\boldsymbol{m} = 2\mu\boldsymbol{S}$ ($2\mu\hbar \equiv -g\mu_B$ [14] in previous notation), then the above considerations tell us that the Hamiltonian describing an (nn) exchange coupled spin system is

$$\mathcal{H} = -J\sum\nolimits_{j\Delta} \boldsymbol{S}_j \cdot \boldsymbol{S}_{j+\Delta} - 2\mu_0 \mu H \sum\nolimits_j S_{jz} . \tag{7.138}$$

j runs over all atoms, and δ runs over the nearest neighbors of j, and also we may redefine J so as to write (7.138) as $\mathcal{H} = (J/2)\sum\ldots$. (We do this sometimes to emphasize that (7.138) double counts each interaction.) From now on it will be assumed that there exist real solids for which (7.138) is applicable. The first term in this equation is the Heisenberg Hamiltonian and the second term is the Zeeman energy.

Let

$$S^2 = (\sum\nolimits_j S_j)^2 , \tag{7.139}$$

and

$$S_z = \sum\nolimits_j S_{jz} . \tag{7.140}$$

Then it is possible to show that the total spin and the total z component of spin are constants of the motion. In other words,

$$[\mathcal{H}, S^2] = 0 , \tag{7.141}$$

and

$$[\mathcal{H}, S_z] = 0 . \tag{7.142}$$

[14] The minus sign comes from the negative charge on the electron.

Spin Waves in a Classical Heisenberg Ferromagnet (B)

We want to calculate the internal energy u (per spin) and the magnetization M. Assuming the magnetization is in the z direction and letting $\langle A \rangle$ stand for the quantum-statistical average of A, we have (if $H = 0$)

$$u = \frac{1}{N}\langle \mathcal{H} \rangle = -\frac{1}{2N}\sum\nolimits_{i,j} J_{ij}\langle \boldsymbol{S}_i \cdot \boldsymbol{S}_j \rangle, \tag{7.143}$$

and

$$M = -\frac{g\mu_B}{V}\sum\nolimits_{iz}\langle \boldsymbol{S}_{iz} \rangle, \tag{7.144}$$

(with the $\boldsymbol{S}$ written in units of $\hbar$ and V is the volume of the crystal and J_{ij} absorbs an $\hbar^2$) where the Heisenberg Hamiltonian is written in the form

$$\mathcal{H} = -\frac{1}{2}\sum\nolimits_{i,j} J_{ij}\boldsymbol{S}_i \cdot \boldsymbol{S}_j .$$

Using the fact that

$$S^2 = S_x^2 + S_y^2 + S_z^2,$$

assuming a ferromagnetic ground state, and very low temperatures (where spin wave theory is valid) so that S_x and S_y are very small,

$$S_z = -\sqrt{S^2 - S_x^2 - S_y^2},$$

(negative so $M > 0$) and thus

$$S_z \cong -S\left(1 - \frac{S_x^2 + S_y^2}{2S^2}\right), \tag{7.145}$$

which can be substituted in (7.144). Then by (7.143)

$$u \cong -\frac{1}{2N}\sum\nolimits_{i,j} S^2 J_{ij}\left\langle\left(1 - \frac{S_{ix}^2 + S_{iy}^2}{2S^2}\right)\left(1 - \frac{S_{jx}^2 + S_{jy}^2}{2S^2}\right)\right\rangle$$
$$-\frac{1}{2N}\sum\nolimits_{i,j} J_{ij}\langle S_{ix}S_{jx} + S_{iy}S_{jy}\rangle.$$

We obtain

$$M = \frac{N}{V}g\mu_B S - \frac{g\mu_B}{2SV}\sum\nolimits_i \langle S_{ix}^2 + S_{iy}^2 \rangle, \tag{7.146}$$

$$u = -\frac{S^2 Jz}{2} + \frac{1}{2N}\sum\nolimits_{i,j} J_{ij}\langle S_{ix}^2 + S_{iy}^2 - S_{ix}S_{jx} - S_{iy}S_{jy}\rangle, \tag{7.147}$$

where z is the number of nearest neighbors. It is now convenient to Fourier transform the spins and the exchange integral

$$\boldsymbol{S}_i = \sum_k S_{\boldsymbol{k}} \mathrm{e}^{\mathrm{i}\boldsymbol{k}\cdot\boldsymbol{R}} \tag{7.148}$$

$$J(\boldsymbol{k}) = \sum_{\boldsymbol{R}} J(\boldsymbol{R}) \mathrm{e}^{\mathrm{i}\boldsymbol{q}\cdot\boldsymbol{R}} . \tag{7.149}$$

Using the standard crystal lattice mathematics and $S_{-kx} = S^*_{kx}$, we find:

$$M = \frac{N}{V} g\mu_B S \left\{ 1 - \frac{1}{2S} \sum_k \left\langle S_{kx} S^*_{kx} + S_{ky} S^*_{ky} \right\rangle \right\} \tag{7.150}$$

$$u = -\frac{S^2 Jz}{2} + \frac{1}{2} \sum_k \left(J(0) - J(k) \right) \left\langle S_{kx} S^*_{kx} + S_{ky} S^*_{ky} \right\rangle . \tag{7.151}$$

We still have to evaluate the thermal averages. To do this, it is convenient to exploit the analogy of the spin waves to a set of uncoupled harmonic oscillators whose energy is proportional to the amplitude squared. We do this by deriving the equations of motion and showing in our low-temperature "spin-wave" approximation that they are harmonic oscillators. We can write the Heisenberg Hamiltonian equation as

$$\mathcal{H} = -\frac{1}{2} \sum_j \left\{ \sum_i J_{ij} \frac{\boldsymbol{S}_i}{-g\mu_B} \right\} (-g\mu_B \boldsymbol{S}_j) , \tag{7.152}$$

where $-g\mu_B S_j$ is the magnetic moment. The 1/2 takes into account the double counting and we therefore identify the effective field acting on $\boldsymbol{S}_j$ as

$$\boldsymbol{B}_{M_j} = -\frac{1}{g\mu_B} \sum_i J_{ij} \boldsymbol{S}_i . \tag{7.153}$$

Treating the $\boldsymbol{S}_i$ as dimensionless so $\hbar \boldsymbol{S}_i$ is the angular momentum, and using the fact that torque is the rate of change of angular momentum and is the moment crossed into field, we have for the equations of motion

$$\hbar \frac{\mathrm{d}\boldsymbol{S}_j}{\mathrm{d}t} = \sum_i J_{ij} \boldsymbol{S}_j \times \boldsymbol{S}_i . \tag{7.154}$$

We leave as a problem to show that after Fourier transformation the equations of motion can be written:

$$\hbar \frac{\mathrm{d}\boldsymbol{S}_k}{\mathrm{d}t} = \sum_{k''} J(\boldsymbol{k}'') \boldsymbol{S}_{k-k''} \times \boldsymbol{S}_{k''} . \tag{7.155}$$

For the ferromagnetic ground state at low temperature, we assume that

$$|\boldsymbol{S}_{k=0}| >> |\boldsymbol{S}_{k\neq 0}| ,$$

since

$$\boldsymbol{S}_{k=0} = \frac{1}{N}\sum\nolimits_R S_{\boldsymbol{R}} ,$$

and at absolute zero,

$$\boldsymbol{S}_{k=0} = S\hat{\boldsymbol{k}} \ , \ S_{k\neq 0} = 0 .$$

Even with small excitations, we assume $S_{0z} = S$, $S_{0x} = S_{0y} = 0$ and S_{kx}, S_{ky} are of first order. Retaining only quantities of first order, we have

$$\hbar\frac{\mathrm{d}S_{kx}}{\mathrm{d}t} = S[J(0) - J(k)]\, S_{ky} \tag{7.156a}$$

$$\hbar\frac{\mathrm{d}S_{ky}}{\mathrm{d}t} = -S[J(0) - J(k)]\, S_{kx} \tag{7.156b}$$

$$\hbar\frac{\mathrm{d}S_{kz}}{\mathrm{d}t} = 0 . \tag{7.156c}$$

Combining (7.156a) and (7.156b), we obtain harmonic-oscillator-type equations with frequencies $\omega(k)$ and energies $\varepsilon(k)$ given by

$$\varepsilon(k) = \hbar\omega(k) = S[J(0) - J(k)] . \tag{7.157}$$

Combining this result with (7.151), we have for the average energy per oscillator,

$$u = -\frac{S^2 Jz}{2} + \frac{1}{2}\sum\nolimits_k \frac{\varepsilon(k)}{S}\left\langle |\, S_{kx}\,|^2 + |\, S_{ky}\,|^2 \right\rangle$$

for z nearest neighbors. For quantized harmonic oscillators, up to an additive term, the average energy per oscillator would be

$$\frac{1}{N}\sum\nolimits_k \varepsilon(k)\langle n_k\rangle .$$

Thus, we identify $\langle n_k\rangle$ as

$$\left\langle \frac{|\, S_{kx}\,|^2 + |\, S_{ky}\,|^2}{2S} \right\rangle N ,$$

and we write (7.150) and (7.151) as

$$M = \frac{N}{V} g\mu_B S\left\{1 - \frac{1}{NS}\sum\nolimits_k \langle n_k\rangle\right\} \tag{7.158}$$

$$u = -\frac{S^2 Jz}{2} + \frac{1}{N}\sum\nolimits_k \varepsilon(k)\langle n_k\rangle . \tag{7.159}$$

Now $\langle n_k \rangle$ is the average number of excitations in mode k (magnons) at temperature T.

By analogy with phonons (which represent quanta of harmonic oscillators) we say

$$\langle n_k \rangle = \frac{1}{e^{\varepsilon(k)/kT} - 1} . \tag{7.160}$$

As an example, we work out the consequences of this for simple cubic lattices with $Z = 6$ and nearest-neighbor coupling.

$$J(k) = \sum J(R) e^{i\boldsymbol{k} \cdot \boldsymbol{R}} = 2J(\cos k_x a + \cos k_y a + \cos k_z a).$$

At low temperatures where only small k are important, we find

$$\varepsilon(k) = S[J(0) - J(k)] \cong SJk^2a^2 . \tag{7.161}$$

We will evaluate (7.158) and (7.159) using (7.160) and (7.161) later after treating spin waves quantum mechanically from the beginning.

The name "spin-waves" comes from the following picture. In Fig. 7.10, suppose

$$S_{kx} = S\sin(\theta)\exp[i\omega(k)t] ,$$

then

$$\hbar\dot{S}_{kx} = i\omega(q)\hbar S_{kx} = \omega(k)\hbar S_{ky}$$

by the equation of motion. So,

$$iS_{kx} = S_{ky} .$$

Therefore, if we had one spin-wave mode q in the x direction, e.g., then

$$S_{Rx} = \exp(i\boldsymbol{k} \cdot \boldsymbol{R})S_{kx} = S\sin(\theta)\exp[i(kR_x + \omega t)],$$
$$S_{Ry} = S\sin(\theta)\exp[i(kR_x + \omega t - \pi/2)] .$$

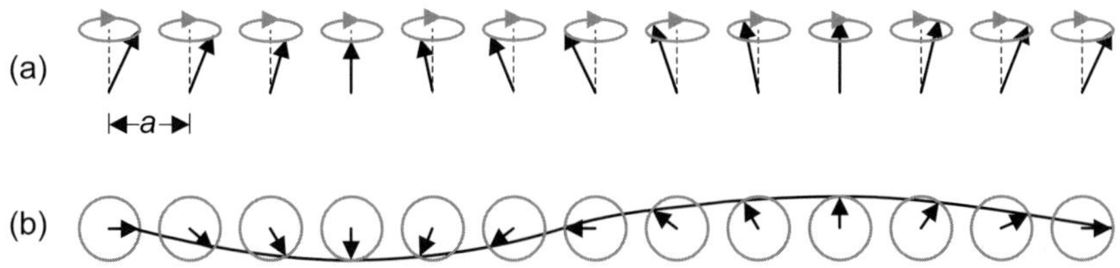

Fig. 7.10. Classical representation of a spin wave in one dimension (**a**) viewed from side and (**b**) viewed from top (along $-z$). The phase angle from spin to spin changes by ka. Adapted from Kittel C, *Introduction to Solid State Physics*, 7th edn, Copyright © 1996 John Wiley and Sons, Inc. This material is used by permission of John Wiley and Sons, Inc

Thus, if we take the real part, we find

$$S_{Rx} = S\sin(\theta)\cos(kR_x + \omega t),$$
$$S_{Ry} = S\sin(\theta)\sin(kR_x + \omega t),$$

and the spins all spin with the same frequency but with the phase changing by ka, which is the change in kR_x, as we move from spin to spin along the x-axis.

As we have seen, spin waves are collective excitations in ordered spin systems. The collective excitations consist in the propagation of a spin deviation, θ. A localized spin at a site is said to undergo a deviation when its direction deviates from the direction of magnetization of the solid below the critical temperature. Classically, we can think of spin waves as vibrations in the magnetic moment density. As mentioned, quanta of the spin waves are called *magnons*. The concept of spin waves was originally introduced by F. Bloch, who used it to explain the temperature dependence of the magnetization of a ferromagnet at low temperatures. The existence of spin waves has now been definitely proved by experiment. Thus the concept has more validity than its derivation from the Heisenberg Hamiltonian might suggest. We will only discuss spin waves in ferromagnets but it is possible to make similar comments about them in any ordered magnetic structure. The differences between the ferromagnetic case and the antiferromagnetic case, for example, are not entirely trivial [60, p 61].

Spin Waves in a Quantum Heisenberg Ferromagnet (A)

The aim of this section is rather simple. We want to show that the quantum Heisenberg Hamiltonian can be recast, in a suitable approximation, so that its energy excitations are harmonic-oscillator-like, just as we found classically (7.161).

Here we make two transformations and a long-wavelength, low-temperature approximation. One transformation takes the Hamiltonian to a localized excitation description and the other to an unlocalized (magnon) description. However, the algebra can get a little complex.

Equation (7.138) (with $\hbar = 1$ or $2\mu = -g\mu_B$) is our starting point for the three-dimensional case, but it is convenient to transform this equation to another form for calculation. From our previous discussion, we believe that magnons are similar to phonons (insofar as their mathematical description goes), and so we might guess that some sort of second quantization notation would be appropriate. We have already indicated that the squared total spin and the z component of total spin give good quantum numbers. We can also show that S_j^2 commutes with the Heisenberg Hamiltonian so that its eigenvalues $S(S + 1)$ are good quantum numbers. This makes sense because it just says that the total spin of each atom remains constant. We assume that the spin S of every ion is the same. Although each atom has three components of each spin vector, only two of the components are independent.

The Holstein and Primakoff Transformation (A) Holstein and Primakoff[15] have developed a transformation that not only has two independent variables, but also utilizes the very convenient second quantization notation. The Holstein–Primakoff transformation is also very useful for obtaining terms that describe magnon–magnon interactions.[16] This transformation is (with $\hbar = 1$ or S representing $S/\hbar$):

$$S_j^+ \equiv S_{jx} + \mathrm{i}S_{jy} = \sqrt{2S}\left[1 - \frac{a_j^\dagger a_j}{2S}\right]^{1/2} a_j, \tag{7.162}$$

$$S_j^- \equiv S_{jx} - \mathrm{i}S_{jy} = \sqrt{2S}\, a_j^\dagger\left[1 - \frac{a_j^\dagger a_j}{2S}\right]^{1/2}, \tag{7.163}$$

$$S_{jz} \equiv S - a_j^\dagger a_j. \tag{7.164}$$

We could use these transformation equations to attempt to determine what properties a_j and $a_j^\dagger$ must have. However, it is much simpler to define the properties of the a_j and $a_j^\dagger$ and show that with these definitions the known properties of the S_j operators are obtained. We will assume that the $a^\dagger$ and a are *boson* creation and annihilation operators (see Appendix G) and hence they satisfy the commutation relations

$$[a_j, a_l^\dagger] = \delta_j^l. \tag{7.165}$$

We first show that (7.164) is consistent with (7.162) and (7.163). This amounts to showing that the Holstein–Primakoff transformation automatically puts in the constraint that there are only two independent components of spin for each atom. We start by dropping the subscript j for a particular atom and by using the fact that S_j^2 has a good quantum number so we can substitute $S(S+1)$ for S_j^2 (with $\hbar = 1$). We can then write

$$S(S+1) = S_x^2 + S_y^2 + S_z^2 = S_z^2 + \tfrac{1}{2}(S^+S^- + S^-S^+). \tag{7.166}$$

By use of (7.162) and (7.163) we can use (7.166) to calculate S_z^2. That is,

$$S_z^2 = S(S+1) - S\left[\left(1 - \frac{a^\dagger a}{2S}\right)^{1/2}(1 + a^\dagger a)\left(1 - \frac{a^\dagger a}{2S}\right)^{1/2} + a^\dagger\left(1 - \frac{a^\dagger a}{2S}\right)a\right]. \tag{7.167}$$

---

[15] See, for example, [7.38].

[16] At least for high magnetic fields; see Dyson [7.18].

Remember that we define a function of operators in terms of a power series for the function, and therefore it is clear that $a^\dagger a$ will commute with any function of $a^\dagger a$. Also note that $[a^\dagger a, a] = a^\dagger aa - aa^\dagger a = a^\dagger aa - (1 + a^\dagger a)a = -a$, and so we can transform (7.167) to give after several algebraic steps:

$$S_z^2 = (S - a^\dagger a)^2 . \tag{7.168}$$

Equation (7.168) is consistent with (7.164), which was to be shown.

We still need to show that S_j^+ and S_j^- defined in terms of the annihilation and creation operators act as ladder operators should act. Let us define an eigenket of S_j^2 and S_{jz}, by (still with $\hbar = 1$)

$$S_j^2 \left| S, m_s \right\rangle = S(S+1) \left| S, m_s \right\rangle , \tag{7.169}$$

and

$$S_{jz} \left| S, m_s \right\rangle = m_s \left| S, m_s \right\rangle . \tag{7.170}$$

Let us further define a spin-deviation eigenvalue by

$$n = S - m_S , \tag{7.171}$$

and for convenience let us shorten our notation by defining

$$\left| n \right\rangle = \left| S, m_s \right\rangle . \tag{7.172}$$

By (7.162) we can write

$$S_j^+ \left| n \right\rangle = \sqrt{2S} \left(1 - \frac{a_j^\dagger a_j}{2S} \right)^{1/2} a_j \left| n \right\rangle = \sqrt{2S} \left(1 - \frac{n-1}{2S} \right)^{1/2} \sqrt{n} \left| n-1 \right\rangle , \tag{7.173}$$

where we have used $a_j|n\rangle = n^{1/2}|n - 1\rangle$ and also the fact that

$$a_j^\dagger a_j \left| n \right\rangle = (S - S_{jz}) \left| n \right\rangle = n \left| n \right\rangle . \tag{7.174}$$

By converting back to the $|S, m_s\rangle$ notation, we see that (7.173) can be written

$$S_j^+ \left| S, m_s \right\rangle = \sqrt{(S - m_s)(S + m_s + 1)} \left| S, m_s + 1 \right\rangle . \tag{7.175}$$

Therefore S_j^+ does have the characteristic property of a ladder operator, which is what we wanted to show. We can similarly show that the S_j^- has the step-down ladder properties.

Note that since (7.175) is true, we must have that

$$S^+ \left| S, m_s = S \right\rangle = 0 . \tag{7.176}$$

A similar calculation shows that

$$S^{-}\left|S,-m_s = S\right\rangle = 0\,. \tag{7.177}$$

We needed to assure ourselves that this property still held even though we defined the S^+ and S^- in terms of the $a_j^\dagger$ and a_j. This is because we normally think of the a as operating on $|n\rangle$, where $0 \le n \le \infty$. In our situation we see that $0 \le n < 2S + 1$. We have now completed the verification of the consistency of the Holstein–Primakoff transformation. It is time to recast the Heisenberg Hamiltonian in this new notation.

Combining the results of Problem 7.10 and the Holstein–Primakoff transformation, we can write

$$\mathcal{H} = -J\sum_{j\Delta}\left\{(S-a_j^\dagger a_j)(S-a_{j+\Delta}^\dagger a_{j+\Delta}) + S\left[a_j^\dagger\left(1-\frac{a_j^\dagger a_j}{2S}\right)^{1/2}\left(1-\frac{a_{j+\Delta}^\dagger a_{j+\Delta}}{2S}\right)^{1/2} a_{j+\delta} + \left(1-\frac{a_j^\dagger a_j}{2S}\right)^{1/2} a_j a_{j+\Delta}^\dagger\left(1-\frac{a_{j+\Delta}^\dagger a_{j+\Delta}}{2S}\right)^{1/2}\right]\right\} + g\mu_B(\mu_0 H)\sum_j(S-a_j^\dagger a_j)\,. \tag{7.178}$$

Equation (7.178) is the Heisenberg Hamiltonian (plus a term for an external magnetic field) expressed in second quantization notation. It seems as if the problem has been complicated rather than simplified by the Holstein–Primakoff transformation. Actually both (7.138) and (7.178) are equally impossible to solve exactly. Both are many-body problems. The point is that (7.178) is in a form that can be approximated fairly easily. The approximation that will be made is to expand the square roots and concentrate on low-order terms. Before this is done, it is convenient to take full advantage of translational symmetry. This will be done in the next section.

Magnons (A) The $a_j^\dagger$ create localized spin deviations at a single site (one atom per unit cell is assumed). What we need (in order to take translational symmetry into account) is creation operators that create Bloch-like nonlocalized excitations. A transformation that will do this is

$$B_{\boldsymbol{k}} = \frac{1}{\sqrt{N}}\sum_j \exp(\mathrm{i}\boldsymbol{k}\cdot\boldsymbol{R}_j)a_j\,, \tag{7.179a}$$

and

$$B_{\boldsymbol{k}}^\dagger = \frac{1}{\sqrt{N}}\sum_j \exp(-\mathrm{i}\boldsymbol{k}\cdot\boldsymbol{R}_j)a_j^\dagger\,, \tag{7.179b}$$

where $\boldsymbol{R}_j$ is defined by (2.171) and cyclic boundary conditions are used so that the $\boldsymbol{k}$ are defined by (2.175). $N = N_1N_2N_3$ and so the delta function relations (2.178) to (2.184) are valid. $\boldsymbol{k}$ will be assumed to be restricted to the first Brillouin zone. Using all these results, we can derive the inverse transformation

$$a_j = \frac{1}{\sqrt{N}} \sum\nolimits_{\boldsymbol{k}} \exp(-\mathrm{i}\boldsymbol{k} \cdot \boldsymbol{R}_j) B_{\boldsymbol{k}} , \tag{7.180a}$$

and

$$a_j^\dagger = \frac{1}{\sqrt{N}} \sum\nolimits_{\boldsymbol{k}} \exp(\mathrm{i}\boldsymbol{k} \cdot \boldsymbol{R}_j) B_{\boldsymbol{k}}^\dagger . \tag{7.180b}$$

So far we have not shown that the B are boson creation and annihilation operators. To show this, we merely need to show that the B satisfy the appropriate commutation relations. The calculation is straightforward, and is left as a problem to show that the B_k obey the same commutation relations as the a_j.

We can give a very precise definition to the word *magnon.* First let us review some physical principles. Exchange coupled spin systems (e.g. ferromagnets and antiferromagnets) have low-energy states that are wave-like. These wave-like energy states are called spin waves. A spin wave is quantized into units called magnons. We may have spin waves in any structure that is magnetically ordered. Since in the low-temperature region there are only a few spin waves that are excited and thus their complicated interactions are not so important, this is the best temperature region to examine spin waves. Mathematically, precisely whatever is created by $B_k^\dagger$ and annihilated by B_k is called a *magnon.*

There is a nice theorem about the number of magnons. The total number of magnons equals the total spin deviation quantum number. This theorem is easily proved as shown below:

$$\begin{aligned}
\Delta_S &= \sum\nolimits_j (S - S_{jz}) = \sum\nolimits_j a^\dagger a_j \\
&= \frac{1}{N} \sum\nolimits_{i,\boldsymbol{k},\boldsymbol{k}'} \exp[\mathrm{i}(\boldsymbol{k} - \boldsymbol{k}') \cdot \boldsymbol{R}_j] B_{\boldsymbol{k}}^\dagger B_{\boldsymbol{k}'} \\
&= \sum\nolimits_{\boldsymbol{k},\boldsymbol{k}'} \delta_{\boldsymbol{k}}^{\boldsymbol{k}'} B_{\boldsymbol{k}}^\dagger B_{\boldsymbol{k}'} \\
&= \sum\nolimits_{\boldsymbol{k}} B_{\boldsymbol{k}}^\dagger B_{\boldsymbol{k}} .
\end{aligned}$$

This proves the theorem, since $B_k^\dagger B_k$ is the occupation number operator for the number of magnons in mode $\boldsymbol{k}$.

The Hamiltonian defined by (7.178) will now be approximated. The spin-wave variables B_k will also be substituted.

At low temperatures we may expect the spin-deviation quantum number to be rather small. Thus we have approximately

$$\left\langle a_j^\dagger a_j \right\rangle << S . \tag{7.181}$$

This implies that the relation between the S and a can be approximated by

$$S_j^- \cong \sqrt{2S}\left(a_j^\dagger - \frac{a_j^\dagger a_j^\dagger a_j}{4S}\right), \tag{7.182a}$$

$$S_j^+ \cong \sqrt{2S}\left(a_j - \frac{a_j^\dagger a_j a_j}{4S}\right), \tag{7.182b}$$

and

$$S_{jz} = S - a_j^\dagger a_j . \tag{7.182c}$$

Expressing these results in terms of the B, we find

$$\begin{aligned} S_j^+ \cong \sqrt{\frac{2S}{N}}\Big\{ & \textstyle\sum_{\boldsymbol{k}} \exp(-\mathrm{i}\boldsymbol{k}\cdot\boldsymbol{R}_j) B_{\boldsymbol{k}} \\ & - \frac{1}{4SN} \textstyle\sum_{\boldsymbol{k},\boldsymbol{k}',\boldsymbol{k}''} \exp[\mathrm{i}(\boldsymbol{k}-\boldsymbol{k}'-\boldsymbol{k}'')\cdot\boldsymbol{R}_j]\, B_{\boldsymbol{k}}^\dagger B_{\boldsymbol{k}'} B_{\boldsymbol{k}''} \Big\}, \end{aligned} \tag{7.183a}$$

$$\begin{aligned} S_j^- \cong \sqrt{\frac{2S}{N}}\Big\{ & \textstyle\sum_{\boldsymbol{k}} \exp(\mathrm{i}\boldsymbol{k}\cdot\boldsymbol{R}_j) B_{\boldsymbol{k}} \\ & - \frac{1}{4SN} \textstyle\sum_{\boldsymbol{k},\boldsymbol{k}',\boldsymbol{k}''} \exp[\mathrm{i}(\boldsymbol{k}+\boldsymbol{k}'-\boldsymbol{k}'')\cdot\boldsymbol{R}_j]\, B_{\boldsymbol{k}}^\dagger B_{\boldsymbol{k}'}^\dagger B_{\boldsymbol{k}''} \Big\}, \end{aligned} \tag{7.183b}$$

and

$$S_{jz} = S - \frac{1}{N} \textstyle\sum_{\boldsymbol{k},\boldsymbol{k}'} \exp[\mathrm{i}(\boldsymbol{k}-\boldsymbol{k}')\cdot\boldsymbol{R}_j]\, B_{\boldsymbol{k}}^\dagger B_{\boldsymbol{k}'} . \tag{7.183c}$$

The details of the calculation begin to get rather long at about this stage. The approximate Hamiltonian in terms of spin-wave variables is obtained by substituting (7.183) into (7.178). Considerable simplification results from the delta function relations. Terms of order $(\langle a_i^\dagger a_i\rangle/\mathrm{S})^2$ are to be neglected for consistency. The final result is

$$\mathcal{H} = \mathcal{H}_0 + \mathcal{H}_{\text{ex}} , \tag{7.184}$$

neglecting a constant term, where Z is the number of nearest neighbors, $\mathcal{H}_0$ is the term that is bilinear in the spin wave variables and is given by

$$\begin{aligned} \mathcal{H}_0 = & -JSZ[\textstyle\sum_{\boldsymbol{k}} (\alpha_{\boldsymbol{k}}(1 + B_{\boldsymbol{k}}^\dagger B_{\boldsymbol{k}}) + \alpha_{-\boldsymbol{k}} B_{\boldsymbol{k}}^\dagger B_{\boldsymbol{k}} - 2B_{\boldsymbol{k}}^\dagger B_{\boldsymbol{k}})] \\ & + g\mu_B(\mu_0 H) \textstyle\sum_{\boldsymbol{k}} B_{\boldsymbol{k}}^\dagger B_{\boldsymbol{k}} , \end{aligned} \tag{7.185}$$

$$\alpha_{\boldsymbol{k}} = \frac{1}{Z} \textstyle\sum_{\boldsymbol{\Delta}} \exp(\mathrm{i}\boldsymbol{k}\cdot\boldsymbol{\Delta}) , \tag{7.186}$$

and $\mathcal{H}_{ex}$ is called the *exchange interaction* Hamiltonian and is biquadratic in the spin-wave variables. It is given by

$$\mathcal{H}_{ex} \propto Z\frac{J}{N}\sum\nolimits_{k_1k_2k_3k_4}\delta_{k_1+k_4}^{k_2+k_3}(B_{k_1}B_{k_2}^{\dagger}-\delta_{k_1}^{k_2})B_{k_3}^{\dagger}B_{k_4}(\alpha_{k_1}-\alpha_{k_1-k_2})\,. \quad (7.187)$$

Note that $\mathcal{H}_0$ describes magnons without interactions and $\mathcal{H}_{ex}$ includes terms that describe the effect of interactions. Mathematically, we do not want to consider interactions. Physically, it makes sense to believe that interactions should not be important at low temperatures. We can show that $\mathcal{H}_{ex}$ can be neglected for long-wavelength magnons, which should be the only important magnons at low temperature. We will therefore neglect $\mathcal{H}_{ex}$ in all discussions below.

$\mathcal{H}_0$ can be somewhat simplified. Incidentally, the formalism that is being used assumes only one atom per unit cell and that all atoms are equally spaced and identical. Among other things, this precludes the possibility of having "optical magnons." This is analogous to the lattice vibration problem where we do not have optical phonons in lattices with one atom per unit cell.

$\mathcal{H}_0$ can be simplified by noting that if the crystal has a center of symmetry, then $\alpha_k = \alpha_{-k}$, and also

$$\sum\nolimits_k \alpha_k = \frac{1}{Z}\sum\nolimits_{\Delta}\sum\nolimits_k \exp(i\boldsymbol{k}\cdot\boldsymbol{\Delta}) = \frac{N}{Z}\sum\nolimits_{\Delta}\delta_{\Delta}^{0} = 0\,,$$

where the last term is zero because $\boldsymbol{\Delta}$, being the vector to nearest-neighbor atoms, can never be zero. Also note that $BB^{\dagger} - 1 = B^{\dagger}B$. Using these results and defining (with $H = 0$)

$$\hbar\omega_{\boldsymbol{k}} = 2JSZ(1-\alpha_{\boldsymbol{k}})\,, \quad (7.188)$$

we find

$$H_0 = \sum\nolimits_{\boldsymbol{k}}\hbar\omega_{\boldsymbol{k}}n_{\boldsymbol{k}}\,, \quad (7.189)$$

where n_k is the occupation number operator for the magnons in mode $\boldsymbol{k}$.

If the wavelength of the spin waves is much greater than the lattice spacing, so that atomic details are not of much interest, then we are in a classical region. In this region, it makes sense to assume that $\boldsymbol{k}\cdot\boldsymbol{\Delta} \ll 1$, which is also the long-wavelength approximation made in neglecting $\mathcal{H}_{ex}$. Thus we find

$$\hbar\omega_{\boldsymbol{k}} \cong JS\sum\nolimits_{\Delta}(\boldsymbol{k}\cdot\boldsymbol{\Delta})^2\,. \quad (7.190)$$

If further we have a simple cubic, bcc, or fcc lattice, then

$$\hbar\omega_{\boldsymbol{k}} = \frac{\hbar^2\boldsymbol{k}^2}{2m^*}\,, \quad (7.191)$$

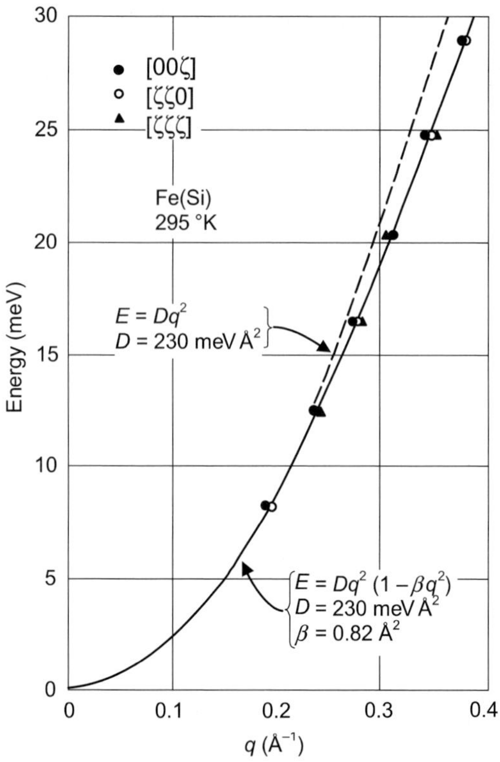

Fig. 7.11. Fe (12 at. % Si) room-temperature spin-wave dispersion relations at low energy. Reprinted with permission from Lynn JW, *Phys Rev B* **11**(7), 2624 (1975). Copyright 1975 by the American Physical Society

where

$$m^* \propto (2ZJSa^2)^{-1}, \tag{7.192}$$

and a is the lattice spacing. The reality of spin-wave dispersion has been shown by inelastic neutron scattering. See Fig. 7.11.

Specific Heat of Spin Waves (A) With

$$\frac{\left\langle a_i^\dagger a_i \right\rangle}{S} << 1, \quad ka << 1, \quad H = 0,$$

and assuming we have a monatomic lattice, the magnons were found to have the energies

$$\hbar\omega_{\boldsymbol{k}} = Ck^2, \tag{7.193}$$

where C is a constant. Thus apart from notation (7.161) and (7.193) are identical. We also know that the magnons behave as bosons. We can return to (7.158), (7.159), (7.160), and (7.161) to evaluate the magnetization as well as the internal energy due to spin waves.

Now in (7.158) we can replace a sum with an integral because for large N the number of states is fairly dense and in d$\boldsymbol{k}$ per unit volume is d$\boldsymbol{k}/(2\pi)^3$. So

$$\sum_k \frac{1}{\exp(JSk^2a^2/k_BT)-1} \to \frac{V}{(2\pi)^3}\int \frac{\mathrm{d}\boldsymbol{k}}{\exp(JSk^2a^2/k_BT)-1}$$
$$\to \frac{V}{(2\pi)^3}\int_0^\infty \frac{k^2\mathrm{d}k}{\exp(JSk^2a^2/k_BT)-1}.$$

Also we have used that at low T the upper limit can be set to infinity without appreciable error. Changing the integration variable to $x = (JS/k_BT)^{1/2}ka$, we find at low temperature

$$\sum_k \frac{1}{\exp(JSk^2a^2/k_BT)-1} \to \frac{V}{(2\pi)^3}\left(\sqrt{\frac{k_BT}{JS}}\frac{1}{a}\right)^3 N_1,$$

where

$$N_1 = \int_0^\infty \frac{x^2\mathrm{d}x}{\exp(x^2)-1}.$$

Similarly

$$\sum_k \frac{JSk^2a^2}{\exp(JSk^2a^2/k_BT)-1} \to \frac{V}{(2\pi)^3}\left(\sqrt{\frac{k_BT}{JS}}\frac{1}{a}\right)^5 N_2,$$

where

$$N_2 = \int_0^\infty \frac{x^4\mathrm{d}x}{\exp(x^2)-1}.$$

N_1 and N_2 are numbers that can be evaluated in terms of gamma functions and Riemann zeta functions. We thus find

$$M = \frac{N}{V}g\mu_B S\left\{1 - \frac{V}{2\pi^2 SN}\left(\frac{k_B}{JSa^2}\right)^{3/2} N_1 T^{3/2}\right\}, \tag{7.194}$$

and

$$u = -\frac{S^2 Jz}{2} + \frac{V}{2\pi^2 N}\left(\frac{k_B}{JSa^2}\right)^{5/2} N_2 T^{5/2}. \tag{7.195}$$

Thus, from (7.195) by taking the temperature derivative we find the low-temperature magnon specific heat, as first shown by Bloch, is

$$C_V \propto T^{3/2} . \tag{7.196}$$

Similarly, by (7.194) the low-temperature deviation from saturation goes as $T^{3/2}$. these results only depend on low-energy excitations going as k^2.

Also at low T, we have a lattice specific heat that goes as T^3. So at low T we have

$$C_V = aT^{3/2} + bT^3 ,$$

where a and b are constants. Thus

$$C_V T^{-3/2} = a + bT^{3/2} ,$$

so theoretically, plotting $CT^{-3/2}$ vs $T^{3/2}$ will yield a straight line at low T. Experimental verification is shown in Fig. 7.12 (note this is for a ferrimagnet for which the low-energy $\hbar\omega_k$ is also proportional to k^2).

At higher temperatures there are deviations from the $^3/_2$ power law and it is necessary to make refinements in the above theory. One source of deviations is spin-wave interactions. We also have to be careful that we do not approximate away the kinematical part, i.e. the part that requires the spin-deviation quantum number on a given site not to exceed $(2S_j + 1)$. Then, of course, in a more careful analysis we would have to pay more attention to the geometrical shape of the Brillouin zone. Perhaps our worst error involves (7.191), which leads to an approximate density of states and hence to an approximate form for the integral in the calculation of C_V and ΔM.

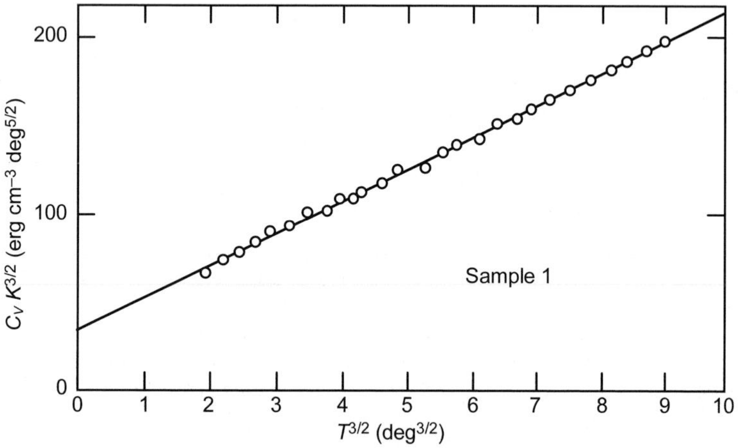

Fig. 7.12. C_V at low T for ferrimagnet YIG. After Elliott RJ and Gibson AF, *An Introduction to Solid State Physics and Applications*, Macmillan, 1974, p 461. Original data from Shinozaki SS, *Phys Rev* **122**, 388 (1961))

Table 7.1. Summary of spin-wave properties (low energy and low temperature)

	Dispersion relation	$\Delta M = M_s - M$ magnetization	C magnetic Sp. Ht.
Ferromagnet	$\omega = A_1 k^2$	$B_1 T^{3/2}$	$B_2 T^{3/2}$
Antiferromagnet	$\omega = A_2 k$	$B_2 T^2$ (sublattice)	$C_2 T^3$

A_i and B_i are constants. For discussion of spin waves in more complicated structures see, e.g., Cooper [7.13].

Equation (7.193) predicts that the density of states (up to cutoff) is proportional to the magnon energy to the $^1/_2$ power. A similar simple development for antiferromagnets [it turns out that the analog of (7.193) only involves the first power of $|k|$ for antiferromagnets] also leads to a relatively smooth dependence of the density of states on energy. In any case, a determination from analyzing the neutron diffraction of an actual magnetic substance will show a result that is not so smooth (see Fig. 7.13). Comparison of spin-wave calculations to experiment for the specific heat for EuS is shown in Fig. 7.14.[17] EuS is an ideal Heisenberg ferromagnet.

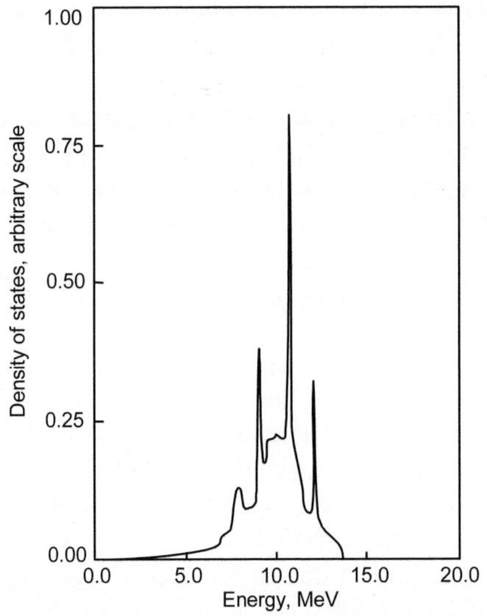

Fig. 7.13. Density of states for magnons in Tb at 90 K. The curve is a smoothed computer plot. [Reprinted with permission from Moller HB, Houmann JCG, and Mackintosh AR, *Journal of Applied Physics*, **39**(2), 807 (1968). Copyright 1968, American Institute of Physics.]

[17] A good reference for the material in this chapter on spin waves is an article by Kittel [7.38]

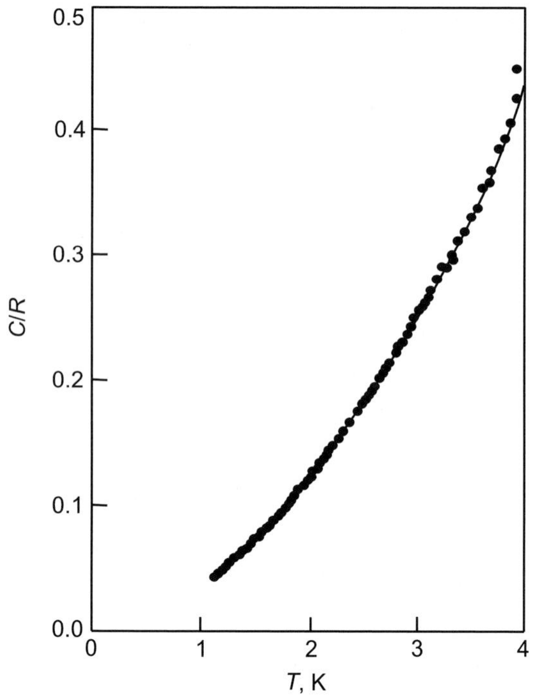

Fig. 7.14. Spin wave specific heat of EuS. An equation of the form $C/R = aT^{3/2} + bT^{5/2}$ is needed to fit this curve. For an evaluation of b, see Dyson FJ, *Physical Review*, **102**, 1230 (1956). [Reprinted with permission from McCollum, Jr. DC, and Callaway J, *Physical Review Letters*, **9** (9), 376 (1962). Copyright 1962 by the American Physical Society.]

Magnetostatic Spin Waves (MSW) (A)

For very large wavelengths, the exchange interaction between spins no longer can be assumed to be dominant. In this limit, we need to look instead at the effect of dipole–dipole interactions (which dominate the exchange interactions) as well as external magnetic fields. In this case spin-wave excitations are still possible but they are called magnetostatic waves. Magnetostatic waves can be excited by inhomogeneous magnetic fields. MSW look like spin waves of very long wavelength, but the spin coupling is due to the dipole–dipole interaction. There are many device applications of MSW (e.g. delay lines) but a discussion of them would take us too far afield. See, e.g., Auld [7.3], and Ibach and Luth [7.33]. Also see Kittel [7.38 p471ff], and Walker [7.65]. There are also surface or Damon–Eshbach wave solutions.[18]

[18] Damon and Eshbach [7.17].

7.2.4 Band Ferromagnetism (B)

Despite the obvious lack of rigor, we have justified qualitatively a Heisenberg Hamiltonian for insulators and rare earths. But what can we do when we have ferromagnetism in metals? It seems to be necessary to take into account the band structure. This topic is very complicated, and only limited comments will be made here. See Mattis [7.48], Morrish [68] and Yosida [7.72] for more discussion.

In a metal, one might hope that the electrons in unfilled core levels would interact by the Heisenberg mechanism and thus produce ferromagnetism. We might expect that the conduction process would be due to electrons in a much higher band and that there would be little interaction between the ferromagnetic electrons and conduction electrons. This is not always the case. The core levels may give rise to a band that is so wide that the associated electrons must participate in the conduction process. Alternatively, the core levels may be very tightly bound and have very narrow bands. The core wave functions may interact so little that they could not directly have the Heisenberg exchange between them. That such materials may still be ferromagnetic indicates that other electrons such as the conduction electrons must play some role (we have discussed an example in Sect. 7.2.1 under *RKKY Interaction*). Obviously, a localized spin model cannot be good for all types of ferromagnetism. If it were, the saturation magnetization per atom would be an integral number of Bohr magnetons. This does not happen in Ni, Fe, and Co, where the number of electrons per atom contributing to magnetic effects is not an integer.

Despite the fact that one must use a band picture in describing the magnetic properties of metals, it still appears that a Heisenberg Hamiltonian often leads to predictions that are approximately experimentally verified. It is for this reason that many believe the Heisenberg Hamiltonian description of magnetic materials is much more general than the original derivation would suggest.

As an approach to a theory of ferromagnetism in metals it is worthwhile to present one very simple band theory of ferromagnetism. We will discuss *Stoner's theory*, which is also known as the *theory of collective electron ferromagnetism*. See Mattis [7.48 Vol. I p250ff] and Herring [7.56 p256ff]. The two basic assumptions of Stoner's theory are:

1. The ferromagnetic electrons or holes are free-electron-like (at least near the Fermi energy); hence their density of states has the form of a constant times $E^{1/2}$, and the energy is

$$E = \frac{\hbar^2 k^2}{2m^*} . \tag{7.197a}$$

2. There is still assumed to be some sort of exchange interaction between the (free) electrons. This interaction is assumed to be representable by a molecular field M. If γ is the molecular field constant, then the exchange interaction energy of the electrons is (SI)

$$E = \pm \mu_0 \gamma M \mu , \tag{7.197b}$$

where μ represents the magnetic moment of the electrons, + indicates electrons with spin parallel, and − indicates electrons with spin antiparallel to $\boldsymbol{M}$.

The magnetization equals μ (here the magnitude of the magnetic moment of the electron $= \mu_B$) times the magnitude of the number of parallel spin electrons per unit volume minus the number of antiparallel spin electrons per unit volume. Using the ideas of Sect. 3.2.2, we can write

$$M = \left| \mu \int [f(E - \mu_0 \gamma M \mu) - f(E + \mu_0 \gamma M \mu)] \frac{K\sqrt{E}}{2V} \mathrm{d}E \right|, \tag{7.198}$$

where f is the Fermi function. The above is the basic equation of Stoner's theory, with the sum of the parallel and antiparallel electrons being constant. For $T = 0$ and sufficiently strong exchange coupling the magnetization has as its saturation value $M = N\mu$. For sufficiently weak exchange coupling the magnetization vanishes. For intermediate values of the exchange coupling the magnetization has intermediate values. Deriving M as a function of temperature from the above equation is a little tedious. The essential result is that the Stoner theory also allows the possibility of a phase transition. The qualitative details of the M versus T curves do not differ enormously from the Stoner theory to the Weiss theory. We develop one version of the Stoner theory below.

The Hubbard Model and the Mean-Field Approximation (A)

So far, except for Pauli paramagnetism, we have not considered the possibility of nonlocalized electrons carrying a moment, which may contribute to the magnetization. Consistent with the above, starting with the ideas of Pauli paramagnetism and adding an exchange interaction leads us to the type of band ferromagnetism called the Stoner model. Stoner's model for band ferromagnetism is the nonlocalized mean field counterpart of Weiss' model for localized ferromagnetism. However, Stoner's model has neither the simplicity, nor the wide applicability of the Weiss approach.

Just as a mean-field approximation to the Heisenberg Hamiltonian gives us the Weiss model, there exists another Hamiltonian called the Hubbard Hamiltonian, whose mean-field approximation gives rise to a Stoner model. Also, just as the Heisenberg Hamiltonian gives good insight to the origin of the Weiss molecular field. So, the Hubbard model gives some physical insight concerning the exchange field for the Stoner model.

The Hubbard Hamiltonian as originally introduced was intended to bridge the gap between a localized and a mobile electron point of view. In general, in a suitable limit, it can describe either case. If one does not go to the limit, it can (in a sense) describe all cases in between. However, we will make a mean-field approximation and this displays the band properties most effectively.

One can give a derivation, of sorts, of the Hubbard Hamiltonian. However, so many assumptions are involved that it is often clearer just to write the Hamiltonian down as an assumption. This is what we will do, but even so, one cannot solve it exactly for cases that approach realism. Here we will solve it within the mean-field approximation, and get, as we have mentioned, the Stoner model of itinerant ferromagnetism.

In a common representation, the Hubbard Hamiltonian is

$$\mathcal{H} = \sum_{k,\sigma} \varepsilon_k a_{k\sigma}^{\dagger} a_{k\sigma} + \frac{I}{2} \sum_{\alpha,\sigma} n_{\alpha\sigma} n_{\alpha,-\sigma} , \tag{7.199}$$

where σ labels the spin (up or down), $\boldsymbol{k}$ labels the band energies, and α labels the lattice sites (we have assumed only one band—say an s-band—with ε_k being the band energy for wave vector $\boldsymbol{k}$). The $a^{\dagger}{}_{k\sigma}$ and $a_{k\sigma}$ are creation and annihilation operators and I defines the interaction between electrons on the same site.

It is important to notice that the Hubbard Hamiltonian (as written above) assumes the electron–electron interactions are only large when the electrons are on the same site. A narrow band corresponds to localization of electrons. Thus, the Hubbard Hamiltonian is often said to be a narrow s-band model. The $n_{\alpha\sigma}$ are Wannier site-occupation numbers. The relation between band and Wannier (site localized) wave functions is given by the use of Fourier relations:

$$\psi_k = \frac{1}{\sqrt{N}} \sum_{R_\alpha} \exp(-\mathrm{i}\boldsymbol{k} \cdot \boldsymbol{R}_\alpha) W(\boldsymbol{r} - \boldsymbol{R}_\alpha) , \tag{7.200a}$$

$$W(\boldsymbol{r} - \boldsymbol{R}_\alpha) = \frac{1}{\sqrt{N}} \sum_{k} \exp(\mathrm{i}\boldsymbol{k} \cdot \boldsymbol{R}_\alpha) \psi_k(\boldsymbol{r}) . \tag{7.200b}$$

Since the Bloch (or band) wave functions ψ_k are orthogonal, it is straightforward to show that the Wannier functions $W(\boldsymbol{r} - \boldsymbol{R}_\alpha)$ are also orthogonal. The Wannier functions $W(\boldsymbol{r} - \boldsymbol{R}_\alpha)$ are localized about site α and, at least for narrow bands, are well approximated by atomic wave functions.

Just as $a^{\dagger}{}_{k\sigma}$ creates an electron in the state ψ_k [with spin σ either + or ↑ (up) or − ↓ (down)], so $c^{\dagger}{}_{\alpha\sigma}$ (the site creation operator) creates an electron in the state $W(\boldsymbol{r} - \boldsymbol{R}_\alpha)$, again with the spin either up or down. Thus, occupation number operators for the localized Wannier states are $n^{\dagger}{}_{\alpha\sigma} = c^{\dagger}{}_{\alpha\sigma} c_{\alpha\sigma}$ and consistent with (7.200a) the two sets of annihilation operators are related by the Fourier transform

$$a_{k\sigma} = \frac{1}{\sqrt{N}} \sum_{R_\alpha} \exp(\mathrm{i}\boldsymbol{k} \cdot \boldsymbol{R}_\alpha) c_{\alpha\sigma} . \tag{7.201}$$

Substituting this into the Hubbard Hamiltonian and defining

$$T_{\alpha\beta} = \frac{1}{N} \sum_{k} \varepsilon_k \exp[\mathrm{i}\boldsymbol{k} \cdot (\boldsymbol{R}_\alpha - \boldsymbol{R}_\beta)] , \tag{7.202}$$

we find

$$\mathcal{H} = \sum_{\alpha,\beta,\sigma} T_{\alpha\beta} c_{\beta\sigma}^{+} c_{\alpha\sigma} + \frac{I}{2} \sum_{\alpha,\sigma} n_{\alpha\sigma}^{+} n_{\alpha-\sigma} . \tag{7.203}$$

This is the most common form for the Hubbard Hamiltonian. It is often further assumed that $T_{\alpha\beta}$ is only nonzero when α and β are nearest neighbors. The first term then represents nearest-neighbor hopping.

Since the Hamiltonian is a many-electron Hamiltonian, it is not exactly solvable for a general lattice. We solve it in the mean-field approximation and thus replace

$$\frac{I}{2}\sum_{\alpha,\sigma} n_{\alpha\sigma} n_{\alpha,-\sigma} ,$$

with

$$I\sum_{\alpha,\sigma} n_{\alpha\sigma}\left\langle n_{\alpha,-\sigma}\right\rangle ,$$

where $\langle n_{\alpha},-\sigma\rangle$ is the thermal average of $n_{\alpha},-\sigma$. We also assume $\langle n_{\alpha},-\sigma\rangle$ is independent of site and so write it down as $n_{-\sigma}$ in (7.204).

Itinerant Ferromagnetism and the Stoner Model (B)

The mean-field approximation has been criticized on the basis that it builds in the possibility of an ordered ferromagnetic ground state regardless of whether the Hubbard Hamiltonian exact solution for a given lattice would predict this. Nevertheless, we continue, as we are more interested in the model we will eventually reach (the Stoner model) than in whether the theoretical underpinnings from the Hubbard model are physical. The mean-field approximation to the Hubbard model gives

$$\mathcal{H} = \sum_{\alpha,\beta,\sigma} T_{\alpha\beta} c^{\dagger}_{\beta\sigma} c_{\alpha\sigma} + I\sum_{\alpha,\sigma} n_{-\sigma} n_{\alpha\sigma} . \tag{7.204}$$

Actually, in the mean-field approximation, the band picture is more convenient to use. Since we can show

$$\sum_{\alpha} n_{\alpha\sigma} = \sum_{\boldsymbol{k}} n_{\boldsymbol{k}\sigma} ,$$

the Hubbard model in the mean field can then be written as

$$\mathcal{H} = \sum_{\boldsymbol{k},\sigma} (\varepsilon_{\boldsymbol{k}} + I n_{-\sigma}) n_{\boldsymbol{k}\sigma} . \tag{7.205}$$

The single-particle energies are given by

$$E_{\boldsymbol{k},\sigma} = \varepsilon_{\boldsymbol{k}} + I n_{-\sigma} . \tag{7.206}$$

The average number of electrons per site n is less than or equal to 2 and $n = n_+ + n_-$, while the magnetization per site n is $M = (n_+ - n_-)\mu_B$, where μ_B is the Bohr magneton.

Note: In order not to introduce another "–" sign, we will say "spin up" for now. This really means "moment up" or spin down, since the electron has a negative charge.

Note $n + (M/\mu_B) = 2n_+$ and $n - (M/\mu_B) = 2n_-$. Thus, up to an additive constant

$$E_{\boldsymbol{k}\pm} = \varepsilon_{\boldsymbol{k}} + I\left(\mp \frac{M}{2\mu_B}\right). \tag{7.207}$$

Note (7.207) is consistent with (7.197b). If we then define $H_{\text{eff}} = IM/2\mu_B^2$, we write the following basic equations for the Stoner model:

$$M = \mu_B(n_\uparrow - n_\downarrow)\,, \tag{7.208}$$

$$E_{\boldsymbol{k},\sigma} = \varepsilon_{\boldsymbol{k}} \mp \mu_B H_{\text{eff}}\,, \tag{7.209}$$

$$H_{\text{eff}} = \frac{IM}{2\mu_B^2}\,, \tag{7.210}$$

$$n_\sigma = \frac{1}{N}\sum\nolimits_{\boldsymbol{k}} \frac{1}{\exp[(E_{\boldsymbol{k}\sigma} - M\mu)/kT] + 1}\,, \tag{7.211}$$

$$n_\uparrow + n_\downarrow = n\,. \tag{7.212}$$

Although these equations are easy to write down, it is not easy to obtain simple convenient solutions from them. As already noted, the Stoner model contains two basic assumptions: (1) The electronic energy band in the metal is described by a known ε_k. By standard means, one can then derive a density of states. For free electrons, $N(E) \propto (E)^{1/2}$. (2) A molecular field approximately describes the effects of the interactions and we assume Fermi–Dirac statistics can be used for the spin-up and spin-down states. Much of the detail and even standard notation has been presented by Wohlfarth [7.69]. See also references to Stoner's work in the works by Wohlfarth.

The only consistent way to determine ε_k and, hence, $N(E)$ is to derive it from the Hubbard Hamiltonian. However, following the usual Stoner model we will just use an $N(E)$ for free electrons.

The maximum saturation magnetization (moment per site) is $M_0 = \mu_B n$ and the actual magnetization is $M = \mu_B(n_\uparrow - n_\downarrow)$. For the Stoner model, a relative magnetization is defined below:

$$\xi = \frac{M}{M_0} = \frac{n_\uparrow - n_\downarrow}{n}\,. \tag{7.213}$$

Using (7.212) and (7.213), we have

$$n_+ = n_\uparrow = (1+\xi)\frac{n}{2}\,, \tag{7.214a}$$

$$n_- = n_\downarrow = (1-\xi)\frac{n}{2}\,. \tag{7.214b}$$

It is also convenient to define a temperature θ', which measures the strength of the exchange interaction

$$k\theta'\xi = \mu_B H_{\text{eff}} \,. \tag{7.215}$$

We now suppose that the exchange energy is strong enough to cause an imbalance in the number of spin-up and spin-down electrons. We can picture the situation with constant Fermi energy $\mu = E_F$ (at $T = 0$) and a rigid shifting of the up N_+ and the down N_- density states as shown in Fig. 7.15.

The ↑ represents the "spin-up" (moment up actually) band and the ↓ the "spin-down" band. The shading represents states filled with electrons. The exchange energy causes the splitting of the two bands. We have pictured the density of states by a curve that goes to zero at the top and bottom of the band unlike a free-electron density of states that goes to zero only at the bottom.

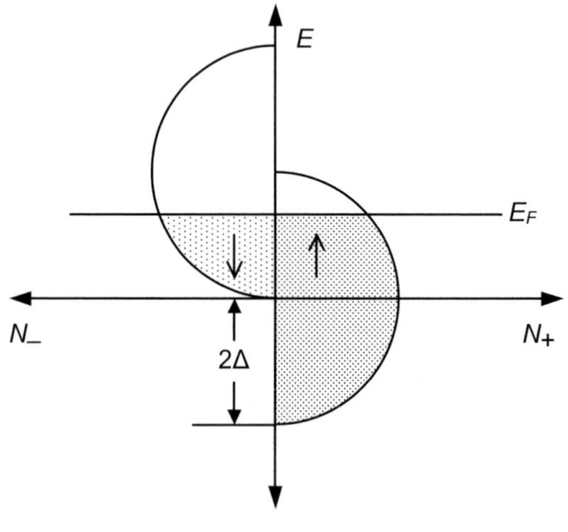

Fig. 7.15. Density states imbalanced by exchange energy

At $T = 0$, we have

$$n_+ = (1+\xi)\frac{n}{2} = \int_{\text{occ. states}} N_+(E)\mathrm{d}E \,, \tag{7.216a}$$

$$n_- = (1-\xi)\frac{n}{2} = \int_{\text{occ. states}} N_-(E)\mathrm{d}E \,. \tag{7.216b}$$

This can be easily worked out for free electrons if $E = 0$ at the bottom of both bands,

$$N_\pm(E) = \frac{1}{2}N_{\text{total}}(E) = \frac{1}{4\pi^2}\left(\frac{2m}{\hbar^2}\right)^{3/2}\sqrt{E} \equiv N(E) \,. \tag{7.217}$$

We now derive conditions for which the magnetized state is stable at $T = 0$. If we just use a single-electron picture and add up the single-electron energies, we find, with the (–) band shifted up by Δ and the (+) band shifted down by Δ, for the energy per site

$$E = n_-\Delta + \int_0^{E_F^-} EN(E)\mathrm{d}E - n_+\Delta + \int_0^{E_F^+} EN(E)\mathrm{d}E \,.$$

The terms involving Δ are the exchange energy. We can rewrite it from (7.208), (7.213), and (7.215) as

$$-\frac{M}{\mu_B}\Delta = -nk\theta'\xi^2 \,.$$

However, just as in the Hartree–Fock analysis, this exchange term has double counted the interaction energies (once as a source of the field and once as interaction with the field). Putting in a factor of 1/2, we finally have for the total energy

$$E = \int_0^{E_F^+} EN(E)\mathrm{d}E + \int_0^{E_F^-} EN(E)\mathrm{d}E - \frac{1}{2}nk\theta'\xi^2 \,. \tag{7.218}$$

Differentiating (d/dξ) (7.216) and (7.218) and combining the results, we can show

$$\frac{1}{n}\frac{\mathrm{d}E}{\mathrm{d}\xi} = \frac{1}{2}(E_F^+ - E_F^-) - k\theta'\xi \,. \tag{7.219}$$

Differentiating (7.219) a second time and again using (7.216), we have

$$\frac{1}{n}\frac{\mathrm{d}^2E}{\mathrm{d}\xi^2} = \frac{n}{4}\left(\frac{1}{N(E_F^+)} + \frac{1}{N(E_F^-)}\right) - k\theta' \,. \tag{7.220}$$

Setting $\mathrm{d}E/\mathrm{d}\xi = 0$, just gives the result that we already know

$$2k\theta'\xi = (E_F^+ - E_F^-) = 2\mu_B H_{\mathrm{eff}} = 2\Delta \,.$$

Note if $\xi = 0$ (paramagnetism) and $\mathrm{d}E/\mathrm{d}\xi = 0$, while $\mathrm{d}^2E/\mathrm{d}\xi^2 < 0$ the paramagnetism is unstable with respect to ferromagnetism. $\xi = 0$, $\mathrm{d}E/\mathrm{d}\xi = 0$ implies $E_F^+ = E_F^-$ and $N(E_F^-) = N(E_F^+) = N(E_F)$. So by (7.220) with $\mathrm{d}^2E/\mathrm{d}\xi^2 \le 0$ we have

$$k\theta' \ge \frac{n}{2N(E_F)} \,. \tag{7.221}$$

For a parabolic band with $N(E) \propto E^{1/2}$, this implies

$$\frac{k\theta'}{E_F} \ge \frac{2}{3} \,. \tag{7.222}$$

We now calculate the relative magnetization (ξ_0) at absolute zero for a parabolic band where $N(E) = K(E)^{1/2}$ where K is a constant. From (7.216)

$$(1+\xi_0)\frac{n}{2} = \frac{2}{3}K(E_F^+)^{3/2},$$

$$(1-\xi_0)\frac{n}{2} = \frac{2}{3}K(E_F^-)^{3/2}.$$

Also

$$n = \frac{4}{3}KE_F^{3/2}.$$

Eliminating K and using $E_F^+ - E_F^- = 2k\theta'\xi_0$, we have

$$\frac{k\theta'}{E_F} = \frac{1}{2\xi_0}[(1+\xi_0)^{2/3} - (1-\xi_0)^{2/3}], \tag{7.223}$$

which is valid for $0 \le \xi_0 \le 1$. The maximum ξ_0 can be is 1 for which $k\theta'/E_F = 2^{-1/3}$, and at the threshold for ferromagnetism ξ_0 is 0. So, $k\theta'/E_F = 2/3$ as already predicted by the Stoner criterion.

Summary of Results at Absolute Zero

We have three ranges:

$$\frac{k\theta'}{E_F} < \frac{2}{3} = 0.667 \quad \text{and} \quad \xi_0 = \frac{M}{n\mu_B} = 0,$$

$$\frac{2}{3} < \frac{k\theta'}{E_F} < \frac{1}{2^{1/3}} = 0.794, \quad 0 < \xi_0 = \frac{M}{n\mu_B} < 1,$$

$$\frac{k\theta'}{E_F} > \frac{1}{2^{1/3}} \quad \text{and} \quad \xi_0 = \frac{M}{n\mu_B} = 1.$$

The middle range, where $0 < \xi_0 < 1$ is special to Stoner ferromagnetism and not to be found in the Weiss theory. This middle range is called "unstructured" or "weak" ferromagnetism. It corresponds to having electrons in both ↑ and ↓ bands. For very low, but not zero, temperatures, one can show for weak ferromagnetism that

$$M = M_0 - CT^2, \tag{7.224}$$

where C is a constant. This is particularly easy to show for very weak ferromagnetism, where $\xi_0 << 1$ and is left as an exercise for the reader.

We now discuss the case of strong ferromagnetism where $k\theta'/E_F > 2^{-1/3}$. For this case, $\xi_0 = 1$, and $n_\uparrow = n$, $n_\downarrow = 0$. There is now a gap E_g between E_F^+ and the bottom of the spin-down band. For this case, by considering thermal excitations to the $n_\downarrow$ band, one can show at low temperature that

$$M = M_0 - K'' T^{3/2} \exp(-E_g / kT)\,, \tag{7.225}$$

where K'' is a constant. However, spin-wave theory says $M = M_0 - C'T^{3/2}$, where C' is a constant, which agrees with low-temperature experiments. So, at best, (7.225) is part of a correction to low-temperature spin-wave theory.

Within the context of the Stoner model, we also need to talk about exchange enhancement of the paramagnetic susceptibility χ_P (*gaussian units with* $\mu_0 = 1$)

$$M = \chi_P B_{\mathrm{eff}}^{\mathrm{Total}}\,, \tag{7.226}$$

where M is the magnetization and χ_P the Pauli susceptibility, which for low temperatures, has a very small αT^2 term. It can be written

$$\chi_P = 2\mu_B^2 N(E_F)(1 + \alpha T^2)\,, \tag{7.227}$$

where $N(E)$ is the density of states for one subband. Since

$$B_{\mathrm{eff}}^{\mathrm{Total}} = H_{\mathrm{eff}} + B = \gamma B + B,$$

it is easy to show that (gaussian with $B = H$)

$$\chi = \frac{M}{B} = \frac{\chi_P}{1 - \gamma\chi_P}\,, \tag{7.228}$$

where $1/(1 - \gamma\chi_P)$ is the exchange enhancement factor.

We can recover the Stoner criteria from this at $T = 0$ by noting that paramagnetism is unstable if

$$\chi_P^0 \gamma \geq 1\,. \tag{7.229}$$

By using $\gamma = k\theta'/n\mu_B^2$ and $\chi_P^0 = 2\mu_B^2 N(E_F)$, (7.229) just gives the Stoner criteria. At finite, but low temperatures where ($\alpha = -|a|$)

$$\chi_P = \chi_P^0(1 - |a|T^2)\,,$$

if we define

$$\theta^2 = \frac{\gamma\chi_P^0 - 1}{\gamma\chi_P^0 |a|}\,,$$

and suppose $|a|T^2 << 1$, it is easy to show

$$\chi = \frac{1}{\gamma |a|} \frac{1}{T^2 - \theta^2}\,.$$

Thus, as long as $T \cong \theta$, we have a Curie–Weiss-like law:

$$\chi = \frac{1}{2\theta\gamma \mid a \mid} \frac{1}{T-\theta}. \tag{7.230}$$

At very high temperatures, one can also show that an ordinary Curie–Weiss-like law is obtained:

$$\chi = \frac{n\mu_B^2}{k} \frac{1}{T-\theta}. \tag{7.231}$$

Summary Comments About the Stoner Model

1. The low-temperature results need to be augmented with spin waves. Although in this book we only derive the results of spin waves for the localized model, it turns out that spin waves can also be derived within the context of the itinerant electron model.

2. Results near the Curie temperature are never qualitatively good in a mean-field approximation because the mean-field approximation does not properly treat fluctuations.

3. The Stoner model gives a simple explanation of why one can have a fractional number of electrons contributing to the magnetization (the case of weak ferromagnetism where $\xi_0 = M_{T=0}/n\mu_B$ is between 0 and 1).

4. To apply these results to real materials, one usually needs to consider that there are overlapping bands (e.g. both s and d bands), and not all bands necessarily split into subbands. However, the Stoner model does seem to work for $ZrZn_2$.

7.2.5 Magnetic Phase Transitions (A)

Simple ideas about spin waves break down as T_c is approached. We indicate here one way of viewing magnetic phenomena near the $T = T_c$ region. In this Section we will discuss magnetic phase transitions in which the magnetization (for ferromagnets with $H = 0$) goes continuously to zero as the critical temperature is approached from below. Thus at the critical temperature (Curie temperature for a ferromagnet) the ordered (ferromagnetic) phase goes over to the disordered (paramagnetic) phase. This "smooth" transition from one phase (or more than one phase in more general cases) to another is characteristic of the behavior of many substances near their critical temperature. In such continuous phase transitions there is no latent heat and these phase transitions are called second-order phase transitions. All second-order phase transitions show many similarities. We shall consider only phase transitions in which there is no latent heat.

No complete explanation of the equilibrium properties of ferromagnets near the magnetic critical temperature (T_c) has yet been given, although the renormalization technique, referred to later, comes close. At temperatures well below T_c we know that the method of spin waves often yields good results for describing the magnetic behavior of the system. We know that high-temperature expansions of the partition function yield good results. The Green function method provides results for interesting physical quantities at all temperatures. However, the Green function results (in a usable approximation) are not valid near T_c. Two methods (which are not as straightforward as one might like) have been used. These are the use of scaling laws[19] and the use of the Padé approximant.[20] These methods often appear to give good quantitative results without offering much in the way of qualitative insight. Therefore we will not discuss them here. The renormalization group, referenced later, in some ways is a generalization of scaling laws. It seems to offer the most in the way of understanding.

Since the region of lack of knowledge (around the phase transition) is only near $\tau = 1$ ($\tau = T/T_c$, where T_c is the critical temperature) we could forget about the region entirely (perhaps) if it were not for the fact that very unusual and surprising results happen here. These results have to do with the behavior of the various quantities as a function of temperature. For example, the Weiss theory predicts for the (zero field) magnetization that $M \propto (T_c - T)^{+1/2}$ as $T \to T_c^-$ (the minus sign means that we approach T_c from below), but experiment often seems to agree better with $M \propto (T_c - T)^{+1/3}$. Similarly, the Weiss theory predicts for $T > T_c$ that the zero-field susceptibility behaves as $\chi \propto (T - T_c)^{-1}$, whereas experiment for many materials agrees with $\chi \propto (T - T_c)^{-4/3}$ as $T \to T_c^+$. In fact, the Weiss theory fails very seriously above T_c because it leaves out the short-range ordering of the spins. Thus it predicts that the (magnetic contribution to the) specific heat should vanish above T_c, whereas the zero-field magnetic specific heat does not so vanish. Using an improved theory that puts in some short-range order above T_c modifies the specific heat somewhat, but even these improved theories [92] do not fit experiment well near T_c. Experiment appears to suggest (although this is not settled yet) that for many materials $C \cong \ln |(T - T_c)|$ as $T \to T_c^+$ (the exact solution of the specific heat of the two-dimensional Ising ferromagnet shows this type of divergence), and the concept of short-range order is just not enough to account for this logarithmic or near logarithmic divergence. Something must be missing. It appears that the missing concept that is needed to correctly predict the "critical exponents" and/or "critical divergences" is the concept of (anomalous) fluctuations. [The exponents $^1/_3$ and $^4/_3$ above are critical exponents, and it is possible to set up the formalism in such a way that the logarithmic divergence is consistent with a certain critical exponent being zero.] Fluctuations away from the thermodynamic equilibrium appear to play a very dominant role in the behavior of thermodynamic functions near the phase transition. Critical-point behavior is discussed in more detail in the next section.

Additional insight into this behavior is given by the Landau theory.[19] The Landau theory appears to be qualitatively correct but it does not predict correctly the critical exponents.

---

[19] See Kadanoff et al [7.35].

[20] See Patterson et al [7.54] and references cited therein.

Critical Exponents and Failures of Mean-Field Theory (B)

Although mean-field theory has been extraordinarily useful and in fact, is still the "workhorse" of theories of magnetism (as well as theories of the thermodynamics behavior of other types of systems that show phase transitions), it does suffer from several problems. Some of these problems have become better understood in recent years through studies of critical phenomena, particularly in magnetic materials, although the studies of "critical exponents" relates to a much broader set of materials than just magnets as referred to above. It is helpful now to define some quantities and to introduce some concepts.

A sensitive test of mean-field theory is in predicting critical exponents, which define the nature of the singularities of thermodynamic variables at critical points of second-order phase transitions. For example,

$$\phi \sim \left|\frac{T_c - T}{T_c}\right|^{\beta} \quad \text{and} \quad \xi = \left|\frac{T_c - T}{T_c}\right|^{-\nu},$$

for $T < T_c$, where β, ν are critical exponents, ϕ is the order parameter, which for ferromagnets is the average magnetization M and ξ is the correlation length. In magnetic systems, the correlation length measures the characteristic length over which the spins are ordered, and we note that it diverges as the Curie temperature T_c is approached. In general, the order parameter ϕ is just some quantity whose value changes from disordered phases (where it may be zero) to ordered phases (where it is nonzero). Note for ferromagnets that ϕ is zero in the disordered paramagnetic phase and nonzero in the ordered ferromagnetic situation.

Mean-field theory can be quite good above an upper critical (spatial) dimension where by definition it gives the correct value of the critical exponents. Below the upper critical dimension (UCD), thermodynamic fluctuations become very important, and mean-field theory has problems. In particular, it gives incorrect critical exponents. There also exists a lower critical dimension (LCD) for which these fluctuations become so important that the system does not even order (by definition of the LCD). Here, mean-field theory can give qualitatively incorrect results by predicting the existence of an ordered phase. The lower critical dimension is the largest dimension for which long-range order is not possible. In connection with these ideas, the notion of a universality class has also been recognized. Systems with the same spatial dimension d and the same dimension of the order parameter D are usually in the same universality class. Range and symmetry of the interaction potential can also play a role in determining the universality class. Quite dissimilar systems in the same universality class will, by definition, exhibit the same critical exponents. Of course, the order parameter itself as well as the critical temperature T_c, may be quite different for systems in the same universality class. In this connection, one also needs to discuss concepts like the renormalization group, but this would take us too far afield. Reference can be made to excellent statistical mechanics books like the one by Huang.[21]

[21] See Huang [7.32, p441ff].

Critical exponents for magnetic systems have been defined in the following way. First, we define a dimensionless temperature that is small when we are near the critical temperature.

$$t = (T - T_c)/T_c \, .$$

We assume $B = 0$ and define critical exponents by the behavior of physical quantities such as M:

Magnetization (order parameter): $M \sim |t|^{\beta}$.

Magnetic susceptibility: $\chi \sim |t|^{-\gamma}$.

Specific heat: $C \sim |t|^{-\alpha}$.

There are other critical exponents, such as the one for correlation length (as noted above), but this is all we wish to consider here. Similar critical exponents are defined for other systems, such as fluid systems. When proper analogies are made, if one stays within the same universality class, the critical exponents have the same value. Under rather general conditions, several inequalities have been derived for critical exponents. For example, the Rushbrooke inequality is

$$\alpha + 2\beta + \gamma \geq 2 \, .$$

It has been proposed that this relation also holds as an equality. For mean-field theory $\alpha = 0$, $\beta = {}^1/_2$, and $\gamma = 1$. Thus, the Rushbrooke relation is satisfied as an equality. However, except for α being zero, the critical exponents are wrong. For ferromagnets belonging to the most common universality class, experiment, as well as better calculations than mean field, suggest, as we have mentioned (Sect. 7.2.5), $\beta = {}^1/_3$, and $\gamma = {}^4/_3$. Note that the Rushbrooke equality is still satisfied with $\alpha = 0$. The most basic problem mean-field theory has is that it just does not properly treat fluctuations nor does it properly treat a related aspect concerning short-range order. It must include these for agreement with experiment. As already indicated, short-range correlation gives a tail on the specific heat above T_c, while the mean-field approximation gives none.

The mean-field approximation also fails as $T \to 0$ as we have discussed. An elementary calculation from the properties of the Brillouin function shows that ($s = 1/2$)

$$M = M_0[1 - 2\exp(-2T_c/T)] \, ,$$

whereas for typical ferromagnets, experiment agrees better with

$$M = M_0(1 - aT^{3/2}) \, .$$

As we have discussed, this dependence on temperature can be derived from spin wave theory.

Although considerable calculation progress has been made by high-temperature series expansions plus Padé Approximants, by scaling, and renormalization group arguments, most of this is beyond the scope of this book. Again,

Huang's excellent text can be consulted.[21] Tables 7.2 and 7.3 summarize some of the results.

Table 7.2. Summary of mean-field theory

Failures	Successes
Neglects spin-wave excitations near absolute zero.	Often used to predict the type of magnetic structure to be expected above the lower critical dimension (ferromagnetism, ferrimagnetism, antiferromagnetism, heliomagnetism, etc.).
Near the critical temperature, it does not give proper critical exponents if it is below the upper critical dimension.	Predicts a phase transition, which certainly will occur if above the lower critical dimension.
May predict a phase transition where there is none if below the lower critical dimension. For example, a one-dimension isotropic Heisenberg magnet would be predicted to order at a finite temperature, which it does not.	Gives at least a qualitative estimate of the values of thermodynamic quantities, as well as the critical exponents – when used appropriately.
Predicts no tail in the specific heat for typical magnets.	Serves as the basis for improved calculations.
	The higher the spatial dimension, the better it is.

Table 7.3. Critical exponents (calculated)

	α	β	γ
Mean field	0	0.5	1
Ising (3D)	0.11	0.32	1.24
Heisenberg (3D)	–0.12	0.36	1.39

Adapted with permission from Chaikin PM and Lubensky TC, *Principles of Condensed Matter Physics*, Cambridge University Press, 1995, p. 231.

Two-Dimensional Structures (A)

Lower-dimensional structures are no longer of purely theoretical interest. One way to realize two dimensions is with thin films. Suppose the thin film is of thickness t and suppose the correlation length of the quantity of interest is c. When the thickness is much less than the correlation length ($t \ll c$), the film will behave two dimensionally and when $t \gg c$ the film will behave as a bulk three-dimensional material. If there is a critical point, since c grows without bound as the critical point is approached, a thin film will behave two-dimensionally near the two-dimensional critical point. Another way to have two-dimensional behavior is in layered magnetic materials in which the coupling between magnetic layers, of spacing d, is weak. Then when $c \ll d$, all coupling between the layers can be neglected and one sees 2D behavior, whereas if $c >> d$, then interlayer coupling can no longer be neglected. This means with magnetic layers, a two-dimensional critical point will be modified by 3D behavior near the critical temperature.

In this chapter we are mainly concerned with materials for which the three-dimensional isotropic systems are a fairly good or at least qualitative model. However, it is interesting that two-dimensional isotropic Heisenberg systems can be shown to have no spontaneous (sublattice – for antiferromagnets) magnetization [7.49]. On the other hand, it can be shown [7.26] that the highly anisotropic two-dimensional Ising ferromagnet (defined by the Hamiltonian $\mathcal{H} \propto \sum_{i,j(\text{nn.})} \sigma_i^z \sigma_j^z$, where the σs refer to Pauli spin matrices, the i and j refer to lattice sites) *must* show spontaneous magnetization.

We have just mentioned the two-dimensional Heisenberg model in connection with the Mermin–Wagner theorem. The planar Heisenberg model is in some ways even more interesting. It serves as a model for superfluid helium films and predicts the long-range order is destroyed by formation of vortices [7.40].

Another common way to produce two-dimensional behavior is in an electronic inversion layer in a semiconductor. This is important in semiconductor devices.

Spontaneously Broken Symmetry (A)

A Heisenberg Hamiltonian is invariant under rotations, so the ensemble average of the magnetization is zero. For every M there is a $-M$ of the same energy. Physically this answer is not correct since magnets do magnetize. The symmetry is spontaneously broken when the ground state does not have the same symmetry as the Hamiltonian, The symmetry is recovered by having degenerate ground states whose totality recovers the rotational symmetry. Once the magnet magnetizes, however, it does not go to another degenerate state because all the magnets would have to rotate spontaneously by the same amount. The probability for this to happen is negligible for a realistic system. Quantum mechanically in the infinite limit, each ground state generates a separate Hilbert space and transitions between them are forbidden—a super selection rule. Because of the symmetry there are excited states that are wave-like in the sense that the local ground state changes slowly over space (as in a wave). These are the Goldstone excitations and they are orthogonal to any ground state. Actually each of the (infinite) number of ground states is orthogonal to each other: The concept of spontaneously broken symmetry is much more general than just for magnets. For ferromagnets the rotational symmetry is broken and spin waves or magnons appear. Other examples include crystals (translation symmetry is broken and phonons appear), and superconductors (local gauge symmetry is broken and a Higgs mode appears—this is related to the Meissner effect – see Chap. 8).[22]

[22] See Weinberg [7.67].

7.3 Magnetic Domains and Magnetic Materials (B)

7.3.1 Origin of Domains and General Comments[23] (B)

Because of their great practical importance, a short discussion of domains is merited even though we are primarily interested in what happens in a single domain.

We want to address the following questions: What are the domains? Why do they form? Why are they important? What are domain walls? How can we analyze the structure of domains, and domain walls? Is there more than one kind of domain wall?

Magnetic domains are small regions in which the atomic magnetic moments are lined up. For a given temperature, the magnetization is saturated in a single domain, but ferromagnets are normally divided into regions with different domains magnetized in different directions.

When a ferromagnet splits into domains, it does so in order to lower its free energy. However, the free energy and the internal energy differ by TS and if T is well below the Curie temperature, TS is small since also the entropy S is small because the order is high. Here we will neglect the difference between the internal energy and the free energy. There are several contributions to the internal energy that we will discuss presently.

Magnetic domains can explain why the overall magnetization can vanish even if we are well below the Curie temperature T_c. In a single domain the M vs. T curve looks somewhat like Fig. 7.16.

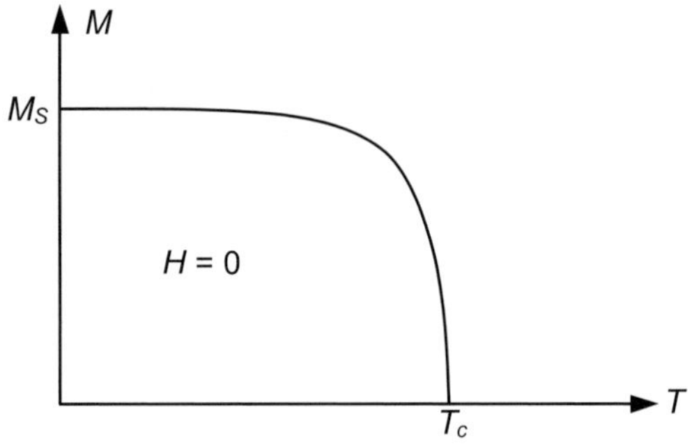

Fig. 7.16. M vs. T curve for a single magnetic domain

For reference, the Curie temperature of iron is 1043 K and its saturation magnetization M_S is 1707 G. But when there are several domains, they can point in different directions so the overall magnetization can attain any value from zero up to saturation magnetization. In a magnetic field, the domains can change in size (with those that are energetically preferred growing). Thus the phenomena of hysteresis, which we sketch in Fig. 7.17 starting from the ideal demagnetized state, can be understood (see Section *Hysteresis, Remanence, and Coercive Force*).

[23] More details can be found in Morrish [68] and Chikazumi [7.11].

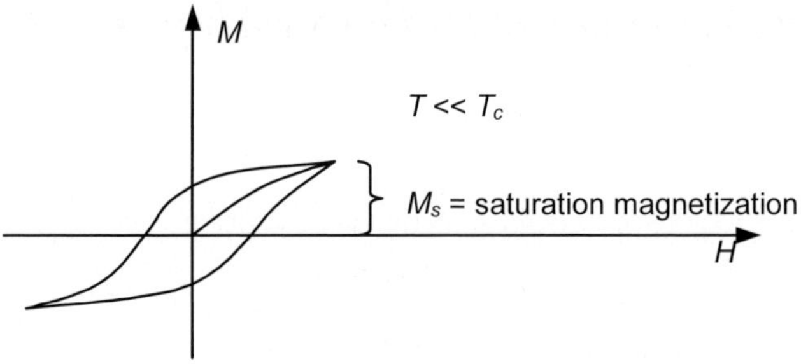

Fig. 7.17. *M* vs. *H* curve showing magnetic hysteresis

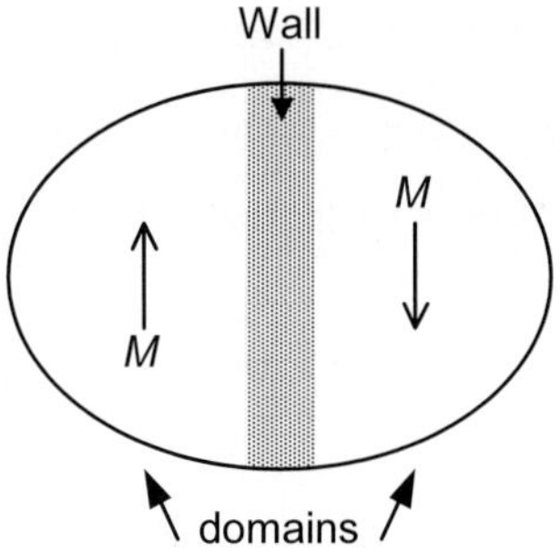

Fig. 7.18. Two magnetic regions (domains) separated by a *domain wall*, where size is exaggerated

In order for some domains to grow at the expense of others, the domain walls separating the two regions must move. Domain walls are transition regions that separate adjacent regions magnetized in different directions. The idea is shown in Fig. 7.18.

We now want to analyze the four types of energy involved in domain formation. We consider (1) exchange energy, (2) magnetostatic energy, (3) anisotropy energy, and (4) magnetostrictive energy. Domain structures with the lower sum of these energies are the most stable.

Exchange Energy (B)

We have seen (see Section *The Heisenberg Hamiltonian and its Relationship to the Weiss Mean-Field Theory*) that quantum mechanics indicates that there may be an interaction energy between atomic spins $\boldsymbol{S}_i$ that is proportional to the scalar product of the spins. From this, one obtains the Heisenberg Hamiltonian describing the interaction energy. Assuming J is the proportionality constant (called the exchange integral) and that only nearest-neighbor (nn) interactions need be considered, the Heisenberg Hamiltonian becomes

$$\mathcal{H} = -J\sum_{\substack{i,j \\ (\mathrm{nn})}} \mathbf{S}_i \cdot \mathbf{S}_j \,, \tag{7.232}$$

where the spin S_i for atom i when averaged over many neighboring spins gives us the local magnetization. We now make a classical continuum approximation. For the interaction energy of two spins we write:

$$U_{ij} = -2J\boldsymbol{S}_i \cdot \boldsymbol{S}_j \,. \tag{7.233}$$

Assuming $\boldsymbol{u}_i$ is a unit vector in the direction of $\boldsymbol{S}_i$ we have since $\boldsymbol{S}_i = S\boldsymbol{u}_i$:

$$U_{ij} = -2JS^2\boldsymbol{u}_i \cdot \boldsymbol{u}_j \,. \tag{7.234}$$

If $\boldsymbol{r}_{ji}$ is the vector connecting spins i and j, then

$$\boldsymbol{u}_j = \boldsymbol{u}_i + \boldsymbol{r}_{ji} \cdot (\nabla \boldsymbol{u})_i \,, \tag{7.235}$$

treating $\boldsymbol{u}$ as a continuous function $\boldsymbol{r}$, $\boldsymbol{u} = \boldsymbol{u}(\boldsymbol{r})$. Then since

$$(\boldsymbol{u}_j - \boldsymbol{u}_i)^2 = u_j^2 + u_i^2 - 2\boldsymbol{u}_i \cdot \boldsymbol{u}_j = 2(1 - \boldsymbol{u}_i \cdot \boldsymbol{u}_j) \,, \tag{7.236}$$

we have, neglecting an additive constant that is independent of the directions of $\boldsymbol{u}_i$ and $\boldsymbol{u}_j$,

$$U_{ij} = +JS^2(\boldsymbol{u}_j - \boldsymbol{u}_i)^2 \,.$$

So

$$U_{ij} = +JS^2(\boldsymbol{r}_{ji} \cdot \nabla \boldsymbol{u})^2 \,. \tag{7.237}$$

Thus the total interaction energy is

$$U = \tfrac{1}{2}\sum U_{ij} = \frac{JS^2}{2}\sum\nolimits_{i,j}(\boldsymbol{r}_{ji} \cdot \nabla \boldsymbol{u})^2 \,, \tag{7.238}$$

where we have inserted a 1/2 so as not to count bonds twice. If

$$\boldsymbol{u} = \alpha_1\boldsymbol{i} + \alpha_2\boldsymbol{j} + \alpha_3\boldsymbol{k} \,,$$

where the α_i are the direction cosines, for $\boldsymbol{r}_{ji} = a\boldsymbol{i}$, for example:

$$\begin{aligned} \sum_{\pm a\boldsymbol{i}}(\boldsymbol{r}_{ji} \cdot \nabla \boldsymbol{u})^2 &= 2a^2\left(\frac{\partial \alpha_1}{\partial x}\boldsymbol{i} + \frac{\partial \alpha_2}{\partial x}\boldsymbol{j} + \frac{\partial \alpha_3}{\partial x}\boldsymbol{k}\right)^2 \\ &= 2a^2\left[\left(\frac{\partial \alpha_1}{\partial x}\right)^2 + \left(\frac{\partial \alpha_2}{\partial x}\right)^2 + \left(\frac{\partial \alpha_3}{\partial x}\right)^2\right]. \end{aligned} \tag{7.239}$$

For a simple cubic lattice where we must also include neighbors at $\boldsymbol{r}_{ji} = \pm a\boldsymbol{j}$ and $\pm a\boldsymbol{k}$, we have:[24]

$$U = \frac{JS^2}{a}\sum_{i\,(\text{all spins})}[(\nabla\alpha_1)^2 + (\nabla\alpha_2)^2 + (\nabla\alpha_3)^2]_i a^3 \,, \tag{7.240}$$

[24] An alternative derivation is based on writing $U \propto \sum \mu_i B_i$, where μ_i is the magnetic moment $\propto \boldsymbol{S}_i$ and B_i is the effective exchange field $\propto \sum_{j(\text{nn})} J_{ij}\boldsymbol{S}_j$, treating the $\boldsymbol{S}_j$ in a continuum spatial approximation and expanding $\boldsymbol{S}_j$ in a Taylor series ($\boldsymbol{S}_j = \boldsymbol{S}_i + a\partial S_i/\partial x$ + etc. to 2nd order). See (7.275) and following.

or in the continuum approximation:

$$U = \frac{JS^2}{a} \int [(\nabla \alpha_1)^2 + (\nabla \alpha_2)^2 + (\nabla \alpha_3)^2] \mathrm{d}V \,. \tag{7.241}$$

For variation of $\boldsymbol{M}$ only in the y direction, and using spherical coordinates r, θ, φ, a little algebra shows that ($\boldsymbol{M} = \boldsymbol{M}(r, \theta, \varphi)$)

$$\frac{\text{Energy}}{\text{Volume}} = A\left\{\left(\frac{\partial \theta}{\partial y}\right)^2 + \sin^2\theta \left(\frac{\partial \varphi}{\partial y}\right)^2\right\}, \tag{7.242}$$

where $A = JS^2/a$ and has the following values for other cubic structures ($A_{\text{fcc}} = 4A$, and $A_{\text{bcc}} = 2A$). We have treated the exchange energy first because it is this interaction that causes the material to magnetize.

Magnetostatic Energy (B)

We have already discussed magnetostatics in Sect. 7.2.2. Here we want to mention that along with the exchange interaction it is one of the two primary interactions of interest in magnetism. It is the driving mechanism for the formation of domains. Also, at very long wavelengths, as we have mentioned, it can be the causative factor in spin-wave motion (magnetostatic spin waves). A review of magnetostatic fields of relevance for applications is given by Bertram [7.6].

Anisotropy (B)

Because of various energy-coupling mechanisms, certain magnetic directions are favored over others. As discussed in Sect. 7.2.2, the physical origin of crystalline anisotropy is a rather complicated subject. As discussed there, a partial understanding, in some materials, relates it to spin-orbit coupling in which the orbital motion is coupled to the lattice. Anisotropy can also be caused by the shape of the sample or the stress it is subjected to, but these two types are not called crystalline anisotropy. Regardless of the physical origin, a ferromagnetic material will have preferred (least energy) directions of magnetization. For uniaxial symmetry, we can write

$$H_{\text{anis}} = -D_a \sum_i (\boldsymbol{k} \cdot \boldsymbol{S}_i)^2 \,, \tag{7.243}$$

where $\boldsymbol{k}$ is the unit vector along the axis of symmetry. If we let $K_1 = D_a S^2/a^3$, where a is the atom–atom spacing, then since $\sin^2\theta = 1 - \cos^2\theta$ and neglecting unimportant additive terms, the anisotropy energy per unit volume is

$$u_{\text{anis}} = K_1 \sin^2\theta \,. \tag{7.244}$$

Also, for proper choice of K_1, this may describe hexagonal crystals, e.g. cobalt (hcp) where θ is the angle between $\boldsymbol{M}$ and the hexagonal axis. Figure 7.19 shows some data related to anisotropy. Note Fe with a bcc structure has easy directions in $\langle 100 \rangle$ and Ni with fcc has easy directions in $\langle 111 \rangle$.

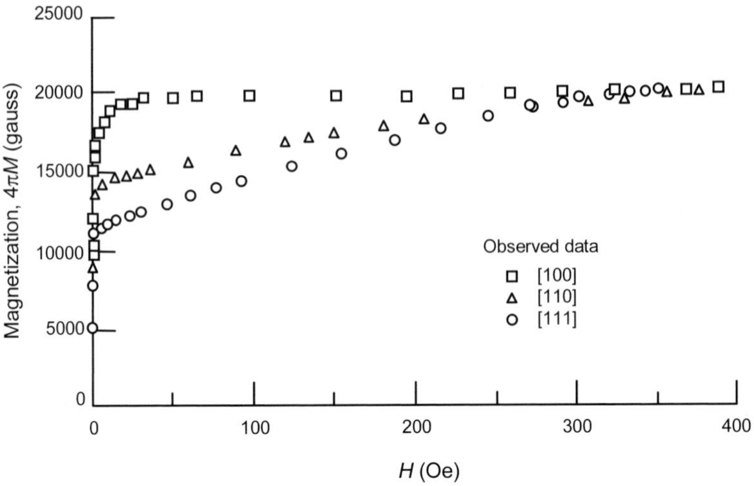

Fig. 7.19. Magnetization curves showing anisotropy for single crystals of iron with 3.85% silicon [Reprinted with permission from Williams HJ, *Phys Rev* **52**, 1 (1937). Copyright 1937 by the American Physical Society.]

Wall Energy (B)

The wall energy is an additive combination of exchange and anisotropy energy, which are independent. Exchange favors parallel moments and a wide wall. Anisotropy prefers moments along an easy direction and a narrow wall. Minimizing the sum of the two determines the width of the wall. Consider a uniaxial ferromagnet with the magnetization varying only in the y direction. If the energy per unit volume is (using spherical coordinates, see, e.g., (7.242) and Fig. 7.25)

$$w = A\left[\left(\frac{\partial\theta}{\partial y}\right)^2 + \left(\sin\theta\frac{\partial\varphi}{\partial y}\right)^2\right] + K_1 \sin^2\theta , \tag{7.245}$$

where

$$A = \alpha_1 \frac{JS^2}{a} \quad \text{and} \quad K_1 = \kappa_1 \frac{D_a S^2}{a^3} , \tag{7.246}$$

and α_1, κ_1 differ for different crystal structures, but both are approximately unity. For simplicity in what follows we will set α_1 and κ_1 equal to one.

Using $\delta\int w\mathrm{d}y = 0$ we get two Euler–Lagrange equations. Inserting (7.245) in the Euler–Lagrange equations, we get the results indicated by the arrows.

$$\frac{\partial w}{\partial \theta} - \frac{\mathrm{d}}{\mathrm{d}y}\frac{\partial w}{\partial \frac{\partial \theta}{\partial y}} = 0 \quad \rightarrow \quad \frac{\mathrm{d}}{\mathrm{d}\theta} K_1 \sin^2\theta = 2A\frac{\mathrm{d}}{\mathrm{d}y}\frac{\partial \theta}{\partial y}, \tag{7.247}$$

$$\frac{\partial w}{\partial \varphi} - \frac{\mathrm{d}}{\mathrm{d}y}\frac{\partial w}{\partial \frac{\partial \varphi}{\partial y}} = 0 \quad \rightarrow \quad 2\frac{\mathrm{d}}{\mathrm{d}y}\left(\sin^2\theta\frac{\partial \varphi}{\partial y}\right) = 0\,. \tag{7.248}$$

For Bloch walls by definition, $\varphi = 0$, which is a possible solution. The first equation (7.247) has a first integral of

$$\sqrt{\frac{A}{K_1}}\frac{\mathrm{d}\theta}{\mathrm{d}y} = \sin\theta\,, \tag{7.249}$$

which integrates in turn to

$$\theta = 2\arctan(\mathrm{e}^{y/\Delta_0})\,, \quad \Delta_0 = \sqrt{\frac{A}{K_1}}\,. \tag{7.250}$$

The effective wall width is obtained by approximating $\mathrm{d}\theta/\mathrm{d}y$ by its value at the midpoint of the wall, where $\theta = \pi/2$.

$$\frac{\mathrm{d}\theta}{\mathrm{d}y} = \sqrt{\frac{K_1}{A}} \cong \frac{1}{a}\sqrt{\frac{D}{J}}\,, \tag{7.251}$$

so the wall width/a is

$$\frac{\text{wall width}}{a} = \pi\sqrt{\frac{D}{J}}\,.$$

One can also show the wall width per unit area (perpendicular to the y-axis in Fig. 7.25) is $4(AK_1)^{1/2}$. For Iron, the wall energy per unit area is of order 1 erg/cm^2, and the wall width is of order 500 Å.

Magnetostrictive Energy (B)

Magnetostriction is the variation of size of a magnetic material when its magnetization varies. Magnetostriction implies a coupling between elastic and magnetic effects caused by the interaction of atomic magnetic moments and the lattice. The magnetostrictive coefficient λ is $\delta l/l$, where δl is the change in length associated with the magnetization change. In general λ can be either sign and is typically of the order of 10^{-5} or so. There may also be a change in volume due to changing magnetization. In any case the deformation is caused by a lowering of the energy.

Magnetostriction is a very complex matter and a detailed description is really outside the scope of this book. We needed to mention it because it has a bearing on domains. See, e.g., Gibbs [7.24].

Formation of Magnetic Domains (B)

We now give a qualitative account of the formation of domains. Consider a cubic material, originally magnetized along an easy direction as shown in Fig. 7.20. Because the magnetization $\boldsymbol{M}$ and demagnetizing fields have opposite directions (7.136), this configuration has large magnetostatic energy. The magnetostatic energy can be reduced if the material splits into domains as shown in Fig. 7.21

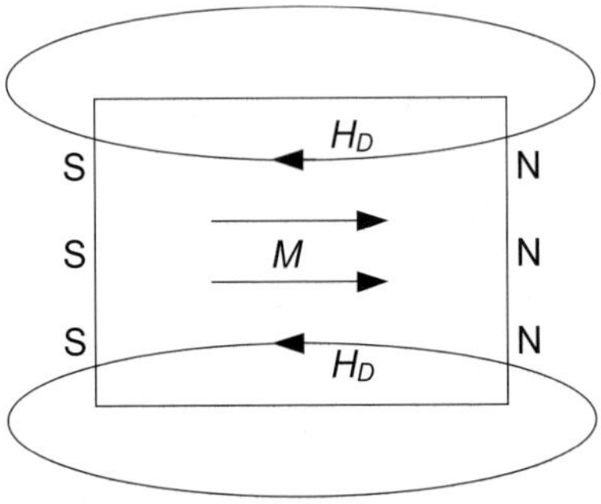

Fig. 7.20. Magnetic domain formation within a material

Since the density of surface poles is $+\boldsymbol{M}\cdot\boldsymbol{n}$ where $\boldsymbol{n}_M$ is the *outward* normal, at an interface the net *magnetic charge* per unit area is

$$(\boldsymbol{M}_2 - \boldsymbol{M}_1)\cdot\boldsymbol{n}_{M_2} ,$$

where $\boldsymbol{n}_{M2}$ is a unit vector pointing from region 1 to region 2. Thus when $\boldsymbol{M}\cdot\boldsymbol{n}$ is continuous, there are no demagnetizing fields (assuming also M is uniform in the interior). Thus (for typical magnetic materials with cubic symmetry) the magnetostatic energy can be further reduced by forming domains of closure, as shown in Fig. 7.22. The overall magnetostrictive and strain energy can be reduced by the formation of more domains of closure (see Fig. 7.23). That is, this splitting into smaller domains reduces the extra energy caused by the internal strain brought about by the spontaneous strain in the direction of magnetization. This process will not continue forever because of the increase in the wall energy (due to exchange and anisotropy). An actual material will of course have many imperfections as well as other complications that will cause irregularities in the domain structure.

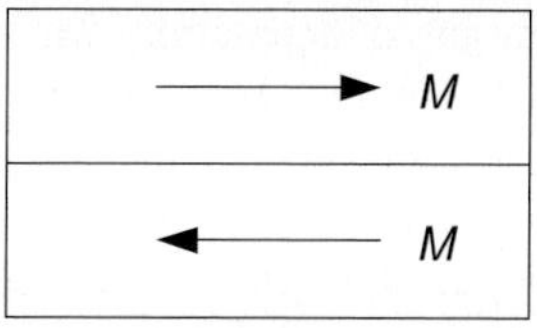

Fig. 7.21. Magnetic-domain splitting within a material

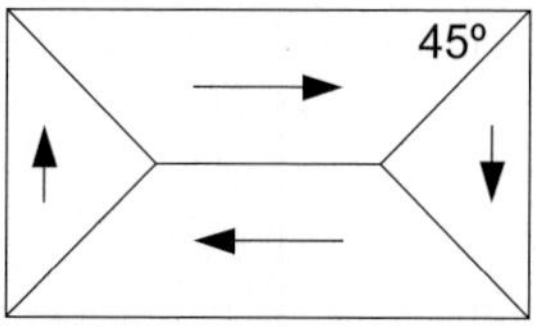

Fig. 7.22. Formation of magnetic domains of closure

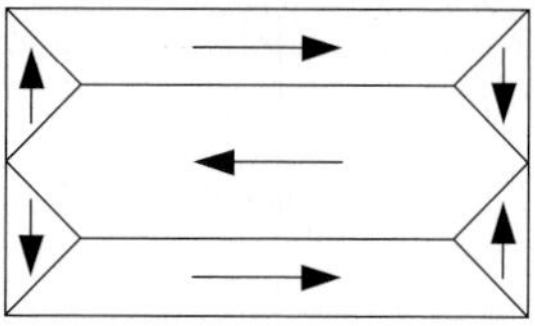

Fig. 7.23. Formation of more magnetic domains of closure

Hysteresis, Remanence, and Coercive Force (B)

Consider an unmagnetized ferromagnet well below its Curie temperature. We can understand the material being unmagnetized if it consists of a large number of domains, each of which is spontaneously magnetized, but that have different directions of magnetization so the net magnetization averages to zero.

The magnetization changes from one domain to another through thin but finite-width domain walls. Typically, domain walls are of thickness of about 10^{-7} meters or some hundreds of atomic spacings, while the sides of the domains are a few micrometers and larger.

The hysteresis loop can be visualized by plotting M vs. H or $B = \mu_0(H + M)$ (in SI) $= H + 4\pi M$ (in Gaussian units) (see Fig. 7.24). The *virgin curve* is obtained by starting in an ideal demagnetized state in which one is at the absolute minimum of energy.

When an external field is turned on, "favorable" domains have lower energy than "unfavorable" ones, and thus the favorable ones grow at the expense of the unfavorable ones.

Imperfections determine the properties of the hysteresis loop. Moving a domain wall generally increases the energy of a ferromagnetic material due to a complex combination of interactions of the domain wall with dislocations, grain boundaries, or other kinds of defects. Generally the first part of the virgin curve is reversible, but as the walls sweep past defects one enters an irreversible region, then in the final approach to saturation, one generally has some rotation of domains. As H is reduced to zero, one is left with a remanent magnetization (in a metastable state with a "local" rather than absolute minimum of energy) at $H = 0$ and B only goes to zero at $-H_c$, the coercive "force".[25] For permanent magnetic materials, M_R and H_c should be as large as possible. On the other hand, soft magnets will have very low coercivity. The hysteresis and domain properties of magnetic materials are of vast technological importance, but a detailed discussion would take us too far afield. See Cullity [7.16].

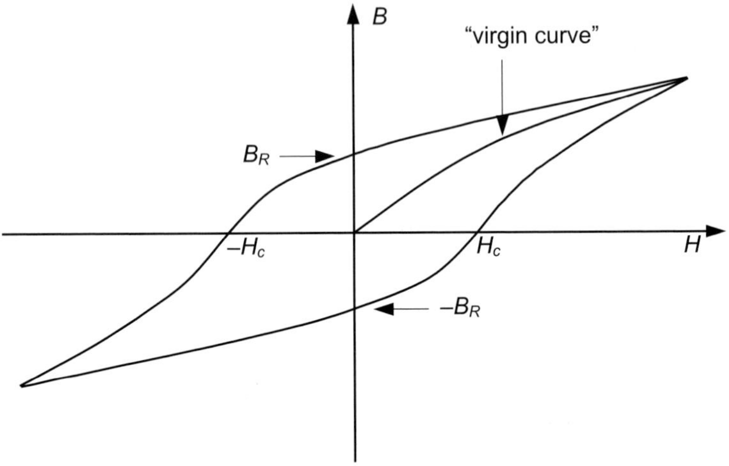

Fig. 7.24. Magnetic hysteresis loop identifying the virgin curve
$H_c \equiv$ coercive "force".
B_R = remanence.
$M_s = [(B - H)/4\pi]_{H\to\infty}$ = saturation magnetization.
$M_R = B_R/4\pi$ = remanent magnetization.

Néel and Bloch Walls (B)

Figure 7.25 provides a convenient way to distinguish Bloch and Néel walls. Bloch walls have $\varphi = 0$, while Néel walls have $\varphi = \pi/2$. Néel walls occur in thin films of materials such as permalloy in order to reduce surface magnetostatic energy as suggested by Fig. 7.26. There are many other complexities involved in domain-wall structures. See, e.g., Malozemoff and Slonczewski [7.44].

[25] Some authors define H_c as the field that reduces M to zero.

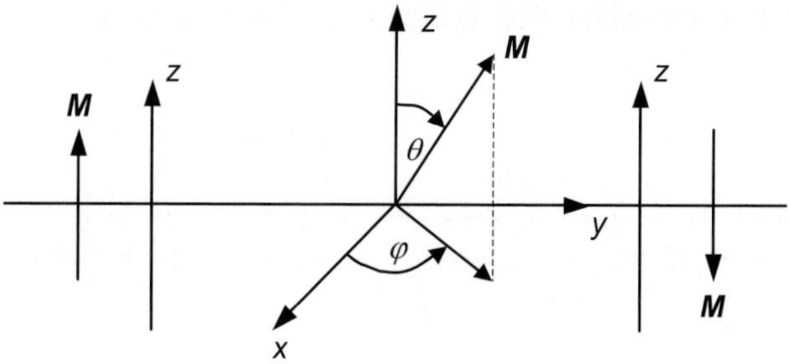

Fig. 7.25. Bloch wall: $\varphi = 0$; Néel wall: $\varphi = \pi/2$

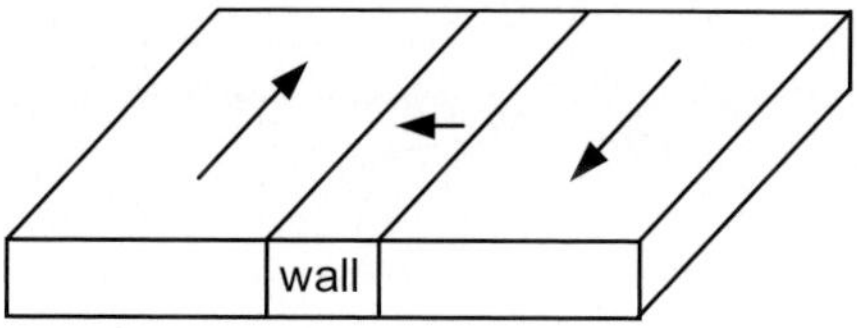

Fig. 7.26. Néel wall in thin film

Methods of Observing Domains (EE, MS)

We briefly summarize five methods.

1. Bitter patterns–a colloidal suspension of particles of magnetite is placed on a polished surface of the magnetic material to be examined. The particles are attracted to regions of nonuniform magnetization (the walls) and hence the walls are readily seen by a microscope.
2. Faraday and Kerr effects–these involve rotation of the plane of polarization on transmission and reflection (respectively) from magnetic substances.
3. Neutrons–since neutrons have magnetic moments they experience interaction with the internal magnetization and its direction, see Bacon GE, "Neutron Diffraction," Oxford 1962 (p355ff).
4. Transmission electron microscopy (TEM)–Moving electrons are influenced by forces due to internal magnetic fields.
5. Scanning electron microscopy (SEM)–Moving secondary electrons sample internal magnetic fields.

7.3.2 Magnetic Materials (EE, MS)

Some Representative Magnetic Materials (EE, MS)

Table 7.4. Ferromagnets

Ferromagnets	T_c (K)	M_s ($T = 0$ K, Gauss)
Fe	1043	1752
Ni	631	510
Co	1394	1446
EuO	77	1910
Gd	293	1980

From Parker SP (ed), *Solid State Physics Sourcebook*, McGraw-Hill Book Co., New York, 1987, p. 225.

Table 7.5. Antiferromagnets

Antiferromagnets	T_N (K)
MnO	122
NiO	523
CoO	293

From Cullity BD, *Introduction to Magnetic Materials*, Addison-Wesley Publ Co, Reading, Mass, 1972, p. 157.

Table 7.6. Ferrimagnets

Ferrimagnets	T_c (K)	M_s ($T = 0$ K, Gauss)
YIG ($Y_3Fe_5O_{12}$)	560	195 a garnet
Magnetite (Fe_3O_4)	858	510 a spinel

(From *Solid State Physics Sourcebook*, op cit p. 225)

We should emphasize that these classes do not exhaust the types of magnetic order that one can find. At suitably low temperatures the heavy rare earths, may show helical or conical order. and there are other types of order, as for example, spin glass order. Amorphous ferromagnets show many kinds of order such as speromagnetic and asperomagnetic. (See, e.g., *Solid State Physics Source Book*, op cit p 89.)

Ferrites are perhaps the most common type of ferrimagnets. Magnetite, the oldest magnetic material that is known, is a ferrite also called lodestone. In general, ferrites are double oxides of iron and another metal such as Ni or Ba (e.g. nickel ferrite: $NiOFe_2O_3$ and barium ferrite: $BaO{\cdot}6Fe_2O_3$). Many ferrites find application in high-frequency devices because they have high resistivity and hence do not have appreciable eddy currents. They are used in microwave devices, radar, etc. Barium ferrite, a hard magnet, is one of the materials used for magnetic recording that is a very large area of application of magnets (see, e.g., Craik [7.15 p. 379]).

Hard and Soft Magnetic Materials (EE, MS) The clearest way to distinguish between hard and soft magnetic materials is by a hysteresis loop (see Fig. 7.27). Hard permanent magnets are hard to magnetize and demagnetize, and their coercive forces can be of the order of 10^6 A/m or larger. For a soft magnetic material, the coercive force can be of order 1 A/m or smaller. For conversions: 1 A/m is $4\pi \times 10^{-3}$ Oersted, 1 kJ/m^3 converts to MGOe (mega Gauss Oersted) if we multiply by 0.04π, 1Tesla = 10^4 G.

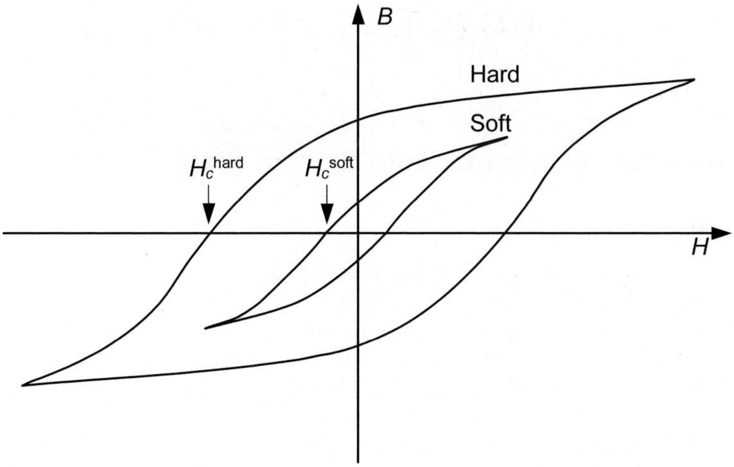

Fig. 7.27. Hard and soft magnetic material hysteresis loops (schematic)

Permanent Magnets (EE, MS) There are many examples of permanent magnetic materials. The largest class of magnets used for applications are permanent magnets. They are used in electric motors, in speakers for audio systems, as wiggler magnets in synchrotrons, etc. We tabulate here only two examples that have among the highest energy products $(BH)_{max}$.

Table 7.7. Permanent Magnets

	T_c (K)	M_s (kA m^{-1})	H_c (kA m^{-1})	$(BH)_{max}$ (kJ m^{-3})
(1) $SmCo_5$	997	768	700–800	183
(2) $Nd_2Fe_{14}B$	~583	—	~880	~290

(1) Craik [7.15 pp. 385, 387]. Sm_2Co_{17} is in some respects better, see [7.15 p. 388].
(2) *Solid State Physics Source Book* op cit p 232. Many other hard magnetic materials are mentioned here such as the AlNiCos, barium ferrite, etc. See also Herbst [7.29].

Soft Magnetic Materials (EE, MS) There are also many kinds of soft magnetic materials. They find application in communication materials, motors, generators, transformers, etc. Permalloys form a very common class of soft magnets. These are Ni-Fe alloys with sometimes small additions of other elements. 78 Permalloy means, e.g., 78% Ni and 22% Fe.

Table 7.8. Soft Magnet

	T_c (K)	H_c (A m^{-1})	B_s (T)
78 Permalloy	873	4	1.08

See *Solid State Physics Source Book* op cit, p. 231. There are several other examples such as high-purity iron.

7.4 Magnetic Resonance and Crystal Field Theory

7.4.1 Simple Ideas About Magnetic Resonance (B)

This Section is the first of several that discuss magnetic resonance. For further details on magnetic resonance than we will present, see Slichter [91]. The technique of magnetic resonance can be used to investigate very small energy differences between individual energy levels in magnetic systems. The energy levels of interest arise from the orientation of magnetic moments of the system in, for example, an external magnetic field. The magnetic moments can arise from either electrons or nuclei.

Consider a particle with magnetic moment $\boldsymbol{\mu}$ and total angular momentum $\boldsymbol{J}$ and assume that the two are proportional so that we can write

$$\boldsymbol{\mu} = \gamma \boldsymbol{J} , \tag{7.252}$$

where the proportionality constant γ is called the *gyromagnetic ratio* and equals $-g\mu_B/\hbar$ (for electrons, it would be + for protons) in previous notation. We will then suppose that we apply a magnetic induction B in the z direction so that the Hamiltonian of the particle with magnetic moment becomes

$$\mathcal{H}_0 = -\gamma\mu_0 H J_z , \tag{7.253}$$

where we have used (7.252), and $B = \mu_0 H$, where H is the magnetic field. If we define j (which are either integers or half-integers) so that the eigenvalues of $\boldsymbol{J}^2$ are $j(j+1)\hbar^2$, then we know that the eigenvalues of $\mathcal{H}_0$ are

$$E_m = -\gamma\hbar\mu_0 H m , \tag{7.254}$$

where $-j \le m \le j$.

From (7.254) we see that the difference between adjacent energy levels is determined by the magnetic field and the gyromagnetic ratio. We can induce transitions between these energy levels by applying an alternating magnetic field (perpendicular to the z direction) of frequency ω, where

$$\hbar\omega = |\gamma| \hbar\mu_0 H \quad \text{or} \quad \omega = |\gamma| \mu_0 H . \tag{7.255}$$

These results follow directly from energy conservation and they will be discussed further in the next section. It is worthwhile to estimate typical frequencies that are involved in resonance experiments for a convenient size magnetic field. For an electron with charge e and mass m, if the gyromagnetic ratio γ is defined as the ratio of magnetic moment to *orbital angular* momentum, it is given by

$$\gamma = e/2m , \quad \text{for } e < 0 . \tag{7.256}$$

For an electron with spin but no orbital angular momentum, the ratio of magnetic moment to spin angular momentum is $2\gamma = e/m$. For an electron with both orbital and spin angular momentum, the contributions to the magnetic moment are as described and are additive. If we use (7.255) and (7.256) with magnetic fields of order 8000 G, we find that the resonance frequency for electrons is in the *microwave* part of the spectrum. Since nuclei have much greater mass, the resonance frequency for nuclei lies in the *radio frequency* part of the spectrum. This change in frequency results in a considerable change in the type of equipment that is used in observing electron or nuclear resonance.

Abbreviations that are often used are NMR for nuclear magnetic resonance and EPR or ESR for electron paramagnetic resonance or electron spin resonance.

7.4.2 A Classical Picture of Resonance (B)

Except for the concepts of spin-lattice and spin-spin relaxation times (to be discussed in the Section on the Bloch equations) we have already introduced many of the most basic ideas connected with magnetic resonance. It is useful to present a classical description of magnetic resonance [7.39]. This description is more pictorial than the quantum description. Further, it is true (with a suitable definition of the time derivative of the magnetic moment operator) that the classical magnetic moment in an external magnetic field obeys the same equations of motion as the magnetic moment operator. We shall not prove this theorem here, but it is because of it that the classical picture of resonance has considerable use. The simplest way of presenting the classical picture of resonance is by use of the concept of the rotating coordinate system. It also should be pointed out that we will leave out of our discussion any relaxation phenomena until we get to the Section on the Bloch equations.

As before, let a magnetic system have angular momentum $\boldsymbol{J}$ and magnetic moment $\boldsymbol{\mu}$, where $\boldsymbol{\mu} = \gamma\boldsymbol{J}$. By classical mechanics, we know that the time rate of change of angular momentum equals the external torque. Therefore we can write for a magnetic moment in an external field $\boldsymbol{H}$,

$$\frac{\mathrm{d}\boldsymbol{J}}{\mathrm{d}t} = \boldsymbol{\mu} \times \mu_0 \boldsymbol{H} . \tag{7.257}$$

Since $\boldsymbol{\mu} = \gamma \boldsymbol{J}$ ($\gamma < 0$ for electrons), we can write

$$\frac{\mathrm{d}\boldsymbol{\mu}}{\mathrm{d}t} = \boldsymbol{\mu} \times (\gamma\mu_0)\boldsymbol{H} . \tag{7.258}$$

This is the general equation for the motion of the magnetic moment in an external magnetic field.

To obtain the solution to (7.258) and especially in order to picture this solution, it is convenient to use the concept of the rotating coordinate system. Let

$$\boldsymbol{A} = \hat{\boldsymbol{i}}A_x + \hat{\boldsymbol{j}}A_y + \hat{\boldsymbol{k}}A_z$$

be any vector, and let $\hat{\boldsymbol{i}}, \hat{\boldsymbol{j}}, \hat{\boldsymbol{k}}$ be unit vectors in a rotating coordinate system. If Ω is the angular velocity of the rotating coordinate system relative to a fixed coordinate system, then relative to a fixed coordinate system we can show that

$$\frac{\mathrm{d}\hat{\boldsymbol{i}}}{\mathrm{d}t} = \boldsymbol{\Omega} \times \hat{\boldsymbol{i}} . \tag{7.259}$$

This implies that

$$\frac{\mathrm{d}\boldsymbol{A}}{\mathrm{d}t} = \frac{\delta \boldsymbol{A}}{\delta t} + \boldsymbol{\Omega} \times \boldsymbol{A} , \tag{7.260}$$

where $\delta\boldsymbol{A}/\delta t$ is the rate of change of $\boldsymbol{A}$ relative to the rotating coordinate system and $\mathrm{d}\boldsymbol{A}/\mathrm{d}t$ is the rate of change of $\boldsymbol{A}$ relative to the fixed coordinate system.

By using (7.260), we can write (7.258) in a rotating coordinate system. The result is

$$\frac{\delta \boldsymbol{\mu}}{\delta t} = \boldsymbol{\mu} \times (\boldsymbol{\Omega} + \gamma\mu_0 \boldsymbol{H}) . \tag{7.261}$$

Equation (7.261) is the same as (7.258). The only difference is that in the rotating coordinate system the effective magnetic field is

$$\boldsymbol{H}_{\text{eff}} = \boldsymbol{H} + \frac{\boldsymbol{\Omega}}{\gamma\mu_0} . \tag{7.262}$$

If $\boldsymbol{H}$ is constant and $\boldsymbol{\Omega}$ is chosen to have the constant value $\boldsymbol{\Omega} = -\gamma\mu_0\boldsymbol{H}$, then $\delta\boldsymbol{\mu}/\delta t = 0$. This means that the spin precesses about $\boldsymbol{H}$ with angular velocity $\gamma\mu_0\boldsymbol{H}$. Note that this is the same as the frequency for magnetic resonance absorption. We will return to this point below.

It is convenient to get a little closer to the magnetic resonance experiment by supposing that we have a static magnetic field H_0 along the z direction and an alternating magnetic field $H_x(t) = 2H'\cos(\omega t)$ along the x-axis. We can resolve the alternating field into two rotating magnetic fields (one clockwise, one counterclockwise) as shown in Fig. 7.28. Simple vector addition shows that the two fields add up to $H_x(t)$ along the x-axis.

With the static magnetic field along the z direction, the magnetic moment will precess about the z-axis. The moment will precess in the same sense as one of the rotating magnetic fields. Now that we have both constant and alternating magnetic

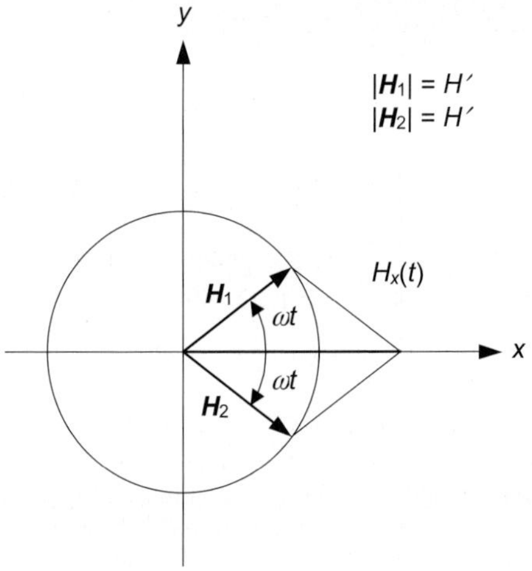

Fig. 7.28. Decomposition of an alternating magnetic field into two rotating magnetic fields

fields, something interesting begins to happen. The component of the alternating magnetic field that rotates in the same direction as the magnetic moment is the important component [91]. Near resonance, the magnetic moment and one of the circularly polarized components of the alternating magnetic field rotate with almost the same angular velocity. In this situation the rotating magnetic field exerts an almost constant torque on the magnetic moment and tends to tip it over. Physically, this is what happens in resonance absorption.

Let us be a little more quantitative about this problem. If we include only one component of the rotating magnetic field and if we assume that Ω is the cyclic frequency of the alternating magnetic field, then we can write

$$\frac{\delta \boldsymbol{\mu}}{\delta t} = \boldsymbol{\mu} \times [\boldsymbol{\Omega} + \gamma\mu_0(\hat{\boldsymbol{i}}H' + \hat{\boldsymbol{k}}H_0)]\,. \tag{7.263}$$

This can be further written as

$$\frac{\delta \boldsymbol{\mu}}{\delta t} = \boldsymbol{\mu} \times \boldsymbol{H}_{\text{eff}}\,, \tag{7.264}$$

where now

$$\boldsymbol{H}_{\text{eff}} \equiv \hat{\boldsymbol{k}}\left(H_0 + \frac{\Omega}{\gamma\mu_0}\right) + \hat{\boldsymbol{i}}H'\,.$$

Since in the rotating coordinate system $\boldsymbol{\mu}$ precesses about $\boldsymbol{H}_{\text{eff}}$, we have the picture shown in Fig. 7.29. If we adjust the static magnetic field so that

$$H_0 = -\frac{\Omega}{\gamma\mu_0}\,,$$

then we have satisfied the conditions of resonance. In this situation $\boldsymbol{H}_{\text{eff}}$ is along the x-axis (in the rotating coordinate system) and the magnetic moment flops up and down with frequency $\gamma\mu_0 H'$.

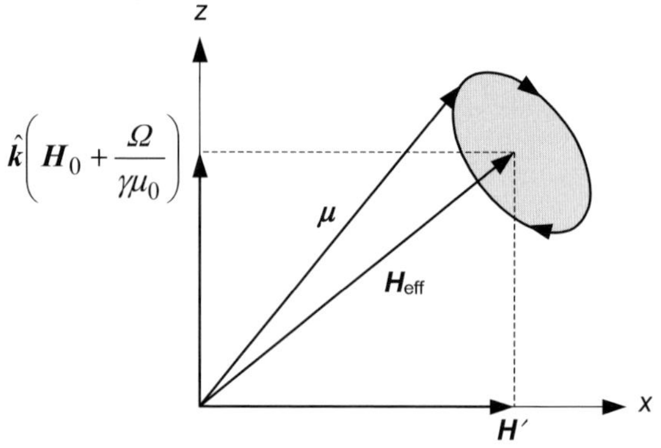

Fig. 7.29. Precession of the magnetic moment $\boldsymbol{\mu}$ about the effective magnetic field $\boldsymbol{H}_{\mathrm{eff}}$ in a coordinate system rotating with angular velocity Ω about the z-axis

Similar quantum-mechanical calculations can be done in a rotating coordinate system, but we shall not do them as they do not add much that is new. What we have done so far is useful in forming a pictorial image of magnetic resonance, but it is not easy to see how to put in spin-lattice interactions, or other important interactions. In order to make progress in interpreting experiments, it is necessary to generalize our formalism somewhat.

7.4.3 The Bloch Equations and Magnetic Resonance (B)

These equations are used for a qualitative and phenomenological discussion of NMR and EPR. In general, however, it is easier to describe NMR than EPR. This is because the nuclei do not interact nearly so strongly with their surroundings as do the electrons. We shall later devote a Section to discussing how the electrons interact with their surroundings.

The Bloch equations are equations that describe precessing magnetic moments, and various relaxation mechanisms. They are almost purely phenomenological, but they do provide us a means of calculating the power absorbed versus the frequency. Without the interactions responsible for the relaxation times, this plot would be a delta function. Such a situation would not be very interesting. It is the relaxation times that give us information about what is going on in the solid.[26]

Definition of Bloch Equations and Relaxation Times (B)

The theory of the resonance of free spins in a magnetic field is simple but it holds little inherent interest. To relate to more physically interesting phenomena it is necessary to include the interactions of the spins with their environment. The Bloch equations include these interactions in a phenomenological way.

---

[26] See Manenkov and Orbach (eds) [7.45].

When we include a relaxation time (or an interaction process), we find that the time rate of change of the magnetization (along the field) is proportional to the deviation of the magnetization from its equilibrium value. This guarantees a relaxation of magnetization along the field. If we add an alternating magnetic field along the x- or y-axes, it is also necessary to add a term $(\boldsymbol{M} \times \boldsymbol{H})_z$ that is proportional to the torque. Thus for the component of magnetization along the constant external magnetic field, it is reasonable to write

$$\frac{\mathrm{d}M_z}{\mathrm{d}t} = \frac{M_0 - M_z}{T_1} + (\gamma\mu_0)(\boldsymbol{M} \times \boldsymbol{H})_z \,. \tag{7.265}$$

As noted, (7.265) has a built-in relaxation process of M_z to M_0, the spin-lattice relaxation time T_1. However, as we approach equilibrium in a static magnetic field $H_0\hat{\boldsymbol{k}}$, we will want both M_x and M_y to tend to zero. For this purpose, a new term with a relaxation time T_2 is often introduced. We write

$$\frac{\mathrm{d}M_x}{\mathrm{d}t} = \gamma\mu_0(\boldsymbol{M} \times \boldsymbol{H})_x - \frac{M_x}{T_2} \,, \tag{7.266}$$

and

$$\frac{\mathrm{d}M_y}{\mathrm{d}t} = \gamma\mu_0(\boldsymbol{M} \times \boldsymbol{H})_y - \frac{M_y}{T_2} \,. \tag{7.267}$$

Equations (7.265), (7.266), and (7.267) are called the *Bloch equations*. T_2 is often called the *spin-spin relaxation time*. The idea is that the term involving T_1 is caused by the interaction of the spin system with the lattice or phonons, while the term involving T_2 is caused by something else. The physical origin of T_2 is somewhat complicated. Consider, for example, two nuclei precessing in an external static magnetic field. The precession of one nucleus produces a varying magnetic field at the second nucleus and hence tends to "flip" the spin of the second nucleus (and vice versa). Waller[27] first pointed out that there are two different types of spin relaxation processes.

Ferromagnetic Resonance (B)

Using a simple quantum picture, for an atomic system, we have already argued (see (7.258))

$$\frac{d\boldsymbol{\mu}}{dt} = \gamma\boldsymbol{\mu} \times \boldsymbol{B}_a \,, \tag{7.268}$$

where $\boldsymbol{B}_a = \mu_0\boldsymbol{H}$. This implies a precession of $\boldsymbol{\mu}$ and $\boldsymbol{M}$ about the constant magnetic field $\boldsymbol{B}_a$ with frequency $\omega = \gamma B_a$ the Larmor frequency, as already noted. For

[27] See Waller [7.66]. Discussion of ways to calculate T_1 and T_2 is contained in White [7.68 p124ff and 135ff].

ferromagnetic resonance (FMR) all spins precess together and $M = N\mu$, where N is the number of spins per unit volume. Thus by (7.268)

$$\frac{\mathrm{d}\boldsymbol{M}}{\mathrm{d}t} = \gamma \boldsymbol{M} \times \boldsymbol{B}_a . \tag{7.269}$$

Several comments can be made. The above equation is valid also for $\boldsymbol{M} = \boldsymbol{M}(\boldsymbol{r})$ varying slowly in space. We will also use this equation for spin-wave resonance when the wavelengths of the waves are long compared to the atom to atom spacing that allows the classical approach to be valid. One generalizes the above equation by replacing $\boldsymbol{B}_a$ by $\boldsymbol{B}$ where

$\boldsymbol{B} = \boldsymbol{B}_a$ (applied)
$+ \boldsymbol{B}_{\mathrm{rf}}$ (due to a radio-frequency applied field)
$+ \boldsymbol{B}_{\mathrm{demag}}$ (from demagnetizing fields that depend on geometry)
$+ \boldsymbol{B}_{\mathrm{exchange}}$ (as derived from the Heisenberg Hamiltonian)
$+ \boldsymbol{B}_{\mathrm{anisotropy}}$ (an effective field arising from interactions producing anisotropy).

We should also include dissipative or damping and relaxation effects.

We start with all fields zero or negligible except for the applied field (note here $\boldsymbol{B}_{\mathrm{exchange}} \propto \boldsymbol{M}$, which is assumed to be uniform, so $\boldsymbol{M} \times \boldsymbol{B}_{\mathrm{exchange}} = 0$). This gives resonance at the natural precessional frequency of the uniform precessional mode. With $\boldsymbol{B} = B_0\hat{\boldsymbol{k}}$ we have

$$\frac{\mathrm{d}M_x}{\mathrm{d}t} = \gamma M_y B_0 , \quad \frac{\mathrm{d}M_y}{\mathrm{d}t} = -\gamma M_x B_0 , \quad \frac{\mathrm{d}M_z}{\mathrm{d}t} = 0. \tag{7.270}$$

We look for solutions with

$$\begin{aligned} M_x &= A_1 \mathrm{e}^{-\mathrm{i}\omega t} , \\ M_y &= A_2 \mathrm{e}^{-\mathrm{i}\omega t} , \\ M_z &= \text{constant}, \end{aligned} \tag{7.271}$$

and so we have a solution provided

$$\begin{vmatrix} -\mathrm{i}\omega & -\gamma B_0 \\ \gamma B_0 & -\mathrm{i}\omega \end{vmatrix} = 0 , \tag{7.272}$$

or

$$|\omega| = |\gamma B_0| , \tag{7.273}$$

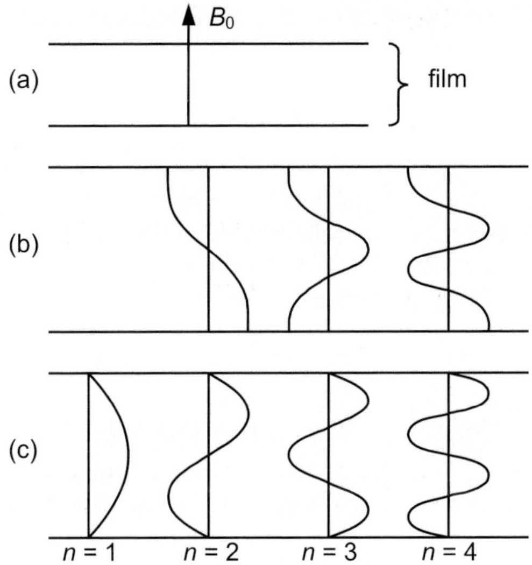

Fig. 7.30. (**a**) Thin film with magnetic field, (**b**) "Unpinned" spin waves, (**c**) "Pinned" spin waves

which as expected is just the Larmor precessional frequency. In actual situations we also need to include demagnetization fields and hence shape effects, which will alter the resonant frequencies. FMR typically occurs at microwave frequencies. Antiferromagnetic resonance (AFMR) has also been studied as a way to determine anisotropy fields.

Spin-Wave Resonance (A)

Spin-wave resonance is a direct way to experimentally prove the existence of spin waves (as is inelastic neutron scattering – see Kittel [7.39 pp456-458]). Consider a thin film with a magnetic field B_0 perpendicular to the film (Fig. 7.30a). In the simplest picture, we view the spin waves as "vibrations" in the spin between the surfaces of the film. Plotting the amplitude versus position, Fig. 7.30b is obtained for unpinned spins. Except for the uniform mode, these have no net interaction (absorption) with the electromagnetic field. The pinned case is a little different (Fig. 7.30c). Here only waves with an even number of half-wavelengths will show no net interaction energy with the field while the ones with an odd number of half-wavelengths (n = 1, 3, etc.) will absorb energy. (Otherwise the induced spin flippings will absorb and emit equal amounts of energy).

We get absorption when

$$n\frac{\lambda}{2} = T \quad \{n \text{ odd, } T \text{ thickness of film}\},$$

$$k = \frac{2\pi}{\lambda} = \frac{n\pi}{T} \quad \text{or} \quad k = (2n+1)\frac{\pi}{T} \quad \{n = 0, 1, 2\ldots\}.$$

With applied field normal to film and with demagnetizing field and exchange $D'k^2$, absorption will occur for

$$\omega_0 = \gamma(B_0 - \mu_0 M) + D'k^2 \quad \text{(SI)},$$

where M is the static magnetization in the direction of B_0. The spin-wave frequency is determined by both the FMR frequency (the first term including demagnetization) and the dispersion relation typical for spin waves.
We now analyze spin-wave resonance in a little more detail. First we develop the Heisenberg Hamiltonian in the continuum approximation,

$$\mathcal{H} = -\sum J_{ij}\boldsymbol{S}_i \cdot \boldsymbol{S}_j = -\tfrac{1}{2}\sum \boldsymbol{\mu}_i \cdot \boldsymbol{B}_i^{\text{ex}} \tag{7.274}$$

defines the effective field $\boldsymbol{B}_i^{\text{ex}}$ acting on the moment at site i, $\boldsymbol{\mu}_i = \gamma\boldsymbol{S}_i$. ($\gamma < 0$ for electrons)

$$\boldsymbol{B}_i^{\text{ex}} = \frac{2}{\gamma}\sum_j J_{ij}\boldsymbol{S}_j. \tag{7.275}$$

Assuming nearest neighbors (nn) at distance a and nn interactions only. We find for a simple cubic (SC) structure after expansion, and using cancellation resulting from symmetry

$$\gamma\boldsymbol{B}_i^{\text{ex}} = 12J\boldsymbol{S}_i + 2Ja^2\nabla^2\boldsymbol{S}_i.$$

Consistent with the classical continuum approximation

$$\frac{\boldsymbol{M}}{M} = \frac{\boldsymbol{S}_i}{S}, \tag{7.276}$$

$$\boldsymbol{B}^{\text{ex}} = \lambda\boldsymbol{M} + K'\nabla^2(\boldsymbol{M}/M), \tag{7.277}$$

where

$$\lambda = \frac{12JS}{\gamma M},\ K' = \frac{2Ja^2S}{\gamma}. \tag{7.278}$$

As an aside we note B^{ex} is consistent with results obtained before (Sect. 7.3.1). Since

$$U = -\tfrac{1}{2}\sum \boldsymbol{\mu}_i \cdot \boldsymbol{B}_i^{\text{ex}}, \tag{7.279}$$

neglecting constant terms (resulting from the magnitude of the magnetization being constant) we have

$$U = -\frac{JS^2}{a}\int \boldsymbol{m}\nabla^2\boldsymbol{m}\,\mathrm{d}V \quad \{\boldsymbol{m} = \boldsymbol{M}/M\}. \tag{7.280}$$

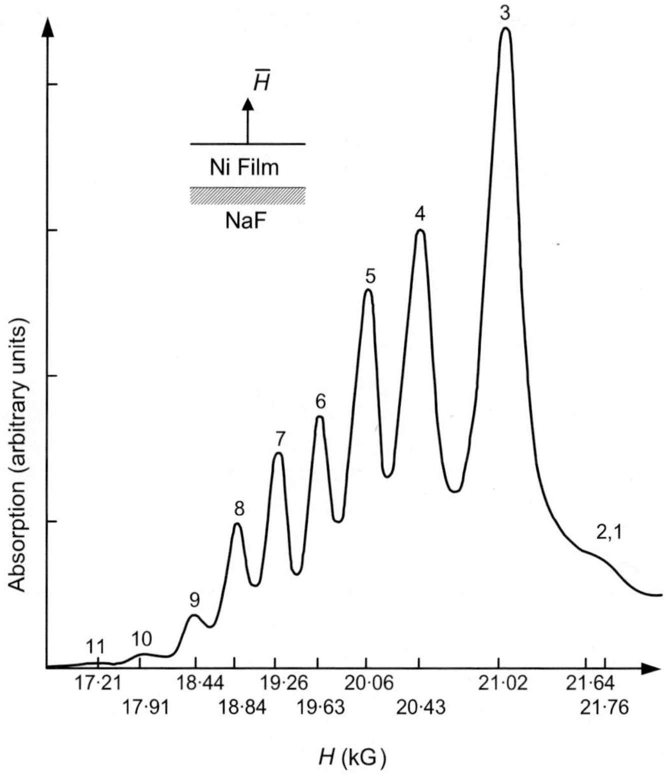

Fig. 7.31. Spin wave resonance spectrum for Ni film, room temperature, 17 GHz. After Puszharski H, "Spin Wave Resonance", *Magnetism in Solids Some Current Topics*, Scottish Universities Summer School in Physics, 1981, p. 287, by permission of SUSSP. Original data in Mitra DP and Whiting JSS, *J Phys F: Metal Physics*, **8**, 2401 (1978)

Assuming $\int m_x \nabla m_x \cdot d\boldsymbol{A}$ etc = 0 for a large surface we can also recast the above as

$$U = \frac{JS^2}{a} \int [(\nabla \alpha_1)^2 + (\nabla \alpha_2)^2 + (\nabla \alpha_3)^2] dV \tag{7.281}$$

which is the same as we obtained before, with a slightly different analysis. The α_i are of course the direction cosines.

The anisotropy energy and effective field can be written in the same way as before, and no further comments need be made about it.

When one generalizes the equation for the time development of M, one has the Landau–Lifshitz equations. Damping causes broadening of the absorption lines. Then

$$\dot{\boldsymbol{M}} = \gamma \boldsymbol{M} \times \boldsymbol{B}^{\text{eff}} + \frac{\alpha}{M} \boldsymbol{M} \times (\boldsymbol{M} \times \boldsymbol{B}^{\text{eff}}), \tag{7.282}$$

where α is a constant characterizing the damping. Spin-wave resonance has been observed as shown in Fig. 7.31. The integers label the modes of excitation. The figure is complicated by surface spin waves that are labeled 2, 1 and not fully resolved. Reference to the original paper must be made for complete details.

We have discussed $\boldsymbol{B}^{\text{eff}}$ in the Section on FMR. Allowing $\boldsymbol{M}$ to vary with $\boldsymbol{r}$ and using the pinned boundary conditions, (7.282) can be used to quantitatively discuss SWR.

7.4.4 Crystal Field Theory and Related Topics (B)

This Section is primarily related to EPR. The general problem is to analyze the effects of neighboring ions on paramagnetic ions in a crystal. This cannot be exactly solved, and so we must seek physically reasonable simplifying assumptions.

Some atoms or ions when placed in a crystal act as if they undergo very little change. When this is so, we can predict the changes by perturbation theory. In order to estimate the perturbing effects of a host crystal on a paramagnetic ion, we ought to be able to treat the host crystal fairly crudely. For example, for an ionic crystal it might be sufficient to treat the ions as point charges. Then it would be fairly simple to estimate the change in the potential at the paramagnetic ion due to the host crystal. This potential energy could serve as a perturbation on the Hamiltonian of the paramagnetic ion.

Another simplification is possible. The crystal potential must have the symmetry of the point group describing the surroundings of the paramagnetic ion. As we will discuss later, group theory is useful in taking this into account.

The effect of the crystal field is to split the energy levels of a paramagnetic ion. In order to show how this comes about, it is useful to know what we mean by the energy levels. The best way to do this is to write down the Hamiltonian (whose eigenvalues are the energy levels) for the electrons. With no external field, the Hamiltonian has a form similar to

$$\begin{aligned}\mathcal{H} = \sum\nolimits_i \left[\frac{P_i^2}{2m} - \frac{Ze^2}{4\pi\varepsilon_0 r_i} + a_i \boldsymbol{J}_i \cdot \boldsymbol{I} - e\phi_c(r_i)\right] \\ + \sum\nolimits_{i,j}' \frac{1}{2}\frac{e^2}{4\pi\varepsilon_0 r_{ij}} + \sum\nolimits_{i,j} \lambda_{ij} \boldsymbol{L}_i \cdot \boldsymbol{S}_j\,.\end{aligned} \tag{7.283}$$

The origin of the coordinate system for (7.283) is the nucleus of the paramagnetic ion. The sum over i and j is a sum over electronic coordinates. The first term is the kinetic energy. The second term is the potential energy of the electrons in the field of the nucleus. The third term is the hyperfine interaction of the electron (with total angular momentum $\boldsymbol{J}_i$) with the nucleus that has angular momentum $\boldsymbol{I}$. The fourth term is the crystal field energy. The fifth term is the potential energy of the electrons interacting with themselves. The last term is the spin ($\boldsymbol{S}_j$)-orbit (angular momentum $\boldsymbol{L}_i$) interaction (see Appendix F) of the electrons. By the unperturbed energy levels of the paramagnetic ion, one often means the energy eigenstates of the first, second, and fifth terms obtained perhaps by Hartree–Fock calculations. The rest of the terms are usually thought of as perturbations. In the discussion that follows, the hyperfine interaction will be neglected.

To avoid complicated many-body effects, we will assume that the sources of the crystal field ($\boldsymbol{E}_c \equiv -\nabla\phi_c$) are external to the paramagnetic ion. Thus in the vicinity of the paramagnetic ion, it can be assumed that $\nabla^2\phi_c = 0$.

Weak, Medium, and Strong Crystal Fields (B)

In discussing the effect of the crystal field on the energy levels, which is important to EPR, three cases can be distinguished [47].

Weak crystal fields are by definition those for which the spin-orbit interaction is stronger than the crystal field interaction. This is often realized when the electrons of the paramagnetic shell of the ion lie "fairly deep" within the ion, and hence are shielded from the crystalline field by the outer electrons. This may happen in ionic compounds of the rare earths. Rare earths have atomic numbers (Z) from 58 to 71. Examples are Ce, Pr, and Ne, which have incomplete 4f shells.

By a *medium crystal field* we mean that the crystal field is stronger than the spin-orbit interaction. This happens when the paramagnetic electrons of the ion are mainly distributed over the outer portions of the ion and hence are not well shielded. In this situation something else may occur. The potential that the paramagnetic ions move in is no longer even approximately spherically symmetric, and hence the orbital angular momentum is not conserved. We say that the orbital angular momentum is (at least partially) "quenched" (this means $\langle\psi|\boldsymbol{L}|\psi\rangle = 0$, $\langle\psi|\boldsymbol{L}^2|\psi\rangle \neq 0$). Paramagnetic crystals that have iron group elements ($Z = 21$ to 29, e.g., Cr, Mn, and Fe that have an incomplete 3d shell) are typical examples of the medium-field case.

Strong crystal field by definition means covalent bonding. In this situation, the wave functions for the paramagnetic ion electrons overlap considerably with the wave functions of the other electrons of the crystal. Crystal field theory does not work here. This type of situation will not be discussed in this chapter.

As we will see, group theory can be an aid in understanding how energy levels are split by perturbations.

Miscellaneous Theorems and Facts (In Relation to Crystal Field Theory) (B)

The theorems below will not be proved. They are stated because they are useful in carrying out actual crystal field calculations.

The Equivalent Operator Theorem. This theorem is used in calculating needed matrix elements in crystal field calculations. The theorem states that within a manifold of states for which l is constant, there are simple relations between the matrix elements of the crystal-field potential and appropriate angular momentum operators. For constant l, the rule says to replace the x by L_x (operator, in this case L_x is the x operator equivalent) and so forth for other coordinates. If the result is a product in which the order of the factors is important, then we must use all possible different permutations. There is a similar rule for manifolds of constant J

(where we include both the orbital angular momentum and the spin angular momentum).

There is a straightforward way of generating operator equivalents (OpEq) by using

$$[L^+, Op\ Eq\ Y_l^M] \propto Op\ Eq\ Y_l^{M+1},$$

and

$$[L^-, Op\ Eq\ Y_l^M] \propto Op\ Eq\ Y_l^{M-1}. \tag{7.284}$$

The constants of proportionality can be computed from a knowledge of the Clebsch–Gordon coefficients.

Table 7.4. Effective magneton number for some representative trivalent lanthanide ions

Ion	Configuration	Ground state	$g\sqrt{J(J+1)}$ *
Pr (3+)	...$4f^2\ 5s^2\ 5p^6$	$^3H_4$	3.58
Nd (3+)	...$4f^3\ 5s^2\ 5p^6$	$^4I_{9/2}$	3.62
Gd (3+)	...$4f^7\ 5s^2\ 5p^6$	$^8S_{7/2}$	7.94
Dy (3+)	...$4f^9\ 5s^2\ 5p^6$	$^6H_{15/2}$	10.63

$$* \; g = g(\text{Lande}) = 1 + \frac{J(J+1) + S(S+1) - L(L+1)}{2J(J+1)}$$

Table 7.5. Effective magneton number for some representative iron group ions*

Ion	Configuration	Ground state	$2\sqrt{S(S+1)}$
Fe (3+)	...$3d^5$	$^6S_{5/2}$	5.92
Fe (2+)	...$3d^6$	$^5D_4$	4.90
Co (2+)	...$3d^7$	$^4F_{9/2}$	3.87
Ni (2+)	...$3d^8$	$^3F_4$	2.83

* Quenching with $J = S$, $L = 0$ (so $g = 2$) is assumed for better agreement with experiment

Kramers' Theorem. This theorem tells us about systems that must have a degeneracy. The theorem says that the systems with an odd number of electrons on which a purely electrostatic field is acting can have no energy levels that are less than two-fold degenerate. If a magnetic field is imposed, this two-fold degeneracy can be lifted.

Jahn–Teller effect. This effect tells us that high degeneracy may be unlikely. The theorem states that a nonlinear molecule that has a (orbitally) degenerate ground state is unstable, and tends to distort itself so as to lift the degeneracy. Because of the Jahn–Teller effect, the symmetry of a given atomic environment in a solid is frequently slightly different from what one might expect. Of course, the Jahn–Teller effect does not remove the fundamental Kramers' degeneracy.

Hund's rules. Assuming Russel–Sanders coupling, these rules tell us what the ground state of an atomic system is. Hund's rules were originally obtained from spectroscopic evidence, but they have been confirmed by atomic calculations that include the Coulomb interactions between electrons. The rules state that in figuring out how electrons fill a shell in the ground state we should (1) assign a maximum S allowed by the Pauli principle, (2) assign maximum L allowed by S, (3) assign $J = L - S$ when the shell is not half-full, and $J = L + S$ when the shell is over half-full. See Problems 7.17 and 7.18. Results from the use of Hund's rules are shown in Tables 7.9 and 7.10.

Energy-Level Splitting in Crystal Fields by Group Theory (A)

In this Section we introduce enough group theory to be able to discuss the relation between degeneracies (in the energies of atoms) and symmetries (of the environment of the atoms). The fundamental work in the field was done by H. A. Bethe (see, e.g., Von der Lage and Bethe [7.64]). For additional material see Knox and Gold [61, in particular see Table 1-2 pp. 5-8 for definitions].

We have already discussed some of the more elementary ideas related to groups in Chap. 1 (see Sect. 1.2.1). The most important new concept that we will introduce here is the concept of group representations. A group representation starts with a set of nonsingular square matrices. For each group element g_i there is a matrix R_i such that $g_i g_j = g_k$ implies that $R_i R_j = R_k$. Briefly stated, a representation of a group is a set of matrices with the same multiplication table as the original group.

Two representations (R', R) of g that are related by

$$R'(g) = S^{-1}R(g)S \tag{7.285}$$

are said to be *equivalent*. In (7.285), S is any nonsingular matrix.

We define

$$R(g) \equiv R^{(1)}(g) \oplus R^{(2)}(g) \equiv \begin{pmatrix} R^{(1)}(g) & 0 \\ 0 & R^{(2)}(g) \end{pmatrix}. \tag{7.286}$$

In (7.286) we say that the representation $R(g)$ is reducible because it can be reduced to the direct sum of at least two representations. If $R(g)$ is of the form (7.286), it is said to be in *block diagonal form*. If a matrix representation can be brought into block diagonal form by a similarity transformation, then the representation is *reducible*. If no matrix representation reduces the representation to block diagonal form, then the matrix representation is *irreducible*. In considering any representation that is reducible, the most interesting information is to find out what irreducible representations are contained in the given reducible representation. We should emphasize that when we say a given representation $R(g)$ is reducible, we mean that a single S in (7.285) will put $R'(g)$ in block diagonal form for all g in the group.

In a typical problem in crystal field theory, a reducible representation (with respect to some group) of interest might be the irreducible representation $R^{(l)}$ of the three-dimensional rotation group. That is, we would like to know what irreducible representations of a group of interest is contained in a given irreducible representation of $R^{(l)}$ for some l. As we will see later, this can tell us a good deal about what happens to the electronic energy levels of a spherical atom in a crystal field.

It is worthwhile to give an explicit example of the irreducible representations of a group. Let us consider the group D_3 already defined in Chap. 1 (see Table 1.2).

In Table 7.6 note that $R^{(1)}$ and $R^{(2)}$ are *unfaithful* (many elements of the group correspond to the same matrix) representations while $R^{(3)}$ is a *faithful* (there is a one-to-one correspondence between group elements and matrices) representation. $R^{(1)}$ is, of course, the trivial representation.

Table 7.6. The irreducible representations of D_3

D_3	g_1	g_2	g_3	g_4	g_5	g_6
$R^{(1)}$	1	1	1	1	1	1
$R^{(2)}$	1	1	1	−1	−1	−1
$R^{(3)}$	$\begin{pmatrix} 1, & 0 \\ 0, & 1 \end{pmatrix}$	$\frac{1}{2}\begin{pmatrix} -1, & +\sqrt{3} \\ -\sqrt{3}, & -1 \end{pmatrix}$	$\frac{1}{2}\begin{pmatrix} -1, & -\sqrt{3} \\ \sqrt{3}, & -1 \end{pmatrix}$	$\frac{1}{2}\begin{pmatrix} 1, & -\sqrt{3} \\ -\sqrt{3}, & -1 \end{pmatrix}$	$\frac{1}{2}\begin{pmatrix} 1, & \sqrt{3} \\ \sqrt{3}, & -1 \end{pmatrix}$	$\begin{pmatrix} -1, & 0 \\ 0, & 1 \end{pmatrix}$

Since a similarity transformation will induce so many equivalent irreducible representations, a quantity that is invariant to similarity transformation might be (and in fact is) of considerable interest. Such a quantity is the *character*. The character of a group element is the trace of the matrix representing that group element.

It is elementary to show that the trace is invariant to similarity transformation. A similar argument shows that all group elements in the same class[28] have the same character. The argument goes as indicated below:

$$\mathrm{Tr}(R(g)) = \mathrm{Tr}(R(g)SS^{-1}) = \mathrm{Tr}(S^{-1}R(g)S) = \mathrm{Tr}(R'(g)),$$

if $R'(g)$ is defined by (7.285).

In summary the characters are defined by

$$\chi^{(i)}(g) = \sum\nolimits_{\alpha} R^{(i)}_{\alpha\alpha}(g) . \tag{7.287}$$

Equation (7.287) defines the character of the group element g in the ith representation. The characters still serve to distinguish various representations. As an example, the character table for the irreducible representation of D_3 is shown in Table 7.7. In Table 7.7, the top row labels the classes.

---

[28] Elements in the same class are conjugate to each other that means if g_1 and g_2 are in the same class there exists a $g \in G \ni g_1 = g^{-1} g_2 g$.

Table 7.7. The character table of D_3

C_1	C_2		C_3		
g_1	g_2	g_3	g_4	g_5	g_6
$\chi^{(1)}$ 1	1	1	1	1	1
$\chi^{(2)}$ 1	1	1	−1	−1	−1
$\chi^{(3)}$ 2	−1	−1	0	0	0

Below we summarize some important rules for constructing the character table for the irreducible representations. These results will not be proved, since they are readily available.[29] These rules are:

1. The number of classes s in the group is equal to the number of irreducible representations of the group.

2. If n_i is the dimension of the ith irreducible representation, then $n_i = \chi_i(E)$, where E is the identity of the group and $\sum_l^s n_i^2 = h$, where h is the order of the group G. For small finite groups, this rule obviously greatly restricts what the n_i can be.

3. If B_k is the number of group elements in the class C_k, then the characters for each class obey the relationship

$$\sum_{k=1}^{s} B_k \chi^{(l)^*}(C_k)\chi^{(j)}(C_k) = h\delta_l^j ,$$

where δ_l^j is the Kronecker delta. This relation is often called the *orthogonality relation for characters.*

4. Suppose the order of a group element g is m (i.e. suppose $g^m = E$). Further suppose that the dimension of a representation (which need not be irreducible) is n. It then follows that $\chi(g)$ equals the sum of n, mth roots of unity.

5. The one-dimensional representation is always present.

Finally it is worth giving the criterion for determining the irreducible representations in a given reducible representation. The rule is if

$$R = \sum_i C_i' R^i , \tag{7.288}$$

then C_i' (which is the number of times that irreducible representation i appears in the reducible representation R) is given by

$$C_i' = (1/h)\sum_k B_k \chi^{(i)}(C_k)^* \chi(C_k) , \tag{7.289}$$

where χ denotes character relative to R and the sum over k is a sum over classes. When a reducible representation is expressible in the form (7.288) it is said to be in, *reduced form*. Putting it into such a form as (7.288) is called *reduction.*

---

[29] See Mathews and Walker [7.47].

A frequent use of these results occurs when the representation R is formed by taking *direct products* (see Sect. 1.2.1 for a definition) of the representations $R^{(i)}$. We can then evaluate (7.289) by remembering that the trace of a direct product is the product of the traces.

There are many ways that group theory has been used as an aid in actual calculations. No doubt there remain other ways that have not yet been discovered. The basic ideas that we will use in our physical calculations involve:

1. The physical system determines a symmetry group with irreducible representations that can be found by group theory.
2. Except for what is called by definition "accidental degeneracy" we have a distinct eigenvalue for each (occurrence of an) irreducible representation. (It is possible for the same irreducible representation to occur many times. The meaning of the word "occur" will be given later.)
3. The dimension of the irreducible representation is the degeneracy of each corresponding eigenvalue.

For a brief insight into the above, let the eigenfunctions of $\mathcal{H}$ corresponding to the eigenvalue E_n be labeled $\psi_{ni}(i \rightarrow 1 \text{ to } d)$. E_n is thus d-fold degenerate. Thus

$$\mathcal{H}\psi_{ni} = E_n \psi_{ni} . \tag{7.290}$$

If g is an element of the symmetry group G, it follows that

$$[g, \mathcal{H}] = 0 . \tag{7.291}$$

From this,

$$\mathcal{H}(g\psi_{ni}) = E_n (g\psi_{ni}) . \tag{7.292}$$

Comparing (7.290) and (7.291), we see that

$$g\psi_{ni} = \sum_{i=1}^{d} C_i^n \psi_{in} . \tag{7.293}$$

It can be shown that C_i^n matrices are a representation of the group G. We thus have the desired connection between energy levels, degeneracy, and representations.

Let us consider the physically interesting problem of an atom with one 4f electron. Let us place this atom in a potential with trigonal symmetry. The group appropriate to trigonal symmetry is our old friend D_3. We want to neglect spin and discover what happens (or may happen) to the 4f energy levels when the atom is placed in a trigonal field. This is a problem that could be directly attacked by perturbation theory, but it is interesting to see what type of statements can be made by the group theory.

If you think a little about the ideas we have introduced and about our problem, you should come to the conclusion that what we have to find is the irreducible representations of D_3 generated by (in previous notation) $\psi_{(4f)i}$. Here i runs from 1 to 7. This problem can be solved by using (7.289).

The first thing we need to know is the character of our rotation group. This is given by [61] for the lth irreducible representation

$$\chi^{(l)}(\phi) = \frac{\sin[(l+\frac{1}{2})\phi]}{\sin(\phi/2)}. \tag{7.294}$$

In (7.294), ϕ is an appropriate rotation angle. Since we are dealing with a 4f level we are interested only in the case $l = 3$:

$$\chi^{(3)}(\phi) = \frac{\sin(7\phi/2)}{\sin(\phi/2)}. \tag{7.295}$$

By (7.289), we need to evaluate (7.294) only for ϕ in each of the three classes of D_3. Since the classes of D_3 correspond to the identity, three-fold rotations and two-fold rotations, we have

$$\chi^{(3)}(0) = 7, \tag{7.296a}$$

$$\chi^{(3)}(2\pi/3) = +1, \tag{7.296b}$$

$$\chi^{(3)}(\pi) = -1. \tag{7.296c}$$

Table 7.13. Character table for calculating C_i'

	C_1	C_2	C_3
B_k	1	2	3
$\chi^{(1)}$	1	1	1
D_3 $\chi^{(2)}$	1	1	−1
$\chi^{(3)}$	2	−1	0
Rotation Group $\chi^{(3)}$	7	+1	−1

We can now construct **Fehler! Verweisquelle konnte nicht gefunden werden.**. Applying (7.289) we have

$$C_1' = \tfrac{1}{6}[(7)(1)(1) + 2(+1)(1) + 3(-1)(1)] = 1,$$
$$C_2' = \tfrac{1}{6}[7(1)(1) + 2(+1)(1) + 3(-1)(-1)] = 2,$$
$$C_3' = \tfrac{1}{6}[7(1)(2) + 2(+1)(-1) + 3(-1)(0)] = 2.$$

Thus

$$R = 2R^{(3)} + R^{(1)} + 2R^{(2)}. \tag{7.297}$$

By (7.297) we expect the 4f level to split into two doubly degenerate levels plus three nondegenerate levels. The two levels corresponding to $R^{(3)}$ that occur twice and the two $R^{(2)}$ levels will probably not have the same energy.

7.5 Brief Mention of Other Topics

7.5.1 Spintronics or Magnetoelectronics (EE)[30]

We are concerned here with spin-polarized transport for which the name spintronics is sometimes used. We need to think back to the ideas of band ferromagnetism as contained for example in the Stoner model. Here one assumes that an exchange interaction can cause the spin-up and spin-down density of states to split apart as shown in the schematic diagram (for simplicity we consider that the majority spin-up band is completely filled). Thus, the number of electrons at the Fermi level with spin up (N_{up}) can differ considerably from the number with spin down (N_{down}). See (7.242) and Fig. 7.32 (spins and moments have opposite directions due to the negative charge of the electron – the spins are drawn in the bands). This results in two phenomena: (a) a net magnetic moment, and (b) a net spin polarization in transport defined by

$$P = \frac{N_{\mathrm{up}} - N_{\mathrm{down}}}{N_{\mathrm{up}} + N_{\mathrm{down}}}. \tag{7.298}$$

Fe, Ni, and Co typically have P of order 50%.

In the figures, the $D(E)$ describe the density of states of up and down spins. As shown also in Fig. 7.33 one can use this idea to produce a "spin valve," which preferentially transmits electrons with one spin direction. Spin valves have many possible device applications (see, e.g., Prinz [7.55]). In Fig. 7.33 we show transport from a magnetized metal to a magnetized metal through a nonmagnetic metal.

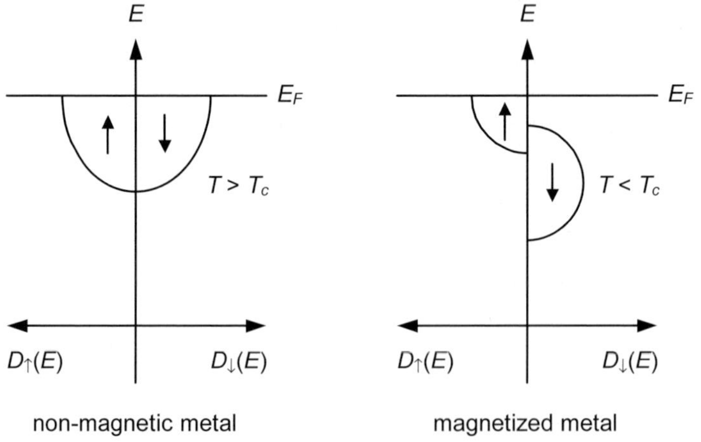

Fig. 7.32. Exchange coupling causes band ferromagnetism. The D are the density of states of the spin-up and spin-down bands. EF is the Fermi energy. Adapted from Prinz [7.55]

30 A comprehensive review has recently appeared, Zutic et al [7.73].

The two ferromagnets are still exchange coupled through the metal separating them. For the case of the second metal being antialigned with the first, the current is reduced and the resistance is high. The electrons with moment up can go from (a) to (b) but are blocked from (b) to (c). The moment-down electrons are inhibited from movement by the gap from (a) to (b). If the second magnetized metal were aligned, the resistance would be low. Since the second ferromagnet's magnetization direction can be controlled by an external magnetic field, this is the principle used in GMR (giant magnetoresistance, discovered by Baibich et al in 1988). See Baibich et al [7.4].

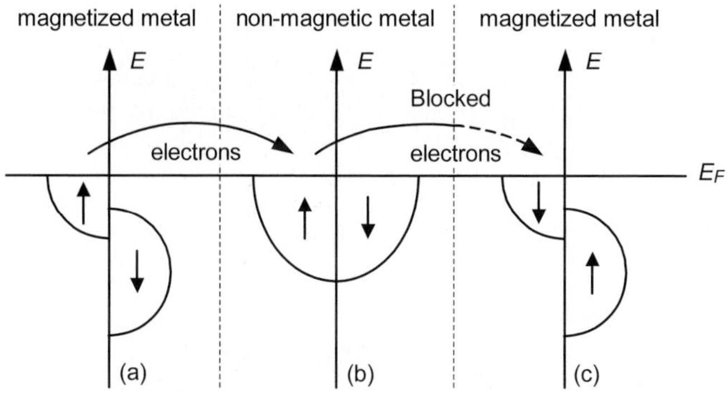

Fig. 7.33. Due to preferential transmission of spin orientation, the resistance is high if the second ferromagnet is antialigned. Adapted from Prinz [7.55]

One should note that spintronic devices are possible because the spin diffusion length that is the square root of the diffusion constant times the spin relaxation time can be fairly large, e.g. 0.1 mm in Al at 40 K. This means that the spin polarization of the transport will typically last over these distances when the polarized current is injected into a nonmagnetized metal or semiconductor. Only in 1988 was it realized that electronic current flowing into an ordinary metal from a ferromagnet could preserve spin, so that spin could be transported just as charge is.

We should also mention that control of spin is important in efforts to achieve quantum computing. Quantum computers perform a series of sequences of unitary transformation on sets of "qubits" – see Bennett [7.5] for a definition. In essence, this holds out the possibility of something like massive parallel computation. Quantum computing is a huge subject; see, e.g., Bennett [7.5].

Hard Drives (EE)

In 1997 IBM introduced another innovation – the giant magnetoresistance (GMR) read head for use in magnetic hard drives – in which magnetic and nonmagnetic materials are layered one in the read head, roughly doubling or tripling its sensitivity. By layering one can design the device with the desired GMR properties. The device works on the quantum-mechanical principle, already mentioned, that when the layers are magnetized in the same direction, the spin-dependent scattering is small, and when the layers are alternatively magnetized in opposite directions, the electrons experience a maximum of spin-dependent scattering (and hence much higher resistivity). Thus, magnetoresistance can be used to read the state of a magnetic bit in a magnetic disk drive. The direction of soft layer in the read head can be switched by the direction of the magnetization in the storing media. The magnetoresistance is thus changed and the direction of storage is then read by the size of the current in the read head. Sandwiches of Co and Cu can be used with the widths of the layers typically of the order of nanometers (a few atoms say) as this is the order of the wavelength of electrons in solids. More generally, magnetic multilayers of ferromagnetic materials (e.g. 3d transition metal ferromagnets) with nonferromagnetic spacers are used. The magnetic coupling between layers can be ferromagnetic or antiferromagnetic depending on spacing. Stuart Parkin of IBM has been a pioneer in the development of the GMR hard disk drive [7.52]

Magnetic Tunnel Junctions (MTJs) (EE)

Here the spacer in a sandwich with two ferromagnetic layers is a thin insulating layer. One difficulty is that it is difficult to make thin uniform insulators. Another difficulty, important for logic devices, is that the ferromagnetic layers need to be ferromagnetic semiconductors (rather than metals with far more mobile electrons than in semiconductors) so that a large fraction of the spin-aligned electrons can get into the rest of the device (made of semiconductors). GaMnAs and $TiCoO_2$ are being considered for use as ferromagnetic semiconductors for these devices.

The tunneling current depends on the relative magnetization directions of the ferromagnetic layers. It should be mentioned here that in the usual GMR structures the current typically flows parallel to the layers (but electrons undergo a random walk, and sample more than one layer so GMR can still operate), while in a MTJ sandwich the current typically flows perpendicular to the layers.

For the typical case, the resistance of the MTJ is lower when the moments of the ferromagnetic layers are aligned parallel and higher when the moments are antiparallel. This produces tunneling magnetic resistance TMR that may be 40% or so larger than GMR. MTJ holds out the possibility of making nonvolatile memories.

Spin-dependent tunneling through the FM-I-FM (ferromagnetic-insulator-ferromagnetic) sandwich had been predicted by Julliere [7.34] and Slonczewski [7.61].

Colossal Magnetoresistance (EE)

Magnetoresistance (MR) can be defined as

$$\mathrm{MR} = \frac{\rho(H) - \rho(0)}{\rho(0)}, \tag{7.299}$$

where ρ is the resistivity and H is the magnetic field.

Typically, MR is a few per cent, while GMR may be a few tens of per cent. Recently, materials with so-called colossal magnetoresistance (CMR) of 100 per cent or more have been discovered. CMR occurs in certain oxides of manganese – manganese perovskites (e.g. $La_{0.75}Ca_{0.25}MnO_3$). Space does not permit further discussion here. See Fontcuberta [7.23]. See also Salamon and Jaime [7.58].

7.5.2 The Kondo Effect (A)

Scattering of conduction electrons by localized moments due to s-d exchange can produce surprising effects as shown by J. Kondo in 1964. Although, this would appear to be a very simple basic phenomena that could be easily understood, at low temperature Kondo carried the calculation beyond the first Born approximation and showed that as the temperature is lowered the scattering is enhanced. This led to an explanation of the old problem of the resistance minimum as it occurred in, e.g., dilute solutions of Mn in Cu.

The Kondo temperature is defined as the temperature at which the Kondo effect clearly appears and for which Kondo's result is valid (see (7.301)). It is given approximately by

$$T_k = E_F \sqrt{J} \exp\left(-\frac{1}{nJ}\right), \tag{7.300}$$

where T_k is the Kondo temperature, E_F is the Fermi energy, J characterizes the strength of the exchange interaction, and n is the density of states. Generally T_k is below the resistance minimum that can be estimated from the approximate expression giving the resistivity ρ,

$$\rho = a - b\ln(T) + cT^5 . \tag{7.301}$$

The $\ln(T)$ term contains the spin-dependent Kondo scattering and cT^5 characterizes the resistivity due to phonon scattering at low temperature (the low temperature is also required for a sharp Fermi surface), and a, b and c are constants with b being proportional to the exchange interaction. This leads to a resistivity minimum at approximately

$$T_M = \left(\frac{b}{5c}\right)^{1/5} . \tag{7.302}$$

In actual practice the Kondo resistivity does not diverge at extremely low temperatures, but rather at temperatures well below the Kondo temperature, the resistivity approaches a constant value as the conduction electrons and impurity spins form a singlet. Wilson has used renormalized group theory to explain this. There are actually three regimes that need to be distinguished. The logarithmic regime is above the Kondo temperature, the crossover region is near the Kondo temperature, and the plateau of the resistivity occurs at the lowest temperatures. To discuss this in detail would take us well beyond the scope of this book. See, e.g., Kirk WP,

"Kondo Effect," pp. 162-165 in [24] and references contained therein. Using quantum dots as artificial atoms and studying them with scanning tunneling microscopes has revived interest in the Kondo effect. See Kouwenhoven and Glazman [7.41].

7.5.3 Spin Glass (A)

Another class of order that may occur in magnetic materials at low temperatures is spin glass. The name is meant to suggest frozen in (long-range) disorder. Experimentally the onset of a spin glass is signaled by a cusp in the magnetic susceptibility at T_f (the freezing temperature) in zero magnetic field. Below T_f there is no long-range order. The classic examples of spin glasses are dilute alloys of iron in gold (Au:Fe, also Cu:Mn, Ag:Mn, Au:Mn and several other examples). The critical ingredients of a spin glass seem to be (a) a competition among interactions as to the preferred direction of a spin (frustration), and (b) a randomness in the interaction between sites (disorder). There are still many questions surrounding spin glasses such as do they have a unique ground state and if the spin glass transition is a true phase transition to a new state (see Bitko [7.7]).

For spin glasses, it is common to define an order parameter by summing over the average spin's squared:

$$q = \frac{1}{N}\sum_{i=1}^{N}\langle \boldsymbol{S}_i \rangle^2 , \tag{7.303}$$

and for $T > T_f$, $q = 0$, while $q \neq 0$ for $T < T_f$. Much further detail can be found in Fischer and Hertz [7.22]. See also the article by Young [7.8 pp. 331-346].

Randomness and frustration (where two paths linking the same pair of spins do not have the same net effective sign of exchange coupling) are shared by many other systems besides spin glasses. Or another way of saying this is that the study of spin glasses fall in the broad category of the study of disordered systems, including random field systems (like diluted antiferromagnets), glasses, neural networks, optimization and decision problems. Other related problems include combinatorial optimization problems, such as the traveling salesman problem, and other problems involving complexity. For the neural network problem see for example, Muller and Reinhardt [7.50]. The book by Fischer and Hertz, already mentioned has a chapter on the physics of complexity with references. Another reference to get started in this general area is Chowdhury [7.12]. Mean-field theories of spin glasses have been promising, but there is no general consensus as to how to model spin glasses.

It is worth looking at a few experimental results to show real spin glass properties. Figure 7.34 shows the cusp in the susceptibility for CuMn. The true T_f occurs as the ac frequency goes to zero. Figure 7.35 shows the temperature dependence of the magnetization and order parameter for CuMn and AuFe. Figure 7.36 shows the magnetic specific heat for two CuMn samples.

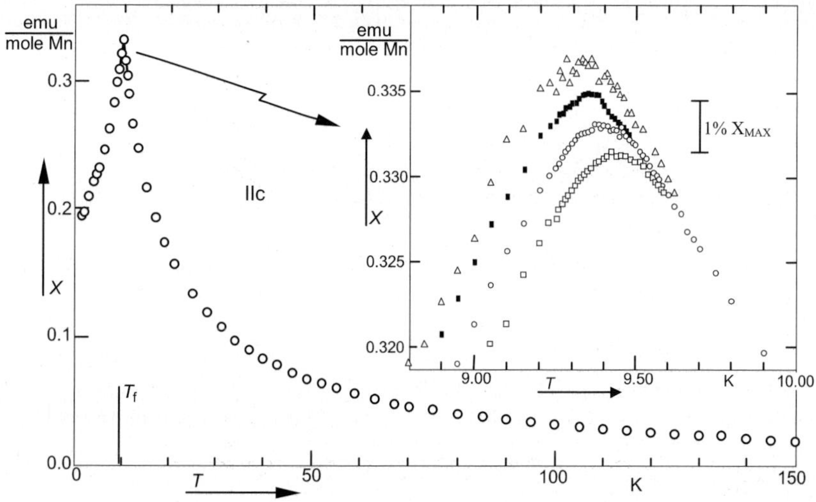

Fig. 7.34. The ac susceptibility as a function of T for CuMn (1 at %). Measuring frequencies: □, 1.33 kHz; **o**, 234 Hz, ■, 10.4 Hz; and Δ, 2.6 Hz. From Mydosh JA, "Spin-Glasses – The Experimental Situation" *Magnetism in Solids Some Current Topics*, Scottish Universities Summer School in Physics, 1981, p. 95, by permission of SUSSP. Data from Mulder CA, van Duyneveldt AJ, and Mydosh JA, *Phys Rev*, **B23**, 1384 (1981)

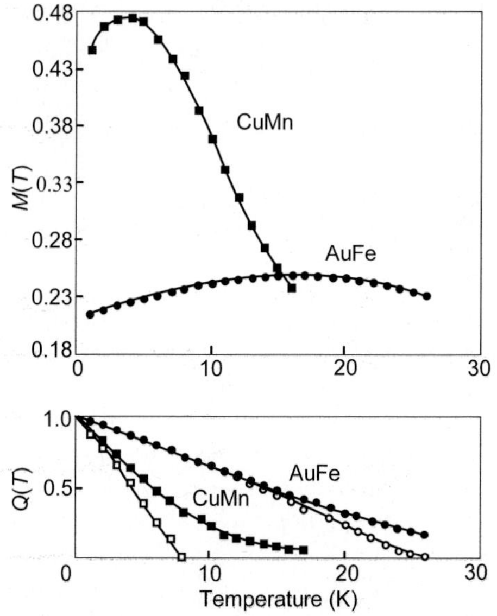

Fig. 7.35. The temperature variation of the magnetization $M(T,H)$ and order parameter $Q(T,H)$ with vanishing field (open symbols) and with 16 kG applied external magnetic field (full symbols) for Cu–0.7 at% Mn and Au–6.6 at% Fe; $M(T,H{=}0)$ is zero. After Mookerjee A and Chowdhury D, *J Physics F, Metal Physics* **13**, 365 (1983), by permission of the Institute of Physics

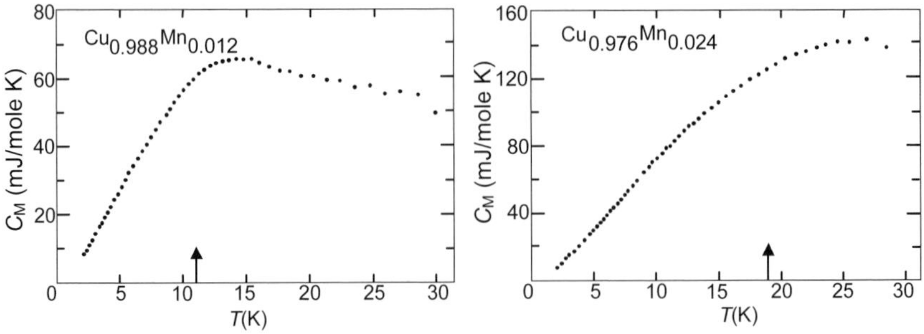

Fig. 7.36. Magnetic specific heat for CuMn spinglasses. The arrows show the freezing temperature (susceptibility peak). Reprinted with permission from Wenger LE and Keesom PH, *Phys Rev B* **13**, 4053 (1976). Copyright 1976 by the American Physical Society

7.5.4 Solitons[31] (A, EE)

Solitary waves are large-amplitude, localized, stable propagating disturbances. If in addition they preserve their identity upon interaction they are called solitons. They are particle-like solutions of nonlinear partial differential equations. They were first written about by John Scott Russell, in 1834, who observed a peculiar stable shallow water wave in a canal. They have been the subject of much interest since the 1960s, partly because of the availability of numerical solutions to relevant partial differential equations. Optical solitons in optical fibers are used to transmit bits of data.

Solitons occur in hydrodynamics (water waves), electrodynamics (plasmas), communication (light pulses in optical fibers), and other areas. In magnetism the steady motion of a domain wall under the influence of a magnetic field is an example of a soliton.[32]

In one dimension, the Korteweg–de Vries equation

$$\frac{\partial u}{\partial t} + A\frac{\partial^3 u}{\partial x^3} + B\frac{\partial}{\partial x}(u^2) = 0$$

(with A and B being positive constants) is used to discuss water waves. In other areas, including magnetism and domain walls, the sine-Gordon equation is encountered

$$A\frac{\partial^2 u}{\partial t^2} - B\frac{\partial^2 u}{\partial x^2} = -\frac{C}{u_0}\sin\left(\frac{2\pi u}{u_0}\right)$$

---

[31] See Fetter and Walecka [7.20] and Steiner, "Linear and non linear modes in 1d magnets," in [7.14, p199ff].

[32] See the article by Krumhansl in [7.8, pp. 3-21] who notes that static solutions are also solitons.

(with A, B, C, and u_0 being positive constants). Generalization to higher dimension have been made. The solitary wave owes its stability to the competition of dispersion and nonlinear effects (such as a tendency to steepen waves). The solitary wave propagates with a velocity that depends on amplitude.

Problems

7.1 Calculate the demagnetization factor of a sphere.

7.2 In the mean-field approximation in dimensionless units for spin 1/2 ferromagnets the magnetization (m) is given by

$$m = \tanh\left(\frac{m}{t}\right),$$

where $t = T/T_c$ and T_c is the Curie temperature. Show that just below the Curie temperature $t < 1$,

$$m = \sqrt{3}\sqrt{1-t}\,.$$

7.3 Evaluate the angular momentum L and magnetic moment μ for a sphere of mass M (mass uniformly distributed through the volume) and charge Q (uniformly distributed over the surface), assuming a radius r and an angular velocity ω. Thereby, obtain the ratio of magnetic moment to angular momentum.

7.4 Derive Curie's law directly from a high-temperature expansion of the partition function. For paramagnets, Curie's law is

$$\chi = \frac{C}{T} \text{ (The magnetic susceptibility)},$$

where Curie's constant is

$$C = \frac{\mu_0 N g^2 \mu_B^2 j(j+1)}{3k}.$$

N is the number of moments per unit volume, g is Lande's g factor, μ_B is the Bohr magneton, and j is the angular momentum quantum number.

7.5 Prove (7.155).

7.6 Prove (7.156).

7.7 In one spatial dimension suppose one assumes the Heisenberg Hamiltonian

$$\mathcal{H} = -\frac{1}{2}\sum\nolimits_{R,R'} J(R-R')\boldsymbol{S}_R \cdot \boldsymbol{S}_{R'}, \quad J(0) = 0,$$

where $\boldsymbol{R} - \boldsymbol{R}' = \pm a$ for nearest neighbor and $J_1 \equiv J(\pm a) > 0$, $J_2 \equiv J(\pm 2a) = -J_1/2$ with the rest of the couplings being zero. Show that the stable ground state is helical and find the turn angle. Assume classical spins. For simplicity, assume the spins are confined to the (x,y)-plane.

7.8 Show in an antiferromagnetic spin wave that the neighboring spins precess in the same direction and with the same angular velocity but have different amplitudes and phases. Assume a one-dimensional array of spins with nearest-neighbor antiferromagnetic coupling and treat the spins classically.

7.9 Show that (7.163) is a consistent transformation in the sense that it obeys a relation like (7.175), but for S_j^-.

7.10 Show that (7.138) can be written as

$$\mathcal{H} = -J\sum\nolimits_{j\delta}[S_{jz}S_{j+\delta,z} + \tfrac{1}{2}(S_j^- S_{j+\delta}^+ + S_j^+ S_{j+\delta}^+)] - 2\mu_0\mu H \sum\nolimits_j S_{jz}\,.$$

7.11 Using the definitions (7.179), show that

$$[b_{\boldsymbol{k}}, b_{\boldsymbol{k}'}^\dagger] = \delta_{\boldsymbol{k}}^{\boldsymbol{k}'},$$
$$[b_{\boldsymbol{k}}, b_{\boldsymbol{k}'}] = 0,$$
$$[b_{\boldsymbol{k}}^\dagger, b_{\boldsymbol{k}'}^\dagger] = 0\,.$$

7.12 (a) Apply Hund's rules to find the ground state of Nd^{3+} $(4f^3 5s^2 p^6)$.
(b) Calculate the Lande g-factor for this case.

7.13 By use of Hund's rules, show that the ground state of Ce^{3+} is ${}^2F_{5/2}$, of Pm^{3+} is ${}^5I_4$, and of Eu^{3+} is ${}^7F_0$.

7.14 Explain what the phrases "$3d^1$ configuration" and "${}^2D$ term" mean.

7.15 Give a rough order of magnitude estimate of the magnetic coupling energy of two magnetic ions in EuO ($T_c \cong 69$ K). How large an external magnetic field would have to be applied so that the magnetic coupling energy of a single ion to the external field would be comparable to the exchange coupling energy (the effective magnetic moment of the magnetic Eu^{2+} ions is 7.94 Bohr magnetons)?

7.16 Estimate the Curie temperature of EuO if the molecular field were caused by magnetic dipole interactions rather than by exchange interactions.

7.17 Prove the Bohr–van Leeuwen theorem that shows the absence of magnetism with purely classical statistics. Hint – look at Chap. 4 of Van Vleck [7.63].

8 Superconductivity

8.1 Introduction and Some Experiments (B)

In 1911 H. Kamerlingh Onnes measured the electrical resistivity of mercury and found that it dropped to zero below 4.15 K. He could do this experiment because he was the first to liquefy helium and thus he could work with the low temperatures required for superconductivity. It took 46 years before Bardeen, Cooper, and Schrieffer (BCS) presented a theory that correctly accounted for a large number of experiments on superconductors. Even today, the theory of superconductivity is rather intricate and so perhaps it is best to start with a qualitative discussion of the experimental properties of superconductors.

Superconductors can be either of type I or type II, whose different properties we will discuss later, but simply put the two types respond differently to external magnetic fields. Type II materials are more resistant, in a sense, to a magnetic field that can cause destruction of the superconducting state. Type II superconductors are more important for applications in permanent magnets. We will introduce the Ginzburg–Landau theory to discuss the differences between type I and type II.

The superconductive state is a macroscopic state. This has led to the development of superconductive quantum interference devices that can be used to measure very weak magnetic fields. We will briefly discuss this after we have laid the foundation by a discussion of tunneling involving superconductors.

We will then discuss the BCS theory and show how the electron–phonon interaction can give rise to an energy gap and a coherent motion of electrons without resistance at sufficiently low temperatures.

Until 1986 the highest temperature that any material stayed superconducting was about 23 K. In 1986, the so-called high-temperature ceramic superconductors were found and by now, materials have been discovered with a transition temperature of about 140 K (and even higher under pressure). Even though these materials are not fully understood, they merit serious discussion. In 2001 MgB_2, an intermetallic material was discovered to superconduct at about 40 K and it was found to have several unusual properties. We will also discuss briefly so-called heavy-electron superconductors.

Besides the existence of superconductivity, Onnes further discovered that a superconducting state could be destroyed by placing the superconductor in a large enough magnetic field. He also noted that sending a large enough current through the superconductor would destroy the superconducting state. Silsbee later suggested that these two phenomena were related. The disruption of the superconductive state is caused by the magnetic field produced by the current at the surface of the wire. However, the critical current that destroys superconductivity is very structure sensitive (see below) so that it can be regarded for some purposes as an independent parameter. The critical magnetic field (that destroys superconductivity) and the critical temperature (at which the material becomes superconducting) are related in the sense that the highest transition temperature occurs when there is no external magnetic field with the transition temperature decreasing as the field increases. We will discuss this a little later when we talk about type I and type II superconductors. Figure 8.1 shows at low temperature the difference in behavior of a normal metal versus a superconductor.

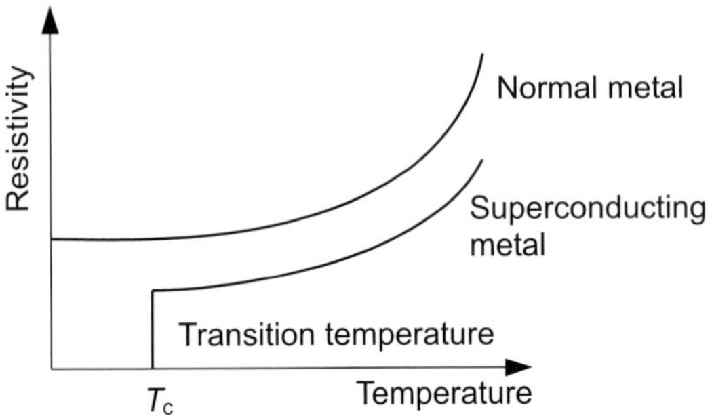

Fig. 8.1. Electrical resistivity in normal and superconducting metals (schematic)

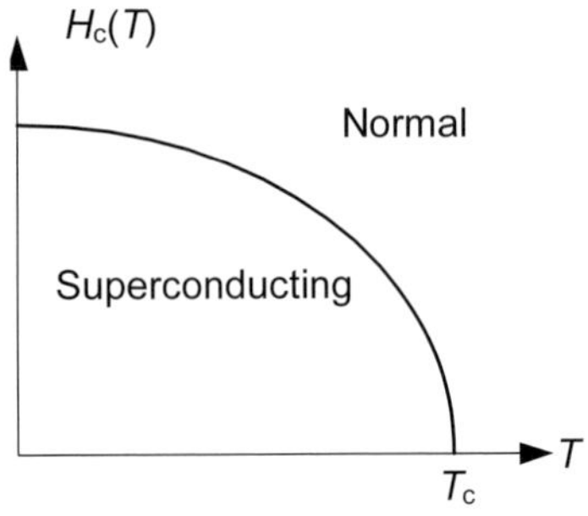

Fig. 8.2. Schematic of critical field vs. temperature for Type I superconductors

In 1933, Meissner and Ochsenfeld made another fundamental discovery. They found that superconductors expelled magnetic flux when they were cooled below the transition temperature. This established that there was more to the superconducting state than perfect conductivity (which would require $\boldsymbol{E} = 0$); it is also a state of perfect diamagnetism or $\boldsymbol{B} = 0$. For a long, thin superconducting specimen, $B = H + 4\pi M$ (cgs). Inside $B = 0$, so $H + 4\pi M = 0$ and $H_{in} = B_a$ (the externally applied B field) by the boundary conditions of H along the length being continuous. Thus, $B_a + 4\pi M = 0$ or $\chi = M/B_a = -1/(4\pi)$, which is the case for a perfect diamagnet. Exclusion of the flux is due to perfect diamagnetism caused by surface currents, which are always induced so as to shield the interior from external magnetic fields. A simple application of Faraday's law for a perfect conductor would lead to a constant flux rather than excluded flux. A plot of critical field *versus* temperature typically (for type I as we will discuss) looks like Fig. 8.2. The equation describing the critical fields dependence on temperature is often empirically found to obey

$$H_c(T) = H_c(0)\left[1 - \left(\frac{T}{T_c}\right)^2\right]. \tag{8.1}$$

In 1950, H. Fröhlich discussed the electron–phonon interaction and considered the possibility that this interaction might be responsible for the formation of the superconducting state. At about the same time, Maxwell and Reynolds, Serin, Wright, and Nesbitt found that the superconducting transition temperature depended on the isotopic mass of the atoms of the superconductor. They found $M^{\alpha}T_c \cong$ constant. This experimental result gave strong support to the idea that the electron–phonon interaction was involved in the superconducting transition. In the simplest model, $\alpha = 1/2$.

In 1957, Bardeen, Cooper, and Schrieffer (BCS) finally developed a formalism that contained the correct explanation of the superconducting state in common superconductors. Their ideas had some similarity to Fröhlich's. A key idea of the BCS theory was developed by Cooper in 1956. Cooper analyzed the electron–phonon interaction in a different way from Fröhlich. Fröhlich had discussed the effect of the lattice vibrations on the self-energy of the electrons. Cooper analyzed the effect of lattice vibrations on the effective interaction between electrons and showed that an attractive interaction between electrons (even a very weak attractive interaction at low enough temperature) would cause pairs of the electrons (the Cooper pairs) to form bound states near the Fermi energy (see Sect. 8.5.3). Later, we will discuss the BCS theory and show the pairing interaction causes a gap in the density of single-electron states.

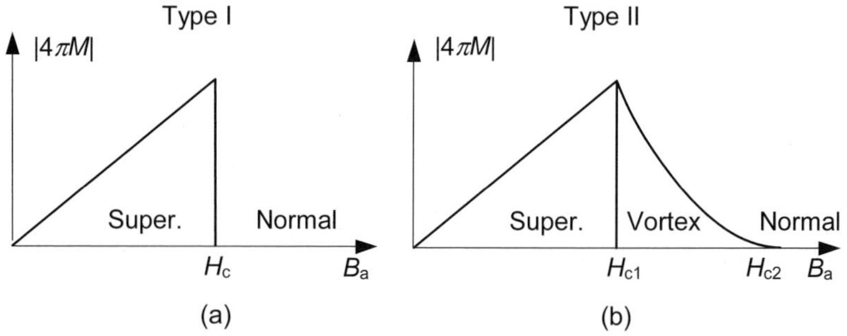

Fig. 8.3. (**a**) Type I and (**b**) Type II superconductors

As we have mentioned a distinction is made between type I and type II superconductors. Type I have only one critical field while type II have two critical fields. The idea is shown in Fig. 8.3a and b. $4\pi M$ is the magnetic field produced by the surface superconducting currents induced when the external field is applied. Type I superconductors either have flux penetration (normal state) or flux exclusion (superconductivity state). For type II superconductors, there is no flux penetration below H_{c1}, the lower critical field, and above the upper critical field H_{c2} the material is normal. But, between H_{c1} and H_{c2} the superconductivity regions are threaded by vortex regions of the flux penetration. The idea is shown in Fig. 8.4.

Type I and type II behavior will be discussed in more detail after we discuss the Ginzburg–Landau equations for superconductivity. We now mention some experiments that support the theories of superconductivity.

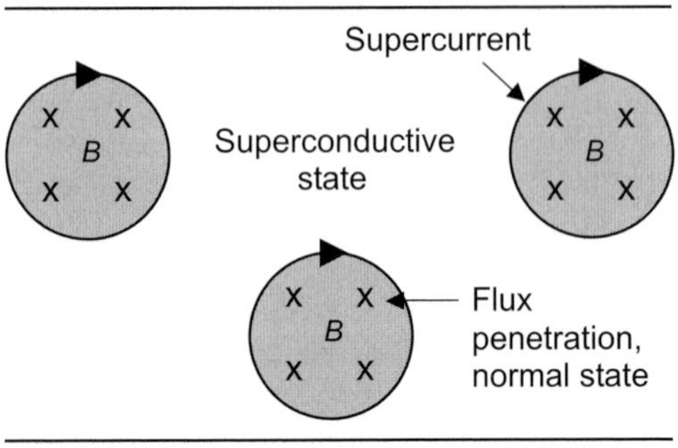

Fig. 8.4. Schematic of flux penetration for type II superconductors. The gray areas represent flux penetration surrounded by supercurrent (vortex). The net effect is that the superconducting regions in between have no flux penetration

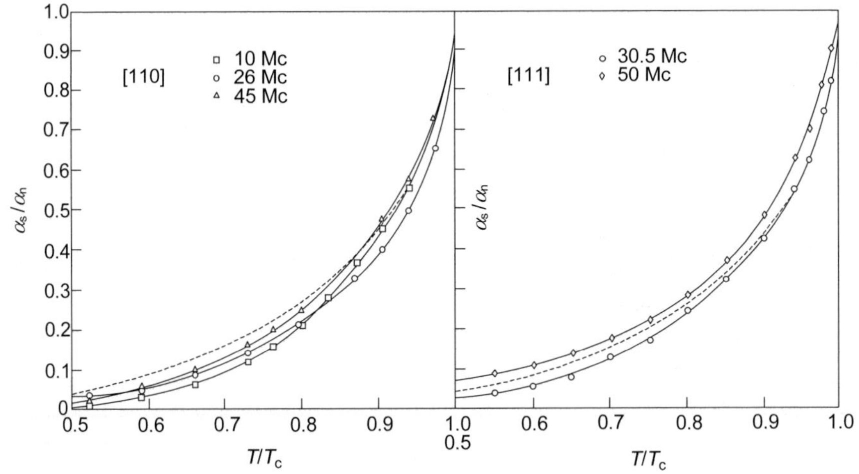

Fig. 8.5. Absorption coefficients ultrasonic attenuation in Pb (α_n refers to the normal state, α_s refers to the superconducting state, and T_c is the transition temperature). The dashed curve is derived from BCS theory and it uses an energy gap of 4.2 kT_c. [Reprinted with permission from Love RE and Shaw RW, *Reviews of Modern Physics* **36**(1) part 1, 260 (1964). Copyright 1964 by the American Physical Society.]

8.1.1 Ultrasonic Attenuation (B)

The BCS theory of the ratio of the normal to the superconducting absorption coefficients (α_n to α_s) as a function of temperature variation of the energy gap (discussed in detail later) can be interpreted in such a way as to give information on the temperature variation of the energy gap. Some experimental results on (α_n/α_s) *versus* temperature are shown in Fig. 8.5. Note the close agreement of experiment and theory, and that the absorption of superconductors is much lower than for the normal case when well below the transition temperature.

8.1.2 Electron Tunneling (B)

There are at least two types of tunneling experiments of interest. One involves tunneling from a superconductor to a superconductor with a thin insulator separating the two superconductors. Here, as will be discussed later, the Josephson effects are caused by the tunneling of pairs of electrons. The other type of tunneling (Giaever) involves tunneling of single quasielectrons from an ordinary metal to a superconducting metal. As will be discussed later, these measurements provide information on the temperature dependence of the energy gap (which is caused by the formation of Cooper pairs in the superconductor), as well as other features.

8.1.3 Infrared Absorption (B)

The measurement of transmission or reflection of infrared radiation through thin films of a superconductor provides direct results for the magnitude of the energy gap in superconductors. The superconductor absorbs a photon when the photon's energy is large enough to raise an electron across the gap.

8.1.4 Flux Quantization (B)

We will discuss this phenomenon in a little more detail later. Flux quantization through superconducting rings of current provides evidence for the existence of paired electrons as predicted by Cooper. It is found that flux is quantized in units of $h/2e$, not h/e.

8.1.5 Nuclear Spin Relaxation (B)

In these experiments, the nuclear spin relaxation time T_1 is measured as a function of temperatures. The time T_1 depends on the exchange of energy between the nuclear spins and the conduction electrons via the hyperfine interaction. The data for T_1 for aluminum looks somewhat as sketched in Fig. 8.6. The rapid change of T_1 near $T = T_c$ can be explained, at least quantitatively, by BCS theory.

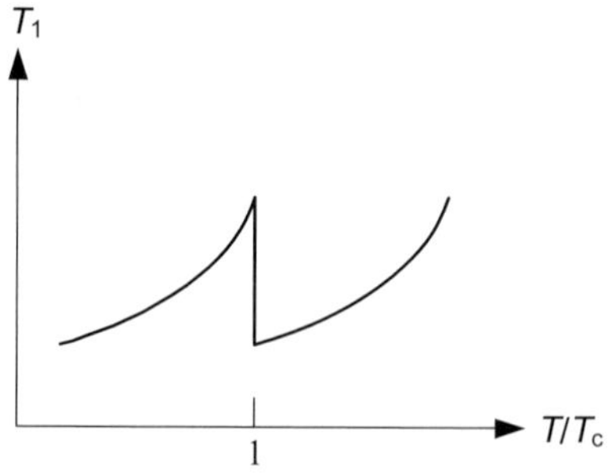

Fig. 8.6. Schematic of nuclear spin relaxation time in a superconductor near T_c

8.1.6 Thermal Conductivity (B)

A sketch of thermal conductivity K versus temperature for a superconductor is shown in Fig. 8.7. Note that if a high enough magnetic field is turned on, the material stays normal—even below T_c. So, a magnetic field can be used to control the thermal conductivity below T_c.

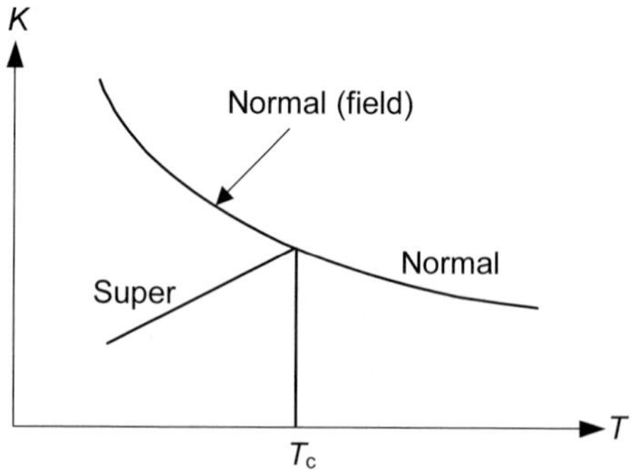

Fig. 8.7. Effect of magnetic field on thermal conductivity K

All of the above experiments have tended to confirm the BCS ideas of the superconducting state. A central topic that needs further elaboration is the criterion for occurrence of superconductivity in any material. We would like to know if the BCS interaction (electrons interacting by the virtual exchange of phonons) is the only interaction. Could there be, for example, superconductivity due to magnetic interactions? Over a thousand superconducting alloys and metals have been found, so superconductivity is not unusual. It is, perhaps, still an open question as to how common it is.

In the chapter on metals, we have mentioned heavy-fermion materials. Superconductivity in these materials seems to involve a pairing mechanism. However, the most probable cause of the pairing is different from the conventional BCS theory. Apparently, the nature of this "exotic" pairing has not been settled as of this writing, and reference needs to be made to the literature (see Sect. 8.7).

For many years, superconducting transition temperatures (well above 20 K) had never been observed. With the discovery of the new classes of high-temperature superconductors, transition temperatures (well above 100 K) have now been observed. We will discuss this later, also. The exact nature of the interaction mechanism is not known for these high-temperature superconductors, either.

8.2 The London and Ginzburg–Landau Equations (B)

We start with a derivation of the Ginzburg–Landau (GL) equations, from which several results will follow, including the London equations. Originally, these equations were proposed on intuitive, phenomenological lines. Later, it was realized they could be derived from the BCS theory. Gor'kov showed the GL theory was a valid description of the BCS theory near T_c. He also showed that the wave function ψ of the GL theory was proportional to the energy gap. Also, the density of superconducting electrons is $|\psi|^2$. Due to spatial inhomogeneities $\psi = \psi(\boldsymbol{r})$, where $\psi(\boldsymbol{r})$ is also called the order parameter. This whole theory was developed further by Abrikosov and is often known as the Ginzburg, Landau, Abrikosov, and Gor'kov theory (for further details, see, e.g., Kuper [8.20]

Near the transition temperature, the free energy density in the phenomenological GL theory is assumed to be (gaussian units)

$$F_S(\boldsymbol{r}) = \alpha|\psi|^2 + \frac{1}{2}\beta|\psi|^4 + \frac{1}{2m_S}\left|\left(\frac{\hbar}{i}\nabla - \frac{q\boldsymbol{A}}{c}\right)\psi\right|^2 + F_N + \frac{h^2}{8\pi}, \tag{8.2}$$

where N and S refer to normal and superconducting phases. The coefficients α and β are phenomenological coefficients to be discussed. $h^2/8\pi$ is the magnetic energy density ($h = h(\boldsymbol{r})$ is local and the magnetic induction $\boldsymbol{B}$ is determined by the spatial average of $h(\boldsymbol{r})$, so $\boldsymbol{A}$ is the vector potential for $\boldsymbol{h}$, $\boldsymbol{h} = \nabla \times \boldsymbol{A}$). $m^* = 2m$ (for pairs of electrons), $q = 2e$ is the charge and is negative for electrons, and ψ is the complex superconductivity wave function. Requiring (in the usual calculus of variations procedure) $\delta F_S/\delta\psi^*$ to be zero ($\delta F_S/\delta\psi = 0$ would yield the complex conjugate of (8.3)), we obtain the first Ginzburg–Landau equation

$$\left[\frac{1}{2m_S^*}\left(\frac{\hbar}{i}\nabla - \frac{q\boldsymbol{A}}{c}\right)^2 + \alpha + \beta|\psi|^2\right]\psi = 0\,. \tag{8.3}$$

F_S can be regarded as a functional of ψ and $\boldsymbol{A}$, so requiring $\delta F_S/\delta\boldsymbol{A} = 0$ we obtain the second GL equation for the current density:

$$\begin{aligned} j &= \frac{c}{4\pi}\nabla\times \mathrm{h} = \frac{q\hbar}{2m^*\mathrm{i}}(\psi^*\nabla\psi - \psi\nabla\psi^*) - \frac{q^2}{m^*c}\psi^*\psi\mathrm{A} \\ &= \frac{q}{m^*}|\psi|^2\left(\hbar\nabla\phi - \frac{q}{c}\mathrm{A}\right), \end{aligned} \tag{8.4}$$

where $\psi = |\psi|e^{i\varphi}$. Note (8.3) is similar to the Schrödinger wave equation (except for the term involving β) and (8.4) is like the usual expression for the current density. Writing $n_S = |\psi|^2$ and neglecting, as we have, any spatial variation in $|\psi|$, we find

$$\nabla \times \boldsymbol{J} = -\frac{q^2 n}{m^* c} \nabla \times \boldsymbol{A} = -\frac{q^2 n_S}{m^* c} \boldsymbol{B},$$

so,

$$\nabla \times \boldsymbol{J} = -\frac{c}{4\pi\lambda_L^2} \boldsymbol{B}, \tag{8.5}$$

where

$$\lambda_L^2 = \frac{m^* c^2}{4\pi n_S q^2}, \tag{8.6}$$

where λ_L is the London penetration depth. Equation (8.5) is London's equation. Note this is the same for a single electron (where $m^* = m$, $q = e$, n_S = ordinary density) or a Cooper pair ($m^* = 2m$, $q = 2e$, $n_S \to n/2$).

Let us show why λ_L is called the London penetration depth. At low frequencies, we can neglect the displacement current in Maxwell's equations and write

$$\nabla \times \boldsymbol{B} = \frac{4\pi}{c} \boldsymbol{J}. \tag{8.7}$$

Combining with (8.5) that we assume to be approximately true, we have

$$\nabla \times (\nabla \times \boldsymbol{B}) = \frac{4\pi}{c} \nabla \times \boldsymbol{J} = -\frac{1}{\lambda_L^2} \boldsymbol{B}, \tag{8.8}$$

or using $\nabla \cdot \boldsymbol{B} = 0$, we have

$$\nabla^2 \boldsymbol{B} = \frac{1}{\lambda_L^2} \boldsymbol{B}. \tag{8.9}$$

For a geometry with a normal material for $x < 0$ and a superconductor for $x > 0$, if the magnetic field at $x = 0$ is B_0, the solution of (8.9) is

$$\boldsymbol{B}(x) = \boldsymbol{B}_0 \exp(-x/\lambda_L). \tag{8.10}$$

Clearly, λ_L is a penetration depth. Thus, if we have a very thin superconducting film (with thickness $<< \lambda_L$), we really do not have a Meissner effect (flux exclusion). Magnetic flux will penetrate the surface of a superconductor over a distance approximately equal to the London penetration depth $\lambda_L \cong 100$ to 1000 Å. Actually, λ_L is temperature dependent and can be well described by

$$\left(\frac{\lambda_L}{\lambda_{L0}}\right)^2 = \frac{1}{1-(T/T_c)^4}, \tag{8.11}$$

where

$$\lambda_{L_0} = \sqrt{\frac{m^* c^2}{4\pi n_S q^2}} \,. \tag{8.12}$$

8.2.1 The Coherence Length (B)

Consider the Ginzburg–Landau equation in the absence of magnetic fields ($\mathbf{A} = 0$). Then, in one dimension from (8.3), we have

$$\left(-\frac{\hbar^2}{2m^*}\frac{\mathrm{d}^2}{\mathrm{d}x^2} + \alpha + \beta|\psi|^2 \right)\psi = 0 \,. \tag{8.13}$$

We define

$$f = \frac{\psi}{\psi_0} \quad \text{where} \quad \psi_0 = \sqrt{-\frac{\alpha}{\beta}} \,. \tag{8.14}$$

Then

$$\left(\frac{-\hbar^2}{2m^*\alpha}\frac{\mathrm{d}^2}{\mathrm{d}x^2} + 1 - |f|^2 \right) f = 0 \,. \tag{8.15}$$

When f has no gradients $|f| = 1$, which would correspond to being well inside the superconductor. We assume a boundary at $x = 0$ between a normal and a superconductor so $f = 1$, $\psi \to \psi_0$ as $x \to \infty$.

Defining

$$\xi^2(T) = \frac{-\hbar^2}{2m^*\alpha} \tag{8.16}$$

and letting $f = 1 + g$, where g is small, then

$$\xi^2(T)\frac{\mathrm{d}^2 g}{\mathrm{d}x^2} + 1 + g - (1+g)(1+g)^2 = 0 \,. \tag{8.17}$$

Keeping only first order in g since it is small,

$$\xi^2(T)\frac{\mathrm{d}^2 g}{\mathrm{d}x^2} - 2g \cong 0 \tag{8.18}$$

$$g(x) \cong \exp[-\sqrt{2}x/\xi(T)] \,. \tag{8.19}$$

Thus, the wave function attains its characteristic value of ψ_0 in a distance $\xi(T)$. $\xi(T)$ is called the coherence length. The coherence length measures the "range" or

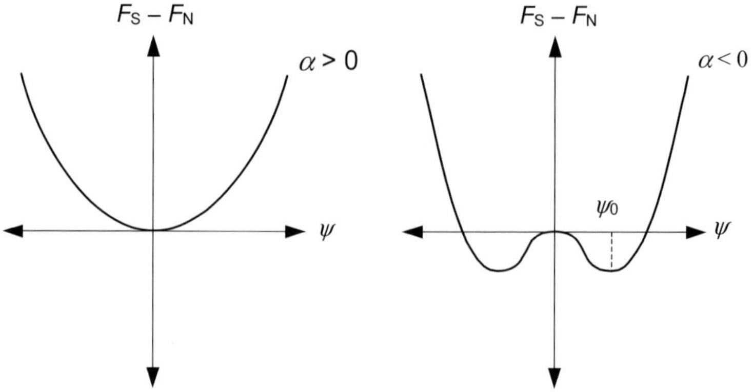

Fig. 8.8. Free energy change at the transition temperature. See (8.20) for how α enters the free energy with no fields or gradients

"size" of Cooper pairs or the distance necessary for the superconducting wave function to change much.

Let us discuss the coherence length further. First, let us review a little about the free energy. The plots for $F_S - F_N$ are shown in Fig. 8.8. The superconducting transition clearly appears at $T = T_c$ or $\alpha = 0$. The free energy for no fields or gradients is

$$F_S - F_N = \alpha\psi^2 + \frac{\beta}{2}\psi^4, \tag{8.20}$$

so

$$\frac{\partial}{\partial\psi}(F_S - F_N) = 0 \tag{8.21}$$

gives

$$\psi^2 = \psi_0^2 = -\frac{\alpha}{\beta} \tag{8.22}$$

at the minimum. Thus at the minimum

$$F_S - F_N = -\frac{\alpha^2}{2\beta}, \tag{8.23}$$

which would also be the stabilization energy. From thermodynamics, if $F = U - TS$, and $\mathrm{d}U = T\mathrm{d}S - M\mathrm{d}H$, then $\mathrm{d}F = -M\mathrm{d}H$ at constant T. For perfect diamagnetism,

$$M = \frac{-1}{4\pi}H. \tag{8.24}$$

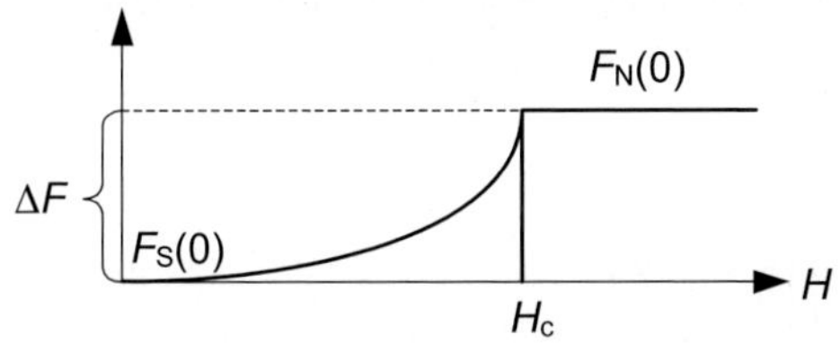

Fig. 8.9. Free energy as a function of field

So,

$$\mathrm{d}F_S = \frac{1}{4\pi} H\mathrm{d}H \,, \tag{8.25}$$

and

$$F_S(H) - F_S(0) = \frac{H^2}{8\pi} \,. \tag{8.26}$$

At $H = H_c$, the critical field that destroys superconductivity,

$$F_N(H_c) = F_S(H_c) \,. \tag{8.27}$$

F_N is almost independent of H so,

$$F_N(H_c) = F_N(0) \,. \tag{8.28}$$

We show the idea schematically in Fig. 8.9. Therefore,

$$F_S(0) - F_N(0) = -\frac{H_c^2}{8\pi} = -\Delta F \tag{8.29}$$

would also be the negative of the stabilization energy, or using (8.23)

$$\frac{H_c^2}{8\pi} = \frac{\alpha^2}{2\beta} \tag{8.30}$$

$$\alpha = \sqrt{\beta \frac{H_c^2}{4\pi}} \,. \tag{8.31}$$

Notice deep in the superconductor

$$\psi_S^2 = -\frac{\alpha}{\beta} = n_S = \frac{n}{2} \,, \tag{8.32}$$

and

$$\lambda^2 = \frac{m^* c^2}{4\pi q^2 n_S} \,. \tag{8.33}$$

So,

$$\alpha = -\beta\frac{n}{2}, \tag{8.34}$$

by (8.32) and thus by (8.33) and (8.31)

$$\alpha = -\frac{2e^2\lambda^2 H_c^2}{m^* c^2}. \tag{8.35}$$

Combining (8.16) and (8.35), an expression for the coherence length can be derived.

We wish to estimate the upper critical field. In Chap. 2 on electrons, we found the allowed energy levels in a constant $\boldsymbol{B}$ field were (in an approximation) free-electron-like parallel to the field and harmonic-oscillator-like in a plane perpendicular to the field. The harmonic energy levels were (dropping * on m)

$$\hbar\omega_c\left(n+\frac{1}{2}\right) = E_{n,k_z} - \frac{\hbar^2 k_z^2}{2m}; \quad \omega_c = \left|\frac{eB}{mc}\right|,$$

or

$$\hbar\frac{|e|B}{mc}\left(n+\frac{1}{2}\right) = -\alpha - \frac{\hbar^2 k_z^2}{2m}, \tag{8.36}$$

for the (linearized) Ginzburg–Landau equation (with $-\alpha$ acting as the eigenvalue in (8.3)). The largest value of B for which solutions of the GL equation exists is ($n = 0$ and letting $k_z = 0$)

$$B_{\max} \equiv H_{c2} = -\frac{2mc\alpha}{\hbar|e|}. \tag{8.37}$$

The two lengths λ and ξ can be defined as a dimensionless ratio K (the GL parameter). H_{c2} can now be described in terms of K and H_c. Using (8.6), (8.16), (8.32), (8.30) and

$$K = \frac{\lambda}{\xi}, \tag{8.38}$$

we find

$$H_{c2} = \sqrt{2}KH_c. \tag{8.39}$$

If $K = \lambda/\xi > 2^{-1/2}$, then $H_{c2} > H_c$. This results in a type II superconductor. The regime of $K > 2^{-1/2}$ is a regime of negative surface energy.

8.2.2 Flux Quantization and Fluxoids (B)

We have for the superconducting current density (by (8.4) with $|\psi|^2$ as spatially constant = n)

$$\boldsymbol{J} = \frac{q}{m^*} n \left(\hbar \nabla \varphi - \frac{q}{c} \boldsymbol{A} \right). \tag{8.40}$$

Well inside a superconductor $\boldsymbol{J} = 0$, so

$$\hbar \nabla \varphi = \frac{q}{c} \boldsymbol{A}. \tag{8.41}$$

Applying this to Fig. 8.10 and integrating around the loop gives

$$\hbar \oint \nabla \varphi \cdot \mathrm{d}l = \frac{q}{c} \oint \boldsymbol{A} \cdot \mathrm{d}l = \frac{q}{c} \int_S \boldsymbol{B} \cdot \mathrm{d}\boldsymbol{S} = \frac{q\Phi}{c} \tag{8.42}$$

$$\hbar (2\pi m) = q \frac{\Phi}{c}; \; m = \text{integer} \tag{8.43}$$

$$\Phi = \frac{hc}{q} m; \quad q = 2e. \tag{8.44}$$

This also applies to Fig. 8.11, so

$$\Phi_0 = \frac{hc}{2e} \equiv \text{the unit of flux of a fluxoid.} \tag{8.45}$$

In the vortex state of the type II superconductors, the minimum current produces the flux Φ_0. In the intermediate state there can be flux tubes threading through the superconductors as shown in Fig. 8.4 and Fig. 8.11. Below, in Fig. 8.12, is a sketch of the penetration depth and coherence length in a superconductor starting with a region of flux penetration. Note $\lambda/\xi < 1$ for type I superconductors.

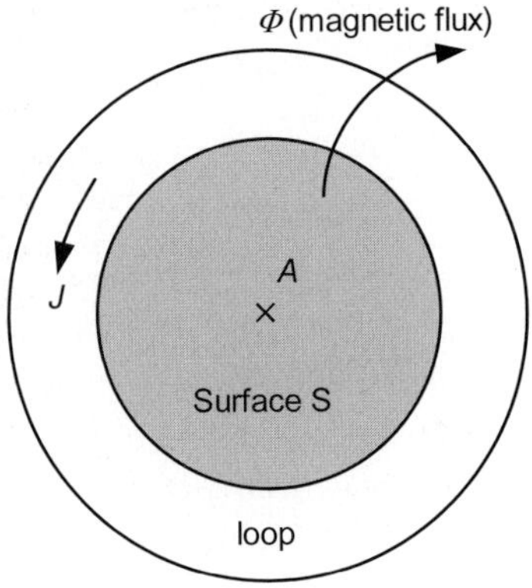

Fig. 8.10. Super current in a ring

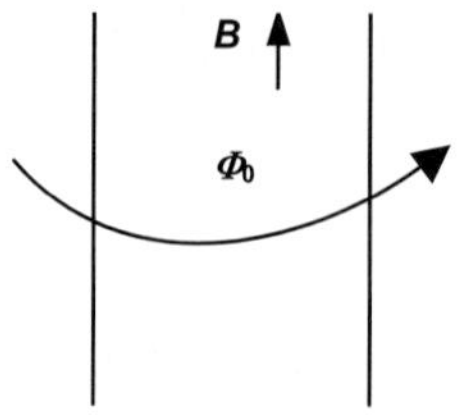

Fig. 8.11. Flux tubes in type II superconductors

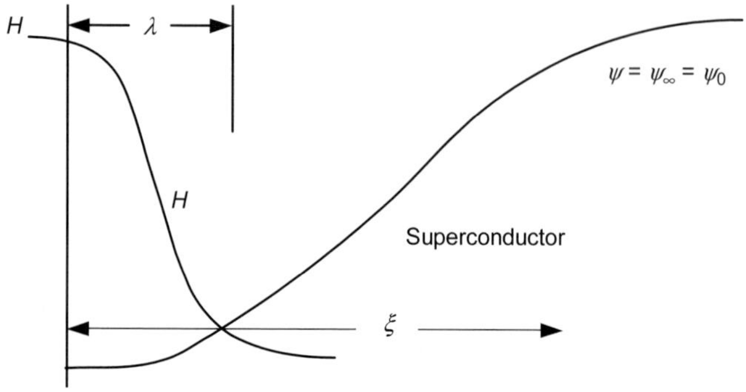

Fig. 8.12. Decay of H and asymptotic value of superconducting wave function

8.2.3 Order of Magnitude for Coherence Length (B)

For type II superconductors, there is a lower critical field H_{c1} for which the flux just begins to penetrate, so

$$H_{c1}\pi\lambda^2 \sim \Phi_0 \tag{8.46}$$

for a single fluxoid. At the upper critical field,

$$H_{c2}\pi\xi_0^2 \sim \Phi_0 , \tag{8.47}$$

so that, by (8.44) with $m = 1$,

$$\xi_0^2 \cong \frac{hc}{2e}\frac{1}{\pi}\frac{1}{H_{c2}} \tag{8.48}$$

for fluxoids packed as closely as possible. ξ_0 is the intrinsic coherence length, to be distinguished from the actual coherence length when the superconductor is "dirty" or possessed of appreciable impurities. A better estimate, based on fundamental parameters is[1]

$$\xi_0 = \frac{2\hbar v_F}{\pi E_g} , \tag{8.49}$$

[1] See Kuper [8.20 p221].

where v_F is the velocity of the Fermi surface and E_g is the energy gap. The coherence length changes in the presence of scattering. If the electron mean free path is l we have

$$\xi = \xi_0 , \tag{8.50}$$

as given by (8.49) for clean superconductors when $\xi_0 < l$ and

$$\xi \cong \sqrt{l \xi_0} , \tag{8.51}$$

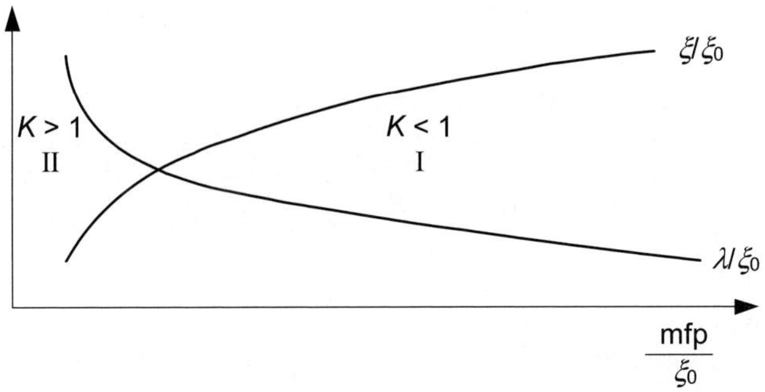

Fig. 8.13. Type I and type II superconductors depending on mfp

for dirty superconductors when $l \ll \xi_0$.[2] That is, dirty superconductors have decreased ξ and increased $K = \lambda/\xi$. The penetration depth can also depend on structure. The idea is schematically shown in Fig. 8.13 where typically the more impure the superconductor the lower the mean free path (mfp) leading to type II behavior.

8.3 Tunneling (B, EE)

8.3.1 Single-Particle or Giaever Tunneling

We anticipate some results of the BCS theory, which we will discuss later. As we will show, when electrons are well separated the electron–lattice interaction can lead to an effective attractive interaction between the electrons. An effective attractive interaction between electrons can cause there to be an energy gap in the single-particle density of states, as we also show later. This energy gap separates the ground state from the excited states and is responsible for most of the unique properties of superconductors.

---

[2] See Saint-James et al [8.27 p. 141]

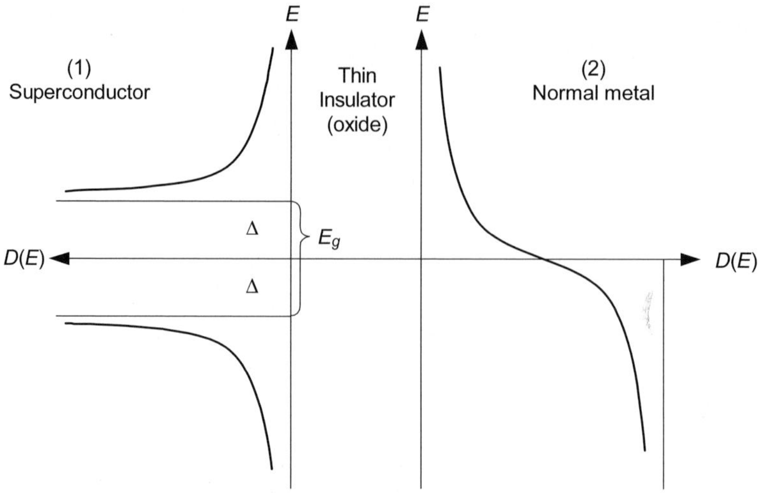

Fig. 8.14. Diagram of energy gap in a superconductor. D(E) is the density of states

Suppose we form a structure as given in Fig. 8.14. Let T be a tunneling matrix element. For the tunneling current we can write (with an applied voltage V)[3]

$$I_{1\to 2} = K' \int_{-\infty}^{\infty} |T|^2 D_1(E) f(E) D_2(E+eV)[1-f(E+eV)]\mathrm{d}E \qquad (8.52)$$

$$I_{2\to 1} = K' \int_{-\infty}^{\infty} |T|^2 D_1(E) D_2(E+eV) f(E+eV)[1-f(E)]\mathrm{d}E \qquad (8.53)$$

$$I = I_{1\to 2} - I_{2\to 1} = K' \int_{-\infty}^{\infty} |T|^2 D_1(E) D_2(E+eV)[f(E)-f(E+eV)]\mathrm{d}E \qquad (8.54)$$

$$I \cong K' D_2(0)|T|^2 \int_{-\infty}^{\infty} D_{1S}(E)\left[-\frac{\partial f}{\partial E} eV\right]\mathrm{d}E\,. \qquad (8.55)$$

[4] In the above, K' is a constant, D_i represents density of states, and f is the Fermi function. If we raise the voltage V by $eV = \Delta$, we get the following (see Fig. 8.15) for the net current, and thus, the energy gap can be determined.

[3] Note this is actually an oversimplified semiconductor-like picture of a complicated many-body effect [8.14 p. 247], but the picture works well for certain aspects and certainly is the simplest way to get a feel for the experiment.

[4] For the superconducting density of states see Problem 8.2.

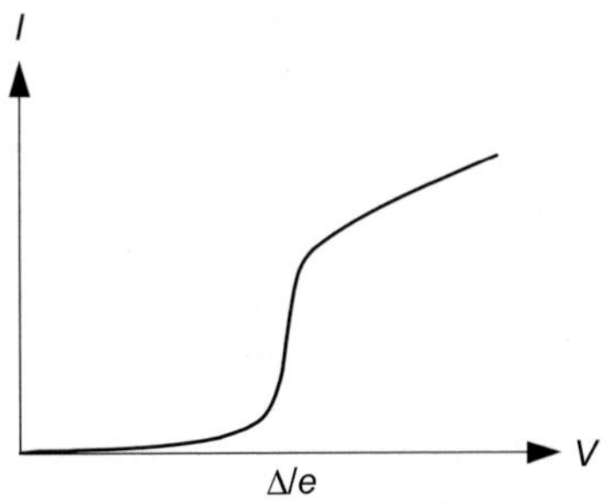

Fig. 8.15. Schematic of Giaever (single-particle) tunneling

8.3.2 Josephson Junction Tunneling

Josephson [8.18] predicted that when two superconductors were separated by an insulator there could be tunneling of Cooper pairs from one to the other provided the insulator was thinner than the coherence length, see Fig. 8.16.

The main concept used to discuss the Josephson effects is that of the phase of the paired electrons. We have already considered this idea in our discussion of flux quantization. F. London had the idea of something like a phase associated with superconducting electrons in that he believed that the motions of electrons in superconductors are correlated over large distances. We now associate the idea of spatial correlation of electrons with the idea of the existence of Cooper pairs. Cooper pairs are sets of two electrons that are attracted to one another (in spite of their Coulomb repulsion) because an electron attracts positive ions. As alluded to earlier, the positive ions in a crystal are much more massive and have, in general, less freedom of movement than the conduction electrons. This means that when an electron has attracted a positive ion to a displaced position, we can imagine the electron as moving out of the area while the positive ion remains displaced for a time. In the region of the crystal where the positive ion(s) is (are) displaced, the crystal has a more positive charge than usual and so this region can attract another electron. We could generalize this argument to consider that the displaced positive ion would be undergoing some sort of motion but still an electron with suitable phase could be attracted to the region of the displaced positive ion. Anyway, the argument seems to make it plausible that there can be an effective attractive interaction between electrons due to the presence of the positive ion lattice. The rather qualitative picture that we have given seems to be the physical content of the statement "Cooper pairs of electrons are formed because of the virtual exchange

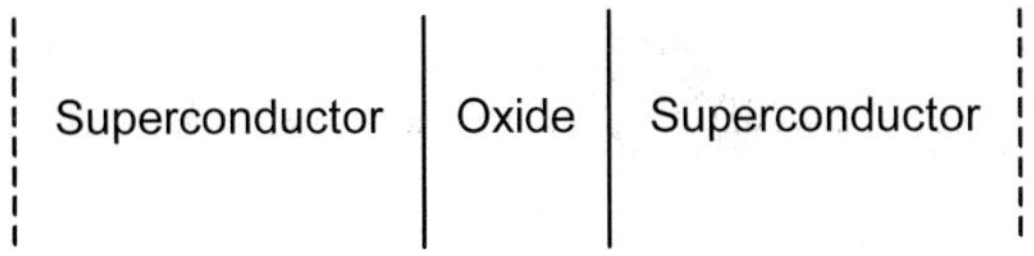

Fig. 8.16. Schematic of Josephson junction

of phonons between the electrons." We also see that only lattices in which the electron–lattice vibration coupling is strong will be good superconductors. Thus, we are led to an understanding of the almost paradoxical fact that good conductors of electricity (with low resistance and hence low electron–phonon coupling) often make poor superconductors. We give details including the role of spin later.

Due to the nature of the attractive mechanism between electrons in a Cooper pair, we should not be surprised that the binding of the electrons is very weak. This means that we have to think of the Cooper pairs as being very large (of the order of many, many lattice spacings) and hence the Cooper pairs overlap with each other a great deal in the solid. As we will see, further analysis of the pairs shows that the electrons in pairs have equal and opposite momentum (in the ground state) and equal and opposite spin. However, the Cooper pairs can accept momentum in such a way that they are still "stable" systems, but so that their center of mass moves. When this happens, the motion of the pairs is influenced by the fact that they are so large many of them must overlap. The Cooper pairs are composed of electrons, and the way electronic wave functions can overlap is limited by the Pauli principle. We now know that overlapping together with the constraint of the Pauli principle causes all Cooper pairs to have the same phase and the same momentum (i.e. the momentum of the center of mass of the Cooper pairs). The pairs are like bosons, in a sense, and condense into a lowest quantum state producing a wave function with phase.

Returning to the coupling of superconductors through an oxide layer, we write a sort of time-dependent "Ginzburg–Landau equations," that allow for coupling,[5]

$$\mathrm{i}\hbar\frac{\partial\psi_1}{\partial t} = H_1\psi_1 + \hbar U\psi_2\,, \tag{8.56}$$

$$\mathrm{i}\hbar\frac{\partial\psi_2}{\partial t} = H_2\psi_2 + \hbar U\psi_1\,. \tag{8.57}$$

If no voltage or magnetic field is applied, we can assume $H_1 = H_2 = 0$. Then

$$\mathrm{i}\hbar\frac{\partial\psi_1}{\partial t} = \hbar U\psi_2\,, \tag{8.58}$$

$$\mathrm{i}\hbar\frac{\partial\psi_2}{\partial t} = \hbar U\psi_1\,. \tag{8.59}$$

[5] See, e.g., Feynman et al [8.13], and Josephson [8.18], this was Josephson's Nobel Prize address. See also Dalven [8.11] and Kittel [23 Chap. 12].

We seek solutions of the form (any complex function can always be written as a product of amplitude ρ and $e^{i\varphi}$ where φ is the phase)

$$\psi_1 = \rho_1 \exp(i\varphi_1)\,, \tag{8.60}$$

$$\psi_2 = \rho_2 \exp(i\varphi_2)\,. \tag{8.61}$$

So, using (8.58) and (8.59) we get

$$i\dot{\rho}_1 - \rho_1\dot{\varphi}_1 = U\rho_2 \exp(i\Delta\varphi)\,, \tag{8.62}$$

$$i\dot{\rho}_2 - \rho_2\dot{\varphi}_2 = U\rho_1 \exp(-i\Delta\varphi)\,, \tag{8.63}$$

where

$$\Delta\varphi = (\varphi_2 - \varphi_1) \tag{8.64}$$

is the phase difference between the electrons on the two sides. Separating real and imaginary parts,

$$\dot{\rho}_1 = U\rho_2 \sin\Delta\varphi\,, \tag{8.65}$$

$$\rho_1\dot{\varphi}_1 = -U\rho_2 \cos\Delta\varphi\,, \tag{8.66}$$

$$\dot{\rho}_2 = -U\rho_1 \sin\Delta\varphi\,, \tag{8.67}$$

$$\rho_2\dot{\varphi}_2 = -U\rho_1 \cos\Delta\varphi\,. \tag{8.68}$$

Assume $\rho_1 \cong \rho_2 \cong \rho$ for identical superconductors, then

$$\frac{d}{dt}(\varphi_2 - \varphi_1) = 0\,, \tag{8.69}$$

$$\varphi_2 - \varphi_1 \cong \text{constant}\,, \tag{8.70}$$

$$\dot{\rho}_1 \cong -\dot{\rho}_2\,. \tag{8.71}$$

The current density J can be written as

$$J \propto \frac{d}{dt}\rho_2^2 = 2\rho_2\dot{\rho}_2\,, \tag{8.72}$$

so

$$J = J_0 \sin(\varphi_2 - \varphi_1)\,. \tag{8.73}$$

This predicts a dc current with no applied voltage. This is the dc Josephson effect. Another more rigorous derivation of (8.73) is given in Kuper [8.20 p 141]. J_0 is the critical current density or the maximum J that can be carried by Cooper pairs.

The ac Josephson effect occurs if we apply a voltage difference V across the junction, so that $\hbar qV$ with $q = 2e$ is the energy change across the junction. The relevant equations become

$$\mathrm{i}\hbar\frac{\partial\psi_1}{\partial t} = \hbar U\psi_2 - eV\hbar\psi_1 , \tag{8.74}$$

$$\mathrm{i}\hbar\frac{\partial\psi_2}{\partial t} = \hbar U\psi_1 + eV\hbar\psi_2 . \tag{8.75}$$

Again,

$$\mathrm{i}\dot{\rho}_1 - \rho_1\dot{\varphi}_1 = U\rho_2 \exp(\mathrm{i}\Delta\varphi) - eV\rho_1 , \tag{8.76}$$

$$\mathrm{i}\dot{\rho}_2 - \rho_2\dot{\varphi}_2 = U\rho_1 \exp(-\mathrm{i}\Delta\varphi) + eV\rho_2 . \tag{8.77}$$

So, separating real and imaginary parts

$$\dot{\rho}_1 = U\rho_2 \sin\Delta\varphi \tag{8.78}$$

$$\dot{\rho}_2 = -U\rho_1 \sin\Delta\varphi \tag{8.79}$$

$$\dot{\rho}_1 \cong -\dot{\rho}_2 \tag{8.80}$$

$$\rho_1\dot{\varphi}_1 = -U\rho_2 \cos\Delta\varphi + eV \tag{8.81}$$

$$\rho_2\dot{\varphi}_2 = -U\rho_1 \cos\Delta\varphi - eV . \tag{8.82}$$

Remembering $\rho_1 \cong \rho_2 \cong \rho$, so

$$\dot{\varphi}_2 - \dot{\varphi}_1 \cong -2eV . \tag{8.83}$$

Therefore

$$\Delta\varphi \cong (\Delta\varphi)_0 - 2eVt , \tag{8.84}$$

and

$$J = J_0 \sin[(\Delta\varphi)_0 - 2eVt] . \tag{8.85}$$

Again, J_0 is the maximum current carried by Cooper pairs. Additional current is carried by single-particle excitations producing the voltage V. The idea is shown later in Fig. 8.18. Therefore, since V is voltage in units of $\hbar$, the current oscillates with frequency (see (8.85))

$$\omega_J = 2eV = 2e\frac{\text{Voltage}}{\hbar} . \tag{8.86}$$

For the dc Josephson effect one can say that for low enough currents there is a current across the insulator in the absence of applied voltage. In effect because of the coherence of Cooper pairs, the insulator becomes a superconductor. Above a critical voltage, V_c, one has single electrons and the material becomes ohmic rather than superconducting. The junction then has resistance, but the current also has a component that oscillates with frequency ω_J as above. One understands this by saying that above V_c one has single particles as well as Cooper pairs. The

Cooper pairs change their energy by $2eV = \hbar\omega_J$ as they cross the energy gap causing radiation at this frequency. The ac Josephson effect, which occurs when

$$\omega = \frac{q}{\hbar}\text{Voltage} \tag{8.87}$$

is satisfied, is even more interesting. With $q = 2e$ (for a Cooper pair), (8.87) is believed to be exact. Thus, the ac Josephson effect can be used for a precise determination of $e/\hbar$. Parker, Taylor, and Langenberg[6] have done this. They used their new value of $e/\hbar$ to determine a new and better value of the fine structure constant α. Their new value of α removed a discrepancy between the quantum-electrodynamics calculation and the experimental value of the hyperfine splitting of atomic hydrogen in the ground state. These experiments have also contributed to better accuracy in the determination of the fundamental constants. There have been many other important developments connected with the Josephson effects, but they will not be presented here. Reference [8.20] is a good source for further discussion. See also Fig. 8.18 for a summary.

Finally, it is worth pointing out another reason why the Josephson effects are so interesting. They represent a quantum effect operating on a macroscopic scale. We can play with words a little, and perhaps convince ourselves that we understand this statement. In order to see quantum effects on a macroscopic scale, we must have many particles in the same state. For example, photons are bosons, and so, we can obtain a large number of them in the same state (which is necessary to see the quantum effects of electrons on a large scale). Electrons are fermions and must obey the Pauli principle. It would appear, then, to be impossible to see the quantum effects of electrons on a macroscopic scale. However, in a certain sense, the Cooper pairs having total spin zero, do act like bosons (but not entirely; the Cooper pairs overlap so much that their motion is highly correlated, and this causes their motion to be different from bosons interacting by a two-boson potential). Hence, we can obtain many electrons in the same state, and we can see the quantum effects of superconductivity on a macroscopic scale.

8.4 SQUID: Superconducting Quantum Interference (EE)

A Josephson junction is shown in Fig. 8.17 below. It is basically a superconductor–insulator–superconductor or a superconducting "sandwich". We now show how flux, due to B, threading the circuit can have profound effects. Using (8.4) with φ the Ginzburg–Landau phase, we have

$$\boldsymbol{J} = \frac{nq}{m}\left[\hbar\nabla\varphi - \frac{q}{c}\boldsymbol{A}\right]. \tag{8.88}$$

[6] See [8.24].

Integrating along the upper path gives

$$\hbar\Delta\varphi_1 = \int_\alpha^\beta \frac{m}{nq} \boldsymbol{J} \cdot \mathrm{d}\boldsymbol{l}_1 + \frac{q}{c} \int_\alpha^\beta \boldsymbol{A} \cdot \mathrm{d}\boldsymbol{l}_1 , \tag{8.89}$$

while integrating along the lower path gives

$$\hbar\Delta\varphi_2 = \int_\alpha^\beta \frac{m}{nq} \boldsymbol{J} \cdot \mathrm{d}\boldsymbol{l}_2 + \frac{q}{c} \int_\alpha^\beta \boldsymbol{A} \cdot \mathrm{d}\boldsymbol{l}_2 . \tag{8.90}$$

Subtracting, we have

$$\hbar(\Delta\varphi_1 - \Delta\varphi_2) = \frac{m}{nq} \oint \boldsymbol{J} \cdot \mathrm{d}\boldsymbol{l} + \frac{q}{c} \oint \boldsymbol{A} \cdot \mathrm{d}\boldsymbol{l} , \tag{8.91}$$

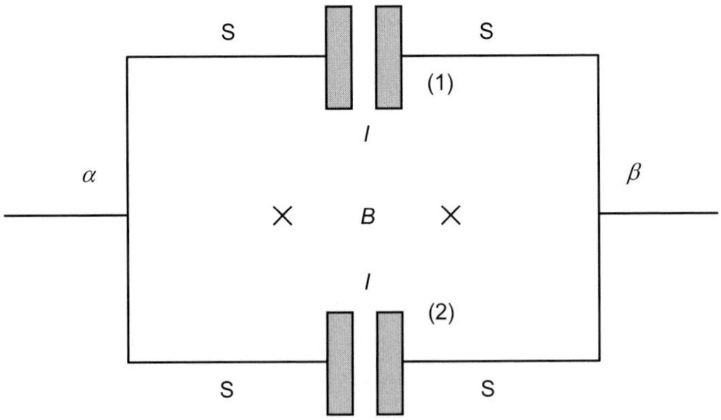

Fig. 8.17. A Josephson junction

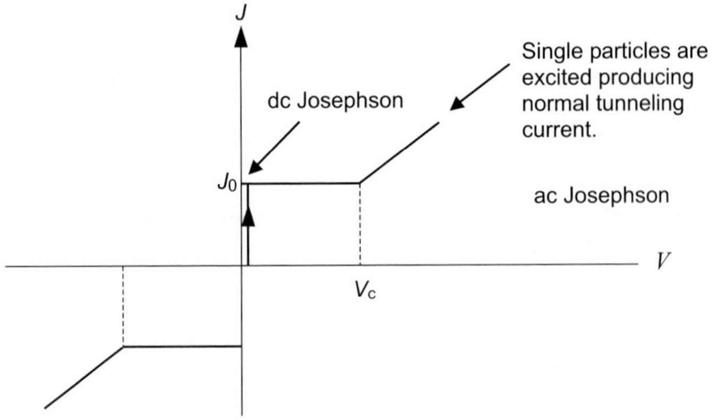

Fig. 8.18. Schematic of current density across junction versus V. The Josephson current $0 < J < J_0$ occurs with no voltage. When $J > J_0$ at $V_c \cong E_g/e$, where E_g is the energy gap, one also has single-particle current

where the first term on the right is zero or negligible. So, using Stokes Theorem and $\boldsymbol{B} = \nabla \times \boldsymbol{A}$ (and choosing a path where $J \cong 0$)

$$\Delta\varphi_1 - \Delta\varphi_2 = \frac{q}{\hbar c}\oint \boldsymbol{A}\cdot \mathrm{d}\boldsymbol{l} = \frac{q}{\hbar c}\int \boldsymbol{B}\cdot \mathrm{d}A = \frac{q\Phi}{\hbar c}\,. \tag{8.92}$$

Defining $\Phi_0 = \hbar c/q$ as per (8.45), we have

$$\Delta\varphi_1 - \Delta\varphi_2 = \frac{\Phi}{\Phi_0}, \tag{8.93}$$

so when $\Phi = 0$, and $\Delta\varphi_1 = \Delta\varphi_2$. We assume the junctions are identical so defining $\varphi_0 = (\Delta\varphi_1 + \Delta\varphi_2)/2$, then

$$\Delta\varphi_1 = \varphi_0 + \frac{\Phi}{2\Phi_0}, \tag{8.94}$$

$$\Delta\varphi_2 = \varphi_0 - \frac{\Phi}{2\Phi_0} \tag{8.95}$$

is a solution. By (8.73)

$$\begin{aligned} J_T &= J_1 + J_2 = J_0(\sin\Delta\varphi_1 + \sin\Delta\varphi_2) \\ &= 2J_0 \sin\left(\frac{\Delta\varphi_1 + \Delta\varphi_2}{2}\right)\cos\left(\frac{\Delta\varphi_1 - \Delta\varphi_2}{2}\right). \end{aligned} \tag{8.96}$$

So

$$J_T = 2J_0 \sin(\varphi_0)\cos\left(\frac{\Phi}{2\Phi_0}\right), \tag{8.97}$$

and

$$J_{T_{\max}} = 2J_0\left|\cos\left(\frac{\Phi}{2\Phi_0}\right)\right|. \tag{8.98}$$

The maximum occurs when $\Phi = 2n\pi\Phi_0$. Thus, quantum interference can be used to measure small magnetic field changes. The maximum current is a periodic function of Φ and, hence, measures changes in the field. Sensitive magnetometers have been constructed in this way. See the original paper about SQUIDS by Silver and Zimmerman [8.31].

8.4.1 Questions and Answers (B)

Q1. What is the simplest way to understand the dc Josephson effect (a current with no voltage in a super–insulator–super sandwich or SIS)?

A. If the insulator is much thinner than the coherence length, the superconducting pairs of electrons tunnel right through, and the insulator does not interfere with them–it is just one superconductor.

Q2. What is the simplest way to understand the ac Josephson effect (a current with a component of frequency $2eV/\hbar$, where V is the applied voltage)?

A. The Cooper pairs have charge $q = 2e$, and when they tunnel across the insulator, they drop in energy by qV. Thus they radiate with frequency $qV/\hbar$. This radiation is linked to the ac current.

8.5 The Theory of Superconductivity[7] (A)

8.5.1 Assumed Second Quantized Hamiltonian for Electrons and Phonons in Interaction (A)

As has already been mentioned, in many materials the superconducting state can be accounted for by an attractive electron–electron interaction due to the virtual exchange of phonons. See, e.g., Fig. 8.19. Thus, if we are going to try to understand the theory of superconductivity from a microscopic viewpoint, then we must examine, in detail, the nature of the electron–phonon interaction. There is no completely rigorous road to the BCS Hamiltonian. The arguments given below are intended to show how the physical origins of the BCS Hamiltonian could arise. It is not claimed that this is the way it *must* arise. However, given the BCS Hamiltonian, it is fair to say that the way it describes superconductivity is well understood.

One could draw an analogy to the Heisenberg Hamiltonian. The road to this Hamiltonian is also not rigorous for real materials, but there seems to be no doubt that it well describes magnetic phenomena in at least certain materials. The phenomena of superconductivity and ferromagnetism are exact, but the road to a quantitative description is not.

We thus start out with the Hamiltonian, which represents the interaction of electrons and phonons. As before, an intuitive approach suggests

$$\mathcal{H}_{\text{ep}} = \sum_{l,b} \boldsymbol{x}_{lb} \cdot [\nabla_{x_{lb}} U(\boldsymbol{r}_i)]_{x=0} \,. \tag{8.99}$$

We have already discussed this Hamiltonian in Chap. 4, which the reader should refer to, if needed. By the theory of lattice vibrations, we also know that (see Chaps. 2 and 4)

$$x_{l,b} = -\sum_{q,p} \sqrt{\frac{\hbar}{2Nm_b \omega_{q,p}}} [\exp(-\mathrm{i}\boldsymbol{q} \cdot \boldsymbol{l}) e^{*}_{q,b,p} (a^{\dagger}_{q,p} - a_{-q,p})] \,. \tag{8.100}$$

[7] See Bardeen et al [8.6].

In the above equation, the a are, of course, phonon creation and annihilation operators.

By a second quantization representation of the terms involving electron coordinates (see Appendix G), we can write

$$\frac{\partial U(\boldsymbol{r}_i)}{\partial x_{l,b}} = \sum_{\boldsymbol{k},\boldsymbol{k}'} \langle \psi_{\boldsymbol{k}} | \nabla_{x_{l,b}} U(r_i) | \psi_{\boldsymbol{k}'} \rangle C_{\boldsymbol{k}}^{\dagger} C_{\boldsymbol{k}'} , \tag{8.101}$$

where the C are electron creation and annihilation operators. The only quantities that we will want to calculate involve matrix elements of the operator $\mathcal{H}_{\text{ep}}$. As we have already shown, these matrix elements will vanish unless the selection rule $\boldsymbol{q} = \boldsymbol{k}' - \boldsymbol{k} - \boldsymbol{G}_n$ is obeyed. Neglecting umklapp processes (assuming $\boldsymbol{G}_n = 0$ the *first major approximation*), we can write

$$\begin{aligned} \mathcal{H}_{\text{ep}} = -\mathrm{i} \sum_{\boldsymbol{l},b} \sum_{\boldsymbol{q},p} \sum_{\boldsymbol{k},\boldsymbol{k}'} & \sqrt{\frac{\hbar}{2Nm_b\omega_{\boldsymbol{q},p}}} \exp(-\mathrm{i}\boldsymbol{q}\cdot\boldsymbol{l}) \\ & \times \langle \psi_{\boldsymbol{k}} | e^{*}_{\boldsymbol{q},b,p} \nabla_{x_{lb}} U(\boldsymbol{r}_i) | \psi_{\boldsymbol{k}'} \rangle \delta_{\boldsymbol{q}}^{\boldsymbol{k}'-\boldsymbol{k}} (a_{\boldsymbol{q},p}^{\dagger} - a_{-\boldsymbol{q},p}) C_{\boldsymbol{k}}^{\dagger} C_{\boldsymbol{k}'}, \end{aligned} \tag{8.102}$$

or

$$\begin{aligned} \mathcal{H}_{\text{ep}} = -\mathrm{i} \sum_{\boldsymbol{l},b} \sum_{\boldsymbol{q},p} \sum_{\boldsymbol{k},\boldsymbol{k}'} & \sqrt{\frac{\hbar}{2Nm_b\omega_{\boldsymbol{q},p}}} \exp(-\mathrm{i}\boldsymbol{q}\cdot\boldsymbol{l}) \\ & \times \langle \psi_{\boldsymbol{k}'-\boldsymbol{q}} | e^{*}_{\boldsymbol{q},b,p} \nabla_{x_{lb}} U(\boldsymbol{r}_i) | \psi_{\boldsymbol{k}'} \rangle (a_{\boldsymbol{q},p}^{\dagger} - a_{-\boldsymbol{q},p}) C_{\boldsymbol{k}'-\boldsymbol{q}}^{\dagger} C_{\boldsymbol{k}'}. \end{aligned} \tag{8.103}$$

Making the dummy variable changes $\boldsymbol{k}' \to \boldsymbol{k}$, $\boldsymbol{q} \to -\boldsymbol{q}$, and dropping the sum over p (assuming, for example, that only longitudinal acoustic phonons are effective in the interaction—this is the *second major approximation*), we find

$$\mathcal{H}_{\text{ep}} = \mathrm{i} \sum_{\boldsymbol{k},\boldsymbol{q}} B_{\boldsymbol{q}} C_{\boldsymbol{k}+\boldsymbol{q}}^{\dagger} C_{\boldsymbol{k}} (a_{\boldsymbol{q}} - a_{-\boldsymbol{q}}^{\dagger}) \tag{8.104}$$

where

$$B_{\boldsymbol{q}} = \sum_{\boldsymbol{l},b} \sqrt{\frac{\hbar}{2Nm_b\omega_{\boldsymbol{q}}}} \exp(\mathrm{i}\boldsymbol{q}\cdot\boldsymbol{l}) \langle \psi_{\boldsymbol{k}+\boldsymbol{q}} | e^{*}_{-\boldsymbol{q},b} \nabla_{x_{l,b}} U(\boldsymbol{r}_i) | \psi_{\boldsymbol{k}} \rangle . \tag{8.105}$$

The only property of $B_{\boldsymbol{q}}$ that we will use from the above equation is $B_{\boldsymbol{q}} = B_{-\boldsymbol{q}}^{*}$. From any reasonable, practical viewpoint, it would be impossible to evaluate the above equation directly and obtain $B_{\boldsymbol{q}}$. Thus, $B_{\boldsymbol{q}}$ will be treated as a parameter to be evaluated from experiment. Note that so far we have not made any approximations that are specifically restricted to superconductivity. The same Hamiltonian could be used in certain electrical-resistivity calculations.

We can now write the total Hamiltonian for interacting electrons and phonons (with $\hbar = 1$, and neglecting the zero-point energy of the lattice vibrations):

$$\mathcal{H} = \mathcal{H}_0 + \mathcal{H}_{\text{ep}} = \sum_q \omega_q a_q^\dagger a_q + \sum_k \varepsilon_k C_k^\dagger C_k + \mathrm{i} \sum_{q,k} B_q C_{k+q}^\dagger C_k (a_q - a_{-q}^\dagger) , \quad (8.106)$$

where the first two terms are the unperturbed Hamiltonian $\mathcal{H}_0$.

The first term is the Hamiltonian for phonons only (with $n_q = a^\dagger{}_q a_q$ as the phonon occupation number operator). The second term is the Hamiltonian for electrons only (with $n_k = C^\dagger{}_k C_k$ as the electron occupation number operator). The third term represents the interaction of phonons and electrons. We have in mind that the second term really deals with quasielectrons. We can assign an effective mass to the quasielectrons in such a manner as partially to take into account the electron–electron interactions, electron interactions with the lattice, and at least partially any other interactions that may be important but only lead to a "renormalization" of the electron mass. Compare Sects. 3.1.4, 3.2.2, and 4.3, as well as the introduction in Chap. 4. We should also include a screened Coulomb repulsion between electrons (see Sect. 9.5.3), but we neglect this here (or better, absorb it in $V_{k,k'}$—to be defined later).

Various experiments and calculations indicate that the energy per atom between the normal and superconducting states is of order 10^{-7} eV. This energy is very small compared to the accuracy with which we can hope to calculate the absolute energy. Thus, a frontal attack is doomed to failure. So, we will concentrate on those terms leading to the energy difference. The rest of the terms can then be pushed aside. The results are nonrigorous, and their main justification is the agreement we get with experiment. The method for separating the important terms is by no means obvious. It took many years to find. All that will be done here is to present a technique for doing the separation.

The technique for separating out the important terms involves making a canonical transformation to eliminate off-diagonal terms of $O(B_q)$ in the Hamiltonian. Before doing this, however, it is convenient to prove several useful results. First, we derive an expansion for

$$\mathcal{H}_S \equiv (\mathrm{e}^{-S})(\mathcal{H})(\mathrm{e}^{S}) , \quad (8.107)$$

where S is an operator.

$$\begin{aligned}(\mathrm{e}^{-S})(\mathcal{H})(\mathrm{e}^{S}) &= \left(1 - S + \frac{1}{2}S^2 + \ldots\right)\mathcal{H}\left(1 + S + \frac{1}{2}S^2 + \ldots\right) \\ &= \mathcal{H} - S\mathcal{H} + \mathcal{H}S + \frac{1}{2}S^2\mathcal{H} - S\mathcal{H}S + \frac{1}{2}\mathcal{H}S^2 ,\end{aligned} \quad (8.108)$$

but

$$[[\mathcal{H}, S], S] = [\mathcal{H}S - S\mathcal{H}, S] = 2\left[\frac{1}{2}\mathcal{H}S^2 + \frac{1}{2}S^2\mathcal{H} - S\mathcal{H}S\right] , \quad (8.109)$$

so that

$$\mathcal{H}_S = \mathcal{H} + [\mathcal{H}, S] + \frac{1}{2}[[\mathcal{H}, S], S] + \ldots . \tag{8.110}$$

We can treat the next few terms in a similar way.

The second useful result is obtained by $\mathcal{H} = \mathcal{H}_0 + X\mathcal{H}_{ep}$ where X is eventually going to be set to one. In addition, we choose S so that

$$X\mathcal{H}_{ep} + [\mathcal{H}_0, S] = 0 . \tag{8.111}$$

We show that in this case $\mathcal{H}_S$ has no terms of O(X). The result is proved by using (8.110) and substituting $\mathcal{H} = \mathcal{H}_0 + X\mathcal{H}_{ep}$. Then

$$\begin{aligned}\mathcal{H}_S &= \mathcal{H}_0 + X\mathcal{H}_{ep} + [\mathcal{H}_0 + X\mathcal{H}_{ep}, S] + \frac{1}{2}[[\mathcal{H}_0 + X\mathcal{H}_{ep}, S], S] + \ldots \\ &= \mathcal{H}_0 + X\mathcal{H}_{ep} + [\mathcal{H}_0, S] + X[\mathcal{H}_{ep}, S] \\ &\quad + \frac{1}{2}[[\mathcal{H}_0, S], S] + \frac{X}{2}[[\mathcal{H}_{ep}, S], S] + \ldots .\end{aligned} \tag{8.112}$$

Using (8.111), we obtain

$$\mathcal{H}_S = \mathcal{H}_0 + X[\mathcal{H}_{ep}, S] + \frac{X}{2}[[\mathcal{H}_{ep}, S], S] + \frac{1}{2}[[\mathcal{H}_0, S], S] + \ldots . \tag{8.113}$$

Since

$$X\mathcal{H}_{ep} + [\mathcal{H}_0, S] = 0 , \tag{8.114}$$

we have

$$X[\mathcal{H}_{ep}, S] = -[[\mathcal{H}_0, S], S] , \tag{8.115}$$

so that

$$\mathcal{H}_S = \mathcal{H}_0 + X[\mathcal{H}_{ep}, S] + \frac{X}{2}[[\mathcal{H}_{ep}, S], S] - \frac{X}{2}[\mathcal{H}_{ep}, S] , \tag{8.116}$$

or

$$\mathcal{H}_S = \mathcal{H}_0 + \frac{X}{2}[\mathcal{H}_{ep}, S] + \mathrm{O}(X^3) . \tag{8.117}$$

Since O(S) = X the second term is of order X^2, which was to be proved.

The point of this transformation is to push aside terms responsible for ordinary electrical resistivity (*third major transformation*). In the original Hamiltonian, terms in X contribute to ordinary electrical resistivity in first order.

From $X\mathcal{H}_{\text{ep}} + [\mathcal{H}_0, S] = 0$, we can calculate S. This is especially easy if we use a representation in which $\mathcal{H}_0$ is diagonal. In such a representation

$$\langle n|X\mathcal{H}_{\text{ep}}|m\rangle + \langle n|\mathcal{H}_0 S - S\mathcal{H}_0|m\rangle = 0\,, \tag{8.118}$$

or

$$\langle n|X\mathcal{H}_{\text{ep}}|m\rangle + (E_n - E_m)\langle n|S|m\rangle = 0\,, \tag{8.119}$$

or

$$\langle n|S|m\rangle = \frac{\langle n|X\mathcal{H}_{\text{ep}}|m\rangle}{E_m - E_n}\,. \tag{8.120}$$

The above equation determines the matrix elements of S and, hence, defines the operator S (for $E_m \neq E_n$).

8.5.2 Elimination of Phonon Variables and Separation of Electron–Electron Attraction Term Due to Virtual Exchange of Phonons (A)

Let us now connect the results we have just derived with the problem of superconductivity. Let $X\mathcal{H}_{\text{ep}}$ be the interaction Hamiltonian for the electron–phonon system. Any operator that we present for S that satisfies

$$\langle n|S|m\rangle = \frac{\langle n|X\mathcal{H}_{\text{ep}}|m\rangle}{E_m - E_n} \tag{8.121}$$

is good enough. In the above equation, $|m\rangle$ means both electron and phonon states. However, let us take matrix elements with respect to phonon states only and select S so that if we were to take electronic matrix elements, the above equation would be satisfied. This procedure is done because the behavior of phonons, except insofar as it affects the electrons, is of no interest. The point of this Section is then to find an effective Hamiltonian for the electrons.

We begin with these ideas. Taking phonon matrix elements, we have

$$\begin{aligned}
\langle n_{q'}+1|S|n_{q'}\rangle &= \frac{\langle n_{q'}+1|X\mathcal{H}_{\text{ep}}|n_{q'}\rangle}{E(\text{total initial state}) - E(\text{total final state})} \\
&= \mathrm{i}\sum_{k,q} B_q \frac{C^\dagger_{k+q} C_k \langle n_{q'}+1|a_q - a^\dagger_{-q}|n_{q'}\rangle}{E_{q'} + \varepsilon_k - (E_{q'} + \omega_{q'}) - \varepsilon_{k+q}} \\
&= -\mathrm{i}\sum_{k,q} B_q \frac{C^\dagger_{k+q} C_k \langle n_{q'}+1|a^\dagger_{-q}|n_{q'}\rangle}{E_{q'} - (E_{q'} + \omega_{q'}) + \varepsilon_k - \varepsilon_{k+q}}\,,
\end{aligned} \tag{8.122}$$

where ω_q is the energy of the created phonon (with $\hbar = 1$ and $\omega_q = \omega_{-q}$). Using

$$\left\langle n_q + 1 \middle| a_q^\dagger \middle| n_q \right\rangle = \sqrt{n_q + 1}\,, \tag{8.123}$$

we find

$$\begin{aligned}\left\langle n_q + 1 \middle| S \middle| n_q \right\rangle &= -\mathrm{i}\sum_{k,q} B_q C_{k+q}^\dagger C_k \frac{\sqrt{n_{q'}+1}}{\varepsilon_k - \varepsilon_{k+q'} - \omega_{q'}}\delta_{q'}^{-q} \\ &= -\mathrm{i}\sum_{k} B_{-q'} C_{k-q'}^\dagger C_k \frac{\sqrt{n_{q'}+1}}{\varepsilon_k - \varepsilon_{k-q'} - \omega_{q'}}.\end{aligned} \tag{8.124}$$

In a similar way we can show

$$\left\langle n_{q'} \middle| S \middle| n_{q'} + 1 \right\rangle = \mathrm{i}\sum_{k} B_{q'} C_{k+q'}^\dagger C_k \frac{\sqrt{n_{q'}+1}}{\varepsilon_k - \varepsilon_{k+q} + \omega_{q'}}. \tag{8.125}$$

Now, using

$$\mathcal{H}_S = \mathcal{H}_0 + \frac{1}{2}[\mathcal{H}_{\mathrm{ep}}S - S\mathcal{H}_{\mathrm{ep}}] + \dots, \tag{8.126}$$

with

$$\mathcal{H}_{\mathrm{ep}} = \mathrm{i}\sum_{k,q} B_q C_{k+q}^\dagger C_k (a_q - a_{-q}^\dagger) \tag{8.127}$$

(X has now been set equal to 1), and taking phonon expectation values for a particular phonon state, we have

$$\begin{aligned}\langle n|\mathcal{H}_S|n\rangle &= \langle n|\mathcal{H}_0|n\rangle + \frac{1}{2}\sum_m \left[\langle n|\mathcal{H}_{\mathrm{ep}}|m\rangle\langle m|S|n\rangle - \langle n|S|m\rangle\langle m|\mathcal{H}_{\mathrm{ep}}|n\rangle\right] \\ &= \langle n|\mathcal{H}_0|m\rangle + \frac{1}{2}\Big[(\mathcal{H}_{\mathrm{ep}})_{n,n-1}S_{n-1,n} + (\mathcal{H}_{\mathrm{ep}})_{n,n+1}S_{n+1,n} \\ &\qquad - S_{n,n-1}(\mathcal{H}_{\mathrm{ep}})_{n-1,n} - S_{n,n+1}(\mathcal{H}_{\mathrm{ep}})_{n+1,n}\Big].\end{aligned} \tag{8.128}$$

Since we are interested only in electronic coordinates, we will write below $\langle n_q|\mathcal{H}_S|n_q\rangle$ as $\mathcal{H}_S$, and $\langle n_q|\mathcal{H}_0|n_q\rangle$ as $\mathcal{H}_0$, and hope that no confusion in notation will arise. Using

$$(\mathcal{H}_{\mathrm{ep}})_{n_q, n_{q-1}} = -\mathrm{i}\sum_k B_{-q} C_{k-q}^\dagger C_k \sqrt{n_q}\,, \tag{8.129}$$

and

$$(\mathcal{H}_{\mathrm{ep}})_{n_q, n_{q+1}} = \mathrm{i}\sum_k B_q C_{k+q}^\dagger C_k \sqrt{n_q + 1}\,, \tag{8.130}$$

the effective Hamiltonian for electrons is given by combining the above. Thus,

$$
\begin{aligned}
\mathcal{H}_S = \mathcal{H}_0 + \frac{1}{2}\left|B_q\right|^2 \sum_{k,k'} \Bigg[& C^{\dagger}_{k-q} C_k C^{\dagger}_{k'+q} C_{k'} n_q \frac{1}{\varepsilon_{k'} - \varepsilon_{k'+q} + \omega_q} \\
& + C^{\dagger}_{k+q} C_k C^{\dagger}_{k'-q} C_{k'} (n_q + 1) \frac{1}{\varepsilon_{k'} - \varepsilon_{k'-q} - \omega_q} \\
& - C^{\dagger}_{k'-q} C_{k'} C^{\dagger}_{k+q} C_k n_q \frac{1}{\varepsilon_{k'} - \varepsilon_{k'-q} - \omega_q} \\
& - C^{\dagger}_{k'+q} C_{k'} C^{\dagger}_{k-q} C_k (n_q + 1) \frac{1}{\varepsilon_{k'} - \varepsilon_{k'+q} + \omega_q} \Bigg].
\end{aligned} \tag{8.131}
$$

Making dummy variable changes, dropping terms that do not involve the interaction of electrons (i.e. that do not involve both $\boldsymbol{k}$ and $\boldsymbol{k}'$), and using the commutation relations for the C, it is possible to write the above in the form

$$
\begin{aligned}
\mathcal{H}_S = \mathcal{H}_0 + \frac{1}{2}\left|B_q\right|^2 & \sum_{k,k'} C^{\dagger}_{k'+q} C_{k'} C^{\dagger}_{k-q} C_k \\
& \times \left(\frac{1}{\varepsilon_k - \varepsilon_{k-q} - \omega_q} - \frac{1}{\varepsilon_{k'} - \varepsilon_{k'+q} + \omega_q} \right).
\end{aligned} \tag{8.132}
$$

In order to properly interpret Hamiltonians such as the above equation, which are expressed in the second quantization notation, it is necessary to keep in mind the appropriate commutation relations of the C. By Appendix G, these are

$$C_k C^{\dagger}_{k'} + C^{\dagger}_{k'} C_k = \delta^{k'}_k , \tag{8.133}$$

$$C^{\dagger}_k C^{\dagger}_{k'} + C^{\dagger}_{k'} C^{\dagger}_k = 0 , \tag{8.134}$$

and

$$C_k C_{k'} + C_{k'} C_k = 0 . \tag{8.135}$$

The Hamiltonian (8.132) describes a process called a *virtual exchange of a phonon*. It has the diagrammatic representation shown in Fig. 8.19.

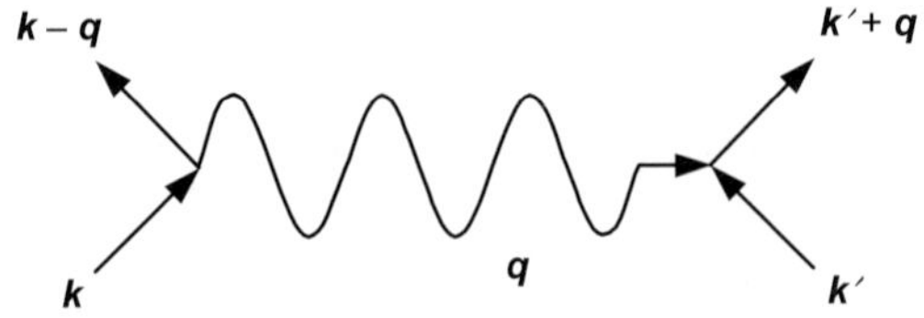

Fig. 8.19. The virtual exchange of a phonon of wave vector $\boldsymbol{q}$. The $\boldsymbol{k}$ are the wave vectors of the electrons. This is the fundamental process of superconductivity

Note that (8.132) is independent of the number of phonons in mode q, and it is the effective electron Hamiltonian with phonons in the single mode q. To get the effective Hamiltonian with phonons in all modes, we merely have to sum over the modes of q. Thus, the total effective interaction Hamiltonian is given by

$$\mathcal{H}_{\mathrm{I}} = \frac{1}{2}\sum_{q}\sum_{k,k'} \left|B_q\right|^2 C^{\dagger}_{k'+q}C_{k'}C^{\dagger}_{k-q}C_k \times\left(\frac{1}{\varepsilon_k - \varepsilon_{k-q} - \omega_q} - \frac{1}{\varepsilon_{k'} - \varepsilon_{k'+q} + \omega_q}\right). \tag{8.136}$$

By dropping further terms that do not involve the interaction of electrons (terms not involving both $\boldsymbol{k}$ and $\boldsymbol{k}'$) and by making variable changes, we can reduce this Hamiltonian to

$$\mathcal{H}_{\mathrm{I}} = \sum_{q}\sum_{k,k'} \left|B_q\right|^2 \frac{\omega_q}{(\varepsilon_k - \varepsilon_{k-q})^2 - \omega_q^2} C^{\dagger}_{k'+q}C^{\dagger}_{k-q}C_k C_{k'} \,. \tag{8.137}$$

From the above equation, we see that there is an attractive electron–electron interaction for $|\varepsilon_k - \varepsilon_{k-q}| < |\omega_q|$. We will assume, for appropriate excitation energies, that the main interaction is attractive. In this connection, most of the electron energies of interest are near the Fermi energy ε_F. A typical phonon energy is the Debye energy $\hbar\omega_D$ (or cutoff frequency with $\hbar = 1$). Many approximations have already been made, and so a very simple criterion for the dominance of the attractive interaction will be assumed. It will be assumed that the interaction is attractive when the electronic energies are in the range of

$$\varepsilon_F - \hbar\omega_D < \varepsilon_k < \varepsilon_F + \hbar\omega_D \quad (\hbar \neq 1 \text{ here}) \,. \tag{8.138}$$

The states that do not satisfy this criterion are not directly involved in the superconducting transition, so their properties are of no particular interest. Hence, the effective Hamiltonian can be written in the following form (*fourth major approximation*):

$$\mathcal{H}_{\mathrm{I}} \equiv -\sum_{q}\sum_{k,k'} V_q C^{\dagger}_{k'+q}C^{\dagger}_{k-q}C_k C_{k'} \,. \tag{8.139}$$

For simplicity, we will assume that V_q is positive and fitted from experiment, that $V_q = V_{-q}$ and $V_q = 0$, unless $\boldsymbol{q}$ is such that (8.138) is satisfied. We assume that any important interactions not included in the above equation can be included by renormalizing (i.e. changing) the quasiparticle mass.

8.5.3 Cooper Pairs and the BCS Hamiltonian (A)

Let us assume that $\varepsilon_k = 0$ at the Fermi level. The total effective Hamiltonian for the electrons is then

$$\mathcal{H} = \sum_{k} \varepsilon_k C^{\dagger}_k C_k - \sum_{k,k',q} V_q C^{\dagger}_{k'+q}C^{\dagger}_{k-q}C_k C_{k'} \,. \tag{8.140}$$

By Appendix G, the Fermion operators satisfy

$$C_j \left| n_1 \dots n_j \dots \right\rangle = (-)^{P_j} n_j \left| n_1 \dots (1 - n_j) \dots \right\rangle , \tag{8.141}$$

$$C_j^\dagger \left| n_1 \dots n_j \dots \right\rangle = (-)^{P_j} (1 - n_j) \left| n_1 \dots (1 + n_j) \dots \right\rangle , \tag{8.142}$$

where

$$P_j = \sum_{P=1}^{j-1} n_P . \tag{8.143}$$

It is essential to notice the alternation in sign defined by (8.142). This alternation is very important for discovering the nature of the lowest-energy state. When we begin to guess a trial wave function, if we pay no attention to this alternation of sign, the presence of the interaction will result in little lowering of the energy. What we need is a way of selecting the trial wave function so that most of the matrix elements of individual terms in the second sum in (8.138) are negative. The way to do this for the ground state is by grouping the electrons into *Cooper pairs*. (These will be precisely defined below.)

There are several assumptions necessary to construct a minimum energy wave function [60, p. 155ff]. For the ground-state wave function, it will be assumed that the Bloch states are occupied only in pairs. In fact, the superconducting ground state is a coherent superposition of Cooper pairs. The Hamiltonian conserves the wave vector, and only pairs with equal total momentum will be considered, i.e.,

$$\boldsymbol{k} + \boldsymbol{k}' = \boldsymbol{K} , \tag{8.144}$$

where $\boldsymbol{K}$ is the same for each pair. It is reasonable to suppose that $\boldsymbol{K}$ is zero for the ground (noncurrent carrying) state of the pairs.

Cooper Pairs[8]

Before proceeding, let us discuss Cooper pairs a little more. A large clue as to the nature of the unusual character of the superconducting state was obtained by L. Cooper in 1956. He showed that the Fermi sea was unstable if electrons interacted by an attractive mechanism—no matter how weak.

Consider the normal Fermi sea of electrons with a well-defined Fermi energy E_F. Now add two more electrons interacting with an attractive interaction $V(1,2)$ and suppose the only interaction with the other electrons is via the Pauli principle.

We write the Schrödinger wave equation for the two electrons as

$$\left[-\frac{\hbar^2}{2m}\nabla_1^2 - \frac{\hbar^2}{2m}\nabla_2^2 + V(1,2) \right] \psi(1,2) = E\psi(1,2) . \tag{8.145}$$

[8] See Cooper [8.10].

We seek a solution of the form

$$\psi(1,2) = \frac{V}{(2\pi)^3} A(1,2) \frac{1}{V} \int e^{i\boldsymbol{k}\cdot(\boldsymbol{r}_1 - \boldsymbol{r}_2)} f(\boldsymbol{k}) d\boldsymbol{k} , \tag{8.146}$$

where $A(1, 2)$ is the antisymmetric spin zero spin wave function

$$A(1,2) = \frac{1}{\sqrt{2}} [\alpha(1)\beta(2) - \alpha(2)\beta(1)] ,$$

with α, β being the usual spin-up and -down wave functions (note $A^\dagger A = 1$) and $f(\boldsymbol{k}) = +f(-\boldsymbol{k})$ so that the spatial wave function is symmetric (it can be shown that the ψ with spin 1 and antisymmetric wave function yields no energy shift, at least in our approximation, and in any case such wave functions correspond to p-state pairs that we are not considering). Note that the spatial wave function pairs off the electrons into $(\boldsymbol{k}, -\boldsymbol{k})$ states.

Inserting (8.146) into (8.145) we have

$$\begin{aligned} &\frac{1}{(2\pi)^3} A(1,2) \int \left[\left(\frac{\hbar^2}{2m} k^2 + \frac{\hbar^2}{2m} k^2 + V \right) f(\boldsymbol{k}) e^{i\boldsymbol{k}\cdot(\boldsymbol{r}_1 - \boldsymbol{r}_2)} \right] d\boldsymbol{k} \\ &= \frac{A(1,2)}{(2\pi)^3} \int E f(\boldsymbol{k}) e^{i\boldsymbol{k}\cdot(\boldsymbol{r}_1 - \boldsymbol{r}_2)} d\boldsymbol{k} . \end{aligned} \tag{8.147}$$

Now multiply by

$$A^\dagger(1,2) \frac{1}{V} e^{-i\boldsymbol{k}'\cdot(\boldsymbol{r}_1 - \boldsymbol{r}_2)} ,$$

and integrate over $\boldsymbol{r}_1$ and $\boldsymbol{r}_2$ and we obtain ($\boldsymbol{r} = \boldsymbol{r}_1 - \boldsymbol{r}_2$, $V(r_1, r_2) = V(\boldsymbol{r}_1 - \boldsymbol{r}_2) = V(\boldsymbol{r})$, and $E_k = \hbar^2 k^2/2m$)

$$\int \int e^{-i\boldsymbol{k}'\cdot\boldsymbol{r}} [2E_k + V(\boldsymbol{r})] f(\boldsymbol{k}) e^{i\boldsymbol{k}\cdot\boldsymbol{r}} d\boldsymbol{r} d\boldsymbol{k} = \int E f(\boldsymbol{k}) e^{i(\boldsymbol{k}-\boldsymbol{k}')\cdot\boldsymbol{r}} d\boldsymbol{k} . \tag{8.148}$$

Using

$$\frac{1}{(2\pi)^3} \int e^{i\boldsymbol{k}\cdot\boldsymbol{r}} d\boldsymbol{k} = \delta(\boldsymbol{k}) ,$$

and

$$V_{\boldsymbol{k}',\boldsymbol{k}} = \frac{1}{V} \int e^{-i\boldsymbol{k}'\cdot\boldsymbol{r}} V(\boldsymbol{r}) e^{i\boldsymbol{k}\cdot\boldsymbol{r}} d\boldsymbol{r} ,$$

we obtain

$$[2E_{k'} - E] f(\boldsymbol{k}') + \frac{V}{(2\pi)^3} \int f(\boldsymbol{k}') V_{\boldsymbol{k}',\boldsymbol{k}} d\boldsymbol{k} = 0 . \tag{8.149}$$

We suppose

$$V_{\boldsymbol{k}',\boldsymbol{k}} = -V_0 < 0 \quad \text{for} \quad E_F < E_{\boldsymbol{k}}\,, \quad E_{\boldsymbol{k}'} < E_F + \hbar\omega_D\,.$$
$$= 0 \quad \text{otherwise.}$$

Notice we are using the ideas that led us to (8.138), divide by $2E_{k'} - E$ and integrate over $\boldsymbol{k}'$ and obtain (after canceling)

$$1 = V_0 \frac{V}{(2\pi)^3} \int \frac{\mathrm{d}\boldsymbol{k}'}{2E_{\boldsymbol{k}'} - E}\,. \tag{8.150}$$

Note that in the limit of large volumes

$$\frac{V/N}{(2\pi)^3} \int \mathrm{d}\boldsymbol{k}'(\,) \leftrightarrow \frac{1}{N} \sum\nolimits_{k'} \leftrightarrow \int N(E')(\,)\mathrm{d}E'\,,$$

where $N(E)$ is the density of state for one spin per unit cell (N unit cells). Thus with $E_{\text{pair}} = E$

$$1 = V_0 \int_{E_F}^{E_F + \hbar\omega_D} \frac{N(E')}{2E' - E_{pair}} \mathrm{d}E'\,. \tag{8.151}$$

Note we can replace $N(E') \cong N(E_F)$ because $\hbar\omega_D << E_F$ so we obtain

$$1 = \frac{V_0 N(E_F)}{2} \ln \frac{2E_F + 2\hbar\omega_D - E_{\text{pair}}}{2E_F - E_{\text{pair}}}\,. \tag{8.152}$$

Let $\delta = 2E_F - E_{\text{pair}}$ so

$$\delta = \hbar\omega_D \frac{\exp\left(\dfrac{-1}{V_0 N(E_F)}\right)}{\sinh \dfrac{1}{V_0 N(E_F)}}\,, \tag{8.153}$$

and in the weak coupling limit

$$\delta = 2\hbar\omega_D \exp\left(\frac{-2}{V_0 N(E_F)}\right). \tag{8.154}$$

We note in particular, the following points:

1. A pair electron wave function that is independent of the direction of $\boldsymbol{r}_1 - \boldsymbol{r}_2$ is said to be an s wave function, which is consistent with an antisymmetric spin wave function.
2. δ is not an analytic function of V_0 so ordinary perturbation theory would not work.
3. In the BCS theory one considers pairing of all electrons.
4. For $\delta > 0$ then the Fermi sea is unstable with respect to the formation of Cooper pairs.

BCS Hamiltonian

Returning to the mainstream of the BCS argument, the above reasoning can be used to pick out the best wave function to use as a trial wave function for evaluating the ground-state energy by variational principle. For mathematical convenience, it is easier to place these assumptions directly in the Hamiltonian. Also, due to exchange, the spins in the Cooper pairs are usually opposite. Thus, the interaction part of the Hamiltonian is now written (*fifth major approximation*) with $\boldsymbol{K} = 0$,

$$\mathcal{H}_{\mathrm{I}} = -\sum_{\boldsymbol{k},\boldsymbol{q}} V_{\boldsymbol{q}} C^{\dagger}_{\boldsymbol{k}+\boldsymbol{q}\uparrow} C^{\dagger}_{-\boldsymbol{k}-\boldsymbol{q}\downarrow} C_{-\boldsymbol{k}\downarrow} C_{\boldsymbol{k}\uparrow} \,. \tag{8.155}$$

Next, assume a "BCS Hamiltonian" for interacting pairs consistent with (8.155), with $\boldsymbol{k}+\boldsymbol{q} \to \boldsymbol{k}$, $\boldsymbol{k} \to \boldsymbol{k}'$, $V_{\boldsymbol{q}} = V_{\boldsymbol{k}-\boldsymbol{k}'} = -V_{\boldsymbol{k},\boldsymbol{k}'}$

$$\mathcal{H} = \sum_{\boldsymbol{k}\sigma} \varepsilon_{\boldsymbol{k}} C^{\dagger}_{\boldsymbol{k}\sigma} C_{\boldsymbol{k}\sigma} + \sum_{\boldsymbol{k},\boldsymbol{k}'} V_{\boldsymbol{k},\boldsymbol{k}'} C^{\dagger}_{\boldsymbol{k}\uparrow} C^{\dagger}_{-\boldsymbol{k}\downarrow} C_{-\boldsymbol{k}'\downarrow} C_{\boldsymbol{k}'\uparrow} \,, \tag{8.156}$$

where

$$\varepsilon_{\boldsymbol{k}} = \frac{\hbar^2 k^2}{2m} - \mu \,, \tag{8.157}$$

and where μ is the chemical potential. Also

$$\mathcal{H} \equiv \mathcal{H}_0 + \mathcal{H}_{\mathrm{I}} \,, \tag{8.158}$$

and note

$$V_{\boldsymbol{k},\boldsymbol{k}'} = V_{\boldsymbol{k}',\boldsymbol{k}} = V^{*}_{\boldsymbol{k},\boldsymbol{k}'} \,. \tag{8.159}$$

As before C are Fermion (electron) annihilation operators, and $C^{\dagger}$ are Fermion (electron) creation operators. Defining the pair creation and annihilation operators

$$b^{\dagger}_{\boldsymbol{k}} = C^{\dagger}_{\boldsymbol{k}\uparrow} C^{\dagger}_{-\boldsymbol{k}\downarrow} \,, \tag{8.160}$$

$$b_{\boldsymbol{k}} = C_{-\boldsymbol{k}\downarrow} C_{\boldsymbol{k}\uparrow} \,; \tag{8.161}$$

and defining

$$\bar{b}_{\boldsymbol{k}} = \frac{Tr(\mathrm{e}^{-\beta H} b_{\boldsymbol{k}})}{Tr(\mathrm{e}^{-\beta H})} \,, \tag{8.162}$$

we can show $\bar{b}_{\boldsymbol{k}}^{\dagger} = \bar{b}_{\boldsymbol{k}}^{*}$ using $\mathrm{Tr}(AB) = \mathrm{Tr}(BA)$. We can also show in the representation we use that $\bar{b}_{\boldsymbol{k}}^{*} = \bar{b}_{\boldsymbol{k}}$. We define

$$\Delta_{\boldsymbol{k}} = -\sum_{\boldsymbol{k}'} V_{\boldsymbol{k},\boldsymbol{k}'} \bar{b}_{\boldsymbol{k}'} = \Delta^{*}_{\boldsymbol{k}} \,. \tag{8.163}$$

As we will demonstrate later, this will turn out to be the gap parameter. We can write the interaction term as

$$\mathcal{H}_\mathrm{I} = \sum_{k,k'} V_{k,k'} b_k^\dagger b_{k'} . \tag{8.164}$$

Note

$$b_{k'} = \bar{b}_{k'} + \delta b_{k'} = \bar{b}_{k'} + (b_{k'} - \bar{b}_{k'}) , \tag{8.165}$$

and

$$b_{k'}^\dagger = \bar{b}_{k'}^* + \delta b_{k'}^\dagger = \bar{b}_{k'}^* + (b_{k'}^\dagger - \bar{b}_{k'}^*) ; \tag{8.166}$$

$$b_k^\dagger = \bar{b}_k^* + \delta b_k^\dagger = \bar{b}_k + \delta b_k^\dagger ; \tag{8.167}$$

and we will neglect $(\delta b_{k'})\times(\delta b_k{}^\dagger)$ terms. (This is sort of a mean-field-like approximation for pairs.) Thus, using (8.166) and (8.167) and neglecting $O(\delta b^2)$ terms, we can write

$$\begin{aligned} b_k^\dagger b_{k'} &= (\bar{b}_k + \delta b_k^\dagger)(\bar{b}_{k'} + \delta b_{k'}) \\ &= \bar{b}_k \bar{b}_{k'} + \bar{b}_{k'} \delta \bar{b}_k^\dagger + \bar{b}_k \delta \bar{b}_{k'} , \\ &= \bar{b}_k b_{k'} + \bar{b}_{k'} b_k^\dagger - \bar{b}_k \bar{b}_{k'} , \end{aligned} \tag{8.168}$$

assuming $\bar{b}_k$ is real. Also,

$$\mathcal{H}_\mathrm{I} = \sum_{k,k'} V_{k,k'} (\bar{b}_{k'} b_k^\dagger + \bar{b}_{k'} b_k - \bar{b}_{k'} \bar{b}_k) . \tag{8.169}$$

Thus,

$$\mathcal{H} = \sum_{k\sigma} \varepsilon_k C_{k\sigma}^\dagger C_{k\sigma} - \sum_k (\Delta_k b_k^\dagger + \Delta_k b_k - \Delta_k \bar{b}_k) . \tag{8.170}$$

We now diagonalize by a Bogoliubov–Valatin transformation:

$$C_{k\uparrow} = u_k \alpha_k + v_k \beta_k^\dagger , \tag{8.171}$$

$$C_{-k\downarrow} = u_k \beta_k - v_k \alpha_k^\dagger ; \tag{8.172}$$

where $u_k{}^2 + v_k{}^2 = 1$ (to preserve anticommutation relations), u_k and v_k are real, and the α and β given by

$$\alpha_k^\dagger = u_k C_{k\uparrow}^\dagger - v_k C_{-k\downarrow} , \tag{8.173}$$

$$\beta_k^\dagger = u_k C_{-k\downarrow}^\dagger + v_k C_{k\uparrow} , \tag{8.174}$$

are Fermion operators obeying the usual anticommutation relations. The $\alpha_k^\dagger$, and $\beta_k^\dagger$ create "bogolons". The algebra gets a bit detailed here and one can *skip* along unless curious,

$$b_k = C_{-k\downarrow}C_{k\uparrow} = (-v_k\alpha_k^\dagger + u_k\beta_k)(u_k\alpha_k + v_k\beta_k^\dagger) \tag{8.175}$$

$$b_k^\dagger = C_{k\uparrow}^\dagger C_{-k\downarrow}^\dagger = (u_k\alpha_k^\dagger + v_k\beta_k)(-v_k\alpha_k + u_k\beta_k^\dagger) \tag{8.176}$$

$$C_{k\uparrow}^\dagger C_{k\uparrow} = (u_k\alpha_k^\dagger + v_k\beta_k)(u_k\alpha_k + v_k\beta_k^\dagger) \tag{8.177}$$

$$C_{-k\downarrow}^\dagger C_{-k\downarrow} = (-v_k\alpha_k + u_k\beta_k^\dagger)(-v_k\alpha_k^\dagger + u_k\beta_k) \tag{8.178}$$

$$b_k^\dagger = -u_k v_k\alpha_k^\dagger\alpha_k + u_k v_k\beta_k\beta_k^\dagger - v_k^2\beta_k\alpha_k + u_k^2\alpha_k^\dagger\beta_k^\dagger \tag{8.179}$$

$$b_k = -v_k u_k\alpha_k^\dagger\alpha_k + u_k v_k\beta_k\beta_k^\dagger - v_k^2\alpha_k^\dagger\beta_k^\dagger + u_k^2\beta_k\alpha_k \tag{8.180}$$

$$C_{k\uparrow}^\dagger C_{k\uparrow} = u_k^2\alpha_k^\dagger\alpha_k + v_k^2\beta_k\beta_k^\dagger + u_k v_k\alpha_k^\dagger\beta_k^\dagger + u_k v_k\beta_k\alpha_k \tag{8.181}$$

$$C_{-k\downarrow}^\dagger C_{-k\downarrow} = u_k^2\beta_k^\dagger\beta_k + v_k^2\alpha_k\alpha_k^\dagger - u_k v_k\beta_k^\dagger\alpha_k^\dagger - v_k u_k\alpha_k\beta_k \tag{8.182}$$

$$\begin{aligned}\mathcal{H} = \textstyle\sum_k [&2\varepsilon_k u_k v_k\alpha_k^\dagger\beta_k^\dagger + 2\varepsilon_k\beta_k\alpha_k u_k v_k \\ &+ \varepsilon_k(u_k^2 - v_k^2)(\alpha_k^\dagger\alpha_k + \beta_k^\dagger\beta_k) + 2\varepsilon_k v_k^2 \\ &+ \Delta_k u_k v_k(\alpha_k^\dagger\alpha_k + \beta_k^\dagger\beta_k) \\ &- \Delta_k u_k v_k + \Delta_k v_k^2\beta_k\alpha_k - \Delta_k u_k^2\alpha_k^\dagger\beta_k^\dagger \\ &+ \Delta_k^* u_k v_k(\alpha_k^\dagger\alpha_k + \beta_k^\dagger\beta_k) - \Delta_k^* u_k v_k \\ &+ \Delta_k v_k^2\alpha_k^\dagger\beta_k^\dagger - \Delta_k^* u_k^2\beta_k\alpha_k] + \textstyle\sum_k \Delta_k\bar{b}_k .\end{aligned} \tag{8.183}$$

Rewriting this we get

$$\begin{aligned}\mathcal{H} = \textstyle\sum_k \{&(2\Delta_k u_k v_k - \varepsilon_k(v_k^2 - u_k^2))(\alpha_k^\dagger\alpha_k + \beta_k^\dagger\beta_k) \\ &+ (2\varepsilon_k u_k v_k + \Delta_k(v_k^2 - u_k^2))(\alpha_k^\dagger\beta_k^\dagger + \beta_k\alpha_k)\} + G,\end{aligned} \tag{8.184}$$

where

$$G = \sum(2\varepsilon_k v_k^2 - 2\Delta_k u_k v_k + \Delta_k\bar{b}_k) .$$

Next, choose

$$2\varepsilon_k u_k v_k + \Delta_k(v_k^2 - u_k^2) = 0 , \tag{8.185}$$

so as to diagonalize the Hamiltonian. Also, using $u_k{}^2 + v_k{}^2 = 1$ let

$$v_k^2 = \frac{1}{2} - a\,, \tag{8.186}$$

$$v_k^2 - u_k^2 = -2a\,, \tag{8.187}$$

$$u_k v_k = \sqrt{\frac{1}{4} - a^2}\,; \tag{8.188}$$

$$2\varepsilon_k \sqrt{\frac{1}{4} - a^2} = \Delta_k 2a\,, \tag{8.189}$$

$$\varepsilon_k^2 \left(\frac{1}{4} - a^2 \right) = \Delta_k^2 a^2\,. \tag{8.190}$$

Thus

$$a = \frac{\varepsilon_k}{2\sqrt{\varepsilon_k^2 + \Delta_k^2}}\,. \tag{8.191}$$

Rewriting,

$$\mathcal{H} = [2\Delta_k u_k v_k - \varepsilon_k (v_k^2 - u_k^2)] \times (\alpha_k^\dagger \alpha_k + \beta_k^\dagger \beta_k) + G\,. \tag{8.192}$$

But, define

$$E_k = \sqrt{\varepsilon_k^2 + \Delta_k^2}\,, \tag{8.193}$$

$$a = \frac{\varepsilon_k}{2E_k}\,, \tag{8.194}$$

and thus

$$2u_k v_k = \frac{\Delta_k}{E_k}\,. \tag{8.195}$$

Thus, after a bit of algebra,

$$2\Delta_k u_k v_k - \varepsilon_k (v_k^2 - u_k^2)] = E_k\,. \tag{8.196}$$

So

$$\mathcal{H} = \sum\nolimits_k E_k (\alpha_k^\dagger \alpha_k + \beta_k^\dagger \beta_k) + G\,, \tag{8.197}$$

and G can be put in the form

$$G = \sum_k (\varepsilon_k - E_k + \Delta_k \bar{b}_k). \tag{8.198}$$

Note by Fig. 8.20 and (8.193) how E_k predicts a gap, for clearly $E_k \geq \Delta_0$. Continuing

$$b_k = -v_k u_k \alpha_k^\dagger \alpha_k + u_k v_k \beta_k \beta_k^\dagger - v_k^2 \alpha_k^\dagger \beta_k^\dagger + u_k^2 \beta_k \alpha_k. \tag{8.199}$$

But $\bar{b}_k$ involves only diagonal terms, so using an appropriate anticommutation relation

$$\bar{b}_k = u_k v_k \left(1 - \overline{\alpha_k^\dagger \alpha_k} - \overline{\beta_k^\dagger \beta_k}\right), \tag{8.200}$$

so

$$\bar{b}_k = u_k v_k (1 - 2n_k), \tag{8.201}$$

where

$$n_k = \frac{1}{\mathrm{e}^{\beta E_k} + 1} = f(E_k). \tag{8.202}$$

$f(E_k)$ is of course the Fermi function but it looks strange without the chemical potential. This is because $\alpha^\dagger$, $\beta^\dagger$ do not change the particle number. See Marder [8.22]. Therefore,

$$\begin{aligned} \Delta_k &= -\sum V_{k,k'} \bar{b}_{k'} \\ &= -\sum V_{k,k'} u_{k'} v_{k'} [1 - 2f(E_{k'})] \\ &= -\sum_{k'} V_{k,k'} \frac{\Delta_{k'}}{2E_{k'}} [1 - 2f(E_{k'})]. \end{aligned} \tag{8.203}$$

Now assume that (not using $\hbar = 1$)

$$\Delta_{k'} = \Delta \quad \text{when} \quad |\varepsilon_k| < \hbar\omega_D, \tag{8.204}$$

$$V_{k,k'} = -V \quad \text{when} \quad |\varepsilon_k| < \hbar\omega_D, \tag{8.205}$$

$$\Delta_k = 0 \quad \text{when} \quad |\varepsilon_k| > \hbar\omega_D, \tag{8.206}$$

and

$$V_{k,k'} = 0 \quad \text{when} \quad |\varepsilon_k| > \hbar\omega_D, \tag{8.207}$$

where ω_D is the Debye frequency (see (8.138)) So,

$$\Delta = V \sum_{|\varepsilon_{k'}| < \hbar\omega_D} \frac{\Delta[1 - 2f(E_{k'})]}{2E_{k'}}. \tag{8.208}$$

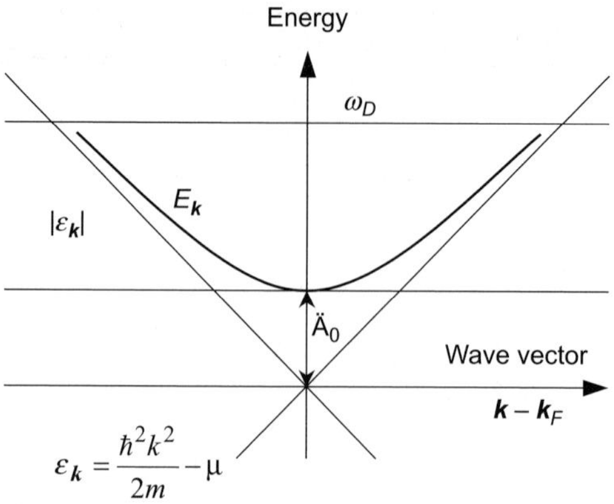

Fig. 8.20. Gap in single-particle excitations near the Fermi energy

For $T = 0$, then

$$\Delta = \sum_{k'} V \frac{\Delta}{2\sqrt{\varepsilon_{k'}^2 + \Delta^2}}, \tag{8.209}$$

and for $T \neq 0$, then

$$\Delta = \sum_{k'} \frac{V\Delta}{2\sqrt{\varepsilon_{k'}^2 + \Delta^2}} \tanh\left(\frac{E_{k'}}{2kT}\right). \tag{8.210}$$

We can then write

$$\Delta \cong \int_0^{\hbar\omega_D} N(E) V \frac{\Delta}{\sqrt{E^2 + \Delta^2}} dE. \tag{8.211}$$

If we further suppose that $N(E) \cong$ constant $\cong N(0) \equiv$ the density of states at the Fermi level, then (8.211) becomes

$$\begin{aligned} \frac{1}{N(0)V} &= \int_0^{\hbar\omega_D} \frac{dE}{\sqrt{E^2 + \Delta^2}} = \ln\left(E + \sqrt{E^2 + \Delta^2}\right)\Bigg|_0^{\hbar\omega_D} \\ &= \ln\left(\frac{\hbar\omega_D + \sqrt{(\hbar\omega_D)^2 + \Delta^2}}{\Delta}\right). \end{aligned} \tag{8.212}$$

This equation can be written as

$$\begin{aligned} \exp\left(-\frac{1}{N(0)V}\right) &= \frac{\Delta}{\hbar\omega_D + \sqrt{(\hbar\omega_D)^2 + \Delta^2}} \\ &= \frac{\Delta}{\hbar\omega_0\left(\sqrt{1 + (\Delta/\hbar\omega_D)^2} + 1\right)} = \frac{\Delta}{2\hbar\omega_D} \end{aligned} \tag{8.213}$$

in the weak coupling limit (when $\Delta << \hbar\omega_D$). Thus, in the weak coupling limit, we obtain

$$\Delta \cong 2\hbar\omega_D \exp\left(-\frac{1}{N(0)V}\right). \tag{8.214}$$

From (8.210) by similar reasoning

$$\frac{1}{N(0)V} = \int_0^{\hbar\omega_D} \frac{\tanh\left(\sqrt{\varepsilon^2+\Delta^2}/2kT\right)}{\sqrt{\varepsilon^2+\Delta^2}} d\varepsilon , \tag{8.215}$$

where, again, N(0) is the density of states at the Fermi energy.

For T greater than some critical temperature there are no solutions for Δ, i.e. the energy gap no longer exists. We can determine T_c by using the fact that at $T = T_c$, $\Delta = 0$. This says that

$$\frac{1}{N(0)V} = \int_0^{\hbar\omega_D} \frac{\tanh(\varepsilon/2kT_c)}{\varepsilon} d\varepsilon . \tag{8.216}$$

In the weak coupling approximation, when $N(0)V << 1$, we obtain from (8.216) that

$$kT_c = 1.14\hbar\omega_D \exp(-1/N(0)V). \tag{8.217}$$

Equation (8.217) is a very important equation. It depends on three material properties:

a) The Debye frequency ω_D

b) V that measures the strength of the electron–phonon coupling and

c) $N(0)$ that measures the number of electrons available at the Fermi energy.

Note that typically $\omega_D \propto (m)^{-1/2}$, where m is the mass of atoms. This leads directly to the isotope effect. Note also the energy gap $E_g = 2\Delta(0)$ at absolute zero.

We can combine this result with our result for the energy gap parameter in the ground state to derive a relation between the energy gap at absolute zero and the critical superconducting transition temperature with no magnetic field. By (8.217) and (8.214), we have that

$$\Delta(0) = 2\hbar\omega_D \exp(-1/N(0)V) = \frac{2}{1.14} kT_c , \tag{8.218}$$

or

$$2\Delta(0) = 3.52kT_c . \tag{8.219}$$

Note that our expression for $\Delta(0)$ and T_c both involve the factor $\exp(-1/N(0)V)$; that is, a power series (in V) expansion for both $\Delta(0)$ and T_c have an essential singu larity in V. We could not have obtained reasonable results if we had tried ordinary perturbation theory because with ordinary perturbation theory, we cannot reproduce the effect of an essential singularity in the perturbation. This is similar to what happened when we discussed a single Cooper pair.

Our discussion has only been valid for weakly coupled superconductors. Roughly speaking, these have $(T_c/\theta_D)^2 \gtrsim (500)^{-2}$. Pb, Hg, and Nb are strongly coupled, and for them $(T_c/\theta_D)^2 \geq (300)^{-2}$. Alternatively, the electron–phonon coupling parameter is about three times larger than is a typical weak coupling superconductor. A result for the strong coupling approximation is given below.

8.5.4 Remarks on the Nambu Formalism and Strong Coupling Superconductivity (A)

The Nambu approach to superconductivity is presented by matrices and diagrams. The Nambu formalism includes the possibility of Cooper pairs in the calculation from the beginning via two component field operators. This approach allows for the treatment of retardation effects that need to be included for the strong (electron lattice) coupling regime. An essential step in the development was taken by Eliashberg and this leads to his equations. The Eliashberg strong coupling calculation of the superconducting transition temperature gives with a computer fit (via McMillan):

$$T_c = \frac{\theta_D}{1.45}\exp\left(\frac{-1.04(1+\lambda)}{\lambda - \mu^*(1+0.62\lambda)}\right).$$

θ_D is the Debye temperature, and for definitions of λ (the coupling constant) and μ^* (the Coulomb pseudopotential term) see Jones and March [8.17]. They also give a nice summary of the calculation. Briefly $\lambda = N(0)V_{\text{phonon}}$, $\mu = N(0)V_{\text{coulomb}}$ where V in (8.218) is $V_{\text{phonon}} - V_{\text{coulomb}}$, and

$$\mu^* = \mu\left(1 + \mu \ln\frac{E_F}{k_B\theta_D}\right)^{-1}.$$

Usually λ empirically turns out to be not much larger than 5/4 (or smaller).

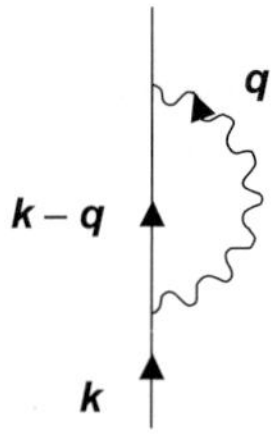

Fig. 8.21. Lowest-order correction to self-energy Feynman diagram (for electrons due to phonons)

The calculation includes the self-energy terms. The lowest-order correction to self-energy for electrons due to phonons is indicated in Fig. 8.21. The BCS theory with the extension of Eliashberg and McMillan has been very successful for many superconductors.

A nice reference to consult is Mattuck [8.23 pp. 267-272].

8.6 Magnesium Diboride (EE, MS, MET)

For a review of the new superconductor magnesium diboride, see, e.g., *Physics Today*, March 2003, p. 34ff. The discovery of the superconductor MgB_2, with a transition temperature of 39 K, was announced by Akimitsu in early 2001. At first sight this might not appear to be a particularly interesting discovery, compared to that of the high-temperature superconductors, but MgB_2 has several interesting properties:

1. It appears to be a conventional BCS superconductor with electron–phonon coupling driving the formation of pairs. It shows a strong isotope effect.
2. It does not appear to have the difficulty that the high-T_c cuprate ceramics have of having grain boundaries that inhibit current.
3. It is a widely available material that comes right off the shelf.
4. MgB_2 is an intermetallic (two metals forming a crystal structure at a well-defined stoichiometry) compound with a transition temperature near double that of Nb_3Ge.

Possibly, the transition temperature can be driven higher by tailoring the properties of magnesium diboride. At this writing, several groups are working intensely on this material, with several interesting results including the fact that it has two superconducting gaps arising from two weakly interacting bands.

8.7 Heavy-Electron Superconductors (EE, MS, MET)

UBe_{13} (T_c = 0.85 K), $CeCu_2Si_2$ (T_c = 0.65 K), and UPt_3 (T_c = 0.54 K) are heavy-electron superconductors. They are characterized by having large low-temperature specific heats due to effective mass being two or three orders of magnitude larger than in normal metals (because of f band electrons). Heavy-electron superconductors do not appear to have a singlet state s-wave pairing, but perhaps can be characterized as d-wave pairing or p-wave pairing (d and p referring to orbital symmetry). It is also questionable whether the pairing is due to the exchange of virtual phonons—it may be due, e.g., to the exchange of virtual magnons. See, e.g., Burns [8.9 p51]. We have already mentioned these in Sect. 5.7.

8.8 High-Temperature Superconductors (EE, MS, MET)

It has been said that Brazil is the country of the future and always has been as well as always will be. A similar comment has been made about superconductors. The problem is that superconductivity applications have been limited by the fact that liquid helium temperatures (of order 4 K) have been necessary to retain superconductivity. Liquid nitrogen (which boils at 77 K) is much cheaper and materials that superconduct at or above the boiling temperature of liquid nitrogen would open a large range of practical applications. Particularly important would be the transport of electrical power.

Just finding a high superconducting transition temperature T_c, however, does not solve all problems. The critical current can be an important limiting factor. Thermally activated creep of fluxoids (due to $\boldsymbol{J} \times \boldsymbol{B}$) can lower J_c (the critical current) as the current interacts with the fluxoid and causes energy loss when the fluxoid becomes unpinned and thus creeps (can move). This is important in the high-T_c superconductors considered in this section.

Until 1986, the highest transition temperature for a superconductor was T_c = 23.2 K for Nb_3Ge. Then Bednortz and Müller found a *ceramic* oxide (product of clay) of lanthanum, barium, and copper became superconducting at about 35 K. For this work they won the Nobel prize for Physics in 1987. Since Bednortz's pioneering work several other high-T_c superconductors have been found.

The "1-2-3" compound $YBa_2Cu_3O_7$, has a T_c of 92 K. The "2-1-4" compound (e.g. $Ba_xLa_{2-x}CuO_{4-y}$) are another class of high-T_c superconductors. $Tl_2Ba_2Ca_2Cu_3O_{10}$ has a remarkably high T_c of 125 K.

The high-T_c materials are type II and typically have a penetration depth to coherence length ratio $K \approx 100$ and typically have a very large upper critical field. As we have mentioned, thermally activated creep of fluxoids due to the $\boldsymbol{J} \times \boldsymbol{B}$ force may cause energy dissipation and limit useable current values. For real materials, the critical current (J_c), critical temperature (T_c), and critical magnetic field (B_c) vary, but can be conveniently represented as shown in Fig. 8.22. As mentioned, the high-temperature superconductors (HTSs) are typically type II and also their J_c parallel to the copper oxide sheets (mentioned below) $\approx 10^7$ A/cm^2, while perpendicular to the sheets J_c can be about 10^7 A/cm^2. A schematic of J, B_c, and T_c is shown in Fig. 8.22 for type I materials. For HTS, the representation of Fig. 8.22 is not complex enough. In Table 8.1 we list selected superconductor elements and compounds along with their transition temperature.

For HTS, we are faced with a puzzle as to what causes some ceramic copper oxide materials to be superconductors at temperatures well above 100 K. In conventional superconductors, we talk about electrons paired into spherically symmetric wave functions (s-waves) due to exchange of virtual phonons. Apparently, lattice vibrations cannot produce a strong enough coupling to produce such high critical temperatures. It appears parallel Cu-O planes in these materials play some very significant but not yet fully understood role. Hole conduction in these planes is important. As mentioned, there is also a strong anisotropy in electrical conduction. Although there seems to be increasing evidence for d-wave pairing, the exchange mechanism necessary to produce the pair is still not clear as of this writing. It could be due to magnetic interactions or there may be new physics. See, e.g., Burns [8.9].

Table 8.1. Superconductors and their transition temperatures

Selected elements*	Transition temperature T_c (K)
Al	1.17
Hg	4.15
Nb	9.25
Sn	3.72
Pb	7.2
Selected compounds*	
Nb_3Ge	23.2
Nb_3Sn	18.
Nb_3Au	10.8
$NbSe_2$	7.2
MgB_2**	39
Copper oxide (HTS)*	
$Bi_2Sr_2Ca_2Cu_3O_{10}$	~110
$YBa_2Cu_3O_7$	~92
$Tl_2Ba_2Ca_3Cu_4O_{11}$	~122
Heavy fermion*	
UBe_{13}	0.85
$CeCu_2Si_2$	0.65
UPt_3	0.54
Fullerenes***	
K_3C_{60}	19.2
$RbCs_2C_{60}$	33

*Reprinted from Burns G, *High Temperature Superconductivity* Table 2-1 p. 8 and Table 3-1 p. 57, Academic Press, Copyright 1992, with permission from Elsevier. On p. 52 Burns also briefly discusses organic superconductors.
Canfield PC and Crabtree GW, *Physics Today* **56(3), 34 (2003).
***Huffmann DR, "Solid C60," *Physics Today* **41**(11), 22 (1991).

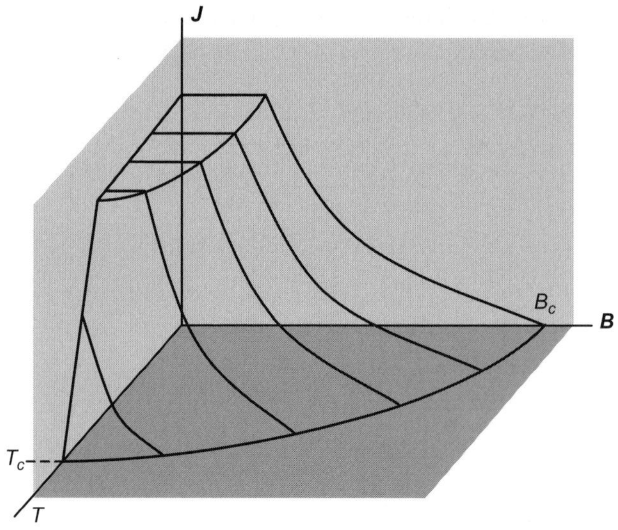

Fig. 8.22. ***J***, ***B***, *T* surface separating superconducting and normal regions

8.9 Summary Comments on Superconductivity (B)

1. In the superconducting state $\boldsymbol{E} = 0$ (superconductivity implies the resistivity ρ vanishes, $\rho \rightarrow 0$).
2. The superconducting state is more than vanishing resistivity since this would imply $\boldsymbol{B}$ was constant, whereas $B = 0$ in the superconducting state (flux is excluded as we drop below the transition temperature).
3. For "normal" BCS theory:

   a) An attractive interaction between electrons can lead to a ground state separated from the excited states by an energy gap. Most of the important properties of superconductors follow from this energy gap.

   b) The electron–lattice interaction, which can lead to an effective attractive interaction, causes the energy gap.

   c) The ideas of the penetration depth (and, hence, the Meissner effect—flux exclusion) and the coherence length follow from the theory of superconductivity.

4. Type II superconductors have upper and lower critical fields and are technically important because of their high upper critical fields. Magnetic flux can penetrate between the upper and lower critical fields, and the penetration is quantized in units of $hc/|2e|$, just as is the magnetic flux through a superconducting ring. Using a unit of charge of $2e$ is consistent with Cooper pairs.

5. In zero magnetic fields, for weak superconductors, superconductivity occurs at the transition temperature:

$$k_B T_c \cong 1.14\hbar\omega_D \exp(-1/N_0V_0)\,, \tag{8.220}$$

where N_0 is half the density of single-electron states, V_0 is the effective interaction between electron pairs near the Fermi surface, and $\hbar\omega_D \cong \hbar\theta_D$, where θ_D is the Debye temperature.

6. The energy gap (2Δ) is determined by (weak coupling):

$$\Delta(0) \cong 2\hbar\omega_D \exp(-1/N_0V_0) = 1.76kT_c \tag{8.221}$$

$$\Delta(T) \cong \Delta(0)\left(1-\frac{T}{T_c}\right)^{1/2} \;;\quad T \le T_c\,. \tag{8.222}$$

7. The critical field is fairly close to the empirical law (for weak coupling):

$$\frac{H_c(T)}{H_c(0)} \approx 1-\left(\frac{T}{T_c}\right)^2 . \tag{8.223}$$

8. The coherent motion of the electrons results in a resistanceless flow because a small perturbation cannot disturb one pair of electrons without disturbing all of them. Thus, even a small energy gap can inhibit scattering.

9. The central properties of superconductors are the penetration depth λ (of magnetic fields) and the coherence length ξ (or "size" of Cooper pairs). Small λ/ξ ratios lead to type I superconductors, and large λ/ξ ratios lead to type II behavior. ξ can be decreased by alloying.

10. The Ginzburg–Landau theory is used for superconductors in a magnetic field where one has inhomogeneities in spatial behavior.

11. We should also mention that one way to think about the superconducting transition is a Bose–Einstein condensation, as modified by their interaction, of bosonic Cooper pairs.

12. See the comment on spontaneously broken symmetry in the chapter on magnetism. Superconductivity can be viewed as a broken symmetry.

13. In the paired electrons of superconductivity, in s and d waves, the spins are antiparallel, and so one understands why ferromagnetism and superconductivity don't appear to coexist, at least normally. However, even p-wave superconductors (e.g. Strontium Ruthenate) with parallel spins the magnetic fields are commonly expelled in the superconducting state. Recently, however, two materials have been discovered in which ferromagnetism and superconductivity coexist. They are UGe_2 (under pressure) and $ZrZn_2$ (at ambient pressure). One idea is that these two materials are p-wave superconductors. The issues about these materials are far from settled, however. See *Physics Today*, p. 16, Sept. 2001.

14. Also, high-T_c (over 100 K) superconductors have been discovered and much work remains to understand them.

In Table 8.2 we give a subjective "Top Ten" of superconductivity research.

Table 8.2 Top 10 of superconductivity (subjective)

Person	Achievement	Date/comments
1. H. Kammerlingh Onnes	Liquefied He Found resistance of Hg $\rightarrow$ 0 at 4.19 K	1908 – Started low-T physics 1911 – Discovered superconducting state 1911 – Nobel Prize
2. W. Meissner and R. Ochsenfeld	Perfect diamagnetism	1933 – Flux exclusion
3. F. and H. London	London equations and flux expulsion	1935 – B proportional to curl of J
4. V.L. Ginzburg and L.D. Landau	Phenomenological equations	1950 – Eventually GLAG equations 1962 – Nobel Prize, Landau 2003 – Nobel Prize, Ginzburg
A. A. Abrikosov	Improvement to GL equations, Type II	1957 – Negative surface energy 2003 – Nobel Prize
L. P. Gor'kov	GL limit of BCS and order parameter	1959 – Order parameter proportional to gap parameter
5. A. B. Pippard	Nonlocal electrodynamics	1953 – x and l dependent on mean free path in alloys
6. J. Bardeen, L. Cooper, and J. Schrieffer	Theory of superconductivity	1957 – e.g. see (8.217) 1972 – Nobel Prize (all three)
7. I. Giaver	Single-particle tunneling	1960 – Get gap energy 1973 – Nobel Prize
8. B. D. Josephson	Pair tunneling	1962 – SQUIDS and metrology 1973 – Nobel Prize
9. Z. Fisk, et al	Heavy fermion "exotic" superconductors	1985 – Pairing different than BCS, probably
10. J.G. Bednorz and K. A. Muller	High-temperature superconductivity	1986 – Now, T_cs are over 100 K 1987 – Nobel Prize (both)

Problems

8.1 Show that the flux in a superconducting ring is quantized in units of h/q, where $q = |2e|$.

8.2 Derive an expression for the single-particle tunneling current between two superconductors separated by an insulator at absolute zero. If E^T is measured from the Fermi energy, you can calculate a density of states as below.

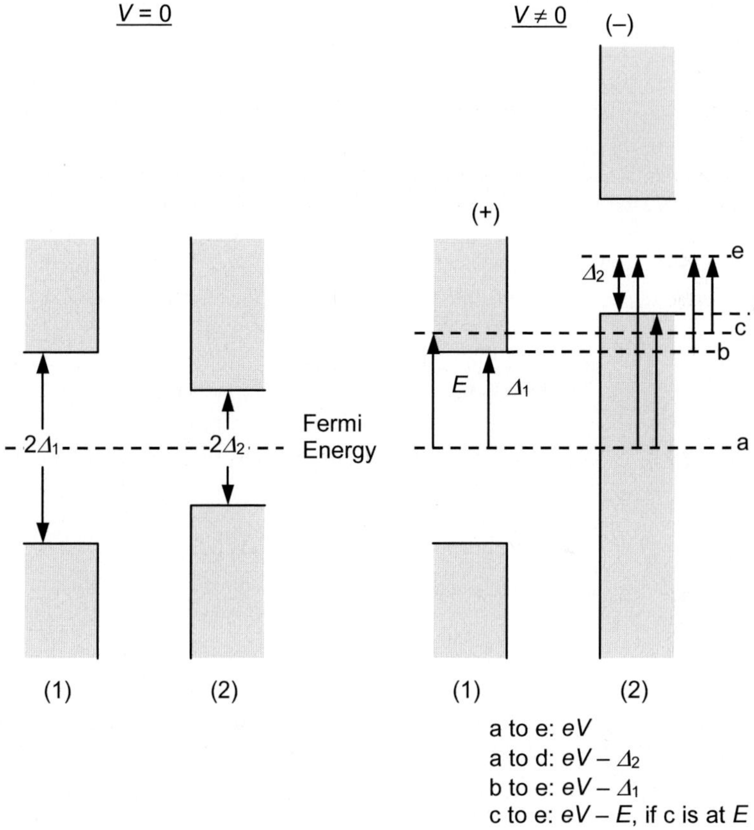

Note:

$$E^T = \sqrt{\varepsilon^2 + \Delta^2} \text{ (compare (8.93))},$$

$$D_S(E^T) = \frac{\mathrm{d}n(E^T)}{\mathrm{d}E^T} = \frac{\dfrac{\mathrm{d}n(E^T)}{\mathrm{d}\varepsilon}}{\dfrac{\mathrm{d}E^T}{\mathrm{d}\varepsilon}} \cong D(0)\frac{E^T}{\sqrt{(E^T)^2 - \Delta^2}},$$

where $D(0)$ is the number of states per unit energy without pairing.

9 Dielectrics and Ferroelectrics

Despite the fact that the concept of the dielectric constant is often taught in introductory physics – because, e.g., of its applications to capacitors – the concept involves subtle physics. The purpose of this chapter is to review the important dielectric properties of solids without glossing over the intrinsic difficulties.

Dielectric properties are important for insulators and semiconductors. When a dielectric insulator is placed in an external field, the field (if weak) induces a polarization that varies linearly with the field. The constant of proportionality determines the dielectric constant. Both static and time-varying external fields are of interest, and the dielectric constant may depend on the frequency of the external field. For typical dielectrics at optical frequencies, there is a simple relation between the index of refraction and the dielectric constant. Thus, there is a close relation between optical and dielectric properties. This will be discussed in more detail in the next chapter.

In some solids, below a critical temperature, the polarization may "freeze in." This is the phenomena of ferroelectricity, which we will also discuss in this chapter. In some ways ferroelectric and ferromagnetic behavior are analogous.

Dielectric behavior also relates to metals particularly by the idea of "dielectric screening" in a quasifree-electron gas. In metals, a generalized definition of the dielectric constant allows us to discuss important aspects of the many-body properties of conduction electrons. We will discuss this in some detail.

Thus, we wish to describe the ways that solids exhibit dielectric behavior. This has practical as well as intrinsic interest and is needed as a basis for the next chapter on optical properties.

9.1 The Four Types of Dielectric Behavior (B)

1. *The polarization of the electronic cloud around the atoms*: When an external electric field is applied, the electronic charge clouds are distorted. The resulting polarization is directly related to the dielectric constant. There are "anomalies" in the dielectric constant or refractive index at frequencies in which the atoms can absorb energies (resonance frequencies, or in the case of solids, interband frequencies). These often occur in the visible or ultraviolet. At lower frequencies, the dielectric constant is practically independent of frequency.

2. *The motion of the charged ions*: This effect is primarily of interest in ionic crystals in which the positive and negative ions can move with respect to one another

and thus polarize the crystal. In an ionic crystal, the resonant frequencies associated with the relative motion of the positive and negative ions are in the infrared and will be discussed in the optics chapter in connection with the restrahlen effect.

3. *The rotation of molecules with permanent dipole moments*: This is perhaps the easiest type of dielectric behavior to understand. In an electric field, the dipoles tend to line up with the electric fields, while thermal effects tend to oppose this alignment, and so, the phenomenon is temperature dependent. This type of dielectric behavior is mostly relevant for liquids and gases.

4. *The dielectric screening of a quasifree electron gas*: This is a many-body problem of a gas of electrons interacting via the Coulomb interaction. The technique of using the dielectric constant with frequency and wave-vector dependence will be discussed. This phenomena is of interest for metals.

Perhaps we should mention *electrets* here as a fifth type of dielectric behavior in which the polarization may remain, at least for a very long time after the removal of an electric field. In some ways an electret is analogous to a magnet. The behavior of electrets appears to be complex and as yet they have not found wide applications. Electrets occur in organic waxes due to frozen in disorder that is long lived but probably metastable.[1]

9.2 Electronic Polarization and the Dielectric Constant (B)

The ideas in this Section link up closely with optical properties of solids. In the chapter on the optical properties of solids, we will relate the complex index of refraction to the absorption and reflection of electromagnetic radiation. Now, we remind the reader of a simple picture, which relates the complex index of refraction to the dynamics of electron motion. We will include damping.

Our model considers matters only from a classical point of view. We limit discussion to electrons in bound states, but for some solids we may want to consider quasifree electrons or both bound and quasifree electrons. For electrons bound by Hooke's law forces, the equation describing their motion in an alternating electric field $E = E_0\exp(-\mathrm{i}\omega t)$ may be written ($e > 0$)

$$m\frac{\mathrm{d}^2x}{\mathrm{d}x^2} + \frac{m}{\tau}\frac{\mathrm{d}x}{\mathrm{d}t} + m\omega_0^2 x = -eE_0\exp(-\mathrm{i}\omega t)\,. \tag{9.1}$$

[1] See Gutmann [9.9]. See also Bauer et al [9.1].

The term containing τ is the damping term, which can be due to the emission of radiation or the other frictional processes. ω_0 is the natural oscillation frequency of the elastically bound electron of charge e and mass m. The steady-state solution is

$$x(t) = -\frac{e}{m}\frac{E_0 \exp(-\mathrm{i}\omega t)}{\omega_0^2 - \omega^2 - \mathrm{i}\omega/\tau}. \tag{9.2}$$

Below, we will assume that the field at the electronic site is the same as the average internal field. This completely neglects local field effects. However, we will follow this discussion with a discussion of local field effects, and in any case, much of the basic physics can be done without them. In effect, we are looking at atomic effects while excluding some interactions.

If N is the number of charges per unit volume, with the above assumptions, we write:

$$P = -Nex = \left(\frac{\varepsilon}{\varepsilon_0} - 1\right)\varepsilon_0 E = N\alpha E, \tag{9.3}$$

where ε is the dielectric constant and α is the polarizability. Using $E = E_0\exp(-\mathrm{i}\omega t)$,

$$\alpha = -\frac{ex}{E} = \frac{e^2}{m}\frac{1}{\omega_0^2 - \omega^2 - \mathrm{i}\omega/\tau}. \tag{9.4}$$

The complex dielectric constant is then given by

$$\frac{\varepsilon}{\varepsilon_0} = 1 + \frac{N}{\varepsilon_0}\frac{e^2}{m}\frac{1}{\omega_0^2 - \omega^2 - \mathrm{i}\omega/\tau} \equiv \varepsilon_r + \mathrm{i}\varepsilon_i, \tag{9.5}$$

where we have absorbed the ε_0 into ε_r and ε_i for convenience. The real and the imaginary parts of the dielectric constant are then given by:

$$\varepsilon_r = 1 + \frac{Ne^2}{m\varepsilon_0}\frac{\omega_0^2 - \omega^2}{(\omega_0^2 - \omega^2)^2 + \omega^2/\tau^2}, \tag{9.6}$$

$$\varepsilon_i = \frac{Ne^2}{m\varepsilon_0}\frac{\omega/\tau}{(\omega_0^2 - \omega^2)^2 + \omega^2/\tau^2}. \tag{9.7}$$

In the chapter on optical properties, we will note that the connection (10.8) between the complex refractive index and the complex dielectric constant is:

$$n_c^2 = (n + \mathrm{i}n_i)^2 = (\varepsilon_r + \mathrm{i}\varepsilon_i). \tag{9.8}$$

Therefore,

$$n^2 - n_i^2 = \varepsilon_r, \tag{9.9}$$

$$2nn_i = \varepsilon_i. \tag{9.10}$$

Thus, explicit equations for fundamental optical constants n and n_i are:

$$n^2 - n_i^2 = 1 + \frac{Ne^2}{m\varepsilon_0} \frac{\omega_0^2 - \omega^2}{(\omega_0^2 - \omega^2)^2 + \omega^2 / \tau^2} \tag{9.11}$$

$$2nn_i = \frac{Ne^2}{m\varepsilon_0} \frac{\omega / \tau}{(\omega_0^2 - \omega^2)^2 + \omega^2 / \tau^2}. \tag{9.12}$$

Quantum mechanics produces very similar equations. The results as given by Moss[2] are

$$n^2 - n_i^2 = 1 + \sum_j \frac{(Ne^2 f_{ij} / m\varepsilon_o)(\omega_{ij}^2 - \omega^2)}{(\omega_{ij}^2 - \omega^2) + \omega^2 / \tau_j^2}, \tag{9.13}$$

$$2nn_i = \sum_j \frac{(Ne^2 f_{ij} / m\varepsilon_o)\omega / \tau_j}{(\omega_{ij}^2 - \omega^2) + \omega^2 / \tau_j^2}, \tag{9.14}$$

where the f_{ij} are called oscillator strengths and are defined by

$$f_{ij} = 2\omega_{ij} \frac{m |\langle \psi_i | x | \psi_i \rangle|^2}{\hbar}, \tag{9.15}$$

where

$$\omega_{ij} = \frac{E_i - E_j}{\hbar}, \tag{9.16}$$

with E_i and E_j being the energies corresponding to the wave functions ψ_i and ψ_j. In a solid, because of the presence of neighboring dipoles, the local electric field does not equal the applied electric field.

Clearly, dielectric and optical properties are not easy to separate. Further discussion of optical-related dielectric properties comes in the next chapter.

We now want to examine some consequences of local fields. We also want to keep in mind that we will be talking about total dielectric constants and total polarizability. Thus in an ionic crystal, there are contributions to the polarizabilities and dielectric constants from both electronic and ionic motion.

The first question we must answer is, "If an external field, E, is applied to a crystal, what electric field acts on an atom in the crystal?" See Fig. 9.1. The slab is maintained between two plates that are connected to a battery of constant voltage V. Fringing fields are neglected. Thus, the electric field, E_0, between the plates before the slab is inserted, is the same as the electric field in the solid-state after insertion

---

[2] See Moss [9.13]. Note n_i refers to the imaginary part of the dielectric constant on the left of these equations and in f_{ij}, i refers to the initial state, while j refers to the final state.

(so, $E_0d = V$). This is also the same as the electric field in a needle-shaped cavity in the slab. The electric field acting on the atom is

$$E_{\text{loc}} = E_0' + E_a + E_b + E_c\,, \tag{9.17}$$

where, E'_0 is the electric field due to charge on the plates after the slab is inserted, E_a is the electric field due to the polarization charges on the faces of the slab, and E_b is the electric field due to polarization charges on the surface of the spherical cavity (which exists in our imagination), and E_c is the polarization due to charges interior to the cavity that we assume (in total) sums to zero.

It is, of course, an approximation to write E_{loc} in the above form. Strictly speaking, to find the field at any particular atom, we should sum over the contributions to this field from all other atoms. Since this is an impossible task, we treat macroscopically all atoms that are sufficiently far from A (and outside the cavity).

By Gauss' law, we know the electric field due to two plates with a uniform charge density ($\pm\sigma$) is $E = \sigma/\varepsilon$. Further, σ due to P ending on the boundary of a slab is $\sigma = P$ (from electrostatics). Since the polarization charges on the surface of the slabs will oppose the electric field of the plate and since charge will flow to maintain constant voltage.

$$\varepsilon_0 E_0 = \varepsilon_0 E_0' - P\,, \tag{9.18}$$

or

$$E_0' = E_0 + \frac{P}{\varepsilon_0}\,. \tag{9.19}$$

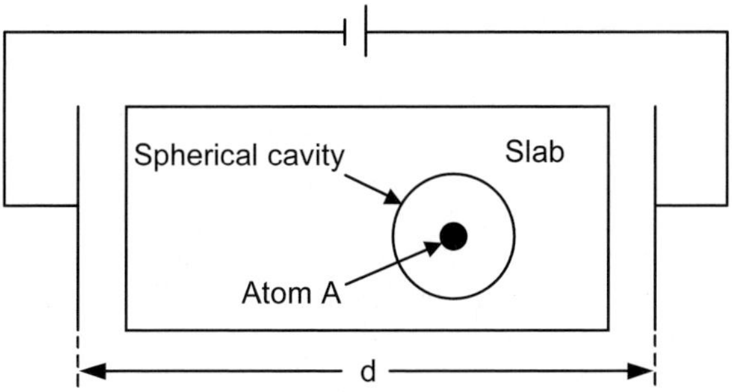

Fig. 9.1. Geometry for local field

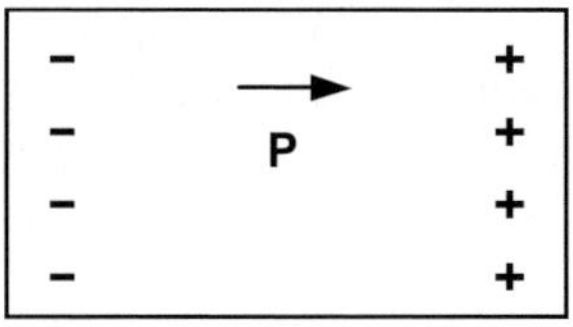

Fig. 9.2. The polarized slab

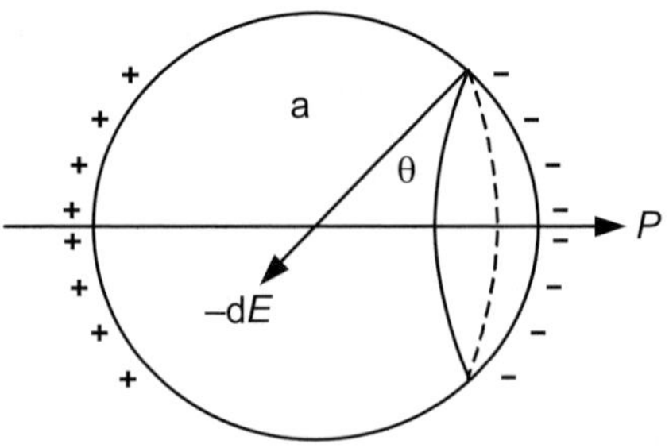

Fig. 9.3. Polarized charges around the cavity

Clearly, $E_a = -P/\varepsilon_0$ (see Fig. 9.2), and for all cubic crystals, $E_c = 0$. So,

$$E_{\text{loc}} = E_0 + E_b \,. \tag{9.20}$$

Using Fig. 9.3, since $\sigma_\rho = \boldsymbol{P}\cdot\boldsymbol{n}$ ($\boldsymbol{n}$ is outward normal), the charge on an annular region of the surface of the cavity is

$$\mathrm{d}q = -P\cos\theta \cdot 2\pi a \sin\theta \cdot a\mathrm{d}\theta \,, \tag{9.21}$$

$$\mathrm{d}E_b = \frac{1}{4\pi\varepsilon_0}\frac{\mathrm{d}q}{a^2}\cdot\cos\theta \,, \tag{9.22}$$

$$E_b = -\frac{P}{2\varepsilon_0}\int_0^\pi \cos^2\theta \cdot \mathrm{d}\cos\theta \,. \tag{9.23}$$

Thus $E_b = P/3\varepsilon_0$, and so we find

$$E_{\text{loc}} = E_0 + \frac{P}{3\varepsilon_0} \,. \tag{9.24}$$

Since E_0 is also the average electric field in the solid, the dielectric constant is defined as

$$\varepsilon = \frac{D}{E_0} = \frac{\varepsilon_0 E_0 + P}{E_0} = \varepsilon_0 + \frac{P}{E_0} \,. \tag{9.25}$$

The polarization is the dipole moment per unit volume, and so, it is given by

$$P = \sum\nolimits_{i(\text{atoms})} E_{\text{loc}}^i N_i \alpha_i \,, \tag{9.26}$$

where N_i is the number of atoms per unit volume of type i, and α_i is the appropriate polarizability (which can include ionic, as well as electronic motions). Thus,

$$P = \left(E_0 + \frac{P}{3\varepsilon_0}\right)\sum\nolimits_i N_i \alpha_i \,, \tag{9.27}$$

or

$$\frac{P}{E_0} = \frac{\sum_i N_i \alpha_i}{1 - \frac{1}{3\varepsilon_0}\sum_i N_i \alpha_i}, \tag{9.28}$$

or

$$\frac{\varepsilon}{\varepsilon_0} = 1 + \frac{1}{\varepsilon_0}\frac{\sum_i N_i \alpha_i}{\left(1 - \frac{1}{3\varepsilon_0}\sum_i N_i \alpha_i\right)}, \tag{9.29}$$

which can be arranged to give the Clausius–Mossotti equation

$$\frac{(\varepsilon/\varepsilon_0) - 1}{(\varepsilon/\varepsilon_0) + 2} = \frac{1}{3\varepsilon_0}\sum_i N_i \alpha_i . \tag{9.30}$$

In the optical range of frequencies (the order of but less than 10^{15} cps), $n^2 = \varepsilon/\varepsilon_0$, and the equation becomes the Lorentz–Lorenz equation

$$\frac{n^2 - 1}{n^2 + 2} = \frac{1}{3\varepsilon_0}\sum_i N_i \alpha_i . \tag{9.31}$$

Finally, we show that when one resonant peak dominates, the only effect of the local field is to shift the dormant resonant (natural) frequency. From

$$\frac{\varepsilon}{\varepsilon_0} = 1 + \frac{1}{\varepsilon_0}\frac{N\alpha}{(1 - N\alpha/3\varepsilon_0)}, \tag{9.32}$$

and

$$\alpha = \frac{e^2}{m}\frac{1}{\omega_0^2 - \omega^2 - \mathrm{i}\omega/\tau}, \tag{9.33}$$

we have

$$\frac{(\varepsilon/\varepsilon_0) - 1}{(\varepsilon/\varepsilon_0) + 2} = \frac{N\alpha}{3\varepsilon_0} = \frac{\omega_p^2}{3}\frac{1}{\omega_o^2 - \omega^2 - \mathrm{i}\omega/\tau}, \tag{9.34}$$

where

$$\omega_p = \sqrt{Ne^2/m\varepsilon_0} \tag{9.35}$$

is the plasma frequency. From this, we easily show

$$\frac{\varepsilon}{\varepsilon_0} = 1 + \frac{\omega_p^2}{\omega_0'^2 - \omega^2 - \mathrm{i}\omega/\tau}, \tag{9.36}$$

where

$$\omega_0'^2 = \omega_0^2 - \frac{1}{3}\omega_p^2 , \tag{9.37}$$

which is exactly what we would have obtained in the beginning (from (9.32) and (9.33)) if $\omega_0 \to \omega'_0$, and if the term $N\alpha/3\varepsilon_0$ had been neglected.

9.3 Ferroelectric Crystals (B)

All ferroelectric crystals are polar crystals.[3] Because of their structure, polar crystals have a permanent electric dipole moment. If $\rho(\boldsymbol{r})$ is the total charge density, we know for polar crystals

$$\int \mathbf{r}\rho(\mathbf{r})\mathrm{d}V \neq 0 . \tag{9.38}$$

Pyroelectric crystals have a polarization that changes with temperature. All polar crystals are pyroelectric, but not all polar crystals are ferroelectric. Ferroelectric crystals are polar crystals whose polarization can be reversed by an electric field. All ferroelectric crystals are also piezoelectric, in which stress changes the polarization. Piezoelectric crystals are suited for making electromechanical transducers with a variety of applications.

Ferroelectric crystals often have unusual properties. Rochelle salt $NaKC_4H_4O_6 \cdot 4H_2O$, which was the first ferroelectric crystal discovered, has both an upper and lower transition temperature. The crystal is only polarized between the two transition temperatures. The "TGS" type of ferroelectric, including triglycine sulfate and triglycine selenate, is another common class of ferroelectrics and has found application to IR detectors due to its pyroelectric properties. Ferroelectric crystals with hydrogen bonds (e.g. KH_2PO_4, which was the second ferroelectric crystal discovered) undergo an appreciable change in transition temperature when the crystal is deuterated (with deuterons replacing the H nuclei). $BaTiO_3$ was the first mechanically hard ferroelectric crystal that was discovered. Ferroelectric crystals are often classified as *displacive*, involving a lattice distortion (i.e. barium titinate, $BaTiO_3$, see Fig. 9.4), or *order–disorder* (i.e. potassium dihydrogen phosphate, KH_2PO_4, which involves the ordering of protons).

[3] Ferroelectrics: The term ferro is used but iron has nothing to do with it. Low symmetry causes spontaneous polarization.

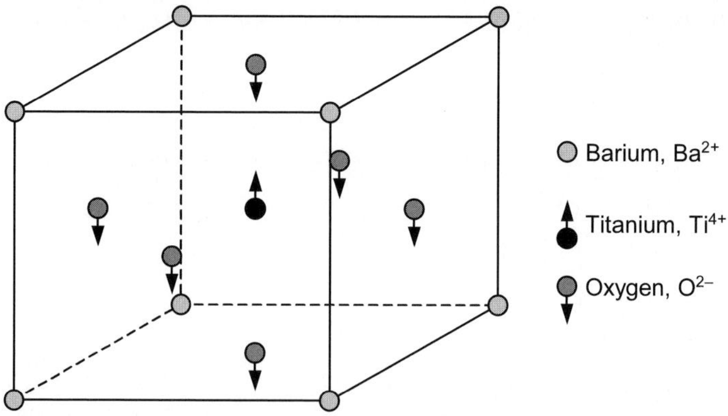

Fig. 9.4. Unit cell of barium titanate. The displacive transition is indicated by the direction of the arrows

In a little more detail, displacive ferroelectrics involve transitions associated with the displacement of a whole sublattice. How this could arises is discussed in Sect. 9.3.3 where we talk about the soft mode model. The soft mode theory, introduced in 1960, has turned out to be a unifying principle in ferroelectricity (see Lines and Glass [9.12]). Order–disorder ferroelectrics have transitions associated with the ordering of ions. We have mentioned in this regard KH_2PO_4 as a crystal with hydrogen bonds in which the motion of protons is important. Ferroelectrics have found application as memories, their high dielectric constant is exploited in making capacitors, and ferroelectric cooling is another area of application.

Other examples include ferroelectric cubic perovskite (PZT) $PbZr_{(x)}Ti_{(1-x)}O_3$, $T_c = 670$ K. The ferroelectric $BaMgF_4$ (BMF) does not show a Curie T even up to melting. These are other familiar ferroelectrics as given below.

The central problem of ferroelectricity is to be able to describe the onset of spontaneous polarization. Spontaneous polarization is said to exist if, in the absence of an electric field, the free energy is minimum for a finite value of the polarization. There may be some ordering involved in a ferroelectric transition, as in a ferromagnetic transition, but the two differ by the fact that the ferroelectric transition in a solid always involves the creation of dipoles.

Just as for ferromagnets, a ferroelectric crystal undergoes a phase transition from the paraelectric phase to the ferroelectric phase, typically, as the temperature is lowered. The transition can be either first order (with a latent heat, i.e. $BaTiO_3$) or second order (without latent heat, i.e. $LiTaO_3$). Just as for ferromagnets, the ferroelectric will typically split into domains of varying size and orientation of polarization. The domain structure forms to reduce the energy. Ferroelectrics show hysteresis effects just like ferromagnets. Although we will not discuss it here, it is also possible to have antiferroelectrics that one can think of as arising from antiparallel orientation of neighboring unit cells. A simple model of spontaneous polarization is obtained if we use the Clausius–Mossotti equation and assume (unrealistically for solids) that polarization arises from orientation effects. This is discussed briefly in a later section.

9.3.1 Thermodynamics of Ferroelectricity by Landau Theory (B)

For both first-order ($\gamma < 0$, latent heat, G continuous) and second-order ($\gamma > 0$, no latent heat, G' (first derivatives) are continuous and we can choose $\delta = 0$), we assume for the Gibbs free energy G' [9.6 Chap. 3, generally assumed for displacive transitions],

$$G = G_0 + \frac{1}{2}\beta(T - T_0)P^2 + \frac{1}{4}\gamma P^4 + \frac{1}{6}\delta P^6 \ ; \ \beta, \delta > 0 \, . \tag{9.39}$$

(By symmetry, only even powers are possible. Also, in a second-order transition, P is continuous at the transition temperature T_c, whereas in a first-order one it is not.) From this we can calculate

$$E = \frac{\partial G}{\partial P} = \beta(T - T_0)P + \gamma P^3 + \delta P^5 \, , \tag{9.40}$$

$$\frac{1}{\chi} = \frac{\partial E}{\partial P} = \beta(T - T_0) + 3\gamma P^2 + 5\delta P^4 \, . \tag{9.41}$$

Notice in the paraelectric phase, $P = 0$ so $E = 0$ and $\chi = 1/\beta(T - T_0)$, and therefore Curie–Weiss behavior is included in (9.39). For $T < T_c$ and $E = 0$ for second order where $\delta = 0$, $\beta(T - T_0)P + \gamma P^3 = 0$, so

$$P^2 = -\frac{\beta}{\gamma}(T - T_0) \, , \tag{9.42}$$

or

$$P = \pm\sqrt{\frac{\beta}{\gamma}(T_0 - T)} \, , \tag{9.43}$$

which again is Curie–Weiss behavior (we assume $\gamma > 0$). For $T = T_c = T_0$, we can show the stable solution is the polarized one.

For first order set $E = 0$, solve for P and exclude the solution for which the free energy is a maximum. We find (where we assume $\gamma < 0$)

$$P_S = \pm\left[-\frac{\gamma}{2\delta}\left(1 + \sqrt{1 - \frac{4\delta\beta}{\gamma^2}(T - T_0)}\right)\right]^{1/2} .$$

Now, $G(P_{SC}) = G_{\text{polar}} = G_{\text{nonpolar}} = G_0$ at the transition temperature. Using the expression for G (9.39) and the expression that results from setting $E = 0$ (9.40), we find

$$T_c = T_0 - \frac{\gamma}{4\beta} P_{SC}^2 \, . \tag{9.44}$$

By $E = 0$, we find (using (9.44))

$$\frac{3\gamma}{4} P_{\mathrm{SC}}^3 + \delta P_{\mathrm{SC}}^5 = 0\,, \tag{9.45}$$

so

$$P_{\mathrm{SC}}^2 = -\frac{3\gamma}{4\delta}\,. \tag{9.46}$$

Putting (9.46) into (9.44) gives

$$T_c = T_0 + \frac{3\gamma^2}{16\beta\delta}\,. \tag{9.47}$$

Figures 9.5, 9.6, and 9.7 give further insight into first- and second-order transitions.

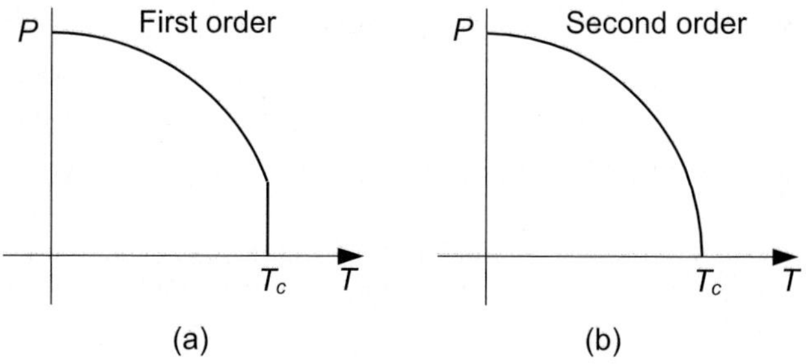

Fig. 9.5. Sketch of (**a**) first-order and (**b**) second-order ferroelectric transitions

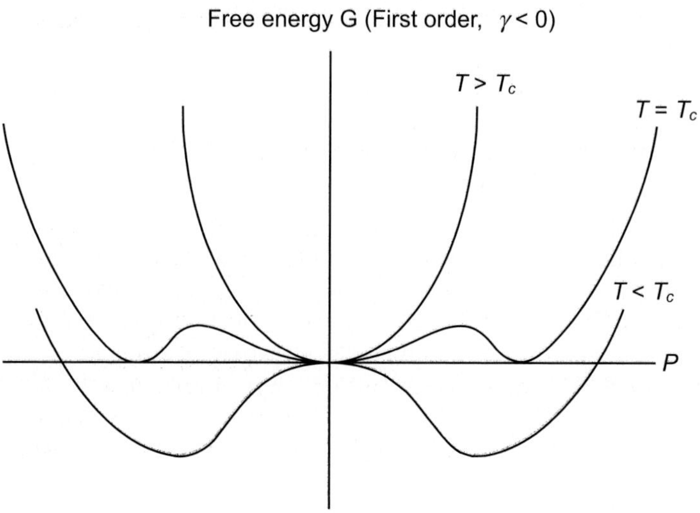

Fig. 9.6. Sketch of variation of Gibbs free energy $G(T, p)$ for first-order transitions

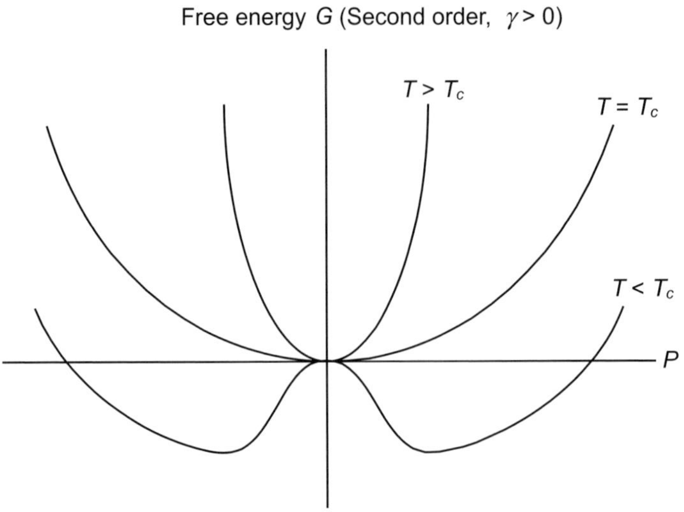

Fig. 9.7. Sketch of variations of Gibbs free energy $G(T,p)$ for second-order transitions

9.3.2 Further Comment on the Ferroelectric Transition (B, ME)

Suppose we have N permanent, noninteracting dipoles P per unit volume, at temperature T, in an electric field E. At high temperature, simple statistical mechanics shows that the polarizability per molecule is

$$\alpha = \frac{p^2}{3kT} . \tag{9.48}$$

Combining this with the Clausius–Mossotti equation (9.29) gives

$$\frac{\varepsilon}{\varepsilon_0} = 1 + \frac{Np^2}{3k\varepsilon_0 (T - T_c)} . \tag{9.49}$$

As $T \to T_c$, we obtain the "polarization catastrophe". For a real crystal, even if this were a reasonable approach, the equation would break down well before $T = T_c$, and at $T = T_c$, we would assume that permanent polarization had set in. Near $T = T_c$, the 1 is negligible, and we have essentially a Curie–Weiss type of behavior. However, this derivation should not be taken too seriously, even though the result is reasonable.

Another way of viewing the ferroelectric transition is by the Lyddane–Sachs–Teller (LST) relation. This is developed in the next chapter, see (10.204). Here an infinite dielectric constant implies a zero-frequency optical mode. This leads to Cochran's theory of ferroelectricity arising from "soft" optic modes. The LST relation can be written

$$\frac{\omega_T^2}{\omega_L^2} = \frac{\varepsilon(\infty)}{\varepsilon(0)}, \tag{9.50}$$

where ω_T is the transverse optical frequency, ω_L is the longitudinal optical frequency (both at low wave vector), $\varepsilon(\infty)$ is the high-frequency limit of the dielectric constant and $\varepsilon(0)$ is the low-frequency (static) limit. Thus a Curie–Weiss behavior for $\varepsilon(0)$ as

$$\frac{1}{\varepsilon(0)} \propto (T - T_c) \tag{9.51}$$

is consistent with

$$\omega_T^2 \propto (T - T_c). \tag{9.52}$$

Table 9.1. Selected ferroelectric crystals

Type	Crystal	T_c (K)
KDP	KH_2PO_4	123
TGS	Triglycine sulfate	322
Perovskites	$BaTiO_3$	406
	$PbTiO_3$	765
	$LiNbO_3$	1483

From Anderson HL (ed), *A Physicists Desk Reference* 2nd edn, American Institute of Physics, Article 20: Frederikse HPR, p.314, Table 20.02.C.1., 1989, with permission of Springer-Verlag. Original data from Kittel C, *Introduction to Solid State Physics*, 4th edn, p.476, Wiley, NY, 1971.

Cochran has pioneered the approach to a microscopic theory of the onset of spontaneous polarization by the soft mode or "freezing out" (frequency going to zero) of an optic mode of zero wave vector. The vanishing frequency appears to result from a canceling of short-range and long-range (Coulomb) forces between ions. Not all ferroelectric transitions are easily associated with phonon modes. For example, the order–disorder transition is associated with the ordering of protons in potential wells with double minima above the transition. Transition temperatures for some typical ferroelectrics are given in Table 9.1.

9.3.3 One-Dimensional Model of the Soft Mode of Ferroelectric Transitions (A)

In order to get a better picture of what the soft mode theory involves, we present a one-dimensional model below that is designed to show ferroelectric behavior.

Anderson and Cochran have suggested that the phase transition in certain ferroelectrics results from an instability of one of the normal vibrational modes of the lattice. Suppose that at some temperature T_c

a) An infinite-wavelength optical mode is accompanied by the condition that the vibrational frequency ω for that mode is zero.

b) The effective restoring force for this mode for the ion displacements equals zero. This condition has prompted the terminology, "soft" mode ferroelectrics.

If these conditions are satisfied, it is seen that the static ion displacements would give rise to a "frozen-in" electric dipole moment–that is, spontaneous polarization. The idea is shown in Fig. 9.8.

We now consider a one-dimensional lattice consisting of two atoms per unit cell, see Fig. 9.9. The atoms (ions) have, respectively, mass m_1 and m_2 with charge $e_1 = e$ and $e_2 = -e$. The equilibrium separation distance between atoms is the distance $a/2$.

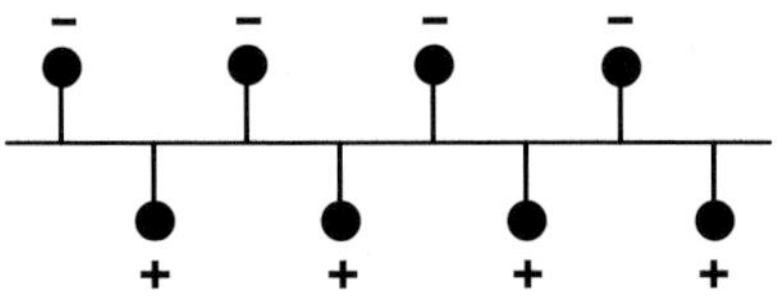

Fig. 9.8. Schematic for ferroelectric mode in one dimension

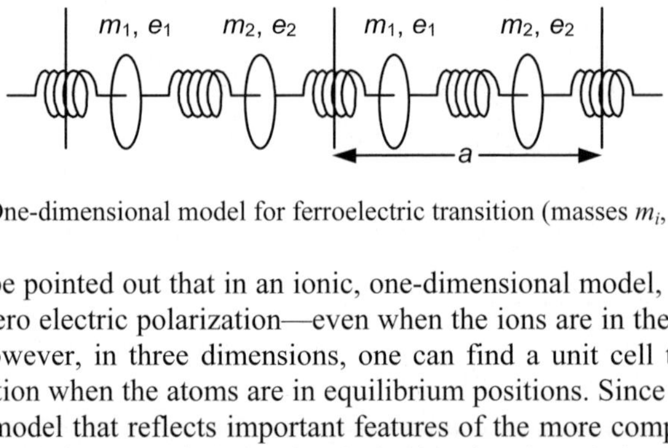

Fig. 9.9. One-dimensional model for ferroelectric transition (masses m_i, charges e_i)

It should be pointed out that in an ionic, one-dimensional model, a unit cell exhibits a nonzero electric polarization—even when the ions are in their equilibrium positions. However, in three dimensions, one can find a unit cell that possesses zero polarization when the atoms are in equilibrium positions. Since our interest is to present a model that reflects important features of the more complicated three-dimensional model, we are interested only in the electric polarization that arises because of displacements away from equilibrium positions. We could propose for the one-dimensional model the existence of fixed charges that will cancel the equilibrium position polarization but that have no other effect. At any rate, we will disregard equilibrium position polarization.

We define x_{kb} as the displacement from its equilibrium position of the bth atom ($b = 1, 2$) in the kth unit cell. For N atoms, we assume that the displacements of the atoms from equilibrium give rise to a polarization, P, where

$$P = \frac{1}{N} \sum_{k',b'} x_{k'b'} e_b \,. \tag{9.53}$$

The equation of motion of the bth atom in the kth unit cell can be written

$$m_b \ddot{x}_{kb} + \sum_{k',b'} J_{bb'}(k - k') x_{k'b'} = c e_b P \,, \tag{9.54}$$

where

$$J_{bb'}(k - k') = \frac{\partial^2 V}{\partial x_{kb} \partial x_{k'b'}} \,. \tag{9.55}$$

This equation is, of course, Newton's second law, $F = ma$, applied to a particular ion. The second term on the left-hand side represents a "spring-like" interaction obtained from a power series expansion to the second order of the potential energy, V, of the crystal. The right-hand side represents a long-range electrical force represented by a local electric field that is proportional to the local electric field $E_{\text{loc}} = cP$, where c is a constant.

As a further approximation, we assume the spring-like interactions are nearest neighbors, so

$$V = \frac{\gamma}{2} \sum_{k''} (x_{k''2} - x_{k''1})^2 + \frac{\gamma}{2} \sum_{k''} (x_{k''+1,1} - x_{k''2})^2 \,, \tag{9.56}$$

where γ is the spring constant. By direct calculation, we find for the $J_{bb'}$

$$\begin{aligned} J_{11}(k' - k) &= 2\gamma \delta_k^{k'} = J_{22}(k' - k) \,, \\ J_{12}(k' - k) &= -\gamma(\delta_k^{k'} + \delta_k^{k'+1}) \,, \\ J_{21}(k' - k) &= -\gamma(\delta_k^{k'} + \delta_k^{k'-1}) \,. \end{aligned} \tag{9.57}$$

We rewrite our dynamical equation in terms of $h = k' - k$

$$m_b \ddot{x}_{kb} + \sum_{h,b'} J_{bb'}(h) x_{h+k,b'} = \frac{c e_b}{N} \sum_{h,b'} x_{h+k,b'} e_{b'} \,. \tag{9.58}$$

Since this equation is translationally invariant, it has solutions that satisfy Bloch's theorem. Thus, there exists a wave vector k such that

$$x_{kb} = \exp(\mathrm{i}kqa) x_{ob} \,, \tag{9.59}$$

where x_{ob} is the displacement of the bth atom in the cell chosen as the origin for the lattice vectors. Substituting, we find

$$m_b \ddot{x}_{kb} + \sum_{h,b'} J_{bb'}(k) \exp(\mathrm{i}hqa) x_{ob'} = \frac{c e_b}{N} \sum_{h,b'} \exp(\mathrm{i}hqa) x_{ob'} e_{b'} \,. \tag{9.60}$$

We simplify by defining

$$G_{bb'}(q) = \sum_h J_{bb'}(h) \exp(ihqa) \,. \tag{9.61}$$

Using the results for $J_{bb'}$, we find

$$\begin{aligned} G_{11} &= 2\gamma = G_{22} \,, \\ G_{12} &= -\gamma[1+\exp(iqa)] \,, \\ G_{21} &= -\gamma[1-\exp(-iqa)] \,. \end{aligned} \tag{9.62}$$

In addition, since

$$\sum_h \exp(ihqa) = N\delta_q^0 \,, \tag{9.63}$$

we finally obtain,

$$m_b \ddot{x}_{ob} + \sum_{b'} G_{bb'}(q) x_{ob'} = ce_b \sum_{b'} \delta_q^0 x_{ob'} e_{b'} \,. \tag{9.64}$$

As in the ordinary theory of vibrations, we assume x_{ob} contains a time factor $\exp(i\omega t)$, so

$$\ddot{x}_{ob} = -\omega^2 x_{ob} \,. \tag{9.65}$$

The polarization term only affects the $q \to 0$ solution, which we look at now. Letting $q = 0$, and $e_1 = -e_2 = e$, we obtain the following two equations:

$$-m_1\omega^2 x_{o1} + 2\gamma x_{o1} - 2\gamma x_{o2} = ce(x_{o1}e - x_{o2}e) \,, \tag{9.66}$$

and

$$-m_2\omega^2 x_{o2} - 2\gamma x_{o1} + 2\gamma x_{o2} = -ce(x_{o1}e - x_{o2}e) \,. \tag{9.67}$$

These two equations can be written in matrix form:

$$\begin{bmatrix} -m_1\omega^2 + d & -d \\ -d & -m_2\omega^2 + d \end{bmatrix} \begin{bmatrix} x_{o1} \\ x_{o2} \end{bmatrix} = 0 \,, \tag{9.68}$$

where $d = 2\gamma - ce^2$. From the secular equation, we obtain the following:

$$\omega^2[m_1 m_2 \omega^2 - (m_1 + m_2)d] = 0 \,. \tag{9.69}$$

The solution $\omega = 0$ is the long-wavelength acoustic mode frequency. The other solution, $\omega^2 = d/\mu$ with $1/\mu = 1/m_1 + 1/m_2$, is the optic mode long-wavelength frequency. For this frequency

$$-m_1 x_{o1} = m_2 x_{o2} \,. \tag{9.70}$$

So,

$$P = x_{o1}e\left(1 + \frac{m_1}{m_2}\right), \tag{9.71}$$

and $P \neq 0$ if $x_{o1} \neq 0$. Suppose

$$\lim_{T \to T_c} [2\gamma(T) - ce^2] = 0, \tag{9.72}$$

then

$$\omega^2 = \frac{d}{\mu} \to 0 \quad \text{at} \quad T = T_c, \tag{9.73}$$

and

$$F_1 = m_1 \ddot{x}_{o1} = d(x_{o1} + x_{o2}) \to 0 \text{ as } T \to T_c. \tag{9.74}$$

So, a solution is x_{o1} = constant $\neq 0$. That is, the model shows a ferroelectric solution for $T \to T_c$.

9.4 Dielectric Screening and Plasma Oscillations (B)

We begin now to discuss more complex issues. We want to discuss the nature of a gas of interacting electrons. This topic is closely related to the occurrence of oscillations in gas-discharge plasmas and is linked to earlier work of Langmuir and Tonks.[4] We begin by considering the subject of plasma oscillations. The general idea can be presented from a classical viewpoint, so we start by assuming the simultaneous validity of Newton's laws and Maxwell's equations.

Let n_0 be the number density of electrons in equilibrium. We assume an equal distribution of positive charge that remains uniform and, thus, supplies a constant background. We will consider one dimension only and, thus, consider only longitudinal plasma oscillations.

[4] See Tonks and Langmuir [9.19].

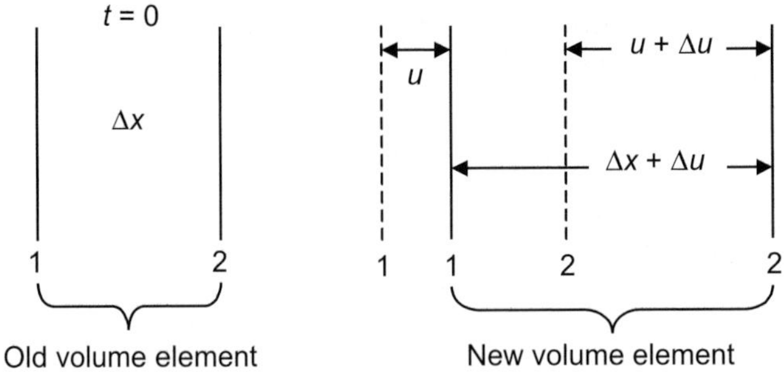

Fig. 9.10. Schematic used to discuss plasma vibration

Let $u(x, t)$ represent the displacement of electrons whose equilibrium position is x and refer to Fig. 9.10 to compute the change in density Let e represent the magnitude of the electronic charge. Since the positive charge remains at rest, the total charge density is given by $\rho = -(n - n_0)e$. Since the same number of electrons is contained in the new volume as the old volume.

$$n = \frac{n_0 \Delta x}{(\Delta x + \Delta u)} \cong n_0 \left(1 - \frac{\mathrm{d}u}{\mathrm{d}x}\right). \tag{9.75}$$

Thus,

$$\rho = n_0 e \frac{\mathrm{d}u}{\mathrm{d}x}. \tag{9.76}$$

In one-dimension, Gauss' law is

$$\frac{\mathrm{d}E_x}{\mathrm{d}x} = \frac{\rho}{\varepsilon_0} = \frac{n_0 e}{\varepsilon_0} \frac{\mathrm{d}u}{\mathrm{d}x}. \tag{9.77}$$

Integrating and using the boundary condition that $(E_x)_{n=0} = 0$, we have

$$E_x = \frac{n_0 e}{\varepsilon_0} u. \tag{9.78}$$

A simpler derivation is discussed in the optics chapter (see Sect. 10.9). Using Newton's second law with force $-eE_x$, we have

$$m \frac{\mathrm{d}^2 u}{\mathrm{d}t^2} = -\frac{n_0 e^2}{\varepsilon_0} u, \tag{9.79}$$

with solution

$$u = u_0 \cos(\omega_p t + \mathrm{const.}), \tag{9.80}$$

where

$$\omega_p = \sqrt{n_0 e^2 / m\varepsilon_0} \tag{9.81}$$

is the plasma frequency of electron oscillation. The quanta associated with this type of excitation are called plasmons. For a typical gas in a discharge tube, $\omega_P \cong 10^{10}\ \mathrm{s}^{-1}$, while for a typical metal, $\omega_P \cong 10^{16}\ \mathrm{s}^{-1}$.

More detailed discussions of plasma effects and electrons can be made by using frequency- and wave-vector-dependent dielectric constants. See Sect. 9.5.3 for further details where we will discuss screening in some detail. We define $\varepsilon(q, \omega)$ as the proportionality constant between the space and time Fourier transform components of the electric field and electric displacement vectors. We generally assume $\varepsilon(\omega) = \varepsilon(q = 0, \omega)$ provides an adequate description of dielectric properties when $q^{-1} >> a$, where a is the lattice spacing. It is necessary to use $\varepsilon(q, \omega)$ when spatial variations not too much larger than the lattice constant are important.

The basic idea is contained in (9.82) and (9.83). For electrical interactions, if the actual perturbation of the potential is of the form

$$V' = \iint v'(q,\omega)\exp(\mathrm{i}\boldsymbol{q}\cdot\boldsymbol{r})\exp(\mathrm{i}\omega t)\mathrm{d}q\cdot\mathrm{d}\omega\,. \tag{9.82}$$

Then, the perturbation of the energy is given by

$$\varepsilon' = \iint \frac{v'(q,\omega)}{\varepsilon(q,\omega)}\exp(\mathrm{i}\boldsymbol{q}\cdot\boldsymbol{r})\exp(\mathrm{i}\omega t)\mathrm{d}q\cdot\mathrm{d}\omega\,. \tag{9.83}$$

$\varepsilon(q,\omega)$ is used to discuss (a) plasmons, (b) the ground-state energy of a many-electron system, (c) screening and Friedel oscillation in charge around a charged impurity in a sea of electrons, (d) the Kohn effect (a singularity in the dielectric constant that implies a change in phonon frequency.), and (e) even other elementary energy excitations, provided enough physics is included in $\varepsilon(q,\omega)$. Some of this is elaborated in Sect. 9.5.

We now discuss two kinds of waves that can occur in plasmas. The first kind concerns waves that propagate in a region with only one type of charge carrier, and in the second we consider both signs of charge carrier. In both cases we assume overall charge neutrality. Both cases deal with electromagnetic waves propagating in a charged media in the direction of a constant magnetic field. Both cases only relate somewhat indirectly to dielectric properties through the Coulomb interaction. They seem to be worth discussing as an aside.

9.4.1 Helicons (EE)

Here we consider electrons as the charge carriers. The helicons are low-frequency (much lower than the cyclotron frequency) waves of circularly polarized electromagnetic radiation that propagate, with little attenuation, along the direction of the external magnetic field. They have been observed in sodium at high field (~ 2.5 T) and low temperatures (~ 4 K). The existence of these waves was predicted by

P. Aigrain in 1960. Since their frequency depends on the Hall coefficient, they have been used to measure it in solids. Their dispersion relation shows that lower frequencies have lower velocities. When high-frequency helicons are observed in the ionosphere, they are called whistlers (because of the way their signal sounds when converted to audio).

For electrons (charge $-e$) in $\boldsymbol{E}$ and $\boldsymbol{B}$ fields with drift velocity $\boldsymbol{v}$, relaxation time τ, and effective mass m, we have

$$m\left(\frac{\mathrm{d}}{\mathrm{d}t}+\frac{1}{\tau}\right)\boldsymbol{v}=-e(\boldsymbol{E}+\boldsymbol{v}\times\boldsymbol{B}). \tag{9.84}$$

Assuming $\boldsymbol{B} = B\hat{\boldsymbol{k}}$ and low frequencies so $\omega\tau << 1$, we can neglect the time derivatives and so

$$\begin{aligned} v_x &= -\frac{e\tau E_x}{m} - \omega_c \tau v_y, \\ v_y &= -\frac{e\tau E_y}{m} + \omega_c \tau v_x, \\ v_z &= -\frac{e\tau E_z}{m}, \end{aligned} \tag{9.85}$$

where $\omega_c = eB/m$ is the cyclotron frequency. Letting, $\sigma_0 = m/ne^2\tau$, where n is the number of charges per unit volume, and the Hall coefficient $R_H = -1/ne$, we can write (noting $\boldsymbol{j} = -ne\boldsymbol{v}, \boldsymbol{j} = \boldsymbol{v}/R_H$):

$$v_x = \sigma_0 R_H (E_x + Bv_y), \tag{9.86}$$

$$v_y = \sigma_0 R_H (E_y - Bv_x). \tag{9.87}$$

Neglecting the displacement current, from Maxwell's equations we have:

$$\begin{aligned} \nabla \times \boldsymbol{B} &= \mu_0 \boldsymbol{j}, \\ \nabla \times \boldsymbol{E} &= -\frac{\partial \boldsymbol{B}}{\partial t}. \end{aligned}$$

Assuming $\nabla \cdot \mathrm{E} = 0$ (overall neutrality), these give

$$\nabla^2 \boldsymbol{E} = \mu_0 \frac{\partial \boldsymbol{j}}{\partial t}. \tag{9.88}$$

If solutions of the form $\boldsymbol{E} = \boldsymbol{E}_0\exp[\mathrm{i}(kx - \omega t)]$ and $v = v_0\exp[\mathrm{i}(kx - \omega t)]$ are sought, we require:

$$-k^2 \boldsymbol{E} = -\mathrm{i}\omega\mu_0 \frac{\boldsymbol{v}}{R_H},$$

$$E_x = \mathrm{i}\frac{\omega\mu_0}{k^2}\frac{v_x}{R_H},$$

$$E_y = \mathrm{i}\frac{\omega\mu_0}{k^2}\frac{v_y}{R_H}.$$

Thus

$$\left(1-\mathrm{i}\sigma_0\frac{\omega\mu_0}{k^2}\right)v_x - \sigma_0 R_H B v_y = 0,$$

$$\sigma_0 R_H B v_x + \left(1-\mathrm{i}\sigma_0\frac{\omega\mu_0}{k^2}\right)v_y = 0. \tag{9.89}$$

Assuming large conductivity, $\sigma_0\omega\mu_0/k^2 >> 1$, and large B, we find:

$$\omega = \frac{k^2}{\mu_0}|R_H|B = \frac{k^2}{\mu_0 ne}B, \tag{9.90}$$

or the phase velocity is

$$v_p = \frac{\omega}{k} = \sqrt{\frac{\omega B}{\mu_0 ne}}, \tag{9.91}$$

independent of m. Note the group velocity is just twice the phase velocity. Since the plasma frequency ω_p is $(ne^2/m\varepsilon_0)^{1/2}$, we can write also

$$v_p = c\sqrt{\frac{\omega\omega_c}{\omega_p^2}}. \tag{9.92}$$

Typically v_p is of the order of sound velocities.

9.4.2 Alfvén Waves (EE)

Alfvén waves occur in a material with two kinds of charge carriers (say electrons and holes). As for helicon waves, we assume a large magnetic field with electromagnetic radiation propagating along the field. Alfvén waves have been observed in Bi, a semimetal at 4 K. The basic assumptions and equations are:

1. $\nabla \times \boldsymbol{B} = \mu_0 \boldsymbol{j}$, neglecting displacement current.
2. $\nabla \times \boldsymbol{E} = -\partial \boldsymbol{B}/\partial t$, Faraday's law.
3. $\rho\dot{\boldsymbol{v}} = \boldsymbol{j} \times \boldsymbol{B}$, where $\boldsymbol{v}$ is the fluid velocity, and the force per unit volume is dominated by magnetic forces.
4. $\boldsymbol{E} = -(\boldsymbol{v} \times \boldsymbol{B})$, from the generalized Ohm's law $j/\sigma = \boldsymbol{E} + \boldsymbol{v} \times \boldsymbol{B}$ with infinite conductivity.
5. $\boldsymbol{B} = B_x\hat{\boldsymbol{i}} + B_y\hat{\boldsymbol{j}}$, where $B_x = B_0$ and is constant while $B_y = B_1(t)$.

6. Only the j_x, E_x, and v_y components need be considered (v_y is the velocity of the plasma in the y direction and oscillates with time).
7. $\dot{\boldsymbol{v}} = \partial \boldsymbol{v}/\partial t$, as we neglect $(\boldsymbol{v}\cdot\nabla)\boldsymbol{v}$ by assuming small hydrodynamic motion. Also we assume the density ρ is constant in time.

Combining (1), (3), and (7) we have

$$\mu_0 \rho \frac{\partial v_y}{\partial t} = [(\nabla \times \boldsymbol{B}) \times \boldsymbol{B}]_y \cong \frac{\partial B_1}{\partial x} B_0. \tag{9.93}$$

By (4)

$$E_z = -B_0 v_y,$$

so

$$-\frac{\mu_0 \rho}{B_0} \frac{\partial E_z}{\partial t} = B_0 \frac{\partial B_1}{\partial x}. \tag{9.94}$$

By (2)

$$\frac{\partial E_z}{\partial x} = -\frac{\partial B_1}{\partial t},$$

so

$$\frac{\partial^2 E_z}{\partial t^2} = -\frac{B_0^2}{\mu_0 \rho} \frac{\partial^2 B_1}{\partial x \partial t} = +\frac{B_0^2}{\mu_0 \rho} \frac{\partial^2 E_z}{\partial x^2}. \tag{9.95}$$

This is the equation of a wave with velocity

$$v_A = \frac{B}{\sqrt{\mu_0 \rho}}, \tag{9.96}$$

the Alfvén velocity. For electrons and holes of equal number density n and effective masses m_e and m_h,

$$v_A = \frac{B}{\sqrt{\mu_0 n (m_e + m_h)}}, \tag{9.97}$$

Notice that $v_A = (B^2/\mu_0 \rho)^{1/2}$ is the velocity in a string of tension B^2/μ_0 and density ρ. In some sense, the media behaves as if the charges and magnetic flux lines move together.

A unified treatment of helicon and Alfvén waves can be found in Elliot and Gibson [9.5] and Platzman and Wolff [9.15]. Alfvén waves are also discussed in space physics, e.g. in connection with the solar wind.

9.5 Free-Electron Screening

9.5.1 Introduction (B)

If you place one charge in the midst of other charges, they will redistribute themselves in such a way as to "damp out" the long-range effects of the original charge. This long-range damping is an aspect of screening. Its origin resides in the Coulomb interactions of charges. This phenomenon was originally treated classically by the Debye–Huckel theory. A semiclassical form is called the Thomas–Fermi Approximation, which also assumes a free-electron gas. Neither the Debye–Huckel Theory nor the Thomas–Fermi model treats screening accurately at small distances. To do this, it is necessary to use the Lindhard theory.

We begin with the linearized Thomas–Fermi and Debye–Huckel methods and show how to use them to calculate the screening due to a single charged impurity. Perhaps the best way to derive this material is through the dielectric function and derive the Lindhard expression for it for a free-electron gas. The Lindhard expression for $\varepsilon(\omega \to 0, q)$ for small q then gives us the Thomas–Fermi expression. Generalization of the dielectric function to band electrons can also be made. The Lindhard approach follows in Sect. 9.5.3.

9.5.2 The Thomas–Fermi and Debye–Huckel Methods (A, EE)

We assume an electron gas with a uniform background charge (jellium). We assume a point charge of charge Ze ($e > 0$) is placed in the jellium. This will produce a potential $\varphi(r)$, which we assume to be weak and to vary slowly over a distance of order $1/k_F$ where k_F is the wave vector of the electrons whose energy equals the Fermi energy. For distances close to the impurity, where the potential is neither weak nor slowly varying our results will not be a very good approximation. Consistent with the slowly varying potential approximation, we assume it is valid to think of the electron energy as a function of position.

$$E_{\boldsymbol{k}} = \frac{\hbar^2 k^2}{2m} - e\varphi(\boldsymbol{r}) , \tag{9.98}$$

where $\hbar$ is Planck's constant (divided by 2π), $\boldsymbol{k}$ is the wave vector, and m is the electronic effective mass.

In order to exhibit the effects of screening, we need to solve for the potential φ. We assume the static dielectric constant is ε and ρ is the charge density. Poisson's equation is

$$\nabla^2 \varphi = \frac{-\rho}{\varepsilon} , \tag{9.99}$$

where the charge density is

$$\rho = eZ\delta(\mathbf{r}) + n_0 e - ne\,, \tag{9.100}$$

where $eZ\delta(\boldsymbol{r})$ is the charge density of the added charge. For the spin $^1/_2$ electrons obeying Fermi–Dirac statistics, the number density (assuming local spatial equilibrium) is

$$n = \int \frac{1}{\exp[\beta(E_{\boldsymbol{k}} - \mu)] + 1} \frac{\mathrm{d}\boldsymbol{k}}{4\pi^3}\,, \tag{9.101}$$

where $\beta = 1/k_B T$ and k_B is the Boltzmann constant. When $\varphi = 0$, then $n = n_0$, so

$$n_0 = n_0(\mu) = \int \frac{1}{\exp[\beta((\hbar^2 k^2/2m) - \mu)] + 1} \frac{\mathrm{d}\boldsymbol{k}}{4\pi^3}\,. \tag{9.102}$$

Note by (9.98) and (9.102), we also have

$$n = n_0[\mu + e\varphi(\boldsymbol{r})]\,. \tag{9.103}$$

This means the charge density can be written

$$\rho = eZ\delta(\boldsymbol{r}) + \rho^{\text{ind}}(\boldsymbol{r})\,, \tag{9.104}$$

where

$$\rho^{\text{ind}}(\boldsymbol{r}) = -e[n_0(\mu + e\varphi(\boldsymbol{r})) - n_0(\mu)]\,. \tag{9.105}$$

We limit ourselves to weak potentials. We can then expand n_0 in powers of φ and obtain:

$$\rho^{\text{ind}}(\boldsymbol{r}) = -e^2 \frac{\partial n_0}{\partial \mu} \varphi(\boldsymbol{r})\,. \tag{9.106}$$

The Poisson equation then becomes

$$\nabla^2 \varphi = -\frac{1}{\varepsilon}\left[Ze\delta(r) - e^2 \frac{\partial n_0}{\partial \mu} \varphi(\boldsymbol{r}) \right]. \tag{9.107}$$

A convenient way to solve this equation is by the use of Fourier transforms. The Fourier transform of the potential can be written

$$\varphi(\boldsymbol{q}) = \int \varphi(\boldsymbol{r}) \exp(-i\boldsymbol{q} \cdot \boldsymbol{r}) \mathrm{d}\boldsymbol{r}\,, \tag{9.108}$$

with inverse

$$\varphi(\boldsymbol{r}) = \frac{1}{(2\pi)^3} \int \varphi(\boldsymbol{q}) \exp(\mathrm{i}\boldsymbol{q} \cdot \boldsymbol{r}) \mathrm{d}\boldsymbol{q} , \tag{9.109}$$

and the Dirac delta function can be represented by

$$\delta(\boldsymbol{r}) = \frac{1}{(2\pi)^3} \int \exp(\mathrm{i}\boldsymbol{q} \cdot \boldsymbol{r}) \mathrm{d}\boldsymbol{q} . \tag{9.110}$$

Taking the Fourier transform of (9.107), we have

$$q^2 \varphi(\boldsymbol{q}) = \frac{1}{\varepsilon} \left[Ze - e^2 \frac{\partial n_0}{\partial \mu} \varphi(\boldsymbol{q}) \right] . \tag{9.111}$$

Defining the screening parameter as

$$k_S^2 = \frac{e^2}{\varepsilon} \frac{\partial n_0}{\partial \mu} , \tag{9.112}$$

we find from (9.111) that

$$\varphi(\boldsymbol{q}) = \frac{Ze}{\varepsilon} \frac{1}{q^2 + k_S^2} . \tag{9.113}$$

Then, using (9.109), we find from (9.113) that

$$\varphi(\boldsymbol{r}) = \frac{Ze}{4\pi\varepsilon r} \exp(-k_S r) . \tag{9.114}$$

Equations (9.112) and (9.114) are the basic equations for screening.

For the classical nondegenerate case, we have from (9.102)

$$n_0(\mu) = \exp(\beta\mu) \int \exp(-\beta\hbar^2 k^2 / 2m) \frac{\mathrm{d}\boldsymbol{k}}{4\pi^3} , \tag{9.115}$$

so that by (9.112)

$$k_S^2 = \frac{e^2}{\varepsilon} \frac{n_0}{k_B T} , \tag{9.116}$$

we get the classical Debye–Huckel result. For the degenerate case, it is convenient to rewrite (9.102) as

$$n_0(\mu) = \int D(E) f(E) \mathrm{d}E , \tag{9.117}$$

so

$$\frac{\partial n_0}{\partial \mu} = \int D(E)\frac{\partial f}{\partial \mu}\mathrm{d}E\,, \tag{9.118}$$

where $D(E)$ is the density of states per unit volume and $f(E)$ is the Fermi function

$$f(E) = \frac{1}{\exp[\beta(E-\mu)]+1}\,. \tag{9.119}$$

since

$$\frac{\partial f(E)}{\partial \mu} \cong \delta(E-\mu)\,, \tag{9.120}$$

at low temperatures when compared with the Fermi temperature; so we have

$$\frac{\partial n_0}{\partial \mu} \cong D(\mu)\,. \tag{9.121}$$

Since the free-electron density of states per unit volume is

$$D(E) = \frac{1}{2\pi^2}\left(\frac{2m}{\hbar^2}\right)^{3/2}\sqrt{E}\,, \tag{9.122}$$

and the Fermi energy at absolute zero is

$$\mu = \frac{\hbar^2}{2m}(3\pi^2 n_0)^{2/3}\,, \tag{9.123}$$

where $n_0 = N/V$, we find

$$D(\mu) = \frac{3n_0}{2\mu}\,, \tag{9.124}$$

which by (9.121) and (9.112) gives the linearized Thomas–Fermi approximation. If we further use

$$\mu = \frac{3}{2}k_B T_F\,, \tag{9.125}$$

we find

$$k_S^2 = \frac{e^2}{\varepsilon}\frac{n_0}{k_B T_F}\,, \tag{9.126}$$

which looks just like the Debye–Huckel result except T is replaced by the Fermi temperature T_F. In general, by (9.112), (9.118), (9.119), and (9.122), we have for free-electrons,

$$k_S^2 = \frac{e^2 n_0}{\varepsilon k_B T_F} \frac{F'_{1/2}(\eta)}{F_{1/2}(\eta)}, \tag{9.127}$$

where $\eta = \mu/k_B T$ and

$$F_{1/2}(\eta) = \int_0^\infty \frac{\sqrt{x}dx}{\exp(x-\eta)+1}, \tag{9.128}$$

is the Fermi integral. Typical screening lengths $1/k_S$ for good metals are of order 1 Å, whereas for typical semiconductors 60 Å is more appropriate. For $\eta << -1$, $F'_{1/2}(\eta)/F_{1/2}(\eta) \approx 1$, which corresponds to the classical Debye–Huckel theory, and for $\eta >> 1$, $F'_{1/2}(\eta)/F_{1/2}(\eta) = 3/(2\eta)$ is the Thomas–Fermi result.

9.5.3 The Lindhard Theory of Screening (A)

Here we do a more general discussion that is self-consistent.[5] We start with the idea of an external potential that determines a set of electronic states. Electronic states in turn give rise to a charge density from which a potential can be determined. We wish to show how we can determine a charge density and a potential in a self-consistent way by using the concept of a frequency- and wave-vector-dependent dielectric constant.

The specific problem we wish to solve is that of the self-consistent response to an applied field. We will assume small applied fields and linear responses. The electronic response to the applied field is called screening, and it arises from the interaction of the electrons with each other and with the external field. Only screening by a free-electron gas will be considered.

Let a charge ρ^{ext} be placed in jellium, and let it produce a potential φ^{ext} (by itself). Let φ be the potential caused by the extra charge, the free-electrons, and the uniform background charge (i.e. extra charge plus jellium). We also let ρ be the corresponding charge density. Then

$$\nabla^2 \varphi^{\text{ext}} = -\frac{\rho^{\text{ext}}}{\varepsilon}, \tag{9.129}$$

$$\nabla^2 \varphi = -\frac{\rho}{\varepsilon}. \tag{9.130}$$

The induced charge density ρ^{ind} is then defined by

$$\rho^{\text{ind}} = \rho - \rho^{\text{ext}}. \tag{9.131}$$

---

5 This topic is also treated in Ziman JM [25, Chap. 5], and Grosso and Paravicini [55 p245ff].

We Fourier analyze the equations in both the space and time domains:

$$q^2\varphi^{\text{ext}}(\boldsymbol{q},\omega) = \frac{\rho^{\text{ext}}(\boldsymbol{q},\omega)}{\varepsilon}, \tag{9.132a}$$

$$q^2\varphi(\boldsymbol{q},\omega) = \frac{\rho(\boldsymbol{q},\omega)}{\varepsilon}, \tag{9.132b}$$

$$\rho(\boldsymbol{q},\omega) = \rho^{\text{ext}}(\boldsymbol{q},\omega) + \rho^{\text{ind}}(\boldsymbol{q},\omega)\,. \tag{9.132c}$$

Subtracting (9.132a) from (9.132b) and using (9.132c) yields:

$$\varepsilon q^2[\varphi(\boldsymbol{q},\omega) - \varphi^{\text{ext}}(\boldsymbol{q},\omega)] = \rho^{\text{ind}}(\boldsymbol{q},\omega)\,. \tag{9.133}$$

We have assumed weak field and linear responses, so we write

$$\rho^{ind}(\boldsymbol{q},\omega) = g(\boldsymbol{q},\omega)\varphi(\boldsymbol{q},\omega)\,, \tag{9.134}$$

which defines $g(\boldsymbol{q}, \omega)$. Thus, (9.133) and (9.134) give this as

$$\varepsilon q^2[\varphi(\boldsymbol{q},\omega) - \varphi^{\text{ext}}(\boldsymbol{q},\omega)] = g(\boldsymbol{q},\omega)\varphi(\boldsymbol{q},\omega)\,. \tag{9.135}$$

Thus,

$$\varphi(\boldsymbol{q},\omega) = \frac{\varphi^{\text{ext}}(\boldsymbol{q},\omega)}{\varepsilon(\boldsymbol{q},\omega)}, \tag{9.136}$$

where

$$\varepsilon(\boldsymbol{q},\omega) = 1 - \frac{g(\boldsymbol{q},\omega)}{\varepsilon q^2}\,. \tag{9.137}$$

To proceed further, we need to calculate $\varepsilon(\boldsymbol{q}, \omega)$ directly. In the process of doing this, we will verify the correctness of the linear response assumption. We write the Schrödinger equation as

$$\mathcal{H}_0|\boldsymbol{k}\rangle = E_{\boldsymbol{k}}|\boldsymbol{k}\rangle\,. \tag{9.138}$$

We assume an external perturbation of the form

$$\delta V(\boldsymbol{r},t) = [V\exp(\mathrm{i}(\boldsymbol{q}\cdot\boldsymbol{r}+\omega t)) + V\exp(-\mathrm{i}(\boldsymbol{q}\cdot\boldsymbol{r}+\omega t))]\exp(\alpha t)\,. \tag{9.139}$$

The factor $\exp(\alpha t)$ has been introduced so that the perturbation vanishes as $t = -\infty$, or in other words, as the perturbation is slowly turned on. V is assumed real. Let

$$\mathcal{H} = \mathcal{H}_0 + \delta V\,. \tag{9.140}$$

We then seek an approximate solution of the time-dependent Schrödinger wave equation

$$\mathcal{H}\psi = \mathrm{i}\hbar\frac{\partial \psi}{\partial t}. \tag{9.141}$$

We seek solutions of the form

$$|\psi\rangle = \sum_{\mathbf{k}'} C_{\mathbf{k}'}(t)\exp(-\mathrm{i}E_{\mathbf{k}'}t/\hbar)|\mathbf{k}'\rangle. \tag{9.142}$$

Substituting,

$$\begin{aligned}&\sum_{\mathbf{k}'}(\mathcal{H}_0 + \delta V)C_{\boldsymbol{k}'}(t)\exp(-iE_{\boldsymbol{k}'}t/\hbar)|\boldsymbol{k}'\rangle\\&= \mathrm{i}\hbar\frac{\partial}{\partial t}\sum_{\boldsymbol{k}'}C_{\boldsymbol{k}'}(t)\exp(-\mathrm{i}E_{\boldsymbol{k}'}t/\hbar)|\boldsymbol{k}'\rangle.\end{aligned} \tag{9.143}$$

Using (9.138) to cancel two terms in (9.143), we have

$$\sum_{\boldsymbol{k}'}\delta V C_{\boldsymbol{k}'}(t)\exp(-\mathrm{i}E_{\boldsymbol{k}'}t/\hbar)|\boldsymbol{k}'\rangle = \mathrm{i}\hbar\sum_{\boldsymbol{k}'}\dot{C}_{\boldsymbol{k}'}(t)\exp(-\mathrm{i}E_{\boldsymbol{k}'}t/\hbar)|\boldsymbol{k}'\rangle. \tag{9.144}$$

Using

$$\langle \boldsymbol{k}''|\boldsymbol{k}'\rangle = \delta_{\boldsymbol{k}'}^{\boldsymbol{k}''}, \tag{9.145}$$

$$\langle \boldsymbol{k}''|\delta V|\boldsymbol{k}'\rangle = \langle \boldsymbol{k}''|\delta V|\boldsymbol{k}'\pm\boldsymbol{q}\rangle\delta_{\boldsymbol{k}'}^{\boldsymbol{k}''\pm\boldsymbol{q}}, \tag{9.146}$$

$$\begin{aligned}\dot{C}_{\boldsymbol{k}''}(t) &= \frac{1}{\mathrm{i}\hbar}C_{\boldsymbol{k}''+\boldsymbol{q}}\exp(-\mathrm{i}E_{\boldsymbol{k}''+\boldsymbol{q}}t/\hbar)\langle\boldsymbol{k}''|\delta V|\boldsymbol{k}''+\boldsymbol{q}\rangle\exp(\mathrm{i}E_{\boldsymbol{k}''}t/\hbar)\\&+\frac{1}{\mathrm{i}\hbar}C_{\boldsymbol{k}''-\boldsymbol{q}}\exp(-\mathrm{i}E_{\boldsymbol{k}''-\boldsymbol{q}}t/\hbar)\langle\boldsymbol{k}''|\delta V|\boldsymbol{k}''-\boldsymbol{q}\rangle\exp(\mathrm{i}E_{\boldsymbol{k}''}t/\hbar).\end{aligned} \tag{9.147}$$

Using (9.139), we have

$$\begin{aligned}\dot{C}_{\boldsymbol{k}''}(t) &= \frac{1}{\mathrm{i}\hbar}C_{\boldsymbol{k}''+\boldsymbol{q}}\exp(-\mathrm{i}(E_{\boldsymbol{k}''+\boldsymbol{q}} - E_{\boldsymbol{k}''})t/\hbar)V\exp(-\mathrm{i}\omega t)\exp(\alpha t)\\&+\frac{1}{\mathrm{i}\hbar}C_{\boldsymbol{k}''-\boldsymbol{q}}\exp(-\mathrm{i}(E_{\boldsymbol{k}''-\boldsymbol{q}} - E_{\boldsymbol{k}''})t/\hbar)V\exp(\mathrm{i}\omega t)\exp(\alpha t).\end{aligned} \tag{9.148}$$

We assume a weak perturbation, and we begin in the state $\boldsymbol{k}$ with probability $f_0(\boldsymbol{k})$, so we have

$$C_{\boldsymbol{k}''}(t) = \sqrt{f_0(\boldsymbol{k})}\delta_{\boldsymbol{k}'',\boldsymbol{k}} + \lambda C_{\boldsymbol{k}''}^{(1)}(t). \tag{9.149}$$

We write out (9.147) to first order for two interesting cases:

$$\begin{aligned}\dot{C}_{k+q}(t) &= \lambda \dot{C}^{(1)}_{k+q}(t) \\ &= \left(\frac{1}{\mathrm{i}\hbar}\right)\sqrt{f_0(\boldsymbol{k})}\exp(-\mathrm{i}(E_{\boldsymbol{k}} - E_{\boldsymbol{k}+\boldsymbol{q}})t/\hbar)V\exp(\mathrm{i}\omega t)\exp(\alpha t),\end{aligned} \tag{9.150}$$

$$\begin{aligned}\dot{C}_{k-q}(t) &= \lambda \dot{C}^{(1)}_{k-q}(t) \\ &= \left(\frac{1}{\mathrm{i}\hbar}\right)\sqrt{f_0(\boldsymbol{k})}\exp(-\mathrm{i}(E_{\boldsymbol{k}} - E_{\boldsymbol{k}-\boldsymbol{q}})t/\hbar)V\exp(-\mathrm{i}\omega t)\exp(\alpha t).\end{aligned} \tag{9.151}$$

Integrating, we find, since $C_{k\pm q}(\infty) = 0$

$$C_{k+q}(t) = \sqrt{f_0(\boldsymbol{k})}\frac{\exp(-\mathrm{i}(E_{\boldsymbol{k}} - E_{\boldsymbol{k}+\boldsymbol{q}})t/\hbar)V\exp(\mathrm{i}\omega t)\exp(\alpha t)}{E_{\boldsymbol{k}} - E_{\boldsymbol{k}+\boldsymbol{q}} - \hbar\omega + \mathrm{i}\hbar\alpha}, \tag{9.152}$$

$$C_{k-q}(t) = \sqrt{f_0(\boldsymbol{k})}\frac{\exp(-\mathrm{i}(E_{\boldsymbol{k}} - E_{\boldsymbol{k}-\boldsymbol{q}})t/\hbar)V\exp(-\mathrm{i}\omega t)\exp(\alpha t)}{E_{\boldsymbol{k}} - E_{\boldsymbol{k}-\boldsymbol{q}} + \hbar\omega + \mathrm{i}\hbar\alpha}. \tag{9.153}$$

We write (9.142) as

$$\psi^{(\boldsymbol{k})} = \sum\nolimits_{\boldsymbol{k}'} C_{\boldsymbol{k}'}(t)\exp(-\mathrm{i}E_{\boldsymbol{k}'}t/\hbar)\psi_{\boldsymbol{k}'}, \tag{9.154}$$

where

$$\psi_{\boldsymbol{k}'}(\boldsymbol{r}) = \frac{1}{\sqrt{\Omega}}\mathrm{e}^{\mathrm{i}\boldsymbol{k}'\cdot\boldsymbol{r}}, \tag{9.155}$$

and Ω is the volume. We put a superscript on ψ because we assume we start in the state $\boldsymbol{k}$. More specifically, (9.153) can be written as

$$\begin{aligned}\psi^{(\boldsymbol{k})} &= \exp(-\mathrm{i}E_{\boldsymbol{k}}t/\hbar)\sqrt{f_0(\boldsymbol{k})}\psi_{\boldsymbol{k}} \\ &\quad + C_{\boldsymbol{k}+\boldsymbol{q}}(t)\exp(-\mathrm{i}E_{\boldsymbol{k}+\boldsymbol{q}}t/\hbar)\psi_{\boldsymbol{k}+\boldsymbol{q}} + C_{\boldsymbol{k}-\boldsymbol{q}}(t)\exp(-\mathrm{i}E_{\boldsymbol{k}-\boldsymbol{q}}t/\hbar)\psi_{\boldsymbol{k}-\boldsymbol{q}}.\end{aligned} \tag{9.156}$$

Any charge density in jellium is an induced charge density (in equilibrium, jellium is uniform and has a net density of zero). Thus,

$$\rho^{\mathrm{ind}} = \frac{eN}{\Omega} - e\sum\nolimits_{\boldsymbol{k}}\left|\psi^{(\boldsymbol{k})}\right|^2. \tag{9.157}$$

Now, note

$$\left|\psi^{(\boldsymbol{k})}\right|^2 = \frac{1}{\Omega} \quad \text{and} \quad \sum\nolimits_{\text{all}\,\boldsymbol{k}} f_0(\boldsymbol{k}) = N, \tag{9.158}$$

so putting (9.155) into (9.156) and retaining no terms beyond first order,

$$\rho^{\text{ind}} = \frac{eN}{\Omega} - \frac{e}{\Omega}\sum\nolimits_{\mathrm{k}} f_0(\mathrm{k})\left\{1 + \frac{V\,exp(\mathrm{iq}\cdot\mathrm{r})exp(\mathrm{i}\omega t)exp(\alpha t)}{E_{\mathrm{k}} - E_{\mathrm{k+q}} - \hbar\omega + \mathrm{i}\hbar\alpha} + \frac{V\,exp(-\mathrm{iq}\cdot\mathrm{r})exp(-\mathrm{i}\omega t)exp(\alpha t)}{E_{\mathrm{k}} - E_{\mathrm{k+q}} + \hbar\omega + \mathrm{i}\hbar\alpha} + \mathrm{c.c.}\right\}, \tag{9.159}$$

or

$$\rho^{\text{ind}} = -\frac{e}{\Omega}\sum\nolimits_{\mathbf{k}}\left\{\frac{[f_0(\boldsymbol{k}) - f_0(\boldsymbol{k}+\boldsymbol{q})]V\exp(\mathrm{i}\boldsymbol{q}\cdot\boldsymbol{r})\exp(\mathrm{i}\omega t)\exp(\alpha t)}{E_{\boldsymbol{k}} - E_{\boldsymbol{k}+\boldsymbol{q}} - \hbar\omega + \mathrm{i}\hbar\alpha} + c.c.\right\}. \tag{9.160}$$

Using

$$V(\boldsymbol{q},\omega) = -e\varphi(\boldsymbol{q},\omega)\,, \tag{9.161}$$

and identifying $\rho^{\text{ind}}(\boldsymbol{q}, \omega)$ as the coefficient of $\exp(\mathrm{i}\boldsymbol{q}\cdot\boldsymbol{r})\exp(\mathrm{i}\omega t)$, we have

$$\rho^{\text{ind}}(\boldsymbol{q},\omega) = -\frac{e^2}{\Omega}\sum\nolimits_{\boldsymbol{k}}\left\{\frac{f_0(\boldsymbol{k}) - f_0(\boldsymbol{k}-\boldsymbol{q})}{E_{\boldsymbol{k}-\boldsymbol{q}} - E_{\boldsymbol{k}} - \hbar\omega + \mathrm{i}\hbar\alpha}\right\}\varphi(\boldsymbol{q},\omega)\,. \tag{9.162}$$

By (9.134) we find $g(\boldsymbol{q}, \omega)$ and by (9.137), we thus find

$$\varepsilon(\boldsymbol{q},\omega) = 1 + \frac{e^2}{\varepsilon\Omega q^2}\sum\nolimits_{\boldsymbol{k}}\frac{f_0(\boldsymbol{k}) - f_0(\boldsymbol{k}-\boldsymbol{q})}{E_{\boldsymbol{k}-\boldsymbol{q}} - E_{\boldsymbol{k}} - \hbar\omega + \mathrm{i}\hbar\alpha}\,. \tag{9.163}$$

Finally, a few notes are provided on notation. We can redefine the Fourier components so as to change the sign of $\boldsymbol{q}$. For example, we can say

$$\varphi(\boldsymbol{r}) = \frac{1}{(2\pi)^2}\int\exp(-\mathrm{i}\boldsymbol{q}\cdot\boldsymbol{r})\varphi(\boldsymbol{q})\mathrm{d}\boldsymbol{q}\,. \tag{9.164}$$

Then defining

$$v_q = \frac{e^2}{\varepsilon\,\Omega\,q^2}, \tag{9.165}$$

gives $\varepsilon(\boldsymbol{q}, \omega)$ in the form given in many textbooks:

$$\varepsilon(\boldsymbol{q},\omega) = 1 - v_q\sum\nolimits_{\boldsymbol{k}}\frac{f_0(\boldsymbol{k}+\boldsymbol{q}) - f_0(\boldsymbol{k})}{E_{\boldsymbol{k}+\boldsymbol{q}} - E_{\boldsymbol{k}} - \hbar\omega + \mathrm{i}\hbar\alpha}\,. \tag{9.166}$$

The limit as $\alpha \to 0$ is tacitly implied in (9.166). In the limit as q becomes small, (9.165) gives, as we will show below, the Thomas–Fermi approximation (when $\omega = 0$). Two notable effects follow from (9.165), but they are not included in the small q limit. An expression for $\varepsilon(q,0)$ at large q is readily obtained for our free-electron case. The result for $\omega = 0$ is

$$\varepsilon(\boldsymbol{q},\omega) = 1 + (\text{constant})D(E_F)\left[\frac{1}{2} + \frac{1-x^2}{4x}\ln\left|\frac{1+x}{1-x}\right|\right], \tag{9.167}$$

where $D(E_F)$ is the density of states at the Fermi energy and $x = q/2k_F$ with k_F being the wave vector at the Fermi energy. This expression has a singularity at $q = 2k_F$, which causes the screening of a charged impurity to have a weakly decaying oscillating term (beyond the Fermi–Thomas potential). This is the origin of *Friedel oscillations*. The Friedel oscillations damp out with distance due to electron scattering. At finite temperature, the singularity disappears causing the Friedel oscillation to damp out.

Further, since ion–ion interactions are screening by $\varepsilon(q)$, the singularity at $q = 2k_F$ is reflected in the phonon spectrum. Kinks in the phonon spectrum due to the singularity in $\varepsilon(q)$ are called *Kohn anomalies*.

Finally, we look at (9.165) for small q, $\omega = 0$ and $\alpha = 0$. We find

$$\begin{aligned}\varepsilon(\boldsymbol{q},\omega) &= 1 - \frac{e^2}{\varepsilon\,\Omega\,q^2}\sum\nolimits_{\boldsymbol{k}}\frac{\partial f_0/\partial\boldsymbol{k}}{\partial E_{\boldsymbol{k}}/\partial\boldsymbol{k}} \\ &= 1 - \frac{e^2}{\varepsilon q^2}\sum\nolimits_{\mathbf{k}} D(E)\frac{\partial f}{\partial E}dE = 1 + \frac{k_S^2}{q^2}.\end{aligned} \tag{9.168}$$

and hence comparing to previous work, we get exactly the Thomas–Fermi approximation.

Problems

9.1 Show that $E'_0 = E_0 + P/\varepsilon_0$, where E_0 is the electric field between the plates before the slab is inserted (9.19).

9.2 Show that $E_1 = -P/\varepsilon_0$ (see Fig. 9.2).

9.3 Show that $E_2 = P/3\varepsilon_0$ (9.23).

9.4 Show for cubic crystals that $E_3 = 0$ (chapter notation is used).

9.5 If we have N permanent free dipoles p per unit volume in an electric field E, find an expression for the polarization. At high temperatures show that the polarizability (per molecule) is $\alpha = p^2/3kT$. What magnetic situation is this analogous to?

9.6 Use (9.30) and (9.48) to show (9.49)

$$\frac{\varepsilon}{\varepsilon_0} = 1 + \frac{Np^2}{3k\varepsilon_0(T - T_c)}.$$

Find T_c. How likely is this to apply to any real material?

9.7 Use the trial wave function $\psi = \psi_{100}\,(1 + pz)$ (where p is the variational parameter) for a hydrogen atom (in an external electric field in the z direction) to show that we obtain for the polarizability $16\pi\varepsilon_0 a_0^3$. (ψ_{100} is the ground-state wave function of the unperturbed hydrogen atom, a_0 is the radius of the first Bohr orbit of the hydrogen atom, and the exact polarizability is $18\pi\varepsilon_0 a_0^3$.)

9.8 (a) Given the Gibbs free energy

$$G = G_0 + \frac{1}{2}\beta(T - T_0)P^2 + \frac{1}{4}\gamma P^4 + \frac{1}{6}\delta P^6\,;$$
$$\beta, \delta > 0\,, \gamma < 0 \quad \text{(first order)},$$

derive an expression for T_c in terms of P_{sc} where $G(P_{sc}) = G_0$ and $E = 0$.

(b) Put the expression for T_c in terms without P_{sc}. That is, fill in the details of Sect. 9.3.1.

10 Optical Properties of Solids

10.1 Introduction (B)

The organization of a solid-state course may vary towards its middle or end. Logical beginnings are fairly easy. One defines the solid-state universe, and this is done with a Section on crystal structures and how they are determined. Then one introduces the main players, and so there are sections on lattice vibrations, phonons, band structure, and electrons. Following this, one can present topics based on the interaction of electrons and phonons and hence discuss, for example, transport. After that come specific materials (semiconductors, magnetic materials, metals, and superconductors) and properties (dielectric, optic, defect, surface, etc). The problem is that some of these categories overlap so that a clean separation is not possible. Optical properties, in particular, seem to spread into many areas, so a well-focused segment on the optical properties of solids can be somewhat tricky to put together.[1]

By optical properties of solids, we mean those properties that relate to the interaction of solids with electromagnetic radiation whose wavelength is in the infrared to the ultraviolet. There are several aspects to optical properties of solids and looking at the subject in full generality can often lead to complexity, whereas treating each part as a separate case often leads to confusion. We will try to keep to a middle ground between these, by emphasizing only one topic (absorption) but treating it in some detail. Although we will concentrate on absorption, we will mention other optical phenomena including emission, reflection, scattering, and photoemission of electrons.

There are several processes involved in absorption, but the main five seem to be:

(a) Absorption due to electronic transitions between bands that involve wavelengths typically less than ten micrometers;

(b) Absorption by excitons at wavelengths with energies just below the absorption edge due to valence–conduction band transitions (in semiconductors);

(c) Excitation and ionization of impurities that involve wavelengths ranging from about one micrometer to one thousand micrometers;

---

[1] A good treatment is Fox [10.12].

(d) Excitation of lattice vibrations (optical phonons) in polar solids for which the usual wavelengths are ten to fifty micrometers;

(e) Free-carrier absorption for frequencies up to the plasma edge. Free-carrier absorption is particularly important in metals, of course. By gathering data about any optical process, we can gain information about the inner workings of the solid.

10.2 Macroscopic Properties (B)

We start by relating the dielectric properties to optical properties, particularly those involving absorption and reflection. The complex dielectric constant, and the relation of its two components by the Kronig–Kramers relation, is particularly important. The imaginary part relates to the absorption coefficient. We assume the total charge density $\rho_{\text{total}} = 0, j = \sigma E$, and $\mu = \mu_0$ (no internal magnetic effects, all in the usual notation). We assume a wavelength large compared with atomic dimensions but small compared with the dimensions of the sample. We start with Maxwell's equations and the constitutive relations in SI in the usual notation:

$$\begin{aligned} &\nabla \cdot E = 0 \qquad \nabla \times E = -\mu_0 \frac{\partial H}{\partial t} = -\frac{\partial B}{\partial t} \\ &\nabla \cdot B = 0 \qquad \nabla \times H = \mathrm{j} + \frac{\partial D}{\partial t} \\ &D = \varepsilon_0 E + P = \varepsilon E \\ &B = \mu_0 (H + M) = \mu H . \end{aligned} \tag{10.1}$$

One then finds

$$\nabla^2 \boldsymbol{E} = \mu_0 \sigma \frac{\partial \boldsymbol{E}}{\partial t} + \mu_0 \frac{\partial^2}{\partial t^2} (\varepsilon \boldsymbol{E}) . \tag{10.2}$$

We look for solutions for each Fourier component

$$\boldsymbol{E}(k,\omega) = \boldsymbol{E}_0 \exp[\mathrm{i}(\boldsymbol{k} \cdot \boldsymbol{r} - \omega t)] , \tag{10.3}$$

and keep in mind that ε should be written $\varepsilon(\boldsymbol{k}, \omega)$. Substituting, one finds

$$k^2 = \mu_0 (\varepsilon \omega^2 + \mathrm{i} \sigma \omega) . \tag{10.4}$$

Or, since $c = 1/(\mu_0 \varepsilon_0)^{1/2}$,

$$k = \frac{\omega}{c} \left(\frac{\varepsilon}{\varepsilon_0} + \mathrm{i} \frac{\sigma}{\varepsilon_0 \omega} \right)^{1/2} . \tag{10.5}$$

For an insulator, $\sigma = 0$ so,

$$k = \frac{\omega}{v} = \frac{\omega n}{c} = \frac{\omega}{c} \sqrt{\frac{\varepsilon}{\varepsilon_0}} . \tag{10.6}$$

where n is the index of refraction. It is then natural to define a complex dielectric constant ε_c and a complex index of refraction n_c so,

$$k = \frac{\omega}{c} n_c , \tag{10.7}$$

where

$$n_c = \sqrt{\frac{\varepsilon}{\varepsilon_0} + \mathrm{i}\frac{\sigma}{\varepsilon_0 \omega}} = n + \mathrm{i}n_i . \tag{10.8}$$

Letting

$$\varepsilon_c = \varepsilon_r + \mathrm{i}\varepsilon_i = \frac{\varepsilon(k,\omega)}{\varepsilon_0} + \mathrm{i}\frac{\sigma}{\varepsilon_0 \omega} , \tag{10.9}$$

squaring both sides and equating real and imaginary parts, we find

$$\varepsilon_r(k,\omega) = n^2 - n_i^2 , \tag{10.10}$$

and

$$\varepsilon_i(k,\omega) = 2nn_i . \tag{10.11}$$

Now, assuming the wave propagates in the z direction, if we substitute

$$k = \frac{\omega}{c}(n + \mathrm{i}n_i) , \tag{10.12}$$

we have

$$\boldsymbol{E} = \boldsymbol{E}_0 \exp\left[\mathrm{i}\omega\left(\frac{nz}{c} - t\right)\right] \exp\left(-\frac{\omega}{c} n_i z\right) . \tag{10.13}$$

So, since energy in the wave is proportional to $|\boldsymbol{E}|^2$, we have that the absorption coefficient is given by

$$\alpha = \frac{2n_i \omega}{c} . \tag{10.14}$$

Another readily measured quantity can be related to n and n_i. If we apply appropriate boundary conditions to a solid surface, we can show as noted below that the reflection coefficient for normal incidence is given by

$$R = \frac{(n-1)^2 + n_i^2}{(n+1)^2 + n_i^2} . \tag{10.15}$$

This relation follows directly from the Maxwell relations. From Faraday's law, we can show that the tangential component of E is continuous, and from Ampere's Law we can show the tangential component of H is continuous. Further manipulation leads to the desired relation. Let us work this out. For normal incidence from the vacuum on a surface at $z = 0$, the incident and reflected waves can be written as

$$E_{i+r} = E_1 \exp\left[\mathrm{i}\omega\left(\frac{z}{c} - t\right)\right] + E_2 \exp\left[-\mathrm{i}\omega\left(\frac{z}{c} + t\right)\right], \tag{10.16}$$

and the refracted wave is given by

$$E_{rf} = E_0 \exp\left[\mathrm{i}\omega\left(n_c \frac{z}{c} - t\right)\right], \tag{10.17}$$

where n_c is the complex index of refraction. Since

$$\nabla \times \boldsymbol{E} + \frac{\partial \boldsymbol{B}}{\partial t} = 0\,, \tag{10.18}$$

we can use the loop of Fig. 10.1 to write

$$\int_A (\nabla \times \boldsymbol{E}) \cdot \mathrm{d}\boldsymbol{s} + \int \frac{\partial \boldsymbol{B}}{\partial t} \cdot \mathrm{d}\boldsymbol{s} = 0\,, \tag{10.19}$$

or

$$\oint_C \boldsymbol{E} \cdot \mathrm{d}\boldsymbol{r} = -\frac{\mathrm{d}}{\mathrm{d}t} \int \boldsymbol{B} \cdot \mathrm{d}\boldsymbol{s}\,, \tag{10.20}$$

as

$$(E_T^1 - E_T^2)l + \mathrm{O}(\varepsilon) = -\frac{\mathrm{d}}{\mathrm{d}t}(B_\perp 2l\varepsilon)\,, \tag{10.21}$$

where the subscript T means the tangential component of the electric field, and the subscript $\perp$ means perpendicular to the page of the paper. Taking the limit as $\varepsilon \to 0$, we obtain

$$E_T^1 - E_T^2 = 0\,, \tag{10.22}$$

or the tangential component of E is continuous. In a similar way we can use

$$\nabla \times \boldsymbol{H} = \boldsymbol{j} + \frac{\partial \boldsymbol{D}}{\partial t} \tag{10.23}$$

to show that

$$(H_T^1 - H_T^2)l + \mathrm{O}(\varepsilon) = \int_A \boldsymbol{j} \cdot \mathrm{d}\boldsymbol{s} + \frac{\mathrm{d}}{\mathrm{d}t}(D_\perp 2l\varepsilon) = j_\perp(2l\varepsilon) + \frac{\mathrm{d}}{\mathrm{d}t}(D_\perp 2l\varepsilon)\,. \tag{10.24}$$

Again taking the limit as $\varepsilon \to 0$, we find

$$(H_T^1 - H_T^2) = 0\,, \tag{10.25}$$

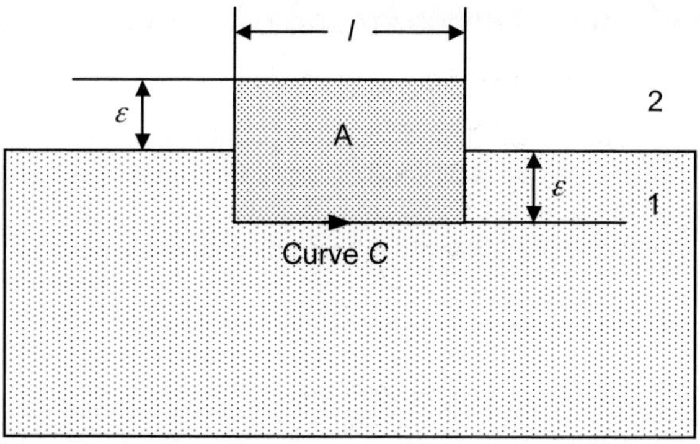

Fig. 10.1. Loop used for deriving field boundary conditions (notice this ε is a distance)

or that the tangential component of H is also continuous. Continuity of the tangential component of H requires (using $\nabla\times\boldsymbol{E} = -\partial\boldsymbol{B}/\partial t$, proper constitutive relations, and (10.16) and (10.17))

$$n_c E_0 = E_1 - E_2 \, . \tag{10.26}$$

Continuity of the tangential component of E requires ((10.16) and (10.17))

$$E_0 = E_1 + E_2 \, . \tag{10.27}$$

Adding these two equations gives

$$E_1 = \frac{E_0(n_c + 1)}{2} \, . \tag{10.28}$$

Subtracting these equations gives

$$E_2 = \frac{E_0(-n_c + 1)}{2} \, . \tag{10.29}$$

Thus, the reflection coefficient is given by

$$R = \left|\frac{E_2}{E_1}\right|^2 = \left|\frac{1-n_c}{1+n_c}\right|^2 = \left|\frac{(n-1)^2 + n_i^2}{(n+1)^2 + n_i^2}\right| . \tag{10.30}$$

Enough has been said to indicate that the theory of the optical properties of solids is intimately related to the complex index of refraction of solids. The complex dielectric constant equals the square of the complex index of refraction. Thus, the optical properties of solids are intimately related to the study of the dielectric properties of solids, and the measurement of the absorptivity and reflectivity determine n and n_i, and hence, ε_r and ε_i.

10.2.1 Kronig–Kramers Relations (A)

We will give a quantum description of the absorption of radiation, but first it is helpful to derive the Kronig–Kramers equations, which give a relation between the real and imaginary parts of the dielectric constant. Let ε be a complex function of ω that converges in the upper half-plane. We need to define the Cauchy principal value P with a real for the following equations and diagrams:

$$P\int_{-\infty}^{\infty}\frac{\varepsilon(\omega)\mathrm{d}\omega}{\omega-a}=\frac{1}{2}\left[\oint_{C'}\frac{\varepsilon(\omega)\mathrm{d}\omega}{\omega-a}+\oint_{C''}\frac{\varepsilon(\omega)\mathrm{d}\omega}{\omega-a}\right], \tag{10.31}$$

as shown by Fig. 10.2. It is assumed that the integral over the large semicircles is zero. From complex variables, we know that if C encloses a and if f has no singularity in C, then

$$\oint_C \frac{f(Z)\mathrm{d}Z}{Z-a}=2\pi_i f(a)\,. \tag{10.32}$$

Using the definition of Cauchy principal value, since we have the integral

$$\int_{C''}=0\,,\quad P\rightarrow\frac{1}{2}\left(\int_{C'}-\int_{C''}\right). \tag{10.33}$$

Thus,

$$P\int\frac{\varepsilon(\omega)\mathrm{d}\omega}{\omega-a}=\frac{1}{2}\int_{\substack{\text{smallcircle}\\ \rho\to 0}}\frac{\varepsilon(\omega)\mathrm{d}\omega}{\omega-a}=\mathrm{i}\pi\varepsilon(a)\,, \tag{10.34}$$

and we have used that $\varepsilon(\omega)$ on the big circle is zero (actually, to achieve this we should use that $\varepsilon(\omega) = [\varepsilon_r(\omega) - 1] + \mathrm{i}\varepsilon_i(\omega)$, which we will put in explicitly at the end). Taking real and imaginary parts we then have,

$$P\int_{-\infty}^{\infty}\frac{\mathrm{Re}[\varepsilon(\omega)]\mathrm{d}\omega}{\omega-a}=-\pi\,\mathrm{Im}[\varepsilon(a)]\,, \tag{10.35}$$

and

$$P\int_{-\infty}^{\infty}\frac{\mathrm{Im}[\varepsilon(\omega)]\mathrm{d}\omega}{\omega-a}=+\pi\,\mathrm{Re}[\varepsilon(a)]\,. \tag{10.36}$$

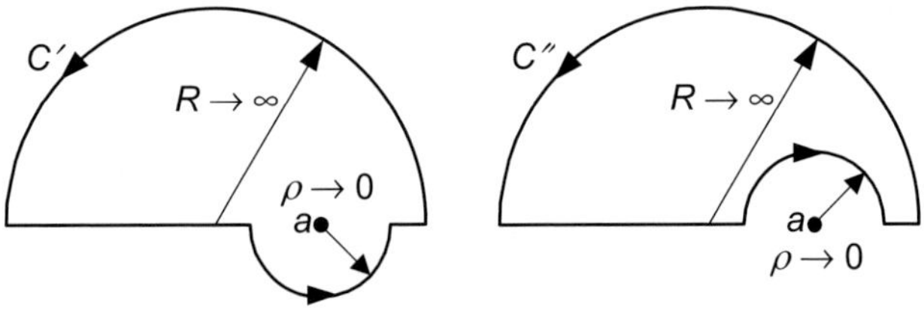

Fig. 10.2. Contours used for Cauchy principal value

There are some other ways to write these relationships,

$$\varepsilon_r(\omega) = \frac{1}{\pi} P \int_{-\infty}^{\infty} \frac{\varepsilon_i(\omega)}{\omega - a} \mathrm{d}\omega = \frac{1}{\pi} P \left[\int_0^{\infty} \frac{\varepsilon_i(\omega)}{\omega - a} \mathrm{d}\omega + \int_{-\infty}^{0} \frac{\varepsilon_i(\omega)}{\omega - a} \mathrm{d}\omega \right]. \quad (10.37)$$

But, the second term can be written

$$-\int_0^{-\infty} \frac{\varepsilon_i(\omega)}{\omega - a} d\omega = \int_0^{-\infty} \frac{[-\varepsilon_i(\omega)]}{-\omega + a} \mathrm{d}(-\omega) = \int_0^{\infty} \frac{[-\varepsilon_i(-\omega)]}{\omega + a} \mathrm{d}\omega\,, \quad (10.38)$$

and $\varepsilon^*(\boldsymbol{r}, t) = \varepsilon(\boldsymbol{r}, t)$, so $\varepsilon(-q,-\omega) = \varepsilon^*(q, \omega)$. Therefore,

$$\varepsilon^*(-\omega) = \varepsilon(\omega)\,; \quad (10.39)$$

or

$$\varepsilon_r(-\omega) = \varepsilon_r(\omega)\,, \quad (10.40)$$

and

$$\varepsilon_i(-\omega) = -\varepsilon_i(\omega)\,. \quad (10.41)$$

We get

$$\frac{1}{\pi} P \int_{-\infty}^{0} \frac{\varepsilon_i(\omega)}{\omega - a} \mathrm{d}\omega = \int_0^{\infty} \frac{\varepsilon_i(\omega)}{\omega + a} \mathrm{d}\omega\,. \quad (10.42)$$

We can thus write the real component of the dielectric constant as

$$\varepsilon_r(a) = \frac{P}{\pi} \left[\int_0^{\infty} \frac{\varepsilon_i(\omega)\mathrm{d}\omega}{\omega - a} + \int_0^{\infty} \frac{\varepsilon_i(\omega)\mathrm{d}\omega}{\omega + a} \right] = \frac{2P}{\pi} \int_0^{\infty} \frac{\omega \varepsilon_i(\omega)\mathrm{d}\omega}{\omega^2 - a^2}, \quad (10.43)$$

and similarly the imaginary component can be written

$$\begin{aligned} \varepsilon_i(a) &= -\frac{P}{\pi} \int_{-\infty}^{\infty} \frac{\varepsilon_r \mathrm{d}\omega}{\omega - a} = -\frac{P}{\pi} \left[\int_0^{\infty} \frac{\varepsilon_r \mathrm{d}\omega}{\omega - a} + \int_{-\infty}^{0} \frac{\varepsilon_r \mathrm{d}\omega}{\omega - a} \right] \\ &= -\frac{P}{\pi} \left[\int_0^{\infty} \frac{\varepsilon_r \mathrm{d}\omega}{\omega - a} + \int_{-\infty}^{0} \frac{\varepsilon_r(-\omega)\mathrm{d}(-\omega)}{-\omega + a} \right] \\ &= -\frac{P}{\pi} \left[\int_0^{\infty} \frac{\varepsilon_r(\omega)\mathrm{d}\omega}{\omega - a} + \int_{\infty}^{0} \frac{\varepsilon_r(\omega)\mathrm{d}\omega}{\omega + a} \right] \\ &= -\frac{P}{\pi} \left[\int_0^{\infty} \frac{\varepsilon_r(\omega)\mathrm{d}\omega}{\omega - a} - \int_0^{\infty} \frac{\varepsilon_r(\omega)\mathrm{d}\omega}{\omega + a} \right] = -\frac{2aP}{\pi} \int_0^{\infty} \frac{\varepsilon_r(\omega)\mathrm{d}\omega}{\omega^2 - a^2}. \end{aligned} \quad (10.44)$$

In summary, the Kronig–Kramers relations can be written, where $\varepsilon_r(\omega) \to \varepsilon_r(\omega) - 1$ should be substituted

$$\varepsilon_r(a) = \frac{P}{\pi} \int_{-\infty}^{\infty} \frac{\mathrm{Im}[\varepsilon(\omega)]\mathrm{d}\omega}{\omega - a} = \frac{2P}{\pi} \int_0^{\infty} \frac{\omega \varepsilon_i(\omega)\mathrm{d}\omega}{\omega^2 - a^2}, \quad (10.45)$$

$$\varepsilon_i(a) = -\frac{P}{\pi} \int_{-\infty}^{\infty} \frac{\mathrm{Re}[\varepsilon(\omega)]\mathrm{d}\omega}{\omega - a} = -\frac{2Pa}{\pi} \int_0^{\infty} \frac{\varepsilon_r(\omega)\mathrm{d}\omega}{\omega^2 - a^2}. \quad (10.46)$$

10.3 Absorption of Electromagnetic Radiation–General (B)

We now give a fairly general discussion of the absorption process by quantum mechanics (see also Yu and Cardona [10.27 Chap. 6] as well as Fox op. cit. Chap. 3). Although much of the discussion is more general, we have in mind the absorption due to transitions between the valence and conduction bands of semiconductors. If $-e$ is the electronic charge, and if we assume the electromagnetic field is described by a vector potential A and a scalar potential ϕ, the Hamiltonian describing the electron in the field is in SI

$$\mathcal{H} = \frac{1}{2m}[\mathbf{p} + e\mathbf{A}]^2 - e(\phi + V), \tag{10.47}$$

where V is the potential in the absence of an electromagnetic field; V would be a periodic potential if the electron were in a solid. We will use the Coulomb gauge to describe the electromagnetic field so $\phi = 0$, $\nabla \cdot A = 0$ and the fields are given by

$$\boldsymbol{E} = -\frac{\partial A}{\partial t}, \quad \boldsymbol{B} = \nabla \times A. \tag{10.48}$$

The Hamiltonian can then be written

$$\mathcal{H} = \frac{1}{2m}[p^2 + eA \cdot \boldsymbol{p} + e\boldsymbol{p} \cdot A + e^2 A^2] - eV. \tag{10.49}$$

The terms quadratic in A will be ignored as they are normally small compared to the terms linear in A. Further in the Coulomb gauge, we can write

$$\boldsymbol{p} \cdot A\psi \propto \nabla \cdot (A\psi) = (\nabla \cdot A)\psi + (A \cdot \nabla)\psi = A \cdot \nabla\psi, \tag{10.50}$$

so that the Hamiltonian can be written

$$\mathcal{H} = \mathcal{H}_0 + \mathcal{H}', \quad \mathcal{H}_0 = \frac{p^2}{2m} - eV;$$

where the perturbation is

$$\mathcal{H}' = \frac{e}{m} A \cdot \boldsymbol{p}. \tag{10.51}$$

We assume the matrix element responsible for electronic transitions will be in the form $\langle f|H|i\rangle$, where i and f refer to the initial and final electron states and $\mathcal{H}'$ is the perturbing Hamiltonian. We assume the vector potential is given by

$$A(\boldsymbol{r}, t) = a\boldsymbol{e}\{\exp[\mathrm{i}(\boldsymbol{k} \cdot \boldsymbol{r} - \omega t)] + \exp[-\mathrm{i}(\boldsymbol{k} \cdot \boldsymbol{r} - \omega t)]\}, \tag{10.52}$$

where $\boldsymbol{e} \cdot \boldsymbol{k} = 0$ and a^2 is given by

$$a = \sqrt{E^2/2\omega^2}, \tag{10.53}$$

where $\overline{E^2}$ is the averaged squared electric field. Then,

$$P_{i\to f}^{\text{absorption}} = \frac{2\pi}{\hbar} a^2 \frac{e^2}{m^2} \left| \langle f | \exp(i\boldsymbol{k}\cdot\boldsymbol{r})\boldsymbol{e}\cdot\boldsymbol{p} | i \rangle \right|^2 , \tag{10.54}$$

and for emission

$$P_{i\to f}^{\text{emission}} = \frac{2\pi}{\hbar} a^2 \frac{e^2}{m^2} \left| \langle f | \exp(-i\boldsymbol{k}\cdot\boldsymbol{r})\boldsymbol{e}\cdot\boldsymbol{p} | i \rangle \right|^2 . \tag{10.55}$$

10.4 Direct and Indirect Absorption Coefficients (B)

Let us now look at the absorption coefficient. Using Bloch wave functions ($\psi_k = \mathrm{e}^{i\boldsymbol{k}\cdot\boldsymbol{r}} u(\boldsymbol{r})$), we have

$$\begin{aligned} \langle f | \exp(i\boldsymbol{k}\cdot\boldsymbol{r})\boldsymbol{e}\cdot\boldsymbol{p} | i \rangle &= \int u_f^* \exp[\mathrm{i}(\boldsymbol{k} - \boldsymbol{k}_f + \boldsymbol{k}_i)\cdot\boldsymbol{r}]\hbar\boldsymbol{e}\cdot\boldsymbol{k}_i u_i \mathrm{d}\Omega \\ &+ \int u_f^* \exp[\mathrm{i}(\boldsymbol{k} - \boldsymbol{k}_f + \boldsymbol{k}_i)\cdot\boldsymbol{r}]\hbar\boldsymbol{e}\cdot\boldsymbol{p} u_i \mathrm{d}\Omega . \end{aligned} \tag{10.56}$$

The first integral can be written as proportional to

$$\begin{aligned} \int \psi_f^* \psi_i \exp(\mathrm{i}\boldsymbol{k}\cdot\boldsymbol{r})\mathrm{d}\Omega &= \sum\nolimits_j \exp[\mathrm{i}(\boldsymbol{k} - \boldsymbol{k}_f + \boldsymbol{k}_i)\cdot\boldsymbol{R}_j] \\ &\times \int_{\Omega_c} u_f^* \exp[\mathrm{i}(\boldsymbol{k} - \boldsymbol{k}_f + \boldsymbol{k}_i)\cdot\boldsymbol{r}] u_i \mathrm{d}\Omega \\ &\cong N \int_{\Omega_c} \psi_f \psi_i \mathrm{d}\Omega \cong 0 , \end{aligned} \tag{10.57}$$

by orthogonality and assuming k is approximately zero, where we have also used

$$\sum\nolimits_j \exp[\mathrm{i}(\boldsymbol{k} - \boldsymbol{k}_f + \boldsymbol{k}_i)\cdot\boldsymbol{R}_j] = \delta_{\boldsymbol{k}}^{\boldsymbol{k}_f - \boldsymbol{k}_i}(N) , \tag{10.58}$$

and Ω_c is the volume of a unit cell. The neglect of all terms but the $k = 0$ terms (called the electric dipole approximation) allows a similar description of the emission term. Following a similar procedure for the second term in (10.56), we obtain for absorption,

$$\langle f | \exp(\mathrm{i}\boldsymbol{k}\cdot\boldsymbol{r})\boldsymbol{e}\cdot\boldsymbol{p} | i \rangle = N \int_{\Omega_c} u_f^* \boldsymbol{e}\cdot\boldsymbol{p} u_i \mathrm{d}\Omega , \tag{10.59}$$

with $k = 0$ and $\boldsymbol{k}_i = \boldsymbol{k}_f$.

Notice in the electric dipole approximation since $k_i = k_f$, we have what are called *direct optical transitions*. If something else such as phonons is involved, direct transitions are not required but the whole discussion must be modified to include this new physical ingredient. The electric dipole transition probability for photon absorption per unit time is

$$P_{i\to f}^{\text{abs}} = \frac{2\pi}{\hbar} \sum_{\boldsymbol{k}} \frac{\overline{E^2}}{2\omega^2} \frac{e^2 N^2}{m^2} \left| \int_{\Omega_c} u_f^* \boldsymbol{e}\cdot\boldsymbol{p} u_i \mathrm{d}\Omega \right|^2 \delta[E_c(\boldsymbol{k}) - E_v(\boldsymbol{k}) - \hbar\omega] . \tag{10.60}$$

The power (per unit volume) lost by the field due to absorption in the medium is the transition probability per unit volume P multiplied by the energy of each photon (where in carrying out the sum over $\boldsymbol{k}$ in (10.60), we will assume we are summing over $\boldsymbol{k}$ states per unit volume). Carrying out the manipulations below, we finally find an expression for the absorption coefficient and, hence, the imaginary part of the dielectric constant. The power lost equals

$$P\hbar\omega = -\frac{\mathrm{d}I}{\mathrm{d}t}, \tag{10.61}$$

where I is the energy/volume. But,

$$-\frac{\mathrm{d}I}{\mathrm{d}t} = -\frac{\mathrm{d}I}{\mathrm{d}x}\frac{\mathrm{d}x}{\mathrm{d}t} = \alpha I\left(\frac{c}{n}\right), \tag{10.62}$$

where $\alpha = 2n_i\omega/c$, and $n_i = \varepsilon_i/2n$. Thus,

$$-\frac{\mathrm{d}I}{\mathrm{d}t} = \frac{\varepsilon_i \omega I}{n^2} = P\hbar\omega . \tag{10.63}$$

Using

$$I = \frac{1}{2}n^2\varepsilon_0\overline{E^2}\times 2, \tag{10.64}$$

where $n = (\varepsilon/\varepsilon_0)^{1/2}$ if $\mu = \mu_0$ and the factor of 2 comes from both magnetic and electric fields carrying current, we find

$$\varepsilon_i(\omega) = \frac{P\hbar}{\varepsilon_0}\frac{1}{\overline{E^2}} . \tag{10.65}$$

Using the Kronig–Kramers relations, we can also derive an expression for the real part of the dielectric constant. Defining

$$\left|M_{vc}\right| = \left|\int_\Omega u_f^* \boldsymbol{e}\cdot\boldsymbol{p}u_i \mathrm{d}\Omega\right|, \tag{10.66}$$

we have (using (10.65), (10.66), and (10.60))

$$\varepsilon_i = \frac{\pi}{\varepsilon_0}\left(\frac{e}{m\omega}\right)^2 \sum\nolimits_{\boldsymbol{k}}\left|M_{vc}\right|^2 \delta(E_c - E_v - \hbar\omega), \tag{10.67}$$

and by (10.45)

$$\varepsilon_r = 1 + \frac{e^2}{m\varepsilon_0}\sum\nolimits_{\boldsymbol{k}}\left(\frac{2}{m\hbar\omega_{cv}}\frac{\left|M_{vc}\right|^2}{\omega_{cv}^2 - \omega^2}\right) \tag{10.68}$$

(where $E_c - E_v \equiv \hbar\omega_{cv}$ and $\delta(ax) = \delta(x)/a$ has been used). Recall that the Σ_k has to be per unit volume and the oscillator strength is defined by

$$f_{vc} = \frac{2|M_{vc}|^2}{m\hbar\omega_{cv}}. \tag{10.69}$$

Classically, the oscillator strength is the number of oscillators per unit volume with frequency ω_{cv}. Thus, the real part of the dielectric constant can be written

$$\varepsilon_r = 1 + \frac{e^2}{m\varepsilon_0}\Sigma_k\left(\frac{f_{vc}}{\omega_{cv}^2 - \omega^2}\right). \tag{10.70}$$

We want to work this out in a little more detail for direct absorption edges. For direct transitions between parabolic valence and conduction bands, effective mass concepts enter because one has to deal with both the valence band and conduction band. For parabolic bands we write

$$E_{vc} = E_g + \frac{\hbar^2 k^2}{2\mu}, \tag{10.71}$$

where

$$\frac{1}{\mu} = \frac{1}{m_c} + \frac{1}{m_v}. \tag{10.72}$$

The joint density of states per unit volume (see (10.94)) is then given by

$$D_j = \left[\frac{\sqrt{2}\mu^{3/2}}{\pi^2\hbar^3}\right]\sqrt{E_{vc} - E_g} \quad \text{where} \quad E_{vc} > E_g, \tag{10.73}$$

and

$$D_j = 0 \quad \text{where} \quad E_{vc} < E_g. \tag{10.74}$$

Thus, we obtain that the imaginary part of the dielectric constant is given by

$$\varepsilon_i(\omega) = \begin{cases} K\sqrt{\Omega - 1}, & \Omega > 1 \\ 0 & , \Omega < 1 \end{cases}, \tag{10.75}$$

where

$$K = \frac{2e^2(2\mu)^{3/2}|M_{vc}|^2\sqrt{E_g}}{m^2\omega^2\hbar^3}, \tag{10.76}$$

and

$$\Omega = \frac{\hbar\omega}{E_g}. \tag{10.77}$$

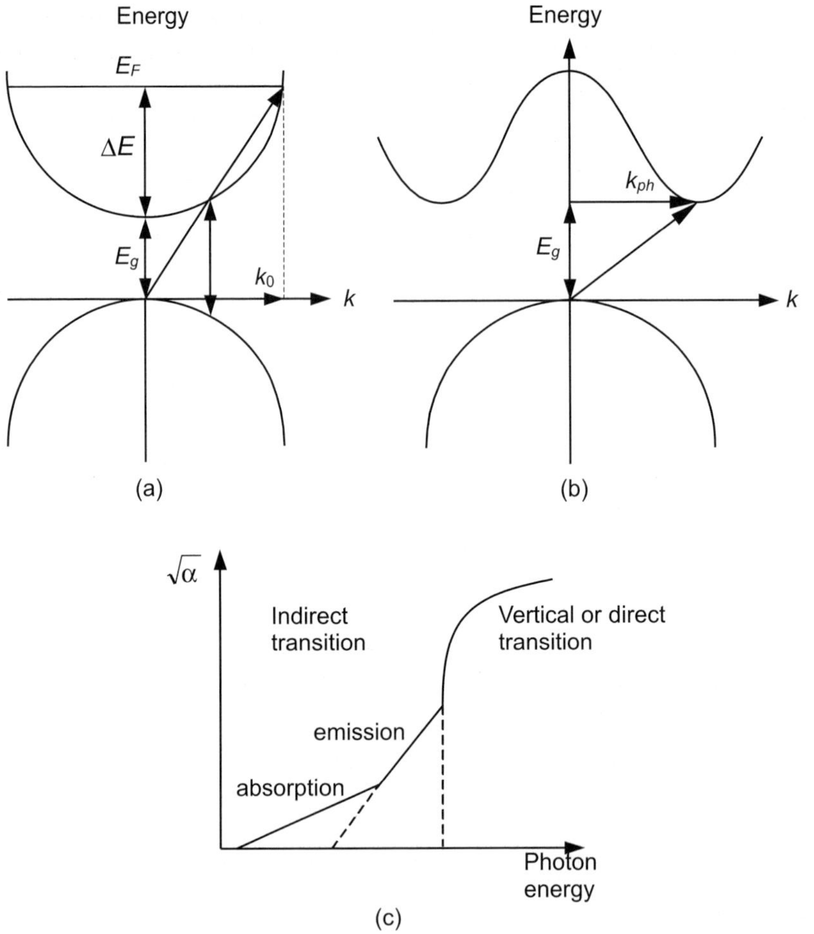

Fig. 10.3. (**a**) Direct transitions and indirect transitions due to band filling; (**b**) Indirect transitions, where k_{ph} is the phonon wave vector; (**c**) Vertical transitions dominate indirect transitions when energy is sufficient to cause them. Emission and absorption refer to phonons in all sketches

From this, one then has an expression for the absorption coefficient (since $\alpha = \omega\varepsilon_i/nc$). Thus for direct transitions and parabolic bands, a plot of the square of the absorption coefficient as a function of the photon energy should be a straight line, at least over a limited frequency. Figure 10.3 illustrates direct and indirect transitions and absorption. Indirect transitions are discussed below.

The fundamental absorption edge due to the bandgap determines the apparent color of semiconductors as seen by transmission.

We now want to discuss indirect transitions. So far, our analysis has assumed a direct bandgap. This means that the k of the initial and final electronic states defining the absorption edge are almost the same (as has been mentioned, the k of the photon causing the absorption is negligible, compared to the Brillouin zone width, for visible wavelengths). This is not true for the two most common semiconductors Si and Ge. For these semiconductors, the maximum energy of the valence band and the minimum energy of the conduction band do not occur at the same k vectors, one has what is called an indirect bandgap semiconductor. For a minimum energy transition across the bandgap, something else, typically a phonon, must be involved in order to conserve wave vector. The requirement of having, for example, a phonon being involved reduces the probability of the event; see Fig. 10.3b, c, Fig. 10.4, (10.82), and consider also Fermi's Golden Rule.

Even in a direct bandgap semiconductor, processes can cause the fundamental absorption edge to shift from direct to indirect, see Fig. 10.3a. For degenerate semiconductors, the optical absorption edge may be a function of the carrier density. In simple models, the location of the Fermi energy in the conduction band can be estimated on the free-electron model. When the Fermi energy is above the bottom of the conduction band, the k vector of the minimum energy that can cause a transition has also shifted from the k of the conduction band minimum. Now direct transitions will originate from deeper states in the valence band, they will be stronger than the threshold energy transitions, but of higher energy.

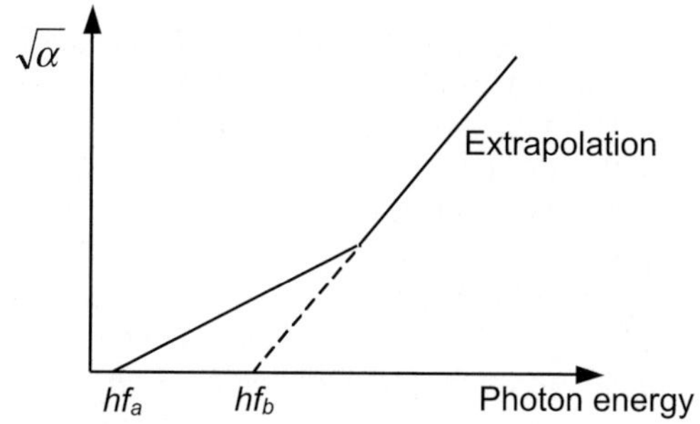

Fig. 10.4. Indirect transitions: $hf_a = E_g - E_{\text{phonon}}$, $hf_b = E_g + E_{\text{phonon}}$, $E_g = (hf_a + hf_b)/2$, sketch

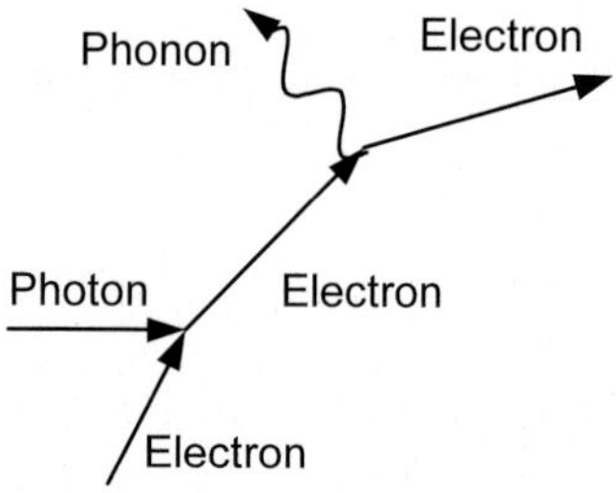

Fig. 10.5. An indirect process viewed in two steps

For indirect transitions, we can write the energy and momentum conservation conditions as follows:

$$\boldsymbol{k}' = \boldsymbol{k} + \boldsymbol{K} \pm \boldsymbol{q}\,, \tag{10.78}$$

where $\boldsymbol{K}$ = photon ≅ 0 and, $\boldsymbol{q}$ = phonon (=k_{ph} in Fig. 10.3b). Also

$$E(\boldsymbol{k}') = E(\boldsymbol{k}) + \hbar\omega \pm \hbar\omega_{\boldsymbol{q}}, \tag{10.79}$$

where $\hbar\omega$ = photon, and $\hbar\omega_q$ = phonon. Note: although the photon makes the main contribution to the transition energy, the phonon carries the burden of insuring that momentum is conserved. Now the Hamiltonian for the process would look like

$$\mathcal{H}' = \mathcal{H}'_{\text{photon}} + \mathcal{H}'_{\text{phonon}}, \tag{10.80}$$

where

$$\mathcal{H}'_{\text{photon}} = \frac{e}{m}\boldsymbol{p}\cdot\boldsymbol{A}, \tag{10.81}$$

and

$$\mathcal{H}'_{\text{phonon}} = \sum\nolimits_{\sigma kq} M_{\boldsymbol{kq}}(a^{+}_{-\boldsymbol{q}} + a_{\boldsymbol{q}})c^{+}_{\boldsymbol{k}+\boldsymbol{q},\sigma}c_{\boldsymbol{k},\sigma}\,. \tag{10.82}$$

One can sketch the indirect process as a two-step process in which the electron absorbs a photon and changes state then absorbs or emits a phonon. See Fig. 10.5.

We mention as an aside another topic of considerable interest. We discuss briefly optical absorption in an electric field. The interesting feature of this phenomenon is that in an electric field, optical absorption can occur for photon energies lower than the normal bandgap energies. The increased optical absorption due to an electric field can be qualitatively understood by thinking about pictures such as in Fig. 10.6. This figure does not present a rigorous concept, but it is helpful.

Very simply, we can think of the triangular area in the figure as a potential barrier that electrons can "tunnel" through. From this point of view, one perhaps believes than an electric field can cause electronic transitions from band 2 to band 1 (This is called the Zener effect). Obviously, the process of tunneling would be greatly enhanced if the electron "picked up some energy from a photon before it began to tunnel." Further details are given by Kane [10.15].

It is not hard to see why the Zener effect (or "Zener breakdown") can be considered as a tunneling effect. The horizontal line corresponds to the motion of an electron (if we describe electrons in terms of wave packets, then we can speak of where they are at various times and we can label positions in terms of distances in the bands). Actually, we should realize that this horizontal line corresponds to the electric field causing the electron to make transitions to higher and higher stationary states in the crystal. When the electron reaches the top of the lower band, we normally think of the electron as being Bragg reflected. However, we should remember what we mean by the energy gap.

The energy gap, E_g, does not represent an absolutely forbidden gap. It simply represents energies corresponding to attenuated, nonpropagating wave functions. The attenuation will be of the form e^{-Kx}, where x represents the distance traveled (K is real and greater than zero) and K is actually a function of x, but this will be ignored here. The electron gains energy from the electric field E as $|eEx|$. When

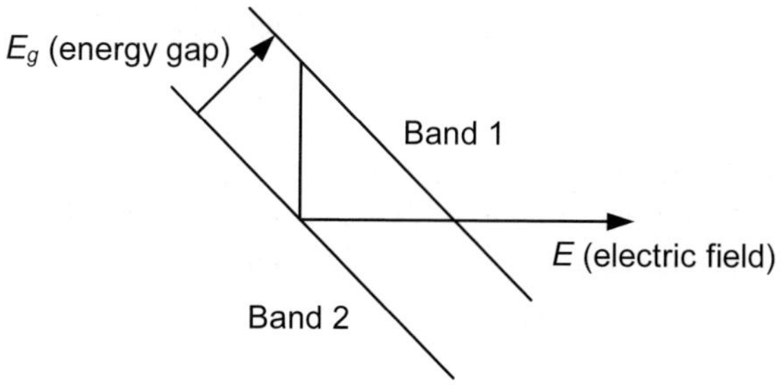

Fig. 10.6. Qualitative effect of an electric field on the energy bands in a solid

the electron has traveled $x = |E_g/eE|$, it has gained sufficient energy to get into the bottom of the upper band if it started at the top of the lower band. In order for the process to occur, we must require that the electron's wave function not be too strongly attenuated, i.e. Zener breakdown will occur if $1/K >> |E_g/eE|$. To see the analogy to tunneling, we observe that the electron's wave function in the triangular region also behaves as e^{-Kx} from a tunneling viewpoint (also with K a function of x), and that the larger we make the electric field, the thinner the area we have to tunnel across, so the greater a band-to-band transition. A more quantitative discussion of this effect is obtained by evaluating K not from the picture, but directly from the Schrödinger equation. The x dependence on K turns out to be fairly easy to handle in the WKB approximation.

Finally, we can summarize the results for many cases in Table 10.1. Absorption coefficients α for various cases (parabolic bands) can be written

$$\alpha = \left(\frac{A}{hf}\right)(hf + \beta - E_g)^{\gamma}, \tag{10.83}$$

where γ, β depend on the process as shown in the table. When phonons are involved we need to add both the absorption and emission ($\pm$) possibilities to get the total absorption coefficient.[2] A very clean example of optical absorption is given in Fig. 10.7. Good optical absorption experiments on InSb were done in the early days by Gobeli and Fan [10.15]. In general, one also needs to take into account the effect of temperature. For example, the indirect allowed term should be written

$$\alpha = \left(\frac{A'}{hf}\right)\left[\frac{(hf + \beta - E_g)^2}{\exp(\beta/kT) - 1} + \frac{(hf - \beta - E_g)^2}{\exp(\beta/kT) - 1}\exp(\beta/kT)\right], \tag{10.84}$$

where A' is a constant independent of the temperature, see, e.g., Bube [10.4] and Pankove [10.22].

[2] An additional very useful reference is Greenaway and Harbeke [10.16]. See also Yu and Cardona [10.27].

Table 10.1. Absorption coefficients

	γ	β
Direct, allowed	1/2	0 See (10.75)
Direct, forbidden	3/2	0
Indirect, allowed	2	$\pm hf_q$ (phonons)
Indirect, forbidden	3	$\pm hf_q$ (phonons)

γ and β are defined by (10.83).

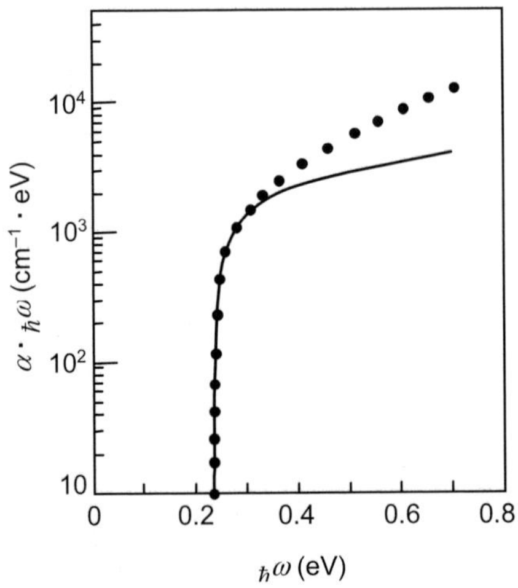

Fig. 10.7. Optical absorption in indium antimonide, InSb at 5 K. The transition is direct because both conduction and valence band edges are at the center of the Brillouin zone, $k = 0$. Notice the sharp threshold. The dots are measurements and the solid line is $(\hbar\omega - E_g)^{1/2}$. (Reprinted with permission from Sapoval B and Hermann C, *Physics of Semiconductors*, Fig. 6.3 p. 154, Copyright 1988 Springer Verlag, New York.)

10.5 Oscillator Strengths and Sum Rules (A)

Let us define the oscillator strength by

$$f_{ij} = b\omega_{ij} \left| \langle i | \boldsymbol{e} \cdot \boldsymbol{r} | j \rangle \right|^2 .$$

We will show this is equivalent to the previous definition with the proper choice of b by using commutation relations to cast it in another form. From $[x, p_x] = \mathrm{i}\hbar$ we can show

$$[\mathcal{H}, x] = -\frac{\mathrm{i}\hbar}{m} p_x . \tag{10.85}$$

Also,

$$[\mathcal{H}, \boldsymbol{e}\cdot\boldsymbol{r}] = -\left(\frac{i\hbar}{m}\right)\boldsymbol{e}\cdot\boldsymbol{p}\,, \tag{10.86}$$

therefore

$$\langle i|\boldsymbol{e}\cdot\boldsymbol{p}|j\rangle = im\omega_{ij}\langle i|\boldsymbol{e}\cdot\boldsymbol{r}|j\rangle\,. \tag{10.87}$$

Thus we can write the oscillator strength as,

$$f_{ij} = b\frac{|\langle i|\boldsymbol{e}\cdot\boldsymbol{p}|j\rangle|^2}{m^2\omega_{ij}}\,, \tag{10.88}$$

which is consistent with how we wrote it before, if $b = -2m/\hbar$ (see (10.69), (10.66)). It is also interesting to show that the oscillator strength obeys a sum rule. If $\boldsymbol{e} = \boldsymbol{i}$, then

$$\begin{aligned}\sum\nolimits_j f_{ij} &= b\sum\nolimits_j \omega_{ij}|\langle i|\boldsymbol{e}\cdot\boldsymbol{r}|j\rangle|^2\\ &= \frac{b}{2}\sum\nolimits_j \frac{1}{im}\{\langle i|\boldsymbol{e}\cdot\boldsymbol{p}|j\rangle\langle j|\boldsymbol{e}\cdot\boldsymbol{r}|i\rangle - \langle i|\boldsymbol{e}\cdot\boldsymbol{r}|j\rangle\langle j|\boldsymbol{e}\cdot\boldsymbol{p}|i\rangle\}\\ &= \frac{b}{2}\frac{1}{im}[\langle i|\boldsymbol{e}\cdot\boldsymbol{p}, \boldsymbol{e}\cdot\boldsymbol{r}|i\rangle] = \frac{b}{2im}[-i\hbar] = 1\,.\end{aligned} \tag{10.89}$$

Classically, for bound states with no damping, we can derive the dielectric constant. Assume N states with frequency ω_0. The result is

$$\frac{\varepsilon}{\varepsilon_0} = 1 + \frac{Ne^2}{m\varepsilon_0}\frac{1}{\omega_0^2 - \omega^2}\,, \tag{10.90}$$

which follows from (9.6) with $\tau \to \infty$ and ω_0 used for several bound states labeled with i. Note that it is just the same as the quantum result (10.70) provided the oscillator strength from one oscillator is one. From this we have the index of refraction, and it is given by $n^2 = \varepsilon/\varepsilon_0$, since ε is real with $\tau \to \infty$. When $\varepsilon/\varepsilon_0$ as the preceding, the resulting equation is often called Sellmeier's equation.

10.6 Critical Points and Joint Density of States (A)

Optical absorption spectra give many details about the band structure. This can be explained by the Van Hove singularities, which appear in the joint density of states as mentioned below. In the integral for the imaginary part of the dielectric constant, we had an expression of the form (10.67):

$$\varepsilon_i \propto \frac{2}{\omega^2}\int\frac{\mathrm{d}^3\boldsymbol{k}}{(2\pi)^3}|M_{vc}|^2\,\delta(E_c - E_v)\,. \tag{10.91}$$

A property of delta functions can be written as

$$\int_a^b g(x)\delta[f(x)]\mathrm{d}x = \sum_{x_p} g(x_p)\frac{1}{\left.\frac{\partial f}{\partial x}\right|_{x=x_p}}, \tag{10.92}$$

where x_p are the zeros of $f(x)$. From which we conclude that the imaginary part of the dielectric constant can be written as

$$\varepsilon_i \propto \frac{2}{\omega^2}\frac{1}{(2\pi)^3}\int_S \frac{\mathrm{d}S|M_{vc}|^2}{|\nabla_{\boldsymbol{k}}(E_c - E_v)|_{E_c - E_v = \hbar\omega}}, \tag{10.93}$$

where dS is a surface of constant $\hbar\omega = E_c - E_v$. The joint density of states is defined as (Yu and Cardona [10.27 p251])

$$J_{vc} = \int \frac{2}{(2\pi)^3}\frac{\mathrm{d}S}{|\nabla_{\boldsymbol{k}}(E_c - E_v)|_{E_c - E_v = \hbar\omega}}, \tag{10.94}$$

and typically the matrix element M_{vc} is a slowly varying function compared with the joint density of states. Now the joint density of states is a strongly varying function of k where the denominator is zero, i.e. where

$$\nabla_{\boldsymbol{k}}(E_c - E_v) = 0 . \tag{10.95}$$

Both valence and conduction band energies must be periodic functions in reciprocal space and so must their difference and from this it follows that there must be a point for which the denominator vanishes (smooth periodic functions have analytic maxima and minima). These critical points lead to singularities in the density of states, the Van Hove singularities. At very highly symmetrical points in the Brillouin zone, we can have critical points due to the gradient of both conduction and valence energies vanishing, at other critical points only the gradient of the difference vanishes. Critical points are defined by the band structure, and in turn, they help determine the absorption coefficient. Reversing the process, studying the absorption coefficient gives information on the band structure.

10.7 Exciton Absorption (A)

In semiconductors, one may detect absorption for energies just below the energy gap where one might have initially expected transparency. This could be due to absorption by bound clectron–hole pairs or excitons. The binding energy of the excitons lowers their absorption below the bandgap energy. It is interesting that one can only think of bound electron–hole pairs if electron and holes move with the same group velocity, in other words the energy gradients of valence and electronic energies need to be the same. That is, excitons form at the critical points of the joint density of states.

One generally talks of two kinds of excitons, the Frenkel excitons and Wannier excitons. The Frenkel excitons are tightly bound and can be described by a variant of tight binding theory. Another way to view Frenkel excitons is as a propagating excited state of a single atom. Thus, we describe it with the Hamiltonian where the states are the localized excited states of each atom. For the Frenkel case let

$$\mathcal{H} = \sum_i \varepsilon |i\rangle\langle i| + \sum_{i,j} V_{ij} |i\rangle\langle j|, \tag{10.96}$$

where with one-dimensional nearest-neighbor hopping

$$V_{ij} = V\delta_i^{j+1} + V\delta_i^{j-1}. \tag{10.97}$$

This can be diagonalized by the substitution:

$$|k\rangle = \sum_j \exp(\mathrm{i}jka)|j\rangle, \tag{10.98}$$

which leads to the energy eigenvalues

$$\mathcal{H}|k\rangle = \varepsilon_k |k\rangle, \tag{10.99}$$

where, $\varepsilon_k = \varepsilon + 2V\cos(ka)$. These Frenkel types of excitons are found in the alkali halides.

In semiconductors, the important types of excitons are the Wannier excitons, which have size much larger than typical interatomic dimensions. The Wannier excitons can be analyzed much as a hydrogen atom with reduced mass defined by the electron and hole masses and with the binding Coulomb potential reduced by the appropriate dielectric constant. That is, the energy eigenvalues are

$$E_n = E_g - \frac{\mu e^4}{2\hbar^2(4\pi\varepsilon)^2 n^2}, \quad n = 1, 2, \ldots, \tag{10.100}$$

where

$$\frac{1}{\mu} = \frac{1}{m_e} + \frac{1}{m_h}. \tag{10.101}$$

Optical absorption in GaAs is shown in Fig. 10.8.

10.8 Imperfections (B, MS, MET)

We will only give a brief discussion here. Reference should be made also to the chapters on semiconductors and defects. Imperfections may produce resonant energy levels in the bands or energy levels that are in the bandgap. Donors and acceptors in semiconductors produce energy levels that may be detected by optical absorption when the thermal energy is much less than their ionization energy. Similarly, deep defects produced in a variety of ways may produce energy levels in the gap, often near the center. Deep defects tend to be very localized in space and therefore to contain a large range of k vectors. Thus, it is possible to have

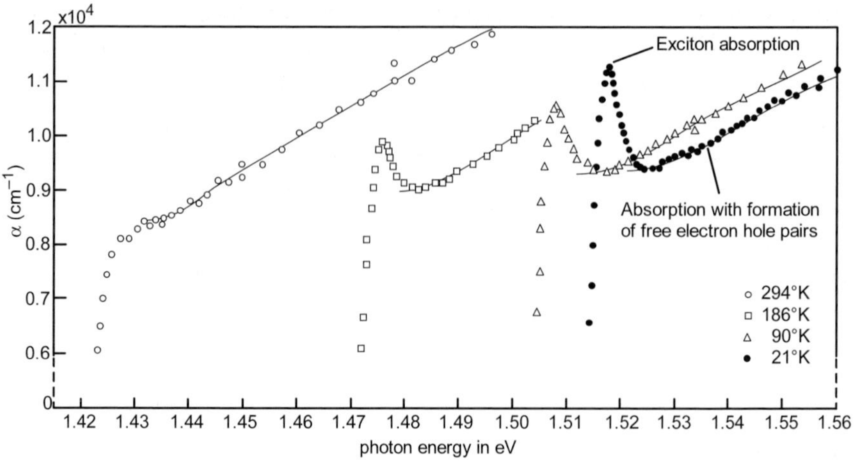

Fig. 10.8. Absorption coefficient near the band edge of GaAs. Note the exciton absorption level below the bandgap E_g [Reprinted with permission from Sturge MD, *Phys Rev* **127**, 771 (1962). Copyright 1962 by the American Physical Society.]

a direct transition from a deep defect to a large range of k values in the conduction band, for example. A shallow level, on the other hand, is well spread out in space and therefore restricted in k value and so direct transitions from it to a band go to quite a restricted range of values. Color centers in alkali halides are examples of other kinds of optically important defects.

Suppose we have some generic defect with energy level in the gap. One could have absorption due to transitions from the valence or conduction band to the level. There could even be absorption between levels due to the defect or different defects. Several types of optical processes are suggested in Fig. 10.9.

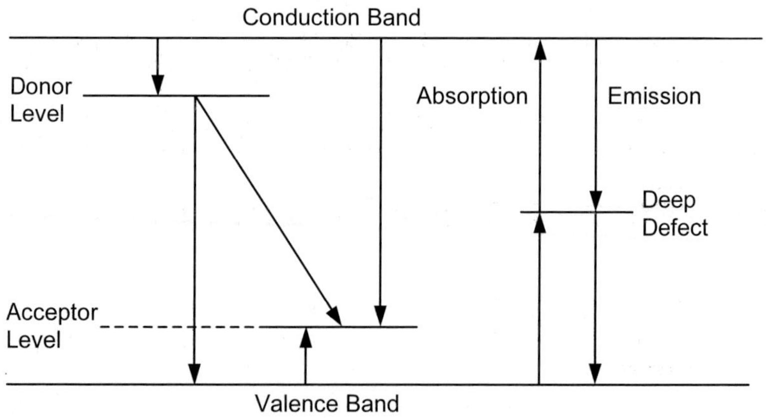

Fig. 10.9. Some typical radiative transitions in semiconductors. Nonradiative (Auger) transitions are also possible

10.9 Optical Properties of Metals (B, EE, MS)

Free-carrier absorption can be viewed as intraband absorption–the electron absorbing the photon remains in the same band.[3] Free-carrier absorption is obviously important for metals, and is often of importance for semiconductors. The electron is accelerated by the photon and gains energy, but since the wave vector of the photon is negligible, something else such as a phonon needs to be involved. For many purposes, the process can be viewed classically by Drude theory with a relaxation time of $\tau \equiv 1/\omega_0$. This relaxation time defines a frictional force constant m^*/τ, where the viscous like frictional force is proportional to the velocity.

We will use classical theory here, but it is worthwhile to make a few comments. It is common to deal with a semiclassical picture of radiation. There we treat the radiation classically, but the underlying electronic systems that absorb and emit the radiation we treat quantum mechanically. Radiation can be treated classically when it is intense enough to have many photons in each mode. Free-electronic systems can be treated classically when their de Broglie wavelengths are small compared to the average interparticle separations.

The de Broglie wavelength can be estimated from the momentum as estimated from equipartition. In practice, this means that for temperatures that are not too low and densities that are not too high, then classical mechanics should be valid. Bound systems are more complicated, but in general, classical mechanics works at higher quantum numbers (higher bound-state energies). In any case, classical and quantum results often overlap in validity well beyond where one might naively expect.

The classical theory can be written, assuming a sinusoidal electric field $E = E_0\exp(-\mathrm{i}\omega t)$ (note these are for free-electrons ($e > 0$) with damping). We also generalize by using an effective mass m^* rather than m:

$$m^*\ddot{x} + \frac{m^*}{\tau}\dot{x} = -eE_0\exp(-\mathrm{i}\omega\tau) . \qquad (10.102)$$

Note this is just (9.1) with $\omega_0 = 0$, as appropriate for free charges. Seeking a steady-state solution of the form $x = x_0\exp(-\mathrm{i}\omega\tau)$, we find

$$x = \frac{-\mathrm{i}eE\tau}{m^*\omega(1-\mathrm{i}\omega\tau)} , \qquad (10.103)$$

which is (9.2) with $\omega_0 = 0$. Thus, the polarization is given by

$$P = -Nex = (\varepsilon - \varepsilon_L)E , \qquad (10.104)$$

where ε_L is the contribution to the dielectric constant of everything except the free carriers (generalizing (9.3)). The frequency-dependent dielectric constant is

$$\varepsilon(\omega) = \varepsilon_L + \mathrm{i}\frac{Ne^2\tau}{m^*\omega}\frac{1}{1-\mathrm{i}\omega\tau} , \qquad (10.105)$$

[3] See also, e.g., Ziman [25, Chap. 8] and Born and Wolf [10.1].

where N is the number of electrons per unit volume. From the real and imaginary parts of ε we find, similar to Sect. 9.2,

$$\varepsilon_r = n^2 - n_i^2 = \frac{\varepsilon_L}{\varepsilon_0} - \frac{\sigma_0\tau/\varepsilon_0}{1+\omega^2\tau^2}, \quad \sigma_0 = \frac{Ne^2\tau}{m^*}, \tag{10.106}$$

and

$$\varepsilon_i = 2nn_i = \frac{\sigma_0}{\varepsilon_0\omega}\left(\frac{1}{1+\omega^2\tau^2}\right). \tag{10.107}$$

It is convenient to write this in terms of the plasma frequency

$$\omega_p^2 = \frac{Ne^2}{m\varepsilon_0} = \frac{\sigma_0}{\tau\varepsilon_0} \equiv \frac{\sigma_0\omega_0}{\varepsilon_0}, \tag{10.108}$$

and so,

$$\varepsilon_r = \frac{\varepsilon_L}{\varepsilon_0} - \frac{\omega_p^2}{\omega_0^2 + \omega^2}, \tag{10.109}$$

and

$$\varepsilon_i = \frac{\omega_0}{\omega}\frac{\omega_p^2}{\omega_0^2 + \omega^2}. \tag{10.110}$$

From here onwards for simplicity we assume $\varepsilon_L \cong \varepsilon_0$. We have three important ω. The plasma frequency ω_p is proportional to the free-carrier concentration, ω_0 measures the electron–phonon coupling and ω is the frequency of light.

We now want to show what these equations predict in three different frequency regions.

(i) $\omega\tau << 1$, the low-frequency region. We obtain by (10.109) with $\omega_0 = 1/\tau$

$$n^2 - n_i^2 = 1 - \omega_p^2\tau^2, \tag{10.111}$$

which is small, and by (10.110)

$$2nn_i = \omega_p^2\frac{\tau}{\omega} = \frac{(\omega_p\tau)^2}{\omega\tau}, \tag{10.112}$$

which is large. Here the imaginary part (of the *dielectric constant*) is much greater than the real part and we have high reflectivity. In this approximation

$$n^2 - n_i^2 = 1 - \omega\tau(2nn_i) \cong 1, \tag{10.113}$$

but neither n nor n_i are small, so $n \cong n_i$, and $n^2 \cong \sigma_0/2\omega\varepsilon_0$. The reflectivity then becomes

$$R = \frac{(n-1)^2 + n_i^2}{(n+1)^2 + n_i^2} \cong 1 - \frac{2}{n} \cong 1 - 2\sqrt{\frac{2\omega\varepsilon_0}{\sigma_0}}. \tag{10.114}$$

This is the Hagen–Rubens relation [10.17].

(ii) $1/\tau << \omega << \omega_p$, the relaxation region. The basic relations become

$$n^2 - n_i^2 = 1 - \frac{\omega_p^2}{\omega^2}, \tag{10.115}$$

which is large and negative, and

$$2nn_i = \left(\frac{\omega_p}{\omega}\right)^2 \frac{1}{\omega\tau}, \tag{10.116}$$

which is smaller than $n^2 - n_i^2$. However, this predicts the metal is still strongly reflecting as we now show. Since $\omega\tau >> 1$ and $\omega_p/\omega >> 1$, we see

$$(n - n_i)(n + n_i) \cong -\left(\frac{\omega_p}{\omega}\right)^2 << 1, \tag{10.117}$$

or

$$(n_i - n)(n + n_i) = \left(\frac{\omega_p}{\omega}\right)^2 >> 1. \tag{10.118}$$

Therefore,

$$n_i >> n, \tag{10.119}$$

$$n_i^2 \cong \left(\frac{\omega_p}{\omega}\right)^2, \tag{10.120}$$

$$n_i \cong \frac{\omega_p}{\omega}, \tag{10.121}$$

$$2nn_i \cong n_i^2 \frac{1}{\omega\tau}, \tag{10.122}$$

$$R = \frac{(n-1)^2 + n_i^2}{(n+1)^2 + n_i^2} = \frac{1 + [(n-1)/n_i]^2}{1 + [(n+1)/n_i]^2} \cong 1 + \left(\frac{n-1}{n_i}\right)^2 - \left(\frac{n+1}{n_i}\right)^2, \tag{10.123}$$

and

$$R = 1 - \frac{4n}{n_i^2} = 1 - \frac{2}{\omega_p\tau}. \tag{10.124}$$

Since $\omega_p\tau >> 1$, the metal is still strongly reflecting.

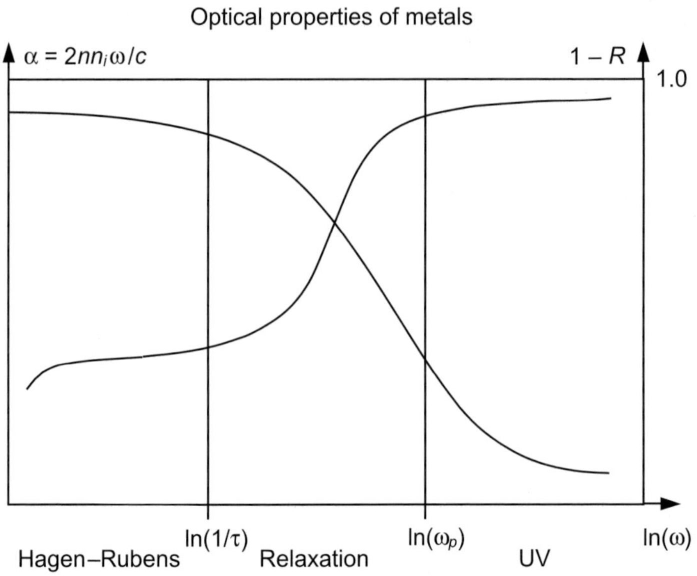

Fig. 10.10. Sketch of absorption and reflection in metals

(iii) $\omega_p << \omega$ or $\omega_p/\omega << 1$. This is the ultraviolet region where we also assume $\omega >> \omega_0$.

$$(n^2 - n_i^2) \cong 1, \tag{10.125}$$

so

$$(n - n_i)(n + n_i) = 1. \tag{10.126}$$

$2nn_i = (\omega_p/\omega)^2(1/\omega\tau)$ is very small. Both n and n_i are not very small, therefore n_i is very small. Therefore,

$$n >> n_i, n \cong 1. \tag{10.127}$$

Therefore,

$$n_i \cong \frac{1}{2}\left(\frac{\omega_p}{\omega}\right)^2 \frac{1}{\omega\tau}. \tag{10.128}$$

So,

$$R \cong \frac{n_i^2}{n_i^2 + 4} \cong \frac{n_i^2}{4} \cong \frac{1}{16}\left(\frac{\omega_p}{\omega}\right)^4 \frac{1}{(\omega\tau)^2} \tag{10.129}$$

is very small. There is little reflectance since this is the ultraviolet transparency region. We summarize our results in Fig. 10.10. See also Seitz [82, p 639], Ziman [25, 1st edn, p 240], and Fox [10.12].

The plasma edge, or the region around the plasma frequency deserves a little more attention. Using Maxwell's equations we have

$$\nabla \times \boldsymbol{E} = -\frac{\partial \boldsymbol{B}}{\partial t}, \quad \nabla \cdot \boldsymbol{E} = \nabla \cdot \boldsymbol{B} = 0 \quad (\rho = 0), \tag{10.130}$$

and

$$\nabla \times \boldsymbol{B} = \mu_0 \frac{\partial^2 \boldsymbol{D}}{\partial t^2} \quad (\boldsymbol{j} = 0), \tag{10.131}$$

and we will include any charge motion in P. Therefore,

$$\nabla^2 E = \mu_0 \frac{\partial^2 \boldsymbol{D}}{\partial t^2}, \quad \boldsymbol{D} = \varepsilon_0 \varepsilon \boldsymbol{E}. \tag{10.132}$$

Note here $\varepsilon \to \varepsilon/\varepsilon_0$. Assume $E = E_0\exp(-\mathrm{i}\omega t)\exp(\mathrm{i}\boldsymbol{k}\cdot\boldsymbol{r})$. We obtain, as shown below ((10.142), (10.143)), for the wave vector

$$k^2 = \varepsilon(\omega)\mu_0\varepsilon_0\omega^2 \Rightarrow (kc)^2 = \varepsilon(\infty)(\omega^2 - \tilde{\omega}_p^2). \tag{10.133}$$

For a free-electron in an electric field we have already derived the plasma frequency in Sect. 9.4. We give here an alternative simple derivation and bring out a few new features,

$$m\frac{\mathrm{d}^2 x}{\mathrm{d}t^2} = -eE, \tag{10.134}$$

$$x = x_0 \exp(-\mathrm{i}\omega t), \tag{10.135}$$

$$E = E_0 \exp(-\mathrm{i}\omega t), \tag{10.136}$$

$$x = \frac{eE}{m\omega^2}. \tag{10.137}$$

Also,

$$P = -Nex = -\frac{Ne^2}{m\omega^2}E, \tag{10.138}$$

$$\varepsilon(\omega) = 1 + \frac{P(\omega)}{\varepsilon_0 E(\omega)} = 1 - \frac{Ne^2}{\varepsilon_0 m\omega^2}, \tag{10.139}$$

$$\omega_p^2 = \frac{Ne^2}{\varepsilon_0 m}, \tag{10.140}$$

$$\varepsilon(\omega) = 1 - \frac{\omega_p^2}{\omega^2}. \tag{10.141}$$

If the positive ion core background has a dielectric constant of $\varepsilon(\infty)$ that is about constant, then (10.141) is modified

$$\varepsilon(\omega) = \varepsilon(\infty)\left[1 - \frac{\tilde{\omega}_p^2}{\omega^2}\right], \tag{10.142}$$

where

$$\tilde{\omega}_p = \frac{\omega_p}{\sqrt{\varepsilon(\infty)}}. \tag{10.143}$$

When the frequency is less than the plasma frequency the squared wave vector is negative (10.133) and gives us total reflection. Above the plasma frequency, the wave vector squared is positive and the material is transparent. That is, simple metals should reflect in the visible and be transparent in the ultraviolet, as we have already seen.

It is also good to remember that at the plasma frequency the electrons undergo low-frequency longitudinal oscillations. See Sect. 9.4. Specifically, note that setting $\varepsilon(\omega) = 0$ defines a frequency $\omega = \omega_L$ corresponding to longitudinal plasma oscillations.

$$\varepsilon(\omega) = 1 - \frac{\omega_p^2}{\omega^2}, \text{ so } \varepsilon(\omega) = 0 \text{ implies } \omega = \omega_p. \tag{10.144}$$

Here we have neglected the dielectric constant of the positive ion cores.

The plasma frequency is also a free longitudinal oscillation. If we have a doped semiconductor with the plasma frequency less than the bandgap over Planck's constant, one can detect the plasma edge, as illustrated in Fig. 10.11. See also Fox op cit, p156. Hence, we can determine the electron concentration.

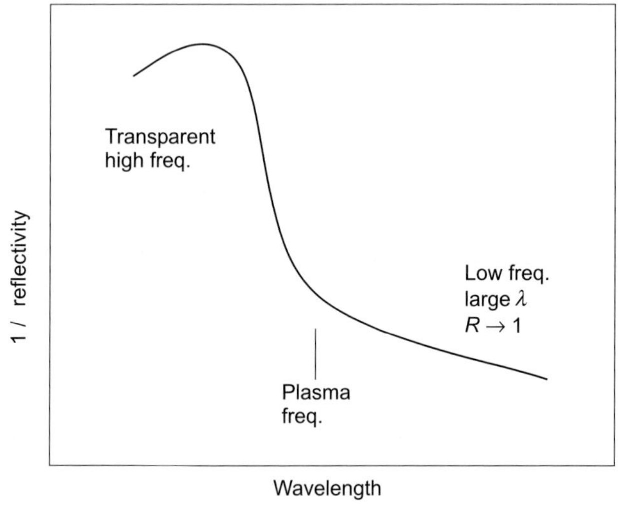

Fig. 10.11. Reflectivity of doped semiconductor, sketch

10.10 Lattice Absorption, Restrahlen, and Polaritons (B)

10.10.1 General Results (A)

Polar solids carry lattice polarization waves and hence can interact with electromagnetic waves (only transverse optical phonons couple to electromagnetic waves by selection rules and conservation laws). The dispersion relations for photons and the phonons of the polarization waves can cross. When these dispersion relations cross, the resulting quanta turn out to be neither photons nor phonons but mixtures called polaritons. One way to view this is shown in Fig. 10.12. We now discuss this process in more detail. We start by considering lattice vibrations in a polar solid. We will later add in a coupling with electromagnetic waves. The displacement of the tth ion in the $\boldsymbol{l}$th cell for the jth component, satisfies

$$M_t \ddot{v}_{t\boldsymbol{l}}^j = -\sum\nolimits_{t'\boldsymbol{h}} G_{tt'}^{jj'}(h) v_{t',\boldsymbol{l}+\boldsymbol{h}}^j , \tag{10.145}$$

where

$$G_{t\boldsymbol{l},t',\boldsymbol{l}'=\boldsymbol{l}+\boldsymbol{h}}^{jj'} = \frac{\partial^2 U}{\partial v_{t\boldsymbol{l}}^j \partial v_{t'\boldsymbol{l}'}^{j'}} , \tag{10.146}$$

and U describes the potential of interaction of the ions. If v_{tl} is a constant,

$$\sum\nolimits_{t'\boldsymbol{h}} G_{tt'}(\boldsymbol{h}) = 0 . \tag{10.147}$$

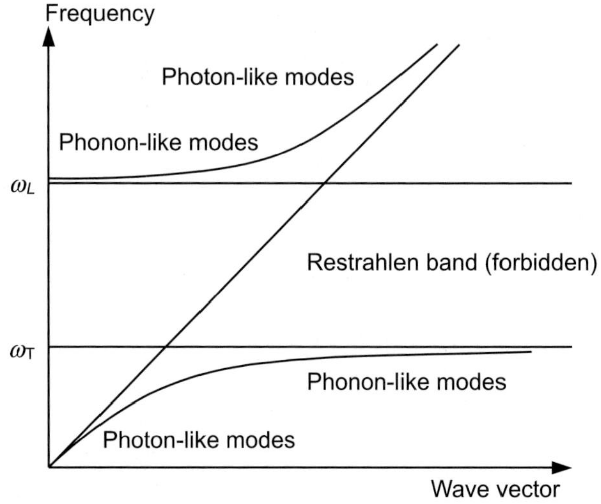

Fig. 10.12. Polaritons as mixtures of photons and transverse phonons. The mathematics of this model is developed in the text

We will add an electromagnetic wave that couples to the system through the force term.

$$e_t E_0 \exp[\mathrm{i}(\boldsymbol{q} \cdot \boldsymbol{l} - \omega t)], \tag{10.148}$$

where e_t is the charge of the tth ion in the cell. We seek solutions of the form

$$v_{sl}(t) = \exp(\mathrm{i}\boldsymbol{q} \cdot \boldsymbol{l}) v_{s,\boldsymbol{q}}(t), \tag{10.149}$$

(now s labels ions) with $\boldsymbol{q} = K$ (dropping the vector notation of $\boldsymbol{q}$, $\boldsymbol{h}$, and $\boldsymbol{l}$ for simplicity from here on) and t is the time. Defining

$$G_{ss'}(K) = \sum\nolimits_h G_{ss'}(h) \exp(\mathrm{i}Kh), \tag{10.150}$$

we have (for one component in field direction)

$$M_s \ddot{v}_{sK} = -\sum\nolimits_{s'} G_{ss'}(K) v_{s'K} + e_s E_0 \exp(-\mathrm{i}\omega t). \tag{10.151}$$

Note that

$$G_{ss'}(K = 0) = \sum\nolimits_h G_{ss'}(h). \tag{10.152}$$

Using the above we find

$$\sum\nolimits_{s'} G_{ss'}(K = 0) = 0. \tag{10.153}$$

Assuming $e_1 = |e|$ and $e_1 = -|e|$ (to build in the polarity of the ions), the following equations can be written (where long wavelengths, $K \cong 0$, and one component of ion location is assumed)

$$M_s \ddot{v}_s = -\sum\nolimits_{s'} G_{ss'} v_{s'} + e_s E_0 \exp(-\mathrm{i}\omega t), \tag{10.154}$$

where

$$G_{ss'} = \sum\nolimits_h G_{ss'}(h). \tag{10.155}$$

If we assume that

$$U = \sum\nolimits_{l',h} \frac{G}{4} (v_{1l'} - v_{2l'+h'})^2, \tag{10.156}$$

where $h' = -1, 0, 1$ (does not range beyond nearest neighbors), then

$$G_{11}(h) = \frac{\partial^2 U}{\partial v_{1l} \partial v_{1l+h}} = G\delta_h^0. \tag{10.157}$$

Similarly,

$$G_{22}(h') = G\delta_{h'}^0, \tag{10.158}$$

$$G_{11}(h) = -G_{12}(h), \tag{10.159}$$

and

$$G_{22}(h) = -G_{21}(h)\,. \tag{10.160}$$

Therefore we can write

$$M_1\ddot{v}_1 = G_{11}(v_2 - v_1) + eE_0 \exp(-\mathrm{i}\omega t)\,, \tag{10.161}$$

and

$$M_2\ddot{v}_2 = G_{22}(v_1 - v_2) - eE_0 \exp(-\mathrm{i}\omega t)\,. \tag{10.162}$$

We now apply this to a dielectric where

$$\varepsilon = \varepsilon_0 + P/E\,, \tag{10.163}$$

and

$$P = \textstyle\sum_i N_i \alpha_i E_{\mathrm{loc},i}\,, \tag{10.164}$$

with N_i = the number of ions/vol of type i and α_i is the polarizability. For cubic crystals as derived in the chapter on dielectrics,

$$E_{\mathrm{loc},i} = E + \frac{P}{3\varepsilon_0}\,. \tag{10.165}$$

Then,

$$\varepsilon = \varepsilon_0 + \frac{1}{E}\sum N_i \alpha_i \left(E + \frac{P}{3\varepsilon_0} \right). \tag{10.166}$$

Let[4]

$$B = \frac{1}{3\varepsilon_0}\sum N_i \alpha_i\,, \tag{10.167}$$

so

$$\varepsilon = \varepsilon_0 + 3\varepsilon_0 B + B(\varepsilon - \varepsilon_0)\,, \tag{10.168}$$

$$\varepsilon(1 - B) = \varepsilon_0 + 2\varepsilon_0 B\,, \tag{10.169}$$

and

$$\varepsilon = \varepsilon_0 \frac{1 + 2B}{1 - B}\,. \tag{10.170}$$

For the diatomic case, define

$$B_{\mathrm{el}} = \frac{1}{3\varepsilon_0} N(\alpha_+ + \alpha_-)\,, \tag{10.171}$$

[4] Grosso and Paravicini [55 p342] also introduce B as a parameter and refer to its effects as a "renormalization" due to local field effects.

$$B_{\text{ion}} = \frac{1}{3\varepsilon_0} N\alpha_{\text{ion}} . \tag{10.172}$$

Then the static dielectric constant is given by

$$\frac{\varepsilon(0)}{\varepsilon_0} = \frac{1 + 2[B_{\text{el}} + B_{\text{ion}}(0)]}{1 - [B_{\text{el}} + B_{\text{ion}}(0)]} , \tag{10.173}$$

while for high frequency

$$\frac{\varepsilon(\infty)}{\varepsilon_0} = \frac{1 + 2B_{\text{el}}}{1 - B_{\text{el}}} . \tag{10.174}$$

We return to the equations of motion of the ions in the electric field—which in fact is a local electric field, and it should be so written. After a little manipulation we can write

$$\mu\ddot{v}_1 = \frac{\mu G}{M_1}(v_2 - v_1) + \frac{\mu}{M_1} eE_{\text{loc}} , \tag{10.175}$$

$$\mu\ddot{v}_2 = \frac{\mu G}{M_2}(v_1 - v_2) - \frac{\mu}{M_2} eE_{\text{loc}} . \tag{10.176}$$

Using

$$\frac{\mu}{M_1} + \frac{\mu}{M_2} = 1 , \tag{10.177}$$

we can write

$$\mu(\ddot{v}_1 - \ddot{v}_2) + G(v_1 - v_2) = eE_{\text{loc}} . \tag{10.178}$$

We first discuss this for transverse optical phonons.[5] Here, the polarization is perpendicular to the direction of travel, so

$$E_{\text{loc}} = \frac{P}{3\varepsilon_0} \tag{10.179}$$

in the absence of an external field. Now the polarization can be written as

$$P = P_{\text{el}} + P_{\text{ion}} = N(\alpha_- + \alpha_+)E_{\text{loc}} + Nev , \; v = v_1 - v_2 , \tag{10.180}$$

$$P = N(\alpha_+ + \alpha_-)\frac{P}{3\varepsilon_0} + Nev , \tag{10.181}$$

and

$$P = \frac{Nev}{1 - B_{\text{el}}} , \tag{10.182}$$

[5] A nice picture of transverse and longitudinal waves is given by Cochran [10.7].

so the local field becomes

$$E_{\text{loc}} = \frac{1}{3\varepsilon_0}\frac{Nev}{1-B_{\text{el}}}. \tag{10.183}$$

The equation of motion can be written

$$\mu\ddot{v} + Gv = \frac{1}{3\varepsilon_0}\frac{Ne^2 v}{1-B_{\text{el}}}. \tag{10.184}$$

Seeking sinusoidal solutions of the form $v = v_0\exp(-i\omega_T t)$ of the same frequency dependence as the local field, then

$$\omega_T^2 = \frac{G}{\mu}\left[1 - \frac{(1/3\varepsilon_0)(Ne^2/G)}{1-B_{\text{el}}}\right]. \tag{10.185}$$

We suppose α_{ion} is the static polarizability so that

$$\frac{ev}{E_{\text{loc}}} = \alpha_{\text{ion}} = \frac{e^2}{G} \tag{10.186}$$

form the equations of motion. So,

$$B_{\text{ion}}(0) = \frac{1}{3\varepsilon_0}N\alpha_{\text{ion}} = \left(\frac{1}{3\varepsilon_0}\right)\left(\frac{Ne^2}{G}\right), \tag{10.187}$$

or

$$\omega_T^2 = \frac{G}{\mu}\left[1 - \frac{B_{\text{ion}}(0)}{1-B_{\text{el}}}\right]. \tag{10.188}$$

For the longitudinal case with $\boldsymbol{q} \parallel P$ we have

$$E_{\text{loc}} = -\frac{P}{\varepsilon_0} + \frac{1}{3}\frac{P}{\varepsilon_0} = -\frac{2}{3}\frac{P}{\varepsilon_0}. \tag{10.189}$$

So,

$$P = P_{\text{el}} + P_{\text{ion}} = N(\alpha_+ + \alpha_-)\left(-\frac{2}{3}\frac{P}{\varepsilon_0}\right) + Nev = -2B_{\text{el}}P + Nev. \tag{10.190}$$

Then, we obtain the equation of motion,

$$\mu\ddot{v} + Gv = -\frac{2}{3\varepsilon_0}\frac{Ne^2 v}{1+2B_{\text{el}}}, \tag{10.191}$$

so

$$\omega_L^2 = \frac{G}{\mu}\left[1 + \frac{(2/3\varepsilon_0)(Ne^2/G)}{1+2B_{\mathrm{el}}}\right]. \tag{10.192}$$

By the same reasoning as before, we obtain

$$\omega_L^2 = \frac{G}{\mu}\left[1 + \frac{2B_{\mathrm{ion}}(0)}{1+2B_{\mathrm{el}}}\right]. \tag{10.193}$$

Thus, we have shown that, in general

$$\omega_L^2 = \frac{G}{\mu}\left[1 + \frac{2[B_{\mathrm{el}} + B_{\mathrm{ion}}(0)]}{1+2B_{\mathrm{el}}}\right], \tag{10.194}$$

and

$$\omega_T^2 = \frac{G}{\mu}\left[1 - \frac{B_{\mathrm{el}} + B_{\mathrm{ion}}(0)}{1-B_{\mathrm{el}}}\right]. \tag{10.195}$$

Therefore, using (10.173), (10.174), (10.194), and (10.195) we find

$$\frac{\varepsilon(\infty)}{\varepsilon(0)} = \frac{\omega_T^2}{\omega_L^2}. \tag{10.196}$$

This is the Lyddane–Sachs–Teller Relation, which was mentioned in Sect. 9.3.2, and also derived in Section 4.3.3 (see 4.79) as an aside in the development of polarons. Compare also Kittel [59, 3rd edn, 1966, p393ff] who gives a table showing experimental confirmation of the LST relation. The original paper is Lyddane et al [10.20]. An equivalent derivation is given by Born and Huang [10.2 p80ff].

For intermediate frequencies $\omega_T < \omega < \omega_L$,

$$\frac{\varepsilon(\omega) - \varepsilon(\infty)}{\varepsilon_0} = \frac{1 + 2[B_{\mathrm{el}} + B_i(\omega)]}{1 - [B_{\mathrm{el}} + B_i(\omega)]} - \frac{1+2B_{\mathrm{el}}}{1-B_{\mathrm{el}}}, \tag{10.197}$$

or

$$\frac{\varepsilon(\omega)}{\varepsilon_0} = \frac{\varepsilon(\infty)}{\varepsilon_0} + \frac{3}{[1 - B_{\mathrm{el}} - B_i(\omega)](1 - B_{\mathrm{el}})} B_i(\omega). \tag{10.198}$$

We need an expression for $B_i(\omega)$. With an external field since only transverse phonons are strongly interacting

$$\mu\ddot{v} + Gv = \frac{1}{3\varepsilon_0}\frac{Ne^2}{1 - B_{\mathrm{el}}} v + eE, \tag{10.199}$$

so

$$B_i(0) = \frac{1}{3\varepsilon_0} N\alpha_i(0) = \frac{1}{3\varepsilon_0} N \frac{ev}{E_{\mathrm{loc}}} = \frac{1}{3\varepsilon_0} N \frac{e^2}{G} . \tag{10.200}$$

Seeking a solution of the form $v = v_0\exp(-\mathrm{i}\omega t)$ we get

$$\omega_T^2 v\mu + Gv - \frac{Gv}{1 - B_{\mathrm{el}}} B_i(0) = eE . \tag{10.201}$$

So,

$$\omega_T^2 = \left(\frac{G}{\mu}\right)\left(1 - \frac{B_i(0)}{1 - B_{\mathrm{el}}}\right), \tag{10.202}$$

or

$$\mu(\omega_T^2 - \omega^2)v = eE . \tag{10.203}$$

So,

$$\alpha_i(\omega) = \frac{ev}{E_{\mathrm{loc}}} = \frac{e}{E_{\mathrm{loc}}} \frac{eE}{\mu(\omega_T^2 - \omega^2)} . \tag{10.204}$$

Using the local field relations, we have

$$\begin{aligned} E_{\mathrm{loc}} &= E + \frac{P}{3\varepsilon_0} = E + \frac{1}{3\varepsilon_0} \frac{Nev}{1 - B_{\mathrm{el}}} \\ &= E + \frac{1}{3\varepsilon_0} \frac{Ne}{1 - B_{\mathrm{el}}} \frac{eE}{\mu} \frac{1}{(\omega_T^2 - \omega^2)} , \end{aligned} \tag{10.205}$$

so,

$$\frac{E}{E_{\mathrm{loc}}} = \left(\frac{1}{1+F}\right)\left(\frac{e^2}{\mu} \frac{1}{\omega_T^2 - \omega^2}\right), \tag{10.206}$$

where,

$$F = \frac{G}{\mu} \frac{B_i(0)}{(1 - B_{\mathrm{el}})(\omega_T^2 - \omega^2)} . \tag{10.207}$$

Or,

$$B_i(\omega) = \frac{1}{3\varepsilon_0} N\alpha_i(\omega) = (1 - B_{\mathrm{el}}) \frac{F}{1+F} , \tag{10.208}$$

or

$$\frac{\varepsilon(\omega)}{\varepsilon_0} = \frac{\varepsilon(\infty)}{\varepsilon_0} + \frac{3(1-B_{\text{el}})F/(1+F)}{(1-B_{\text{el}})[(1-B_{\text{el}})-(1-B_{\text{el}})F/(1+F)]}, \tag{10.209}$$

or

$$\frac{\varepsilon(\omega)}{\varepsilon_0} = \frac{\varepsilon(\infty)}{\varepsilon_0} + \frac{3}{1-B_{\text{el}}}\frac{G}{\mu}\frac{B_i(0)}{(1-B_{\text{el}})(\omega_T^2-\omega^2)}. \tag{10.210}$$

Defining

$$c = 3\frac{G}{\mu}\frac{B_i(0)}{(1-B_{\text{el}})^2}, \tag{10.211}$$

after some algebra we also find

$$\omega_T^2 + \frac{c\varepsilon_0}{\varepsilon(\infty)} = \omega_L^2. \tag{10.212}$$

10.10.2 Summary of the Properties of $\varepsilon(q, \omega)$ (B)

Since $n = \varepsilon^{1/2}$ with $\sigma = 0$ (see (10.8)), if $\varepsilon < 0$, one gets high reflectivity (by (10.15) with n_c pure imaginary). Note if

$$\omega_T^2 < \omega^2 < \omega_T^2 + \frac{c\varepsilon_0}{\varepsilon(\infty)}, \tag{10.213}$$

then $\varepsilon(\omega) < 0$, since by (10.210), (10.211), and (10.212) we can also write

$$\varepsilon(\omega) = \varepsilon(\infty)\frac{\omega_L^2-\omega^2}{\omega_T^2-\omega^2},$$

and one has high reflectivity ($R \to 1$). Thus, one expects a whole band of forbidden nonpropagating electromagnetic waves. ω_T is called the Restrahl frequency and the forbidden gap extends from ω_T to ω_L. We only get Restrahl absorption in semiconductors that show ionic character; it will not happen in Ge and Si. We give some typical values in Table 10.2. See also Born and Huang [2, p118].

Table 10.2. Selected lattice frequencies and dielectric constants

Crystal	ω_T (cm^{-1})	ω_L (cm^{-1})	$\varepsilon(0)$ (cgs)	$\varepsilon(\infty)$ (cgs)
InSb	185	197	17.88	15.68
GaAs	269	292	12.9	10.9
NaCl	164	264	5.9	2.25
KBr	113	165	4.9	2.33
LiF	306	659	8.8	1.92
AgBr	79	138	13.1	4.6

From Anderson HL (ed), *A Physicists Desk Reference* 2nd edn, American Institute of Physics, Article 20: Frederikse HPR, Table 20.02.B.1 p.312, 1989, with permission of Springer-Verlag. Original data from Mitra SS, *Handbook on Semiconductors*, Vol 1, Paul W (ed), North-Holland, Amsterdam, 1982, and from *Handbook of Optical Constants of Solids*, Palik ED (ed), Academic Press, Orlando, FL, 1985.

10.10.3 Summary of Absorption Processes: General Equations (B)

Much of what we have discussed can be summed up in Fig. 10.13. Summary expressions for the dielectric constants are given in (10.67) and (10.68). See also Yu and Cardona [10.27, p. 251], and Cohen [10.8] as well as Cohen and Chelikowsky [10.9, p31].

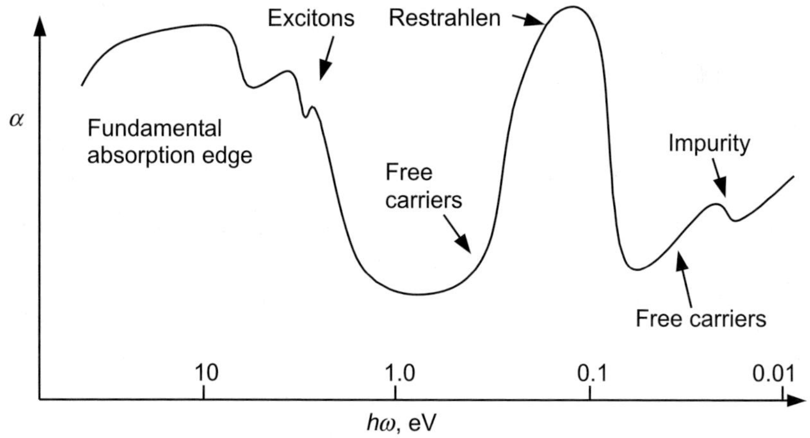

Fig. 10.13. Sketch of absorption coefficient of a typical semiconductor such as GaAs. Adapted from Elliott and Gibson [10.11, p. 208]

10.11 Optical Emission, Optical Scattering and Photoemission (B)

10.11.1 Emission (B)

We will only tread lightly on these topics, but they are important to mention. For example, photoemission (the ejection of electrons from the solid due to photons) can often give information that is not readily available otherwise, and it may be easier to measure than absorption. Photoemission can be used to study electron structure. Two important kinds are XPS – X-ray photoemission from solids, and UPS ultraviolet photoemission. Both can be compared directly with the valence-band density of states. See Table 10.3. A related discussion is given in Sect. 12.2.

Table 10.3. Some optical experiments on solids

High-energy reflectivity	The low-energy range below about 10 eV is good for investigating transitions between valence and conduction bands. The use of synchrotron radiation allows one to consider much higher energies that can be used to probe transitions between the conduction-band and core states. Since core levels tend to be well defined, such measurements provide direct data about conduction band states including critical point structure. The penetration depth is large compared to the depth of surface irregularities and thus this measurement is not particularly sensitive to surface properties. Only relative energy values are measured.
Modulation spectroscopy	This involves measuring derivatives of the dielectric function to eliminate background and enhance critical point structure. The modulation can be of the wavelength, temperature, stress, etc. See Cohen and Chelikowsky p. 52.
Photoemission	Can provide absolute energies, not just relative ones. Can use to study both surface and bulk states. Use of synchrotron radiation is extremely helpful here as it provides a continuous (from infrared to X-ray) and intense bombarding spectrum.
XPS and UPS	X-ray photoemission spectroscopy and ultraviolet photoemission spectroscopy. Both can now use synchrotron radiation as a source. In both cases, one measures the intensity of emitted electrons versus their energy. At low energy this can provide good checks on band-structure calculations.
ARPES	Angle-resolved photoemission spectroscopy. This uses the wave-vector conservation rule for wave vectors parallel to the surface. Provided certain other bits of information are available (see Cohen and Chelikowsky, p.68), information about the band structure can be obtained (see also Sect. 3.2.2).

Reference: Cohen and Chelikowsky [10.8]. See also Brown [10.3].

Also, the topic of emission is important because it involves applications—fluorescent lighting and television are obviously important and based on emission not on absorption. There are perhaps four principal aspects of optical emission. First, there are many types of transitions allowed. A second aspect is the excitation mechanism that positions the electron for emission. Third are the mechanisms that delay emission and give rise to luminescence. Finally, there are those combinations of mechanisms that produce laser action. Luminescence is often defined as light emission that is not due just to the temperature of the emitting body (that is, it is not black-body emission). There are several different kinds of luminescence depending on the source of the energy. For example, one uses the term photoluminescent if the energy comes from IR, visible, or UV light. Although there seems to be no universal agreement on the terms phosphorescence and fluorescence, phosphorescence is used for delayed light emission and fluorescence sometimes just means the light emitted due to excitation. Metals have high absorption at most optical frequencies, and so when we deal with photoemission, we normally deal with semiconductors and insulators.

10.11.2 Einstein *A* and *B* Coefficients (B, EE, MS)

We give now a brief discussion of emission as it pertains to the lasers and masers. The MASER (microwave amplification by stimulated emission of radiation) was developed by C. H. Townes in 1951, also independently by N. G. Basov and A. M. Prokhorov at about the same time). The first working LASER (light amplification by stimulated emission of radiation) was achieved by T. H. Maiman in 1960 using a ruby crystal. Ruby is sapphire (Al_2O_3) with a small amount of chromium impurities.

The Einstein *A* and *B* coefficients are easiest to discuss in terms of discrete levels, and exhibit a main idea of lasers. See Fig. 10.14. For a complete discussion of how lasers produce intense, coherent, and monochromatic beams of light see the references on applied physics [32–35]. Let the spontaneous emission and the induced transition rates be defined as follows:

Spontaneous emission	$n \to m$	A_{nm}
Induced emission	$n \to m$	B_{nm}
Induced absorption	$m \to n$	B_{mn}

From the Planck distribution we have for the density of photons

$$\rho(\nu) = \frac{8\pi h^3 \nu^2}{c^3} \frac{1}{\exp(h\nu/kT) - 1} . \tag{10.214}$$

Thus, generalizing to band-to-band transitions, we can write the generation rate as

$$G_{mn} = B_{mn} N_m f_m N_n (1 - f_n) \rho(\nu_{mn}) , \tag{10.215}$$

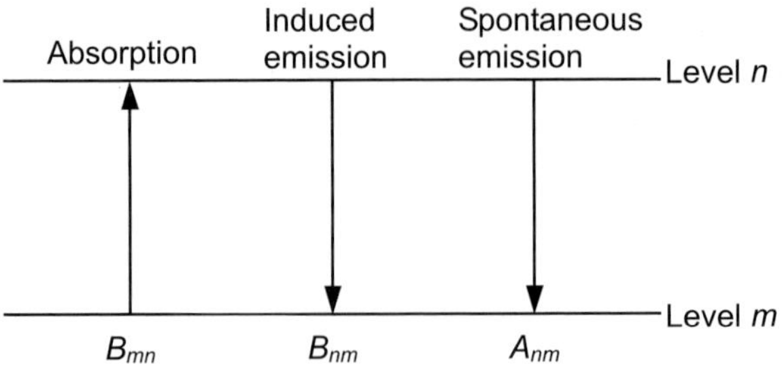

Fig. 10.14. The Einstein A and B coefficients

where N represents the number and f is the Fermi function. Also, we can write the recombination rate as

$$R_{nm} = B_{nm}N_n f_n N_m(1-f_m)\rho(\nu_{mn}) + A_{nm}N_n f_n N_m(1-f_m)\,. \quad (10.216)$$

In steady state, $G_{mn} = R_{nm}$. From the Fermi function we can show

$$\frac{f_m(1-f_n)}{f_n(1-f_m)} = \exp\left(\frac{E_n - E_m}{kT}\right). \quad (10.217)$$

Thus, since $B_{nm} = B_{mn}$, we have from (10.215) and (10.216)

$$B_{nm}\rho(\nu)\left[\exp\left(\frac{E_n - E_m}{kT}\right) - 1\right] = A_{nm}\,, \quad (10.218)$$

and

$$E_n - E_m = h\nu_{mn}\,, \quad (10.219)$$

we find for the ratio between the A and B coefficients,

$$\frac{A}{B} = \frac{8\pi n^3 \nu^2}{c^3}\,. \quad (10.220)$$

10.11.3 Raman and Brillouin Scattering (B, MS)

The laser has facilitated many optical experiments such as, for example, Raman scattering. We now discuss briefly Raman and Brillouin scattering. One refers to the inelastic scattering of light by phonons as Raman scattering if optical phonons are involved, and Brillouin scattering if acoustic phonons are. If phonons are emitted one speaks of the Stokes line and if absorbed as the anti-Stokes line. Note that these processes are two-photon processes (there is one photon in and one out). Raman and Brillouin scattering are made possible by the strain dependence of the electronic polarization. The relevant conservation equations can be written:

$$\omega_k = \omega_{k'} \pm \omega_K\,, \quad (10.221)$$

$$k = k' \pm K\,, \quad (10.222)$$

where ω and k refer to photons and ω_K and K to phonons. Since the value of the wave vector of photons is very small, the phonon wave vector can be at most twice that of the photon, and hence is very small compared to the Brillouin zone width. Hence, the energy of the optical phonons is very nearly constant at the optical phonon energy of zero wave vectors.

Brillouin scattering from longitudinal acoustic waves can be viewed as scattering from a density grating that moves at the speed of sound. Raman scattering can be used to determine the frequency of the zone-center phonon modes. Since the processes depend on phonons, a temperature dependence of the relative intensity of the Stokes and anti-Stokes lines can be predicted.

A simple idea as to the temperature dependence of the Stokes and the anti-Stokes lines is as follows [23, p. 323]. (For a more complete analysis see [10.2, p272]. See also Fox op. cit. p222.)

Stokes:
$$Intensity \propto \left|\langle n_K + 1|a_k^\dagger|n_K\rangle\right|^2 \propto n_K + 1\,, \tag{10.223}$$

Anti-Stokes:
$$Intensity \propto \left|\langle n_K - 1|a_k|n_K\rangle\right|^2 \propto n_K \tag{10.224}$$

$$\frac{I(\omega+\omega_K)}{I(\omega-\omega_K)} = \frac{n_K}{n_K+1} = \exp(-\hbar\beta\Omega)\,. \tag{10.225}$$

A diagram of Raman/Brillouin scattering involving absorption of a phonon (anti-Stokes) is shown in Fig. 10.15. As we have shown above, the intensity of the anti-Stokes line goes to zero at absolute zero, simply because there are no phonons available to absorb.

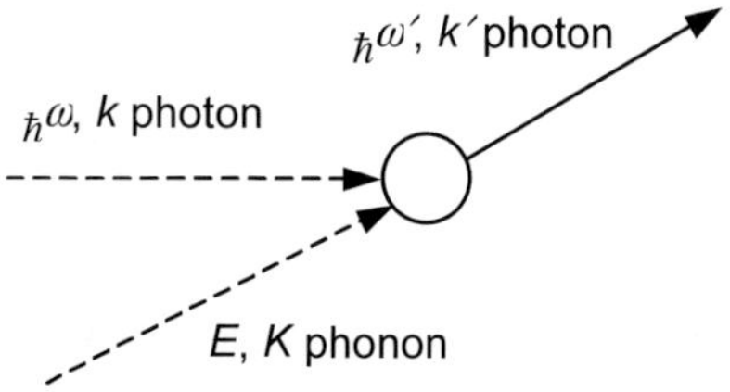

Fig. 10.15. Raman and Brillouin scattering. The diagram shows absorption. Acoustic phonons are involved for Brillouin scattering, and optical phonons for Raman

An expression for the frequency shift of both of these processes is now given. For absorption

$$k + K = k'\,, \tag{10.226}$$

and

$$\omega_k + \omega_K = \omega_{k'}\,. \tag{10.227}$$

Assuming the wavelength of the phonon is much greater than the wavelength of light, we have $k \cong k'$. If we let θ be the angle between k and k', then it is easy to see that

$$K = 2k\sin\left(\frac{\theta}{2}\right). \tag{10.228}$$

The shift in frequency of the scattered light is ω_K. For Brillouin scattering, with $V \cong \omega_K/K$ being the phonon velocity and n being the index of refraction, one finds

$$\omega_K = \left(\frac{2n\omega_k V}{c}\right)\sin\left(\frac{\theta}{2}\right), \tag{10.229}$$

and thus n can be determined. When phonons are absorbed, the photons are shifted up in frequency by ω_K, and when phonons are emitted, they are shifted down in frequency by this amount.

10.12 Magneto-Optic Effects: The Faraday Effect (B, EE, MS)

The rotation of the plane of polarization of plane-polarized light, which is propagating along an external magnetic field, is called the Faraday effect.[6] Substances for which this occurs naturally without an applied field are said to be optically active. One way of understanding this effect is to resolve the plane-polarized light into counterrotating circularly polarized components. Each component will have (see below) a different index of refraction and so propagates at a different speed, thus when they are recombined, the plane of polarization has been rotated. The two components behave differently because they interact with electrons via the two rotating electric fields. The magnetic field in effect causes a different radial force depending on the direction of rotation, and this modifies the effective spring constant. Both free and bound carriers can contribute to this effect. A major use of the Faraday effect is as an isolator that allows electromagnetic waves to propagate only in one direction. If the wave is polarized, and then rotated by 45 degrees by the Faraday rotator, any wave reflected back through the rotator will be rotated another 45 degrees in the same direction and hence be at 90 degrees to the polarizer and so cannot travel that way.

A simple classical picture of the effect works fairly well. We assume an electron bound by an isotropic Hooke's law spring in an electric and a magnetic field. By Newton's second law ($e > 0$):

$$m\ddot{\boldsymbol{r}} = -k\boldsymbol{r} - e(\boldsymbol{E} + \dot{\boldsymbol{r}} \times \boldsymbol{B}). \tag{10.230}$$

[6] A comprehensive treatment has been given by Caldwell [10.5].

Defining $\omega_0{}^2 = k/m$ (a different use of ω_0 from that in (10.108)!), letting $\boldsymbol{B} = B\boldsymbol{k}$, and assuming the electric field is in the (x,y)-plane, if we write out the x and y components of the above equation we have

$$\ddot{x} + \frac{e}{m}\dot{y}B + \omega_0^2 x = -\frac{e}{m}E_x, \tag{10.231}$$

$$\ddot{y} - \frac{e}{m}\dot{x}B + \omega_0^2 y = -\frac{e}{m}E_y. \tag{10.232}$$

We define $w_\pm = x \pm \mathrm{i}y$ and $E_\pm = E_x \pm \mathrm{i}E_y$. Note that the real and imaginary parts of E_+ correspond to "right-hand waves" (thumb along z) and the real and imaginary parts of E_- correspond to "left-hand waves".

We assume for the two circularly polarized components,

$$E_\pm = E_0 \exp[\pm\mathrm{i}(\omega t - k_\pm z)], \tag{10.233}$$

which when added together gives a plane-polarized beam along x at z =0. We seek steady-state solutions for which

$$w_\pm = \exp[\pm\mathrm{i}(\omega t - k_\pm z)]. \tag{10.234}$$

Substituting we find

$$w_\pm = \frac{(-e/m)E_\pm}{(\omega_0^2 - \omega^2) \pm (e/m)B\omega}. \tag{10.235}$$

The polarization P is given by

$$\boldsymbol{P} = -Ne\boldsymbol{r}, \tag{10.236}$$

where N is the number of electrons/volume:

$$P_\pm = \frac{(Ne^2/m)E_\pm}{(\omega_0^2 - \omega^2) \pm (e/m)B\omega}. \tag{10.237}$$

It is convenient to write this in terms of two special frequencies. The cyclotron frequency is

$$\omega_c = \frac{eB}{m}, \tag{10.238}$$

and the plasma frequency is

$$\omega_p = \sqrt{\frac{Ne^2}{m\varepsilon_0}}. \tag{10.239}$$

Thus (10.237) can be written

$$P_\pm = \frac{\varepsilon_0 \omega_p^2 E_\pm}{(\omega_0^2 - \omega^2) \pm \omega_c \omega}. \tag{10.240}$$

As usual we write

$$D_\pm = \varepsilon_0 E_\pm + P_\pm, \tag{10.241}$$

or

$$D_{\pm} = \varepsilon_{\pm} E_{\pm} . \tag{10.242}$$

Using (10.240), (10.241), (10.242), and

$$n_{\pm}^2 = \frac{\varepsilon_{\pm}}{\varepsilon_0} , \tag{10.243}$$

we find

$$n_{\pm}^2 = 1 + \frac{\omega_p^2}{(\omega_0^2 - \omega^2) \pm \omega_c \omega} . \tag{10.244}$$

The total angle that the polarization turns through is

$$\Theta = \frac{1}{2}(\Theta_{+} - \Theta_{-}) , \tag{10.245}$$

where in a distance l (and with period of rotation T)

$$\Theta_{\pm} = 2\pi \frac{l}{v_{\pm} T} = \frac{\omega l}{v_{\pm}} = \frac{\omega l}{c} n_{\pm} . \tag{10.246}$$

Thus,

$$\Theta = \frac{1}{2} \frac{\omega l}{c} (n_{+} - n_{-}) . \tag{10.247}$$

If

$$\omega_p^2 << (\omega_0^2 - \omega^2) \pm \omega_c \omega , \tag{10.248}$$

then

$$n_{\pm} \cong 1 + \frac{1}{2} \frac{\omega_p^2}{(\omega_0^2 - \omega^2) \pm \omega_c \omega} . \tag{10.249}$$

So, combining (10.247) and (10.249)

$$\Theta = -\frac{\omega_c \omega_p^2 \omega^2 l}{2c} \frac{1}{(\omega_0^2 - \omega^2)^2 - \omega_c^2 \omega^2} . \tag{10.250}$$

For free carriers $\omega_0 = 0$, we find if $\omega_c << \omega$,

$$\Theta = -\frac{l \omega_p^2 \omega_c}{2c\omega^2} . \tag{10.251}$$

Note a positive B (along z) with propagation along z will give a negative Verdet constant (the proportionality between the angle and the product of the field and path length) and a clockwise Θ when it is viewed along (i.e. in the direction of) $-z$.

Problems

10.1 In a short paragraph explain what photoconductivity is, and describe any photoconductivity experiment.

10.2 Describe, very briefly, the following magneto-optical effects: (a) Zeeman effect, (b) inverse Zeeman effect, (c) Voigt effect, (d) Cotton–Mouton effect, (e) Faraday effect, (f) Kerr magneto-optic effect.

Describe briefly the following electro-optic effects: (g) Stark effect, (h) inverse Stark effect, (i) electric double refraction, (j) Kerr electro-optic effect.

Descriptions of these effects can be found in any good optics text.

10.3 Given a plane wave $E = E_0\exp[i(\boldsymbol{k}\cdot\boldsymbol{r} - \omega t)]$ normally incident on a surface, detail the assumptions, conditions and steps to show $n_c E_0 = E_1 - E_2$, (cf. (10.26)).

10.4 (a) From $[x, p_x] = i\hbar$, show that

$$[H, \hat{\boldsymbol{e}} \cdot \boldsymbol{r}] = -\frac{i\hbar}{m} \hat{\boldsymbol{e}} \cdot \boldsymbol{p},$$

(b) For $\hat{\boldsymbol{e}} = \hat{\boldsymbol{\imath}}$, show the oscillator strength f_{ij} obeys the sum rule $\sum_j f_{ij} = 1$.

10.5 For intermediate frequencies $\omega_T < \omega < \omega_L$, given (by (10.198))

$$\frac{\varepsilon(\omega)}{\varepsilon_0} - \frac{\varepsilon(\infty)}{\varepsilon_0} = \frac{3B_{\text{ion}}(\omega)}{[1 - B_{\text{el}} - B_{\text{ion}}(\omega)](1 - B_{\text{el}})},$$

and the equation of motion (by (10.199))

$$\mu\ddot{v} + Gv = \frac{1}{3\varepsilon_0} \frac{Ne^2}{1 - B_{\text{el}}} v + eE,$$

derive the equation

$$\omega_T^2 + \frac{c\varepsilon_0}{\varepsilon(\infty)} = \omega_L^2,$$

where c is a defined as constant within the derivation. In this process, show intermediate derivations for the following equations defining constants as necessary:

$$\mu(\omega_T^2 - \omega^2)v = eE,$$

$$\frac{E}{E_{\text{loc}}} = \frac{1}{1 + F},$$

$$\varepsilon(\omega) = \varepsilon(\infty) + \frac{c\varepsilon_0}{\omega_T^2 - \omega^2}.$$

10.6 This problem fills in the details of Sect. 10.11.2.

(a) Describe the factors that make up the generation rate

$$G_{mn} = B_{mn} N_m f_m N_n (1 - f_n) \rho(\nu_{mn}) .$$

(b) Show from the Fermi function that

$$\frac{f_n(1 - f_m)}{f_m(1 - f_n)} = \exp\left(\frac{E_m - E_n}{kT}\right) .$$

(c) Starting from $G_{mn} = R_{nm}$, show that

$$\frac{A}{B} = \frac{8\pi n^3 \nu^2}{c^3} .$$

11 Defects in Solids

11.1 Summary About Important Defects (B)

A defect in a solid is any deviation from periodicity in the solid. All solids have defects, but for some applications, they can be neglected, while for others, the defects can be very important. By now, simple defects are well understood, but for more complex defects, a considerable amount of fundamental work remains to be accomplished for a thorough understanding.

Some discussion of defects has already been made. In Chap. 2, the effects of defects on the phonon spectrum of a one-dimensional lattice were discussed, whereas in Chap. 3 the effects of defects on the electronic states in a one-dimensional lattice were considered. In the semiconductor chapter, donor and acceptor states were used, but some details were postponed until this chapter.

There is only one way to be perfect, but there are numerous ways to be imperfect. Thus, we should not be surprised that there are many kinds of defects. The mere fact that no crystal is infinite is enough to introduce *surface* defects, which could be electronic or vibrational. Electronic surface states are classified as Tamm states (if they are due to a different potential in the last unit cell at the surface edge with atoms far apart) or Shockley states (the cells remain perfectly repetitive right up to the edge, but with atoms close enough so as to have band crossing[1]). Whether or not Tamm and Shockley states should be distinguished has been the subject of debate that we do not wish to enter into here. In any case, the atoms on the surface are not in the same environment as interior atoms, and so, their contribution to the properties of the solid must be different. The surface also acts to scatter both electrons and phonons. The properties of surfaces are of considerable practical importance. All input and output to solids goes through the surfaces. Thermionic and cold field emission from surfaces is discussed in Sects. 11.7 and 11.8. Surface reconstruction is discussed in Chap. 12. Another important application of surface physics is to better understand corrosion.

Besides surfaces, we briefly review other ways crystals can have defects, starting with *point defects* (see Crawford and Slifkin [11.7]). When a crystal is grown, it is not likely to be pure. Foreign impurity atoms will be present, leading to *substitutional* or *interstitial* defects (see Fig. 11.1). Interstitial atoms can originate from atoms of the crystal as well as foreign atoms. These may be caused by thermal effects (see below) or may be introduced artificially by *radiation damage*. Radiation damage (or thermal effects) may also cause vacancies. Also,

---

1 See, e.g., Davison and Steslicka [11.8].

when a crystal is composed of more than one element, these elements may not be exactly in their proper chemical proportions. The stoichiometric derivations can result in vacancies as well as *antisite defects* (an atom of type A occupying a site normally occupied by an atom of type B in an AB compound material).

Vacancies are always present in any real crystal. Two sorts of point defects involving vacancies are so common that they are given names. These are the Schottky and Frenkel defects, shown for an ionic crystal in Fig. 11.2. Defects such as *Schottky* and *Frenkel* defects are always present in any real crystal at a finite temperature in equilibrium. The argument is simple. Suppose we assume that the free energy $F = U - TS$ has a minimum in equilibrium. The defects will increase U, but they cause disorder, so they also cause an increase in the entropy S. At high enough temperatures, the increase in U can be more than compensated by the decrease in $-TS$. Thus, the stable situation is the situation with defects.

Mass transport is largely possible because of defects. Vacancies can be quite important in controlling diffusion (discussed later in Sect. 11.5). Ionic conductivity studies are important in studying the motion of lattice defects in ionic crystals. *Color centers* are another type of point defect (or complex of point defects). We will discuss them in a little more detail later (Sect. 11.4). Color centers are formed by defects and their surrounding potential, which trap electrons (or holes).

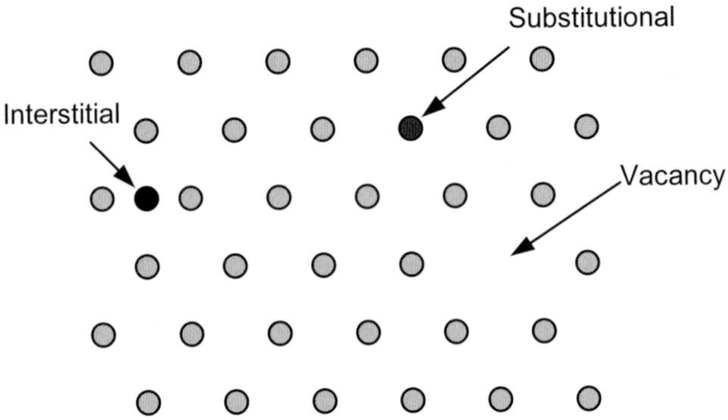

Fig. 11.1. Point defects

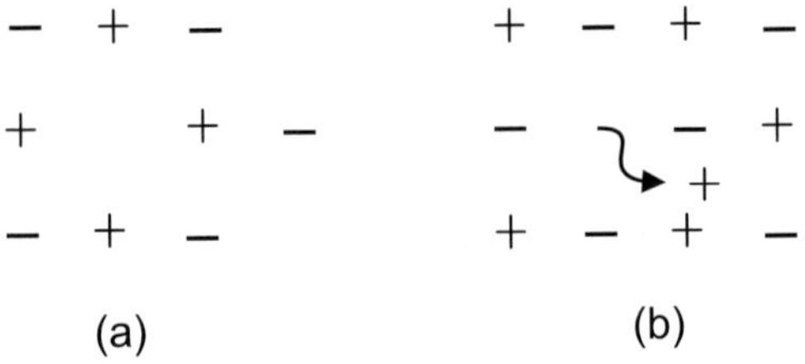

Fig. 11.2. (**a**) Schottky and (**b**) Frenkel defects

Table 11.1. Summary of common crystal lattice defects

Point defects	*Comments*
Foreign atoms	Substitutional or interstitial
Vacancies	Schottky defect is vacancy with atom transferred to surface
Antisite	Example: A on a B site in an AB compound
Frenkel	Vacancy with foreign atom transferred to interstice
Color centers	Several types – F is vacancy with trapped electron (ionic crystals – see Sect. 11.4
Donors and acceptors	Main example are shallow defects in semiconductors – see Sects. 11.2 and 11.3
Deep levels in semiconductors	See Sects. 11.2 and 11.3
Line defects	*Comments*
Dislocations	Edge and screw – see Sect. 11.6 – General dislocation is a combination of these two
Surface defects	*Comments*
External	
Tamm and Shockley electronic states	See Sect. 11.1
Reconstruction	See Sect. 12.2
Internal	
Stacking fault	Example: a result of an error in growth[2]
Grain boundaries	Tilt between adjacent crystallites – can include low angle (with angle, in radians, being the ratio of the Burgers vector (magnitude) to the dislocation spacing) to large angle (which includes twin boundaries)
Heteroboundary	Between different crystals
Volume defects	*Comments*
Many examples	Three-dimensional precipitates and complexes of defects

See, e.g., Henderson [11.16].

Vacancies, substitutional atoms, and interstitial atoms are all point defects. Surfaces are planar defects. There is another class of defects called *line defects*. *Dislocations* are important examples of line defects, and they will be discussed

[2] A fcc lattice along (1,1,1) is composed of planes ABCABC etc. If an A plane is missing then we have ABCBCABC, etc. This introduces a local change of symmetry. See, e.g., Kittel [23, p. 18].

later (Sect. 11.6). They are important for determining how easily crystals deform and may also relate to crystal growth.

Finally, there are defects that occur over a whole volume. It is usually hard to grow a *single crystal*. In a single crystal, the lattice planes are all arranged as expected–in a perfectly regular manner. When we are presented with a chunk of material, it is usually in a polycrystal form. That is, many little crystals are stuck together in a somewhat random way. The boundary between crystals is also a two-dimensional defect called a *grain boundary*. We have summarized these ideas in Table 11.1.

11.2 Shallow and Deep Impurity Levels in Semiconductors (EE)

We start by considering a simple chemical model of shallow donor and acceptor defects. We will give a better definition later, but for now, by "shallow", we will mean energy levels near the bottom of the conduction band for donor level and near the top of the valence band for acceptors.

Consider Si^{14} as the prototype semiconductor. In the usual one-electron shell notation, its electron structure is denoted by

$$1s^2 2s^2 2p^6 3s^2 3p^2 .$$

There are four valence electrons in the $3s^2 3p^2$ shell, which requires eight to be filled. We think of neighboring Si atoms sharing electrons to fill the shells. This sharing lowers energy and binds the electrons. We speak of covalent bonds. Schematically, in two dimensions, we picture this occurring as in Fig. 11.3. Each line represents a shared electron. By sharing, each Si in the outer shell has eight electrons. This is of course like the discussion we gave in Chap. 1 of the bonding of C to form diamond.

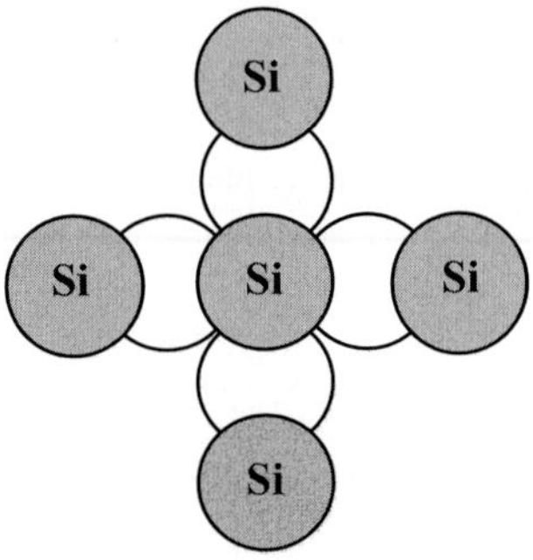

Fig. 11.3. Chemical model of covalent bond in Si

Now, suppose we have an atom, say As, which substitutionally replaces a Si. The sp shell of As has five electrons ($4s^2 4p^3$) and only four are needed to "fill the shell". Thus, As acts as a donor with an additional loosely bound electron (with a large orbit encompassing many atoms), which can be easily excited into the conduction band at room temperature.

An acceptor like In (with three electrons in its outer sp shell ($5s^2 5p$)) needs four electrons to complete its covalent bonds. Thus, In can accept an electron from the valence band, leaving behind a hole. The combined effects of effective mass and dielectric constant cause the carrier to be bound much less tightly than in an analogous hydrogen atom. The result is that donors introduce energy levels just below the conduction-band minimum and acceptors introduce levels just above the top of the valence band. We discuss this in more detail below.

In brief, it turns out that the ground-state donor energy level is given by (atomic units, see the appendix)

$$E_n = -\frac{m^* / m}{2n^2 \varepsilon^2}, \tag{11.1}$$

where m^*/m is the effective mass ratio typically about 0.25 for Si and ε is the dielectric constant (about 11.7 in Si). E_n in (11.1) is measured from the bottom of the conduction band. Except for the use of the dielectric constant and the effective mass, this is the same result as obtained from the theory of the energy levels of hydrogen. A similar, remarkably simple result holds for acceptor states. These results arise from pioneering work by Kohn and Luttinger as discussed in [11.17], and we develop the basics below.

11.3 Effective Mass Theory, Shallow Defects, and Superlattices (A)

11.3.1 Envelope Functions (A)

The basic model we will use here is called the envelope approximation.[3] It will allow us to justify our treatments of effective mass theory and of shallow defects in semiconductors. With a few more comments, we will then be able to relate it to a simple approach to superlattices, which will be discussed in more detail in Chap. 12.

Let

$$\mathcal{H}_0 = -\frac{\hbar^2}{2m}\nabla^2 + V(\boldsymbol{r}), \tag{11.2}$$

[3] Besides [11.17], see also Luttinger and Kohn [11.22] and Madelung [11.23].

where $V(\boldsymbol{r})$ is the periodic potential. Let $\mathcal{H} = \mathcal{H}_0 + U$ where $U = V_D(\boldsymbol{r})$ is the extra defect potential. Now, $\mathcal{H}_0\psi_n(\boldsymbol{k}, \boldsymbol{r}) = E_n\psi_n(\boldsymbol{k}, \boldsymbol{r})$ and $\mathcal{H}\psi = E\psi$.

We expand the wave function in Bloch functions

$$\psi = \sum\nolimits_{n,\boldsymbol{k}} a_n(\boldsymbol{k})\psi_n(\boldsymbol{k},\boldsymbol{r}), \tag{11.3}$$

where n is the band index. Also, since $E_n(\boldsymbol{k})$ is a periodic function in $\boldsymbol{k}$-space, we can expand it in a Fourier series with the sum restricted to lattice points

$$E_n(\boldsymbol{k}) = \sum\nolimits_m F_{nm}\mathrm{e}^{\mathrm{i}\boldsymbol{k}\cdot\boldsymbol{R}_m}. \tag{11.4}$$

We define an operator $E_n(-\mathrm{i}\nabla)$ by substituting $-\mathrm{i}\nabla$ for $\boldsymbol{k}$:

$$\begin{aligned} E_n(-\mathrm{i}\nabla)\psi_n(\boldsymbol{k},\boldsymbol{r}) &= \sum\nolimits_m F_{nm}e^{\boldsymbol{R}_m\cdot\nabla}\psi_n(\boldsymbol{k},\boldsymbol{r}) \\ &= \sum\nolimits_m F_{nm}[1+\boldsymbol{R}_m\cdot\nabla+\tfrac{1}{2}(\boldsymbol{R}_m\cdot\nabla)^2+\ldots]\psi_n(\boldsymbol{k},\boldsymbol{r}) \\ &= \sum\nolimits_m F_{nm}\psi_n(\boldsymbol{k},\boldsymbol{r}+\boldsymbol{R}_m), \end{aligned} \tag{11.5}$$

by the properties of Taylor's series. Then using Bloch's theorem

$$E_n(-\mathrm{i}\nabla)\psi_n(\boldsymbol{k},\boldsymbol{r}) = \sum\nolimits_m F_{nm}e^{\mathrm{i}\boldsymbol{k}\cdot\boldsymbol{R}_m}\psi_n(\boldsymbol{k},\boldsymbol{r}), \tag{11.6}$$

and by (11.4)

$$E_n(-\mathrm{i}\nabla)\psi_n(\boldsymbol{k},\boldsymbol{r}) = E_n(\boldsymbol{k})\psi_n(\boldsymbol{k},\boldsymbol{r}). \tag{11.7}$$

Substituting (11.3) into $\mathcal{H}\psi = E\psi$, we have (using the fact that ψ_n is an eigenfunction of $\mathcal{H}_0$ with eigenvalue $E_n(\boldsymbol{k})$)

$$\begin{aligned} \sum\nolimits_{n,\boldsymbol{k}} E_n(\boldsymbol{k})a_n(\boldsymbol{k})\psi_n(\boldsymbol{k},\boldsymbol{r}) + \sum\nolimits_{n,\boldsymbol{k}} V_D a_n(\boldsymbol{k})\psi_n(\boldsymbol{k},\boldsymbol{r}) \\ = \sum\nolimits_{n,\boldsymbol{k}} E a_n(\boldsymbol{k})\psi_n(\boldsymbol{k},\boldsymbol{r}). \end{aligned} \tag{11.8}$$

If we use (11.4) and (11.6), this becomes

$$\sum\nolimits_{n,\boldsymbol{k}} a_n(\boldsymbol{k})[E_n(-\mathrm{i}\nabla)+V_D]\psi_n(\boldsymbol{k},\boldsymbol{r}) = E\psi. \tag{11.9}$$

11.3.2 First Approximation (A)

We neglect band-to-band interactions and hence, neglect the summation over n. Dropping n entirely from (11.9), we have

$$\psi = \sum\nolimits_{\boldsymbol{k}} a(\boldsymbol{k})\psi(\boldsymbol{k},\boldsymbol{r}), \tag{11.10}$$

and

$$[E(-\mathrm{i}\nabla)+V_D]\psi(\boldsymbol{k},\boldsymbol{r}) = E\psi. \tag{11.11}$$

11.3.3 Second Approximation (A)

We assume a large extension in real space that means that only a small range of k values are important – say the ones near a parabolic (assumed for simplicity) minimum at $k = 0$ (Madelung op. cit. Chap. 9).

We assume, then,

$$\psi(\boldsymbol{k},\boldsymbol{r}) = \mathrm{e}^{\mathrm{i}\boldsymbol{k}\cdot\boldsymbol{r}}u(\boldsymbol{k},\boldsymbol{r}) \cong \mathrm{e}^{\mathrm{i}\boldsymbol{k}\cdot\boldsymbol{r}}u(0,\boldsymbol{r}) = \mathrm{e}^{\mathrm{i}\boldsymbol{k}\cdot\boldsymbol{r}}\psi(0,\boldsymbol{r}) \tag{11.12}$$

so using (11.10) and (11.12),

$$\psi = F(\boldsymbol{r})\psi(0,\boldsymbol{r})\,, \tag{11.13}$$

where

$$F(\boldsymbol{r}) = \sum\nolimits_{\boldsymbol{k}} a(\boldsymbol{k})\mathrm{e}^{\mathrm{i}\boldsymbol{k}\cdot\boldsymbol{r}}\,. \tag{11.14}$$

So, we have by (11.11)

$$[E(-\mathrm{i}\nabla) + V_D]F(\boldsymbol{r})\psi(0,\boldsymbol{r}) = EF(\boldsymbol{r})\psi(0,\boldsymbol{r})\,. \tag{11.15}$$

Using the definition of $E(-\mathrm{i}\nabla)$ as in (11.5) we have with n suppressed,

$$\sum\nolimits_m F_m F(\boldsymbol{r} + \boldsymbol{R}_m)\psi(0,\boldsymbol{r} + \boldsymbol{R}_m) = (E - V_D)F(\boldsymbol{r})\psi(0,\boldsymbol{r})\,. \tag{11.16}$$

But, $\psi(0, r + R_m) = \psi(0, r)$, so it can be cancelled. Thus retracing our steps, we have

$$[E(-\mathrm{i}\nabla) + V_D]F(\boldsymbol{r}) = EF(\boldsymbol{r})\,. \tag{11.17}$$

This simply means that a rapidly varying function has been replaced by a slowly varying function $F(r)$ called the "envelope" function. This immediately leads to the concept of shallow donors. Consider the bottom of a parabolic conductor band near $k = 0$ and expand about $k = 0$;

$$E(-\mathrm{i}\nabla) = E_c + \left.\frac{\mathrm{d}E}{\mathrm{d}k}\right|_{k=0}(-i\nabla) + \frac{1}{2}\frac{\mathrm{d}^2E}{\mathrm{d}k^2}(-\mathrm{i}\nabla)^2\,. \tag{11.18}$$

Also,

$$\left.\frac{\mathrm{d}E}{\mathrm{d}k}\right|_{k=0} = 0\,, \tag{11.19}$$

and

$$\left.\frac{\mathrm{d}^2E}{\mathrm{d}k^2}\right|_{k=0} = \frac{\hbar^2}{m^*}\,, \tag{11.20}$$

where m^* is the effective mass. Thus, we find

$$E(-\mathrm{i}\nabla) = E_c - \frac{\hbar^2}{2m^*}\nabla^2\,. \tag{11.21}$$

And, if $V_D = e^2/4\pi\varepsilon r$, our resulting equation is

$$\left(-\frac{\hbar^2}{2m^*}\nabla^2 - \frac{e^2}{4\pi\varepsilon r}\right)F(\boldsymbol{r}) = (E - E_c)F(\boldsymbol{r}) . \tag{11.22}$$

Except for the use of ε and m^*, these solutions are just hydrogenic wave functions and energies, and so our use of the hydrogenic solution (11.1) is justified.

Now let us discuss briefly electron and hole motion in a perfect crystal. If $U = 0$, we simply write

$$\left(-\frac{\hbar^2}{2m_e^*}\nabla^2 + E_c\right)F = EF . \tag{11.23}$$

On the other hand, suppose U is still 0, but consider a valence band with a maximum at $k = 0$. We then can expand about that point with the following result:

$$E(-\mathrm{i}\nabla) = E_v + \frac{1}{2}\frac{\mathrm{d}^2 E}{\mathrm{d}k^2}(-\mathrm{i}\nabla)^2 . \tag{11.24}$$

Using the hole mass, which has the opposite sign for the electron mass ($m_h = -m_e$), we can write

$$E(-\mathrm{i}\nabla) = E_v + \frac{\hbar^2}{2m_h^*}(\nabla^2) , \tag{11.25}$$

so the relevant Schrödinger equation becomes

$$\left[\frac{-\hbar^2}{2m_h^*}(\nabla^2) - E_v\right]F = -EF . \tag{11.26}$$

Looking at (11.23) and (11.26), we see how discontinuities in band energies can result in effective changes in the potential for the carriers, and we see why the hole energies are inverted from the electron energies.

Now let us consider superlattices with a set of layers so there is both a lattice periodicity in each layer and a periodicity on a larger scale due to layers (see Sect. 12.6). The layers A and B could for example be laid down as ABABAB…

There are several more considerations, however, before we can apply these results to superlattices. First, we have to consider that if we are to move from a region of one band structure (layer) to another (layer), the effective mass changes since adjacent layers are different. With the possibility of change in effective mass, the Hamiltonian is often written as

$$H = -\frac{\hbar^2}{2}\frac{\partial}{\partial z}\left(\frac{1}{m^*(z)}\frac{\partial}{\partial z}\right) + V(z) , \tag{11.27}$$

rather than in the more conventional way. This allows the Hamiltonian to remain Hermitian, even with varying m^*, and it leads to a probability current density of

$$j_z(z) = \frac{\hbar}{2i}\left[\frac{\psi^*}{m^*}\cdot\frac{\partial\psi}{\partial z} - \frac{\psi}{m^*}\cdot\frac{\partial\psi^*}{\partial z}\right], \tag{11.28}$$

from which we apply the requirement of continuity on ψ and $\partial\psi/(m^*\partial z)$ rather than ψ and $\partial\psi/\partial z$.

We have assumed the thickness of each layer is sufficient that the band structure of the material can be established in this thickness. Basically, we will need both layers to be several monolayers thick. Also, we assume in each layer that the electron wave function is an envelope function (different for different monolayers) times a Bloch function (see (11.13)). Finally, we assume that in each layer $U = U_0$ (a constant appropriate to the layer) and the carrier motion perpendicular to the layers is free-electron-like so,

$$F(\boldsymbol{r}) = \varphi(z)\mathrm{e}^{\mathrm{i}(k_x x + k_y y)}, \tag{11.29}$$

$$\left[-\frac{\hbar^2}{2m^*}(k_x^2 + k_y^2) - \frac{\hbar^2 \mathrm{d}^2}{2m^* \mathrm{d}z^2} + U_0\right]\varphi(z)\mathrm{e}^{\mathrm{i}(k_x x + k_y y)} = EF, \tag{11.30}$$

which means

$$\left[-\frac{\hbar^2}{2m^*}\frac{\mathrm{d}^2}{\mathrm{d}z^2} + U_0\right]\varphi(z) = E_z\varphi(z), \tag{11.31}$$

where for each layer

$$E = E_z + \frac{\hbar^2}{2m^*}(k_x^2 + k_y^2). \tag{11.32}$$

There are many, many complications to the above. We have assumed, e.g., that $m^*_{xy} = m^*_z$ which may not be so in all cases. The book by Bastard [11.1] can be consulted. See also, Mitin et al [11.25].

In semiconductors, shallow levels are often defined as being near a band edge and deep levels as being near the center of the forbidden energy gap. In more recent years, a different definition has been applied based on the nature of the causing agent. Shallow levels are now defined as defect levels produced by the long-range Coulomb potential of the defect and deep levels[4] are defined as being produced by the central cell potential of the defect, which is short ranged. Since

---

[4] See, e.g., Li and Patterson [11.20, 11.21] and references cited therein.

the potential is short range, a modification of the Slater-Koster model, already discussed in Chap. 2, is a convenient starting point for discussing deep defects. Some reasons for the significance of shallow and deep defects are given in Table 11.2. Deep defects are commonly formed by substitutional, interstitial, and antisite atoms and by vacancies.

Table 11.2. Definition and significance of deep and shallow levels

Shallow levels are defect levels produced by the long-range Coulomb potential of defects. Deep Levels are defect levels produced by the central cell potential of defect.

	Deep level	Shallow level
Energy	May or may not be near band edge. Spectrum is not hydrogen-like.	Near band edge. Spectrum is hydrogen-like.
Typical properties	Recombination centers. Compensators. Electron–hole generators.	Suppliers of carriers.

11.4 Color Centers (B)

The study of color centers arose out of the curiosity as to what caused the yellow coloration of rock salt (NaCl) and other coloration in similar crystals. This yellow color was particularly noted in salt just removed from a mine. Becquerel found that NaCl could be colored by placing the crystal near a discharge tube. From a fundamental point of view, NaCl should have an infrared absorption due to vibrations of its ions and an ultraviolet absorption due to excitation of the electrons. A perfect NaCl crystal should not absorb visible light, and should be uncolored. The coloration of NaCl must be due, then, to defects in the crystal. The main absorption band in NaCl occurs at about 4650 Å (the "F"-band). This blue absorption is responsible for the yellow color that the NaCl crystal can have. A further clue to the nature of the absorption is provided by the fact that exposure of a colored crystal to white light can result in the bleaching out of the color. Further experiments show that during the bleaching, the crystal becomes photoconductive, which means that electrons have been promoted to the conduction band. It has also been found that NaCl could be colored by heating it in the presence of Na vapor. Some of the Na atoms become part of the NaCl crystal, resulting in a deficiency of Cl and, hence, Cl^- vacancies. Since photoconductive experiments show that F-band defects can release electrons, and since Cl^- vacancies can trap electrons, it seems very suggestive that the defects responsible for the F-band (called "F-centers") are electrons trapped at Cl^- vacancies. (Note: the "F" comes from the German *farbe*, meaning "color".) This is the explanation accepted today. Of course, since some Cl^- vacancies are always present in a NaCl crystal in thermodynamic equilibrium, any sort of radiation that causes electrons to be knocked into the Cl^- vacancies will form F-centers. Thus, we have an explanation of Becquerel's early results as mentioned above.

More generally, color centers are formed when point defects in crystals trap electrons with the resultant electronic energy levels at optical frequencies. Color centers usually form "deep" traps for electrons, rather than "shallow" traps, as donor impurities in semiconductors do, and, their theoretical analysis is complex. Except for relatively simple centers such as F-centers, the analysis is still relatively rudimentary.

Typical experiments that yield information about color centers involve optical absorption, paramagnetic resonance and photoconductivity. The absorption experiments give information about the transition energies and other properties of the transition. Paramagnetic resonance gives wave function information about the trapped electron, while photoconductivity yields information on the quantum efficiency (number of free electrons produced per incident photon) of the color centers.

Mostly by interpretation of experiment, but partly by theoretical analysis, several different color centers have been identified. Some of these are listed below. The notation is

$$[\text{missing ion} \mid \text{trapped electron} \mid \text{added ion}],$$

where our notation is p ≡ proton, e ≡ electron, − ≡ halide ion, + ≡ alkaline ion, and M^{++} ≡ doubly charged positive ion. The usual place to find color centers is in ionic crystals.

$$\begin{aligned}
[-|e|] &= \text{F-center} \\
[-|2e|] &= \text{F}'\text{-center} \\
[\mp-|2e|] &= \text{M-center (?)} \\
[|e|p] &= \text{U}_2\text{-center} \\
[+|e|M^{++}] &= \text{Z}_1\text{-center (?)}.
\end{aligned}$$

In Figs 11.4 and 11.5 we give models for two of the less well-known color centers. In these two figures, ions enclosed by boxes indicate missing ions, a dot means an added electron, and a circle includes a substitutionally added ion. We include several references to color centers. See, e.g., Fowler [11.12] or Schulman and Compton [11.28].

Color centers turn out to be surprisingly difficult to treat theoretically with precision. But success has been obtained using modern techniques on, e.g., F centers in LiCl. See, e.g., Louie p. 94, in Chelikowsky and Louie [11.4]. In recent years tunable solid-state lasers have been made using color centers at low temperatures.

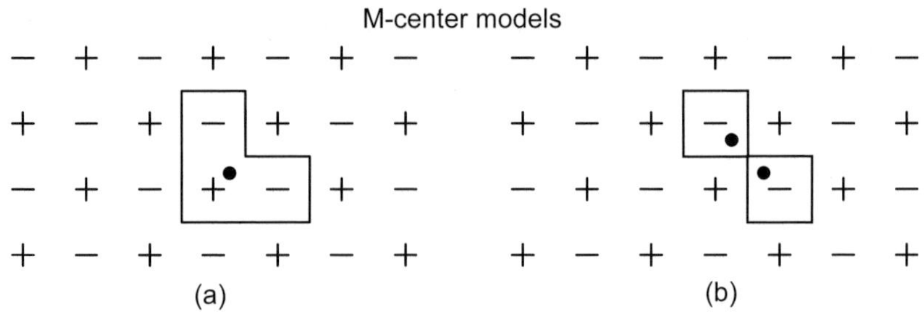

Fig. 11.4. Models of the M-center: (**a**) Seitz, (**b**) Van Doorn and Haven. [Reprinted with permission from Rhyner CR and Cameron JR, *Phys Rev* **169**(3), 710 (1968). Copyright 1968 by the American Physical Society.]

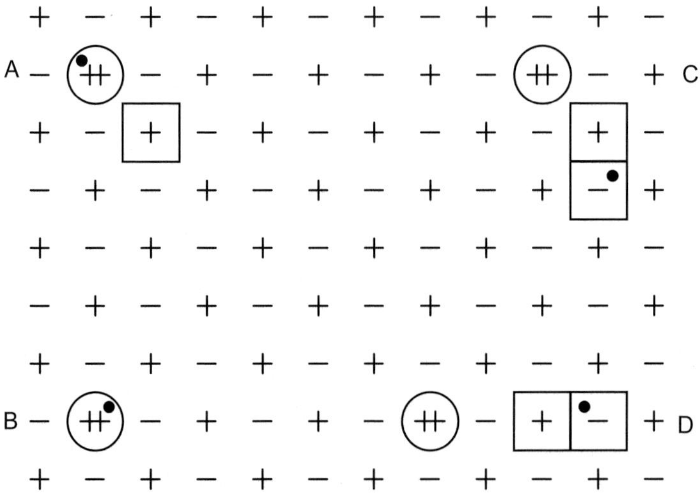

Fig. 11.5. Four proposed models for Z_1-centers [Reprinted with permission from Paus H and Lüty F, *Phys Rev Lett* **20**(2), 57 (1968). Copyright 1968 by the American Physical Society.]

11.5 Diffusion (MET, MS)

Point defects may diffuse through the lattice, while vacancies may provide a mechanism to facilitate diffusion. Diffusion and defects are intimately related, so we give a brief discussion of diffusion. If C is the concentration of the diffusing quantity, Fick's Law says the flux of diffusing quantities is given by

$$J = -D\frac{\partial C}{\partial x}, \tag{11.33}$$

where D is, by definition, the diffusion constant. Combining this with the equation of continuity

$$\frac{\partial J}{\partial x} + \frac{\partial C}{\partial \tau} = 0 , \tag{11.34}$$

leads to the diffusion equation

$$\frac{\partial C}{\partial \tau} = D \frac{\partial^2 C}{\partial x^2} . \tag{11.35}$$

For solution of this equation, we refer to several well-known treatises as referred to in Borg and Dienes [11.2]. Typically, the diffusion constant is a function of temperature via

$$D = D_0 \exp(-E_0 / kT) , \tag{11.36}$$

where E_0 is the activation energy that depends on the process. Interstitial defects moving from one site to an adjacent one typically have much less E_0 than say, vacancy motion. Obviously, the thermal variation of defect diffusion rates has wide application.

11.6 Edge and Screw Dislocation (MET, MS)

Any general dislocation is a combination of two basic types: the *edge* and the *screw* dislocations. The edge dislocation is perhaps the easiest to describe. If we imagine the pages in a book as being crystal planes, then we can visualize an edge dislocation as a book with half a page (representing a plane of atoms) missing. The edge dislocation is formed by the missing half-plane of atoms. The idea is depicted in Fig. 11.6. The motion of edge dislocations greatly reduces the shear strength of crystals. Originally, the shear strength of a crystal was expected to be much greater than it was actually found to be for real crystals. However, all large crystals have dislocations, and the movement of a dislocation can greatly aid the shearing of a crystal. The idea involves similar reasoning as to why it is easier to move a rug by moving a wrinkle through it rather than moving the whole rug. The force required to move the wrinkle is much less.

Crystals can be strengthened by introducing impurity atoms (or anything else), which will block the motion of dislocations. Dislocations themselves can interfere with each other's motions and bending crystals can generate dislocations, which then causes *work hardening*. Long, but thin, crystals called *whiskers* have been grown with few dislocations (perhaps one screw dislocation to aid growth – see below). Whiskers can have the full theoretical strength of ideal, perfect crystals.

The other type of dislocation is called a screw dislocation. *Screw* dislocations can be visualized by cutting a book along A (see Fig. 11.7), then moving the upper half of the book a distance of one page and taping the book into a spiral staircase. Another view of the dislocation is shown in Fig. 11.8 where successive atomic planes are joined together to form one surface similar to the way a kind of Riemann surface can be defined. Screw dislocations greatly aid crystal growth. During the growth, a wandering atom finds two surfaces to "stick" to at the growth edge (or jog) (see Fig. 11.8) rather than only one flat plane. Actual crystals have shown little spirals on their surface corresponding to this type of growth.

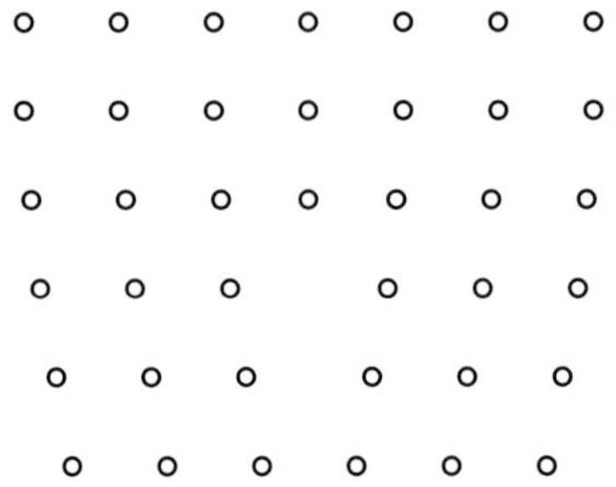

Fig. 11.6. An edge dislocation

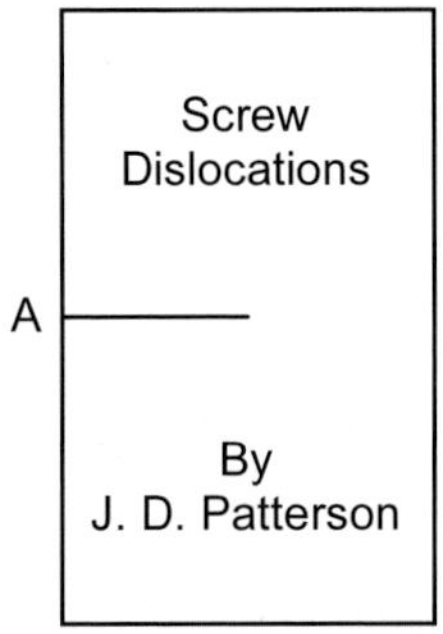

Fig. 11.7. A book can be used to visualize screw dislocations

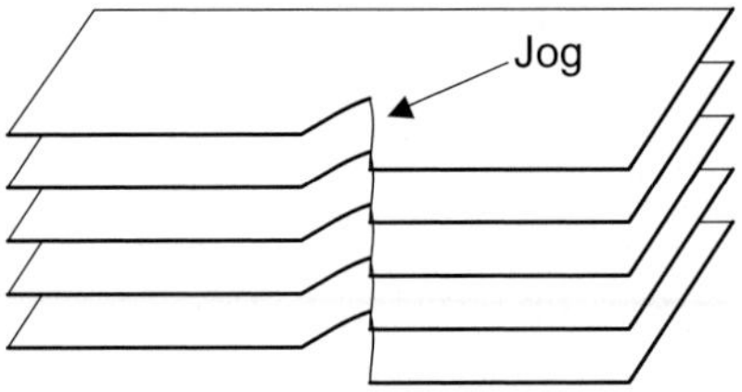

Fig. 11.8. A screw dislocation

We have already mentioned that any general dislocation is a combination of the edge and screw. It is well at this point to make the idea more precise by the use of the *Burgers vector*, which is depicted in Fig. 11.9. We take an atom-to-atom path around a dislocation line. The path is drawn in such a way that it would close on itself as if there were no dislocations. The additional vector needed to close the

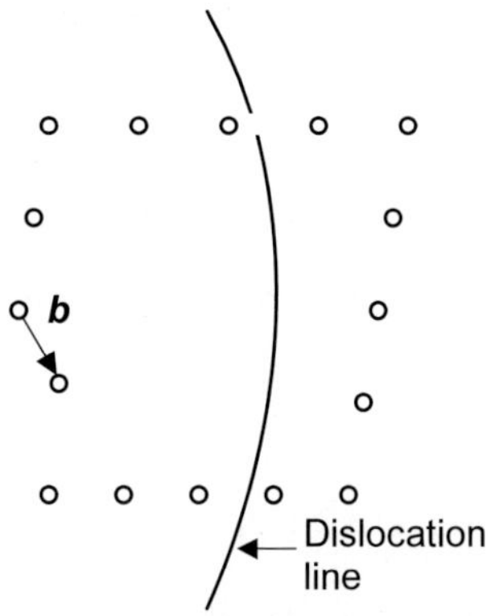

Fig. 11.9. Diagram used for the definition of the Burgers vector ***b***

path is the Burgers vector. For a pure edge dislocation, the Burgers vector is perpendicular to the dislocation line; for a pure screw dislocation, it is parallel. In general, the Burgers vector can make any angle with the dislocation line, which is allowed by crystal symmetry. The book by Cottrell [11.6] is a good source of further details about dislocations. See also deWit [11.9].

11.7 Thermionic Emission (B)

We now discuss two very classic and important properties of the surfaces of metals – in this Section thermionic emission and in the next Section cold-field emission.

So far, we have mentioned the role of Fermi–Dirac statistics in calculating the specific heat, Pauli paramagnetism, and Landau diamagnetism. In this Section we will apply Fermi–Dirac statistics to the emission of electrons by heated metals. It will turn out that the fact that electrons obey Fermi–Dirac statistics is relatively secondary in this situation.

It is also possible to have cold (no heating) emission of electrons. Cold emission of electrons is obtained by applying an electric field and allowing the electrons to tunnel out of the metal. This was one of the earliest triumphs of quantum mechanics in explaining hitherto unexplained phenomena. It will be explained in the next section.[5]

For the purpose of the calculation in this section, the surface of the metal will be pictured as in Fig. 11.10. In Fig. 11.10, E_F is the Fermi energy, ϕ is the work function, and E_0 is the barrier height of the potential. The barrier can at least be partially understood by an image charge calculation.

---

[5] A comprehensive review of many types of surface phenomena is contained in Gundry and Tompkins [11.14].

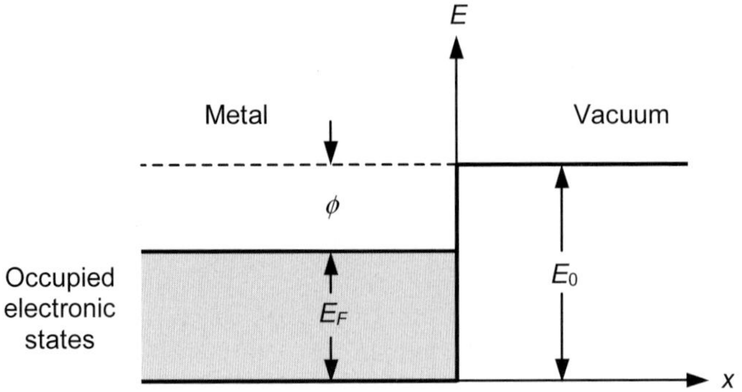

Fig. 11.10. Model of the surface of a metal used to explain thermionic emission

We wish to calculate the current density as a function of temperature for the heated metal. If $n(\boldsymbol{p})\,\mathrm{d}^3\boldsymbol{p}$ is the number of electrons per unit volume in $\boldsymbol{p}$ to $\boldsymbol{p} + \mathrm{d}^3\boldsymbol{p}$ and if v_x is the x component of velocity of the electrons with momentum $\boldsymbol{p}$, we can write the rate at which electrons with momentum from $\boldsymbol{p}$ to $\boldsymbol{p} + \mathrm{d}^3\boldsymbol{p}$ hit a unit area in the (x,y)-plane as

$$v_x n(\boldsymbol{p})\mathrm{d}^3\boldsymbol{p} = n(\boldsymbol{p})(p_x/m)\mathrm{d}p_x\mathrm{d}p_y\mathrm{d}p_z\,. \tag{11.37}$$

Now

$$n(\boldsymbol{p})\mathrm{d}^3\boldsymbol{p} = n(\boldsymbol{k})\mathrm{d}^3\boldsymbol{k} = 2f(E)\frac{\mathrm{d}^3\boldsymbol{k}}{(2\pi)^3} \tag{11.38}$$

$$\begin{aligned} &= 2f(E)\frac{\mathrm{d}^3\boldsymbol{p}}{(2\pi\hbar)^3} \\ &= \frac{2f(E)}{h^3}\mathrm{d}^3\boldsymbol{p}, \end{aligned} \tag{11.39}$$

so that

$$n(\boldsymbol{p}) = 2f(E)/h^3\,. \tag{11.40}$$

In (11.40), $f(E)$ is the Fermi function and the factor 2 takes the spin degeneracy of the electronic states into account. Finally, we need to consider that only electrons whose x component of momentum p_x satisfies

$$p_x^2/2m > \phi + E_F \tag{11.41}$$

will escape from the metal.

If we assume the probability of reflection at the surface of the metal is R and is constant (or represents an average value), the emission current density j is e (the electronic charge, here $e > 0$) times the rate at which electrons of sufficient energy strike unit area of the surface times $T_r \equiv 1 - R$. Thus, the emission current density is given by

$$j = -\frac{2e}{h^3} T_r \int_{-\infty}^{\infty} \left(\int_{-\infty}^{\infty} \left\{ \int_{\phi + E_F}^{\infty} \frac{\mathrm{d}(p_x^2/2m)}{\exp[(E - E_F)/kT] + 1} \right\} \mathrm{d}p_z \right) \mathrm{d}p_y . \quad (11.42)$$

Since $E = (1/2m)(p_x^2 + p_y^2 + p_z^2)$, we can write this expression as

$$j = -kT \frac{2e}{h^3} T_r \int_{-\infty}^{\infty} \left[\int_{-\infty}^{\infty} \left(\int_0^{\infty} \frac{\mathrm{d}E'}{e^{E'/kT} \exp\{[\phi + (p_x^2 + p_y^2)/2m]/kT\} + 1} \right) \mathrm{d}p_z \right] \mathrm{d}p_y ,$$

where $E' = (p_x^2/2m) - \phi - E_F$. But

$$\int_0^{\infty} \frac{\mathrm{d}n}{ae^n + 1} = \ln(1 + a^{-1}) ,$$

so that

$$j = -(T_r) \frac{2kTe}{h^3} \int_{-\infty}^{\infty} \left[\int_{-\infty}^{\infty} \ln(1 + e^{-G}) \mathrm{d}p_z \right] \mathrm{d}p_y , \quad (11.43)$$

where

$$G = \frac{\phi + (1/2m)(p_x^2 + p_y^2)}{kT} . \quad (11.44)$$

At common operating temperatures, $G >> 1$, so since $\ln(1 + \varepsilon) \approx \varepsilon$ (for small ε) we can write (this approximation amounts to replacing Fermi–Dirac statistics by Boltzmann statistics for all electrons that get out of the metal)

$$j = -\frac{T_r 2kTe}{h^3} \int_{-\infty}^{\infty} \left\{ \int_{-\infty}^{\infty} \exp(-\phi/kT) \cdot \exp\left[\frac{-1}{2mkT} (p_x^2 + p_y^2) \right] \mathrm{d}p_z \right\} \mathrm{d}p_y .$$

Thus, so far as the temperature dependence goes, we can write

$$j = AT^2 \mathrm{e}^{-\phi/kT} , \quad (11.45)$$

where A is a quantity that can be determined from the above expressions. In actual practice there is little point to making this evaluation. Our A depends on having an idealized surface, which is never realized. Typical work functions ϕ, as determined from thermionic emission data, are of the order of 5 eV, see Table 11.3.

Equation (11.45) is often referred to as the *Richardson–Dushmann* equation. It agrees with experimental results at least qualitatively. Account must be taken, however, of adsorbates that can lower the effective work function.[6]

---

[6] See Zanquill [11.33].

Table 11.3. Work functions

Element	φ (eV)
Ag	
100	4.64
110	4.52
111	4.74
poly	4.26
Co poly	5
Cu poly	4.65
Fe poly	4.5
K poly	2.3
Na poly	2.75
Ni poly	5.15
W poly	4.55

From Anderson HL (ed), *A Physicists Desk Reference* 2nd edn, Article 21: Hagstum HD, Surface Physics, p. 330, American Institute of Physics, (1989) by permission of Springer-Verlag. Original data from Michaelson HB, *J Appl Phys* **48**, 4729 (1977).

11.8 Cold-Field Emission (B)

To have a detectable cold-field emission it is necessary to apply a strong electric field. The strong electric field can be obtained by using a sharp point, for example. We shall assume that we have applied an electric field $\boldsymbol{E}_1$ in the $-x$ direction to the metal so that the electron's potential energy (with $-e$ the charge of the electron) produced by the electric field is $V = E_0 - eE_1x$. The form of the potential function near the surface of the metal will be assumed to be as in Fig. 11.11.

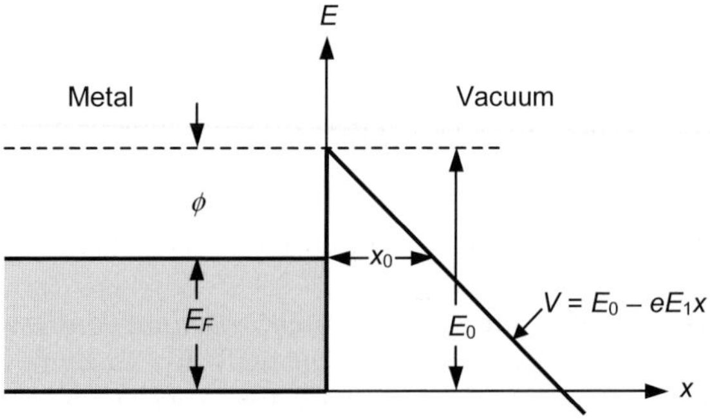

Fig. 11.11. Potential energy for tunneling from a metal in the presence of an applied electric field

To calculate the current density, which is emitted by the metal when the electric field is applied, it is necessary to have the transmission coefficient for tunneling through the barrier. This transmission coefficient can perhaps be adequately evaluated by use of the WKB approximation. For a high and broad barrier, the WKB approximation gives for the transmission coefficient

$$T = \exp\left(-2\int_{x=0}^{x_0} K(x)\mathrm{d}x\right), \tag{11.46}$$

where

$$K(x) = \sqrt{2m/\hbar^2}\sqrt{V(x)-E}\ , \tag{11.47}$$

x_0 is the second classical turning point, and E is, of course, the energy.

The upper limit of the integral is determined (for an electron of energy E) from

$$E = E_0 - eE_1x_0 \quad \text{or} \quad x_0 = \frac{E_0 - E}{eE_1}.$$

Therefore

$$\int_{x=0}^{x_0} K(x)\mathrm{d}x = \tfrac{2}{3}\sqrt{2m(E_0-E)^3/(\hbar eE_1)^2}\ .$$

Since $(E_0 - E_F) = \phi$, the transmission coefficient for electrons with the Fermi energy is given by

$$T = \exp\left(-\tfrac{4}{3}\sqrt{\frac{2m}{\hbar^2}\frac{\phi^3}{(eE_1)^2}}\right). \tag{11.48}$$

Further analysis shows that the current density for field-emitted electrons is given approximately by $J \propto E_1{}^2 T$ so,

$$J = aE_1^2 e^{-b/E_1}, \tag{11.49}$$

where a and b are different constants for different materials. Equation (11.49), where b is commonly proportional to $\phi^{3/2}$, is often referred to as the *Fowler–Nordheim* equation. The ideas of Fowler–Nordheim tunneling are also used for the tunneling of electrons in a metal-oxide-semiconductor (MOS) structure. See also Sarid [11.27].

There is another type of electron emission that is present when an electric field is applied. When an electric field is applied, the height of the potential barrier is slightly lowered. Thus more electrons can be classically emitted (without tunneling) by thermionic emission than previously. This additional emission due to the lowering of the barrier is called *Schottky emission*. If we imagine the barrier is caused by image charge attraction, it is fairly easy to see why the maximum barrier height should decrease with field strength. Simple analysis predicts the barrier lowering to be proportional to the square root of the magnitude of the electric field. The idea is shown in Fig. 11.12. See Problem 11.7.

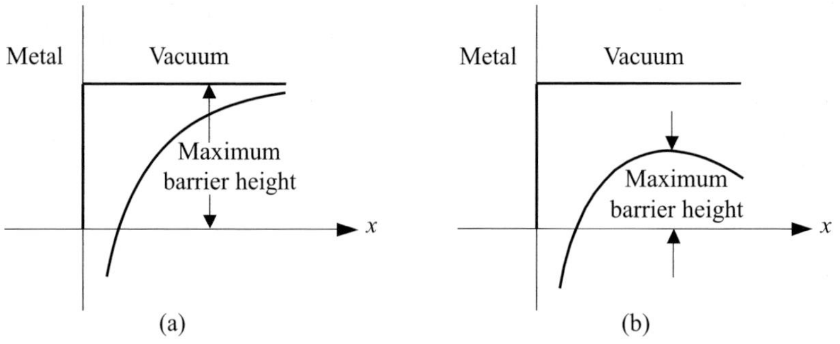

Fig. 11.12. The effect of an electric field on the surface barrier of a metal: (**a**) with no field, and (**b**) with a field

11.9 Microgravity (MS)

It is believed that crystals grown in microgravity will often be more perfect. S. Lehoczky of Marshall Space Flight Center has been experimenting for years with growing mercury cadmium telluride in microgravity (on the Space Shuttle) with the idea of producing more perfect crystals that would yield better infrared detectors. See, e.g., Lehoczky et al [11.19].

First, we should talk about what microgravity is and what it is not. It is not the absence of gravity, or even a region where gravity is very small. Unless one goes very far from massive bodies, this is impossible. Even at a Shuttle orbit of 300 km above the Earth, the force of gravity is about 90% the value experienced on the Earth.

Newton himself understood the principle. If one mounts a cannon on a large mountain on an otherwise flat Earth and fires the cannonball horizontally, it will land some distance away from the base of the mountain. Adding more powder will cause the ball to go further. Finally, a point will be reached when the ball falls exactly the same amount that the earth curves. The ball will then be in free-fall and in orbit. The effects of gravity for objects inside the ball will be very small. In an orbiting satellite, there will be exactly one surface where the effects of gravity are negligible. At other places inside, one has "microgravity".

There are many ways to produce microgravity; all you have to do is arrange to be in free-fall. Drop towers and drop tubes offer two ways of accomplishing this. The first commercial use of microgravity was probably the drop tower used in 1785 in England to make spherical lead balls. Marshall Space Flight Center had both a drop tower and a drop tube 100 meters high–this alone allowed free-fall, or microgravity, for about 5 s. In a drop tower, the entire experimental package is dropped. For crystal growth experiments, this means the furnace as well as the instrumentation and the specimen are all placed in a special canister and dropped.

In a drop tube, there is an enclosure in which, for example, only the molten sample would be dropped. Special aerodynamic design, vacuum, or other means is used to reduce air drag and, hence, obtain real free-fall. For slightly longer times (20 s or so), the KC 135 aircraft can be put into a parabolic path to produce microgravity. Extending this idea, rockets have been used to produce microgravity for periods of about 400 s.

Problems

11.1 Give a simple derivation of Ivey's law. Ivey's law states that fa^2 = constant where f is the frequency of absorption in the F-band and a is the lattice spacing in the colored crystal. Use as a model for the F-center an electron in a box and assume that the absorption is due to a transition between the ground and first excited energy states of the electron in the box.

11.2 The F-center absorption energy in NaCl is about 2.7 eV. For a particle in a box of side $a_{\text{NaCl}} = 5.63 \times 10^{-10}$ m, find the excitation energy of an electron from the ground to the first excited state.

11.3 A low-angle grain boundary is found with a tilt angle of about 20 s on a (100) surface of Ge. What is the prediction for the linear dislocation density of etch pits predicted?

11.4 Find the allowed energies of a hydrogen atom in two dimensions. The answer you should get is [12.54]

$$E_n = -\frac{R}{(n - \frac{1}{2})^2},$$

where n is a nonzero integer. R is the Rydberg constant that can be written as

$$R = -\frac{me^4}{2(4\pi\varepsilon_0 K\hbar)^2},$$

where $K = \varepsilon/\varepsilon_0$ with ε the appropriate dielectric constant. Since the Bohr radius is

$$a_B = \frac{4\pi\varepsilon_0 K\hbar^2}{me^2},$$

one can also write

$$R = -\frac{\hbar^2}{2ma_B^2}.$$

Note that the result is the same as the three-dimensional hydrogen atom if one replaces n by $n - ½$.

11.5 Quantum wells will be discussed in Chap. 12. Find the allowed energies of a donor atom, represented by a hydrogen atom with electron mass m and in a region of dielectric constant as above. Suppose the quantum well is of width w and with infinite sides with potential energy $V(z)$. Also suppose $w \ll a_B$. In this case the wave function for a donor in a quantum well is

$$\left[-\frac{\hbar^2}{2m}\nabla^2 - \frac{e^2}{4\pi\varepsilon_0 K\sqrt{x^2+y^2}} + V(z)\right]\psi = E\psi ,$$

where $V(z) = 0$ for $0 < z < w$ and is infinite otherwise. The answer is

$$E = E_{p,n} = \frac{\hbar^2\pi^2 p^2}{2mw^2} - \frac{R}{(n-\frac{1}{2})^2},$$

p, n are nonzero integers and R is the Rydberg constant [12.54].

11.6 a) Show that a solution of the one-dimensional diffusion equation is

$$C(x,t) = \frac{A}{\sqrt{t}}\exp\left(-\frac{x^2}{4Dt}\right).$$

b) If $\int_{-\infty}^{\infty} C(x,t)\mathrm{d}x = Q$, show that

$$A = \frac{Q}{2\sqrt{\pi D}}.$$

11.7 This problem illustrates the Schottky effect. See Fig. 11.11 and Fig. 11.12. Suppose the attraction outside the metal is caused by an image charge.

a) Show that in the absence of an electric field we can write the potential energy as

$$V(x) = E_0 - \frac{e^2}{16\pi\varepsilon_0 x},$$

so that with an external field

$$V(x) = E_0 - eE_1 x - \frac{e^2}{16\pi\varepsilon_0 x}.$$

b) Thus show that with the electric field E_1, the barrier height is reduced from E_0 to $E_0 - \Delta$, where

$$\Delta = \frac{1}{2}\sqrt{\frac{e^3 E_1}{\pi\varepsilon_0}}.$$

12 Current Topics in Solid Condensed–Matter Physics

This chapter is concerned with some of the newer areas of solid condensed-matter physics and so contains a variety of topics in nanophysics, surfaces, interfaces, amorphous materials, and soft condensed matter.

There was a time when the living room radio stood on the floor and people gathered around in the evening and "watched" the radio. Radios have become smaller and smaller and thus, increasingly cheaper. Eventually, of course, there will be a limit in smallness of size to electronic devices. Fundamental physics places constraints on how small the device can be and still operate in a "conventional way". Recently people have realized that a limit for one kind of device is an opportunity for another. This leads to the topic of new ways of using materials, particularly semiconductors, for new devices.

Of course, the subject of electronic technology, particularly semiconductor technology, is too vast to consider here. One main concern is the fact that quantum mechanics places basic limits on the size of devices. This arises because quantum mechanics associates a wavelength with the electrons that carry current and electrical signals. Quantum effects become important when electron wavelength becomes comparable to component size. In particular, the phenomenon of tunneling, which is often assumed to be of no importance for most ordinary microelectronic devices becomes important in this limit. We will discuss some of the basic physics needed to understand these devices, in which tunneling and related phenomena are important. Here we get into the area of bandgap engineering to attain structures that have desired properties not attainable with homostructures. Generally, these structures are *nanostructures*. A nanostructure is a condensed-matter structure having at least one minimum dimension between about 1 nm to 10 nm.

We will start by discussing surfaces and then consider how to form nanostructures on surfaces by molecular beam epitaxy. Nanostructures may be two dimensional (quantum wells), one dimensional (quantum wires), or "zero" dimensional (quantum dots). We will discuss all of these and also talk about heterostructures, superlattices, quantum conductance, Coulomb blockade, and single-electron devices.

Another reduced-dimensionality effect is the quantum Hall effect, which arises when electrons in a magnetic field are confined two dimensionally. As we will see, the ideas and phenomena involved are quite novel.

Carbon, carbon nanotubes, and fullerene nanotechnology may lead to entirely new kinds of devices and they are also included in this chapter, as the nanotubes are certainly nanostructures.

Amorphous, noncrystalline disordered solids have become important and we discuss them as examples of new materials if not reduced dimensionality.

Finally, the new area of soft condensed-matter physics is touched on. This area includes liquid crystals, polymers, and other materials that may be "soft" to the touch. The unifying idea here is the ease with which the materials deform due to external forces.

12.1 Surface Reconstruction (MET, MS)

As already mentioned, the input and output of a device go through the surface, so physical understanding of surfaces is critical. Of course, the nature of the surface also affects crystal growth, chemical reactions, thermionic emission, semiconducting properties, etc.

One generally thinks of the surface of a material as being the top two or three layers. The rest can be called the bulk or substrate. The distortion near the surface can be both perpendicular (stretching or contracting) as well as parallel. Below we concentrate on that which is parallel.

If we project the bulk with its periodicity on the surface and if no reconstruction occurs we say the surface is 1×1. More likely the lack of bonding forces on the surface side will cause the surface atoms to find new locations of minimum energy. Then the projection of the bulk on the surface is different in symmetry from the surface. For the special case where the projection defines primitive surface vectors $\boldsymbol{a}$ and $\boldsymbol{b}$, while the actual surface has primitive vectors $\boldsymbol{a}_S = N\boldsymbol{a}$ and $\boldsymbol{b}_S = M\boldsymbol{b}$ then one says one has an $N \times M$ reconstruction. If there also is a rotation R of β associated with $\boldsymbol{a}_S$ and $\boldsymbol{b}_S$ primitive cell compared to the $\boldsymbol{a}$, $\boldsymbol{b}$ primitive cell we write the reconstruction as

$$\left(\frac{|\boldsymbol{a}_S|}{|\boldsymbol{a}|} \times \frac{|\boldsymbol{b}_S|}{|\boldsymbol{b}|}\right) R\beta\,.$$

Note that the vectors $\boldsymbol{a}$ and $\boldsymbol{b}$ depend on whether the original (unreconstructed or unrelaxed) surface is (1, 1, 1) or (1, 0, 0), or in general (h, k, l). For a complete description the surface involved would also have to be included. The reciprocal lattice vectors $\boldsymbol{A}$, $\boldsymbol{B}$ associated with the surface are defined in the usual way as

$$\boldsymbol{A} \cdot \boldsymbol{a}_S = \boldsymbol{B} \cdot \boldsymbol{b}_S = 2\pi\,, \tag{12.1a}$$

and

$$\boldsymbol{A} \cdot \boldsymbol{b}_S = \boldsymbol{B} \cdot \boldsymbol{a}_S = 0\,, \tag{12.1b}$$

where the 2π now inserted in an alternative convention for reciprocal lattice vectors. One uses these to discuss two-dimensional diffraction.

Low-energy electron diffraction (LEED, see Sects. 1.2.7 and 12.2) is commonly used to examine the structure of surfaces. This is because electrons, unlike photons, have charge and thus, do not penetrate too far into materials. There are theoretical techniques, including those using the pseudopotential, which are available. See Chen and Ho [12.12].

Since surfaces are so important for solid-state properties we briefly review techniques for their characterization in the next section.

12.2 Some Surface Characterization Techniques (MET, MS, EE)

AFM: Atomic Force Microscopy–This instrument detects images of surfaces on an atomic scale by sensing atomic forces between the sample and a cantilevered tip (in one kind of mode, there are various modes of operation). Unlike STMs (see below), this instrument can be used for nonconductors as well as conductors.

AES: Auger Electron Spectroscopy–uses an alternative (to X-ray emission) decay scheme for an excited core hole. The core hole is often produced by the impact of energetic electrons. An electron from a higher level makes a transition to fill the hole, and another bound electron escapes with the left-over energy. The Auger process leaves two final-state holes. The energy of the escaping electron is related to the characteristic energies of the atom from which it came, and therefore chemical analysis is possible.

EDX: Energy Dispersive X-ray Spectroscopy–electrons are incident at a grazing angle and the energy of the grazing X-rays that are produced, are detected and analyzed. This technique has sensitivities comparable to Auger electron spectroscopy.

Ellipsometry–study of the reflection of plane-polarized light from the surface of materials to determine the properties of these materials by measuring the ellipticity of the reflected light.

EELS: Electron Energy Loss Spectroscopy–electrons scattered from surface atoms may lose amounts of energy dependent on surface excitations. This can be used to examine surface vibrational modes. It is also used to detect surface plasmons.

EXAFS: Extended X-ray Absorption Spectroscopy–photoelectrons caused to be emitted by X-rays are backscattered from surrounding atoms. They interfere with the emitted photoelectrons and give information about the geometry of the atoms that surround the original absorbing atom. When this technique is surface specific, as for detecting Auger electrons, it is called *SEXAFS*.

FIM: Field Ion Microscopy–this can be used to detect individual atoms. Ions of the surrounding imaging gas are produced by field ionization at a tip and are detected on a fluorescent screen placed at a distance, to which ions are repelled.

LEED: Low-Energy Electron Diffraction–due to their charge, electrons do not penetrate deeply into a surface. *LEED* is the coherent reflection or diffraction of electrons typically with energy less than hundreds of electron volts from the surface layers of a solid. Since it is from the surface, the diffraction is two-dimensional and can be used to examine surface reconstruction.

RHEED: Reflection High-Energy Electron Diffraction–high-energy electrons can also be diffracted from the surface, provided they are at grazing incidence and so do not greatly penetrate.

SEM: Scanning Electron Microscopy–a focused electron beam is scanned across a surface. The emitted secondary electrons are used as a signal that, in a synchronous manner, is displayed on the surface of an oscilloscope. An electron spectrometer can be used to only display electrons whose energies correspond to an Auger peak, in which case the instrument is called a scanning Auger microscope (SAM).

SIMS: Secondary Ion Mass Spectrometry–a destructive but sensitive surface technique. Kiloelectron-volt ions bombard a surface and knock off or sputter ions, which are analyzed by a mass spectrometer and thus can be chemically analyzed.

TEM: Transmission Electron Microscopy–this is like SEM except that the electrons transmitted through a thin specimen are examined. Both elastically and inelastically scattered electrons can be examined, and high contrast is possible.

STM: Scanning Tunneling Microscopy–A sample (metal or semiconductor) has a sharp metal tip placed within 10 Å or less of its surface. A small voltage of order 1 V is established between the two. Since the wave functions of the atoms on the surface of the sample and the tip overlap, in equilibrium the Fermi energies of the sample and tip equalize and under the voltage difference a tunneling current of order nanoamperes will flow between the two. Since the current flow is due to tunneling, it depends exponentially on the distance from the sample to the tip. The exponential dependence makes the tunneling sensitive to sub-angstrom changes in distance, and hence it is possible to use this technique to detect and image individual atoms. The current depends on the local density of states (LDOS) at the surface of the sample and hence is used for LDOS mapping. The position of the tip is controlled by piezoelectric transducers. The apparatus is operated in either the constant-distance or constant-current mode.

UPS: Ultraviolet Photoelectron (or Photoemission) Spectroscopy–the binding energy of a core electron is measured by measuring the energy of the core electron ejected by the ultraviolet photon. For energies not too high, the energy distribution of emitted electrons is dominated by the joint density of initial and final states. An angle-resolved mode is often used since the parallel (to the surface) component of the k vector as well as the energy is conserved. This allows experimental determination of the energy of the initial occupied state for which k parallel is thus known (see Sect. 3.2.2). See also Table 10.3.

XPS: X-ray Photoelectron (or Photoemission) Spectroscopy–the binding energy of a core electron is measured by measuring the energy of the core electron ejected by the X-ray photon–also called *ESCA*. See also Table 10.3

There are of course many other characterization techniques that we could discuss. There are many kinds of scanning probe microscopes, for example. There are many kinds of characterization techniques that are not primarily related to surface properties. Some ideas have already been discussed. Elastic and inelastic X-ray and neutron scattering come immediately to mind. Electrical conductivity and other electrical measurements can often yield much information, as can the many kinds of magnetic resonance techniques. Optical techniques can yield important information (see, e.g., Perkowitz [12.49], as well as Chap. 10 on optical properties in this book). Raman scattering spectroscopy is often important in the infrared. Spectroscopic data involves information about intensity versus frequency. In Raman scattering, the incident photon is inelastically scattered by phonons. Commercial instruments are available, as they are for FTIR (Fourier transform infrared spectroscopy), which use a Michelson interferometer to increase the signal-to-noise ratio and get the Fourier transform of the intensity versus frequency. A FFT (fast Fourier transform) algorithm is then used to get the intensity versus frequency in real time. Perkowitz also mentions photoluminescence spectroscopy, where in general after photon excitation an electron returns to its initial state. Commercial instruments are also available. This technique gives fingerprints of excited states. Considerable additional information about characterization can be found in Bullis et al [12.5]. For a general treatment see Prutton [12.52] and Marder (preface ref. 6, pp73-82).

12.3 Molecular Beam Epitaxy (MET, MS)

Molecular beam epitaxy (MBE) was developed in the 1970s and is by now a common technology for use in making low-dimensional solid-state structure. MBE is an ultrathin film vacuum technique in which several atomic and/or molecular beams collide with and stick to the substrate. Epitaxy means that at the interface of two materials, there is a common crystal orientation and registry of atoms. The substrates are heated to temperature T and mounted suitably. Each effusion cell, from which the molecular beams originate, are held at appropriate temperatures to maintain a suitable flux. The effusion cells also have shutters so that the growth of layers due to the molecular beams can be controlled (see Fig. 12.1). MBE produces high-purity layers in ultrahigh vacuum. Abrupt transitions on an atomic scale can be grown at a rate of a few (tens of) nanometers per second. See, e.g., Joyce [12.31]. Other techniques for producing layered structures include chemical vapor deposition and electrochemical deposition.

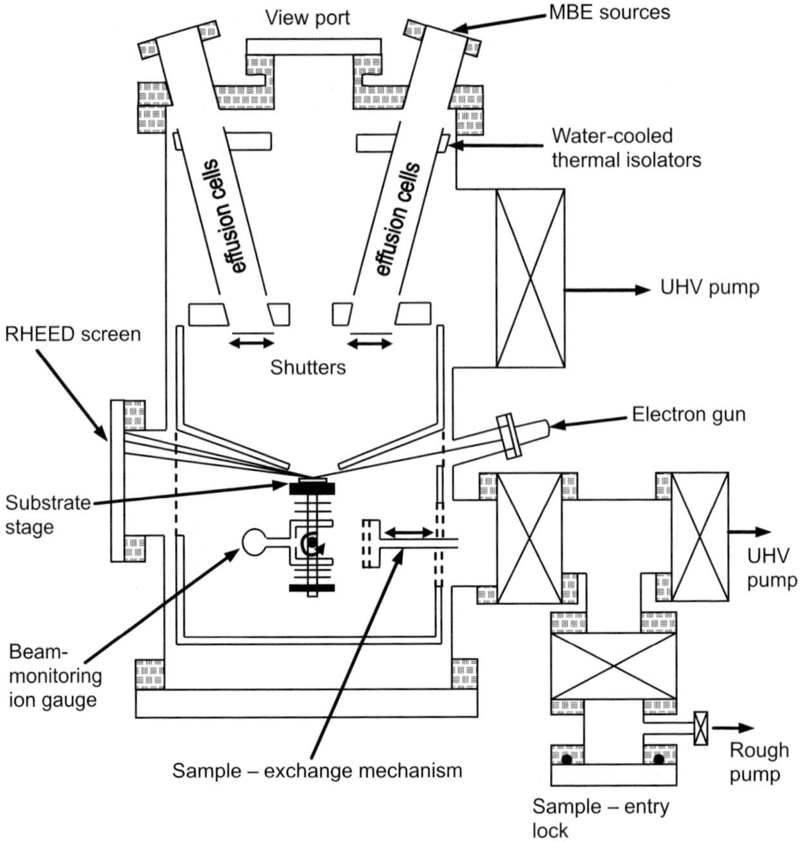

Fig. 12.1. Schematic diagram of an ultrahigh vacuum, molecular beam growth system (adapted from Joyce BA, *Rep Prog Physics* **48**, 1637 (1985), by permission of the Institute of Physics). Reflection high-energy electron diffraction (RHEED) is used for monitoring the growth

12.4 Heterostructures and Quantum Wells

By use of MBE or other related techniques, heterostructures and quantum wells can be formed. Heterostructures are layers of semiconductors with the same crystal structure, grown coherently, but with different bandgaps. Their properties depend heavily on their type. Two types are shown in Fig. 12.2: normal (example GaAlAs-GaAs) and broken (example GaSb-InAs). There are also other types. See Butcher et al [12.6 p. 15]. ΔE_c is the conduction-band offset.

Two-dimensional quantum wells are formed by sandwiching a small-bandgap material between two large-bandgap materials. Energy barriers are formed that quantize the motion in one direction. These can be used to form resonant tunneling devices (e.g. by depositing small-bandgap – large-bandgap – small-bandgap – large – small, etc. See applications of superlattices in Sect. 12.6.1). A quantum well can show increased tunneling currents due to resonance at allowed energy levels in the well. The current versus voltage can even show a decrease with voltage for certain values of voltage. See Fig. 12.11. Diodes and transistors have been constructed with these devices.

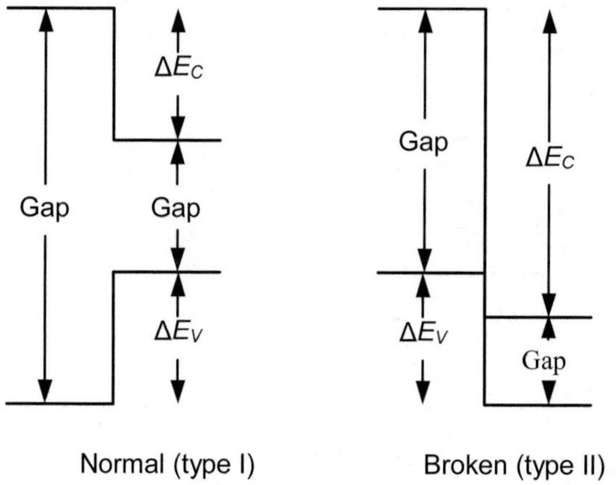

Fig. 12.2. Normal and broken heterostructures

12.5 Quantum Structures and Single-Electron Devices (EE)

Dimension is an important aspect of small electronic devices. Dimensionality can be controlled by sandwiching. If the center of the sandwich is bordered by planar materials for which the electronic states are higher (wider bandgaps), then three-dimensional motion can be reduced to two, producing quantum wells. Similarly one can make linear one-dimensional "quantum wires" and nearly zero-dimensional or "quantum dot" materials. That is, a quantum wire is made by laying down a line of narrow-gap semiconductors surrounded by a wide-gap one with the carriers confined in two dimensions, while a quantum dot involves only a small volume of narrow-gap material surrounded by wide-gap material and the carriers are confined in all three dimensions. With the quantum-dot structure, one may confine or exchange one electron at a time and develop single-electron transistors that would be fast, low power, and have essentially error-free signals. These three types of quantum structures are summarized in Table 12.1.

Table 12.1. Summary of three types of quantum structures

Nanostructures	Comments
Quantum wells	Superlattices can be regarded as quantum well layers – alternating layers of different crystals (when the wells are not too far apart)
Quantum wires	A crystal enclosed on two sides by another crystalline material, with appropriate wider bandgaps
Quantum dots	A crystal enclosed on three sides by another crystalline material – sometimes descriptively called a quantum box

Note: Nanostructures have a least one dimension of a size between approximately one to ten nanometers. See Sects. 12.6 and 7.4.
References: 1. Bastard [12.2]
2. Weisbuch and Vinter [12.65]
3. Mitin et al [12.47]

12.5.1 Coulomb Blockade[1] (EE)

The Coulomb blockade model shows how electron–electron interactions can give rise to effects that in certain circumstances are very easy to understand. It relates to the ideas of single-electron transistors, quantum dots, charge quantization leading to an energy gap in the density of states for tunneling, and is sometimes even qualitatively likened to a dripping faucet. For purposes of illustration, we consider a simple model of an artificial atom represented by the metal particle shown in Fig. 12.3.

Experimentally, the conductance (current per voltage bias) from source to drain shows large changes with gate voltage. We wish to analyze this with the Coulomb blockade model. Let C be the total capacitance between the metal particle and the rest of the system, which we will assume is approximately the capacitance between the metal particle and the gate. Let V_g be the gate voltage, relative to source, and assume the source, particle, and drain voltages are close (but sufficiently different to have the possibility of drawing current from source to drain). If there is a charge Q on the metal particle, then its electrostatic energy is

$$U = QV_g + \frac{Q^2}{2C} . \tag{12.2}$$

Setting $\partial U/\partial Q = 0$, we find U has a minimum at

$$Q = Q_0 = -CV_g . \tag{12.3}$$

If N is an integer, let $Q_0 = -(N + \eta)e$, where $e > 0$, so

$$CV_g = (N + \eta)e . \tag{12.4}$$

[1] See Kastner [12.32]. See also Kelly [12.33, pp. 300-305].

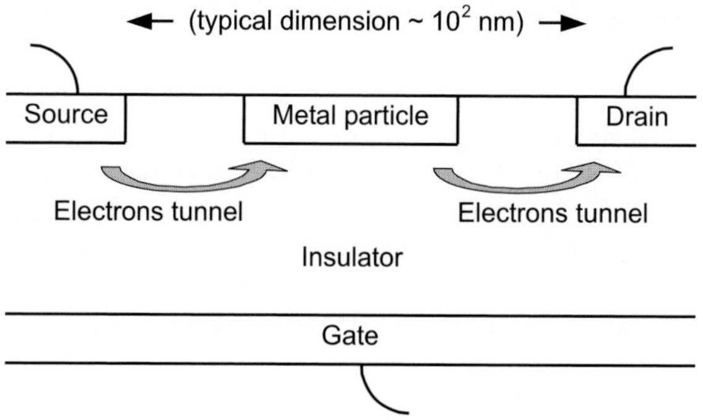

Fig. 12.3. Model of a single-electron transistor

Note that while Q_0 can be any value, the actual physical situation will be determined by the integer number of electrons on the artificial atom (metal particle) that makes U the smallest. This will only be at a mathematical minimum if V_g is an integral multiple of e/C.

For $-1/2 < \eta < 1/2$, and $V_g = (N + \eta)e/C$, the minimum energy is obtained for N electrons on the metal particle. The Coulomb blockade arises because of the energy required to transfer an electron to (or from) the metal particle (you can't transfer less than an electron). We can easily calculate this as follows. Let us consider η between zero and one half. Combining (12.2) and (12.4),

$$U = \frac{1}{C}\left[Q(N+\eta)e + \frac{Q^2}{2}\right]. \tag{12.5}$$

Let the initial charge on the particle be $Q_i = -Ne$ and the final charge be $Q_f = -(N \pm 1)e$. Then for the energy difference,

$$\Delta U^{\pm} = U_f^{\pm} - U_i \ ,$$

we find

$$\Delta U^{+} = \frac{e^2}{C}\left(\frac{1}{2} - \eta\right), \quad \Delta U^{-} = \frac{e^2}{C}\left(\frac{1}{2} + \eta\right).$$

We see that for $\eta < 1/2$ there is an energy gap for tunneling: the Coulomb blockade. For $\eta = 1/2$ the energies for the metal particle having N and $N + 1$ electrons are the same and the gap disappears. Since the source and the drain have approximately the same Fermi energy, one can understand this result from Fig. 12.4. Note ΔU^+ is the energy to add an electron and ΔU^- is the energy to take away an electron (or to add a hole). Thus the gap in the allowed states of the particle is e^2/C. Just above $\eta = 1/2$, the number of electrons on the artificial atoms increases by 1 (to $N + 1$) and the process repeats as V_g is increased. It is indeed reminiscent of a dripping faucet.

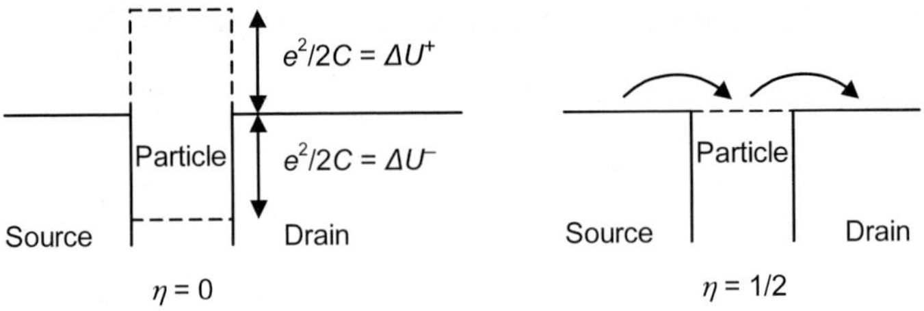

Fig. 12.4. Schematic diagram of the Coulomb blockade at $\eta = 0$. At $\eta = 1/2$ the energy gap ΔE disappears

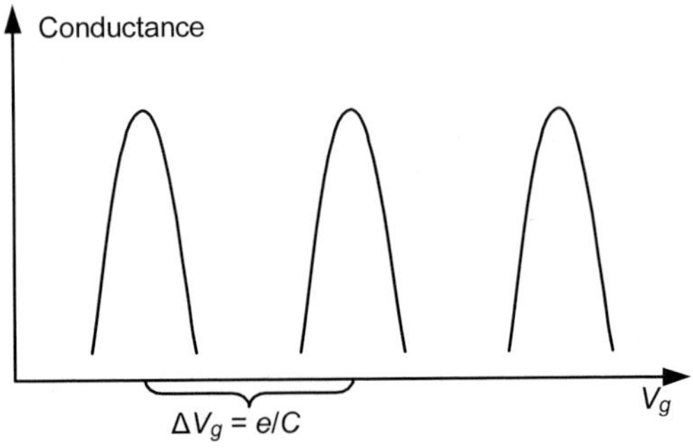

Fig. 12.5. Periodic conductance peaks

The total voltage change from one turn on to the next turn on occurs, e.g., when η goes from 1/2 to 3/2 or

$$\Delta V_g = \frac{e}{C}\left[N + \frac{3}{2} - \left(N + \frac{1}{2}\right)\right] = \frac{e}{C}.$$

A sketch of the conductance versus gate voltage in Fig. 12.5 shows periodic peaks. In order to conduct, an electron must go from source to particle, and then from particle to drain (or a hole from drain to particle, etc.).

Low temperatures are required to see this effect, as one must have

$$kT < \frac{e^2}{2C},$$

so that thermal effects do not wash out the gap. This condition requires small temperatures and small capacitances, such as encountered in nanodevices. In addition the metal particle–artificial atom has discrete energy levels that may be observed in tunneling experiments by fixing V_g and varying the drain-to-source voltage. See Kasner op cit.

12.5.2 Tunneling and the Landauer Equation (EE)

Metal-Barrier-Metal Tunneling (EE)

We start by considering tunneling through a barrier as suggested in Fig. 12.6. We assume each (identical) metal is in local equilibrium with a chemical potential μ. Due to an applied external potential difference φ, we assume the chemical potential is shifted down by $-e\varphi/2$ ($e > 0$) for metal 1 and up by $e\varphi/2$ for metal 2 (see Fig. 12.7).

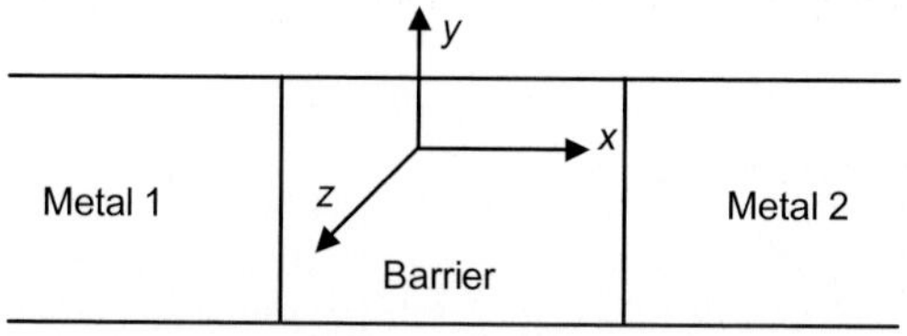

Fig. 12.6. Schematic diagram for barrier tunneling

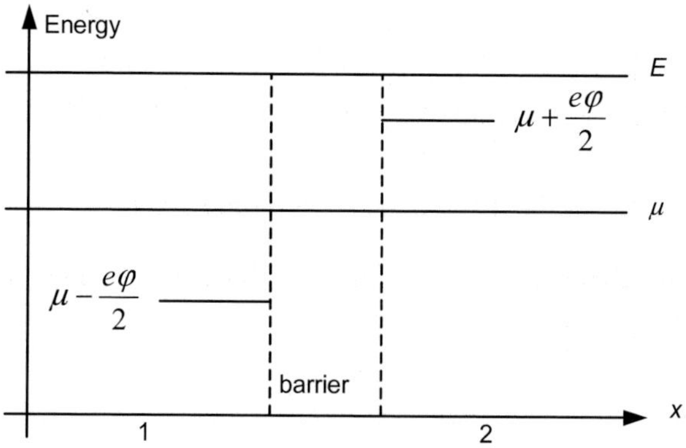

Fig. 12.7. Tunneling sketch

We consider an electron of energy E and assume it tunnels through the barrier without changing energy. We write its energy as (with W defined by the equation and assuming for simplicity the same effective mass in all directions)

$$E = W + E_{\|} + C = \frac{\hbar^2 k_x^2}{2m^*} + \frac{\hbar^2}{2m^*}(k_y^2 + k_z^2) + C\,,$$

where C is a constant that determines the bottom of the conduction band and m^*, assumed constant, is the effective mass. We assume, for this case, that the transmission coefficient T across the barrier depends only on W, $T = T(W)$. We insert a factor of 2 for the spin and consider electron flow in the $\pm x$ directions. With $\varphi = 0$, let the chemical potential in each metal be μ and the Fermi function

$$f(E,\mu) = \frac{1}{\exp[(E-\mu)/kT]+1}\,.$$

Notice $\mu \to \mu - e\varphi/2$ is the same as $E \to E + e\varphi/2$. Then the current density J is (considering current flowing each way, $\pm x$)

$$dJ = -2ev_x[f(E+e\varphi/2,\mu)(1-f(E-e\varphi/2,\mu)) \\ - f(E-e\varphi/2,\mu)(1-f(E+e\varphi/2,\mu))]T(W)\frac{\mathrm{d}^3\boldsymbol{k}}{(2\pi)^3}.$$

Since

$$v_x = \frac{1}{\hbar}\frac{\partial E}{\partial k_x},$$

then

$$v_x \mathrm{d}k_x = \frac{1}{\hbar}\mathrm{d}E .$$

Also $\mathrm{d}^3\boldsymbol{k} = \mathrm{d}k_x\mathrm{d}k_y\mathrm{d}k_z$ and since

$$W = E - \frac{\hbar^2 k_{\|}^2}{2m^*} - C \quad \text{with} \quad (k_{\|}^2 = k_y^2 + k_z^2),$$

we have

$$\mathrm{d}k_y\mathrm{d}k_z = 2\pi k_{\|}\mathrm{d}k_{\|} = \pi \mathrm{d}k_{\|}^2 = -\frac{2\pi m^*}{\hbar^2}\mathrm{d}W ,$$

so substituting we find

$$\mathrm{d}J = \frac{m^* e}{2\pi^2\hbar^3}[f(E+e\varphi/2) - f(E-e\varphi/2)]\mathrm{d}E\, T(W)\mathrm{d}W .$$

When the form of the barrier is known and is suitably simple, the transmission coefficient is often evaluated by the WKB approximation. J can then be calculated by integrating over appropriate limits (W from 0 to $E - C$ and E from C to infinity). This is the standard simple way of looking at tunneling conductance. A different situation is presented below.

Landauer Equation and Quantum Conductance (EE)

In mesoscopic (intermediate between atomic and macroscopic sizes) channels at small sizes, it may be necessary to have a different picture of transport because of quantum effects. In mesoscopic channels at low voltage and low temperatures and few inelastic collisions, Landauer has derived that the electronic conductance is $2e^2/h$ times the number of conductance channels corresponding to all (quantized) transverse energies from zero to the Fermi energy. Transverse energy is defined as the total energy minus the kinetic energy for velocities in the direction of the channel. We derive this result below (see, e.g., Imry I and Landauer R, *More Things in Heaven and Earth*, Bederson B (ed), Springer-Verlag, 1999, p515ff.)

We here write the electron energy as

$$E = \frac{\hbar^2 k_x^2}{2m} + E_{ny,nz},$$

where $E_{ny,nz}$ represents the quantized energy corresponding to the y and z directions. We have replaced the barrier by a device of conductance length L in the x direction and with small size in the y and z directions. We assume this small size is of order of the electron wavelength and thus $E_{ny,nz}$ is clearly quantized. We also regard the two metals as leads to the device and we continue to assume we can treat each lead as essentially in thermal equilibrium.

We assume $T_{ny,nz}(E)$ is the transmission coefficient of the device. Note we have allowed for the possibility that T depends on the quantized motion in the y and z directions. Thus the current is

$$I = -\frac{2e}{L} \sum\nolimits_{ny,nz} L \int \frac{\mathrm{d}k_x}{2\pi} v_x T_{ny,nz}(E)[f(E + e\varphi/2, \mu) - f(E - e\varphi/2, \mu)] .$$

Note that $\mathrm{d}k_x/2\pi$ is the number of states per unit length, so we multiply by L. then we end up with (effectively) the number of electrons, but we want the number per unit length so we divide by L. If φ and the temperature are small then

$$\begin{aligned}[f(E + e\varphi/2, \mu) - f(E - e\varphi/2, \mu)] &= \left.\frac{\partial f}{\partial E}\right)_{\varphi=0} (e\varphi) \\ &= -\delta(E - \mu) e\varphi .\end{aligned}$$

Then using $v_x \mathrm{d}k_x = (1/\hbar)\mathrm{d}E$ as before, we have

$$I = \frac{2e}{h} \sum\nolimits_{ny,nz} T_{ny,nz}(\mu) e\varphi .$$

We thus obtain for the conductance

$$G = \frac{I}{\varphi} = \frac{2e^2}{h} \sum\nolimits_{ny,nz} T_{ny,nz}(\mu) .$$

Note that the sum is only over states with total energy μ so $E_{ny,nz} \le \mu$.

The quantity e^2/h is called the quantum of conductance G_0 so

$$G = 2G_0 \sum_{\substack{ny,nz \\ (E_{ny,nz} \le \mu)}} T_{ny,nz}(\mu) ,$$

which is the Landauer equation. This equation has been verified by experiment. Recently, a similar effect has been seen for thermal conduction by phonons. Here the unit of thermal conduction is $(\pi k_B)^2 T/3h$ (see Schab [12.53].

12.6 Superlattices, Bloch Oscillators, Stark–Wannier Ladders

A superlattice is a set of essentially epitaxial layers (with thickness in nanometers) laid down in a periodic way so as to introduce two periodicities: the lattice periodicity, and the layer periodicity. One can introduce this additional periodicity by doping variations or by compositional variations. A particularly interesting type of superlattice is the strained layer. This is a superlattice in which the lattice constants do not exactly match. It has been found that one can do this without introducing defects provided the layers are sufficiently thin. The resulting strain can be used to productively modify the energy levels.

Minibands can appear in a superlattice. These are caused by quantum wells with discrete levels that are split into minibands due to tunneling between the wells. Some applications of superlattices will be discussed later. For a more quantitative discussion of superlattices, see the sections on Envelope Functions, Effective Mass Theory, Shallow Defects, and Superlattices in Sect. 11.3, and also Mendez and Bastard [12.46].

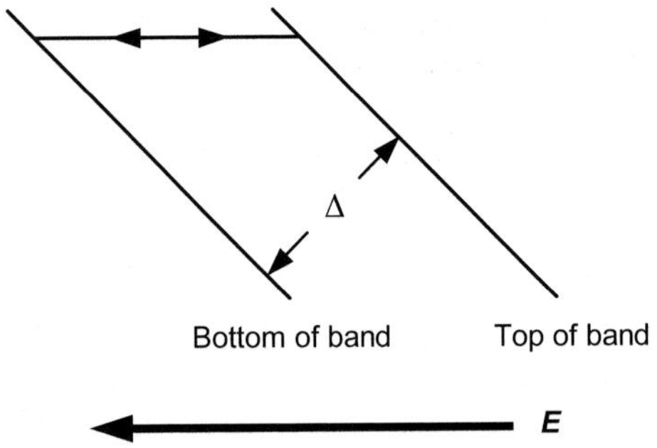

Fig. 12.8. Miniband "tilted" by electric field, and Bloch oscillations

Bloch oscillations can occur in minibands. Consider a portion of a miniband when it is "tilted" by an electric field as shown in Fig. 12.8. An electron in the band will lower its potential energy in the electric field while gaining in kinetic energy, and thus, follow a constant energy path from the bottom of the band to the top, as illustrated above. For very narrow minibands, there is a good chance it will reach the top before phonon emission. In such cases, it could be Bragg reflected. Several reflections between the top and bottom could be possible. These are the Bragg reflections.

We can be slightly more quantitative about Bloch oscillations. The equation of motion of an electron in a lattice is

$$\hbar \frac{\mathrm{d}k}{\mathrm{d}t} = -eE \,, \quad e > 0 \,. \tag{12.6}$$

The width of the Brillouin zone associated with the superlattice is

$$K = \frac{2\pi}{p}, \tag{12.7}$$

where p is the length of the fundamental repeat distance for the superlattice and K is thus a reciprocal lattice vector of the superlattice. Integrating (12.6) from one side of the zone to the other, we find

$$\hbar K = -eEt. \tag{12.8}$$

The Bloch frequency for an oscillation corresponding to the time required to cross the Brillouin zone boundary is given by

$$\omega_B = \frac{2\pi}{t} = \frac{peE}{\hbar}. \tag{12.9}$$

In a tight binding approximation, the energy band structure is given by

$$E_k = A - B\cos(kp), \quad -\frac{\pi}{p} \le k \le \frac{\pi}{p}. \tag{12.10}$$

The group velocity can then be calculated by

$$v_g = \frac{1}{\hbar}\frac{\mathrm{d}E_k}{\mathrm{d}k}. \tag{12.11}$$

In time zero to t_1, the electron moves

$$x_1 = \int_0^{t_1} v_g \mathrm{d}t = \int_0^{-eEt_1/\hbar} v_g \frac{\mathrm{d}t}{\mathrm{d}k}\mathrm{d}k. \tag{12.12}$$

Combining (12.12), (12.11), (12.10), (12.9), and (12.6), we find

$$x_1 = \frac{B}{eE}[\cos(\omega_B t_1) - 1]. \tag{12.13}$$

The electron oscillates in real space with the Bloch frequency ω_B, as expected. In a normal material (nonsuperlattice), the band width is much larger than the miniband width Δ, so that phonon emission before Bloch oscillations set in is overwhelmingly probable. Note that the time required to cross the (superlattice) Brillouin zone is also the time required to go from $k = 0$ to π/p (assuming bottom of band is at 0 and top at $2\pi/p$) then be Bragg reflected to $-\pi/p$ and hence go from $k = -\pi/p$ to 0. So the Bloch oscillation is a complete oscillation of the band to the top and back.

Consider a superlattice of quantum wells producing a narrow miniband. On applying an electric field, the whole drawing "tilts" producing a set of discrete energy levels known as a Stark–Wannier Ladder (see Fig. 12.9). The presence of the (sufficiently strong) electric field may cause the extended wave functions of the miniband to become localized wave functions. If p is the thickness of the period of

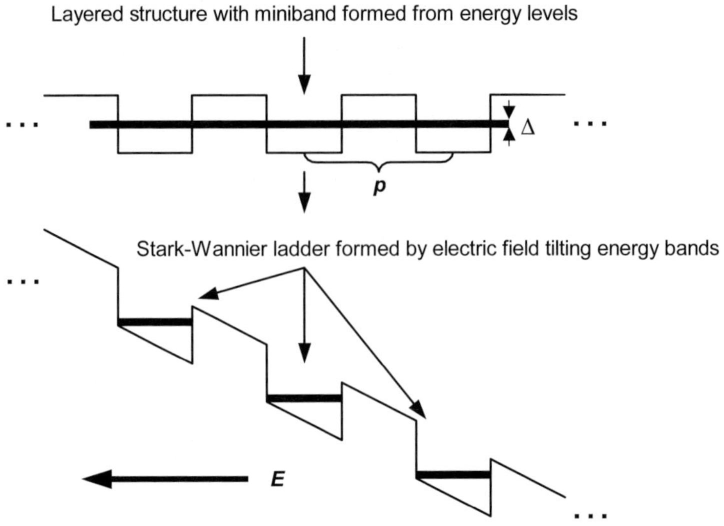

Fig. 12.9. An applied electric field to a superlattice may create a Stark–Wannier ladder when electrons in the discrete levels have no states to easily tunnel to. Only one miniband is shown and the tilt is exaggerated

the superlattice and Δ is the width of the miniband, the Stark–Wannier ladder occurs where $|eEp| \geq \Delta$. Actual realistic calculation gives a set of sharp resonances rather than discrete levels, and the Stark–Wannier ladder has been verified experimentally. Stark–Wannier ladders were predicted by Wannier [12.64]. See also Lyssenko et al [12.45], and [55, p31ff].

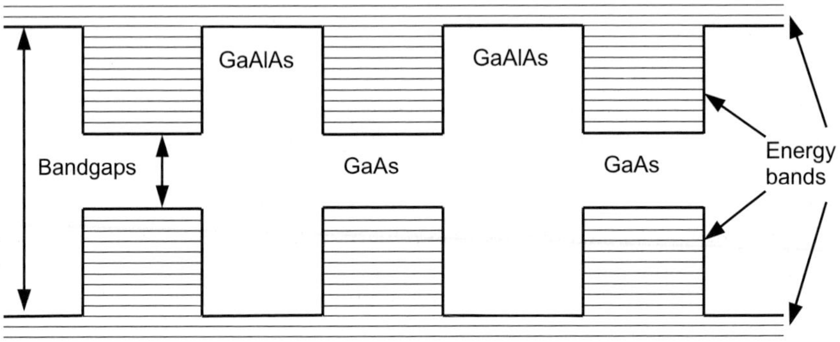

Fig. 12.10. GaAs-GaAlAs superlattice

12.6.1 Applications of Superlattices and Related Nanostructures (EE)

High Mobility (EE)

See Fig. 12.10. Suppose the GaAlAs is heavily donor doped. The donated electrons will fall into the GaAs wells where they would be *separated* from the impurities (donor ions) that furnished them and could scatter them. Thus, high mobility would be created. So, this structure would create high-conductivity semiconductors. Superlattices were proposed by Esaki and Tsu [12.18]. They have since become a very large part of basic and applied research in solid-state physics.

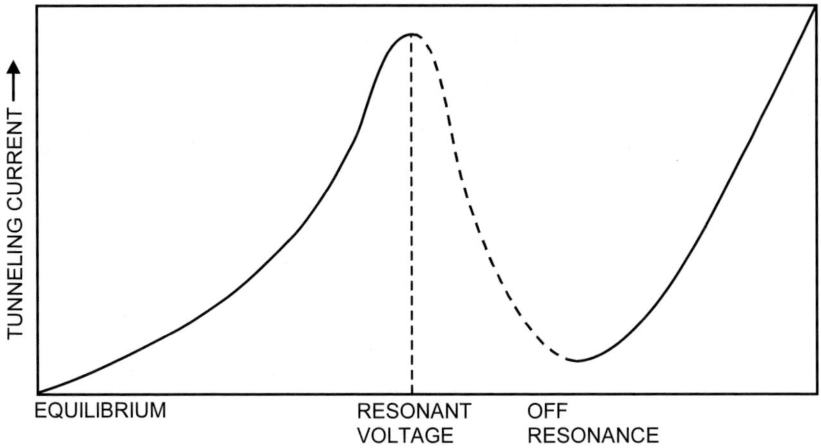

Fig. 12.11. *V-I* curve showing the peak and valley indicating resonant tunneling for a double barrier structure with metals (Fermi energy E_F) on each side

Resonant Tunneling Devices (EE)

A quantum well is formed by layers of wide-bandgap, narrow-bandgap, and wide-bandgap semiconductors. Quantum barriers can be formed from narrow-gap, wide gap, narrow-gap semiconductors. A resonant tunneling device can be formed by surrounding a well with two barriers. Outside the barrier, electrons populate states up to the Fermi energy. If a voltage is applied across the device, the (quasi) Fermi energy on the input side can be moved until it equals the energy of one of the discrete energies within the well.

Typically, the current increases with increasing voltage until a match is obtained, and as the voltage is further increased, the current decreases. The decrease in current with increasing voltage is called negative differential resistance, which can be applied in making high-frequency devices (See Fig. 12.11). See, e.g., Beltram and Capasso in Butcher et al [12.6 Chap. 15]. See also Capasso and Datta [12.8].

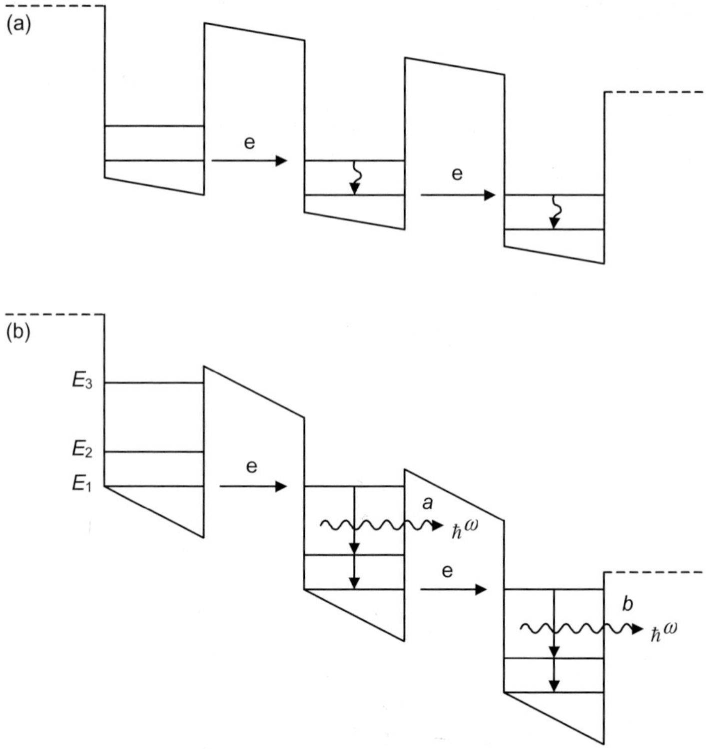

Fig. 12.12. (**a**) Resonant tunneling through a superlattice with a discrete Stark-Wannier "ladder" of states. (**b**) Resonant tunneling laser (emission *a* may trigger emission *b*, etc.). Note that in (**a**) we are considering non radiative transitions while (**b**) has indicated radiative transitions *a* and *b*. Adapted from Capasso F, *Science* **235**, 175 (1987).

Lasers (EE)

We start with a superlattice (or at least a multiple quantum well structure) of alternating wide-gap, narrow-gap materials. The quantum wells form where we have narrow-gap semiconductors and the electrons settle into discrete ground states in the quantum wells. Now, apply an electric field so that the ground state of one level is in resonance with the excited state of the next level. One then gets resonant tunneling between these two states. In effect, one can obtain a population inversion leading to lasing action (see Fig. 12.12). For further details, discussion of relevance of minibands, etc., see Capasso et al [12.9]. Lasers using quantum wells are used in compact disk players.

Infrared Detectors (EE)

This can be made similarly to the way the laser is made, except one deals with excitations to the conduction band and subsequent collection by the electric field. See Fig. 12.13 where the idea is sketched. One assumes the excitation energy is in the infrared.

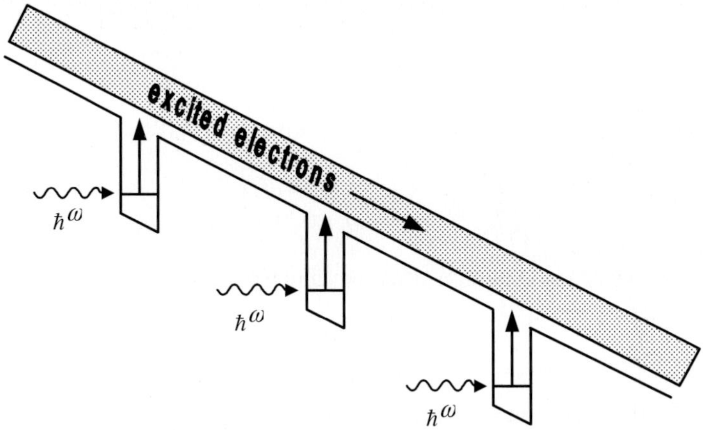

Fig. 12.13. Infrared photodetector made with quantum wells. As shown, the electrons in the wells are excited into the conduction band states and then can be collected and detected. Adapted from Capasso and Datta [12.8, p. 81]

12.7 Classical and Quantum Hall Effect (A)

12.7.1 Classical Hall Effect – CHE (A)

The Hall effect has been important for many reasons. For example, in semiconductors it can be used for determining the sign and the concentration of charge carriers. The fractional quantum Hall effect, in terms of basic physics ideas, may be the most important discovery in solid-state physics in the last quarter of a century. To start, we first reconsider the classical quantum Hall effect for electrons only.

Let electrons move in the (x,y)-plane with a magnetic field in the z direction and an electric field also in the (x,y)-plane. In MKS units and standard notation $(e > 0)$

$$F_x = -eE_x - eV_yB - \frac{mV_x}{\tau}, \tag{12.14}$$

$$F_y = -eE_y + eV_xB - \frac{mV_y}{\tau}, \tag{12.15}$$

where the term involving the relaxation time τ is due to scattering. The current density is given by

$$J_x = -neV_x\,, \tag{12.16}$$

$$J_y = -neV_y\,, \tag{12.17}$$

where n is the number of electrons per unit volume. Letting the dc conductivity be

$$\sigma_0 = \frac{ne^2\tau}{m}\,, \tag{12.18}$$

we can write (in the steady state when F_x, $F_y = 0$ using (12.14)–(12.18))

$$\begin{pmatrix} E_x \\ E_y \end{pmatrix} = \frac{1}{\sigma_0}\begin{pmatrix} 1 & \omega_c\tau \\ -\omega_c\tau & 1 \end{pmatrix}\begin{pmatrix} J_x \\ J_y \end{pmatrix}, \tag{12.19}$$

where $\omega_c = eB/m$ is the cyclotron frequency and we can show (by (12.18))

$$\frac{B}{ne} = \frac{\omega_c\tau}{\sigma_0}\,. \tag{12.20}$$

The inverse to (12.19) can be written

$$\begin{pmatrix} J_x \\ J_y \end{pmatrix} = \frac{\sigma_0}{1+(\omega_c\tau)^2}\begin{pmatrix} 1 & -\omega_c\tau \\ \omega_c\tau & 1 \end{pmatrix}\begin{pmatrix} E_x \\ E_y \end{pmatrix}. \tag{12.21}$$

We will use the geometry as shown in Fig. 12.14. We rederive the Hall coefficient. Setting $J_y = 0$, then

$$E_y = |\boldsymbol{V}\times\boldsymbol{B}| = -\frac{\omega_c\tau}{\sigma_0}J_x = -\frac{B}{ne}J_x\,, \tag{12.22}$$

where $\boldsymbol{V} = V_x = J_x/ne$ from (12.16). The Hall coefficient is defined as

$$R_{\mathrm{H}} = \frac{E_y}{J_x B_z} = -\frac{1}{ne} \tag{12.23}$$

as usual. The Hall voltage over the length w would then be

$$V_{\mathrm{H}} = -E_y w = \frac{BJ_x w}{ne}\,. \tag{12.24}$$

The current through the segment of area tw is

$$I_x = J_x t w\,, \tag{12.25}$$

so

$$V_{\mathrm{H}} = \frac{BI_x}{nte}\,. \tag{12.26}$$

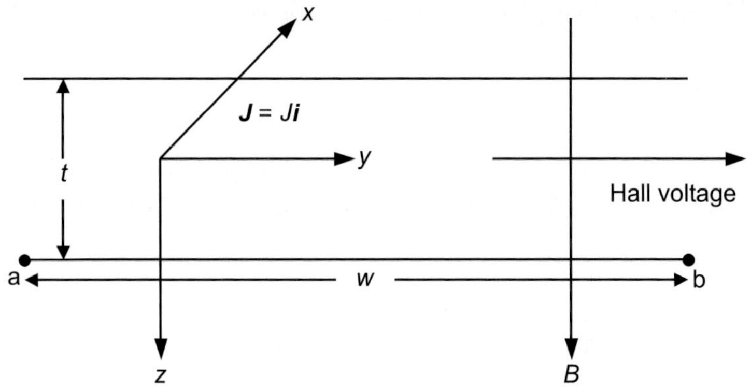

Fig. 12.14. Schematic diagram of classical Hall effect

Define n_a as the number of electrons per unit area (projected into the (x,y)-plane) so the Hall voltage can be written

$$V_{\mathrm{H}} = \frac{I_x B}{n_a e}. \tag{12.27}$$

The Hall conductance $1/R_{xy}$ is

$$\frac{1}{R_{xy}} = \frac{I_x}{V_{\mathrm{H}}} = \frac{n_a e}{B}. \tag{12.28}$$

Longitudinally over a length L, the voltage change is

$$V_L = E_x L = \frac{J_x L}{\sigma_0} = \frac{I_x L}{tw\sigma_0}, \tag{12.29}$$

which we find to be independent of B. This is the usual Drude result. However, this result is based on the assumption that all electrons are moving with the same velocity. If we allow the electrons to have a distribution of velocities by doing a proper Boltzmann equation calculation, we find there is a magnetoresistance effect. The result is (Blakemore [12.3]).

$$\sigma = \frac{\sigma_0}{1 + (\sigma_0 R_{\mathrm{H}})^2 \dfrac{|\boldsymbol{J} \times \boldsymbol{B}|^2}{|J|^2}}. \tag{12.30}$$

In addition, when band-structure effects are taken into account one finds there also may be a magnetoresistance even when $\boldsymbol{J} \times \boldsymbol{B} = 0$. Classically then we predict behavior for the Hall effect (with I_x = constant) as shown schematically in Fig. 12.15.

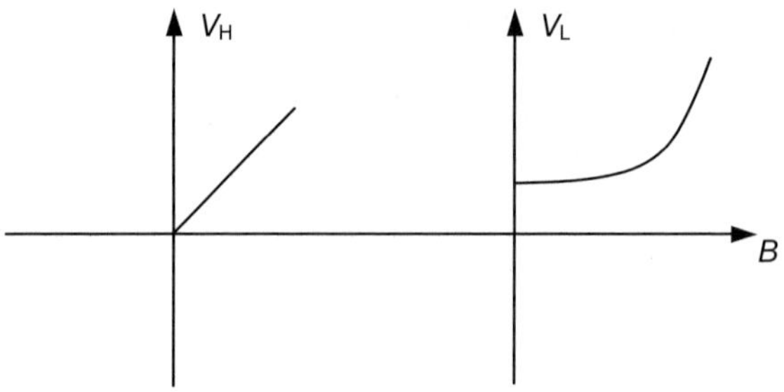

Fig. 12.15. Schematic diagram of classical Hall effect behavior. See (12.27) for V_H and (12.29) for V_L

12.7.2 The Quantum Mechanics of Electrons in a Magnetic Field: The Landau Gauge (A)

We start by solving the problem of electrons moving in two dimensions (x, y) in a magnetic field in the z direction (see, e.g., [12.41, 12.51, 12.56, 12.59]). The essential ideas of the quantum Hall effect can be made clear by ignoring electron spin, and so we do. The limit to two dimensions is necessary for the quantum Hall effect as we will discuss later. The discussion of Landau diamagnetism (Sect. 3.2.2) may be helpful here as a review of the quantum mechanics of electrons in magnetic fields.

For $\boldsymbol{B} = B\boldsymbol{k}$, one choice of $\boldsymbol{A}$ is:

$$\boldsymbol{A} = -\frac{1}{2}\boldsymbol{r}\times\boldsymbol{B}\,, \tag{12.31}$$

which is a cylindrically symmetric gauge. Instead, we use the Landau gauge where $A_x = -yB$, $A_y = 0$, and $A_z = 0$. This yields a simpler solution for the Hall situation that we consider.

The free-electron Hamiltonian can then be written

$$\mathcal{H} = \frac{1}{2m}[\boldsymbol{p} - q\boldsymbol{A}]^2\,, \tag{12.32}$$

where $q = -e < 0$. In two dimensions this becomes (compare Sect. 3.2.2)

$$\mathcal{H} = \frac{1}{2m}\left[\left(\frac{\hbar}{i}\frac{\partial}{\partial x} - eyB\right)^2 - \hbar^2\frac{\partial^2}{\partial y^2}\right]. \tag{12.33}$$

Introducing the "magnetic length"

$$l_\mu = \sqrt{\frac{\hbar}{eB}}\,, \tag{12.34}$$

we can then write the Schrödinger equation as

$$-\frac{\hbar^2}{2m}\left[\frac{\partial^2}{\partial y^2}-\left(\frac{1}{\mathrm{i}}\frac{\partial}{\partial x}-\frac{y}{l_\mu^2}\right)^2\right]\psi = E\psi . \tag{12.35}$$

We seek a solution of the form

$$\psi = A\mathrm{e}^{\mathrm{i}kx}\varphi(y) ,$$

and thus

$$\left[-\frac{\hbar^2}{2m}\frac{\partial^2}{\partial y^2}+\frac{\hbar^2}{2ml_\mu^4}(y-l_\mu^2 k)^2\right]\varphi = E\varphi . \tag{12.36}$$

Since also

$$l_\mu = \sqrt{\frac{\hbar}{m\omega_c}} ,$$

and from (12.34) and from the preceding equation for l_μ, we have

$$\frac{\hbar^2}{2ml_\mu^4} = \frac{1}{2}m\omega_c^2 .$$

This may be recognized as a harmonic oscillator equation with the quantized energies

$$E_n = \hbar\omega_c\left(n+\frac{1}{2}\right), \quad n = 0,1,2\ldots ,$$

and the eigenfunctions are

$$\varphi_n = \left(\frac{m\omega_c}{\pi\hbar}\right)^{1/4}\frac{1}{\sqrt{2^n n!}}H_n\left(\frac{y}{l_\mu}-l_\mu k\right)\exp\left[-\frac{1}{2}\left(\frac{y-l_\mu^2 k}{l_\mu}\right)^2\right] , \tag{12.37}$$

where the $H_n(x)$ are the Hermite polynomials

$$H_0(x)=1, \quad H_1(x)=2x, \quad H_2(x)=4x^2-2, \quad \text{etc.}$$

For the Hall effect we now solve for the case in which there is also an electric field in the y direction (the Hall field). This adds a potential of

$$U = eEy . \tag{12.38}$$

The drift velocity in crossed E and B fields is

$$V = \frac{E}{B} ,$$

so by (12.38), the above, and $\omega_c = eB/m$

$$U = m\omega_c V y \,. \tag{12.39}$$

Thus we can write from (12.38):

$$\left[-\frac{\hbar^2}{2m}\frac{\partial^2}{\partial y^2} + \frac{1}{2}m\omega_c^2(y - l_\mu^2 k)^2 + m\omega_c V y\right]\varphi = E\varphi \,. \tag{12.40}$$

Now since V is very small, we can neglect terms involving the square of V. Then if we define the origin so $y = y' - aV$, with $a = 1/\omega_c$, the Schrödinger equation simplifies to the same form as (12.36):

$$\left[-\frac{\hbar^2}{2m}\frac{\partial^2}{\partial y^2} + \frac{1}{2}m\omega_c^2(y' - l_\mu^2 k)^2\right]\varphi = [E - m\omega_c l_\mu^2 k V]\varphi \,. \tag{12.41}$$

Thus using (12.37) in new notation,

$$\varphi_n \propto H_n\left(\frac{y + V/\omega_c}{l_\mu} - l_\mu k\right)\exp\left[-\frac{1}{2}\left(\frac{y + V/\omega_c - l_\mu^2 k}{l_\mu}\right)^2\right], \tag{12.42}$$

and

$$E_n = \hbar\omega_c\left(n + \frac{1}{2}\right) + m\omega_c l_\mu^2 k V \,. \tag{12.43}$$

Now let us discuss some qualitative results related to these states.

12.7.3 Quantum Hall Effect: General Comments (A)

We first present the basic experimental results of the quantum Hall effect and then indicate how it can be explained. We have already described the Hall geometry. The Hall resistance is V_{H}/I_x, where I_x may be held constant. The longitudinal resistance is V_L/I_x. One finds plateaus at values of $(h/e^2)/\nu$ with e^2/h being called the quantum of conductance and ν is an integer for the integer quantum Hall effect and a fraction for the fractional quantum Hall effect.

As shown in Fig. 12.16, V_L/I_x appears to be zero when the Hall resistance is on a plateau. The figures only schematically illustrate the effect for $\nu = 2$, 1, and 1/3. There are many other plateaus, which we have omitted.

The quantum Hall effect requires two dimensions, low temperatures, electrons, and a large external magnetic field. Two dimensions are necessary so the gaps in between the Landau levels ($E_g = \hbar\omega_c$) are not obliterated by the continuous energy introduced by motion in the third dimension. (The IQHE involves filled or empty Landau levels. Gaps for the FQHE, which involve partially filled Landau levels, are introduced by electron–electron interactions.) Low temperatures are necessary so as not to wipe out the quantization of levels by thermal-broadening effects.

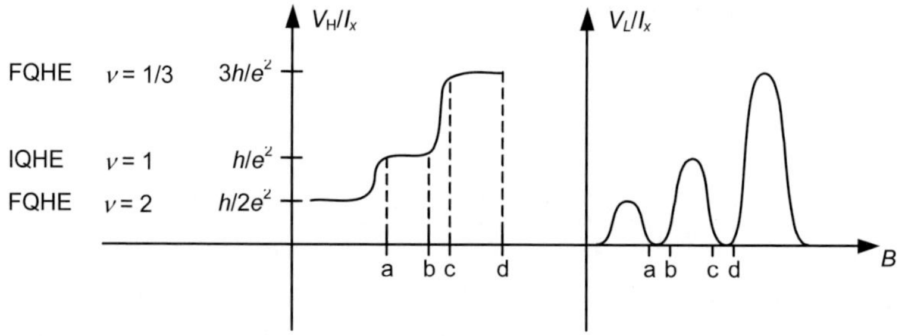

Fig. 12.16. Schematic diagram of quantum Hall effect behavior

There are two convenient ways to produce the two dimensional electron systems (2DES). One way is with MOSFETs. In a MOSFET a positive gate voltage can create a 2DES in an inversion region at the Si and SiO_2 interface. One can also use GaAs and AlGaAs heterostructures with donor doping in the AlGaAs so the electrons go to the GaAs region that has lower potential. This separates the electrons from the donor impurities and hence the electrons can have high mobility due to low scattering of them.

The IQHE was discovered by Klaus von Klitzing in 1980 and for this he was awarded the Nobel prize in 1985. About two years later, Stormer and Tsui discovered the FQHE and they along with Laughlin (for theory) were awarded the 1998 Nobel prize for this effect.

Qualitatively, the IQHE can be fairly simply explained. As each Landau level is filled there is a gap to the next Landau level. The gap is filled by localized non-conducting states, and as the Fermi level moves through this gap, no change in current is observed. The Landau levels themselves are conducting. For the IQHE the electron–electron interactions effects are really not important, but the disorder that causes the localized states in the gap is crucial.

The fractional quantum Hall effect occurs for partially filled Landau levels and electron–electron interaction effects are crucial. They produce an excitation gap reminiscent of the gap produced in the Mott insulating transition. Potential fluctuations cause localized states and plateau formation.

The Integer Quantum Hall Effect – IQHE–Simple Picture (A)

We give an elementary picture of the IQHE. We start with four results.

a. The Landau degeneracy per spin is eB/h. (This follows because the number of states per unit area in k-space (ΔA) and in real space is $(\Delta A)/(2\pi)^2$. Then from (6.29), $(\Delta A) = (2\pi)^2(eB/h)$. Thus, the number of states per unit area in real space is $n_B \equiv eB/h$).

b. The drift velocity perpendicular to E and B field is $V = E/B$.

c. Flux quanta have the value $\Phi_0 = h/e$ (see (8.47)).

d. The number of filled Landau levels $\nu = N/N_\Phi$, where N is the number of electrons and N_Φ is the number of flux quanta. This follows from $\nu = N/(eBLw/h) = N/(\Phi/\Phi_0)$.

Then

$$I_x = J_x wt = neVwt \,, \tag{12.44}$$

where n = the number of electrons per unit volume and

$$n = \frac{N}{wtL} = \nu \frac{eB}{h}(Lw)\frac{1}{wtL} = \nu \frac{eB}{ht} \,. \tag{12.45}$$

So since $V = E/B$,

$$I_x = \nu \frac{eB}{ht} e \frac{E}{B} wt = \nu e^2 \frac{Ew}{h} = \nu e^2 \frac{V_{\mathrm{H}}}{h} \,, \tag{12.46}$$

or

$$\frac{I}{V_{\mathrm{H}}} = \frac{1}{R_{xy}} = \text{the Hall conductance } = \frac{\nu e^2}{h} \,. \tag{12.47}$$

If B changes, as long as the Fermi level stays in the gap, the Landau levels are filled or empty and the current over the voltage remains on a plateau of fixed n. (It can be shown that the total current carried by a full Landau level remains constant even as the number of electrons that fill it varies with the Landau degeneracy.)

Incidentally, when $1/R_{xy} = \nu e^2/h$ then $1/R_{xx} = I/V_L \to \infty$ or $R_{xx} \to 0$. This is because the electrons in conducting states have no available energy states into which they can scatter.

Fractional Quantum Hall Effect – FQHE (A)

One needs to think about the FQHE both by thinking about the Laughlin wave functions and by thinking of their physical interpretation. For example, for the ν = 1/3 case with m = 3 (see general comments, next section), the wave function is (see [12.41]):

$$\psi(z_1,\ldots,z_N) = \prod_{j<k}^{N}(z_j - z_k)^m \exp\left(-\frac{1}{4l_\mu^2}\sum_{j=1}^{N}\left|z_j\right|^2\right), \tag{12.48}$$

where $z_j = x_j + \mathrm{i}y_j$ locates the jth electron in 2D. Positive and negative excitations at $z = z_0$ are given by (see also [12.59])

$$\psi^+ = \exp\left(-\frac{1}{4l_\mu^2}\sum_{j=1}^{N}\left|z_j\right|^2\right)\prod_{j}^{N}(z_j - z_0)\prod_{j<k}^{N}(z_j - z_k)^m \,, \tag{12.49}$$

$$\psi^- = \exp\left(-\frac{1}{4l_\mu^2}\sum_{j=1}^{N}\left|z_j\right|^2\right)\prod_{j}^{N}\left(2l_\mu^2\frac{\partial}{\partial z_j} - z_0^*\right)\prod_{j<k}^{N}(z_j - z_k)^m \,. \tag{12.50}$$

For $m = 3$, these excitations have effective charges of magnitude $e/3$. The ground state of the FQHE is considered to be like an incompressible fluid as the density is determined by the magnetic field and is fairly rigidly locked. The papers by Laughlin should be consulted for full details.

These wave functions have led to the idea of composite particles (CPs). Rather than considering electrons in 2D in a large magnetic field, it turns out to be possible to consider an equivalent system of electrons plus attached field vortices (see Fig. 16, p. 885 in [12.51]). The attached vortices account for most of the magnetic field and the new particles can be viewed as weakly interacting because the vortices minimize the electron–electron interactions.

General Comments (A)

It turns out that the composite particles may behave as either bosons or fermions according to the number of attached flux quanta. Electrons plus an odd number of surrounding flux quanta are Bose CPs and electrons with an even number of attached quanta are Fermi CPs. The $\nu = 1/3$, $m = 3$ case involves electrons with three attached quanta and hence these CPs are bosons that can undergo a Bose–Einstein-like condensation, produce an energy gap, and have a FQHE with plateaus. For the $\nu = 1/2$ case, there are two attached quanta, the systems behaves as a collection of fermions, there is no Bose–Einstein condensation and no FQHE.

In general, when the magnetic field increases, electrons can "absorb" some field and become "anyons." These can be shown to obey fractional statistics and seem to be intermediate between fermions and bosons. This topic takes us too far afield and references should be consulted.[2]

There are different ways to construct CPs to describe the same physical situation, but normally one tries to use the simplest. Also, there are still problems connected with the understanding of some values of ν. A complete description would take us further than we intend to go, but the chapter references listed at the end of this book can be a good starting point for further investigation.

The quantum Hall effects are very rich in physical effects. So far, they are not so rich in applications. However, the experiments do determine e^2/h to three parts in ten million or better, and hence they provide an excellent resistance standard. Also, since the speed of light is a defined quantity, the QHE also determines the fine structure constant $e^2/\hbar c$ to high accuracy. It is interesting that the quantum Hall effect determines e^2/h, while we found earlier that e/h could be determined by the Josephson effect. Thus the two can be used to determine both e and h individually.

[2] See Lee [12.43].

12.8 Carbon – Nanotubes and Fullerene Nanotechnology (EE)

Carbon is very versatile and important both to living tissues and to inanimate materials. Carbon of course forms diamond and graphite. In recent years the ability of carbon to aggregate into fullerenes and nanotubes has been much discussed.

Fullerenes are stable, cage-like molecules of carbon with often a nearly spherical appearance. A C_{60} molecule is also called a Buckyball. Both are named after Buckminster Fuller because of their resemblance to the geodesic domes he designed. Buckyballs were discovered in 1985 as a byproduct of laser-vaporized graphite. Some of the fullerides (salts such as K-C_{60}) can be superconductors (see, e.g., Hebard [12.25]).

Carbon nanotubes are one or more cylindrical and seamless shells of graphitic sheets. Their ends are capped by half of a fullerene molecule. They were discovered in 1993 by Sumio Iijima and mass produced in 1995 by Rick Smalley. For more details see, e.g., Dresselhaus et al [12.17]. While carbon nanotubes are now easy to produce, they are not easy to produce in a controlled fashion.

To form them, start with a single sheet of graphite called graphene whose band structure leads to a semimetal (where the conduction band edge is very close to the valence band edge). A picture of the dispersion relations show a two-dimensional E vs. k relationship where two cones touch at their tips with the same conic axis and in an end-to-end fashion. See Fig. 12.17. Where the cones touch is the Fermi energy, or as it is called, the Fermi point.

Nanotubes can be semiconductors or metals. It depends on the boundary conditions on the wave function as determined by how the sheet is rolled up. Both the circumference and twist are important. This, in turn, affects whether a bandgap is introduced where the Fermi point in graphene was. The semiconducting bandgap can be varied by the circumference. Multiwalled nanotubes are more complex.

Semiconductor nanotubes can be made to act as transistors by using a gate voltage. A negative bias (to the gate) induces holes and makes them conduct. Positive bias makes the conductance shut off. They have even been made to act as simple logic devices. See McEven PL, "Single-Wall Carbon Nanotubes," *Physics World*, pp. 32-36 (June 2000). One interesting feature about nanotubes is that they provide a way around the fundamental size limits of Si devices. This is because they can be made very small and are not plagued with surface states (they have no surface formed by termination of a 3D structure and as cylinders they have no edges).

Carbon nanotubes are a fascinating example of one-dimensional transport in hopefully easy to make structures. They are quantum wires with ballistic electrons – and they show many interesting quantum effects.

An additional feature of interest is that carbon nanotubes show significant mechanical strength. Their strength arises from the carbon bond.[3]

---

[3] Carbon is becoming an increasingly interesting material with the suggestion that under certain circumstances it can even be magnetic. See Coey M and Sanvito S, Physics World, Nov 2004, p33ff.

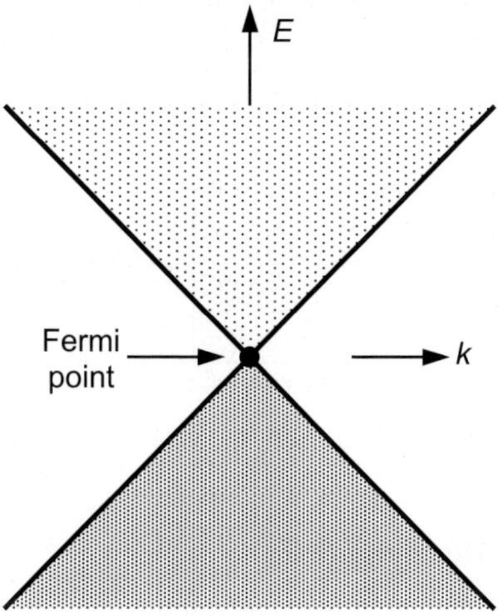

Fig. 12.17. Dispersion relation for graphene

12.9 Amorphous Semiconductors and the Mobility Edge (EE)

By amorphous, we will mean noncrystalline. Here, rather than an energy gap one has a mobility gap separating localized and nonlocalized states. The localization of electron states is an important concept. The electron–electron interaction itself may give rise to localization as shown by Mott [12.48], as we have discussed earlier in the book. In effect, the electron–electron interaction can split the originally partially filled band into a filled band and an empty band separated by a bandgap. We are more interested here in the Anderson localization transition caused by random local field fluctuations due to disorder. In amorphous semiconductors, this can lead to "mobility edges" rather than band edges (see Fig. 12.18).

The dc conductivity of an amorphous semiconductor is of the form

$$\sigma = \sigma_0 \exp\left(-\frac{\Delta E}{kT}\right), \tag{12.51}$$

for charge transport by *extended state carriers*, where ΔE is of the order of the mobility gap and σ_0 is a conductor. For *hopping of localized carriers*

$$\sigma = \sigma_0 \exp\left(-\frac{T_0}{T}\right)^{1/4}, \tag{12.52}$$

where σ_0 and T_0 are constants. Memory and switching devices have been made with amorphous chalcogenide semiconductors. The meaning of (12.52) is amplified in the next section.

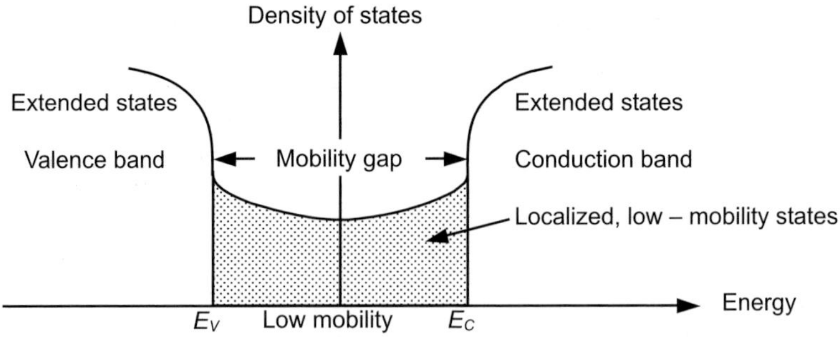

Fig. 12.18. Area of mobility between valence and conduction bands

12.9.1 Hopping Conductivity (EE)

So far, we have discussed band conductivity. Here electrons move along at constant energy, in the steady state the energy they gain from the field is dissipated by collisions. One can even have band conductivity in impurity bands when the impurity wave functions overlap sufficiently to form a band. One usually thinks of impurity states as being localized, and for localized states there is no dc conductivity at absolute zero. However, at nonzero temperatures, an electron in a localized state may make a transition to an empty localized state, getting any necessary energy from a phonon, for example. We say the electron hops from state to state. In general, then, an electron hop is a transition of the electron involving both its position and energy.

The topic of hopping conductivity is very complicated and a thorough treatment would take us too far afield. The books by Shoklovskii and Efros [12.55], and Mott [12.48], together with copious references cited therein, can be consulted. In what is given below, we are primarily concerned with hopping conductivity in lightly doped semiconductors.

Suppose the electron jumps to a state a distance R. We assume very low temperatures with the relevant states localized near the Fermi energy. We assume states just below the Fermi energy hop to states just above gaining the energy E_a (from a phonon). Letting $N(E)$ be the number of states per unit volume, we estimate:

$$\frac{1}{E_a} \approx \frac{4}{3}\pi R^3 N(E_F), \tag{12.53}$$

thus we estimate (see Mott [12.48]) the hopping probability and hence the conductivity is proportional to

$$\exp(-2\alpha R - E_a / kT), \tag{12.54}$$

where α is a constant denoting the rate of exponential decrease of the wave function of the localized state $\exp(-\alpha r)$.

Substituting (12.53) into (12.54) and maximizing the expression with regard to the hopping range R gives:

$$\sigma = \sigma_0 \exp[(-T_0 / T)^{1/4}], \tag{12.55}$$

where

$$T_0 = 1.5\alpha^3 \beta / N(E_F), \tag{12.56}$$

and β is a constant, whose value follows from the derivation, but in fact needs to be more precisely evaluated in a more rigorous presentation.

Maximizing also yields

$$R = \text{constant}(1/T)^{1/4}, \tag{12.57}$$

so the theory is said to be for variable-range hopping (VRH); the lower the temperature, the longer the hopping range and the less energy is involved.

Equation (12.55), known as Mott's law, is by no means a universal expression for the hopping conductivity. This law may only be true near the Mott transition, and even then that is not certain. Electron–electron interactions may cause a Coulomb gap (Coulombic correlations may lead the density of states to vanish at the Fermi level), and lead to a different exponent (from one quarter–actually to 1/2 for low-temperature VRH).

12.10 Amorphous Magnets (MET, MS)

Magnetic effects are typically caused by short-range interactions, and so they are preserved in the amorphous state although the Curie temperature is typically lowered. A rapid quench of a liquid metallic alloy can produce an amorphous alloy. When the alloy is also magnetic, this can produce an amorphous magnet. Such amorphous magnets, if isotropic, may have low anisotropy and hence low coercivities. An example is $Fe_{80}B_{30}$, where the boron is used to lower the melting point, which makes quenching easier. Transition metal amorphous alloys such as $Fe_{75}P_{15}C_{10}$ may also have very small coercive forces in the amorphous state.

On the other hand, amorphous NdFe may have a high coercivity if the quench is slow so as to yield a multicrystalline material. Rare earth alloys (with transition metals) such as $TbFe_2$ in the amorphous state may also have giant coercive fields (~ 3 kOe). For further details, see [12.20, 12.26, 12.36].

We should mention that bulk amorphous steel has been made. It has approximately twice the strength of conventional steel. See Lu et al [12.44].

Nanomagnetism is also of great importance, but is not discussed here. However, see the relevant chapter references at the end of this book.

12.11 Soft Condensed Matter (MET, MS)

12.11.1 General Comments

Soft condensed-matter physics occupies an intermediate place between solids and fluids. We can crudely say that soft materials will not hurt your toe if you kick them.

Generally speaking, hard materials are what solid-state physics discusses and the focus of this book was crystalline solids. Another way of contrasting soft and hard materials is that soft ones are typically not describable by harmonic excitations about the ground-state equilibrium positions. Soft materials are also often complex, as well as flexible. Soft materials have a shape but respond more easily to forces than crystalline solids.

Soft condensed-matter physics concerns itself with *liquid crystals* and *polymers*, which we will discuss, and fluids as well as other materials that feel soft. Also included under the umbrella of soft condensed matter are *colloids*, *emulsions*, and *membranes*. As a reminder, colloids are solutes in a solution where the solute clings together to form 'particles,' and emulsions are two-phase systems with the dissolved phase being minute drops of a liquid. A membrane is a thin, flexible sheet that is often a covering tissue. Membranes are two-dimensional structures built from molecules with a hydrophilic head and a hydrophobic tail. They are important in biology.

For a more extensive coverage the books by Chaikin and Lubensky [12.11], Isihara [12.27], and Jones [12.30] can be consulted.

We will discuss liquid crystals in the next Section and then we have a Section on polymers, including rubbers.

12.11.2 Liquid Crystals (MET, MS)

Liquid crystals involve phases that are intermediate between liquids and crystals. Because of their intermediate character some call them *mesomorphic* phases. Liquid crystals consist of highly anisotropic weakly coupled (often rod-like) molecules. They are liquid-like but also have some anisotropy. The anisotropic properties of some liquid crystals can be changed by an electric field, which affects their optical properties, and thus watch displays and screens for computer monitors have been developed. J. L. Fergason [12.19] has been one of the pioneers in this as well as other applications.

There are two main classes of liquid crystals: nematic and smectic. In nematic liquid crystals the molecules are partly aligned but their position is essentially random. In smectic liquid crystals, the molecules are in planes that can slide over each other. Nematic and smectic liquid crystals are sketched in Fig. 12.19. An associated form of the nematic phase is the cholesteric. Cholesterics have a director (which is a unit vector along the average axis of orientation of the rod-like molecules) that has a helical twist.

Liquid crystals still tend to be somewhat foreign to many physicists because they involve organic molecules, polymers, and associated structures. For more details see deGennes PG and Prost [12.15] and Isihara [12.27 Chap. 12].

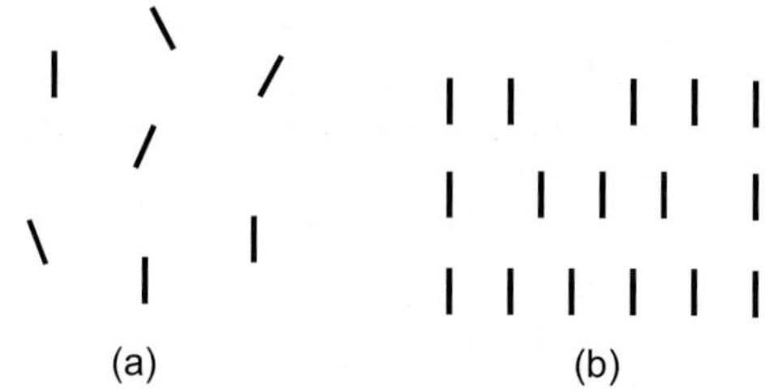

Fig. 12.19. Liquid crystals. (**a**) Nematic (long-range orientational order but no long-range positional order), and (**b**) Smectic (long-range orientational order and in one dimension long-range positional order)

12.11.3 Polymers and Rubbers (MET, MS)

Polymers are a classic example of soft condensed matter. In this section, we will discuss polymers[4] and treat rubber as a particular example.

A monomer is a simple molecule that can join with itself or similar molecules (many times) to form a giant molecule that is referred to as a polymer. (From the Greek, polys – many and meros – parts). A polymer may be either naturally occurring or synthetic. The number of repeating units in the polymer is called the degree of polymerization (which is typically of order 10^3 to 10^5). Most organic substances associated with living matter are polymers, thus examples of polymers are myriad. Plastics, rubbers, fibers, and adhesives are common examples. Bakelite was the first thermosetting plastic found. Rayon, Nylon, and Dacron (polyester) are examples of synthetic fibers. There are crystalline polymer fibers such as cellulose (wood is made of cellulose) that diffract X-rays and by contrast there are amorphous polymers (rubber can be thought of as made of amorphous polymers) that don't show diffraction peaks.

There are many subfields of polymers of which rubber is one of the most important. A rubber consists of many long chains of polymers connected together somewhat randomly. The chains themselves are linear and flexible. The random linking bonds give shape. Rubbers are like liquids in that they have a well-defined volume, but not a well-defined shape. They are like a solid in that they maintain their shape in the absence of forces. The most notable property of rubbers is that

[4] As an aside we mention the connection of polymers with fuel cells, which have been much in the news. In 1839 William R. Grove showed the electrochemical union of hydrogen and oxygen generates electricity—the idea of the fuel cell. Hydrogen can be extracted from say methanol, and stored in, for example, metal hydrides. Fuel cells can run as long as hydrogen and oxygen are available. The only waste is water from the fuel-cell reaction. In 1960 synthetic polymers were introduced as electrolytes.

they have a very long and reversible elasticity. Vulcanizing soft rubber, by adding sulfur and heat treatment makes it harder and increases its strength. The sulfur is involved in linking the chains.

A rubber can be made by repetition of the isoprene group (C_5H_8, see Fig. 12.20).[5] Because the entropy of a polymer is higher for configurations in which the monomers are randomly oriented than for which they are all aligned, one can estimate the length of a long linear polymer in solution by a random-walk analysis. The result for the overall length is the length of the monomer times the square root of their number (see below). The radius of a polymer in a ball is given by a similar law. More complicated analysis treats the problem as a self-avoiding random walk and leads to improved results (such as the radius of the ball being approximately the length of the monomer times their number to the 3/5 power). Another important feature of polymers is their viscosity and diffusion. The concept of reptation (which we will not discuss here, see Doi and Edwards [12.16]), which means snaking, has proved to be very important. It helps explain how one polymer can diffuse through the mass of the others in a melt. One thinks of the Brownian motion of a molecule along its length as aiding in disentangling the polymer.

$$\overset{\displaystyle CH_3}{\underset{\displaystyle +CH_2 - C = CH - CH_2 +}{|}}$$

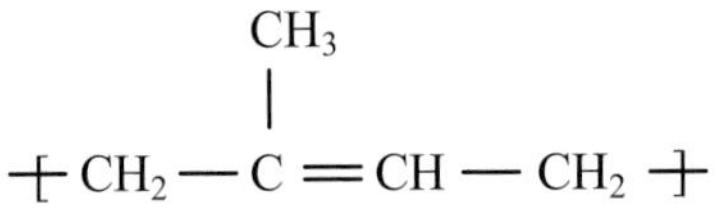

Fig. 12.20. Chemical structure of isoprene (the basic unit for natural rubber)

We first give a one-dimensional model to illustrate how the length of a polymer can be estimated from a random-walk analysis. We will then discuss a model for estimating the elastic constant of a rubber.

We suppose N monomers of length a linked together along the x-axis. We suppose the ith monomer to be in the $+x$ direction with probability of $^1/_2$ and in the $-x$ direction with the same probability. The rms length R of the polymer is calculated below.

Let $x_i = a$ for the monomer in the $+x$ direction and $-a$ for the $-x$ direction. Then the total length is $x = \Sigma x_i$ and the average squared length is

$$\left\langle x^2 \right\rangle = \left\langle \Sigma x_i \right\rangle^2 = \Sigma \left\langle x_i^2 \right\rangle, \qquad (12.58)$$

since the cross terms drop out, so

$$\left\langle x^2 \right\rangle = Na^2, \qquad (12.59)$$

or

$$R = a\sqrt{N}. \qquad (12.60)$$

---

[5] See, e.g., Brown et al [12.4]. See also Strobl [12.57].

We have already noted that a similar scaling law applies to the radius of a N-monomer polymer coiled in a ball in three dimensions.

In a similar way, we can estimate the tension in the polymer. This model or generalizations of it to two or three dimensions (See, e.g., Callen [12.7]) seem to give the basic idea. Let n^+ and n^- represent the links in the + and − directions. The length x is

$$x = (n^+ - n^-)a\,, \tag{12.61}$$

and the total number of monomers is

$$N = (n^+ + n^-)\,. \tag{12.62}$$

Thus

$$n^+ = \frac{1}{2}\left(N + \frac{x}{a}\right),\quad n^- = \frac{1}{2}\left(N - \frac{x}{a}\right). \tag{12.63}$$

The number of ways we can arrange N monomers with n^+ in the $+x$ direction and n^- in the − direction is

$$W = \frac{N!}{n^+!\,n^-!}. \tag{12.64}$$

Using $S = k\ln(W)$ and using Stirling's approximation, we can find the entropy S. Then since $\mathrm{d}U = T\mathrm{d}S + F\mathrm{d}x$, where T is the temperature, U the internal energy and F the tension, we find

$$F = -T\frac{\partial S}{\partial x} + \frac{\partial U}{\partial x}\,, \tag{12.65}$$

so we find (assuming we use a model in which $\partial U/\partial x$ can be neglected)

$$F = \frac{kT}{2a}\ln\left(\frac{1 + x/Na}{1 - x/Na}\right) = \frac{kTx}{Na^2} \quad (\text{if } x \ll Na). \tag{12.66}$$

The tension F comes out to be proportional to both the temperature and the extension x (it becomes stiffer as the temperature is raised!). Another way to look at this is that the polymer contracts on warming. In 3D, we think of the polymer curling up at high temperatures and the entropy increasing.

Problems

12.1 If the periodicity $p = 50$ Å and $E = 5\times10^4$ V/cm, calculate the fundamental frequency for Bloch oscillations. Compare the results to relaxation times τ typical for electrons, i.e. compute $\omega_B\tau$.

12.2 Find the minimum radius of a spherical quantum dot whose electron binding energy is at least 1 eV.

12.3 Discuss how the Kronig–Penny model can be used to help understand the motion of electrons in superlattices. Discuss both transverse and in-plane motion. See, e.g., Mitin et al [12.47 pp. 99-106].

12.4 Consider a quantum well parallel to the (x,y)-plane of width w in the z direction. For simplicity assume the depth of the quantum well is infinite. Assume also for simplicity that the effective mass is a constant m for motion in all directions, See, e.g., Shik [12.54, Chaps. 2 and 4] .

a) Show the energy of an electron can be written

$$E = \frac{\hbar^2\pi^2 n^2}{2mw^2} + \frac{\hbar^2}{2m}(k_x^2 + k_y^2),$$

where $p_x = \hbar k_x$ and $p_y = \hbar k_y$ and n is an integer.

b) Show the density of states can be written

$$D(E) = \frac{m}{\pi\hbar^2}\Sigma_n \theta(E - E_n),$$

where $D(E)$ represents the number of states per unit area per unit energy in the (x,y)-plane and

$$E_n = \frac{\hbar^2\pi^2}{2mw^2} n^2 .$$

$\theta(x)$ is the step function $\theta(x) = 0$ for $x < 0$ and $= 1$ for $x > 0$.

c) Show also $D(E)$ at $E \gtrsim E_3$ is the same as $D_{3D}(E)$ where D_{3D} represents the density of states in 3D without the quantum well (still per unit area in the (x,y)-plane for a width w in the z direction)

d) Make a sketch showing the results of b) and c) in graphic form.

12.5 For the situation of Problem 12.4 impose a magnetic field B in the z direction. Show then that the allowed energies are discrete with values

$$E_{n,p} = \frac{\hbar^2\pi^2 n^2}{2mw^2} + \hbar\omega_c\left(p + \frac{1}{2}\right),$$

where n, p are integers and $\omega_c = |eB/m|$ is the cyclotron frequency. Show also the two-dimensional density of states per spin (and per unit energy and area in (x,y)-plane) is

$$D(E) = \frac{eB}{h} \sum_p \delta\left[E - \hbar\omega\left(p + \frac{1}{2}\right)\right],$$

when

$$\frac{\hbar^2\pi^2}{2mw^2} < E < \frac{4\hbar^2\pi^2}{2mw^2}.$$

These results are applicable to a 2D Fermi gas, see, e.g., Shik [12.54, Chaps. 7] as well as 12.7.2 and 12.7.3.

Appendices

A Units

The choice of a system of units to use is sometimes regarded as an emotionally charged subject. Although there are many exceptions, experimental papers often use mksa (or SI) units, and theoretical papers may use Gaussian units (or perhaps a system in which several fundamental constants are set equal to one).

All theories of physics must be checked by comparison to experiment before they can be accepted. For this reason, it is convenient to express final equations in the mksa system. Of course, much of the older literature is still in Gaussian units, so one must have some familiarity with it. The main thing to do is to settle on a system of units and stick to it. Anyone who has reached the graduate level in physics can convert units whenever needed. It just may take a little longer than we wish to spend.

In this appendix, no description of the mksa system will be made. An adequate description can be found in practically any sophomore physics book.[1]

In solid-state physics, another unit system is often convenient. These units are called Hartree atomic units. Let e be the charge on the electron, and m be the mass of the electron. The easiest way to get the Hartree system of units is to start from the Gaussian (cgs) formulas, and let $|e|$ = Bohr radius of hydrogen = $|m|$ = 1. The results are summarized in Table A.1. The Hartree unit of energy is 27.2 eV. Expressing your answer in terms of the fundamental physical quantities shown in Table A.1 and then using Hartree atomic units leads to simple numerical answers for solid-state quantities. In such units, the solid-state quantities usually do not differ by too many orders of magnitude from one.

---

[1] Or see "Guide for Metric Practice," by Robert A. Nelson at http://www.physicstoday.org/guide/metric.html.

Table A.1. Fundamental physical quantities*

Quantity	Symbol	Expression / value in mksa units	Expression / value in Gaussian units	Value in Hartree units
Charge on electron	e	1.6×10^{-19} coulomb	4.80×10^{-10} esu	1
Mass of electron	m	0.91×10^{-30} kg	0.91×10^{-27} g	1
Planck's constant	$\hbar$	1.054×10^{-34} joule s	1.054×10^{-27} erg s	1
Compton wavelength of electron	λ_c	$2\pi(\hbar/mc)$ 2.43×10^{-12} m	$2\pi(\hbar/mc)$ 2.43×10^{-10} cm	$(2\pi)\frac{1}{137}$
Bohr radius of hydrogen	a_0	$4\pi\varepsilon_0\hbar^2/me^2$ 0.53×10^{-10} m	$\hbar^2/me^2$ 0.53×10^{-8} cm	1
Fine structure constant	α	$e^2/\hbar c$ $\frac{1}{137}$ (approx.)	$e^2/\hbar c$ $\frac{1}{137}$	$\frac{1}{137}$
Speed of light	c	3×10^8 m s^{-1}	3×10^{10} cm s^{-1}	137
Classical electron radius	r_0	$e^2/4\pi\varepsilon_0 mc^2$ 2.82×10^{-15} m	e^2/mc^2 2.82×10^{-13} cm	$(\frac{1}{137})^2$
Energy of ground state of hydrogen (1 Rydberg)	E_0	$e^4m/32(\pi\varepsilon_0\hbar)^2$ 13.61† eV	$me^4/2\hbar^2$ 13.61† eV	$\frac{1}{2}$
Bohr magneton (calculated from above)	μ_B	$e\hbar/2m$ 0.927×10^{-23} amp meter^2	$e\hbar/2mc$ 0.927×10^{-20} erg gauss^{-1}	$\frac{1}{274}$
Cyclotron frequency (calculated from above)	ω_c, or ω_h	$(\mu_0 e/2m)(2H)$	$(e/2mc)(2H)$	$\frac{1}{274}$ $(2H)$

* The values given are greatly rounded off from the standard values. The list of fundamental constants has been updated and published yearly in part B of the August issue of *Physics Today*. See, e.g., Peter J. Mohr and Barry N. Taylor, "The Fundamental Physical Constants," *Physics Today*, pp. BG6-BG13, August, 2003. Now see http://www.physicstoday.org/guide/fundcon.html.

† 1 eV = 1.6×10^{-12} erg = 1.6×10^{-19} joule.

B Normal Coordinates

The main purpose of this appendix is to review clearly how the normal coordinate transformation arises, and how it leads to a diagonalization of the Hamiltonian. Our development will be made for classical systems, but a similar development can be made for quantum systems. An interesting discussion of normal modes has been given by Starzak.[2] The use of normal coordinates is important for collective excitations such as encountered in the discussion of lattice vibrations.

We will assume that our mechanical system is described by the Hamiltonian

$$\mathcal{H} = \frac{1}{2}\sum\nolimits_{i,j}(\dot{x}_i\dot{x}_j\delta_{ij} + \upsilon_{ij}x_ix_j)\,. \tag{B.1}$$

In (B.1) the first term is the kinetic energy and the second term is the potential energy of interaction among the particles. We consider only the case that each particle has the same mass that has been set equal to one. In (B.1) it is also assumed that $\upsilon_{ij} = \upsilon_{ji}$; and that each of the υ_{ij} is real. The coordinates x_i in (B.2) are measured from equilibrium that is assumed to be stable. For a system of N particles in three dimensions, one would need $3N$ x_i to describe the vibration of the system. The dot of $\dot{x}_i$ of course means differentiation with respect to time, $\dot{x}_i = \mathrm{d}x_i/\mathrm{d}t$.

The Hamiltonian (B.1) implies the following equation of motion for the mechanical system:

$$\sum\nolimits_j(\delta_{ij}\ddot{x}_j + \upsilon_{ij}x_j) = 0\,. \tag{B.2}$$

The normal coordinate transformation is the transformation that takes us from the coordinates x_i to the normal coordinates. A normal coordinate describes the motion of the system in a normal mode. In a normal mode each of the coordinates vibrates with the same frequency. Seeking a normal mode solution is equivalent to seeking solutions of the form

$$x_j = ca_j\mathrm{e}^{-\mathrm{i}\omega t}\,. \tag{B.3}$$

In (B.3), c is a constant that is usually selected so that $\sum_j|x_j|^2 = 1$, and $|ca_j|$ is the amplitude of vibration of x_j in the mode with frequency ω. The different frequencies ω for the different normal modes are yet to be determined.

Equation (B.2) has solutions of the form (B.3) provided that

$$\sum\nolimits_j(\upsilon_{ij}a_j - \omega^2\delta_{ij}a_j) = 0\,. \tag{B.4}$$

Equation (B.4) has nontrivial solutions for the a_j (i.e. solutions in which all the a_j do not vanish) provided that the determinant of the coefficient matrix of the a_j vanishes. This condition determines the different frequencies corresponding to the different normal modes of the mechanical system. If V is the matrix whose

---

2 See Starzak [A.25 Chap. 5].

elements are given by υ_{ij} (in the usual notation), then the eigenvalues of V are ω^2, determined by (B.4). V is a real symmetric matrix; hence it is Hermitian; hence its eigenvalues must be real.

Let us suppose that the eigenvalues ω^2 determined by (B.4) are denoted by Ω_k. There will be the same number of eigenvalues as there are coordinates x_i. Let a_{jk} be the value of a_j, which has a normalization determined by (B.7), when the system is in the mode corresponding to the kth eigenvalue Ω_k. In this situation we can write

$$\sum_j \upsilon_{ij} a_{jk} = \Omega_k \sum_j \delta_{ij} a_{jk} \,. \tag{B.5}$$

Let A stand for the matrix with elements a_{jk} and Ω be the matrix with elements $\Omega_{lk} = \Omega_k \delta_{lk}$. Since $\Omega_k \sum_j \delta_{ij} a_{jk} = \Omega_k a_{ik} = a_{ik}\Omega_k = \sum_l a_{il}\Omega_k \delta_{lk} = \sum_l a_{il}\Omega_{lk}$, we can write (B.5) in matrix notation as

$$VA = A\Omega \,. \tag{B.6}$$

It can be shown [2] that the matrix A that is constructed from the eigenvectors is an orthogonal matrix, so that

$$A\tilde{A} = \tilde{A}A = I \,. \tag{B.7}$$

$\tilde{A}$ means the transpose of A. Combining (B.6) and (B.7) we have

$$\tilde{A}VA = \Omega \,. \tag{B.8}$$

This equation shows how V is diagonalized by the use of the matrix that is constructed from the eigenvectors.

We still must indicate how the new eigenvectors are related to the old coordinates. If a column matrix a is constructed from the a_j as defined by (B.3), then the eigenvectors E (also a column vector, each element of which is an eigenvector) are defined by

$$E = \tilde{A}a \,, \tag{B.9a}$$

or

$$a = AE \,. \tag{B.9b}$$

That (B.9) does define the eigenvectors is easy to see because substituting (B.9b) into the Hamiltonian reduces the Hamiltonian to diagonal form. The kinetic energy is already diagonal, so we need consider only the potential energy

$$\begin{aligned}
\sum \upsilon_{ij} a_i a_j &= \tilde{a}Va = \tilde{E}\tilde{A}VAE = \tilde{E}\Omega E \\
&= \sum_{k,j} (\tilde{E})_j \Omega_{jk} E_k = \sum_{j,k} (\tilde{E})_j \Omega_k \delta_{jk} E_k \\
&= \sum_j (\tilde{E})_j \Omega_j E_j = \sum_{j,k} \omega_j^2 (\tilde{E}_i) E_i \delta_{jk} \,,
\end{aligned}$$

which tells us that the substitution reduces V to diagonal form. For our purposes, the essential thing is to notice that a substitution of the form (B.9) reduces the Hamiltonian to a much simpler form.

An example should clarify these ideas. Suppose the eigenvalue condition yielded

$$\det\begin{pmatrix} 1-\omega^2 & 2 \\ 2 & 3-\omega^2 \end{pmatrix} = 0 . \tag{B.10}$$

This implies the two eigenvalues

$$\omega_1^2 = 2+\sqrt{5} \tag{B.11a}$$

$$\omega_2^2 = 2-\sqrt{5} . \tag{B.11b}$$

Equation (B.4) for each of the eigenvalues gives for

$$\omega = \omega_1^2 : a_1 = \frac{2a_2}{1+\sqrt{5}}, \tag{B.12a}$$

and for

$$\omega = \omega_2^2 : a_1 = \frac{2a_2}{1-\sqrt{5}} . \tag{B.12b}$$

From (B.12) we then obtain the matrix A

$$\widetilde{A} = \begin{pmatrix} \dfrac{2N_1}{1+\sqrt{5}}, & N_1 \\ \dfrac{2N_2}{1-\sqrt{5}}, & N_2 \end{pmatrix}, \tag{B.13}$$

where

$$(N_1)^{-1} = \left[\frac{4}{(\sqrt{5}+1)^2}+1\right]^{1/2}, \tag{B.14a}$$

and

$$(N_2)^{-1} = \left[\frac{4}{(\sqrt{5}-1)^2}+1\right]^{1/2} . \tag{B.14b}$$

The normal coordinates of this system are given by

$$E = \begin{pmatrix} E_1 \\ E_2 \end{pmatrix} = \begin{pmatrix} \dfrac{2N_1}{1+\sqrt{5}}, & N_1 \\ \dfrac{2N_2}{1-\sqrt{5}}, & N_2 \end{pmatrix} \begin{pmatrix} a_1 \\ a_2 \end{pmatrix} . \tag{B.15}$$

Problems

B.1 Show that (B.13) satisfies (B.7)

B.2 Show for A defined by (B.13) that

$$\tilde{A}\begin{pmatrix} 1 & 2 \\ 2 & 3 \end{pmatrix} A = \begin{pmatrix} 2+\sqrt{5}, & 0 \\ 0, & 2-\sqrt{5} \end{pmatrix}.$$

This result checks (B.8).

C Derivations of Bloch's Theorem

Bloch's theorem concerns itself with the classifications of eigenfunctions and eigenvalues of Schrödinger-like equations with a periodic potential. It applies equally well to electrons or lattice vibrations. In fact, Bloch's theorem holds for any wave going through a periodic structure. We start with a simple one-dimensional derivation.

C.1 Simple One-Dimensional Derivation[3–5]

This derivation is particularly applicable to the Kronig–Penney model. We will write the Schrödinger wave equation as

$$\frac{d^2\psi(x)}{dx^2} + U(x)\psi(x) = 0\,, \tag{C.1}$$

where $U(x)$ is periodic with period a, i.e.,

$$U(x+na) = U(x)\,, \tag{C.2}$$

with n an integer. Equation (C.1) is a second-order differential equation, so that there are two linearly independent solutions ψ_1 and ψ_2:

$$\psi_1'' + U\psi_1 = 0\,, \tag{C.3}$$

$$\psi_2'' + U\psi_2 = 0\,. \tag{C.4}$$

---

[3] See Ashcroft and Mermin [A.3].
[4] See Jones [A.10].
[5] See Dekker [A.4].

From (C.3) and (C.4) we can write

$$\psi_2\psi_1'' + U\psi_2\psi_1 = 0\,,$$

$$\psi_1\psi_2'' + U\psi_1\psi_2 = 0\,.$$

Subtracting these last two equations, we obtain

$$\psi_2\psi_1'' - \psi_1\psi_2'' = 0\,. \tag{C.5}$$

This last equation is equivalent to writing

$$\frac{\mathrm{d}W}{\mathrm{d}x} = 0\,, \tag{C.6}$$

where

$$W = \begin{vmatrix} \psi_1 & \psi_2 \\ \psi_1' & \psi_2' \end{vmatrix} \tag{C.7}$$

is called the *Wronskian*. For linearly independent solutions, the Wronskian is a constant not equal to zero.

It is easy to prove one result from the periodicity of the potential. By dummy variable change $(x) \to (x + a)$ in (C.1) we can write

$$\frac{\mathrm{d}^2\psi(x+a)}{\mathrm{d}x^2} + U(x+a)\psi(x+a) = 0\,.$$

The periodicity of the potential implies

$$\frac{\mathrm{d}^2\psi(x+a)}{\mathrm{d}x^2} + U(x)\psi(x+a) = 0\,. \tag{C.8}$$

Equations (C.1) and (C.8) imply that if $\psi(x)$ is a solution, then so is $\psi(x + a)$. Since there are only two linearly independent solutions ψ_1 and ψ_2, we can write

$$\psi_1(x+a) = A\psi_1(x) + B\psi_2(x) \tag{C.9}$$

$$\psi_2(x+a) = C\psi_1(x) + D\psi_2(x)\,. \tag{C.10}$$

The Wronskian W is a constant $\neq 0$, so $W(x + a) = W(x)$, and we can write

$$\begin{vmatrix} A\psi_1 + B\psi_2 & C\psi_1 + D\psi_2 \\ A\psi_1' + B\psi_2' & C\psi_1' + D\psi_2' \end{vmatrix} = \begin{vmatrix} \psi_1 & \psi_2 \\ \psi_1' & \psi_2' \end{vmatrix}\begin{vmatrix} A & C \\ B & D \end{vmatrix} = \begin{vmatrix} \psi_1 & \psi_2 \\ \psi_1' & \psi_2' \end{vmatrix},$$

or

$$\begin{vmatrix} A & C \\ B & D \end{vmatrix} = 1\,,$$

or

$$AD - BC = 1 . \tag{C.11}$$

We can now prove that it is possible to choose solutions $\psi(x)$ so that

$$\psi(x+a) = \Delta\,\psi(x) , \tag{C.12}$$

where Δ is a constant $\neq 0$. We want $\psi(x)$ to be a solution so that

$$\psi(x) = \alpha\psi_1(x) + \beta\psi_2(x) , \tag{C.13a}$$

or

$$\psi(x+a) = \alpha\psi_1(x+a) + \beta\psi_2(x+a) . \tag{C.13b}$$

Using (C.9), (C.10), (C.12), and (C.13), we can write

$$\begin{aligned} \psi(x+a) &= (\alpha A + \beta C)\psi_1(x) + (\alpha B + \beta D)\psi_2(x) \\ &= \Delta\alpha\psi_1(x) + \Delta\beta\psi_2(x). \end{aligned} \tag{C.14}$$

In other words, we have a solution of the form (C.12), provided that

$$\alpha A + \beta C = \Delta\alpha ,$$

and

$$\alpha B + \beta D = \Delta\beta .$$

For nontrivial solutions for α and β, we must have

$$\begin{vmatrix} A - \Delta & C \\ B & D - \Delta \end{vmatrix} = 0 . \tag{C.15}$$

Equation (C.15) is equivalent to, using (C.11),

$$\Delta + \Delta^{-1} = A + D . \tag{C.16}$$

If we let Δ_+ and Δ_- be the eigenvalues of the matrix $\left(\begin{smallmatrix} A & C \\ B & D \end{smallmatrix}\right)$ and use the fact that the trace of a matrix is the sum of the eigenvalues, then we readily find from (C.16) and the trace condition

$$\begin{aligned} \Delta_+ + (\Delta_+)^{-1} &= A + D , \\ \Delta_- + (\Delta_-)^{-1} &= A + D , \end{aligned} \tag{C.17}$$

and

$$\Delta_+ + \Delta_- = A + D .$$

Equations (C.17) imply that we can write

$$\Delta_+ = (\Delta_-)^{-1}. \tag{C.18}$$

If we set

$$\Delta_+ = \mathrm{e}^b, \tag{C.19}$$

and

$$\Delta_- = \mathrm{e}^{-b}, \tag{C.20}$$

the above implies that we can find linearly independent solutions ψ_i^1 that satisfy

$$\psi_1^1(x+a) = e^b \psi_1^1(x), \tag{C.21}$$

and

$$\psi_2^1(x+a) = e^{-b} \psi_2^1(x). \tag{C.22}$$

Real b is ruled out for finite wave functions (as $x \to \pm\infty$), so we can write $b = \mathrm{i}ka$, where k is real. Dropping the superscripts, we can write

$$\psi(x+a) = \mathrm{e}^{\pm \mathrm{i}ka} \psi(x). \tag{C.23}$$

Finally, we note that if

$$\psi(x) = \mathrm{e}^{\mathrm{i}kx} u(x), \tag{C.24}$$

where

$$u(x+a) = u(x), \tag{C.25}$$

then (C.23) is satisfied. (C.23) or (C.24), and (C.25) are different forms of Bloch's theorem.

C.2 Simple Derivation in Three Dimensions

Let

$$\mathcal{H}\psi(x_1 \cdots x_N) = E\psi(x_1 \cdots x_N) \tag{C.26}$$

be the usual Schrödinger wave equation. Let $T_{\boldsymbol{l}}$ be a translation operator that translates the lattice by $l_1\boldsymbol{a}_1 + l_2\boldsymbol{a}_2 + l_3\boldsymbol{a}_3$, where the l_i are integers and the a_i are the primitive translation vectors of the lattice.

Since the Hamiltonian is invariant with respect to translations by $T_{\boldsymbol{l}}$, we have

$$[\mathcal{H}, T_{\boldsymbol{l}}] = 0, \tag{C.27}$$

and

$$[T_l, T_{l}'] = 0 \, . \tag{C.28}$$

Now we know that we can always find simultaneous eigenfunctions of commuting observables. Observables are represented by Hermitian operators. The T_l are unitary. Fortunately, the same theorem applies to them (we shall not prove this here). Thus we can write

$$\mathcal{H}\psi_{E,l} = E\psi_{E,l} \, , \tag{C.29}$$

$$T_l\psi_{E,l} = t_l\psi_{E,l} \, . \tag{C.30}$$

Now certainly we can find a vector k such that

$$t_l = \mathrm{e}^{\mathrm{i}\boldsymbol{k}\cdot\boldsymbol{l}} \, . \tag{C.31}$$

Further

$$\int_{\text{all space}} |\psi(\boldsymbol{r})|^2 \mathrm{d}\tau = \int |\psi(\boldsymbol{r}+\boldsymbol{l})|^2 \mathrm{d}\tau = |t_l|^2 \int |\psi(\boldsymbol{r})|^2 \mathrm{d}\tau \, ,$$

so that

$$|t_l|^2 = 1 \, . \tag{C.32}$$

This implies that $\boldsymbol{k}$ must be a vector over the real field.

We thus arrive at Bloch's theorem

$$T_l\psi(\boldsymbol{r}) = \psi(\boldsymbol{r}+\boldsymbol{l}) = \mathrm{e}^{\mathrm{i}\boldsymbol{k}\cdot\boldsymbol{l}}\psi(\boldsymbol{r}) \, . \tag{C.33}$$

The theorem says we can always choose the eigenfunctions to satisfy (C.33). It does not say the eigenfunction must be of this form. If periodic boundary conditions are applied, the usual restrictions on the $\boldsymbol{k}$ are obtained.

C.3 Derivation of Bloch's Theorem by Group Theory

The derivation here is relatively easy once the appropriate group theoretic knowledge is acquired. We have already discussed in Chaps. 1 and 7 the needed results from group theory. We simply collect together here the needed facts to establish Bloch's theorem.

1. It is clear that the group of the T_l is abelian (i.e. all the T_l commute).
2. In an abelian group each element forms a class by itself. Therefore the number of classes is O(G), the order of the group.
3. The number of irreducible representations (of dimension n_i) is the number of classes.

4. $\sum n_i^2 = O(G)$ and thus by above

$$n_1^2 + n_2^2 + \cdots + n_{0(G)}^2 = 0(G).$$

This can be satisfied only if each $n_i = 1$. Thus the dimensions of the irreducible representations of the T_l are all one.

5. In general

$$T_l \psi_i^k = \sum_j A_{ij}^{l,k} \psi_j^k ,$$

where the $A_{i\,j}^{l,k}$ are the matrix elements of the T_l for the kth representation and the sum over j goes over the dimensionality of the kth representation. The ψ_i^k are the basis functions selected as eigenfunctions of $\mathcal{H}$ (which is possible since $[\mathcal{H}, T^l] = 0$). In our case the sum over j is not necessary and so

$$T^l \psi^k = A^{l,k} \psi^k .$$

As before, the $A^{l,k}$ can be chosen to be $e^{il \cdot k}$. Also in one dimension we could use the fact that $\{T^l\}$ is a cyclic group so that the $A^{l,k}$ are automatically the roots of one.

D Density Matrices and Thermodynamics

A few results will be collected here. The proofs of these results can be found in any of several books on statistical mechanics.

If $\psi^i(x, t)$ is the wave function of system (in an ensemble of N systems where $1 \le i \le N$) and if $|n\rangle$ is a complete orthonormal set, then

$$\left|\psi^i(x,t)\right\rangle = \sum_n c_n^i(t) |n\rangle .$$

The density matrix is defined by

$$\rho_{nm} = \frac{1}{N} \sum_{i=1}^N c_m^{i*}(t) c_m^i(t) \equiv \overline{c_m^* c_n} .$$

It has the following properties:

$$Tr(\rho) \equiv \sum_n \rho_{nn} = 1 ,$$

the ensemble average (denoted by a bar) of the quantum-mechanical expectation value of an operator A is

$$\left\langle \overline{A} \right\rangle \equiv Tr(\rho A) ,$$

and the equation of motion of the density operator ρ is given by

$$-i\hbar\frac{\partial\rho}{\partial t}=[\rho,H],$$

where the density operator is defined in such a way that $\langle n|\rho|m\rangle \equiv \rho_{nm}$. For a canonical ensemble in equilibrium

$$\rho=\exp\left(\frac{F-H}{kT}\right).$$

Thus we can readily link the idea of a density matrix to thermodynamics and hence to measurable quantities. For example, the internal energy for a system in equilibrium is given by

$$U=\langle\overline{H}\rangle=Tr\left[H\exp\left(\frac{F-H}{kT}\right)\right]=\frac{Tr[H\exp(-H/kT)]}{Tr[\exp(-H/kT)]}.$$

Alternatively, the internal energy can be calculated from the free energy F where for a system in equilibrium,

$$F=-kT\ln Tr[\exp(-H/kT)].$$

It is fairly common to leave the bar off $\langle\overline{A}\rangle$ so long as the meaning is clear. For further properties and references see Patterson [A.19], see also Huang [A.8].

E Time-Dependent Perturbation Theory

A common problem in solid-state physics (as in other areas of physics) is to find the transition rate between energy levels of a system induced by a small time-dependent perturbation. More precisely, we want to be able to calculate the time development of a system described by a Hamiltonian that has a small time-dependent part. This is a standard problem in quantum mechanics and is solved by the time-dependent perturbation theory. However, since there are many different aspects of time-dependent perturbation theory, it seems appropriate to give a brief review without derivations. For further details any good quantum mechanics book such as Merzbacher[6] can be consulted.

[6] See Merzbacher [A.15 Chap. 18].

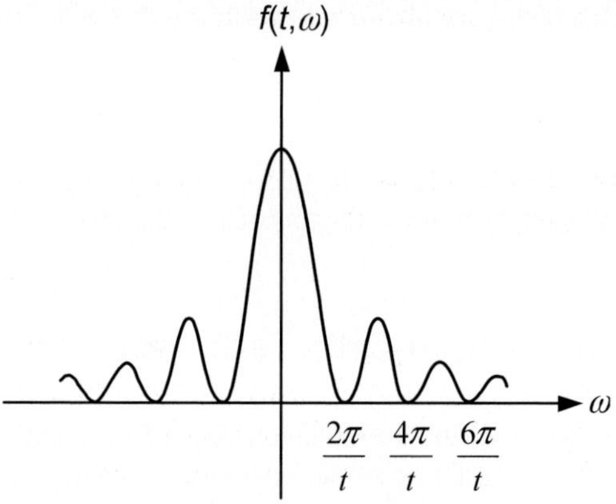

Fig. E.1. $f(t, \omega)$ versus ω. The area under the curve is $2\pi t$

Let

$$\mathcal{H}(t) = \mathcal{H}^0 + V(t), \tag{E.1}$$

$$\mathcal{H}^0 |l\rangle = E_l^0 |l\rangle, \tag{E.2}$$

$$V_{kl}(t) = \langle k | V(t) | l \rangle, \tag{E.3}$$

$$\omega_{kl} = \frac{E_k^0 - E_l^0}{\hbar}. \tag{E.4}$$

In first order in V, for V turned on at $t = 0$ and constant otherwise, the probability per unit time of a discrete $i \to f$ transition for $t > 0$ is

$$P_{i \to f} \cong \frac{2\pi}{\hbar} |V_{fi}|^2 \delta(E_i^0 - E_f^0). \tag{E.5}$$

In deriving (E.5) we have assumed that the $f(t, \omega)$ in Fig. E.1 can be replaced by a Dirac delta function via the equation

$$\lim_{t \to \infty} \frac{1 - \cos(\omega_{if} t)}{(\hbar \omega_{if})^2} = \frac{\pi t}{\hbar} \delta(E_i^0 - E_f^0) = \frac{f(t, \omega)}{2\hbar^2}. \tag{E.6}$$

If we have transitions to a group of states with final density of states $p_f(E_f)$, a similar calculation gives

$$P_{i \to f} = \frac{2\pi}{\hbar} |V_{fi}|^2 p_f(E_f). \tag{E.7}$$

In the same approximation, if we deal with periodic perturbations represented by

$$V(t) = g e^{i\omega t} + g^{\dagger} e^{-i\omega t}, \tag{E.8}$$

which are turned on at $t = 0$, we obtain for transitions between discrete states

$$P_{i\to f} = \frac{2\pi}{\hbar}\left|g_{fi}\right|^2 \delta(E_i^0 - E_f^0 \pm \hbar\omega)\,. \tag{E.9}$$

In the text, we have loosely referred to (E.5), (E.7), or (E.9) as the Golden rule (according to which is appropriate to the physical situation).

F Derivation of The Spin-Orbit Term From Dirac's Equation

In this appendix we will indicate how the concepts of spin and spin-orbit interaction are introduced by use of Dirac's relativistic theory of the electron. For further details, any good quantum mechanics text such as that of Merzbacher[7], or Schiff[8] can be consulted. We will discuss Dirac's equation only for fields described by a potential V. For this situation, Dirac's equation can be written

$$[c(\boldsymbol{\alpha}\cdot\boldsymbol{p}) + m_0c^2\beta + V]\psi = E\psi\,. \tag{F.1}$$

In (F.1), c is the speed of light, $\boldsymbol{\alpha}$ and β are 4×4 matrices defined below, $\boldsymbol{p}$ is the momentum operator, m_0 is the rest mass of the electron, ψ is a four-component column matrix (each element of this matrix may be a function of the spatial position of the electron), and E is the total energy of the electron (including the rest mass energy that is m_0c^2). The $\boldsymbol{\alpha}$ matrices are defined by

$$\boldsymbol{\alpha} = \begin{pmatrix} 0 & \boldsymbol{\sigma} \\ \boldsymbol{\sigma} & 0 \end{pmatrix}, \tag{F.2}$$

where the three components of $\boldsymbol{\sigma}$ are the 2×2 Pauli spin matrices. The definition of β is

$$\beta = \begin{pmatrix} I & 0 \\ 0 & -I \end{pmatrix}, \tag{F.3}$$

where I is a 2×2 unit matrix.

For solid-state purposes we are not concerned with the fully relativistic equation (F.1), but rather we are concerned with the relativistic corrections that (F.1) predicts should be made to the nonrelativistic Schrödinger equation. That is, we want to consider the Dirac equation for the electron in the small velocity limit. More precisely, we will consider the limit of (F.1) when

$$\varepsilon \equiv \frac{(E - m_0c^2) - V}{2m_0c^2} << 1\,, \tag{F.4}$$

---

[7] See Merzbacher [A.15 Chap. 23].
[8] See Schiff [A.23].

and we want results that are valid to first order in ε, i.e. first-order corrections to the completely nonrelativistic limit. To do this, it is convenient to make the following definitions:

$$E = E' + m_0 c^2 , \tag{F.5}$$

and

$$\psi = \begin{pmatrix} \chi \\ \phi \end{pmatrix} , \tag{F.6}$$

where both χ and ϕ are two-component wave functions.

If we substitute (F.5) and (F.6) into (F.1), we obtain an equation for both χ and ϕ. We can combine these two equations into a single equation for χ in which ϕ does not appear. We can then use the small velocity limit (F.4) together with several properties of the Pauli spin matrices to obtain the Schrödinger equation with relativistic corrections

$$E'\chi = \left[\frac{p^2}{2m_0} - \frac{p^4}{8m_0^3c^2} + V - \frac{\hbar^2}{4m_0^2c^2}\nabla V \cdot \nabla + \frac{\hbar^2}{4m_0^2c^2}\boldsymbol{\sigma} \cdot ((\nabla V) \times \boldsymbol{p}) \right] \cdot \chi . \tag{F.7}$$

This is the form that is appropriate to use in solid-state physics calculations. The term

$$\frac{\hbar^2}{4m_0^2c^2}\boldsymbol{\sigma} \cdot [(\nabla V) \times \boldsymbol{p}] \tag{F.8}$$

is called the spin-orbit term. This term is often used by itself as a first-order correction to the nonrelativistic Schrödinger equation. The spin-orbit correction is often applied in band-structure calculations at certain points in the Brillouin zone where bands come together. In the case in which the potential is spherically symmetric (which is important for atomic potentials but not crystalline potentials), the spin-orbit term can be cast into the more familiar form

$$\frac{\hbar^2}{2m_0^2c^2}\frac{1}{r}\frac{\mathrm{d}V}{\mathrm{d}r}\boldsymbol{L} \cdot \boldsymbol{S} , \tag{F.9}$$

where $\boldsymbol{L}$ is the orbital angular momentum operator and $\boldsymbol{S}$ is the spin operator (in units of $\hbar$).

It is also interesting to see how Dirac's theory works out in the (completely) nonrelativistic limit when an external magnetic field $\boldsymbol{B}$ is present. In this case the magnetic moment of the electron is introduced by the term involving $\boldsymbol{S} \cdot \boldsymbol{B}$. This term automatically appears from the nonrelativistic limit of Dirac's equation. In addition, the correct ratio of magnetic moment to spin angular momentum is obtained in this way.

G The Second Quantization Notation for Fermions and Bosons

When the second quantization notation is used in a nonrelativistic context it is simply a notation in which we express the wave functions in occupation-number space and the operators as operators on occupation number space. It is of course of great utility in considering the many-body problem. In this formalism, the symmetry or antisymmetry of the wave functions is automatically built into the formalism. In relativistic physics, annihilation and creation operators (which are the basic operators of the second quantization notation) have physical meaning. However, we will apply the second quantization notation only in nonrelativistic situations. No derivations will be made in this section. (The appropriate results will just be concisely written down.) There are many good treatments of the second quantization or occupation number formalism. One of the most accessible is by Mattuck.[9]

G.1 Bose Particles

For Bose particles we deal with b_i and $b_i^\dagger$ operators (or other letters where convenient): $b_i^\dagger$ *creates* a Bose particle in the state i; b_i *annihilates* a Bose particle in the state f. The b_i operators obey the following commutation relations:

$$[b_i, b_j] \equiv b_i b_j - b_j b_i = 0,$$
$$[b_i^\dagger, b_j^\dagger] = 0,$$
$$[b_i, b_j^\dagger] = \delta_{ij}.$$

The occupation number operator whose eigenvalues are the number of particles in state i is

$$n_i = b_i^\dagger b_i ,$$

and

$$n_i + 1 = b_i b_i^\dagger .$$

The effect of these operators acting on different occupation number kets is

$$b_i \left| n_1, \ldots, n_i, \ldots \right\rangle = \sqrt{n_i} \left| n_1, \ldots, n_i - 1, \ldots \right\rangle,$$
$$b_i^\dagger \left| n_1, \ldots, n_i, \ldots \right\rangle = \sqrt{n_i + 1} \left| n_1, \ldots, n_i + 1, \ldots \right\rangle,$$

where $|n_1,\ldots,n_i,\ldots\rangle$ means the ket appropriate to the state with n_1 particles in state 1, n_2 particles in state 2, and so on.

---

[9] See Mattuck [A.14].

The matrix elements of these operators are given by

$$\langle n_i - 1 | b_i | n_i \rangle = \sqrt{n_i},$$
$$\langle n_i | b_i^\dagger | n_i - 1 \rangle = \sqrt{n_i}.$$

In this notation, any one-particle operator

$$f_{op}^{(1)} = \sum_l f^{(1)}(\boldsymbol{r}_l)$$

can be written in the form

$$f_{op}^{(1)} = \sum_{i,k} \langle i | f^{(1)} | k \rangle b_i^\dagger b_k ,$$

and the $|k\rangle$ are any complete set of one-particle eigenstates.

In a similar fashion any two-particle operator

$$f_{op}^{(2)} = \sum_{l,m} f^{(2)}(\boldsymbol{r}_l - \boldsymbol{r}_m)$$

can be written in the form

$$f_{op}^{(2)} = \sum_{i,k,l,m} \langle i(1)k(2) | f^{(2)} | l(1)m(2) \rangle b_i^\dagger b_k^\dagger b_m b_l .$$

Operators that create or destroy base particles at a given point in space (rather than in a given state) are given by

$$\psi(\boldsymbol{r}) = \sum_\alpha u_\alpha(\boldsymbol{r}) b_\alpha,$$
$$\psi^\dagger(\boldsymbol{r}) = \sum_\alpha u_\alpha^*(\boldsymbol{r}) b_\alpha^\dagger,$$

where $u_\alpha(\boldsymbol{r})$ is the single-particle wave function corresponding to state α. In general, r would refer to *both space* and *spin* variables. These operators obey the commutation relation

$$[\psi(\boldsymbol{r}), \psi^\dagger(\boldsymbol{r})] = \delta(\boldsymbol{r} - \boldsymbol{r}').$$

G.2 Fermi Particles

For Fermi particles, we deal with a_i and $a_i^\dagger$ operators (or other letters where convenient): $a_i^\dagger$ creates a fermion in the state i; a_i annihilates a fermion in the state i. The a_i operators obey the following anticommutation relations:

$$\{a_i, a_j\} \equiv a_i a_j + a_j a_i = 0,$$
$$\{a_i^\dagger, a_j^\dagger\} = 0,$$
$$\{a_i, a_j^\dagger\} = \delta_{ij}.$$

The occupation number operator whose eigenvalues are the number of particles in state i is

$$n_i = a_i^\dagger a_i \ ,$$

and

$$1 - n_i = a_i a_i^\dagger \ .$$

Note that $(n_i)^2 = n_i$, so that the only possible eigenvalues of n_i are 0 and 1 (the Pauli principle is built in!).

The matrix elements of these operators are defined by

$$\langle \cdots n_i = 0 \cdots | a_i | \cdots n_i = 1 \cdots \rangle = (-)^{\Sigma(1,i-1)} ,$$

and

$$\langle \cdots n_i = 1 \cdots | a_i^\dagger | \cdots n_i = 0 \cdots \rangle = (-)^{\Sigma(1,i-1)} ,$$

where $\Sigma(1, i-1)$ equals the sum of the occupation numbers of the states from 1 to $i-1$.

In this notation, any one-particle operator can be written in the form

$$f_0^{(1)} = \Sigma_{i,j} \langle i | f^{(1)} | j \rangle a_i^\dagger a_j \ ,$$

where the $|j\rangle$ are any complete set of one-particle eigenstates. In a similar fashion, any two-particle operator can be written in the form

$$f_{op}^{(2)} = \Sigma_{i,j,k,l} \langle i(1) j(2) | f^{(2)} | k(1) l(2) \rangle a_j^\dagger a_i^\dagger a_k a_l \ .$$

Operators that create or destroy Fermi particles at a given point in space (rather than in a given state) are given by

$$\psi(\boldsymbol{r}) = \Sigma_\alpha u_\alpha(\boldsymbol{r}) a_\alpha \ ,$$

where $u_\alpha(\boldsymbol{r})$ is the single-particle wave function corresponding to state α, and

$$\psi^\dagger(\boldsymbol{r}) = \Sigma_\alpha u_\alpha^*(\boldsymbol{r}) a_\alpha^\dagger \ .$$

These operators obey the anticommutation relations

$$\{\psi(\boldsymbol{r}), \psi^\dagger(\boldsymbol{r})\} = \delta(\boldsymbol{r} - \boldsymbol{r}') \ .$$

The operators also allow a convenient way of writing *Slater* determinants, e.g.,

$$a_\alpha^\dagger a_\beta^\dagger |0\rangle \leftrightarrow \frac{1}{\sqrt{2}} \begin{vmatrix} u_\alpha(1) & u_\alpha(2) \\ u_\beta(1) & u_\beta(2) \end{vmatrix} ;$$

$|0\rangle$ is known as the *vacuum* ket.

The easiest way to see that the second quantization notation is consistent is to show that matrix elements in the second quantization notation have the same values as corresponding matrix elements in the old notation. This demonstration will not be done here.

H The Many-Body Problem

Richard P. Feynman is famous for many things, among which is the invention, in effect, of a new quantum mechanics. Or maybe we should say of a new way of looking at quantum mechanics. His way involves taking a process going from A to B and looking at all possible paths. He then sums the amplitude of the all paths from A to B to find, by the square, the probability of the process.

Related to this is a diagram that defines a process and that contains by implication all the paths, as calculated by appropriate integrals. Going further, one looks at all processes of a certain class, and sums up all diagrams (if possible) belonging to this class. Ideally (but seldom actually) one eventually treats all classes, and hence arrives at an exact description of the interaction.

Thus, in principle, there is not so much to treating interactions by the use of Feynman diagrams. The devil is in the details, however. Certain sums may well be infinite–although hopefully disposable by renormalization. Usually doing a nontrivial calculation of this type is a great technical feat.

We have found that a common way we use Feynman diagrams is to help us understand what we mean by a given approximation. We will note below, for example, that the Hartree approximation involves summing a certain class of diagrams, while the Hartree–Fock approximation involves summing these diagrams along with another class. We believe, the diagrams give us a very precise idea of what these approximations do.

Similarly, the diagram expansion can be a useful way to understand why a perturbation expansion does not work in explaining superconductivity, as well as a way to fix it (the Nambu formalism).

The practical use of diagrams, and diagram summation, may involve great practical skill, but it seems that the great utility of the diagram approach is in clearly stating, and in keeping track of, what we are doing in a given approximation.

One should not think that an expertise in the technicalities of Feynman diagrams solves all problems. Diagrams have to be summed and integrals still have to be done. For some aspects of many-electron physics, density functional theory (DFT) has become the standard approach. Diagrams are usually not used at the beginning of DFT, but even here they may often be helpful in discussing some aspects.

DFT was discussed in Chap. 3, and we briefly review it here, because of its great practical importance in the many-electron problem of solid-state physics. In the beginning of DFT Hohenberg and Kohn showed that the N-electron Schrödinger wave equation in three dimensions could be recast. They showed that an equation for the electron density in three dimensions would suffice to determine ground-state properties. The Hohenberg–Kohn formulation may be regarded as a generalization of the Thomas–Fermi approximation. Then came the famous Kohn–Sham equations that reduced the Hohenberg–Kohn formulation to the problem of noninteracting electrons in an effective potential (somewhat analogous to the Hartree equations, for example). However, part of the potential, the exchange correlation part could only be approximately evaluated, e.g. in the local density approximation (LDA) – which assumed a locally homogeneous electron gas. A problem with DFT-LDA is that it is not necessarily clear what the size of the errors are, however, the DFT is certainly a good way to calculate, ab initio, certain ground-state properties of finite electronic systems, such as the ionization energies of atoms. It is also very useful for computing the electronic ground-state properties of periodic solids, such as cohesion and stability. Excited states, as well as approximations for the exchange correlation term in N-electron systems continue to give problems. For a nice brief summary of DFT see Mattsson [A.13].

For quantum electrodynamics, a brief and useful graphical summary can be found at: http://www2slac.standford.edu/vvc/theory/feynman.html. We now present a brief summary of the use of diagrams in many-body physics.

In some ways, trying to do solid-state physics without Feynman diagrams is a little like doing electricity and magnetism (EM) without resorting to drawing Faraday's lines of electric and magnetic fields. However, just as field lines have limitations in describing EM interactions, so do diagrams for discussing the many-body problem [A.1]. The use of diagrams can certainly augment one's understanding.

The distinction between quasi- or dressed particles and collective excitations is important and perhaps is made clearer from a diagrammatic point of view. Both are 'particles' and are also elementary energy excitations. But after all a polaron (a quasi-particle) is not the same kind of beast as a magnon (a collective excitation). Not everybody makes this distinction. Some call all 'particles' quasiparticles. Bogolons are particles of another type, as are excitons (see below for definitions of both). All are elementary excitations and particles, but not really collective excitations or dressed particles in the usual sense.

H.1 Propagators

These are the basic quantities. Their representation is given in the next section. The single-particle propagator is a sum of probability amplitudes for all the ways of going from r_1, t_1 to r_2, t_2 (adding a particle at 1 and taking out at 2).

The two-particle propagator is the sum of the probability amplitudes for all the ways two particles can enter a system, undergo interactions and emerge again.

H.2 Green Functions

Propagators are represented by Green functions. There are both advanced and retarded propagators. Advanced propagators can describe particles traveling backward in time, i.e. holes. The use of Fourier transforms of time-dependent propagators led to simpler algebraic equations. For a retarded propagator the free propagator is:

$$G_0^+(\boldsymbol{k},\omega) = \frac{1}{\omega - \varepsilon_k + \mathrm{i}\delta}. \tag{H.1}$$

For quasiparticles, the real part of the pole of the Fourier transform of the single-particle propagator gives the energy, and the imaginary part gives the width of the energy level. For collective excitations, one has a similar statement, except that two-particle propagators are needed.

H.3 Feynman Diagrams

Rules for drawing diagrams are found in Economu [A.5 pp. 251-252], Pines [A.22 pp. 49-50] and Schrieffer [A.24 pp. 127-128]. Also, see Mattuck [A.14 p. 165]. There is a one-to-one correspondence between terms in the perturbation expansion of the Green functions and diagrammatic representation. Green functions can also be calculated from a hierarchy of differential equations and an appropriate decoupling scheme. Such approximate decoupling schemes are always equivalent to a partial sum of diagrams.

H.4 Definitions

Here we remind you of some examples. A more complete list is found in Chap. 4.

Quasiparticle – A real particle with a cloud of surrounding disturbed particles with an effective mass and a lifetime. In the usual case it is a dressed fermion. Examples are listed below.

Electrons in a solid – These will be dressed electrons. They can be dressed by interaction with the static lattice, other electrons or interactions with the vibrating lattice. It is represented by a straight line with an arrow to the right ⟶ if time goes that way

Holes in a solid – One can view the ground state of a collection of electrons as a vacuum. A hole is then what results when an electron is removed from a normally occupied state. It is represented by a straight line with an arrow to the left ⟵ .

Polaron – An electron moving through a polarizable medium surrounded by its polarization cloud of virtual phonons.

Photon – Quanta of electromagnetic radiation (e.g. light) – it is represented by a wavy line ∿∿∿.

Collective Excitation – These are elementary energy excitations that involve wave-like motion of all the particles in the systems. Examples are listed below.

Phonon – Quanta of normal mode vibration of a lattice of ions. Also often represented by wavy line.

Magnon – Quanta of low-energy collective excitations in the spins, or quanta of waves in the spins.

Plasmon – Quanta of energy excitation in the density of electrons in an interacting electron gas (viewing, e.g., the positive ions as a uniform background of charge).

Other Elementary Energy Excitations – Excited energy levels of many-particle systems.

Bogolon – Linear combinations of electrons in a state $+k$ with 'up' spin and $-k$ with 'down' spin. Elementary excitations in a superconductor.

Exciton – Bound electron–hole pairs.

Some examples of interactions represented by vertices (time going to the right):

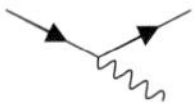
An electron emitting a phonon

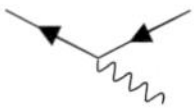
A hole emitting a phonon.

Diagrams are built out of vertices with conservation of momentum satisfied at the vertices. For example

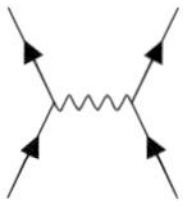

represents a coulomb interaction with time going up.

H.5 Diagrams and the Hartree and Hartree–Fock Approximations

In order to make these concepts clearer it is perhaps better to discuss an example that we have already worked out without diagrams. Here, starting from the Hamiltonian we will discuss briefly how to construct diagrams, then explain how to associate single-particle Green functions with the diagrams and how to do the partial sums representing these approximations. For details, the references must be consulted.

In the second quantization notation, a Hamiltonian for interacting electrons

$$\mathcal{H} = \sum_i V(i) + \frac{1}{2}\sum_{i,j} V(ij)\,, \tag{H.2}$$

with one- and two-body terms can be written as

$$\mathcal{H} = \sum_{i,j} \langle i(1)|V(1)|\,j(1)\rangle a_i^\dagger a_j + \frac{1}{2}\sum_{ijkl} \langle i(1)\,j(2)|V(1,2)|k(1)\,j(2)\rangle a_j^\dagger a_i^\dagger a_k a_l\,, \tag{H.3}$$

where

$$\langle i(1)|V(1)|j(1)\rangle = \int \phi_i^*(\boldsymbol{r}_1)V(\boldsymbol{r}_1)\phi_j(\boldsymbol{r}_1)d^3\boldsymbol{r}_i\,, \tag{H.4}$$

and

$$\langle i(1)j(2)|V(1,2)|k(1)l(2)\rangle = \int \phi_i^*(\boldsymbol{r}_1)\phi_j^*(\boldsymbol{r}_2)V(1,2)\phi_k(\boldsymbol{r}_1)\phi_l(\boldsymbol{r}_2)d^3\boldsymbol{r}_1 d^3\boldsymbol{r}_2\,, \tag{H.5}$$

and the annihilation and creation operators have the usual properties

$$\begin{aligned} a_i a_j^\dagger + a_j^\dagger a_i &= \delta_{ij}\,, \\ a_i a_j + a_j a_i &= 0. \end{aligned}$$

We now consider the Hartree approximation. We assume, following Mattuck [A.14] that the interactions between electrons is mostly given by the forward scattering processes where the interacting electrons have no momentum change in the interaction. We want to get an approximation for the single-particle propagator that includes interactions. In first order the only possible process is given by a bubble diagram where the hole line joins on itself. One thinks of the particle in state k knocking a particle out of and into a state l instantaneously. Since this can happen any number of times, we get the following partial sum for diagrams representing the single-particle propagator. The first diagram on the right-hand side represents the free propagator where nothing happens (Mattuck [A.14 p. 89][10]).

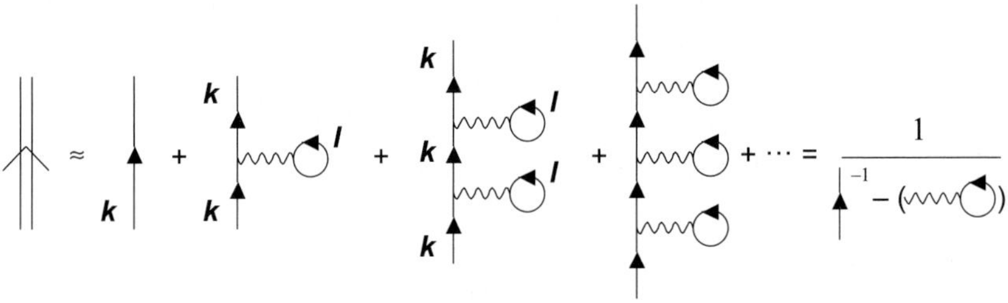

Using the "dictionary" given by Mattuck [A.14 p. 86], we substitute propagators for diagrams and get

$$G^+(\boldsymbol{k},\omega) = \frac{1}{\omega - \varepsilon_k - \sum_{l(\mathrm{occ.})} V_{klkl} + \mathrm{i}\delta}\,. \tag{H.6}$$

Since the poles give the elementary energy excitations we have

$$\varepsilon_k' = \varepsilon_k + \sum_{l(\mathrm{occ.})} V_{klkl}\,, \tag{H.7}$$

[10] Reproduced with permission from Mattuck RD, *A Guide to Feynman Diagrams in the Many-Body Problem*, 2nd edn, (4.67) p. 89, Dover Publications, Inc., 1992.

which is exactly the same as the Hartree approximation (see (3.21)) since

$$\sum_l V_{klkl} = \int \mathrm{d}^3 r_2 \phi_k^*(r_2)\phi_k(r_2) \sum_l \int \phi_l^*(r_1)\phi_l(r_1) V(1,2)\mathrm{d}^3 r_1 \,. \tag{H.8}$$

It is actually very simple to go from here to the Hartree–Fock approximation – all we have to do is to include the exchange terms in the interactions. These are the "open-oyster" diagrams

where a particle not only strikes a particle in l and creates an instantaneous hole, but is exchanged with it. Doing the partial sum of forward scattering and exchange scattering one has (Mattuck [A.14 p. 91][11]):

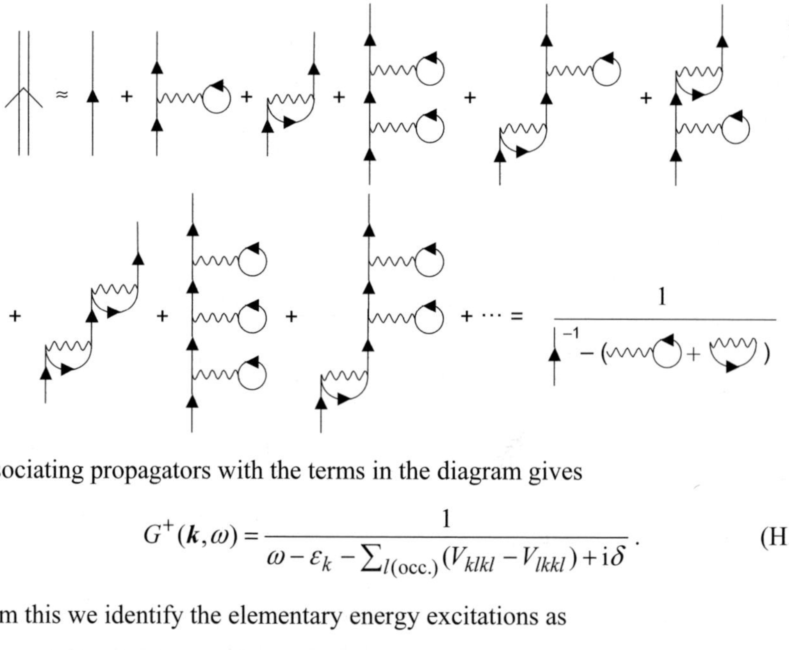

Associating propagators with the terms in the diagram gives

$$G^+(\boldsymbol{k},\omega) = \frac{1}{\omega - \varepsilon_k - \sum_{l(\text{occ.})}(V_{klkl} - V_{lkkl}) + \mathrm{i}\delta} \,. \tag{H.9}$$

From this we identify the elementary energy excitations as

$$\varepsilon_k' = \varepsilon_k + \sum_{l(\text{occ.})}(V_{klkl} - V_{lkkl}) \,, \tag{H.10}$$

which is just what we got for the Hartree–Fock approximation (see (3.50)).

The random-phase approximation [A.14] can also be obtained by a partial summation of diagrams, and it is equivalent to the Lindhard theory of screening.

---

[11] Reproduced with permission from Mattuck RD, *A Guide to Feynman Diagrams in the Many-Body Problem*, 2nd edn, (4.76) p. 91, Dover Publications, Inc., 1992.

H.6 The Dyson Equation

This is the starting point for many approximations both diagrammatic, and algebraic. Dyson's equation can be regarded as a generalization of the partial sum technique used in the Hartree and Hartree–Fock approximations. It is exact. To state Dyson's equation we need a couple of definitions. The self-energy part of a diagram is a diagram that has no incoming or outgoing parts and can be inserted into a particle line. The bubbles of the Hartree method are an example. An irreducible or proper self-energy part is a part that cannot be further reduced into unconnected self-energy parts. It is common to define

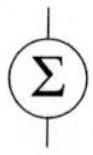

as the sum over all proper self-energy parts. Then one can sum over all repetitions of sigma ($\sum \boldsymbol{k},\omega$) to get

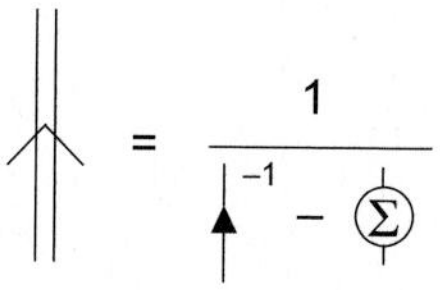

Dyson's equation yields an exact expression for the propagator,

$$G(\boldsymbol{k},\omega) = \frac{1}{\omega - \varepsilon_k - \sum_{l(\text{occ.})}(\boldsymbol{k},\omega) + \mathrm{i}\delta_k}, \tag{H.11}$$

since all diagrams are either proper diagrams or their repetition.

In the Hartree approximation

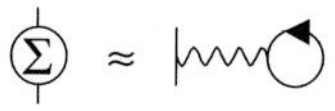

and in the Hartree–Fock approximation

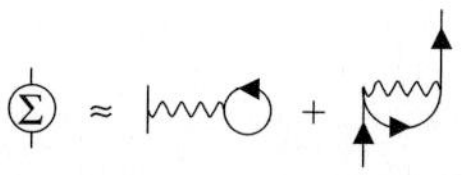

Although the Dyson equation is in principle exact, one still has to evaluate sigma, and this is in general not possible except in some approximation.

We cannot go into more detail here. We have given accurate results for the high and low-density electron gas in Chap. 2. In general, the ideas of Feynman diagrams and the many-body problem merit a book of their own. We have found the book by Mattuck [A.14] to be particularly useful, but note the list of references at the end of this section. We have used some ideas about diagrams when we discussed superconductivity.

Bibliography

Chapter 1

1.1. Anderson PW, *Science* **177**, 393-396 (1972).

1.2. Bacon GE, *Neutron Diffraction*, Clarendon Press, Oxford, 2nd edn (1962).

1.3. Bradley CJ and Cracknell AP, The Mathematical Theory of Symmetry in Solids: Representation Theory for Point Groups and Space Groups, Clarendon Press, Oxford (1972).

1.4. Brown PJ and Forsyth JB, *The Crystal Structure of Solids*, Edward Arnold, London, 1975 (Chap. 3).

1.5. Buerger MJ, *Elementary Crystallography*, John Wiley and Sons, New York, 1956.

1.6. Daw MS, "Model of metallic cohesion: The embedded-atom method," *Phys Rev B* **39**, 7441-7452 (1989)

1.7. de Gennes PG and Prost J, *The Physics of Liquid Crystals*, Clarendon Press, Oxford, 2nd edn (1993).

1.8. Evjan HM, *Physical Review*, **39**, 675 (1932)

1.9. Ghatak AK and Kothari LS, *An Introduction to Lattice Dynamics*, Addison Wesley Publishing Co., Reading, Mass. 1972 p165ff.

1.10. Herzfield CM and Meijer PHE, "Group Theory and Crystal Field Theory," *Solid State Physics, Advances in Research and Applications* **12**, 1-91 (1961).

1.11. Horton GK, "Ideal Rare-Gas Crystals," *American Journal of Physics*, **36**(2), 93 (1968).

1.12. Janot C, *Quasicrystals, A Primer*, Clarendon Press, Oxford (1992).

1.13. Kittel C, *Introduction to Solid State Physics*, John Wiley and Sons, New York, Seventh Edition, 1996, Chaps. 1-3, 19.

1.14. Koster GF, "Space Groups and Their Representations," *Solid State Physics, Advances in Research and Applications* **5**, 174-256 (1957).

1.15. Levine D and Steinhardt PJ, *Phys Rev Lett* **53**, 2477-83 (1984).

1.16. Maradudin AA, Montrol EW, and Weiss GW, *Theory of Lattice Dynamics in the Harmonic Approximation*, Academic Press, N.Y. 1963 p 245.

1.17. Moffatt WG , Pearsall GW, and Wulff J, *The Structure and Properties of Materials, Vol. 1*, John Wiley and Sons, Inc., New York, 1964.

1.18. Pauling L, *Nature of the Chemical Bond*, 2nd edn., Cornell University Press, Ithaca, 1945.

1.19. Phillips FC, *An Introduction to Crystallography*, John Wiley and Sons, Inc., New York, 4th edn., 1971.

1.20. Pollock Daniel D, "Physical Properties of Materials for Engineers," CRC Press, Boca Raton, 1993.

1.21. Shechtman D, Blech I, Gratias D, and Cahn JW, *Phys Rev Lett* **53**, 1951-3 (1984).

1.22. Steinhardt PJ and Ostlund S, *The Physics of Quasicrystals*, World Scientific Publishing, Singapore, 1987.
1.23. Streetman BG, *Solid State Electronic Devices*, Prentice Hall, Englewood Cliffs, NJ, 4th edn. (1995).
1.24. Tinkham M, *Group Theory and Quantum Mechanics*, McGraw-Hill Book Company, New York, 1964.
1.25. Tosi MP, "Cohesion of Ionic Solids in the Born Model," *Solid State Physics, Advances in Research and Applications* **16**, 1-120 (1964).
1.26. Tran HT and Perdew JP, "How Metals Bind: The deformable jellium model with correlated electrons," *Am J Phys* **71**, 1048-1061 (2003).
1.27. Webb MB and Lagally MG, "Elastic Scattering of Low Energy Electrons from Surfaces," *Solid State Physics, Advances in Research and Applications* 28, 301-405 (1973).
1.28. West Anthony R, *Solid State Chemistry and its Properties*, John Wiley and Sons, New York, 1984.
1.29. Wigner EP and Seitz F, "Qualitative Analysis of the Cohesion in Metals," *Solid State Physics, Advances in Research and Applications* **1**, 97-126 (1955).
1.30. Wyckoff RWG, "Crystal Structures," Vols 1-5, John Wiley and Sons, New York 1963-1968.

Chapter 2

2.1. Anderson PW, *Science* **177**, 393-396 (1972).
2.2. Bak TA (ed), *Phonons and Phonon Interactions*, W A Benjamin, New York, 1964.
2.3. Bilz H and Kress W, *Phonon Dispersion Relations in Insulators*, Springer, 1979
2.4. Blackman M, "The Specific Heat of Solids," *Encyclopedia of Physics*, Vol. VII, Part 1, Crystal Physics 1, Springer Verlag, Berlin, 1955, p. 325.
2.5. Born M and Huang K, *Dynamical Theory of Crystal Lattices*, Oxford University Press, New York, 1954.
2.6. Brockhouse BN and Stewart AT, *Rev. Mod Phys* **30**, 236 (1958).
2.7. Brown FC, *The Physics of Solids*, W A Benjamin, Inc., New York, 1967, Chap. 5.
2.8. Choquard P, *The Anharmonic Crystal*, W A Benjamin, Inc., New York (1967).
2.9. Cochran W, "Interpretation of Phonon Dispersion Curves," in Proceedings of the International Conference on Lattice Dynamics, Copenhagen, 1963, Pergamon Press, New York, 1965.
2.10. Cochran W, "Lattice Vibrations," *Reports on Progress in Physics*, Vol. XXVI, The Institute of Physics and the Physical Society, London, 1963, p. 1. See also Cochran W, *The Dynamics of Atoms in Crystals*, Edward Arnold, London, 1973.
2.11. deLauney J, "The Theory of Specific Heats and Lattice Vibrations," *Solid State Physics: Advances in Research and Applications* **2**, 220-303 (1956).
2.12. Dick BG Jr, and Overhauser AW, *Physical Review* **112**, 90 (1958).
2.13. Dorner B, Burkel E, Illini T, and Peisl J, *Z für Physik* **69**, 179-183 (1989)
2.14. Dove MT, *Structure and Dynamics*, Oxford University Press, 2003.
2.15. Elliott RJ and Dawber DG, *Proc. Roy. Soc.* **A223**, 222 (1963)
2.16. Ghatak K and Kothari LS, *An Introduction to Lattice Dynamics*, Addison-Wesley Publ. Co., Reading, MA, 1972, Chap. 4.

2.17. Grosso G and Paravicini GP, *Solid State Physics, Academic Press*, New York, 2000, Chaps. VIII and IX.
2.18. Huntington HB, "The Elastic Constants of Crystals," *Solid State Physics: Advances in Research and Applications* **7**, 214-351 (1958).
2.19. Jensen HH, "Introductory Lectures on the Free Phonon Field," in *Phonons and Phonon Interactions*, Bak TA (ed), W. A. Benjamin, New York, 1964.
2.20. Jones W and March NA, *Theoretical Solid State Physics*, John Wiley and Sons, (1973), Vol I, Chap. 3.
2.21. Joshi SK and Rajagopal AK, "Lattice Dynamics of Metals," *Solid State Physics: Advances in Research and Applications* **22**, 159-312 (1968).
2.22. Kunc K, Balkanski M, and Nusimovici MA, *Phys Stat Sol B* **71**, 341; **72**, 229, 249 (1975).
2.23. Lehman GW, Wolfram T, and DeWames RE, *Physical Review,* **128**(4), 1593 (1962).
2.24. Leibfried G and Ludwig W, "Theory of Anharmonic Effects in Crystals," *Solid State Physics: Advances in Research and Applications* **12**, 276-444 (1961).
2.25. Lifshitz M and Kosevich AM, "The Dynamics of a Crystal Lattice with Defects," *Reports on Progress in Physics*, Vol. XXIX, Part 1, The Institute of Physics and the Physical Society, London, 1966, p. 217.
2.26. Maradudin A, Montroll EW, and Weiss GH, "Theory of Lattice Dynamics in the Harmonic Approximation," *Solid State Physics: Advances in Research and Applications*, Supplement 3 (1963).
2.27. Messiah A, *Quantum Mechanics*, North Holland Publishing Company, Amsterdam, 1961, Vol. 1, p 69.
2.28. Montroll EW, *J. Chem. Phys.* **10**, 218 (1942), 11, 481 (1943).
2.29. Schaefer G, *Journal of Physics and Chemistry of Solids* **12**, 233 (1960).
2.30. Scottish Universities Summer School, *Phonon in Perfect Lattices and in Lattices with Point Imperfections*, 1965, Plenum Press, New York, 1960.
2.31. Shull CG and Wollan EO, "Application of Neutron Diffraction to Solid State Problems," *Solid State Physics, Advances in Research and Applications* **2**, 137 (1956).
2.32. Srivastava GP, *The Physics of Phonons*, Adam Hilger, Bristol, 1990.
2.33. Strauch D, Pavone P, Meyer AP, Karch K, Sterner H, Schmid A, Pleti Th, Bauer R, Schmitt M, "Festorkorperproblem," *Advances in Solid State Physics* **37**, 99-124 (1998), Helbig R (ed), Braunschweig/Weisbaden: Vieweg.
2.34. Toya T, "Lattice Dynamics of Lead," in Proceedings of the International Conference on Lattice Dynamics, Copenhagen, 1963, Pergamon Press, New York, 1965.
2.35. Van Hove L, *Phys Rev* **89**, 1189 (1953).
2.36. Vogelgesang R et al, *Phys Rev* **B54**, 3989 (1996).
2.37. Wallis RF (ed), Proceedings of the International Conference on Lattice Dynamics, Copenhagen, 1963, Pergamon Press, New York, 1965.
2.38. Ziman JM, *Electrons and Phonons*, Oxford, Clarendon Press (1962).
2.39. 1962 Brandeis University Summer Institute Lectures in Theoretical Physics, Vol. 2, W. A. Benjamin, New York, 1963.

Chapter 3

3.1. Altman SL, *Band Theory of Solids*, Clarendon Press, Oxford, 1994. See also Singleton J, *Band Theory and Electronic Structure*, Oxford University Press, 2001.

3.2. Aryasetiawan F and Gunnarson D, *Rep Prog Phys* **61**, 237 (1998).

3.3. Austin BJ et al, *Phys Rev* **127**, 276 (1962)

3.4. Berman R, *Thermal Conduction in Solid*, Clarendon Press, Oxford, 1976, p. 125.

3.5. Blount EI, "Formalisms of Band Theory," *Solid State Physics, Advances in Research and Applications*, **13**, 305-373 (1962).

3.6. Blount EI, *Lectures in Theoretical Physics*, Vol. V, Interscience Publishers, New York, 1963, p422ff.

3.7. Bouckaert LP, Smoluchowski R, and Wigner E, *Physical Review*, **50**, 58 (1936).

3.8. Callaway J and March NH, "Density Functional Methods: Theory and Applications," *Solid State Physics, Advances in Research and Applications*, **38**, 135-221 (1984).

3.9. Ceperley DM and Alder BJ, *Phys Rev Lett* **45**, 566 (1980).

3.10. Chelikowsky JR and Louie SG (eds), *Quantum Theory of Real Materials*, Kluwer Academic Publishers, Dordrecht, 1996.

3.11. Cohen ML, *Physics Today* **33**, 40-44 (1979)

3.12. Cohen ML and Chelikowsky JR, *Electronic Structure and Optical Properties of Semiconductors*, Springer-Verlag, Berlin, 2nd edn. (1989).

3.13. Cohen ML and Heine V, "The Fitting of Pseudopotentials to Experimental Data and their Subsequent Application," *Solid State Physics, Advances in Research and Applications*, **24**, 37-248 (1970).

3.14. Cohen M and Heine V, *Phys Rev* **122**, 1821 (1961).

3.15. Cusack NE, *The Physics of Disordered Matter*, Adam Hilger, Bristol, 1987, see especially Chaps. 7 and 9.

3.16. Dimmock JO, "The Calculation of Electronic Energy Bands by the Augmented Plane Wave Method," *Solid State Physics, Advances in Research and Applications*, **26**, 103-274 (1971).

3.17. Fermi E, *Nuovo Cimento* **2**, 157 (1934)

3.18. Friedman B, *Principles and Techniques of Applied Mathematics*, John Wiley and Sons, New York, 1956.

3.19. Harrison WA, *Pseudopotentials in the theory of Metals*, W A Benjamin, Inc., New York, 1966.

3.20. Heine V, "The Pseudopotential Concept," *Solid State Physics, Advances in Research and Applications*, **24**, 1-36 (1970).

3.21. Herring C, *Phys Rev* **57**, 1169 (1940).

3.22. Herring C, *Phys Rev* **58**, 132 (1940).

3.23. Hohenberg PC and Kohn W, "Inhomogeneous Electron Gas," *Phys Rev*, **136**, B804-871 (1964).

3.24. lzynmov YA, *Advances in Physics* **14**(56), 569 (1965).

3.25. Jones RO and Gunnaisson O, "The Density Functional Formalism, its Applications and Prospects," *Rev Modern Phys*, **61**, 689-746 (1989).

3.26. Jones W and March NH, *Theoretical Solid State Physics*, Vol. 1 and 2, John Wiley and Sons, 1973.

3.27. Kohn W, "Electronic Structure of Matter–Wave Functions and Density Functionals," *Rev Modern Phys*, **71**, 1253-1266 (1999).

3.28. Kohn W and Sham LJ, "Self Consistent Equations Including Exchange and Correlation Effects," *Phys Rev*, **140**, A1133-1138 (1965).
3.29. Kohn W and Sham LJ, *Phys Rev* **145**, 561 (1966).
3.30. Kronig and Penny, *Proceedings of the Royal Society* (London), **A130**, 499 (1931).
3.31. Landau L, *Soviet Physics* JETP, **3**, 920 (1956).
3.32. Loucks TL, *Phys Rev Lett*, **14**, 693 (1965).
3.33. Löwdin PO, *Advances in Physics* **5**, 1 (1956)
3.34. Marder MP, *Condensed Matter Physics*, John Wiley and Sons, Inc., New York, 2000.
3.35. Mattuck RD, *A Guide to Feynman Diagrams in the Many-Body Problem*, McGraw-Hill Book Company, New York, 2nd Ed., 1976. See particularly Chap. 4.
3.36. Negele JW and Orland H, *Quantum Many Particle Systems*, Addison-Wesley Publishing Company, Redwood City, California (1988).
3.37. Nemoshkalenko VV and Antonov VN, *Computational Methods in Solid State Physics*, Gordon and Breach Science Publishers, The Netherlands, 1998.
3.38. Parr RG and Yang W, Oxford Univ. Press, New York, 1989.
3.39. Pewdew JP and Zunger A, *Phys Rev* **B23**, 5048 (1981).
3.40. Phillips JC and Kleinman L, *Phys Rev* **116**, 287-294 (1959).
3.41. Pines D, *The Many-Body Problem*, W A Benjamin, New York, 1961.
3.42. Raimes S, *The Wave Mechanics of Electrons in Metals*, North-Holland Publishing Company, Amsterdam (1961).
3.43. Reitz JR, "Methods of the One-Electron Theory of Solids," *Solid State Physics, Advances in Research and Applications*, **1**, 1-95 (1955).
3.44. Schlüter M and Sham LJ, "Density Functional Techniques," *Physics Today*, Feb. 1982, pp. 36-43.
3.45. Singh DJ, *Plane Waves, Pseudopotentials, and the APW Method*, Kluwer Academic Publishers, Boston (1994).
3.46. Singleton J, *Band Theory and Electronic Properties of Solids*, Oxford University Press (2001).
3.47. Slater JC [88, 89, 90].
3.48. Slater JC, "The Current State of Solid-State and Molecular Theory," *International Journal of Quantum Chemistry*, **I**, 37-102 (1967).
3.49. Slater JC and Koster GF, *Phys Rev* **95**, 1167 (1954).
3.50. Slater JC and Koster GF, *Phys Rev* **96**, 1208 (1954).
3.51. Smith N, "Science with soft x-rays," *Physics Today* **54**(1), 29-54 (2001).
3.52. Spicer WE, *Phys Rev* **112**, p114ff (1958).
3.53. Stern EA, "Rigid-Band Model of Alloys," *Phys Rev*, **157**(3), 544 (1967).
3.54. Thouless DJ, *The Quantum Mechanics of Many-Body Systems*, Academic Press, New York, 1961.
3.55. Tran HT and Pewdew JP, "How metals bind: The deformable-jellium model with correlated electrons," *Am. J. Phys.* **71**(10), 1048-1061 (2003).
3.56. Wannier GH, "The Structure of Electronic Excitation Levels in Insulating Crystals," *Phys Rev* **52**, 191-197, (1937).
3.57. Wigner EP and Seitz F, "Qualitative Analysis of the Cohesion in Metals," *Solid State Physics, Advances in Research and Applications*, **1**, 97-126 (1955).
3.58. Woodruff TO, "The Orthogonalized Plane-Wave Method," *Solid State Physics, Advances in Research and Applications*, **4**, 367-411 (1957).
3.59. Ziman JM, "The Calculation of Bloch Functions," *Solid State Physics, Advances in Research and Applications*, **26**, 1-101 (1971).

Chapter 4

4.1. Anderson HL (ed), *A Physicists Desk Reference*, 2nd edn, Article 20: Frederikse HPR, p. 310, AIP Press, New York, 1989.

4.2. Appel J, "Polarons," *Solid State Physics, Advances in Research and Applications*, **21**, 193-391 (1968). A comprehensive treatment.

4.3. Arajs S, *American Journal of Physics*, **37** (7), 752 (1969).

4.4. Bergmann DJ, *Physics Reports* 43, 377 (1978).

4.5. Brockhouse BN, *Rev Modern Physics* **67**, 735-751 (1995).

4.6. Callaway J, "Model for Lattice Thermal Conductivity at Low Temperatures," *Physical Review*, **113**, 1046 (1959).

4.7. Feynman RP, Statistical Mechanics, Addison-Wesley Publ. Co., Reading MA, 1972, Chap. 8.

4.8. Fisher ME and Langer JS, "Resistive Anomalies at Magnetic Critical Points," *Physical Review Letters*, **20**(13), 665 (1968).

4.9. Garnett M, *Philos. Trans. R. Soc.* (London), **203**, 385 (1904).

4.10. Geiger Jr. FE and Cunningham FG, "Ambipolar Diffusion in Semiconductors," *American Journal of Physics*, **32**, 336 (1964).

4.11. Halperin BI and Hohenberg PC, "Scaling Laws for Dynamical Critical Phenomena," *Physical Review*, **177**(2), 952 (1969).

4.12. Holland MG, "Phonon Scattering in Semiconductors from Thermal Conductivity Studies," *Physical Review*, **134**, A471 (1964).

4.13. Howarth DJ and Sondheimer EH, Proc. Roy. Soc. **A219**, 53 (1953)

4.14. Jan JP, "Galvanomagnetic and Thermomagnetic Effects in Metals," *Solid State Physics, Advances in Research and Applications*, **5**, 1-96 (1957).

4.15. Kadanoff LP, "Transport Coefficients Near Critical Points," *Comments on Solid State Physics*, **1**(1), 5 (1968).

4.16. Katsnelson AA, Stepanyuk VS, Szász AI, and Farberovich DV, *Computational Methods in Condensed Matter: Electronic Structure*, American Institute of Physics, 1992.

4.17. Kawasaki K, "On the Behavior of Thermal Conductivity Near the Magnetic Transition Point," *Progress in Theoretical Physics (Kyoto)*, **29**(6), 801 (1963).

4.18. Klemens PG, "Thermal Conductivity and Lattice Vibration Modes," *Solid State Physics, Advances in Research and Applications*, **7**, 1-98 (1958).

4.19. Kohn W, *Physical Review*, **126**, 1693 (1962).

4.20. Kohn W, "Nobel Lecture: Electronic Structure of Matter–Wave Functions and density Functionals," *Rev. Modern Phys.* **71**, 1253-1266 (1998)

4.21. Kondo J, "Resistance Minimum in Dilute Magnetic Alloys," *Progress in Theoretical Physics (Kyoto)*, **32**, 37 (1964).

4.22. Kothari LS and Singwi KS, "Interaction of Thermal Neutrons with Solids," *Solid State Physics, Advances in Research and Applications*, **8**, 109-190 (1959).

4.23. Kuper CG and Whitfield GD, *POLARONS AND EXCITONS*, Plenum Press, New York, 1962. There are lucid articles by Fröhlich, Pines, and others here, as well as a chapter by F. C. Brown on experimental aspects of the polaron.

4.24. Langer JS and Vosko SH, *Journal of Physics and Chemistry of Solids*, **12**, 196 (1960).

4.25. MacDonald DKC, "Electrical Conductivity of Metals and Alloys at Low Temperatures," *Encyclopedia of Physics*, Vol. XIV, Low Temperature Physics I, Springer-Verlag, Berlin, 1956, p. 137.

4.26. Madelung O, *Introduction to Solid State Theory*, Springer-Verlag, 1978, pp. 153-155, 183-187, 370-373. A relatively simple and clear exposition of both the large and small polaron.

4.27. Mahan GD, *Many Particle Physics*, Plenum Press, New York, 1981, Chaps. 1 and 6. Green's functions and diagrams will be found here.

4.28. Mattuck RD, *A Guide to Feynman Diagrams in the Many-Body Problem*, McGraw-Hill Book Company, New York, 1967.

4.29. McMillan WL and Rowell JM, *Physical Review Letters*, **14** (4), 108 (1965).

4.30. Mendelssohn K and Rosenberg HM, "The Thermal Conductivity of Metals a Low Temperatures," *Solid State Physics, Advances in Research and Applications*, **12**, 223-274 (1961).

4.31. Mott NF, *Metal-Insulator Transitions*, Taylor and Francis, London, 1990 (2nd edn).

4.32. Olsen JL, *Electron Transport in Metals*, Interscience, New York, 1962.

4.33. Patterson JD, "Modern Study of Solids," *Am. J. Phys.* **32**, 269-278 (1964).

4.34. Patterson JD, "Error Analysis and Equations for the Thermal Conductivity of Composites," *Thermal Conductivity* **18**, Ashworth T and Smith DR (eds), Plenum Press, New York, 1985, pp 733-742.

4.35. Pines D, "Electron Interactions in Metals," *Solid State Physics, Advances in Research and Applications*, **1**, 373-450 (1955).

4.36. Reynolds JA and Hough JM, *Proc. Roy. Soc.* (London), **B70**, 769-775 (1957).

4.37. Sham LJ and Ziman JM, "The Electron-Phonon Interaction," *Solid State Physics, Advances in Research and Applications*, **15**, 223-298 (1963).

4.38. Stratton JA, *Electromagnetic Theory*, McGraw Hill, 1941, p. 211ff.

4.39. Ziman JM, *Electrons and Phonons*, Oxford, London, 1962, Chap. 5 and later chapters (esp. p. 497)

Chapter 5

5.1. Alexander, W. and Street A, *Metals in the Service of Man*, 7th edn. Middlesex, England: Penguin, 1979.

5.2. Blatt FJ, *Physics of Electronic Conduction in Solids*, McGraw-Hill (1968).

5.3. Borg RJ and Dienes GJ, *An Introduction to Solid State Diffusion*, Academic Press, San Diego, 1988, p 148-151.

5.4. Cottrell A, *Introduction to the Modern Theory of Metals*, the Institute of Metals, London, 1988.

5.5. Cracknell AP and Wong KC, *The Fermi Surface: Its Concept, Determination, and Use in the Physics of Metals*, Clarendon Press, Oxford, 1973.

5.6. Duke CB, "Tunneling in Solids," in Supplement 10, Solid State Physics, Advances in Research and Applications (1969).

5.7. Fiks VB, *Sov Phys Solid State*, **1**, 14 (1959).

5.8. Fisk Z et al, "The Physics and Chemistry of Heavy Fermions," *Proc Natl Acad Sci USA* **92**, 6663-6667 (1995).

5.9. Gantmakher VF, "Radio Frequency Size Effect in Metals," *Progress in Low Temperature Physics*, Vol. V, Gorter CJ (ed), North-Holland Publishing Company, Amsterdam, 1967, p. 181.

5.10. Harrison WA, *Applied Quantum Mechanics*, World Scientific, Singapore, 2000, Chap. 21.

5.11. Harrison WA and Webb MB (eds), *The Fermi Surface*, John Wiley and Sons, New York, 1960.

5.12. Huang K, Statistical Mechanics, John Wiley and Sons, 2nd edn, 1987, pp 247-255.

5.13. Huntington HB and Grove AR, *J Phys Chem Solids*, **20**, 76 (1961).

5.14. Kahn AH and Frederikse HPR, "Oscillatory Behavior of Magnetic Susceptibility and Electronic Conductivity," *Solid State Physics, Advances in Research and Applications* **9**, 259-291 (1959).

5.15. Kittel C and Kroemer H, *Thermal Physics*, W. H. Freeman and Company, San Francisco, 2nd edn., 1980, Chap. 11.

5.16. Langenberg DN, "Resource Letter OEPM-1 on the Ordinary Electronic Properties of Metals," *American Journal of Physics*, **36** (9), 777 (1968).

5.17. Lax B and Mavroides JG, "Cyclotron Resonance," *Solid State Physics, Advances in Research and Applications* **11**, 261-400 (1960).

5.18. Lloyd JR, "Electromigration in integrated circuit conductors", *J Phys D: Appl Phys* **32**, R109-R118 (1999).

5.19. Mackintosh AR, *Sci. Am.* **209**, 110, (1963).

5.20. Onsager L, Phil. Mag. **93**, 1006-1008 (1952).

5.21. Overhauser AW, "Charge Density Wave," *Solid State Physics Source Book*, Parker SP (ed), McGraw-Hill Book Co., 1987, pp 142-143.

5.22. Overhauser AW, "Spin-Density Wave," *Solid State Physics Source Book*, Parker SP (ed), McGraw-Hill Book Co., 1987, pp. 143-145

5.23. Peierls R, *More Surprises in Theoretical Physics*, Princeton University Press, NJ, 1991, p29.

5.24. Pippard AB, "The Dynamics of Conduction Electrons," *Low Temperature Physics*, deWitt C, Dreyfus B, and deGennes PG (eds), Gordon and Breach, New York, 1962. Also Pippard AB, *Magnetoresistance in Metals*, Cambridge University Press, 1988.

5.25. Radousky HB, *Magnetism in Heavy Fermion Systems*, World Scientific, Singapore, 2000.

5.26. Shapiro SL and Teukolsky SA, *Black Holes, White Dwarfs and Neutron Stars: The Physics of Compact Objects*, John Wiley and Sons, Inc., New York, 1983.

5.27. Shoenberg D, *Magnetic Oscillations in Metals*, Cambridge University Press, 1984.

5.28. Sorbello RS, "Theory of Electromigration," *Solid State Physics, Advances in Research and Applications* **51**, 159-231 (1997).

5.29. Stark RW and Falicov LM, "Magnetic Breakdown in Metals," *Progress in Low Temperature Physics*, Vol. V, Gorter CJ (ed), North-Holland Publishing Company, Amsterdam, 1967, p. 235.

5.30. Stewart GR, "Heavy-Fermion Systems," *Rev Modern Physics*, **56**, 755-787 (1984).

5.31. Thorne RE, "Charge Density Wave Conductors," *Physics Today*, pp 42-47 (May 1996).

5.32. Wigner E and Huntington HB, *J Chem Phys* **3**, 764-770 (1935).

5.33. Wilson AH, *The Theory of Metals*, Cambridge, 1954.

5.34. Zak J, "Dynamics of Electrons in Solids in External Fields," I, *Physical Review*, **168**(3), 686 (1968); II, *Physical Review*, **177**(3), 1151 (1969). Also "The kq-Representation in the Dynamics of Metals" *Solid State Physics, Advances in Research and Applications* **27**, 1-62 (1972).
5.35. Ziman JM, *Principles of Theory of Solids*, 2nd edn., Cambridge, 1972.
5.36. Ziman JM, *Electrons in Metals, A Short Guide to the Fermi Surface*, Taylor and Francis, London, 1963.

Chapter 6

6.1. Alferov ZI, "Nobel Lecture: The Double Heterostructure Concept and its application in Physics, Electronics, and Technology," *Rev. Modern Phys.* **73**(3), 767-782 (2001).
6.2. Ashcroft NW and Mermin ND, *Solid State Physics*, Holt, Rinehart and Winston, New York, 1976, Chapters 28 and 29.
6.3. Bardeen J, "Surface States and Rectification at a Metal-Semiconductor Contact", *Physical Review*, **71**, 717-727 (1947).
6.4. Blakemore JS, *Solid State Physics*, Second Edition, W. B. Saunders Co., Philadelphia, 1974.
6.5. Boer KW, Survey of Semiconductor Physics, Electrons and Other Particles in Bulk Semiconductors, Van Nostrand Reinhold, New York, 1990.
6.6. Bube R, *Electronics in Solids*, Academic Press, Inc., New York, 1992 3rd edn.
6.7. Chen A and Sher A, *Semiconductor Alloys*, Plenum Press, New York, 1995.
6.8. Cohen ML and Chelikowsky JR, *Electronic Structure and Optical Properties of Semiconductors*, Springer-Verlag, Berlin, 2nd edn, 1989.
6.9. Conwell E and Weisskopf VF, *Physical Review*, **77**, 388 (1950).
6.10. Dalven R, *Introduction to Applied Solid State Physics*, Plenum Press, New York, 1980. See also second edition, 1990.
6.11. Dresselhaus G, Kip AF and Kittel C, *Phys Rev* **98**, 368 (1955).
6.12. Einspruch NG, "Ultrasonic Effects in Semiconductors," *Solid State Physics, Advances in Research and Applications* **17**, 217-268 (1965).
6.13. Fan HY, "Valence Semiconductors, Ge and Si," *Solid State Physics, Advances in Research and Applications* **1**, 283-265 (1955).
6.14. Fraser DA, *The Physics of Semiconductor Devices*, Clarendon Press, Oxford, 4th edition, 1986.
6.15. Handler P, "Resource Letter Scr-1 on Semiconductors," *American Journal of Physics*, **32** (5), 329 (1964).
6.16. Kane EO, *J. Phys. Chem. Solids* **1**, 249 (1957).
6.17. Kittel C, *Introduction to Solid State Physics*, Seventh Edition, John Wiley and Sons, New York, 1996, Chap. 8.
6.18. Kohn W, "Shallow Impurity States in Si and Ge," *Solid State Physics, Advances in Research and Applications* **5**, 257-320 (1957).
6.19. Kroemer H, "Nobel Lecture: Quasielectronic Fields and Band Offsets: Teaching Electrons New Tricks," *Rev. Modern Phys.* **73**(3), 783-793 (2001).
6.20. Li M-F, "Modern Semiconductor Quantum Physics," *World Scientific*, Singapore, 1994.

6.21. Long D, *Energy Bands in Semiconductors*, Interscience Publishers, New York, 1968.

6.22. Ludwig GW and Woodbury HH, "Electron Spin Resonance in Semiconductors," *Solid State Physics, Advances in Research and Applications* **13**, 223-304 (1962).

6.23. McKelvey JP, *Solid State and Semiconductor Physics,* Harper and Row Publishers, New York, 1966.

6.24. Merzbacher E, *Quantum Mechanics*, 2nd edn., John Wiley & Sons, Inc., New York, 1970, Chap. 2.

6.25. Moss TS (ed), *Handbook on Semiconductors*, Vol. 1, Landberg PT (ed), North Holland/Elsevier (1992), Amsterdam (There are additional volumes).

6.26. Nakamura S, Pearton S, Fasol G, *The Blue Laser Diode: The Complete Story*, Springer-Verlag, New York, 2000.

6.27. Ovshinsky SR, "Reversible Electrical Switching Phenomena in Disordered Structures," *Physical Review Letters*, **21**, 1450 (1968).

6.28. Pankove JI and Moustaka TD (eds), "Gallium Nitride I," *Semiconductors and Semimetals*, Vol. 50, Academic Press, New York, 1997.

6.29. Pantiledes ST (editor), *Deep Centers in Semiconductors*, Gordon and Breach Publishers, Yverdon, Switzerland, 1992.

6.30. Patterson JD, "Narrow Gap Semiconductors," *Condensed Matter News* **3** (1), 4-11 (1994).

6.31. Perkowitz S, *Optical Charaterization of Semiconductors*, Academic Press, San Diego, 1993.

6.32. Ridley BK, *Quantum Processes in Semiconductors*, Clarendon Press, Oxford, 1988.

6.33. Sapoval B and Hermann C, *Physics of Semiconductors*, Springer-Verlag, New York, 1995.

6.34. Seeger K, *Semiconductor Physics*, Springer-Verlag, Berlin, 4th edn, 1989.

6.35. Seitz F, *Physical Review*, **73**, 549 (1948).

6.36. Shockley W, *Electrons and Holes in Semiconductors*, D. Van Nostrand, New York, 1950.

6.37. Slater JC, *Quantum Theory of Molecules and Solids*, Vol. III, *Insulators, Semiconductors, and Metals,* McGraw-Hill Book Company, New York, 1967.

6.38. Smith RA, *Wave Mechanics of Crystalline Solids*, John Wiley and Sons, 1961, section 8.8 and appendix 1.

6.39. Smith RA (ed), *Semiconductors*, Proceedings of the International School of Physics, "Enrico Fermi" Course XXII, Academic Press, New York, 1963.

6.40. Streetman BG, *Solid State Electronic Devices*, 2nd ed. Prentice Hall, Englewood Cliffs, N.J., 1980. Also see the third edition (1990).

6.41. Sze SM, *Physics of Semiconductor Devices*, 2nd edn, Wiley, New York, 1981.

6.42. Sze SM (ed), *Modern Semiconductor Device Physics*, John Wiley and Sons, Inc., New York, 1998.

6.43. Willardson RK, and Weber ER, "Gallium Nitride II," *Semiconductors and Semimetals*, Vol. 57, Academic Press, 1998.

6.44. Yu PY and Cardona M, *Fundamentals of Semiconductors*, Springer Verlag, Berlin, 1996.

Chapter 7

7.1. Anderson PW, "Theory of Magnetic Exchange Interactions: Exchange in Insulators and Semiconductors," *Solid State Physics, Advances in Research and Applications*, **14**, 99-214 (1963).

7.2. Ashcroft NW and Mermin ND, *Solid State Physics*, Holt, Rinehart and Winston, New York, 1976, Chaps. 31, 32 and 33.

7.3. Auld BA, "Magnetostatic and Magnetoelastic Wave Propagation in Solids", *Applied Solid State Science*, Vol. 2, Wolfe R and Kriessman CJ (eds), Academic Press 1971.

7.4. Baibich MN, Broto JM, Fert A, Nguyen Van Dau F, Petroff F, Eitenne P, Creuzet G, Friederich A, and Chazelas J, *Phys. Rev. Lett.*, **61**, 2472 (1988).

7.5. Bennett C, "Quantum Information and Computation," *Physics Today*, October 1995, pp. 24-30.

7.6. Bertram HN, *Theory of Magnetic Recording*, Cambridge University Press, 1994, Chap. 2.

7.7. Bitko D et.al., *J Research of NIST*, **102**(2), 207-211 (1997)).

7.8. Blackman JA and Tagüeña J, *Disorder in Condensed Matter Physics, A Volume in Honour of Roger Elliott*, Clarendon Press, Oxford, 1991.

7.9. Blundell S, *Magnetism in Condensed Matter*, Oxford University Press, 2001.

7.10. Charap SH and Boyd EL, *Physical Review*, **133**, A811 (1964).

7.11. Chikazumi S, *Physics of Ferromagnetism*, (Translation editor, Graham CD), Oxford at Clarendon Press, 1977.

7.12. Chowdhury D, *Spin Glasses and Other Frustrated Systems*, Princeton University Press, 1986.

7.13. Cooper B, "Magnetic Properties of Rare Earth Metals," *Solid State Physics, Advances in Research and Applications*, **21**, 393-490 (1968).

7.14. Cracknell AP and Vaughn RA, *Magnetism in Solids Some Current Topics*, Scottish Universities Summer School, 1981.

7.15. Craik D, *Magnetism Principles and Applications*, John Wiley and Sons, 1995.

7.16. Cullity BD, *Introduction to Magnetic Materials*, Addison-Wesley, Reading, Mass., 1972.

7.17. Damon R and Eshbach J, *J Phys Chem Solids*, **19**, 308 (1961).

7.18. Dyson FJ, *Physical Review*, **102**, 1217 (1956).

7.19. Elliott RJ, *Magnetic Properties of Rare Earth Metals*, Plenum Press, London, 1972.

7.20. Fetter AL and Walecka JD, *Theoretical Mechanics of Particles and Continua*, McGraw-Hill, pp. 399-402, 1980.

7.21. Fisher ME, "The Theory of Equilibrium Critical Phenomena," *Reports on Progress in Physics*, XXX(II), 615 (1967).

7.22. Fischer KH and Hertz JA, *Spin Glasses*, Cambridge University Press, 1991.

7.23. Fontcuberta J, "Colossal Magnetoresistance," *Physics World*, February 1999, pp. 33-38.

7.24. Gibbs MRJ (ed), *Modern Trends in Magnetostriction Study and Application*, Kluwer Academic Publishers, Dordrecht, 2000.

7.25. Gilbert W, *De Magnete* (originally published in 1600), Translated by P. Fleury Mottelay, Dover, New York (1958).

7.26. Griffiths RB, *Physical Review*, **136**(2), 437 (1964).

7.27. Heitler W, *Elementary Wave Mechanics*, Oxford, 1956, 2nd edn, Chap. IX.

7.28. Heller P, "Experimental Investigations of Critical Phenomena," *Reports on Progress in Physics*, XXX(II), 731 (1967).

7.29. Herbst JF, *Rev Modern Physics* **63**(4), 819-898 (1991).

7.30. Herring C, *Exhange Interactions among Itinerant Electrons in Magnetism*, Rado GT and Suhl H (eds), Academic Press, New York, 1966.

7.31. Herzfield CM and Meijer HE, "Group Theory and Crystal Field Theory," *Solid State Physics, Advances in Research and Applications*, **12**, 1-91 (1961).

7.32. Huang K, *Statistical Mechanics*, 2nd edn, John Wiley and Sons, New York, 1987.

7.33. Ibach H and Luth H, *Solid State Physics*, Springer-Verlag, Berlin, 1991, p 152.

7.34. Julliere M, *Phys Lett* **54A**, 225 (1975).

7.35. Kadanoff LP et al, *Reviews of Modern Physics*, **39** (2), 395 (1967).

7.36. Kasuya T, *Progress in Theoretical Physics (Kyoto)*, **16**, 45 and 58 (1956).

7.37. Keffer F, "Spin Waves," *Encyclopedia of Physics*, Vol. XVIII, Part 2, Ferromagnetism, Springer-Verlag, Berlin, 1966.

7.38. Kittel C, "Magnons," *Low Temperature Physics*, DeWitt C, Dreyfus B, and deGennes PG (eds), Gordon and Breach, New York, 1962.

7.39. Kittel C, *Introduction to Solid State Physics*, 7th edn, John Wiley and Sons, New York, 1996, Chapters 14, 15, and 16.

7.40. Kosterlitz JM and Thouless DJ, *J Phys C* **6**, 1181 (1973).

7.41. Kouwenhoven L and Glazman L, *Physics World*, pp. 33-38, Jan. 2001.

7.42. Langer JS and S. H. Vosko, *J Phys Chem Solids* **12**, 196 (1960).

7.43. Levy RA and Hasegawa R, *Amorphous Magnetism II*, Plenum Press, New York, 1977.

7.44. Malozemoff AP and Slonczewski JC, *Magnetic Domain Walls in Bubble Materials*, Academic Press, New York, 1979.

7.45. Manenkov AA and Orbach R (eds), *Spin-Lattice Relaxation in Ionic Solids,* Harper and Row Publishers, New York, 1966.

7.46. Marshall W (ed), *Theory of Magnetism in Transition Metals,* Proceedings of the International School of Physics, "Enrico Fermi" Course XXXVII, Academic Press, New York, 1967.

7.47. Mathews J and Walker RL, *Mathematical Methods of Physics*, W. A. Benjamin, New York, 1967.

7.48. Mattis DC, The Theory of Magnetism I Statics and Dynamics, Springer-Verlag, 1988 and II Thermodynamics and Statistical Mechanics, Springer-Verlag 1985.

7.49. Mermin ND and Wagner H, *Physical Review Letters*, **17**(22), 1133 (1966).

7.50. Muller B and Reinhardt J, *Neural Networks, An Introduction*, Springer-Verlag, Berlin, 1990.

7.51. Pake GE, "Nuclear Magnetic Resonance," *Solid State Physics, Advances in Research and Applications*, **2**, 1-91 (1956).

7.52. Parkin S, *J App Phys* **85**, 5828 (1999).

7.53. Patterson JD, *Introduction to the Theory of Solid State Physics*, Addison-Wesley Publishing Co., Reading MA, 1971 p176ff.

7.54. Patterson JD et al, *Journal of Applied Physics*, **39** (3), 1629 (1968), and references cited therein.

7.55. Prinz GA, Science Vol. 282, 27 Nov. 1998, p 1660.

7.56. Rado GT and Suhl H (eds), Vol. II Part A, *Statistical Models, Magnetic Symmetry, Hyperfine Interactions, and Metals*, Academic Press, New York, 1965. Vol. IV, *Exchange Interactions among Itinerant Electrons* by Conyers Herring, Academic Press, New York, 1966.

7.57. Ruderman MA and Kittel C, *Physical Review*, **96**, 99 (1954).[1]

7.58. Salamon MB and Jaime M, "The Physics of Manganites Structure and Transport," *Rev. Modern Physics*, **73**, 583-628 (2001).

7.59. Schrieffer JR, "The Kondo Effect–The Link Between Magnetic and NonMagnetic Impurities in Metals?", *Journal of Applied Physics*, **38**(3), 1143 (1967).

7.60. Slichter CP, *Principles of Magnetic Resonance*, Harper and Row, Evanston,1963.

7.61. Slonczewski JC, *Phys Rev* **B39**, 6995 (1989).

7.62. Tyalblikov SV, *Methods in the Quantum Theory of Magnetism*, Plenum Press, New York, 1967.

7.63. Van Vleck JH, *The Theory of Electric and Magnetic Susceptibilities*, Oxford University Press, 1932.

7.64. Von der Lage FC and Bethe HA, *Phys Rev* **71**, 612 (1947).

7.65. Walker LR, *Phys Rev* **105**, 309 (1957).

7.66. Waller I, *Z Physik*, **79**, 370 (1932).

7.67. Weinberg S, *The Quantum Theory of Fields, Vol I Modern Applications*, Cambridge University Press, 1996, pp. 332-352.

7.68. White RM, *Quantum Theory of Magnetism*, McGraw Hill, New York, 1970.

7.69. Wohlfarth EP, *Magnetism Vol. III*, Rado GT and Suhl H, (eds), Academic Press, New York, 1963. Wohlfarth EP, *Rev Mod Phys.* **25**, 211 (1953). Wohlfarth EP, *Handbook of Magnetic Materials*, Elsevier, several volumes (~1993).

7.70. Wojtowicz PJ, *Journal of Applied Physics*, **35**, 991 (1964).

7.71. Yosida K, *Physical Review*, **106**, 893 (1957).[1]

7.72. Yosida K, *Theory of Magnetism, Springer*, Berlin, 1998.

7.73. Zutic I et al, "Spintronics: Fundamentals and applications," *Rev Mod Phys* **76**, 325 (2004).

Chapter 8

8.1. Allen PB and Mitrovic B, "Theory of Superconductivity T_c," *Solid State Physics, Advances in Research and Applications*, **37**, 2-92 (1982).

8.2. Anderson PW, "The Josephson Effect and Quantum Coherence Measurements in Superconductors and Superfluids," *Progress in Low Temperature Physics*, Vol. V, Gorter CJ (ed), North-Holland Publishing Company, Amsterdam, 1967, p. 1. See also Annett JF, *Superconductivity, Superfluids, and Condensates*, Oxford University Press, 2004.

8.3. Annett JF, *Superconductivity, Superfluids, and Condensates*, Oxford University Press, 2004.

[1] These papers deal with the indirect interaction of nuclei by their interaction with the conduction electrons and of the related indirect interaction of ions with atomic magnetic moments by their interaction with the conduction electrons. In the first case the hyperfine interaction is important and in the second the exchange interaction is important.

8.4. Bardeen J, "Superconductivity," *1962 Cargese Lectures in Theoretical Physics*, Levy M (ed), W. A. Benjamin, New York, 1963.

8.5. Bardeen J and Schrieffer JR, "Recent Developments in Superconductivity," *Progress in Low Temperature Physics*, Vol. III, Gorter CJ (ed), North-Holland Publishing Company, Amsterdam, 1961, p. 170.

8.6. Bardeen J, Cooper LN, and Schreiffer JR, "Theory of Superconductivity," *Phys Rev* **108**, 1175-1204 (1957).

8.7. Beyers R and Shaw TM, "The Structure of $Y_1Ba_2Cu_3O_{7-\delta}$ and its Derivatives," *Solid State Physics, Advances in Research and Applications*, **42**, 135-212 (1989).

8.8. Bogoliubov NN (ed), *The Theory of Superconductivity*, Gordon and Breoch, New York, 1962.

8.9. Burns G, *High Temperature Superconductivity an Introduction*, Academic Press, Inc., Boston, 1992.

8.10. Cooper L, "Bound Electron Pairs in a Degenerate Fermi Gas," *Phys Rev* **104**, 1187-1190 (1956).

8.11. Dalven R, *Introduction to Applied Solid State Physics*, Plenum Press, New York, 1990, 2nd edn, Ch. 8.

8.12. deGennes PG, *Superconductivity of Metals and Alloys*, W. A. Benjamin, New York, 1966.

8.13. Feynman RP, Leighton RB, and Sands M, *The Feynman Lectures on Physics*, Vol. III, Addison-Wesley Publishing Co., 1965.

8.14. Giaever I, "Electron Tunneling and Superconductivity," *Rev Modern Phys*, **46**(2), 245-250 (1974).

8.15. Goodman BB, "Type II Superconductors," *Reports on Progress in Physics*, Vol. XXIX, Part 11, The Institute of Physics and The Physical Society, London, 1966, p. 445.

8.16. Hass KC, "Electronics Structure of Copper-Oxide Superconductors," *Solid State Physics, Advances in Research and Applications*, **42**, 213-270 (1989).

8.17. Jones W and March NH, *Theoretical Solid State Physics*, Vol. 2, Dover, 1985, p895ff and p1151ff.

8.18. Josephson B, "The Discovery of Tunneling Supercurrents," *Rev Modern Phys*, **46**(2), 251-254 (1974).

8.19. Kittel C, [60], Chap. 8.

8.20. Kuper CG, *An Introduction to the Theory of Superconductivity*, Clarendon Press, Oxford, 1968.

8.21. Mahan G, *Many-Particle Physics*, Plenum Press, New York, 1981, Ch. 9.

8.22. Marder, *Condensed Matter Physics*, John Wiley and Sons, 2000, p. 819.

8.23. Mattuck RD, *A Guide to Feynman Diagrams in the Many-Body Problem*, 2nd edn, Dover, New York, 1992, Chap. 15.

8.24. Parker WH, Taylor BN and Langenberg DN, *Physical Review Letters*, **18** (8), 287 (1967).

8.25. Parks RD (ed), *Superconductivity*, Vols. 1 and 2, Marcel Dekker, New York, 1969.

8.26. Rickaysen G, *Theory of Superconductivity*, Interscience, New York, 1965.

8.27. Saint-James D, Thomas EJ, and Sarma G, Type II Superconductivity, Pergamon, Oxford, 1969.

8.28. Scalapino DJ, "The Theory of Josephson Tunneling," *Tunneling Phenomena in Solids*, Burstein E and Lundquist S (eds), Plenum Press, New York, 1969.

8.29. Schafroth MR, "Theoretical Aspects of Superconductivity," *Solid State Physics, Advances in Research and Applications*, **10**, 293-498 (1960).
8.30. Schrieffer JR, *Theory of Superconductivity*, W. A. Benjamin, New York, 1964.
8.31. Silver AH and Zimmerman JE, *Phys Rev* **157**, 317 (1967).
8.32. Tinkham M, Introduction to Superconductivity, McGraw-Hill, New York, 2nd edn, 1996.
8.33. Tinkham M and Lobb CJ, "Physical Properties of the New Superconductors," *Solid State Physics, Advances in Research and Applications*, **42**, 91-134 (1989).

Chapter 9

9.1. Bauer S, Gerhard-Multhaupt R, and Sessler GM, "Ferroelectrets: Soft Electroactive Foams for Transducers," *Physics Today* **57**, 39-43 (Feb. 2004).
9.2. Böttcher CJF, *Theory of Electric Polarization*, Elsevier Publishing Company, New York, 1952.
9.3. Brown WF Jr, "Dielectrics," *Encyclopedia of Physics*, Vol. XVII, Flügge S (ed), Springer-Verlag, Berlin, 1956.
9.4. Devonshire AF, "Some Recent Work on Ferroelectrics," *Reports on Progress in Physics*, Vol. XXVII, The Institute of Physics and The Physical Society, London, 1964, p. 1.
9.5. Elliot RJ and Gibson AF, *An Introduction to Solid State Physics and its Applications*, Harper and Row 1974 p277ff.
9.6. Fatuzzo E and Merz WJ, *Ferroelectricity*, John Wiley and Sons, New York, 1967.
9.7. Forsbergh PW Jr., "Piezoelectricity, Electrostriction, and Ferroelectricity," *Encyclopedia of Physics*, Vol. XVII, Flügge S (ed), Springer Verlag, Berlin, 1956.
9.8. Fröhlich H, *Theory of Dielectrics—Dielectric Constant and Dielectric Loss*, Oxford University Press, New York, 1949.
9.9. Gutmann F, *Rev Modern Phys* **70**, 457 (1948).
9.10. Jona F and Shirane G, *Ferroelectric Crystals*, Pergamon Press, New York, 1962.
9.11. Kanzig W, "Ferroelectrics and Antiferroelectrics," *Solid State Physics, Advances in Research and Applications* **4**, 1-97 (1957).
9.12. Lines ME and Glass AM, Principles and Applications of Ferroelectrics and Related Materials, Oxford, 1977.
9.13. Moss TS, *Optical Properties of Semi-Conductors*, Butterworth and Company Pubs., London, 1959.
9.14. Pines D, "Electron Interaction in Metals," *Solid State Physics, Advances in Research and Applications* **1**, 373-450 (1955).
9.15. Platzman PM and Wolff PA, *Waves and Interactions in Solid State Plasmas*, Academic Press, New York, 1973, Chaps. VI and VII.
9.16. Smyth CP, *Dielectric Behavior and Structure*, McGraw-Hill Book Company, New York, 1955.
9.17. Samara GA and Peercy PS, "The Study of Soft-Mode Transitions at High Pressure," *Solid State Physics, Advances in Research and Applications* **36**, 1-118 (1981).
9.18. Steele MC and Vural B, *Wave Interactions in Solid State Plasmas*, McGraw-Hill Book Company, New York, 1969.

9.19. Tonks L and Langmuir I, *Phys Rev* **33**, 195-210 (1929).
9.20. Uehling EA, "Theories of Ferroelectricity in KH_2P0_4," in *Lectures in Theoretical Physics*, Vol. V, Briton WE, Downs BW, and Downs J (eds), Interscience Publishers, New York, 1963.
9.21. Zheludev IS, "Ferroelectricity and Symmetry," *Solid State Physics, Advances in Research and Applications* **26**, 429-464 (1971).

Chapter 10

10.1. Born M and Wolf E, Principles of Optics, 2nd (Revised) edn, MacMillan, 1964, especially Optics of Metal (Chap. XIII) and Optics of Crystals (Chap. XIV).
10.2. Born M and Huang K, *Dynamical Theory of Crystal Lattices*, Oxford at the Clarenden Press, 1954, see especially the optical effects (Chap. VII).
10.3. Brown FC, "Ultraviolet Spectroscopy of Solids with the Use of Synchrotron Radiation," *Solid State Physics, Advances in Research and Applications* **29**, 1-73 (1974).
10.4. Bube RH, *Photoconductivity of Solids*, John Wiley and Sons, New York, 1960.
10.5. Caldwell DJ, "Some Observations of the Faraday Effect," *Proc Natl Acad Sci* **56**, 1391-1398 (1966).
10.6. Callaway J, "Optical Absorption in an Electric-Field," *Physical Review*, **130**(2), 549 (1963).
10.7. Cochran W, *The Dynamics of Atoms in Crystals*, Edward Arnold, London, 1973, p 90.
10.8. Cohen MH, *Phil Mag* **3**, 762 (1958)
10.9. Cohen ML and Chelikowsky JR, *Electronic Structure and Optical Properties of Semiconductors*, 2nd edn, Springer-Verlag, Berlin, 1989.
10.10. Dexter DL, "Theory of the Optical Properties of Imperfections in Nonmetals," *Solid State Physics, Advances in Research and Applications* **6**, 353-411 (1958).
10.11. Elliott RJ and Gibson AF, *An Introduction to Solid State Physics*, Macmillan, 1974, Chap. 6, 7.
10.12. Fox M, *Optical Properties of Solids*, Oxford University Press, 2002.
10.13. Frova A, Handler P, Germano FA, and Aspnes DE, "Electro-Absorption Effects at the Band Edges of Silicon and Germanium," *Physical Review*, **145**(2), 575 (1966).
10.14. Givens MP, "Optical Properties of Metals," *Solid State Physics, Advances in Research and Applications* **6**, 313-352 (1958).
10.15. Gobeli GW and Fan HY, *Phys Rev* **119**(2), 613-620, (1960).
10.16. Greenaway DL and Harbeke G, *Optical Properties and Band Structures of Semiconductors*, Pergamon Press, Oxford, 1968.
10.17. Hagen E and Rubens H, *Ann. d. Physik* (4) **11**, 873 (1903).
10.18. Kane EO, *J Phys Chem Solids* **12**, 181 (1959).
10.19. Knox RS, "Theory of Excitons," *Solid State Physics, Advances in Research and Applications*, Supplement 5, 1963.
10.20. Lyddane RH, Sachs RG, and Teller E, *Phys Rev* **59**, 673 (1941).
10.21. Moss TS, *Optical Properties of Semiconductors*, Butterworth, London, 1961.
10.22. Pankove JL, *Optical Processes in Semiconductors*, Dover, New York, 1975.

10.23. Phillips JC, "The Fundamental Optical Spectra of Solids," *Solid State Physics, Advances in Research and Applications* **18**, 55-164 (1966).
10.24. Stern F, "Elementary Theory of the Optical Properties of Solids," *Solid State Physics, Advances in Research and Applications* **15**, 299-408 (1963).
10.25. Tauc J (ed), *The Optical Properties of Solids*, Proceedings of the International School of Physics, "Enrico Fermi" Course XXXIV, Academic Press, New York, 1966.
10.26. Tauc J, *The Optical Properties of Semiconductors*, Academic Press, New York, 1966.
10.27. Yu PY and Cordona M, *Fundamentals of Semiconductors*, Springer-Verlag, Berlin, 1996, chapters 6,7,8.

Chapter 11

11.1. Bastard G, Wave Mechanics Applied to Semiconductor Heterostructures, Halsted (1988)
11.2. Borg RJ and Dienes GJ, *An Introduction to Solid State Diffusion*, Academic Press, San Diego, 1988.
11.3. Bube RH, "Imperfection Ionization Energies in CdS-Type Materials by Photoelectronic Techniques," *Solid State Physics: Advances in Research and Applications* **11**, 223-260 (1960).
11.4. Chelikowsky JR and Louie SG, *Quantum Theory of Real Materials*, Kluwer Academic Publishers, Dordrecht, 1996.
11.5. Compton WD and Rabin H, "F-Aggregate Centers in Alkali Halide Crystals," *Solid State Physics: Advances in Research and Applications* **16**, 121-226 (1964).
11.6. Cottrell AH, *Dislocations and Plastic Flow in Crystals*, Oxford University Press, New York, 1953.
11.7. Crawford JH Jr. and Slifkin LM, *Point Defects in Solids, Vol. 1 General and Ionic Crystals, Vol. 2 Defects in Semiconductors*, Plenum Press, New York, Vol. 1 (1972), Vol. 2 (1975).
11.8. Davison SG and Steslicka M, *Basic Theory of Surface States*, Clarendon Press, Oxford, 1992, p. 155.
11.9. deWit R, "The Continuum Theory of Stationary Dislocations," *Solid State Physics: Advances in Research and Applications* **10**, 249-292 (1960).
11.10. Dexter DL, "Theory of Optical Properties of Imperfections in Nonmetals," *Solid State Physics: Advances in Research and Applications* **6**, 353-411 (1958).
11.11. Eshelby JD, "The Continuum Theory of Lattice Defects," *Solid State Physics: Advances in Research and Applications* **3**, 79-144 (1956).
11.12. Fowler WB (ed), *Physics of Color Centers*, Academic Press, New York, 1968.
11.13. Gilman JJ and Johnston WG, "Dislocations in Lithium Fluoride Crystals," *Solid State Physics: Advances in Research and Applications* **13**, 147-222 (1962).
11.14. Gundry PM and Tompkins FC, "Surface Potentials," in *Experimental Methods of Catalysis Research*, Anderson RB (ed), Academic Press, New York, 1968, pp. 100-168.

11.15. Gourary BS and Adrian FJ, "Wave Functions for Electron-Excess Color Centers in Alkali Halide Crystals," *Solid State Physics: Advances in Research and Applications* **10**, 127-247 (1960).
11.16. Henderson B, *Defects in Crystalline Solids*, Crane, Russak and Company, Inc., New York, 1972.
11.17. Kohn W, "Shallow Impurity States in Silicon and Germanium," *Solid State Physics: Advances in Research and Applications* **5**, 257-320 (1957).
11.18. Kröger FA and Vink HJ, "Relations between the Concentrations of Imperfections in Crystalline Solids," *Solid State Physics: Advances in Research and Applications* 3, 307-435 (1956).
11.19. Lehoczky SL et al, NASA CR-101598, "Advanced Methods for Preparation and Characterization of Infrared Detector Materials, Part I," July 5, 1981.
11.20. Li W and Patterson JD, "Deep Defects in Narrow-Gap Semiconductors," *Phys Rev* **50**, 14903-14910 (1994).
11.21. Li W and Patterson JD, "Electronic and Formation Energies for Deep Defects in Narrow-Gap Semiconductors," *Phys Rev* **53**, 15622-15630 (1996), and references cited therein.
11.22. Luttinger JM and Kohn W, "Motion of Electrons and Holes in Perturbed Periodic Fields," *Phys Rev* **99**, 869-883 (1955).
11.23. Madelung O, *Introduction to Solid-State Theory*, Springer-Verlag, Berlin (1978), Chaps. 2 and 9.
11.24. Markham JJ, "F-Centers in the Alkali Halides," *Solid State Physics: Advances in Research and Applications*, Supplement 8 (1966).
11.25. Mitin V, Kochelap VA, and Stroscio MA, *Quantum Heterostructures*, Cambridge University Press, 1999.
11.26. Pantelides ST, *Deep Centers in Semiconductors*, 2nd edn, Gordon and Breach, Yverdon, Switzerland, (1992).
11.27. Sarid D, *Exploring Scanning Probe Microscopy with Mathematica*, John Wiley and Sons, Inc., 1997, Chap. 11.
11.28. Schulman JH and Compton WD, *Color Centers in Solids*, The Macmillan Company, New York, 1962.
11.29. Seitz F and Koehler JS, "Displacement of Atoms During Irradiation," *Solid State Physics: Advances in Research and Applications* **2**, 305-448 (1956).
11.30. Stoneham HM, *Theory of Defects in Solids*, Oxford, 1973.
11.31. Wallis RF (ed), *Localized Excitations in Solids*, Plenum Press, New York, 1968.
11.32. West AR, *Solid State Chemistry and its Applications*, John Wiley and Sons, New York, 1984.
11.33. Zanquill A, *Physics at Surfaces*, Cambridge University Press, 1988, p. 293.

Chapter 12

12.1. Barnham K and Vvendensky D, *Low Dimensional Semiconductor Structures*, Cambridge University Press, Cambridge, 2001.
12.2. Bastard G, *Wave Mechanics Applied to Semiconductor Heterostructures*, Halsted Press, New York, 1988.

12.3. Blakemore JS, *Solid State Physics*, 2nd edn, W. B. Saunders Company, Philadelphia, 1974, p. 168.
12.4. Brown TL, LeMay, HE Jr., and Bursten BE, *Chemistry The Central Science*, 6th edn, Prentice Hall, Englewood Cliff, NJ 07632, 1994.
12.5. Bullis WM, Seiler DG, and Diebold AC (eds), *Semiconductor Characterization—Present Status and Future Needs*, AIP Press, Woodbury, New York, 1996.
12.6. Butcher P, March NH, and Tosi MP, *Physics of Low-Dimensional Semiconductor Structures*, Plenum Press, New York, 1993.
12.7. Callen HB, *Thermodynamics and an introduction to Thermostatistics*, John Wiley and Sons, New York, 1985, p339ff.
12.8. Capasso F and Datta S, "Quantum Electron Devices," *Physics Today* **43**, 74-82 (1990).
12.9. Capasso F, Gmachl C, Siveo D, and Cho A, "Quantum Cascade Lasers," *Physics Today* **55**, 34-40 (May 2002).
12.10. Cargill GS, "Structure of Metallic Alloy Glasses," *Solid State Physics, Advances in Research and Applications* **30**, 227-320 (1975).
12.11. Chaikin PM and Lubensky TC, *Principles of Condensed Matter Physics*, Cambridge University Press, 1995.
12.12. Chen CT and Ho KM, Chap. 20 "Metal Surface Reconstructions" in Chelikowsky JR and Louie SG, *Quantum Theory of Real Materials*, Kluwer Academic Publishers, Dordrecht, 1996.
12.13. Davies JH and Long AR (eds), *Nanostructures*, Scottish Universities Summer School and Institute of Physics, Bristol and Philadelphia, 1992.
12.14. Davison SG and Steslika M, *Basic Theory of Surface States*, Clarendon Press, Oxford, 1992.
12.15. deGennes PG and Prost J, *The Physics of Liquid Crystals*, Clarendon Press, Oxford, 2nd edn (1993).
12.16. Doi M and Edwards SF, *The Theory of Polymer Dynamics*, Oxford University, Oxford, 1986.
12.17. Dresselhaus MS, Dresselhaus G, and Avouris P, *Carbon Nanotubes*, Springer-Verlag, 2000.
12.18. Esaki L and Tsu R, *IBM J Res Devel* **14**, 61 (1970).
12.19. Fergason JL, *The Scientific American*, 74 (Aug. 1964).
12.20. Fisher KH and Hertz JA, *Spin Glasses*, Cambridge University Press, 1991, see especially p. 55, pp. 346-353.
12.21. Gaponenko SV, *Optical Properties of Semiconductor Nanocrystals*, Cambridge University Press, 1998.
12.22. Girvin S, "Spin and Isospin: Exotic Order in Quantum Hall Ferromagnets," *Physics Today*, 39-45 (June 2000).
12.23. Grahn HT (ed), *Semiconductor Superlattices–Growth and Electronic Properties*, World Scientific, Singapore, 1998.
12.24. Halperin BI, "The Quantized Hall Effect", *Scientific American*, April 1986, 52-60.
12.25. Hebard A, "Superconductivity in Doped Fullerenes," *Physics Today*, 26-32 (November, 1992).
12.26. Herbst JF, "$R_2Fe_{14}B$ Materials, Intrinsic Properties and Technological Aspects," *Rev Modern Physics* **63**, 819-898 (1991).
12.27. Isihara A, *Condensed Matter Physics*, Oxford University Press, New York, 1991.
12.28. Jacak L, Hawrylak P, Wojs A, *Quantum Dots*, Springer, Berlin 1998.

12.29. Jain JK, "The Composite Fermion: A Quantum Particle and its Quantum Fluid", *Physics Today*, April 2000, 39-45.

12.30. Jones RAL, *Soft Condensed Matter*, Oxford University Press, 2002.

12.31. Joyce BA, *Rep Prog Physics* **48**, 1637 (1985).

12.32. Kastner M, "Artificial Atoms," *Physics Today* **46**(1), 24-31 (Jan. 1993).

12.33. Kelly MJ, Low Dimensional Semiconductors–Materials, Physics, Technology, Devices, Clarendon Press, Oxford, 1995.

12.34. Kivelson S, Lee D-H and Zhang S-C, "Global phase diagram in the quantum Hall effect", *Phys Rev B* **46**, 2223-2238 (1992).

12.35. Kivelson S, Lee D-H and Zhang S-C, "Electrons in Flatland", *Scientific American*, March 1996, 86-91.

12.36. Levy RA and Hasegana R (eds), *Amorphous Magnetism II*, Plenum Press, New York, 1977.

12.37. Lockwood DJ and Pinzuk A (eds), *Optical Phenomena in Semiconductor Structures of Reduced Dimensions*, Kluwer Academic Publishers, Dordrecht, 1993.

12.38. Laughlin RB, "Quantized Hall conductivity in two dimensions", *Phys Rev B* **23**, 5632-5633 (1981).

12.39. Laughlin RB, "Anomalous quantum Hall effect: An incompressible quantum fluid with fractionally charged excitations," *Phys Rev Lett* **50**, 1395-1398 (1983).

12.40. Laughlin RB, *Phys Rev Lett* **80**, 2677 (1988).

12.41. Laughlin RB, *Rev Mod Phys* **71**, 863-874 (1999).

12.42. Lee DH, *Phys Rev Lett* **80** 2677 (1988).

12.43. Lee DH, "Anyon superconductivity and the fractional quantum Hall effect", *International Journal of Modern Physics B* **5**, 1695 (1991).

12.44. Lu ZP et al, *Phys Rev Lett* **92**, 245503 (2004).

12.45. Lyssenko VG, et al, "Direct Measurement of the Spatial Displacement of Bloch-Oscillating Electrons in Semiconductor Superlattices," *Phys Rev Lett* **79**, 301 (1997).

12.46. Mendez EE and Bastard G, "Wannier-Stark Ladders and Bloch Oscillations in Superlattices," *Physics Today* **46**(6), 39-42 (June, 1993).

12.47. Mitin VV, Kochelap VA, and Stroscio MA, *Quantum Heterostructures*, Cambridge University Press, 1999.

12.48. Mott NF, *Metal-Insulator Transitions*, Taylor and Francis, London, 1990, 2nd edn, See especially pp. 50-54.

12.49. Perkowitz S, *Optical Characterization of Semiconductors*, Academic Press, New York, 1993.

12.50. Poon W, McLeish T, and Donald A, "Soft Condensed Matter: Where Physics meets Biology," *Physics Education* **37**(1), 25-33 (2002).

12.51. Prange RE and Girvin SM (eds), *The Quantum Hall Effect*, 2nd edn, Springer New York, 1990.

12.52. Prutton M, *Introduction to Surface Physics*, Clarendon Press, Oxford, 1994.

12.53. Schab K et al, *Nature* **404**, 974 (2000).

12.54. Shik A, *Quantum Wells, Physics and Electronics of Two-Dimensional Systems*, World Scientific, Singapore, 1997.

12.55. Shklovskii BI and Efros AL, *Electronic Properties of Doped Semiconductors*, Springer-Verlag, Berlin, 1984.

12.56. Stormer HL, *Rev Mod Phys* **71**, 875-889 (1999).

12.57. Strobl G, *The Physics of Polymers*, Springer-Verlag, Berlin, 2nd edn, 1997.

12.58. Tarton R, *The Quantum Dot*, Oxford Press, New York, 1995.
12.59. Tsui DC, *Rev Modern Phys* **71**, 891-895 (1999).
12.60. Tsui DC, Stormer HL, and Gossard AC, "Two-dimensional magnetotransport in the extreme quantum limit", *Phys Rev Lett* **48**, 1559-1562 (1982).
12.61. Vasko FT and Kuznetsov AM, *Electronic States and Optical Transitions in Semiconductor Heterostructures*, Springer, Berlin 1993.
12.62. von Klitzing K, Dorda G, and Pepper M, "New Method for high-accuracy determination of the fine-structure constant based on quantum Hall resistance", *Phys Rev Lett* **45**, 1545-1547 (1980).
12.63. von Klitzing K, "The quantized Hall effect", *Rev Modern Phys* **58**, 519-531 (1986).
12.64. Wannier GH, *Phys Rev* **117**, 432 (1969).
12.65. Weisbuch C and Vinter B, *Quantum Semiconductor Structures*, Academic Press, Inc., Boston, 1991.
12.66. Wilczek F, "Anyons," *Scientific American*, May 1991, 58-65.
12.67. Zallen R, *The Physics of Amorphous Solids*, John Wiley, New York, 1983.
12.68. Zargwill A, *Physics at Surfaces*, Cambridge University Press, New York, 1988.
12.69. Zhang SC, "The Chern-Simons-Landau-Ginzburg theory of the fractional quantum Hall effect", *International Journal of Modern Physics B* **6**, 25, 1992.

Additional References on nanophysics, especially nanomagnetism (some of this material also relates to Chap. 7, see section 7.5.1 on spintronics). Thanks to D. J. Sellmyer, Univ. of Nebraska-Lincoln, for this list.

12.70. Hadjipanayis GC and Prinz GA (eds), "Science and Technology of Nanostructured Magnetic Materials," *NATO Proceedings*, Kluwer, Dordrecht (1991)
12.71. Hadjipanayis GC and Siegel RW (eds), *Nanophase Materials: Synthesis - Properties - Applications*, Kluwer, Dortrecht (1994)
12.72. Hernando A (Ed.), "Nanomagnetism," *NATO Proceedings*, Kluwer, Dordrecht (1992)
12.73. Jena P, Khanna SN, and Rao BK (eds), "Cluster and Nanostructure Interfaces," *Proceedings of International Symposium*, World Scientific, Singapore (2000)
12.74. Maekawa S and Shinjo T, *Spin Dependent Transport in Magnetic Nanostructures*, Taylor & Francis, London (2002).
12.75. Nalwa HS (ed), *Magnetic Nanostructures*, American Scientific Publishers, Los Angeles (2001)
12.76. Nedkov I and Ausloos M (eds), *Nano-Crystalline and Thin Film Magnetic Oxides*, Kluwer, Dordrecht, (1999).
12.77. Shi D, Aktas B, Pust L, and Mikallov F (eds), *Nanostructured Magnetic Materials and Their Applications*, Springer, Berlin (2003)
12.78. Wang ZL, Liu Y, and Zhang Z (eds), *Handbook of Nanophase and Nanostructured Materials*, Kluwer, Dortrecht (2002)
12.79. Zhang J et al (eds), *Self-Assembled Nanostructures*, Kluwer, Dordrecht (2002)

Appendices

A.1. Anderson PW, "Brainwashed by Feynman?," *Physics Today*, 53(2), 11-12, (Feb. 2000).

A.2. Anderson PW, *Concepts in Solids*, W. A. Benjamin, New York, 1963.

A.3. Ashcroft NW and Mermin ND, *Solid State Physics*, Holt, Rhinhart, and Wilson, New York, 1976, pp. 133-141.

A.4. Dekker AJ, *Solid State Physics*, Prentice-Hall, Inc., Englewood Cliffs, NJ, 1957, pp. 240-242.

A.5. Economou EN, *Green's Functions in Quantum Physics*, Springer, Berlin 1990.

A.6. Enz CP, "A Course on Many-Body Theory Applied to Solid-State Physics", *World Scientific*, Singapore, 1992.

A.7. Fradkin E, *Field Theories of Condensed Matter Systems*, Addison-Wesley Publishing Co., Redwood City, CA, 1991.

A.8. Huang K, *Statistical Mechanics*, 2nd edn., John Wiley and Sons, New York, 1987, pp. 174-178.

A.9. Huang K, *Quantum Field Theory From Operators to Path Integrals*, John Wiley and Sons, Inc., New York, 1998.

A.10. Jones H, *The Theory of Brillouin Zones and Electronic States in Crystals*, North-Holland Pub. Co., Amsterdam, 1960, Chap. 1.

A.11. Levy M (ed), *1962 Cargese Lectures in Theoretical Physics*, W. A. Benjamin, Inc., New York 1963.

A.12. Mahan GD, *Many-Particle Physics*, Plenum, New York, 1981.

A.13. Mattsson AE, "In Pursuit of the "Divine" Functional," *Science* **298**, 759-760 (25 October 2002).

A.14. Mattuck RD, *A Guide to Feynman Diagrams in the Many-Body Problem*, 2nd edn, Dover edition, New York 1992.

A.15. Merzbacher E, *Quantum Mechanics*, 2nd edn., John Wiley and Sons, Inc., New York, 1970.

A.16. Mills R, *Propagators for Many-particle Systems*, Gordon and Breach Science Publishers, New York, 1969.

A.17. Negele JW and Henri Orland, *Quantum Many-Particle Systems*, Addison-Wesley Publishing Co., Redwood City, CA, 1988.

A.18. Nozieres P, *Theory of Interacting Fermi Systems*, W. A. Benjamin, Inc., New York 1964, see especially pp. 155-167 for rules about Feynman diagrams.

A.19. Patterson JD, *American Journal of Physics*, **30**, 894 (1962).

A.20. Phillips P, *Advanced Solid State Physics,* Westview Press, Boulder. CO, 2003.

A.21. Pines D, *The Many-Body Problem*, W. A. Benjamin, New York 1961.

A.22. Pines D, *Elementary Excitation in Solids*, W. A. Benjamin, New York, 1963.

A.23. Schiff LI, *Quantum Mechanics* 3rd edn, McGraw-Hill Book Company, New York, 1968.

A.24. Schrieffer JR, *Theory of Superconductivity*, W. A. Benjamin, Inc., New York 1964.

A.25. Starzak ME, *Mathematical Methods in Chemistry and Physics*, Plenum Press, New York, 1989, Chap. 5.

A.26. Van Hove L, Hugenholtz NM, and Howland LP, *Quantum Theory of Many-Particle Systems*, W. A. Benjamin, Inc., New York 1961.

A.27. Zagoskin AM, *Quantum Theory of Many-Body Systems*, Springer, Berlin 1998.

Subject References

Solid state, of necessity, draws on many other disciplines. Suggested background reading is listed in this bibliography.

Mechanics

1. Fetter AL and Walecka JD, *Theoretical Mechanics of Particles and Continua*, McGraw-Hill Book Co., New York, 1980. Advanced
2. Goldstein H, *Classical Mechanics*, 2nd edn, Addison-Wesley Publishing Co., Reading, MA 1980. Advanced
3. Marion JB and Thornton ST, *Classical Dynamics of Particles and Systems,* Saunders College Publ. Co., Fort Worth, 1995. Intermediate

Electricity

4. Jackson JD, *Classical Electrodynamics*, John Wiley and Sons, 2nd edn, New York, 1975. Advanced
5. Reitz JD, Milford FJ, and Christy RW, *Foundations of Electromagnetic Theory*, Addison-Wesley Publishing Co., Reading, MA, 1993. Intermediate

Optics

6. Guenther RD, *Modern Optics*, John Wiley and Sons, New York, 1990. Intermediate
7. Klein MV and Furtak TE, *Optics*, 2nd edn, John Wiley and Sons, New York, 1986. Intermediate

Thermodynamics

8. Espinosa TP, *Introduction to Thermophysics*, W. C. Brown, Dubuque, IA 1994. Intermediate
9. Callen HB, *Thermodynamics and an Introduction to Thermostatics*, John Wiley and Sons, New York, 1985. Intermediate to Advanced

Statistical Mechanics

10. Kittel C and Kroemer H, *Thermal Physics*, 2nd edn, W. H. Freeman and Co., San Francisco, 1980. Intermediate
11. Huang K, *Statistical Physics*, 2nd edn, John Wiley and Sons, New York, 1987. Advanced

Critical Phenomena

12. Binney JJ, Dowrick NJ, Fisher AJ, and Newman MEJ, *The Theory of Critical Phenomena*, Clarendon Press, Oxford, 1992. Advanced

Crystal Growth

13. Tiller WA, The Science of Crystallization-Macroscopic Phenomena and Defect Generation, Cambridge U. Press, 1991 and The Science of Crystallization-Microscopic Interfacial Phenomena, Cambridge University Press, Cambridge, 1991. Advanced

Modern Physics

14. Born M, *Atomic Physics*, 7th edn, Hafner Publishing Company, New York, 1962. Intermediate
15. Eisberg R and Resnick R, *Quantum Physics of Atoms, Molecules, Solids, Nuclei, and Particles*, 2nd edn, John Wiley and Sons, 1985. Intermediate

Quantum Mechanics

16. Bjorken JD and Drell SD, *Relativistic Quantum Mechanics*, McGraw-Hill, New York, 1964 and *Relativistic Quantum Fields*, McGraw-Hill, New York, 1965. Advanced
17. Mattuck RD, *A Guide to Feynman Diagrams in the Many-body Problem*, 2nd edn, McGraw-Hill Book Company, New York, 1976. Intermediate to Advanced
18. Merzbacher E, *Quantum Mechanics*, 2nd edn, John Wiley, New York, 1970. Intermediate to Advanced
19. Park D, *Introduction to the Quantum Mechanics*, 3rd edn, McGraw-Hill, Inc., New York, 1992. Intermediate and very readable.

Math Physics

20. Arfken G, *Mathematical Methods for Physicists*, 3rd edn, Academic Press, Orlando, 1980. Intermediate

Solid State

21. Ashcroft NW and Mermin ND, *Solid State Physics*, Holt Reiehart and Winston, New York, 1976. Intermediate to Advanced
22. Jones W and March NH, *Theoretical Solid State Physics*, Vol. 1, *Perfect Lattices in Equilibrium*, Vol. 2, *Non-equilibrium and Disorder*, John Wiley and Sons, London, 1973 (also available in a Dover edition). Advanced
23. Kittel C, *Introduction to Solid State Physics*, 7th edn, John Wiley and Sons, Inc., New York, 1996. Intermediate
24. Parker SP, Editor in Chief, *Solid State Physics Source Book*, McGraw-Hill Book Co., New York, 1987. Intermediate
25. Ziman JM, *Principles of the Theory of Solids*, Second Edition, Cambridge University Press, Cambridge 1972. Advanced

Condensed Matter

26. Chaikin PM and Lubensky TC, *Condensed Matter Physics*, Cambridge University Press, Cambridge, 1995. Advanced

27. Isihara A, *Condensed Matter Physics*, Oxford University Press, Oxford, 1991. Advanced

Computational Physics

28. Koonin SE, *Computational Physics*, Benjamin/Cummings, Menlo Park, CA, 1986. Intermediate to Advanced

29. Press WH, Flannery BP, Teukolsky SA, and Vetterling WT, *Numerical Recipes-The Art of Scientific Computing*, Cambridge University Press, Cambridge, 1986. Advanced

Problems

30. Goldsmid HJ (ed), *Problems in Solid State Physics*, Academic Press, New York, 1968. Intermediate

General Comprehensive Reference

31. Seitz F, Turnbull D, Ehrenreich H (and others depending upon volume), *Solid State Physics, Advances in Research and Applications*, Academic Press, New York, a continuing series at research level.

Applied Physics

32. Dalven R, *Introduction to Applied Solid State Physics*, 2nd edn, Plenum, New York, 1990. Intermediate

33. Fraser DA, *The Physics of Semiconductor Devices*, 4th edn, Oxford University Press, Oxford, 1986. Intermediate

34. Kroemer H, *Quantum Mechanics for Engineering, Materials Science, and Applied Physics,* Prentice-Hall, Englewood Cliffs, NJ, 1994. Intermediate

35. Sze SM, *Semiconductor Devices*, Physics and Technology, John Wiley and Sons, 2nd edn, New York, 1985. Advanced

Rocks

36. Gueguen Y and Palciauskas V, *Introduction to the Physics of Rocks*, Princeton University Press, Princeton, 1994. Intermediate

History of Solid State Physics

37. Seitz F, *On the Frontier-My Life in Science*, AIP Press, New York, 1994. Descriptive

38. Hoddeson L, Braun E, Teichmann J, and Weart S (eds), *Out of the Crystal Maze–Chapters from the History of Solid State Physics*, Oxford University Press, Oxford, 1992. Descriptive plus technical

The Internet

39. (http://xxx.lanl.gov/lanl/), This gets to arXiv which is an e-print source in several fields including physics. It is presently owned by Cornell University.
40. (http://online.itp.ucsb/online/), Institute of Theoretical Physics at the University of California, Santa Barbara, programs and conferences available on line.

Periodic Table

When thinking about solids it is often useful to have a good tabulation of atomic properties handy. The Welch periodic chart of the atoms by Hubbard and Meggers is often useful as a reference tool.

Further Reading

The following mostly older books have also been useful in the preparation of this book, and hence the student may wish to consult some of them from time to time.

41. Anderson PW, *Concepts in Solids*, W. A. Benjamin, New York, 1963. Emphasizes modern and quantum ideas of solids.
42. Bates LF, *Modern Magnetism*, Cambridge University Press, New York, 1961. An experimental point of view.
43. Billington DS, and Crawford JH Jr., *Radiation Damage in* Solids, Princeton University Press; Princeton, New Jersey, 1961. Describes a means for introducing defects in solids.
44. Bloembergen N, *Nuclear Magnetic Relaxation*, W. A. Benjamin, New York, 1961. A reprint volume with a pleasant mixture of theory and experiment.
45. Bloembergen N, *Nonlinear Optics,* W. A. Benjamin, New York, 1965. Describes the types of optics one needs with high intensity laser beams.
46. Born M, and Huang K, *Dynamical Theory of Crystal Lattices,* Oxford University Press, New York, 1954. Useful for the study of lattice vibrations.
47. Brillouin L, *Wave Propagation in Periodic Structures,* McGraw-Hill Book Company, New York, 1946. Gives a unifying treatment of the properties of different kinds of waves in periodic media.
48. Brout R, *Phase Transitions,* W. A. Benjamin, New York, 1965. A very advanced treatment of freezing, ferromagnetism, and superconductivity.
49. Brown FC, *The Physics of Solids—Ionic Crystals, Lattice Vibrations, and Imperfections*, W. A. Benjamin, New York, 1967. A textbook with an unusual emphasis on ionic crystals. The book has a particularly complete chapter on color centers.
50. Buerger MJ, *Elementary Crystallography,* John Wiley and Sons, New York, 1956. A very complete and elementary account of the symmetry properties of solids.

51. Choquard P, *The Anharmonic Crystal*, W. A. Benjamin, New York, 1967. This book is intended mainly for theoreticians, except for a chapter on thermal properties. The book should convince you that there are still many things to do in the field of lattice dynamics.

52. Debye P, *Polar Molecules,* The Chemical Catalog Company, 1929, reprinted by Dover Publications, New York. Among other things this book should aid the student in understanding the concept of the dielectric constant.

53. Dekker AJ, Solid *State Physics,* Prentice-Hall, Engelwood Cliffs, New Jersey, 1957. Has many elementary topics and treats them well.

54. Frauenfelder H, *The Mossbauer Effect*, W. A. Benjamin, New York, 1962. A good example of relationships between solid state and nuclear physics.

55. Grosso G. and Paravicini GP, *Solid State Physics*, Academic Press, 2000, modern.

56. Harrison WA, *Pseudopotentials in the Theory of Metals*, W. A. Benjamin, New York, 1966. The first book-length review of pseudopotentials.

57. Holden A, *The Nature of Solids,* Columbia University Press, New York, 1965. A greatly simplified view of solids. May be quite useful for beginners.

58. Jones H, *The Theory of Brillouin Zones and Electronic States in Crystals,* North-Holland Publishing Company, Amsterdam, 1960. Uses group theory to indicate how the symmetry of crystals determines in large measure the electronic band structure.

59. Kittel C, *Introduction to Solid State Physics,* John Wiley and Sons, New York. All editions have some differences and can be useful. The latest is listed in [23]. The standard introductory text in the field.

60. Kittel C, *Quantum Theory of Solids,* John Wiley and Sons, New York, 1963. Gives a good picture of how the techniques of field theory have been applied to solids. Most of the material is on a high level.

61. Knox RS, and Gold A, *Symmetry in the Solid State,* W. A. Benjamin, New York, 1964. Group theory is vital for solid state physics, and this is a good review and reprint volume.

62. Lieb EH, and Mattis DC, *Mathematical Physics in One Dimension,* Academic Press, New York, 1966. This book is a collection of reprints with an introductory text. Because of mathematical simplicity, many topics in solid state physics can best be introduced in one dimension. This book offers many examples of one dimensional calculations which are of interest to solid state physics.

63. Loucks T, *Augmented Plane Wave Method, W.* A. Benjamin, New York, 1967. With the use of the digital computer, the APW method developed by J. C. Slater in 1937 has been found to be a practical and useful technique for doing electronic band structure calculations. This lecture note and reprint volume is by the man who developed a relativistic generalization of the APW method.

64. *Magnetism and Magnetic Materials Digest*, A Survey of Technical Literature of the Preceding Year, Academic Press, New York. This is a useful continuing series put out by different editors in different years. Mention should also be made of the survey volumes of Bell Telephone Laboratories, called *Index to the Literature of Magnetism.*

65. *Materials, A Scientific American Book*, W. H. Freeman and Company, San Francisco, *1967.* A good, elementary, and modern view of many of the properties of solids. Written in the typical *Scientific American* style.

66. Mattis DC, *The Theory of Magnetism—An Introduction to the Study of Cooperative Phenomena,* Harper and Row Publishers, New York, 1965. A modern authoritative account of magnetism; advanced. See also The Theory of Magnetism I and II, Springer-Verlag, Berlin, 1988 (I), 1985 (II).
67. Mihaly L and Martin MC, *Solid State Physics—Problems and Solutions*, John Wiley, 1996
68. Morrish AH, *The Physical Principles of Magnetism,* John Wiley and Sons, New York, 1965. A rather complete and modern exposition of intermediate level topics in magnetism.
69. Moss TS, *Optical Properties of Semi-Conductors*, Butterworth and Company Publishers, London, 1959. A rather special treatise, but it gives a good picture of the power of optical measurements in determining the properties of solids.
70. Mott NF, and Gurney RW, *Electronic Processes in Ionic Crystals,* Oxford University Press, New York, 1948. A good introduction to the properties of the alkali halides.
71. Mott NF, and Jones H, *Theory of the Properties of Metals and Alloys*, Oxford University Press, New York, 1936. A classic presentation of the free-electron properties of metals and alloys.
72. Nozieres P, *Theory of Interacting Fermi Systems*, W. A. Benjamin, New York, 1964. A good account of Landau's ideas of quasi-particles. Very advanced, but helps to explain why "free-electron theory" seems to work for many metals. In general it discusses the many-body problem, which is a central problem of solid state physics.
73. Olsen JL, *Electron Transport in Metals,* Interscience Publishers, New York, 1962. A simple outline of theory and experiment.
74. Pake GE, *Paramagnetic Resonance*, W. A. Benjamin, New York, 1962. This book is particularly useful for the discussion of crystal field theory.
75. Peierls RE, *Quantum Theory of Solids*, Oxford University Press, New York, 1955. Very useful for physical insight into the basic nature of a wide variety of topics.
76. Pines D, *Elementary Excitations in Solids*, W. A. Benjamin, New York, 1963. The preface states that the course on which the book is based concerns itself with the "view of a solid as a system of interacting particles which, under suitable circumstances, behaves like a collection of nearly independent elementary excitations."
77. Rado, GT and Suhl H (eds), *Magnetism*, Vols. I, IIA, IIB, III, and IV, Academic Press, New York. Good summaries in various fields of magnetism which take one up to the level of current research.
78. Raimes S, *The Wave Mechanics of Electrons in Metals*, North-Holland Publishing Company, Amsterdam, 1961. Gives a fairly simple approach to the applications of quantum mechanics in atoms and metals.
79. Rice FO, and Teller E, *The Structure of Matter*, John Wiley and Sons, New York, 1949. A very simply written book; mostly words and no equations.
80. Schrieffer JR, *Theory of Superconductivity*, W. A. Benjamin, New York, 1964. An account of the Bardeen, Cooper, and Schrieffer theory of superconductivity, by one of the originators of the theory.
81. Schulman JH, and Compton WD, *Color Centers in Solids*, The Macmillan Company, New York, 1962. A nonmathematical account of color center research.

82. Seitz F, *The Modern Theory of Solids*, McGraw-Hill Book Company, New York, 1940. This is still probably the most complete book on the properties of solids, but it may be out of date in certain sections.
83. Seitz F, and Turnbull D (eds) (these are the original editors, later volumes have other editors), *Solid State Physics Advances in Research and Applications*, Academic Press, New York. Several volumes; a continuing series. This series provides excellent detailed reviews of many topics.
84. Shive JN, *Physics of Solid State Electronics*, Charles E. Merrill Books, Columbus, Ohio, 1966. An undergraduate level presentation of some of the solid state topics of interest to electrical engineers.
85. Shockley W, *Electrons and Holes in Semiconductors*, D. van Nostrand Company, Princeton, New Jersey, 1950. An applied point of view.
86. Slater JC, *Quantum Theory of Matter,* McGraw-Hill Book Company, New York, 1951, also 2nd edn, 1968. Good for physical insight.
87. Slater JC, *Atomic Structure*, Vols. I, II, McGraw-Hill Book Company, New York, 1960.
88. Slater JC, *Quantum Theory of Molecules and Solids*, Vol. I, *Electronic Structure of Molecules*, McGraw-Hill Book Company, New York, 1963.
89. Slater JC, *Quantum Theory of Molecules and Solids*, Vol. II, *Symmetry and Energy Bands in Crystals*, McGraw-Hill Book Company, New York, 1965.
90. Slater JC, *Quantum Theory of Molecules and Solids*, Vol. 111, *Insulators, Semiconductors, and Metals*, McGraw-Hill Book Company, New York, 1967. The titles of these books [87 through 90] are self-descriptive. They are all good books. With the advent of computers, Slater's ideas have gained in prominence.
91. Slichter CP, *Principles of Magnetic Resonance with Examples from Solid State Physics*, Harper and Row Publishers, New York, 1963. This is a special topic but the book is very good and it has many transparent applications of quantum mechanics. Also, see 3rd edn, Springer-Verlag, Berlin, 1980.
92. Smart JS, *Effective Field Theories of Magnetism*, W. B. Saunders Company, Philadelphia, 1966. A good summary of Weiss field theory and its generalizations.
93. Smith RA, *Wave Mechanics of Crystalline Solids*, John Wiley and Sons, New York, 1961. Among other things, this book has some good sections on one-dimensional lattice vibrations.
94. Van Vleck JH, *Theory of Electric and Magnetic Susceptibilities*, Oxford University Press, New York, 1932. Old, but still very useful.
95. Wannier GH, *Elements of* Solid *State Theory*, Cambridge University Press, New York, 1959. Has novel points of view on many topics.
96. Weinreich G, *Solids: Elementary Theory for Advanced Students*, John Wiley and Sons, New York, 1965. The title is descriptive of the book. The preface states that the book's "purpose is to give the reader some feeling for what solid state physics is all about, rather than to cover any appreciable fraction" of the theory of solids.
97. Wilson AH, *The Theory of Metals*, Cambridge University Press, New York, 1954, 2nd edn. This book gives an excellent, detailed account of the quasi-free electron picture of metals and its application to transport properties.
98. Wood EA, *Crystals and Light*, D. van Nostrand Company, Princeton, New Jersey, 1964. An elementary viewpoint of this subject.

99. Ziman JM, *Electrons and Phonons,* Oxford University Press, New York, 1960. Has interesting treatments of electrons, phonons, their interactions, and applications to transport processes.

100. Ziman JM, *Electrons in Metals—A Short Guide to the Fermi Surface*, Taylor and Francis, London, 1963. Short, qualitative, and excellent.

101. Ziman JM, *Elements of Advanced Quantum Theory*, Cambridge University Press, New York, 1969. Excellent for gaining an understanding of the many-body techniques now in vogue in solid state physics.

Index

A

B

C

D

E

F

G

M

N

Q

R

S

Printing: Krips bv, Meppel
Binding: Stürtz, Würzburg